Galactic Center Workshop 2006 – From the Center of the Milky Way to Nearby Low-Luminosity Galactic Nuclei

Journal of Physics: Conference Series Volume 54

Bad Honnef, Germany
18-22 April 2006

Editors:

Rainer Schödel
Geoffrey C. Bower
Michael P. Muno
Sergei Nayakshin
Thomas Ott

ISBN: 978-1-61738-260-4

Printed from e-media with permission by:

Curran Associates, Inc.
57 Morehouse Lane
Red Hook, NY 12571
www.proceedings.com

Some format issues inherent in the e-media version may also appear in this print version.

Printed by Curran Associates, Inc. (2010)

Institute of Physics
Dirac House, Temple Back
Bristol BS1 6BE UK

Phone: 44 1 17 929 7481
Fax: 44 1 17 920 0979

techtracking@iop.org

TABLE OF CONTENTS

We are pleased to present the proceedings from the Galactic Center Workshop 2006—*From the Center of the Milky Way to Nearby Low-Luminosity Galactic Nuclei*. The conference took place in the Physikzentrum, Bad Honnef, Germany, on 18 to 22 April 2006. It is the third workshop of this kind, following the Galactic Center Workshops held 1998 in Tucson, Arizona, and 2002 in Kona, Hawaii.

The center of the Milky Way is the only galactic nucleus of a fairly common spiral galaxy that can be observed in great detail. With a distance of roughly 8 kpc, the resolution that can currently be achieved is of the order 40 mpc/8000 AU in the X-ray domain, 2 mpc/400 AU in the near-infrared, and 0.01 mpc/1 AU with VLBI in the millimeter domain. This is two to three orders of magnitude better than for any comparable nearby galaxy, making thus the center of the Milky Way *the*template object for the general physical interpretation of the phenomena that can be observed in galactic nuclei.

We recommend the summary article *News from the year 2006 Galactic Centre workshop*by Mark Morris and Sergei Nayakshin—who also gave the summary talk of the conference—to the reader in order to obtain a first, concise overview of the results presented at the workshop and some of the currently most exciting—and debated—developments in recent GC research.

While the workshops held in 1998 and 2002 were dedicated solely to the center of the Milky Way, the field of view was widened in Bad Honnef to include nearby low-luminosity nuclei. This new feature followed the realization that not only the GC serves as a template for understanding extragalactic nuclei, but that the latter can also provide the context and broader statistical base for understanding the center of our Milky Way. This concerns especially the accretion and emission processes related to the Sagittarius A*, the manifestation of the super massive black hole in the GC, but also the surprising observation of great numbers of massive, young stars in an environment that was previously thought hostile to star formation.

Highlights of the GC Workshop 2002 in Kona were certainly the extraordinary evidence for the black hole nature of Sagittarius A* provided by the observation of orbits of individual stars around the central dark mass of the Milky Way, and the observations of short radiation outbreaks from Sgr A* in the X-ray regime, the so-called *flares*. These events with their variability on the minute time scale provided additional hard evidence for the black hole nature of Sgr A*. Also, the new observational capabilities provided by *Chandra*and *XMM*provided spectacular new insights into the physics of the GC.

With the black hole nature of Sagittarius A* now confirmed beyond reasonable doubt, theory and observation are zooming into ever smaller scales, trying to understand the exact physics behind the extreme under-luminosity (10^{-9}–10^{-10} in terms of Eddington luminosity) of Sagittarius A*. Especially observations of the submm, NIR, and X-ray flares appear to be promising sources to analyze the accretion physics of Sagittarius A* near its event horizon. Some flares may carry the actual imprint of plasma moving at relativistic velocities near the last orbit of the black hole. Another topic that has moved into the focus of attention is star formation near the central black hole. One, possibly two, disks of young massive stars in the central half parsec, generally assumed hostile to star formation, challenge current theories of star formation. Through new infrared integral field spectrometers there is now strong evidence that the stars in the immediate environment of Sagittarius A*, the so-called *S-stars* are B-type main sequence stars. There are many hypothesis for their origin, but no model could yet provide entirely satisfactory explanations for their existence.

A significant number of new telescopes and instruments access ever greater detail at all wavelengths. The continuing development of (sub)-millimeter instrumentation, for example, provides high spatial resolution as well as important spectroscopic data on the complex chemistry of the central tens of parsecs. The new X-ray satellite *Suzaku* delivers unprecedented spectral resolution in the X-ray domain. Instruments such as H.E.S.S. open, for the first time, the window toward observations in the TeV regime with sub-arcminute resolution. A spectacular near-to-mid-infrared survey of the central hundred parsecs of the Milky Way has been conducted with the *Spitzer* infrared satellite. VLBI is moving to ever shorter wavelengths, opening the tantalizing prospect of imaging processes near the event horizon of Sagittarius A* within the next decade. Finally, the brightest near- and mid-infrared sources in the central parsec have been and are continued to be examined with infrared interferometry at the ESO Very Large Telescope Interferometer. These first observations clear the way for future ambitious measurements of relativistic effects in the immediate environment of the black hole.

The articles in this volume can also be accessed on-line in electronic form, including full color figures and multimedia files. The proceedings can be accessed via the internet site of the *Journal of Physics: Conference Series* of Institute of Physics Publishing. A link to this site can be also found on the web site of the GC 2006 Workshop, http://www.ph1.uni-koeln.de/GC06. The complete program of the conference, pictures, and other material can be accessed via this site as well.

We are grateful to everyone who helped with their efforts to make this conference a fruitful and enjoyable one. Special thanks to the LOC and to the employees of the Physikzentrum in Bad Honnef. The location proved to be a perfect choice, providing all the necessary infrastructure on a high level, a fantastic lecture hall, and the very comfortable basement hall for discussing and socializing in the evenings. We also would like to thank all people working as editors for *GCNEWS—The Galactic Center Newsletter*, who helped with their know-how from previous GC workshops. GCNEWS is a newsletter that appears several times each year with articles and news on the Galactic Center, including abstracts of recently published papers. Please see http://www.aoc.nrao.edu/~gcnews for further information. GCNEWS is a communication platform for the steadily increasing community of GC researchers and is the backbone behind the GC Workshops.

We acknowledge financial support for the conference by the *Deutsche Forschungsgemeinschaft (DFG)* Sonderforschungsbereich project number SFB 494.

Finally, we would like to thank the more than 100 participants of the GC Workshop 2006 for their enthusiasm and the numerous active contributions which made this conference such a success.

Rainer Schödel, Geoffrey C Bower, Michael P Muno, Sergei Nayakshin and Thomas Ott
Editors

Institute of Physics Publishing
doi:10.1088/1742-6596/54/1/001
Journal of Physics: Conference Series **54** (2006) 1–9
Galaxy Center Workshop 2006

The Galactic Center Magnetosphere

Mark Morris

Department of Physics & Astronomy, University of California, Los Angeles, CA 90095-1547, USA

E-mail: `morris@astro.ucla.edu`

Abstract. The magnetic field within a few hundred parsecs of the center of the Galaxy is an essential component of any description of that region. The field has several pronounced observational manifestations: 1) morphological structures such as nonthermal radio filaments (NTFs) – magnetic flux tubes illuminated by synchrotron emission from relativistic electrons – and a remarkable, large-scale, helically wound structure, 2) relatively strong polarization of thermal dust emission from molecular clouds, presumably resulting from magnetic alignment of the rotating dust grains, and 3) synchrotron emission from cosmic rays. Because most of the NTFs are roughly perpendicular to the Galactic plane, the implied large-scale geometry of the magnetic field is dipolar. Estimates of the mean field strength vary from tens of microgauss to $\sim$ a milligauss. The merits and weaknesses of the various estimations are discussed here. If the field strength is comparable to a milligauss, then the magnetic field is able to exert a strong influence on the dynamics of molecular clouds, on the collimation of a Galactic wind, and on the lifetimes and bulk motions of relativistic particles. Related to the question of field strength is the question of whether the field is pervasive throughout the central zone of the Galaxy, or whether its manifestations are predominantly localized phenomena. Current evidence favors the pervasive model.

1. Introduction

The magnetic field at the center of the Galaxy (hereafter, the "field") has been studied with a wide variety of techniques for over 20 years, and while there is some consensus that the predominant, global geometry within the central 200 - 300 parsecs is poloidal, the discussion at this workshop has emphasized that there is no universal agreement on the strength of the field and on the extent to which the field strength varies from one place to another. In this review, I summarize the evidence characterizing the various points of view. Earlier reviews of the Galactic center magnetic field have described many of the central points that have been known for some time [1, 2, 3, 4, 5, 6], but recent observations have added considerably to the information that can be brought to bear on this discussion.

The primary probe of the large-scale field has been radio observations of polarized, filamentary structures which, while typically < 0.5 pc in width, are tens of parsecs in length. The strong radio polarization, and the occasional filamentary counterpart at X-ray wavelengths [7] indicate that the emission is synchrotron radiation, and the position angle of the polarization, once corrected for Faraday rotation, confirms that the magnetic field lies along the filaments [8, 9, 10, 11]. The almost invariant curvature of the filaments, and their absence of distortion in spite of clear interactions with the highly turbulent interstellar medium, led Yusef-Zadeh & Morris (1987 [12], see also [5]) to note that the implied rigidity of the filaments requires a field strength on the order of a milligauss, which is surprisingly large, given the scale of these structures.

The orientation of the most prominent NTFs is roughly perpendicular to the Galactic plane, as illustrated in Figure 1, a schematic diagram depicting all filaments identified in the λ20-cm VLA survey by Yusef-Zadeh et al. (2004 [13]). Because the individual filaments define the local field direction, the ensemble of filaments has been interpreted in terms of a predominantly dipolar field, extending at least 200 pc along the Galactic plane [14]. The deviations from perfect verticality of many of the filaments can be ascribed to a global divergence of the field above and below the Galactic plane. The short, nonconforming filaments are discussed in §2.3 (and [15]).

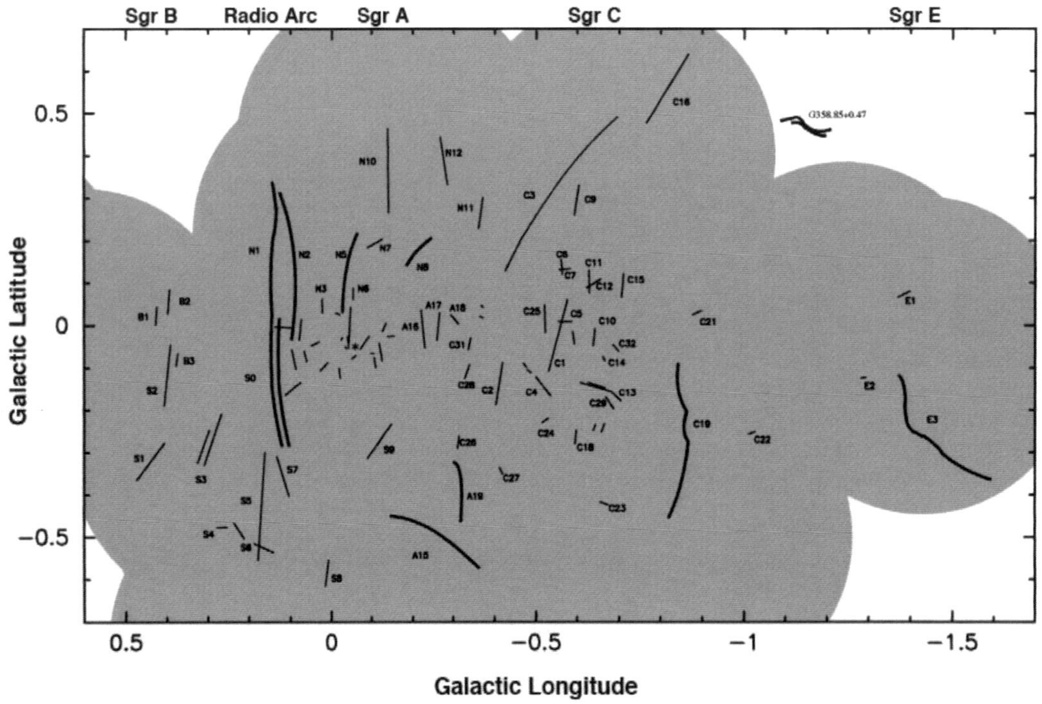

Figure 1. Schematic map showing the radio filaments catalogued by Yusef-Zadeh et al. (2004, [13]) in the course of their λ20-cm survey of the Galactic center.

Quite a different probe of the magnetic field is provided by mid- and far-IR observations of thermal dust emission from magnetically aligned dust grains. The rotation axes of dust grains align with the magnetic field by dissipative torques [16], leading to a net polarization of the thermal emission such that the E-vector is perpendicular to the magnetic field. This probe, however, is strongly dominated by dense, warm clouds, so it is quite different from the NTFs, which sample the field in the intercloud medium occupying most of the volume of the Galactic center. The magnetic field implied by the polarized dust emission is parallel to the Galactic plane [17, 18, 19, 20, 21], and thus perpendicular to the large-scale intercloud field revealed by the NTFs. The perhaps surprising orthogonality of these two systems can be understood in terms of the tidal shear suffered by molecular clouds inhabiting the central molecular zone (CMZ). Any portion of a molecular cloud located a distance R_{gc} pc from the Galactic center, and having a density less than 10^4 cm^{-3} [75 pc/R_{gc}]$^{1.8}$ is subject to such shear [22, 23], so cloud envelopes tend to get stretched into tidal streams that may subtend a large angle at the Galactic center (e.g., [24]). Any magnetic field within the clouds – presumably flux-frozen to the partially ionized molecular gas – will thus be deformed into an azimuthal configuration, with the field lines oriented predominantly along the direction of the shear [17].

There is little evidence that the cloud and inter-cloud environments are magnetically coupled to each other in any significant way, as might have been expected if the field lines were anchored to the cloud layer, and if the rotation of the cloud layer thus imposes a global twist upon the vertical field [25, 26]. The most prominent NTFs show very little deformation where they pass through the Galactic plane and interact with gas in the CMZ (*e.g.*, [12]). Some case can be made that Faraday rotation measurements are consistent with the geometry of a twisted, large-scale field ([6], and references therein), but these data remain too sparse to draw any firm conclusions.

If, as the evidence does indicate, the magnetic field is not anchored in the CMZ, then it is either anchored in the essentially non-rotating Galactic halo or beyond, or it arcs back to the Galactic plane at relatively large radii and is anchored there. In either case, the field lines do not rotate with the CMZ, and the molecular clouds move through the field with a large relative velocity. This gives rise to an induced $\mathbf{v}\times\mathbf{B}$ electric field at cloud surfaces (10^{-4} B(mG) V/cm) which can accelerate particles, drive currents and contribute to the cloud heating [27, 28].

The residence time of clouds in the Galactic center is a few hundred million years as a result of angular momentum loss resulting from both dynamical friction and magnetic drag [29, 2, 30], so it is not clear how clouds forming at the outside edge of the CMZ [31] will retain any magnetic contact with their surroundings as they migrate inwards through the vertical field. Any original connection between the cloud and extra-cloud fields could have pinched off during the inward migration, leaving the clouds magnetically isolated. If typical cloud lifetimes are less than the inspiral times of clouds, presumably because clouds are sheared in the tidal field, then the situation is more complex, but these comments can still apply to sheared cloud streams and the new clouds that reform as the streams interact with each other.

The remainder of this review focuses on several topics of current interest – both observational and theoretical – and culminates in a description of what I think are some of the most important open questions.

2. Uniformity of the Galactic Center Field

2.1. Pressure Confinement of Magnetic Structures

Regardless of the magnetic field strength, the pressure of the interstellar medium in the CMZ is very large compared to the Galactic disk [32]. A hot diffuse gas (T $\sim 10^8$ K, n ~ 0.04 cm^{-3}) that pervades much of the volume of the Galactic center [33, 34, 35] has a pressure of 6 x 10^{-10} dynes cm^{-2}, and is in approximate pressure equilibrium with the warm ($\sim$150 K, low-density molecular medium [36, 37], if the velocity dispersion of $\sim$20 km s^{-1} is used to calculate a turbulent pressure. This pressure is at least two orders of magnitude higher than is characteristic of the Galactic disk. The magnetic field, on the other hand, has a pressure of 4 x 10^{-8} B(mG)^2 dynes cm^{-2}. Consequently, if the magnetic field strength in observed magnetic field structures is $\sim$ a milligauss, then those structures are not confined, and would expand and disappear on a short time scale. This consideration led to the argument that a milligauss magnetic field must be pervasive throughout the CMZ [38]; the strong and extended magnetic field would then provide its own support. In this view, the NTFs are then simply illuminated magnetic flux tubes into which relativistic electrons have been injected, and along which the electrons are constrained to flow [1]. A ring current at the outer edge of the CMZ, or distributed over some range of radii there, is required to generate and confine the overall dipole field [5].

2.2. Models of Localized Magnetic Structures

The alternative to a strong, pervasive field is that the NTFs represent localized peaks in the magnetic field strength. A force-free magnetic field configuration might be considered as a way of tying a local current to a local enhancement of the magnetic field strength [39, 40], but unless the overall configuration is pressure confined, it will be transient and short-lived.

A recent suggestion by Boldyrev & Yusef-Zadeh [41] is that the NTF's are localized structures of milligauss field strength confined by the effective pressure of large-scale turbulence in the Galactic center. In their model, the turbulent cells expulse the field, and concentrate it in regions between the cells. However, while the field will indeed diffuse out of a zone of strong turbulence, the turbulence itself is generally accompanied by the generation of new field at a rate at least as fast as the rate of outward diffusion. Consequently, while this mechanism raises the interesting possibility that the geometry of the boundary field might be different from that within the turbulent zones because of the interactions of the field emanating from the different zones, it is not obvious how this mechanism would lead to a relative enhancement of the field strength at those boundaries. Furthermore, the turbulence in this model must be organized in such a way that the resulting magnetic filaments are predominantly vertical. This places a strong constraint on the overall helicity distribution of plasma motions in the Galactic center. Numerical models that address these concerns are needed to assess this model further.

While other models for localized structures have been proposed [42, 43, 44], they lack the generality needed to account for the population and the orientations of the filaments.

2.3. Significance of the Short Radio Streaks?

One relatively recent finding that has called the notion of a pervasive, uniform field into question is a population of short radio filaments, or streaks, that occupy much of the same Galactic longitude range as the prominent NTFs [14, 45, 15]. These structures are largely included in figure 1. They differ in three ways from the long-known, prominent NTFs:

(i) They are typically quite short, ~0.1 pc.

(ii) Their surface brightness is typically about 1/4 that of the prominent NTFs.

(iii) They appear to be more or less randomly oriented, and thus do not conform to the global verticality of the prominent NTFs. This point has been raised as an argument against a globally ordered, dipole magnetic field.

Given these pronounced differences, one could argue that the radio streaks represent a different population with a separate origin, such as localized oblique shock structures, or strong local deformations of the large-scale field as a result of some local, energetic disturbance. It is premature to conclude that they are inconsistent with a predominantly ordered, large-scale dipole field. Further study of these features is warranted to determine whether they differ systematically from the prominent NTFs in other ways as well, such as in terms of spectral index and polarization properties, and whether they are connected to other interstellar structures in the same way that the prominent NTFs are.

2.4. Dynamical Consequences

As mentioned above, a pervasive, dipole field exerts a magnetic drag force on clouds moving through it, enhancing the rate at which they spiral inwards. If sufficiently strong, the field can also collimate winds and energetic particles that emanate from the center, creating a chimney effect. This is consistent with observations of extended columnar radio features in nearby, radio-bright galactic nuclei [46, 47, 48], although the extent to which the energetic winds in such galaxies have been collimated by the magnetic field, as opposed to the back pressure of their stratified interstellar gas layers, has not been settled.

Recent work by Belmont *et al.* [34] has shown that at least the hydrogen in the hot, diffuse gas at the Galactic center is unbound, so a thermal galactic wind is implied. A dipole magnetic field can collimate this wind to an extent that depends on the field strength, so observations of the large-scale morphology of thermal X-ray emission from the hot gas will be a useful probe of both the wind and the magnetic field.

Cosmic rays will also be confined by a pervasive, vertical field. This has two important consequences: first, the residence time for cosmic rays in the Galactic center will be relatively short (a few $\times$ 10^5 yr) compared to that in the Galactic disk (a few $\times$ 10^6 yr), because the constraint that cosmic rays diffuse primarily along the field lines implies, in the Galactic center, that they diffuse directly away from the Galactic plane, whereas in the Galactic disk, they are largely trapped by the azimuthal field. This relatively short residence time implies a much smaller cosmic ray density than one might infer from the volume rate of supernovae alone. This is consistent with the fact that the high-energy γ-ray emission intensity across the CMZ does not have a peak comparable in its contrast to the peak in the total column density of gas [49, 50]. Second, the longitudinal diffusion of cosmic rays, especially electrons, would be suppressed by a pervasive vertical field. Such diffusion – for protons – is assumed in a recent model for the extended TeV emission observed by HESS invoking a single source of high-energy cosmic rays [51, 52]; this model is probably inconsistent with the presence of a strong, pervasive, vertical field.

3. Comments on Arguments for a Weak Field

3.1. The Minimum Energy Assumption

A number of researchers have estimated the strength of the Galactic center magnetic field using the minimum energy assumption, also referred to as "equipartition", applied to observations of synchrotron emission from relativistic particles (e.g., [60]). This assumption can be applied to a medium in which energy exchange takes place between particles and fields on time scales much less than the energy loss times of particles or the field generation time from macroscopic particle dynamics. This can, for example, describe environments characterized by isotropic turbulence and tangled fields, such as the hot spots in the lobes of double radio source galaxies. However, it is quite generally inapplicable to the Galactic center, except perhaps in very local environments in which energetic events have recently occurred. The striking large-scale order of the Galactic center magnetic field implies that its energy content is not responding in any significant way to local fluid motions or relativistic particle dynamics. The relativistic particles are responding to the field, but the reverse is not true. The energy content of the Galactic center field is far greater than that of the emitting particles, and thus the field strength can be much larger than the equipartition value.

3.2. Zeeman Measures

The most compelling measure of field strength would be a direct measure via the Zeeman effect. Zeeman measures have indeed been made in Galactic center clouds in lines of both H and OH [53, 54, 55, 56, 57], with the result that, where any significant Zeeman signal is seen at all, it implies a field strength on the order of a milligauss or larger. However, there are only a few places where a significant Zeeman signal has been detected. (We do not include in these comments the Zeeman measures deduced from 1720-MHz OH masers around Sgr A East and the circumnuclear disk, which give field strengths of 3-5 mG [58, 59], because such masers presumably arise from locally compressed gas, and may therefore not be representative of the magnetic field on large scales.) One strong selection effect in Zeeman measures is that the extremely broad lines of Galactic center clouds make detection of the Zeeman splitting very difficult unless the field strength exceeds $\sim$1 mG.

Two other points must be considered when interpreting Zeeman measurements: first, they apply largely to the magnetic field within clouds or at the surfaces of clouds. As the above discussion indicates, the magnetic field geometry in clouds is not necessarily related to the large-scale intercloud field. Second, the Zeeman effect measures only the mean line-of-sight component of the field, so if there are field reversals along the line of sight, or if the field direction changes across the radiotelescope beam, then there is significant averaging and dilution of the Zeeman signal. In any case, even if Zeeman measures were able to provide insight into the strength of the intercloud field, the line-of-sight restriction makes it difficult to draw conclusions about a largely vertical dipole field. Further Zeeman measurements, not only of H and OH with improved sensitivity and spatial resolution, but also of other molecules that probe denser regions, will be very important for achieving a more complete understanding of the Galactic center field.

3.3. Synchrotron Lifetimes

One argument that has been raised against a pervasive field of milligauss strength is that the synchrotron lifetime of the electrons responsible for the nonthermal radio emission is relatively short, $\sim 10^5$ years for electrons responsible for the 330-MHz radio emission arising from the central $4° \times 2°$ diffuse nonthermal source [60]. So the supernova rate in the CMZ (or in the Galactic and nuclear bulges above it, since not much less than half of the relativistic electrons created in a supernova will diffuse along the field lines and reach the Galactic plane) must be somewhat larger than 1 per 10^5 yrs if supernovae alone are to account for the uniformity of the synchrotron emission. The rate of only Type Ia supernovae in the Galactic bulge has been estimated at 30 per 10^5 yrs, [61], and in the nuclear bulge (defined in [62, 63]) it is about 20 per 10^5 yrs, so allowing also for core collapse supernovae, the particle production rate seems abundantly sufficient, even if no particles diffuse to the Galactic center from the rest of the Galaxy [64], and if there is no particle reacceleration process operating.

The synchrotron lifetimes of electrons responsible for the 5-GHz radio emission from the NTFs is only $\sim 10^4$ years, so if they diffuse along the field lines at the Alfvén speed, 2200 km s^{-1} B(mG)/n(cm^{-3})$^{1/2}$, then the net distance they can travel before losing an appreciable amount of energy is $\sim$20 pc $\times$ B(mG)/n(cm^{-3})$^{1/2}$, somewhat shorter than the length of the longest filaments (60 pc). (The Alfvén speed is assumed because the diffusion is usually limited by scattering of the streaming particles off of Alfvén waves propagating along the field lines). So far, observations indicate that the radio spectral index has no noticeable variation along the length of the filaments (*e.g.*, [10]). Consequently, if the relativistic electrons are produced at a specific location along them, then the synchrotron lifetime may present a problem unless the field strength is substantially less than a milligauss. Two possible alternatives warrant consideration: first that the diffusion along the field lines is much faster than the relatively slow rate assumed here because the magnetic field is much more rigid and smooth than in most situations where the Alfvén speed is invoked. Second, a reacceleration process may take place along the filaments via shocks, wave dissipation, or reconnection, in analogy with the reacceleration processes needed to account for the persistence of highly relativistic particles in extragalactic jet sources, in spite of their synchrotron and Compton losses.

4. The Double Helix Nebula

A potential new probe of the Galactic center magnetic field was recently revealed at 24 μm with the Spitzer Space Telescope [65]. At a distance of $\sim$100 pc toward positive Galactic latitude from the Galactic center, a nebula having the form of an intertwined double helix extends over at least 50 pc, with its long axis oriented approximately perpendicular to the Galactic plane (Figure 2). This feature was interpreted as a torsional Alfvén wave propagating away from the Galactic center along the magnetic field, and driven by the rotation of the circumnuclear gas disk (CND). The few-parsec scale of the CND matches the width of the nebula, and the wavelength of the torsional wave, 19 pc, corresponds to the $\sim 10^4$-year rotation period of the CND if the Alfvén speed is 10^3 km s^{-1}. This speed, in turn, constrains the magnetic field to have a strength of 0.5 $n^{1/2}$ mG in the context of this hypothesis, where n is the hydrogen density in the medium through which the wave propagates. The density is not known, but for values of the magnetic field ranging from 0.1 to 1 mG, a plausible density is found: n = 0.04 - 4 cm^{-3}. The presence of two strands has been attributed to an apparent "dumbbell" asymmetry of the driving disk (see [65]); the magnetic field threading the disk is concentrated into two diametrically opposed density maxima.

A potential weakness of the torsional wave hypothesis is that the wave cannot yet be followed all the way down to its hypothetical source, the CND. However, this also raises the question of why the double helix is visible in the first place; its mid-infrared emission is most likely thermal emission from dust, so the visibility of the nebula at its present location presumably requires that the wave has levitated charged dust grains. Because of variable conditions at the base of the wave over the past 10^5 years (indeed, the CND is a rather disturbed, non-equilibrated disk [5]), such dust may not have been continuously available to highlight the wave. This may also help explain why a similar nebula is not present on the opposite side of the CND.

An alternative scenario for understanding the Double Helix feature is that it be connected in some way with the linear radio filaments of the Galactic Center Radio Arc. If the Northern extension of the Arc [66] is followed and extrapolated to Galactic latitudes beyond $0.5°$ (see fig 20b of [13]), then it coincides approximately with the long axis of the Double Helix. However, there is no continuous connection in the radio maps between the linear filaments and the Double Helix, and the only radio emission associated with the Double Helix lies outside the mid-IR strands (C. Law, personal communication). There is so far no explanation for how a long bundle of linear, nonthermal filaments could culminate in helically wound, thermal structures. Whether or not the CND hypothesis for the Double Helix is valid, further study of this feature should provide valuable insight into the Galactic center magnetic field.

5. Open Questions

The questions that seem now to be the most compelling for guiding near-future research on the Galactic center magnetic field, besides those already mentioned above, are the following:

- Whether or not the central vertical field is more or less uniform, how and where does it merge with the azimuthal field of the Galactic disk?
- If the Galactic center magnetosphere is defined as the region in which nonthermal radio filaments are observed, then its outer edge roughly coincides with the edge of the CMZ, with the Galaxy's inner inner Lindblad resonance, and with the transition from X1 to X2 gas orbits in the bar. What is the interplay between these phenomena, at this critical juncture in the Galaxy?
- Can high-resolution observations be used to obtain more detail on the points of interaction between cloud and intercloud fields? This may best be done with a combination of radio and far-infrared polarization measurements.
- What process produces the relativistic particles that illuminate the NTFs via their synchrotron emission?
- At the moment, we lack consensus on the power source for the 10^8 K gas occupying much of the volume of the nuclear bulge. Can we appeal to the stirring that takes place as clouds move through the field, leaving magnetosonic and Alfvén waves in their wake? Or can the energy be supplied by magnetic field line annihilation of new vertical field constantly migrating inwards from the rest of the Galaxy?
- What is the origin of the poloidal field? Dynamo models have been hard-pressed to produce a dipole field like that observed, and a promising possibility is that the central field represents protogalactic field that has been concentrated over the history of the Galaxy by mass inflow [67]. Now is a propitious time to take these models to the next stage of sophistication.

Acknowledgments
I gratefully acknowledge stimulating discussions with Steve Cowley.

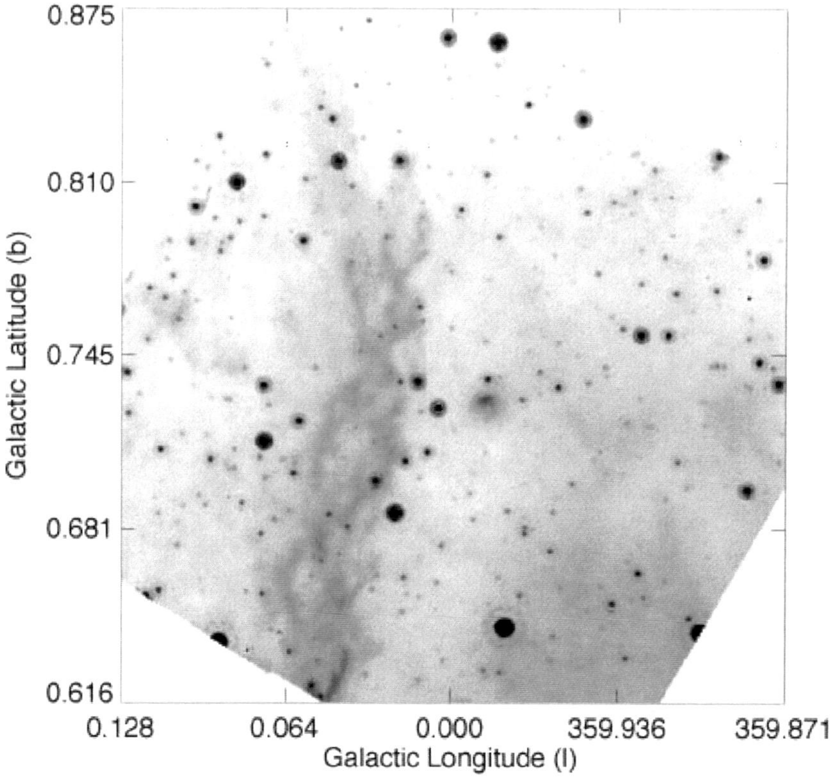

Figure 2. The Double Helix Nebula, observed with the MIPS instrument on the Spitzer Space Telescope [65]. This structure was also seen at shorter wavelengths, 3.6-8 μm, with the IRAC camera on Spitzer.

References

[1] Morris, M. 1990, in IAU Symp. No. 140: *Galactic and Intergalactic Magnetic Fields*, eds: R. Beck, P. Kronberg, and R. Wielebinski, Dordrecht: Kluwer, p. 361

[2] Morris, M. 1994, in *The Nuclei of Normal Galaxies: Lessons from the Galactic Center*, NATO ASI Series C: Vol. 445, eds: R. Genzel and A.I. Harris, Dordrecht: Kluwer, p. 185

[3] Morris, M. 1996, in IAU Symp. No. 169: *Unsolved Problems in the Milky Way*, eds: L. Blitz & PJ Teuben, Dordrecht: Kluwer, p. 247

[4] Morris, M. 1998, in IAU Symp. No. 184: *The Central Regions of the Galaxy and Galaxies*, ed: Y. Sofue, Dordrecht: Kluwer, p. 331

[5] Morris, M. & Serabyn, E. 1996, *Ann. Rev. Astron. Ap.* **34**, 645

[6] Novak, G. 2005, in *Magnetic Fields in the Universe: From Laboratory and Stars to Primordial Structures*, eds: E.M. de Gouveia Dal Pino, G. Lugones, & A. Lazarian, New York: AIP, p 329

[7] Yusef-Zadeh, F., Wardle, M., Muno, M., Law, C. & Pound, M. 2005, *Adv. Spa. Res.* **35**, 1074

[8] Tsuboi, M., Inoue, M., Handa, T., Tabara, H., Kato, T., Sofue, Y. & Kaifu, N. 1986, *AJ* **92**, 818

[9] Yusef-Zadeh, F., Wardle, M. & Parastaran, P. 1997, *ApJL* **475**, L119

[10] Lang, C.C, Morris, M. & Echevarria, L. 1999, *ApJ* **526**, 727.

[11] Lang, C.C. 2007, in this volume.

[12] Yusef-Zadeh, F. & Morris, M. 1987, *AJ* **94**, 1178

[13] Yusef-Zadeh, F., Hewitt, J.W. & Cotton, W. 2004, *ApJS* **155**, 421

[14] Nord, M.E., Lazio, T., Kassim, N., Hyman, S., LaRosa, T., Brogan, C. & Duric, N. 2004, *ApJ* **128**, 1646

[15] LaRosa, T.N. 2007, in this volume.

[16] Hildebrand, R.H. 1988, *QJRAS* **29**, 327
[17] Werner, M.W., Davidson, J.A., Morris, M., Novak, G., Platt, S.R. & Hildebrand, R.H. 1988, *ApJ* **333**, 729
[18] Morris, M., Davidson, J.A., Werner, M., Dotson, J., Figer, D.F., Hildebrand, R. Novak, G. & Platt, S. 1992, *ApJL* **399**, 63
[19] Hildebrand, R.H., Davidson, J.A., Dotson, J., Figer, D., Novak, G., Platt, S. & Tao, L. 1993, *ApJ* **417**, 565
[20] Novak, G., Chuss, D.T., Renbarger, T., Griffin, G.S., Newcomb, M.G., Peterson, J.B., Loewenstein, R.F., Pernic, D. & Dotson, J.L. 2003, *ApJL* **583**, L83
[21] Chuss, D.T., Davidson, J.A., Dotson, J.L., Dowell, C.D., Hildebrand, R.H., Novak, G. & Vaillancourt, J.E. 2003, *ApJ* **599**, 1116
[22] Güsten, R. 1989, in *The Center of the Galaxy*, IAU Symp. No. 136, ed: M. Morris, Dordrecht: Kluwer, p 89
[23] Stark, A.A., Martin, C.L., Walsh, W.M., Xiao, K. & Lane, A.P. 2004, *ApJL* **614**, L41.
[24] Tsuboi, M., Handa, T. & Ukita, N. 1999, *ApJS* , **120**, 1
[25] Uchida, Y., Sofue,Y. & Shibata, K. 1985, *Nature* **317** 699
[26] Shibata, K. & Uchida, Y. 1987, *PASJ* **39**, 559
[27] Benford, G. 1988, *ApJ* **333**, 735
[28] Morris, M. & Yusef-Zadeh, F. 1989, *ApJ* **343**, 703
[29] Stark, A.A., Bally, J., Gerhard, O.E. & Binney, J. 1991, *MNRAS* **248**, 14P
[30] Belmont, R. & Tagger, M. 2006, *A&A* **452**, 15
[31] Binney, J., Gerhard, O.E., Stark, A.A., Bally, J. & Uchida, K.I. 1991, *MNRAS* **252**, 210
[32] Spergel, D.N. & Blitz, L. 1992, *Nature* **357**, 665
[33] Muno, M., Baganoff, F.K., Bautz, M.W., Feigelson, E.D., Garmire, G.P., Morris, M.R., Park, S., Ricker, G.R. & Townsley, L.K. 2004, *ApJ* **613**, 326
[34] Belmont, R., Tagger, M., Muno, M., Morris, M. & Cowley, S. 2005, *ApJL* **631**, L53
[35] Belmont, R. & Tagger, M. 2007, this volume
[36] Dahmen, G., Hüttemeister, S., Wilson, T.L. & Mauersberger, R. 1998, *A&A* **331**, 959
[37] Rodriguez-Fernandez, N.J., Martin-Pintado, J., Fuente, A. & Wilson, T.L. 2004, *A&A* **427**, 217
[38] Morris, M. & Yusef-Zadeh, F. 1989, *ApJ* **343**, 703
[39] Yusef-Zadeh, F., Morris, M. & Chance, D. 1984, *Nature* **310** 557
[40] Yusef-Zadeh, F. & Morris, M. 1987, *ApJ* **322** 721
[41] Boldyrev, S. & Yusef-Zadeh, F. 2006, *ApJL* **637**, L101
[42] Bicknell, G.V. & Li, J. 2001, *ApJ* **548**, L72
[43] Shore, S.N. & LaRosa, T.N. 1999, *ApJ* **593**, 587
[44] Dahlburg, R.B., Einaudi, G., LaRosa, T.N. & Shore, S.N. 2002, *ApJ* **568**, 220
[45] LaRosa, T.N. Nord, M.E., Lazio, T.J.W. & Kassim, N.E. 2004, *ApJ* **607**, 302
[46] Hummel, E., van Gorkom, J.H. & Kotanyi, C.G. 1983, *ApJL* **267**, L5
[47] Duric, N. & Seaquist, E.R. 1988, *ApJ* **326**, 574
[48] Keel, W.C. & Wehrle, A.E. 1993, *AJ* **106**, 236
[49] Hunter, S.D. *et al.* 1997, *ApJ* **481**, 205
[50] Strong, A.W., Moskalenko, I.V. & Reimer, O. 2004, *ApJ* **613**, 962
[51] Aharonian, F. *et al.* 2006, *Nature* **439**, 695
[52] Hinton, J.A. 2007, this volume
[53] Schwarz, U.J. & Lasenby, J. 1990, in *Galactic & Intergalactic Magnetic Fields*, eds: R. Beck, P.P. Kronberg, R. Wielebinski, Dordrecht: Kluwer, p 383
[54] Killeen, N.E.B., Lo, K.Y. & Crutcher, R.M. 1992, *ApJ* **385**, 585
[55] Plante, R.L., Lo, K.Y. & Crutcher, R.M. 1995, *ApJL* **445**, L113
[56] Uchida, K.I. & Güsten, R. 1995, *A&A* **298** 473
[57] Marshall, A.N., Lasenby, A.N. & Yusef-Zadeh, F. 1995, *MNRAS 274*, 519
[58] Yusef-Zadeh, F., Roberts, D.A., Goss, W.M., Frail, D.A. & Green, A.J. 1999, *ApJ* **512**, 230
[59] Sjouwerman, L. 2007, this volume
[60] LaRosa, T.N., Brogan, C.L., Shore, S.N., Lazio, T.J., Kassim, N.E. & Nord, M.E. 2005, *ApJL* **626**, L23
[61] Schanne, S., Cassé, M., Sizun, P., Cordier, B. & Paul, J. 2006, astro-ph/0609566
[62] Serabyn, E. & Morris, M. 1996, *Nature* **382**, 602
[63] Launhardt, R., Zylka, R. & Mezger, P.G. 2002, *A&A* **384**, 112
[64] Cowin, B. & Morris, M. 2007, in preparation
[65] Morris, M., Uchida, K.I. & Do, Tuan 2006, *Nature* **440**, 308
[66] Yusef-Zadeh, F. & Morris, M. 1988, *ApJ* **329**, 729
[67] Chandran, B.D.G., Cowley, S.C. & Morris, M. 2000, *ApJ* **528**, 723

Institute of Physics Publishing
doi:10.1088/1742-6596/54/1/002
Journal of Physics: Conference Series **54** (2006) 10–15
Galaxy Center Workshop 2006

The Strength and Structure of the Galactic Center Magnetic Field

T. N. LaRosa[1], S. N. Shore[2], T. Joseph W. Lazio[3] & Namir E. Kassim[3]

[1] Department of Biological & Physical Sciences, Kennesaw State University, 1000 Chastain Road, Kennesaw, GA 30144 USA,
[2] Dipartimento di Fisica "Enrico Fermi," Università' di Pisa and INFN, Sezione di Pisa, largo B. Pontecorvo 3, I-56127 Pisa, Italy;
[3] Naval Research Laboratory, 4555 Overlook Ave. SW, Washington DC 20375-5351

E-mail: ted@avatar.kennesaw.edu

Abstract. This paper summarizes recently obtained, strong evidence for a weak global field in the Galactic center (GC): the existence of a large-scale region of diffuse, low-frequency, non-thermal emission coincident with the central molecular zone. The overall energetics of this emission, considered along with constraints on GC cosmic ray energy density and diffusion, indicate clearly that the magnetic field pervading this region is $\sim 10\ \mu$G. For completeness, additional points on the orientation of the GC nonthermal filaments, rotation measures of extragalactic sources seen through the GC, and comparison with other normal spiral galaxies are also reviewed.

1. Introduction

Many physical processes in the Galactic center (GC) region depend on the characteristics of the magnetic field. It controls the lifetimes of high energy electrons as well as the energization and transport of cosmic rays. Magnetic fields also affect, for instance, the accretion of ambient gas onto the central engine, star formation, and the dynamics of the local interstellar medium. At present there no consensus on the strength and structure of the magnetic field. The controversey centers on whether there is a strong (of order 1 mG) globally organized magnetic field ([1], [2]) or a weak (of order 10 μG) global field with strong enhancements occurring only in localized regions ([3]; [4]; [5]).

2. Diffuse Non-thermal Source

Recent observations with the Very Large Array (VLA) at 4-meter wavelength (74 MHz) and with the Green Bank Telescope (GBT) at 1-meter wavelength (330 MHz) have revealed a large region ($6^\circ \times 2^\circ$) of diffuse nonthermal emission ([3]; [6]). Assuming equipartition, a minimum energy analysis using the formulation outlined in [7], yields a total energy of 4×10^{51} ergs and a magnetic field of 15 μG in the inner $1.5^\circ \times 0.5^\circ$ of this region. It is within this central region that the GC nonthermal filaments (NTFs) are found. These minimum energy values were found assuming that the ratio of the energy per proton to that of the electrons, is $\mathbf{K} \sim 100$ ([7]).

The corresponding particle energy density is 5.6 eV cm^{-3}. The energy required to power this diffuse non-thermal source is of order 1 supernova (SN) every 10^5 yr, consistent with the rate required to generate the observed soft X-ray emission in this region ([8]). The rate of massive star formation in the inner 50 pc is estimated to exceed that in the disk by a factor of 250 ([9]) and can supply the required production rate for SNe. Thus in [3], we proposed that the diffuse radio source reflects the collective synchrotron emission from residual relativistic electrons of the cumulative SN activity in the GC.

The actual energy density within this diffuse emission region depends upon the estimated radio flux densities at the two wavelengths and the spectral index between them. At 4-meter wavelength, the flux density estimate may be affected by thermal absorption. We obtain an *upper* limit for the spectral index of $\alpha = -0.7$ (for a radio flux density $S \propto \nu^{\alpha}$). This steep spectral index, characteristic of older (aged) electrons, is similar to that of the long-wavelength synchrotron background ([10]) commonly assumed to be produced by electrons originally accelerated in supernova remnant shocks and subsequently diffusing through the Galaxy.

This scenario, however, assumes equipartition, an assumption that would not be justified if these electrons were accelerated in a single energetic event. Let us assume this is the case and that they subsequently propagate in a 1 mG magnetic field. The observed synchrotron emission (approximately 1000 Jy beam^{-1} at 1-meter wavelength, observed with a $39'$ beam at the GBT) would be accounted for by an electron energy density, with energies of order 1 MeV, of 0.04 eV cm^{-3}. Extrapolating to the GeV range, for those particles relevant to the meter-wavelength emission detected, using the observed spectrum gives an electron energy density in the 1 GeV range of only 7×10^{-3} eV cm^{-3}, more than an order of magnitude below the value in the local interstellar medium (0.2 eV cm^{-3}, [11]).

There is no direct measurement of the electron energy density in the GC in the MeV to GeV range. An indirect measure can be obtained from diffuse γ-ray emission. Diffuse Galactic γ-ray emission is produced by the interaction of high-energy cosmic rays with interstellar material, as well as a relatively uncertain contribution from unresolved compact sources (e.g., pulsars, X-ray binaries). Thus, γ-ray emission can be expected to peak where the cosmic ray density is high, the gas density is high, or there is a high concentration of compact objects. A recent re-analysis of all available EGRET data toward the GC ([12]) suggests that the diffuse γ-ray emission is consistent with a cosmic-ray density at least as high as the local value, with a possible additional contribution from the Radio Arc or compact objects.

Although there is a strong GC wind that can advect cosmic rays out of the region at a rate perhaps 100 times faster than in the local disk, the injection rate from SN-generated cosmic rays is also at least 100 times greater in the GC. The near balance between these competing effects yields a particle energy density consistent with the local value. Moreover, a diffuse *TeV* γ-ray emission region has been discovered recently in the central molecular zone ([13]; [14]). While this is not the energy range relevant for synchrotron emission, extrapolation of their results to lower energies also implies an electron energy density comparable to the local value. The upcoming GLAST[1] γ-ray mission will be able to measure diffuse emission at energies around 1 GeV ([14]) and should be able to probe the relevant electron energies directly.

A second indirect constraint on the GC cosmic ray energy density comes from observations of the H_3^+ ion ([15]; [16]). The formation of H_3^+ is thought to be powered by cosmic ray ionization and observed H_3^+ column densities can be used to estimate ζ, the cosmic ray primary ionization rate. For GC clouds, [14] estimate $\zeta \sim 2\text{–}7 \times 10^{-15}$ s^{-1}, substantially higher than the local value, 3×10^{-17} s^{-1} ([17]). Although cloud dynamics may influence these analyses ([18]), again they indicating that the electron energy density within the GC is at least as high as the local value and potentially much higher.

Finally, low energy cosmic rays (tens of keV to MeV) also may generate the X-ray emission

[1] http://www-glast.standford.edu/

from molecular clouds such as Sgr B2. While it was first thought the X-ray emission from these clouds was florescence from an X-ray event associated with the central object ([19]), it is now thought that ionization by low energy cosmic rays can naturally generate this emission. Collectively these results are incompatible with an anomalously low GC cosmic ray energy density.

The diffuse non-thermal source extends about 400 pc on either side of the GC. Assuming that the bulk of high energy electrons responsible for this emission is generated in the inner 50 to 100 pc, they must diffuse at least 300 pc. A number of processes govern the diffusion of high energy particles in the ISM: magnetohydrodynamic (MHD) turbulence in the background medium, self-generated waves, and stochastic magnetic field ([20]). For particles with GeV energies, it is often assumed that scattering from self-generated waves (SGW) (and the resulting pitch angle scattering) is the primary mechanism, and, if there are tangled fields or a high level of background waves, then the diffusion will be even slower. Thus the SGW-dominated diffusion provides a lower limit for the diffusion time. The diffusion coefficient scales as $D_\theta \equiv \langle(\Delta\theta)^2\rangle/\Delta t \propto \Omega(\delta B/B)^2$, where Ω is the electron gyrofrequency in the background magnetic field B and the waves have amplitude δB (e.g., [21]), corresponding to a spatial diffusion coefficient $D_{\rm sp} = c^2/D_\theta$. The observed isotropy of cosmic rays indicates that $\delta B/B \sim 10^{-4}$. Using this value and a magnetic field of 15 μG, the time for a GeV electron to diffuse 300 pc is $\sim 3 \times 10^4$ yr, considerably less than the synchrotron lifetime of 10^7 yr in this field. However, in a 1 mG field, the time required would be $\sim 3 \times 10^6$ yr, orders of magnitude longer than the radiative lifetime of the electrons in such a strong field. The spatial diffusion is faster in the lower field since it takes more time to turn a particle (the rigidity is higher) resulting in a longer mean free path.

In summary, we conclude that the large-scale magnetic field in the GC region is weak, ~ 10 μG. We turn now to assessing the consistency of this conclusion with other evidence taken as indicators of strong fields such as the NTFs and the recently discovered, infrared double helix nebula ([22]).

3. Topology of the NTFs

Prior to the observation of the diffuse nonthermal source, the NTFs (Figure 1; for reviews see [1]; [2]) were used to infer both the strength and topology of the GC magnetic field. More recently, with the discovery of the diffuse nonthermal source and additional NTFs, at issue is whether these are locally generated structures or simply flux tubes that are part of a global field. Rotation measures studies of the NTFs indicate that their magnetic fields are aligned along the filaments (e.g., [23]; [24]). An estimate for the magnetic field strength is inferred from their regidity in the presence of large turbulent pressures from the surrounding gas. Equating the ram pressure with the internal magnetic field pressure gives field strengths of order 1 mG ([25]). To date, a dozen isolated NTFs—as well as a dozen or so bundled NTFs in the Radio Arc—have been identified unambiguously with polarization observations ([2]). There are, however, many tens of additional filamentary candidates that have been discovered with wide-field imaging ([26]; [27]). Their orientations do not appear to be random: most are more or less perpendicular to the Galactic Plane. However, two recently confirmed NTFs lie at large angles with the plane and an explanation of their unexpected orientation is central for understanding the GC magnetic field.

Randomly oriented NTFs were revealed following the development of improved techniques for wide-field, long-wavelength imaging ([28]; [29]; [30]). Earlier, ([31]) had proposed that the perpendicular orientation could be explained by a global, dipolar field centered on the GC. In this scenario the NTFs are interpreted as the flux tubes of a global field that are fortuitously visible only because high energy electrons are locally accelerated. A pervasive 1 mG field was invoked to confine the NTFs and prevent them from expanding rapidly on the Alfvén timescale, since the internal pressure corresponding to a localized field would exceed the ambient thermal pressure

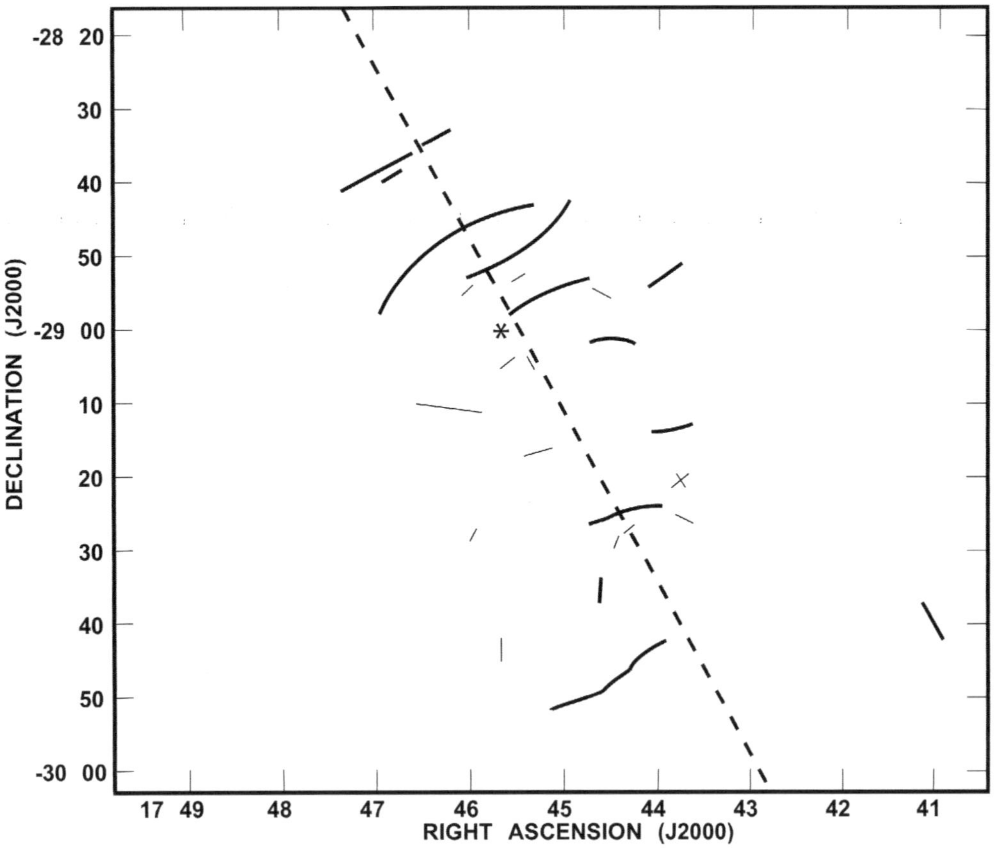

Figure 1. A schematic of the 1-meter wavelength VLA image from [2]). The bold lines depict the confirmed NTFs and the thin lines are the candidates. The Galactic plane is shown as the dashed line.

by orders of magnitude. However, the existence of oblique NTFs contradicts this interpretation. For example, NTF 359.32−0.16 is only 15 pc (in projection) from the Sgr C NTF, and it is not possible to satisfy $\nabla \cdot \mathbf{B} = 0$ if both are part of the same global field. Polarization observations of additional oblique NTF candidates (G359.43+0.13) with the Australia Telescope Compact Array (ATCA) telescope are underway. If confirmed as further examples of oblique NTFs, they would present additional challenges for the global field paradigm. It may be possible to explain them within a global model by asserting that they are a "local" phenomenon generated by a different process than is responsible the longer perpendicular NTFs. However, the surface brightness of the oblique NTFs is generally only a factor of 4 lower than the larger filaments and small changes in the field strength, the particle density or both can explain this difference easily. Aside from their lengths, these objects exhibit the same phenomenology and it seems contrived to propose they are a separate class of NTFs.

Moreover, a pervasive magnetic field is not required to explain the origin of the NTFs. Several dynamical mechanisms have already been suggested. ([32]) suggested the NTFs are (cometary-like) magnetic wakes generated through the interaction of molecular clouds with the GC wind, and are thus necessarily confined laterally by the ram pressure of the strong GC wind. Another attractive idea is that of ([4]) who propose the NTFs are part of a cascade of flux ropes generated by a turbulent dynamo. It is interesting to note that this process may provide the seed flux ropes for the cometary wake model. In both scenarios, the strong magnetic fields within the NTFs are generated locally and do not require a global field.

4. Discussion & Conclusions

Recently, 8-μm *Spitzer* observations of a helical structure some 300 pc from Sgr A have been interpreted as a large-scale torsional Alfven wave ([22]). The estimate of the magnetic field strength in this structure is of order 100 μG. However, there is no direct evidence that this structure is generated magnetically. It also may be interpreted as a dynamical structure formed through a hydrodynamic instability (e.g., the Kelvin-Helmholtz instability). This explanation is especially attractive given its association with the Ω lobe, thought to be part of a large-scale outflow from the GC (e.g., [33]). For instance, a helical-like structure—the Honeycomb Nebula—has been observed in the vicinity of the 30 Dor star-forming region of the Large Magellanic Cloud ([34]; [35]). The structure has a scale of approximately 30 pc, and ([36]) have proposed that it is produced by the blast wave from a younger SN interacting with an older supernova shell that has undergone a Rayleigh-Taylor instability. In this context, additional multiwavelength observations of the Double Helix nebula will be required to confirm whether it is a magnetic or dynamic feature. It is likely that many, if not all, of these processes are operating somewhere at some time in the GC region and are also relevant to other active star forming regions.

Other important observations that constrain the GC magnetic field are the infrared observations of dust polarization ([37]) and rotation measure studies ([38]). The dust polarization traces the magnetic field in the dense molecular gas and reveals clearly a toroidal field in those regions. The relationship between this toroidal field and that in the ionized medium is not clear and requires further theoretical study. ([38]) observed a number of extragalactic sources through the GC region and found that the rotation measures (RMs) towards these sources was significantly smaller than the RMs toward the NTFs ($\sim$ 500 rad m^{-2} compared to $\sim$ 5000 rad m^{-2}). They infer that the mean global field in the GC region must be less than 50 μG. Lastly, there is no evidence for strong magnetic fields in the nuclear regions of other normal spiral galaxies. ([39]) suggest that strong starbursts may have magnetic fields that exceed minimum energy estimates considerably, but, for more normal spiral galaxies, the magnetic field strength appears to scale with the observed surface gas densities, indicating magnetic pressures comparable to the hydrostatic pressures. Since a weak field picture emerges from a number of different lines of evidence, we recommend that both a weak magnetic field value ($\sim$ 10 μG) and a strong field value ($\sim$ 1 mG) be used when making estimates for physical processes and parameters in the GC.

We end with the observation that whatever is happening in the central star forming region of our Galaxy is a pale shadow of the activity seen in starburst systems but likely represents many of the same phenomena. Thus, future work could include assessing the extent to which insights from the GC, especially related to the organization and strength of the magnetic field and the cosmic ray distribution, can be scaled up to the several orders of magnitude greater rates of star formation seen in such systems.

Acknowledgments

TNL thanks the INFN PISA and the Kennesaw State Univ. Foundation for travel support. Basic research in radio astronomy at the NRL is supported by the Office of Naval Research.

References

[1] Morris, M and Serabyn E G 1996, *Ann. Rev. Astron. Astrophys.*, **34**, 645
[2] Morris, this volume
[3] LaRosa, T N, Nord, M E, Lazio, T J W and Kassim, N E 2004, *Astrophys. J.*, **607**, 302
[4] LaRosa, T N, Brogan, C L, Shore, S N, Lazio, T J W, Kassim, N E and Nord, M 2005, *Astrophys. J.*, **626**, L23

[5] Boldyrev, S and Yusef-Zadeh, F 2006, *Astrophys. J.*, **637**, L101
[6] Brogan, C L 2006, *From Clark Lake to the Long Wavelength Array: Bill Erickson's Radio Science*, ASP Conference Series, Vol. 345, eds N Kassim, M Perez, W Junor and P Henning (San Francisco: Astronomical Society of the Pacific) p 187
[7] Beck, R and Krause, M 2005, *Astron. Nach.*, **326**, 414
[8] Muno, M P, et al. 2004, *Astrophys. J.*, **613**, 326
[9] Figer, D F, Rich, R M, Kim, S S, Morris, M and Serabyn, E 2004, *Astrophys. J.*, **601**, 319
[10] Platania, P, Burigana, C, Maino, D, Caserini, E, Bersanelli, M, Cappellini, B and Mennella, A 2003, *Astron. & Astrophys.*, **410**, 847
[11] Mayer-Hasslewander, H A, et al. 1998, *Astron. & Astrophs.*, **335**, 161
[12] Webber, W R 1998, *Astrophys. J.*, **506**, 329
[13] Aharonian F, et al., 2006, *Nature*, **439**, 695
[14] Thompson, D J 2004, *New Astron. Rev.*, **48**, 543
[15] Goto, M, et al. 2002, *Pub. Astron. Soc. Japan*, **54**, 951
[16] Oka, T, Geballe, T R, Goto, M, Usuda, T and McCall, B J 2005, *Astrophs. J.*, **632**, 882
[17] van Dishoeck, E F and Black, J H 1986, *Astrophys. J. Suppl.*, **62**, 109
[18] Lintott, C J and Rawlings, J M C 2006, *Astron. & Astrophs.*, **448**, 425
[19] Murakami, H, Koyama, K, Sakano, M, Tsujimoto, M and Maeda, Y 2000, *Astrophys. J.*, **534**, 283
[20] Scalo, J and Elmegreen, B G 2004, *Ann. Rev. Astron. & Astrophys.*, **42**, 275
[21] Kulsrud, R M 2005, *Plasma Physics for Astrophysics* (Princeton University: Princeton NJ)
[22] Morris, M, Uchida, K and Do, T 2006, *Nature*, **440**, 308
[23] Yusef-Zadeh, F, Wardle, M and Parastaran, P 1997, *Astrophys. J.*, **475**, L119
[24] Lang, C C, Morris, M and Echevarria, L 1999, *Astrophs. J.*, **526**, 727
[25] Yusef-Zadeh, F and Morris, M 1987a, *Astrophs. J.*, **320**, 545
[26] Nord, M E, et al., 2004, *Astron. J.*, **128**, 1646
[27] Yusef-Zadeh, F, Hewitt, J W and Cotton, W 2004, *Astrophys. J. Suppl.*, **155**, 421
[28] Kassim, N E, LaRosa, T N, Lazio, T J W and Hyman, S D 1999, *The Central Parsecs of the Galaxy*, ASP Conference Series, Vol. 186, eds H Falcke, A Cotera, W J Duschl, F Melia and M J Rieke (San Francisco: Astronomical Society of the Pacific) p. 403
[29] Lang, C C, Anantharamaiah, K, Kassim, N E and Lazio, T J W 1999, *Astrophy. J.*, **521**, L41
[30] LaRosa, T N, Kassim, N E, Lazio, T J W and Hyman, S D 2000, *Astron. J.*, **119**, 207
[31] Yusef-Zadeh, F and Morris, M 1987b, *Astrophs. J.*, **322**, 721
[32] Shore, S N and LaRosa, T N 1999, *Astrophys. J.*, **521**, 587
[33] Bland-Hawthorn, J and Cohen, M 2003, *Astrophs. J.*, **582**, 246
[34] Meaburn, J, Wang, L, Palmer, J and Lopez J A 1993, *Mon. Not. R. Astron. Soc.*, **263**, L6
[35] Chu Y H, Dickel J R, Staveley-Smith, L, Osterberg, J and Smith R C 1995, *Astron. J.* **109**, 1729
[36] Redman, M P, Al-Mostafa, Z A, Meaburn, J, Bryce, M and Dyson, J E 1999, *Astron. & Astrophys.*, **345**, 943
[37] Novak, G, et al. 2003, *Astrophys. J.*, **583**, L83
[38] Roy, S., et al. 2006, *Mon. Not. R. Astron. Soc.*, submitted
[39] Thompson, T A, Quataert, E, Waxman, E, Murray, N and Martin, C L 2006, *Astrophys. J.*, submitted

Institute of Physics Publishing
doi:10.1088/1742-6596/54/1/003

Journal of Physics: Conference Series **54** (2006) 16–21
Galaxy Center Workshop 2006

Interaction between the SNR Sagittarius A East and the 50-km s^{-1} Molecular Cloud

Masato Tsuboi[1], Sachiko K. Okumura[1], and Atsushi Miyazaki[2]

[1] Nobeyama Radio Observatory, Minamisaku, Nagano, 384-1305, Japan
[2] Shanghai Astronomical Observatory, Nandan Road, Shanghai,China

E-mail: `tsuboi@nro.nao.ac.jp`

Abstract. We performed high-resolution observations of the Galactic Center 50-km s^{-1} molecular cloud in the CS $J = 1 - 0$ line using the Nobeyama Millimeter Array. The 50-km s^{-1} molecular cloud corresponds to a break in the Sagittarius (Sgr) A east shell. A very broad and negative velocity wing feature is detected at an apparent contact spot between the molecular cloud and the Sgr A east shell. The velocity width of the wing feature is over 50-km s^{-1}. The width is three times wider than those of typical Galactic Center clouds. This strongly suggests that the shell is interacting physically with the molecular cloud. The asymmetric velocity profile of the wing feature indicates that the Sgr A east shell expands and crashes into the far side of the molecular cloud. About 50 clumps are identified in the cloud using CLUMPFIND. The velocity width-size relation and the mass spectrum of clumps in the cloud are similar to those in Central Molecular Zone (CMZ).

1. Introduction

The Galactic Center region is the nearest nucleus of a spiral galaxy. We can observe unique features there related to star formation fed by infalling molecular clouds in detail using present radio telescopes. Such features include young and highly luminous star clusters in the Galactic Center region: the Arches cluster, Quintuplet cluster, and central cluster [1]. It is an open question what mechanism is responsible for the formation of such bright star clusters around the Galactic Center. Molecular clouds in the Galactic Center region are much denser, warmer, and more turbulent than disk clouds, and they have filamentary shapes with several clumps [2] [3] [4]. Detailed observations of the dense molecular clouds of the Galactic Center region should provide unprecedented information about the mechanism of star formation in the Galactic Center region. The "50-km s^{-1} molecular cloud," which is located only $3'$ from Sagittarius (Sgr) A* is a most remarkable Galactic Center molecular cloud (see figure 3 and 4 in [4]), being associated with compact HII regions [5][6]. Moreover, the morphological relationship between the 50-km s^{-1} molecular cloud and the Sgr A east shell, which is presumably a luminous and young SNR located in the Galactic Center region, suggests a physical interaction [7]. The interaction of SNR with the molecular clouds may play an important role in star formation in the Galactic Center region, exactly as in the Galactic Disk region. In addition, the statistical properties of the molecular clouds, for example the mass spectrum of the molecular clumps, could reveal how the clouds are changed before and after interaction with a SNR. We have observed the detailed structure of the 50-km s^{-1} molecular cloud in the CS $J = 1 - 0$ line using the Nobeyama Millimeter Array (NMA). CS emission is expected to be nearly free from the strong contamination near 0 km s^{-1}

from the foreground and background disk molecular clouds because it has a high critical density, $n(H_2) \simeq 10^4$ cm^{-3}. Only the NMA can make high resolution observations in the CS $J = 1 - 0$ line. Throughout this paper, we will adopt 8.5 kpc as the distance of the Galactic Center. At this distance, 1 pc corresponds to about 24".

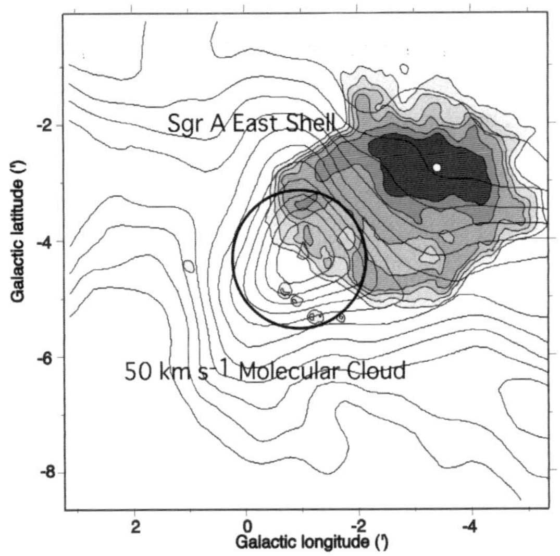

Figure 1. Relative location among the FOV of NMA (thick circle), the 50 km s^{-1} molecular cloud in CS $J = 1 - 0$ line with the 45-m telescope (contours), and Sgr A east at 5 GHz.

2. Observations and Data Analysis

The 50-km s^{-1} molecular cloud, which is centered at $l = 359°58'54.8", b = -0°4'18.8"$, was observed with NMA in the CS $J = 1 - 0$ line (48.990964GHz). The field of view (FOV) was about 140" at 49 GHz, which was the FWHM (Full Width at Half Maximum) of the 10-m antenna elements of the NMA. Figure 1 demonstrates the relative location of the FOV, the 50 km s^{-1} molecular cloud in CS $J = 1 - 0$ line from the 45-m telescope [4], and the Sgr A east shell at 5 GHz [8]. The FOV included the 50-km s^{-1} molecular cloud and the east edge of Sgr A east shell. It was difficult to make a mapping of the field in the vicinity of a strong source such as Sgr A west because the side lobes of a strong source induce artifacts that overwhelm faint structures. We chose the field center so that Sgr A west was settled in the first null of the element antenna response of the NMA. Thus the resultant maps were almost free from the side-lobe effect of Sgr A west. The receivers of the NMA at 49 GHz were SIS mixer receivers. The system noise temperature, including atmospheric noise, was about 400 K during the observations. The back end was a FX-type digital correlator. The total velocity coverage and velocity resolution were 490 km s^{-1} and 0.48 km s^{-1}, respectively. We used NRAO530 as the phase and the gain calibration source, for which the flux density during our observations was 7.0 Jy at 49 GHz. The absolute flux density scale was established using Uranus and Mars.

In order to make maps, we used the CLEAN method in NRAO AIPS package. The size of the synthesized beam is 8.5" × 10"($\phi = 24°$) with a natural weighting, which corresponds to about 0.35 pc ×0.42 pc. Features with spatial extents larger than $1'$ are resolved out. Therefore, the NMA is best used as a spatial filter to find compact components. It should be noted here that all maps in this paper are not corrected for the primary beam attenuation of the element antenna. Finally, a comparison of the $C^{32}S$ $J = 1 - 0$ and $C^{34}S$ $J = 1 - 0$ lines observed by the NRO 45-m telescope indicates that average optical depth of the $C^{32}S$ $J = 1 - 0$ line is about 3 in the central part of the 50-km s^{-1} molecular cloud [4]. We keep in mind in the following analysis that the CS $J = 1 - 0$ line is moderately optically thick.

3. Results

Figure 2 shows the velocity channel map taken with the NMA of the CS $J = 1-0$ emission toward the 50 km s^{-1} molecular cloud. The integrated velocity width of each panel is 3.8 km s^{-1}. The r.m.s. noise levels of the resultant maps are about 0.05 Jy beam^{-1} at the centers of the fields. The molecular cloud can be detected in panels with $V_{LSR} = 10.6 - 83.2$ km s^{-1}. The molecular cloud is resolved into many clumps and/or several filaments in the channel maps. The Sgr A east shell is located northwest of the cloud, but it is already faded out at these frequencies. In the velocity range of $V_{LSR} = 10.6 - 52.7$ km s^{-1}, the molecular cloud is elongated mainly from the lower left to the upper right. The clouds with more positive velocities have more complicated structures. We used CLUMPFIND on FITS data of this channel map in order to find molecular clumps automatically. These will be discussed in the next section.

Figure 3 shows a velocity-integrated map of the 50-km s^{-1} molecular cloud. The velocity range is $V_{LSR} = 8 - 72$ km s^{-1}, and covers almost all of the 50-km s^{-1} molecular cloud. Although the molecular cloud in the lower-resolution map is smooth and has a triangular shape (also see figure 1), the higher-resolution map shows that the CS $J = 1 - 0$ intensity in the cloud is concentrated in three components. The emission detected within the FOV by the NMA is about 40 % of that detected by the NRO 45-m telescope. The circle superimposed on the figure shows the FOV of the NMA. The shaded contours in the figure indicates the continuum emission at 5 GHz of Sgr A east shell. The figure also shows the detailed morphological complementarity between the molecular cloud and Sgr A east shell. The western boundary of the molecular cloud apparently intrudes into the break of the Sgr A east shell, which is darker than other parts of the shell. A component is located at an apparent contact point of the 50-km s^{-1} molecular cloud with the Sgr A east shell, $l = -1.0', b = -4.2'$. This component is elongated along the boundary of Sgr A east shell. This component is identified in the channel maps with $V_{LSR} = 22.1 - 56.76$ km s^{-1} (see figure 2). Another component is located in the break of the shell, $l = -1.3', b = -3.5'$. This is also elongated along the boundary. This component has velocity of $V_{LSR} = 25.9 - 71.8$ km s^{-1}(see figure 2). The lower left component is located at one of the compact HII regions embedded in this cloud [5] [6]. The component has velocity of $V_{LSR} = 14.4 - 41.1$ km s^{-1}.

Figure 4 is the position-velocity diagram along the path through these components. The 50 km s^{-1} molecular cloud is seen as an inclined ridge in the diagram. The velocity width increases up to 50 km s^{-1} at the boundary. This velocity width is three times larger than those of other parts of the 50 km s^{-1} molecular cloud and 10 times wider than those expected from their sizes based on the size-velocity width relation for disk clouds. This wide velocity dispersion presumably is produced by an interaction with the Sgr A east shell. Thus the boundary is a real contact point between the 50 km s^{-1} molecular cloud and Sgr A east shell. In addition, the line profile at the contact point has only a negative velocity wing. H_2 emission [9] and collisionally-excited OH masers (1720 MHz) [10] have been observed around the boundary.

4. Discussion

As mentioned in the previous section, the line profile at the contact point has only a negative velocity wing. Although the poor sensitivity of the NMA may prevent us from detecting the positive velocity wing, any positive velocity wing is much weaker than the negative one. This suggests that the 50 km s^{-1} molecular cloud is located in front of Sgr A east shell, so that the expanding shell crashes into the far side of the molecular cloud and blocks or suppresses the growth of shocked gas with positive velocity. As we know, the radio shadow or silhouette of Sgr A west against Sgr A east shell at low frequency indicates that Sgr A west is located in front of Sgr A east shell [11]. These facts suggest the positional relation among Sgr A west, Sgr A east, and the 50 km s^{-1} cloud along the line of sight that is shown in figure 5. However, the distance between Sgr A west and east is not determined from these observations.

We used CLUMPFIND to find clumps in the 50 km s^{-1} molecular cloud and to derive the

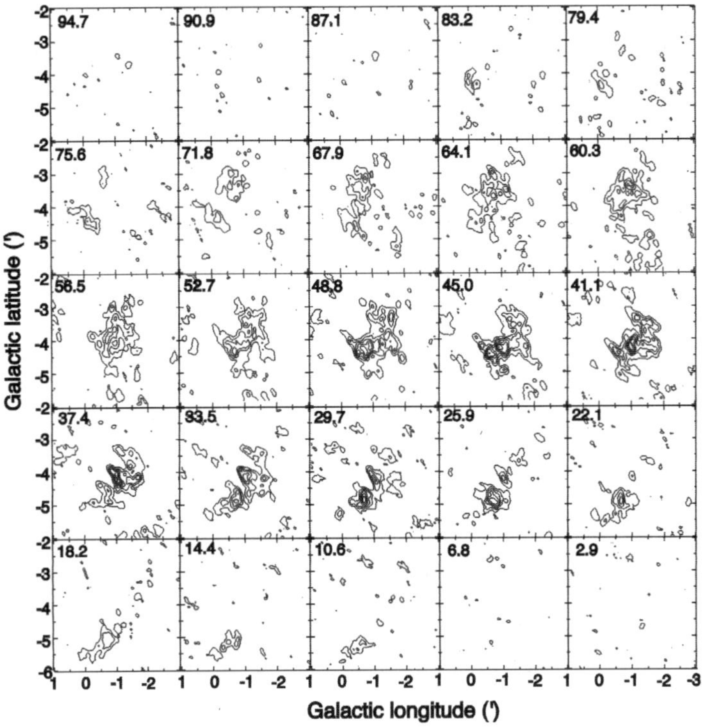

Figure 2. Velocity channel map from the NMA of CS $J = 1 - 0$ emission toward the 50-km s^{-1} molecular cloud. The velocity width of each panel is 3.8 km s^{-1}. The numbers in the top left corners show the central LSR velocities of the each panel. The size of the synthesized beam is 8.5" × 10" ($\phi = 24°$).

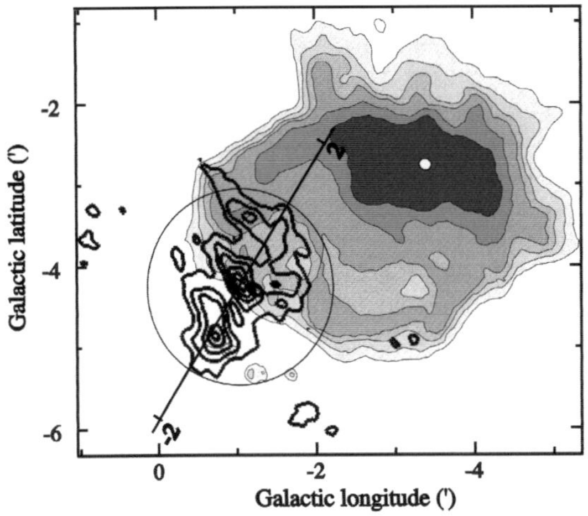

Figure 3. Velocity-integrated map (contours) of the 50 km s^{-1} molecular cloud overlaid on an image of Sgr A east at 5 GHz. The velocity range is $V_{LSR} = 8 - 72$ km s^{-1}.

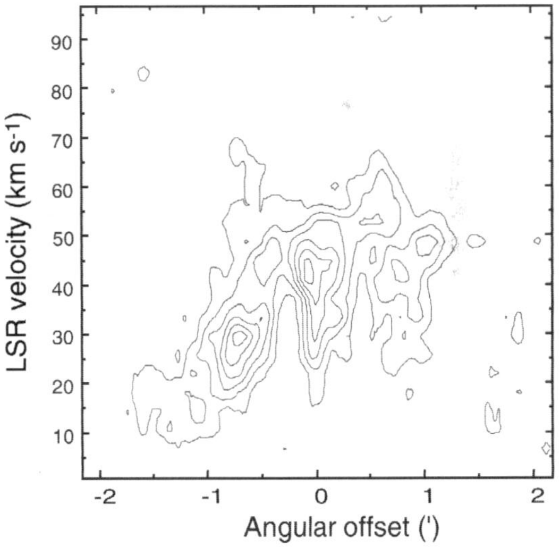

Figure 4. Position-velocity diagram along the thick line in figure 3. The velocity width at the boundary between the 50 km s^{-1} molecular cloud and Sgr A east (angular offset $= 0'$) is three times wider than that of other parts of the cloud

statistical properties of the clumps. About 50 clumps are identified in the cloud. As mentioned in the previous section, the 50 km s^{-1} molecular cloud is located at the east boundary of Sgr A east shell. The ragged limb appears to fit in the break of Sgr A east shell. The components in the cloud are elongated along the boundary. Because the age of the Sgr A east shell is expected to be about 1500 years from X-ray observations [12], the interaction should have started several hundred years ago. However, even molecular gas with a velocity of 100 km s^{-1} moves only 0.1 pc in 1000 years. The time since the beginning of the interaction is too short to make the elongated component by a snowplow effect. Then the appearance may be caused by an increase of the fractional abundance of the CS molecule within the shock. It is unclear whether the statistical properties of the clouds will be changed in such a case.

Figure 6 indicates the velocity width-size relation of clumps in several regions. The velocity width-size relation is generally represented by $\Delta V \simeq aR^{-0.5}$ [13]. The relations of the Galactic Center molecular clouds [3][14][15] and Ori A GMC [16] are $\Delta V \simeq 11R^{-0.5}$ and $\Delta V \simeq 2R^{-0.5}$, respectively. The identified clumps are located on the smaller-size extension of the relation for Galactic Center molecular clouds. Although the clumps in the 50 km s^{-1} molecular cloud would be influenced by the incident shock wave, the velocity width-size relation in Central Molecular Zone (CMZ) is consistent with this cloud. Figure 7 shows the mass spectrum of clumps in the 50 km s^{-1} molecular cloud. The sampling mass of this observation is $10^{2.5} - 10^{3.5} M_\odot$. The power law index in the cloud is $\gamma \simeq -2$ ($dN/dM \propto M^\gamma$). The figure also shows mass spectra in other regions of the CMZ for comparison. The sampling mass of the left and the right parts are $10 - 10^2 M_\odot$ and $10^{4.5} - 10^6 M_\odot$, respectively. For wide mass range of $10 - 10^6 M_\odot$, the index of the spectra is $\gamma \simeq -2$. The power law index of the interacting region is also $\gamma \simeq -2$, which is the same as that of the non-interacting region. There is no significant difference among these mass spectra. The power law index is also similar to that of disk clouds (e.g. Ori A GMC;[16]), even though the magnetic field is very strong (mGauss) and the velocity width is very wide (several 10 km s^{-1}) in the CMZ. It is not clear what causes these relations. However, they should provide information to help the discussion of star formation in the CMZ.

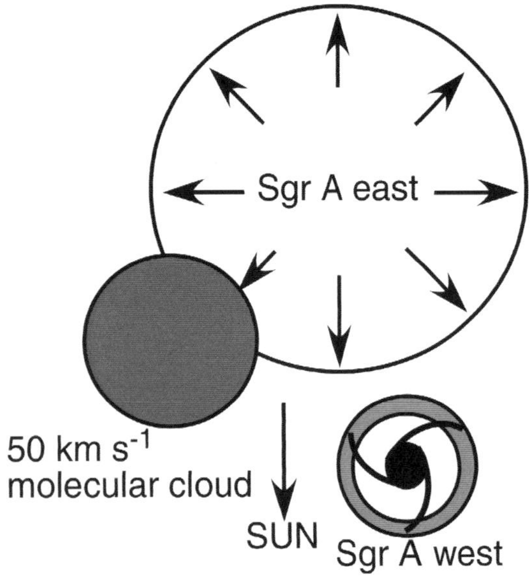

Figure 5. Schematic display of the face-on view of the Galactic Center.

5. Conclusions

We have obtained 8.5" × 10" resolution maps in CS $J = 1 - 0$ emission line of the 50-km s^{-1} molecular cloud in the Galactic Center region with NMA. The molecular cloud is physically interacting with Sgr A east. The velocity width-size relation and the mass spectrum of clumps identified in the cloud are similar to those in CMZ.

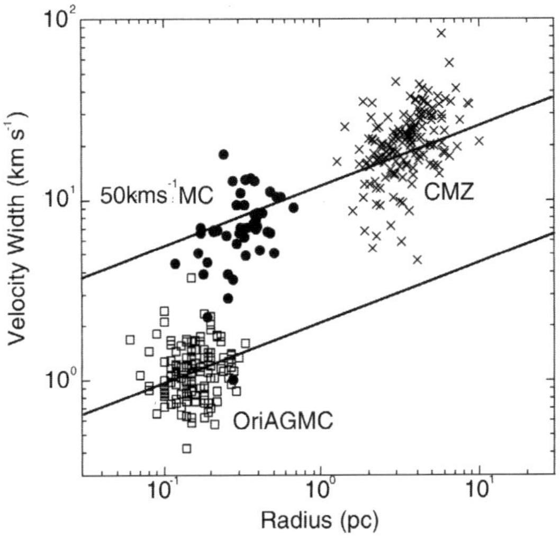

Figure 6. Velocity width-size relation of clumps in the 50 km s^{-1} molecular cloud, CMZ, and Ori A GMC.

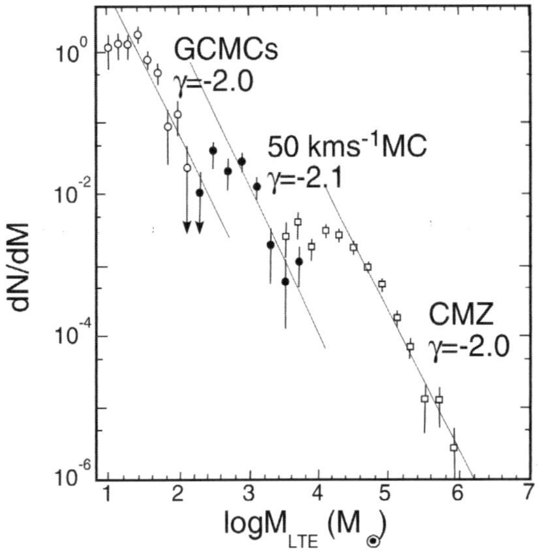

Figure 7. Mass spectra of clumps in the 50 km s^{-1} molecular cloud (center), small molecular clouds in CMZ (left), and CMZ (right)

Acknowledgments

The authors would like to thank the staff of the Nobeyama Radio Observatory for the excellent help during the observations.

References

[1] Figer, D. F. et al. 2002 *Astrophysical Journal* **581** 258
[2] Bally, J. et al. 1987 *Astrophysical Journal Supplement* **65** 13
[3] Oka, T. et al. 1998 *Astrophysical Journal Supplement* **118** 455
[4] Tsuboi, M. et al. 1999 *Astrophysical Journal Supplement* **120** 1
[5] Ekers, R. D. et al. 1983 *Astronomy and Astrophysics* **122** 143
[6] Goss, W. M. et al. 1985 *MNRAS* **215** 69
[7] Ho, P. T. P. et al. 1985 *Astrophysical Journal* **288** 575
[8] Yusef-Zadeh, F., Morris, M. 1987 *Astrophysical Journal* **320** 545
[9] Lee, S. et al. 2003 *MNRAS* **341** 509
[10] Yusef-Zadeh et al. 2001 *Astrophysical Journal* **560** 749
[11] Pedlar, A. et al. 1989 *Astrophysical Journal* **342** 749
[12] Koyama, K. 2006 private communication
[13] Larson, R.B. 1981 *MNRAS* **194** 809
[14] Miyazaki, A., Tsuboi, M. 2000 *Astrophysical Journal* **536** 357
[15] Miyazaki, A., Tsuboi, M. 2001 *Proceedings of IAU Symposium* **205** 356
[16] Tatematsu, K. et al. 1993 *Astrophysical Journal* **404** 643

Institute of Physics Publishing
doi:10.1088/1742-6596/54/1/004

Journal of Physics: Conference Series **54** (2006) 22–28
Galaxy Center Workshop 2006

Shock interactions between Sgr A East and its environments

Sungho Lee[1], Soojong Pak[2], T. R. Geballe[3], Sang-Gak Lee[4], Christopher J. Davis[5], Minho Choi[1], Y. C. Minh[1,6], Robeson M. Herrnstein[7] and Paul T. P. Ho[8,9]

[1]Korea Astronomy and Space Science Institute, 61-1 Hwaam-dong, Yuseong-gu, Daejeon 305-348, South Korea
[2]Department of Astronomy and Space Science, Kyung Hee University, Yongin-si, Gyeonggi-do 446-701, South Korea
[3]Gemini Observatory, 670 N. A'ohoku Place, Hilo, HI 96720, USA
[4]Astronomy Department, School of Physics and Astronomy, Seoul National University, Shillim-dong, Seoul 151-742, South Korea
[5]Joint Astronomy Centre, University Park, 660 North A'ohoku Place, Hilo, HI 96720, USA
[7]Department of Astronomy, Columbia University, 550 West 120th St. New York, NY 10027, USA
[8]Academia Sinica Institute of Astronomy and Astrophysics, P.O. Box 23-141, Taipei, 106 Taiwan
[9]Harvard-Smithsonian Center for Astrophysics, 60 Garden Street, Cambridge, MA 02138, USA

E-mail: leesh@kasi.re.kr

Abstract. We have surveyed the interaction regions between Sgr A East and the surrounding molecular clouds in the H_2 1−0 S(1) (2.1218 μm) line emission. The resulting data cube shows the distribution of the H_2 emission both in space (with a resolution of about 2 arcsec) and in velocity (with a resolution of about 18 km s^{-1}). Our data are compared with NH_3(3,3) data from radio observations to investigate the gas kinematics. Based on the H_2 1−0 S(1) line's intensity, its ratio to H_2 2−1 S(1), and its gas kinematics, we conclude that the emitting H_2 is excited by strong molecular shocks. We estimate shock velocities ($\simeq$ 100 km s^{-1}) by comparing the H_2 line profiles with those of NH_3, and derive an initial explosion energy of 0.2 – 4 $\times 10^{53}$ ergs. Such a high energy excludes a single normal supernova for the origin of Sgr A East. We examine other hypotheses, including the tidal disruption of a star by the supermassive black hole (Sgr A*), multiple supernovae, and a hypernova. We suggest that a hypernova, which is believed to originate from a collapsar or microquasar, is the most probable origin.

1. Introduction

Sgr A East is believed to play an essential role in the large-scale structure and kinematics of the central 10 pc of our Galaxy. However, the nature of Sgr A East is still controversial. The object frequently has been interpreted as a supernova remnant (SNR) because of its shell structure and non-thermal spectrum in the radio [1, 2]. Some recent studies, however, have suggested that the energy, size, and elongated morphology of Sgr A East cannot have been produced by a typical supernova [3, 4, 5]. A theoretical study [6] suggested that Sgr A East may be the

[6] Present address: Academia Sinica Institute of Astronomy and Astrophysics, P.O. Box 23-141, Taipei, Taiwan

remnant of a solar mass star tidally disrupted by a 10^6 $M_\odot$ supermassive black hole (SMBH), of which Sgr A* is the most probable candidate. This model can explain the elongated shape of Sgr A East as well as the extreme energetics ($\sim 4 \times 10^{52}$ ergs). However, observations by the *Chandra X-ray Observatory* [7] suggested that Sgr A East should be classified as a metal-rich 'mixed morphology' remnant of a single Type II supernova explosion with an energy of 10^{51} ergs. Another study using NH_3 data [8] estimated the energy of the progenitor explosion of Sgr A East as $2 - 9 \times 10^{51}$ ergs.

The energy required to make the Sgr A East shell is a key parameter to understanding its origin and can be directly measured by studying the interaction of the shell with the ambient material. Sgr A East is surrounded by various molecular structures including two giant molecular clouds (GMCs) M-0.02-0.07 and M-0.13-0.08 (also known as the '50 km s^{-1} cloud' and the '20 km s^{-1} cloud', respectively), and several filamentary molecular features: the 'molecular ridge', the 'southern streamer', the 'northern ridge', and the 'western streamer' (contours in Figure 1 and refer to [8, 9]). Previous studies of the interaction, however, are mostly based on indirect evidence from morphology, kinematics of neighboring molecular clouds, or absorption of background radiation, rather than on direct, physical interactions involving Sgr A East. To investigate the origin of Sgr A East, we have observed molecular hydrogen (H_2) line emission, which is an excellent tracer of interactions between energetic wind sources such as Sgr A East and dense molecular clouds.

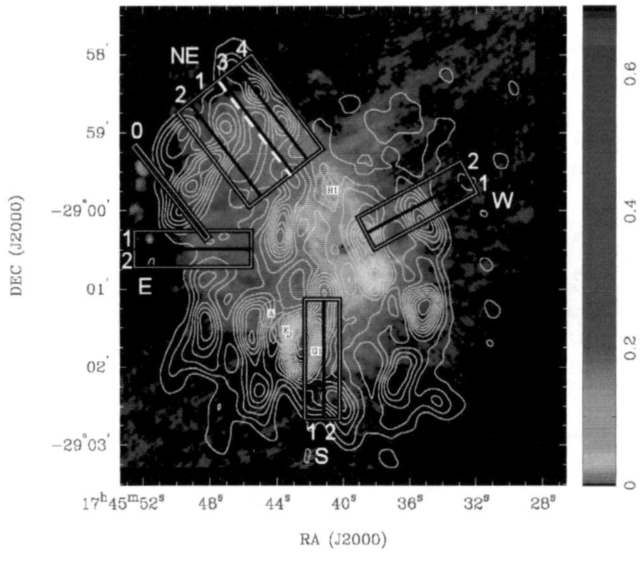

Figure 1. The background image is the 6 cm continuum map of the central 10 pc with $NH_3(3,3)$ emission contours superimposed [9]. Sgr A East surrounds Sgr A* and Sgr A West in the radio continuum. The four fields observed in H_2 1–0 S(1) are indicated by the labelled white boxes. Each field is divided into 2 or 4 scanning blocks labelled with numbers. The narrow box labelled '0' belongs to field E. The white dashed line in field NE-1 indicates the slit position of the H_2 2−1 S(1) observation. Contour levels are in intervals of 4 σ, where $\sigma = 0.33$ Jy beam^{-1} km s^{-1} (with the $\sim$ 15" × 13" beam), and the scale bar ranges from 0 to 0.7 Jy beam^{-1} (where the beam size is 3.4" × 3.0").

2. Observations and data reduction

Four fields around Sgr A East were observed in H_2 $1-0$ S(1) with a slit-scanning technique (see Figure 1). The H_2 $2-1$ S(1) line (2.2477μm) also was observed at the location where the H_2 1–0 S(1) emission is brightest. The data were obtained at the 3.8 m United Kingdom Infrared Telescope (UKIRT) during 2001 and 2003 using the Cooled Grating Spectrometer 4 [10] with its 31 l/mm echelle, 300 mm focal length camera, and two-pixel-wide slit. The pixel scale along the slit was 0.90 arcsec for H_2 $1-0$ S(1) and 0.84 arcsec for H_2 $2-1$ S(1); the slit widths on the sky were 0.83 and 0.89 arcsec, respectively, for these two configurations. The angular resolution was about 2 arcsec ($\sim$ 0.1 pc at the distance to the Galactic centre [11]) and the spectral resolution was ~ 18 km s^{-1} for both lines. The slit length was ~ 90 arcsec, which allowed a large area to be efficiently surveyed.

After the data were flatfielded through the UKIRT reduction pipeline ORAC-DR, we employed standard IRAF[1] routines to perform sky subtraction, interpolate over bad pixels, remove image distortions, and impose an accurate (to ± 1 km s^{-1}) wavelength calibration using telluric OH lines. We also removed stellar continua and residual skylines, and flux-calibrated using the spectra of HR 6496 and HR 6310.

The final stage of data reduction involved the use of MIRIAD [12, 13], a program package generally used for reduction and analysis of radio interferometric data. However, MIRIAD can also be used for the general reduction of continuum and spectral line data such as our near-infrared H_2 data. We employed MIRIAD to stack the 2-dimensional (2-D) spectral images into a single 3-dimensional (3-D) data cube for each of the fields and then combined them into a total data cube (Figure 2) that contains coordinate information for every position along every slit for each slit orientation [14].

3. H_2 excitation mechanism

The near-infrared H_2 emission arises either from thermal excitation by shocks or from non-thermal excitation by absorption of far-ultraviolet (far-UV) photons. The H_2 2–1 S(1) / 1–0 S(1) line ratio, which is the most commonly used indicator of excitation mechanism, can distinguish between UV in low-density ($n(H_2) < 5 \times 10^4$ cm^{-3} [15]) gas, UV in dense gas, jump-shocks (J-shocks) [16], slow J-shocks [17], and continuous-shocks (C-shocks) [17, 18].

From the observed line ratios of 0.3 – 0.5 [19], we can rule out many of the above possibilities, but cannot distinguish between fast J-shocks, a dense PDR, or by a combination of fluorescence and shocks. However, fast J-shocks are excluded since J-shocks typically produce lower H_2 $1-0$ S(1) line intensities ($< 10^{-4}$ ergs s^{-1} cm^{-2} sr^{-1} or $< 3 \times 10^{-18}$ W m^{-2} arcsec^{-2} [16]) than our observed results ($5 - 20 \times 10^{-18}$ W m^{-2} arcsec^{-2}).

We can also consider the kinematic information obtained from the observed line profiles. In a pure PDR or in a J-shock the H_2 line profiles are narrow. In a PDR the line is un-shifted with respect to the cloud velocity, whereas in a J-shock it is shifted by more than ~ 20 km s^{-1}. C-shock profiles are broad extending from the ambient cloud velocity to the shock velocity. To investigate the kinematic properties of both the excited (hot) gas and ambient (cool; < 100 K) gas, we compare the high resolution H_2 1–0 S(1) line profiles with the NH_3(3,3) line profiles [9], at eight representative positions where bright emission lines of both molecules are detected (see Figure 2).

For most spectra in Figure 3, the H_2 line widths are much larger (typically 50 – 100 km s^{-1}) than the NH_3 line widths (typically 20 – 40 km s^{-1}). Because of a blending of five hyperfine lines, the observed NH_3 line widths may be greatly overestimated and the recovered intrinsic line widths are about 15 km s^{-1} [20]. Thus we conclude that the H_2 lines are much broader than the NH_3 lines, by as much as 30 – 80 km s^{-1}, and exclude the pure fluorescence models. Fast J-shocks also seem to be inconsistent with our results, owing to the low peak velocities of the H_2 lines relative to the NH_3 lines. Instead, the gas kinematics indicates C-shocks. However, the high values (up to 0.5) of the line ratio at some positions [19] point to fluorescent contributions. Therefore, a combination of C-shocks and fluorescence [21] is the most reasonable explanation for the H_2 excitation.

[1] IRAF is distributed by the National Optical Astronomy Observatories, which are operated by the Association of Universities for Research in Astronomy, Inc., under cooperative agreement with the National Science Foundation.

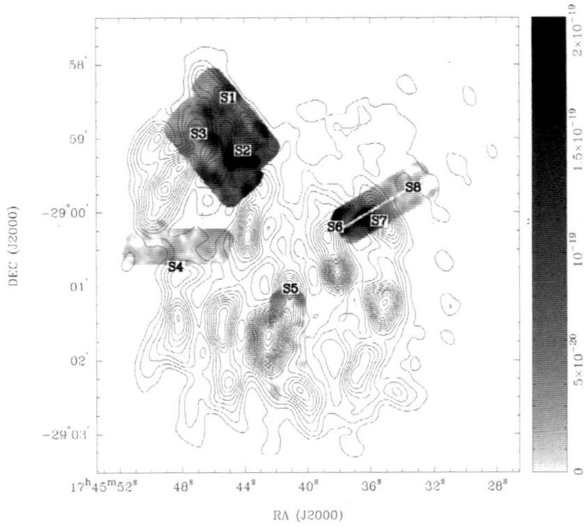

Figure 2. Integrated intensity map of H_2 $1-0$ S(1) line emission, smoothed by a Gaussian with FWHM = 5". The gray-scaled intensity level is indicated by the right hand bar in units of W m^{-2} arcsec^{-2}. The overlaid contours show the velocity-integrated NH_3(3,3) emission map [9]. The contour levels are in intervals of 3σ (the RMS noise $\sigma = 0.33$ Jy beam^{-1} km s^{-1}). The positions of the eight spectra of H_2 $1-0$ S(1) and NH_3(3,3) in Figure 3 are marked on the map.

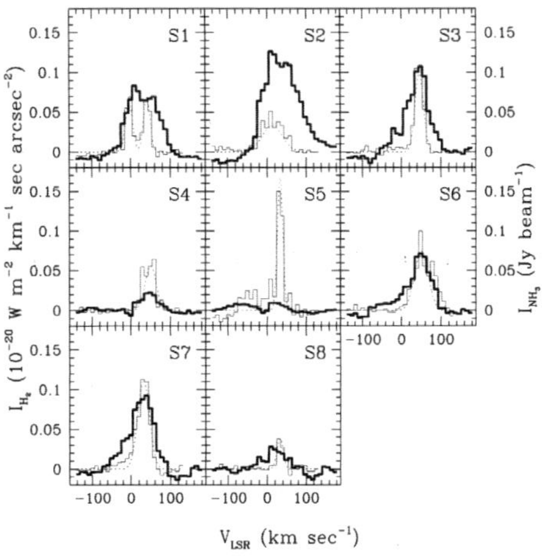

Figure 3. Spectra of H_2 $1-0$ S(1) and NH_3(3,3) in 30" FWHM apertures at the positions marked in Figure 2. Thick solid lines are the H_2 $1-0$ S(1) spectra, for which the intensity scale is indicated on the left-side, and thin solid lines NH_3(3,3), for which the scale is on the right-side. The dotted lines are Gaussian fits to the NH_3(3,3) line profiles. The spectra were obtained by convolving each data cube with a circular Gaussian beam and are not corrected for instrumental broadening in velocity.

4. Shock velocities of Sgr A East from line profiles

Since the radiation is emitted as the gas is being heated and accelerated in the shock front [22], the speed of a C-shock can be determined from the velocity corresponding to the edge of the line profile, hereafter denoted by V_{max}. According to detailed models for the line profiles of C-shocks [23], the full-width-at-zero-intensity (FWZI) of a line is 90 per cent of the shock velocity. Because line profiles from different regions are often superimposed, however, the low velocity edge (the pre-shock velocity) of an individual shock profile is not always obvious. Thus, instead of using the FWZI to determine the speed of the C-shock, we use the difference between V_{max,H_2} of a post-shock H_2 profile and V_{0,NH_3}, the system velocity of the pre-shock gas as traced by the NH_3 emission.

The velocity of NH_3 line center (V_{0,NH_3}) is measured from a Gaussian fit and V_{max,H_2} is determined from a deconvolved profile (using a Gaussian of FWHM = 18 km s^{-1} to approximate the instrumental resolution [14]). The resulting line-of-sight velocities of the C-shocks range from 44 km s^{-1} to 128 km s^{-1} depending on the location. After corrections for geometrical projection effects assuming an oblate shell for the Sgr A East shock front [14], the mean velocity of the shocks is 109 ± 34km s^{-1}. The detailed results are presented elsewhere [14].

The derived shock velocities seem to be too large to be acceptable in typical C-shock models, which have breakdown velocities of < 50 km s^{-1} [24]. The breakdown velocity, however, depends on the density and magnetic field strength in the pre-shock gas [25]. It is generally thought that a magnetic field of a few mG exists throughout the Galactic Center region [26]. In such a strong magnetic field and in clouds of mean density 10^4 cm^{-3} [27], the breakdown velocity is increased to > 200 km s^{-1}.

5. Explosion energy and age of Sgr A East

To estimate the initial energy and the age of Sgr A East we use a standard model [28] of the evolution of a SNR in a medium of uniform density. Sgr A East is thought to be in a pressure-driven radiative phase (or snowplow phase) [4, 7, 8]. Then the energy of the initial explosion, E_0, is given by

$$E_0 = 4.3 \times 10^{51} \left(\frac{E_{shell}}{10^{51}\ \mathrm{ergs}}\right)^{2/3} \left(\frac{R_{shell}}{1\ \mathrm{pc}}\right)^{4/3} \left(\frac{n_0}{10^4\ \mathrm{cm}^{-3}}\right)^{2/3} \quad [\mathrm{ergs}]$$

$$E_{shell} = \frac{2\pi}{3} R_{shell}^3\, n_0\, m_H\, V_{shell}^2$$

where R_{shell} is the current radius of the shell, n_0 is the mean initial H number density of the pre-shock gas, m_H is the proton mass, and V_{shell} is the expansion velocity of the shell material [4], which is assumed to be the same as the mean velocity of the molecular shocks. For n_0 we suppose two extreme cases: a dense giant molecular cloud (scenario I), and a low density halo [7] (scenario II).

In scenario I, shocks with a mean velocity of $V_{shock} = 109$ km s^{-1} are propagating through molecular clouds with mean H_2 densities of $n_{H_2} = 10^4$ cm^{-3} [27]. Thus $V_{shell,I} = V_{shock} = 109$ km s^{-1} and $n_{0,I} = 2\ n_{H_2} = 2 \times 10^4$ cm^{-3}. Since the mean shell radius, $R_{shell} = 4.6$ pc, which is derived from the elliptical boundary of 5.4 pc × 3.8 pc [14], the shell's kinetic energy is calculated to be $E_{shell,I} = 2.3 \times 10^{52}$ ergs and the initial explosion energy $E_{0,I} = 4.2 \times 10^{53}$ ergs.

In scenario II, we assume that the shock front of Sgr A East has just started to contact its surrounding molecular clouds and the shocks are slowed down by dense gas. Since the measured shock velocities from the H_2 lines are for these slowed shocks, we use instead the velocities of fast shocks that propagated through the low density gas halo before contacting the molecular clouds. The velocity of the fast ambient shocks in the halo can be derived using $V_{s,halo}/V_{s,cloud} = \sqrt{n_{cloud}/n_{halo}}$ where $n_{halo} = 10^3$ cm^{-3} [7]. It is reasonable to regard the compressed shell as the synchrotron shell which contacts the inner edge of the dust ring [4]. Hence we use $R_{shell,II} = 3.7$ pc, which is smaller by 80 per cent than the size determined from the H_2 emission. Using $V_{s,cloud} = 109$ km s^{-1} and $n_{cloud} = 2 \times 10^4$ cm^{-3}, we calculate $V_{shell,II} = V_{s,halo} = 487$ km s^{-1} and $n_{0,II} = n_{halo} = 10^3$ cm^{-3}. Then the kinetic shell energy is calculated to be $E_{shell,II} = 1.2 \times 10^{52}$ ergs and the initial explosion energy $E_{0,II} = 2.7 \times 10^{52}$ ergs.

We can calculate the age of Sgr A East from

$$t = 114 \left(\frac{R_{shell}}{0.329\ \mathrm{pc}}\right)^{7/2} \left(\frac{E_0}{10^{51}\ \mathrm{ergs}}\right)^{-3/4} \left(\frac{n_0}{10^4\ \mathrm{cm}^{-3}}\right) \quad [\mathrm{yr}],$$

using the estimated explosion energy, E_0 [28]. Then the estimated age is 2.5×10^4 yr in scenario I and 4.6×10^3 yr in scenario II.

6. Origin of Sgr A East

The explosion energy of 0.2 – 4 $\times 10^{53}$ ergs excludes the hypothesis of a single, typical supernova with a kinetic energy of 10^{51} ergs for the origin of Sgr A East.

The predicted frequency (once every 10^4–10^5 yr) of the tidal disruption of stars by a SMBH and energy released (0.4 – 4 $\times 10^{53}$ ergs) [6] are consistent with our estimates of the age and explosion energy, respectively. On the other hand, to explain the 3-D geometry of the Sgr A complex, this model needs a highly collimated eruption since Sgr A*, which is thought to be the source of explosion (i.e. the SMBH), is located at the edge of Sgr A East along the line of sight [8, 14]. An anisotropic explosion, however, is not likely according to detailed studies of the evolution of stellar debris [29, 30]. In addition, this model cannot reproduce the high metallicity observed in X-ray [7] since the explosion is driven by gravity rather than by nuclear reactions.

The extremely large energy of the Sgr A East explosion may be explained by typical supernovae if several tens of them explode over a short time span, which cannot be longer than the age of Sgr A East ($\sim 10^4$ yr). This might be possible in a stellar cluster containing a huge number of massive stars, in which many stars have similar masses so that their SN-end-lifetimes have a very small dispersion. In the Galactic center, there are three extraordinarily massive and dense young clusters; the Central, Arches, and Quintuplet clusters. Using the estimated initial masses [31] and evolution models of massive stars [32, 33], we calculate the SN-end-lifetimes of the member stars in the Arches cluster as an example. Figure 4 shows that the estimated lifetimes are concentrated around 4500 kyr, which corresponds to $M_{init} \simeq 40$–70 $M_\odot$. The peak at that age contains ~50 stars and covers $\sim 3 \times 10^5$ yr, corresponding to an average supernova rate of slightly more than one per 10^4 yr. Thus at most only a few explosions are expected during the lifetime of Sgr A East, and multiple supernovae cannot account for its energy.

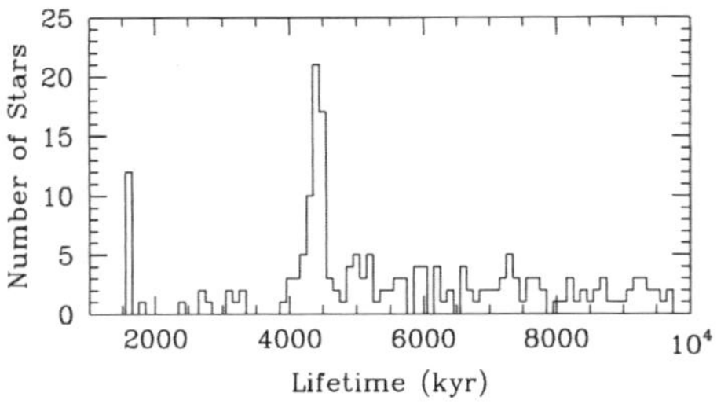

Figure 4. Histogram for the lifetimes of the stars in the Arches cluster. The data are binned by 10^5 yr for the lifetime.

The extremely high energies ($> 10^{52}$ ergs) of hypernova [34, 35] can be reproduced by black hole formation mechanisms such as a "collapsar" or "jet-powered supernova (JetSN)" [33, 36], or the "microquasar" model [34]. Either occurrence satisfies both the energy and metallicity constraints. There are many potential cadidate hypernovae near Sgr A East. JetSNe are possible in the range of $M_{init} \simeq 25$–60 $M_\odot$ (GRBs are also possible when $M_{init} > 35$ $M_\odot$) under the metallicity condition in the Galactic center [36]. In the Arches cluster, for example, there are ~ 130 massive stars in the potential mass range of JetSNe and ~ 80 stars in the range of JetSNe+GRBs. These numbers may be doubled if the Central and Quintuplet clusters, which have similar properties with the Arches cluster, are included. We conclude that only the hypothesis of a hypernova can satisfy all the observational constraints above.

7. Future work

The interstellar medium in the central 10 pc is known to be very clumpy and dense. Thus the evolution of Sgr A East needs to be studied using more realistic, numerical simulations than the standard analytic model. For the black hole origin, only the tidal disruption model has been tested in this work, but disruption is only one possibility related to SMBHs. Other phenomena observed in active galactic nuclei, e.g. bursts or outflows, should also be examined as possible origins of Sgr A East. Note also that a very inhomogeneous environment near the site of the explosion might circumvent the problem of high collimation. Finally, the viability of the hypernova model should be also examined based on its other properties; for example, its radio or X-ray emission.

Acknowledgments

We thank the staff at UKIRT for their excellent support during our observations. S. Lee thanks Luis Ho for the insightful suggestions on future work. Fig. 1 is reproduced from Fig. 10 of [9] by permission of the AAS. The United Kingdom Infrared Telescope is operated by the Joint Astronomy Centre on behalf of the U.K. Particle Physics and Astronomy Council. This work was financially supported by the BK21 Project of the Korean Government. T. R. Geballe's research is supported by the Gemini Observatory, which is operated by the Association of Universities for Research in Astronomy, Inc., on behalf of the international Gemini partnership of Argentina, Australia, Brazil, Canada, Chile, the United Kingdom and the United States of America.

References

[1] Jones T W 1974 *A&A* **30** 37
[2] Goss W M, Schwarz U J, Ekers R D and van Gorkom J H 1983 *Proc. IAU Symp.* vol 101 ed J Danziger and P Gorenstein (Dordrecht: Reidel) p 65
[3] Yusef-Zadeh F and Morris M 1987 *ApJ* **320** 545
[4] Mezger P G, Zylka R, Salter C J, Wink J E, Chini R, Kreysa E and Tuffs R 1989 *A&A* **209** 337
[5] Fatuzzo M, Melia F, Yusef-Zadeh F and Markoff S 1999 *ASP Conf. Ser.* vol 186 ed H Falcke et al (San Francisco: ASP) p 560
[6] Khokhlov A and Melia F 1996 *ApJ* **457** L61
[7] Maeda Y et al 2002 *ApJ* **570** 671
[8] Herrnstein R M and Ho P T P 2005 *ApJ* **620** 287
[9] McGary R S, Coil A L and Ho P T P 2001 *ApJ* **559** 326
[10] Mountain C M, Robertson D J, Lee T J and Wade R 1990 *Proc. SPIE* vol 1235 ed D L Crawford (Bellingham: SPIE) p 25
[11] Reid M J 1993 *ARA&A* **31** 345
[12] Sault R J, Teuben P J and Wright M C H 1995 *ASP Conf. Ser.* vol 77 ed R A Shaw et al (San Francisco: ASP) p 433
[13] Hoffman W, Hudson J, Sharpe R K, Grossman A W, Morgan J A and Teuben P J 1996 *ASP Conf. Ser.* vol 101 ed G H Jacoby and J Barnes (San Francisco: ASP) p 436
[14] Lee S 2005 *PhD thesis* Seoul National University
[15] Black J H and van Dishoeck E F 1987 *ApJ* **322** 412
[16] Hollenbach D and McKee C F 1989 *ApJ* **342** 306
[17] Smith M D 1995 *A&A* **296** 789
[18] Kaufman M J and Neufeld D A 1996 *ApJ* **456** 611
[19] Lee S, Pak S, Davis C J, Herrnstein R M, Geballe T R, Ho P T P and Wheeler J C 2003 *MNRAS* **341** 509
[20] McGary R S and Ho P T P 2002 *ApJ* **577** 757
[21] Fernandes A J L, Brand P W J L and Burton M G 1997 *MNRAS* **290** 216
[22] Hollenbach D J, Chernoff D F and McKee C F 1989 *Proc. of the 22nd Eslab Symp.* vol SP-290 ed B H Kaldeich (European Space Agency) p 245
[23] Smith M D and Brand P W J L 1990 *MNRAS* **243** 498
[24] Draine B T, Roberge W G and Dalgarno A 1983 *ApJ* **264** 485
[25] Smith M D, Brand P W J L and Moorhouse A 1991 *MNRAS* **248** 730
[26] Chuss D T, Davidson J A, Dotson J L, Dowell C D, Hildebrand R H, Novak G and Vaillancourt J E 2003 *ApJ* **599** 1116
[27] Güsten R and Philipp S D 2004 *Proc. of the 4th Cologne-Bonn-Zermatt Symp.* vol 91 ed S Pfalzner et al (Heidelberg: Springer) p 253
[28] Shull J M 1980 *ApJ* **237** 769
[29] Lee H M and Kim S S 1996 *JKAS* **29** 195
[30] Bogdanović T, Eracleous M, Mahadevan S, Sigurdsson S and Laguna P 2004 *ApJ* **610** 707
[31] Figer D F et al 2002 *ApJ* **581** 258
[32] Woosley S E, Langer N and Weaver T A 1993 *ApJ* **411** 823
[33] Woosley S E, Heger A and Weaver T A 2002 *Rev. Mod. Phys.* **74** 1015
[34] Paczyński B 1998 *ApJ* **494** L45
[35] Nakamura T, Umeda H, Iwamoto K, Nomoto K, Hashimoto M, Hix W R and Thielemann F-K 2001 *ApJ* **555** 880
[36] Heger A, Fryer C L, Woosley S E, Langer N and Hartmann D H 2003 *ApJ* **591** 288

Institute of Physics Publishing
doi:10.1088/1742-6596/54/1/005

Journal of Physics: Conference Series **54** (2006) 29–34
Galaxy Center Workshop 2006

Hot molecular gas in the central region around Sgr A*

María Montero-Castaño[1,2], Robeson M. Herrnstein[3] and Paul T.P. Ho[1,4]

[1] Harvard-Smithsonian Center for Astrophysics, 60 Garden St., Cambridge, MA 02138, USA
[2] Departamento de Astrofísica, Facultad de Ciencias Físicas, Universidad Complutense de Madrid, 28040-Madrid, Spain
[3] Department of Astronomy, Columbia University, 550 West 120th St., New York, NY 10027, USA
[4] Academia Sinica Institute of Astronomy and Astrophysics, Taipei, Taiwan

E-mail: `mmontero@cfa.harvard.edu`

Abstract. We present our latest Submillimeter Array[1] (SMA) results on HCN(4-3) in the Sgr A* vicinity. We have produced a 37-pointing mosaic, covering a 2' × 2' area around Sgr A*. The angular resolution is 2.5" and the spectral window covered the velocity range of -180 - 1490 km s^{-1}. We compare our results to the HCN(1-0) observations from the Owens Valley Radio Observatory (OVRO) by [1]. Because of the higher excitation of the (4-3) transition, we find that absorption by lower density foreground material is not important. By comparing the (4-3) and (1-0) results, we find that the excitation is not uniform within the circumnuclear disk (CND). The southern part is stronger in the higher-excitation line. This result was previously suggested by single-dish observations from [2]. We discuss our results in the context of the structure and stability of the CND.

1. Introduction

One of the most fascinating characteristics of the Galactic Center is the type of interstellar medium (ISM) it is composed of. This ISM is very dense, warm and turbulent, much more than in other regions of the Milky Way, and it is mostly concentrated in giant molecular clouds (GMCs) ([3]). Because the temperature is higher, the molecular species detected in this region are highly excited, and even rare species, hardly observed anywhere else in the Galaxy, are present.

Different molecular studies have demonstrated that the central 10 pc of the Galaxy have a very remarkable geometry, as can be seen in figure 1. Many different features have been detected in this region, surrounding the central black hole, Sgr A*. The close proximity of all of these structures implies that powerful interactions are taking place between them. The study of these interactions should shed light on our understanding of the dynamics of the Galactic Center.

Previous studies have addressed the study of this region using different molecular tracers, with both interferometers and single-dish antennas. [4] studied the central 10 pc of the Milky

[1] The Submillimeter Array is a joint project between the Smithsonian Astrophysical Observatory and the Academia Sinica Institute of Astronomy and Astrophysics, and is funded by the Smithsonian Institution and the Academia Sinica

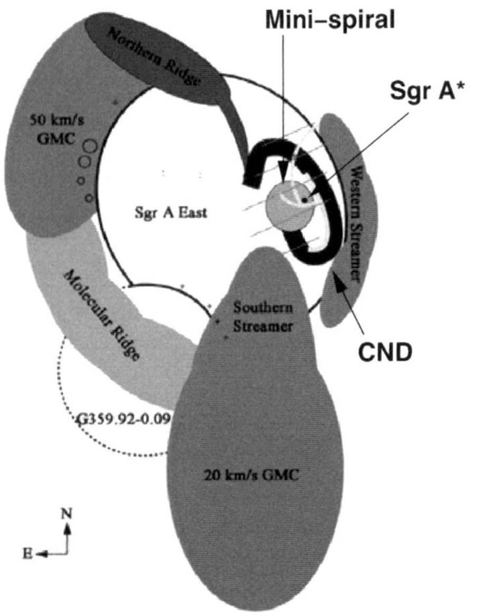

Figure 1. Schematic drawing of the central 10 pc of the Milky Way. Adapted from [4].

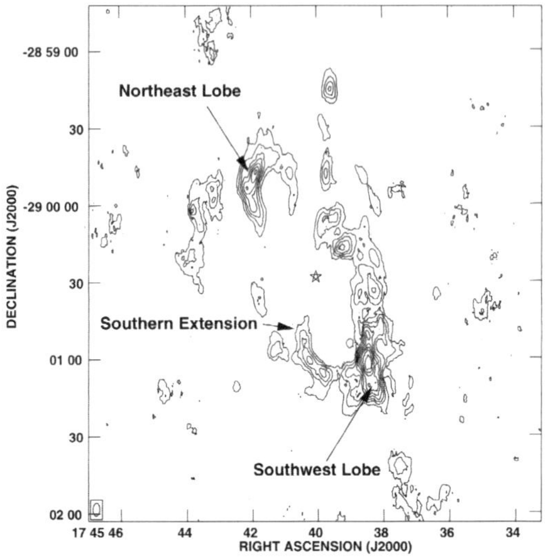

Figure 2. HCN(1-0) integrated intensity from [1]. The contour levels are in steps of 10% of the intensity peak, from 8.3 Jy beam^{-1} km s^{-1} to 74.5 Jy beam^{-1} km s^{-1}. Sgr A* is marked by a star.

Way using a high-density molecular tracer, NH_3 ($\sim 10^5$ cm^{-3}), in four different transitions, (1,1), (2,2), (3,3) and (6,6), using the Very Large Array (VLA). The lower transitions, (1,1) and (2,2), do not show very strong emission and this is mostly extended. (3,3), on the other hand, seems to trace quite well most of the structures in this region marked in figure 1. The explanation of these different behaviors is that (1,1) and (2,2) are suffering absorption from cold material in the line of sight, while the higher transitions are not affected by this problem. (6,6), which traces material at a very high temperature, ~ 400 K, is detected very close to Sgr A*, but it is mostly absent everywhere else. These results confirm that really hot and dense material can be found in the close proximity of Sgr A*. This interesting conclusion brings the necessity to expand the study using another high-density molecular tracer, hence HCN ($\sim 10^6$ cm^{-3}).

[5] used the Hat-Creek Interferometer to observe the Galactic Center in HCN(1-0). Their results showed that (1-0) traces nicely the CND but for a gap in the eastern part, being the southern part stronger than the northern, and even emission to the south, outside of the CND, is seen. A few years later [6] observed the same region with the Berkeley-Illinois-Maryland Association (BIMA), producing a very similar emission map, where the emission coming from the southern part of the CND is stronger than the emission from the northern part. The latest and most improved results in HCN(1-0), in terms of angular resolution, were obtained using OVRO by [1]. In the emission map produced by these results (Fig. 2), we can appreciate the different structures more clearly than in the previous studies, even seeing the southwest lobe and the northeast lobe as separated structures. But HCN(1-0) is affected by self-absorption, as it has been reported by [7]. Therefore, the necessity to find a transition that is not affected by this problem. HCN(4-3) fulfills this requirement. HCN(4-3) was not available before for high-resolution interferometric studies, since no array had a receiver that covered the 354 GHz frequency. Nowadays, the SMA has solved this problem, allowing us to continue our study.

2. Observations

Observations of the central 2 pc of the Milky Way (which cover Sgr A*, the ionizing gas forming the mini-spiral and the CND surrounding them both, see figure 1) in HCN(4-3) (ν = 354.505 GHz) were made with the SMA in the compact configuration on 2004 July 9 and 2005 July 1 and 7 and August 13. The primary beam for the 354 GHz frequency is 36", therefore to cover the central 2 pc ($\sim$ 2' $\times$ 2') we needed a mosaic. We produced a 37-pointing mosaic in 2004 and a 25-pointing mosaic in 2005, both centered on Sgr A*. The data were reduced using the MIR package developed for the SMA calibration and consequently imaged using MIRIAD (Multichannel Image Reconstruction, Image Analysis and Display). The synthesized beam is 2."5 $\times$ 1."7.

3. Results

Figure 3 shows the integrated intensity map obtained combining the four nights of data. There is a very clear detection of the southern part of the CND, but the emission from the northern part is almost non-existent. This result is somehow not surprising since previous single-dish studies by [2] showed the same pattern (Fig. 5): the emission from the southern part of the CND is much stronger than the emission from the northern part.

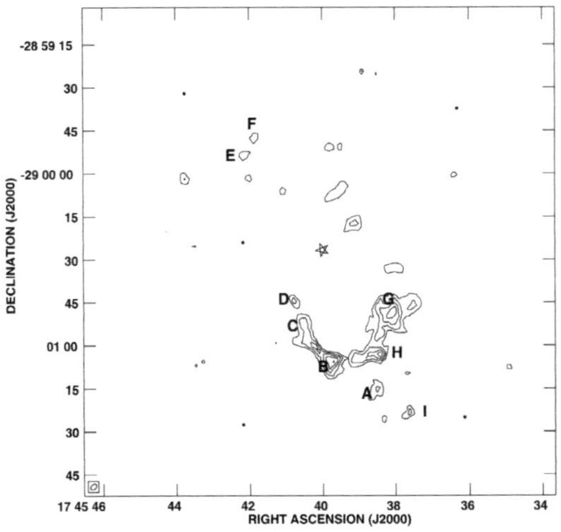

Figure 3. HCN(4-3) integrated intensity in contours. The contour levels are in steps of 3σ, from 3 to 15σ. Sgr A* is marked by a star.

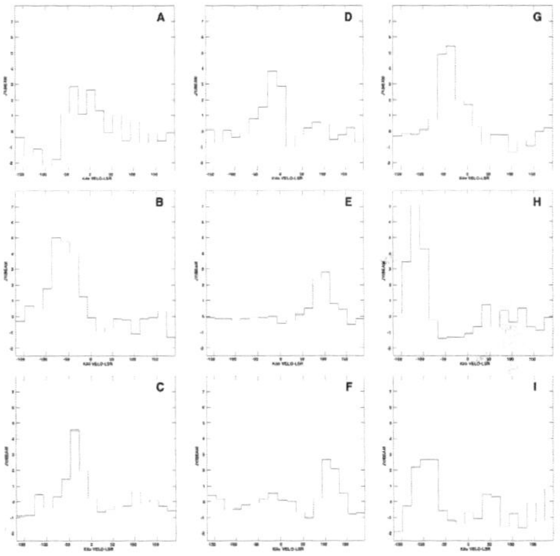

Figure 4. Spectra at the different positions marked in figure 3.

[2] explained this behavior by distinguishing the northern and the southern lobes of the CND as independent structures following the same rotation pattern. While most of the material in the southern part of the CND follows this pattern, only the high velocity material of the north does, since emission at lower velocities seems to be dominated by material that does not belong to the CND, but that it is rather in the line of sight. By plotting the spectra at the different peaks marked in figure 3 (Fig. 4), we can see that the trend followed is certainly that of a rotation around Sgr A*, with the northern part receding and the southern part approaching.

Comparing the (4-3) results with the (1-0) emission map from [1] (Fig. 6) we can see that the emission from the lower-excitation line traces the CND more efficiently, but also that (1-0), as well as (4-3), is stronger in the southern part of the CND (Fig. 2). This result means that not only the northern and the southern parts of the CND are independent structures, but also that they have different excitation levels, being the material composing the southern part of the CND warmer than the material in the northern part.

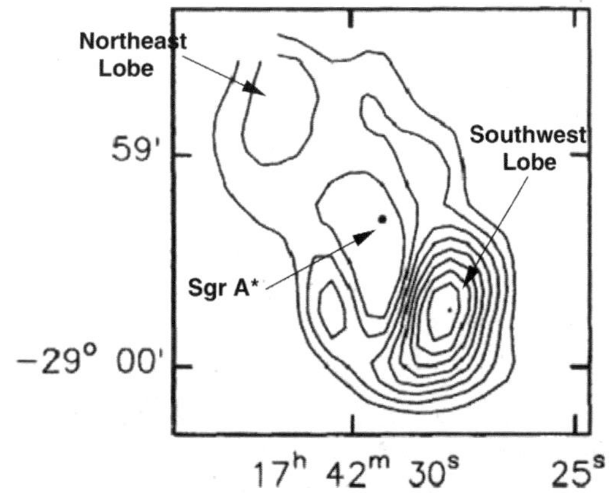

Figure 5. HCN(4-3) integrated intensity adapted from [2]. The contour levels are in steps of 25 K km s^{-1} from 75 K km s^{-1}.

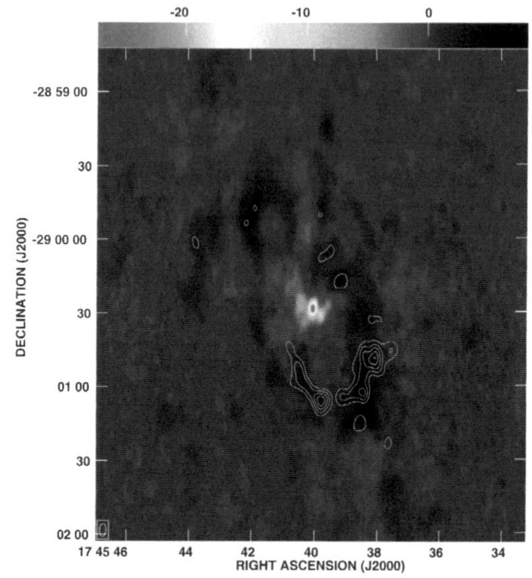

Figure 6. HCN(4-3) integrated intensity in contours and HCN(1-0) integrated intensity in false-color scale from [1]. The HCN(4-3) map has been smoothed to match the HCN(1-0) resolution. The contour levels are in steps of 3σ, from 3 to 12σ, and the last contour level is at 14σ.

We also compare our results with those obtained from another high-density tracer, NH_3, in its two highest mapped transitions, (3,3) and (6,6), ([7] and [8], respectively) observed with the Very Large Array (VLA) (Fig. 7 and 9). NH_3 traces material at higher temperature but less dense than the material traced by HCN. More precisely, NH_3(3,3) traces material at $\sim$ 125 K and (6,6) at $\sim$ 412 K, while HCN(4-3) traces material at $\sim$ 25 K, but with a density one order of magnitude higher than that of the material traced by NH_3. The first remarkable feature observed in both figures (7 and 9), is that NH_3, in both transitions, is mostly absent from the northern part of the CND, a behavior also observed in HCN(4-3). Apart from that similarity, the maps look really different. NH_3(3,3) seems to trace the western part of the CND, but it is not very strong, in fact, it is stronger towards the southeast of the CND. It seems to trace the southern streamer that connects the "20 km s^{-1} GMC" with the CND (see figure 1), but it disappears once it reaches the CND. In order to see if the NH_3(3,3) and the HCN(4-3) are really "connected", we plot the spectra at the same peaks or at least close by. We can observe that the peaks are not exactly coincident in the western peak (named H). The emission peak for NH_3(3,3) is red-shifted, indicating that HCN(4-3) and NH_3(3,3) are not tracing the same material. The spectra corresponding to the eastern peaks (since they are not really coincident for NH_3(3,3) and HCN(4-3), named K and L) show that the NH_3(3,3) does not have a large velocity gradient, unlike the HCN(4-3), and it is also red-shifted. This means that the material traced by both molecular transitions is probably unrelated and what we see as a "connection" between the CND and the southern streamer might be a projection effect.

NH_3(6,6) is mostly absent from the southern lobe of the CND, but it appears on the eastern

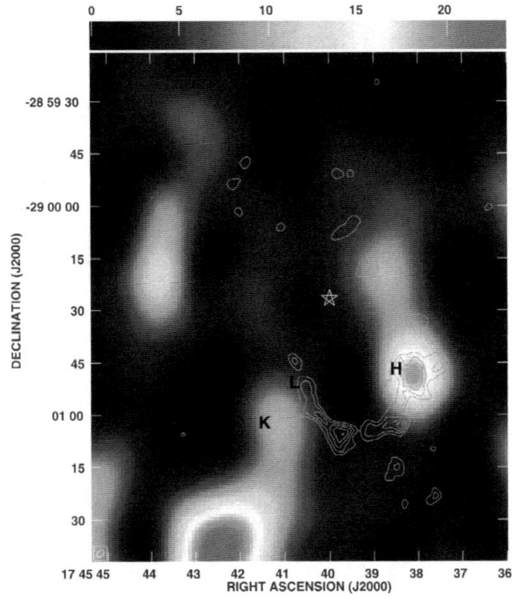

Figure 7. HCN(4-3) integrated intensity in contours and NH_3(3,3) integrated intensity in false-color scale from [7]. Contour levels are as in figure 3. Sgr A* is marked by a star.

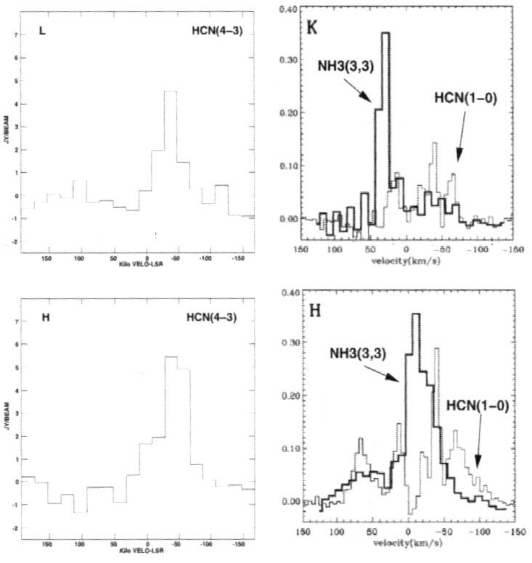

Figure 8. Spectra at the positions marked in figure 7. The left-side spectra correspond to our HCN(4-3) results. The right-side spectra are NH_3(3,3) spectra in thick lines and HCN(1-0) spectra in thin lines adapted from [7] and [6] (the self-absorption suffered by HCN(1-0) is clearly seen in the bottom-right spectrum).

part of the CND, looking as a continuation towards the north of the southern extension of the CND. But, unlike HCN(4-3), NH_3(6,6) is detected inside the CND. This situation could happen because the conditions inside the CND allow gas at 412 K to survive, but not at 25 K. Looking at the spectra, we can observe that the spectra of the western peak (named E), both in HCN(4-3) and NH_3(6,6), have a similar linewidth and are centered at the same velocity, therefore NH_3(6,6) and HCN(4-3) seem to be tracing the same kind of material. The situation is completely different for the eastern peak (named D). The NH_3(6,6) line is very broad, which means that it is being affected by some kind of interaction, most likely related to the central black hole, while the HCN(4-3) seems to be merely following the rotation pattern mentioned previously.

4. Conclusions

We have succesfully detected and mapped the emission corresponding to a high-density molecular tracer in a high-excitation state, HCN(4-3), in the Galactic Center, using the SMA. The detected emission shows that the CND might be composed by material in different excitation states, being the northern part of the CND colder than the southern part. Also, the connection that seems to be present between the southern streamer and the CND is most likely a projection effect. HCN(4-3) is not detected inside the CND, unlike the less dense but warmer NH_3(6,6). The interactions that the material inside the CND might be undergoing could prevent the colder gas from surviving there.

Acknowledgments

During the development of this study M.M.-C. has been supported by an Academia Sinica Institute of Astronomy and Astrophysics fellowship and a Smithsonian Institution visiting

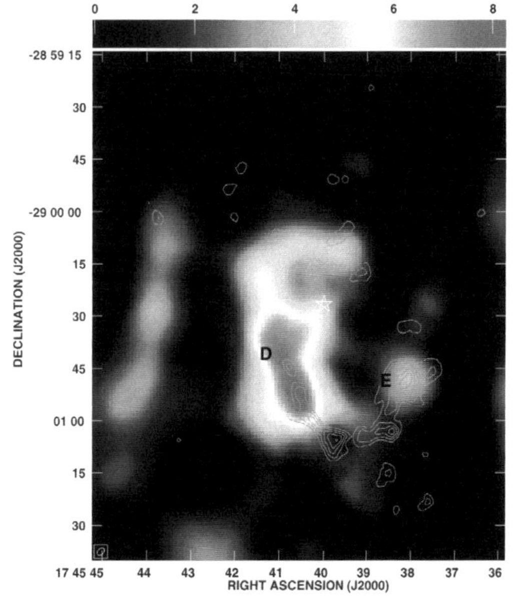

Figure 9. HCN(4-3) integrated intensity in contours and $NH_3(6,6)$ integrated intensity in false-color scale from [8]. Contour levels are as in figure 3. Sgr A* is marked by a star.

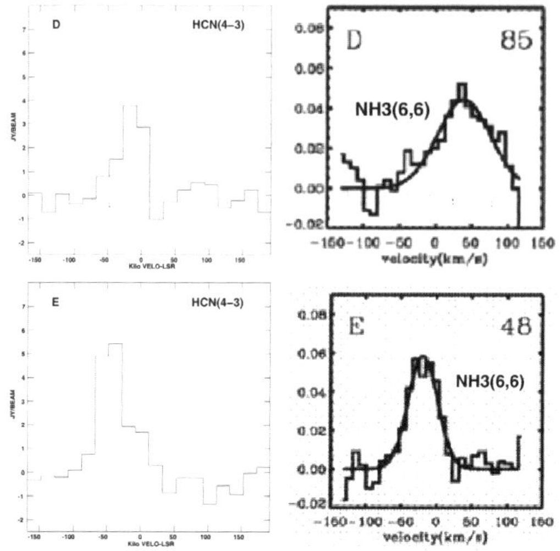

Figure 10. Spectra at the positions marked in figure 9. The left-side spectra correspond to our HCN(4-3) results. The right-side spectra are $NH_3(6,6)$ spectra adapted from [8].

student grant.

References

[1] Christopher, M.H., Scoville, N.Z., Stolovy, S.R. and Yun, M.S. 2005 *ApJ* **622** 346

[2] Marshall, J., Lasenby, A.N. and Harris, A.I. 1995 *MNRAS* **277** 594

[3] Hüttemeister, S., Dahmen, G., Mauersberger, R. Henkel, C., Wilson, T.L. and Martín-Pintado, J. 1998 *A&A* **334** 646

[4] Herrnstein, R.M. and Ho, P.T.P. 2005 *ApJ* **620** 287

[5] Güsten, R., Genzel, R., Wright, M.C.H., Jaffe, D.T., Stutzki, J. and Harris, A.I. 1987 *ApJ* **318** 124

[6] Wright, M.C.H., Coil, A.L., McGary, R.S., Ho, P.T.P. and Harris, A.I. 2001 *ApJ* **551** 254

[7] McGary, R.S., Coil, A.L. and Ho, P.T.P. 2001 *ApJ* **559** 326

[8] Herrnstein, R.M. and Ho, P.T.P. 2002 *ApJ* **579** L83

Institute of Physics Publishing
doi:10.1088/1742-6596/54/1/006
Journal of Physics: Conference Series **54** (2006) 35–41
Galaxy Center Workshop 2006

Physical conditions and molecular chemistry of the Central Molecular Zone

Nemesio J. Rodríguez-Fernández[1]

[1] Observatoire de Bordeaux (L3AB/OASU), CNRS/Université Bordeaux 1, BP 89, 33270 Floirac, France

E-mail: nemesio.rodriguez@obs.u-bordeaux1.fr

Abstract. We discuss the physical conditions of the different kinematical components of the Central Molecular Zone. In particular we compare the properties of the clouds moving with in elongated orbits along the Galactic bar with those of the well-known Galactic center ring (GCR) clouds (Sgr A, Sgr B2,...). We show that all the components contain dense clouds that can withstand the tidal shear. The SiO abundance in the clouds with non-circular velocities is high ($\sim 10^{-8}$), in perfect agreement with that of the GCR clouds. We discuss the role of the UV radiation and C-shocks in the heating of the neutral gas and the high abundances of some molecules like SiO. The SiO emission in the clouds moving in elongated trajectories is probably due to the cloud collisions expected in the inner regions of a bar.

1. Morphology and kinematics of the Central Molecular Zone

Figure 1 shows a schematic view of the longitude-velocity diagram (hereafter lv-diagram) of the Galactic center (GC) as observed in CO [3, 7, 24, 5]. The Central Molecular Zone (CMZ) is the region from $l = -1°$ to $l = 1.5°$ (the $l = 1.3°$-complex) [23]. The important cloud complexes from Sgr C to Sgr D are located approximately along a line passing through the origin. Therefore, in a face-on view of the Galactic center the Sgr complexes would be located in a circular ring (Fig. 2). Hereafter, we refer to this structure as the Galactic center ring (GCR). However, the GCR clouds and the $l = 1.3°$-complex are not the only cloud components in the GC. The lv-diagram also exhibits many structures corresponding to clouds moving in non-circular trajectories like the Clump 2, the Connecting Arm, or the structures that we have labeled as J, K, L, M and N (Fig. 1). The easiest explanation for the non-circular trajectories is the gas response to a bar potential [4, 8].

Numerical simulations of the gas dynamics in the inner region of barred galaxies [1, 8] show that the gas follows elongated orbits along the bar major axis until the contact point where these gas clouds interact with the clouds in the nuclear disk (GCR). The gas is then sprayed and it collides with the material on the opposite side of the bar major axis giving rise to shocks in the molecular cloud component and to the characteristic dust lanes that are observed near the leading edges of the bars in external galaxies like NGC 1300, NGC 1530 or NGC 6951. In the GC, the near-side dust lane has been identified as the Connecting Arm [8]. The structure K, which seems to be linked to the Connecting Arm at $l \sim 3°$ (Fig. 1), represents gas clouds that move in the decelerating section of elongated orbits, i.e, gas that does not fall in the nuclear ring at the first passage around the interaction region (see Fig. 2). The structure J has nearly the same inclination as that of the Connecting Arm (Fig. 1), and it could be a kind of small dust lane, but in contrast with the Connecting Arm and the structure K, the gas in the structure J will easily interact with the material of the nuclear disk in the area of the $l = 1.3°$-complex (see Fig. 2). Thus, the $l = 1.3°$-complex would be the contact point of the dust lane and the GCR as proposed by previous works [11, 8]. The clump at $l = 5°$ and the Clump 2 are gas clouds that are about to enter the dust lane shock. The lv-diagram at $l < 0$ is more difficult to interpret. Due to perspective effects the far side dust lane is elongated along the line of sight and thus it is a vertical structure in the lv-diagram close to the extremity of the nuclear ring (see Fig. 23 of reference [8]). Therefore, the lv-diagram is not expected to be symmetric. The structure M would be part of the gas spray due to the interaction of the far-side dust lane and the GCR.

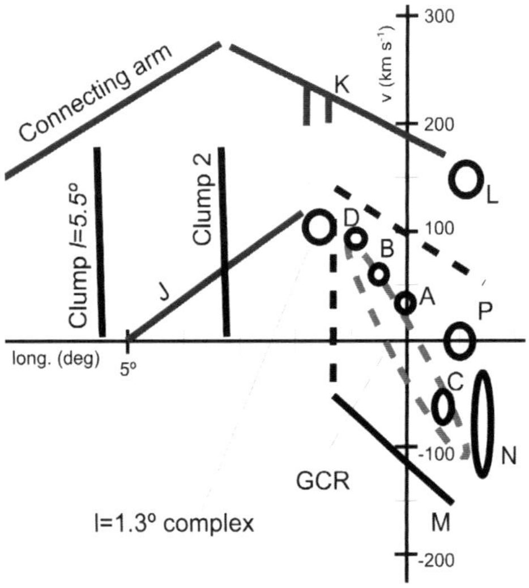

Figure 1. Schema illustrating the main kinematical structures in the longitude-velocity diagram of the inner degrees of the Galaxy. The labels A to D stand for Sgr A to Sgr D complexes. See the text for an explanation of the other labels.

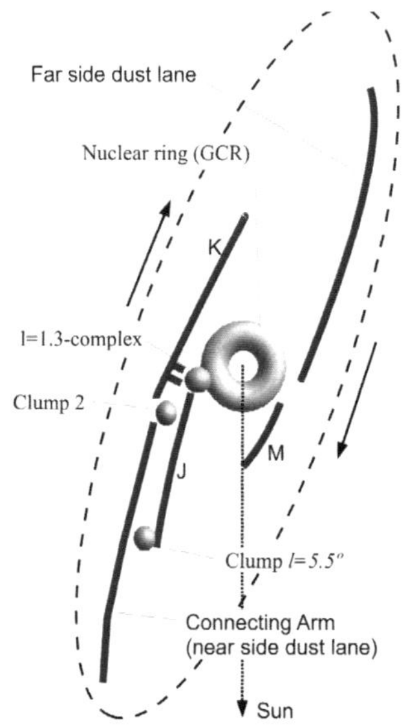

Figure 2. Schema of a plausible face on view of the Galactic center.

2. Physical conditions and molecular chemistry

2.1. The Galactic center ring

The physical conditions of the neutral gas in the GC have been studied by observing high density tracers like CS [3, 39] or HCN [13] and by multi-level studies of NH_3 [11], H_2 [28, 30] or CO [34, 16]. These works, among others, have shown that the clouds in the GCR and the $l = 1.3°$-complex are dense (10^4 cm^{-3}) and warm (up to $\sim$ 150 K). Fine structure lines observations of several elements in different ionization stages have allowed to study in detail the heating mechanisms of the warm molecular gas [30, 31]. The CMZ is permeated by hot star radiation ($\sim$ 35000 K) like that present in the vicinity of the Arches and the Quintuplet clusters in the Radio Arc region [29]. This radiation ionizes the surface of the molecular clouds [31]. In the interface of the ionized gas and the well-shielded molecular gas there are mainly neutral photo-dissociation regions (PDRs). The intensity of the main cooling lines ([CII] 158 μm, [OI] 63 μm) and the temperature of the warm H_2 ($\sim$ 150 K) are in perfect agreement with PDR models with an incident far-UV field 1000 times higher than the local interstellar radiation field and a density of 1000 cm^{-3} [30]. The far-infrared continuum emission from the warm ($\sim$ 35 K) dust component also agrees with this scenario of heating in PDRs. However, only a small fraction (10-30 %) of the total column density of warm H_2 can be explained by the PDR models [30]. One possibility to explain the heating of the large quantity of molecular gas measured in the GC clouds is the presence of magneto-hydrodynamic continuous (C) shocks. An important parameter in C-shock models is $b = B/\sqrt{n}$, where B is the intensity of the magnetic field expressed in μG and n is the hydrogen density in cm^{-3}. We have compared the measured intensities of the lowest H_2 pure-rotational lines with the C-shock models by Pineau des Forets and collaborators [6]. Figure 3 shows that C-shocks models with $b = 10$ (for instance $B = 300$ μG and $n = 1000$ cm^{-3}) can explain the observed intensity of the H_2 lines in the GC (see also [30]).

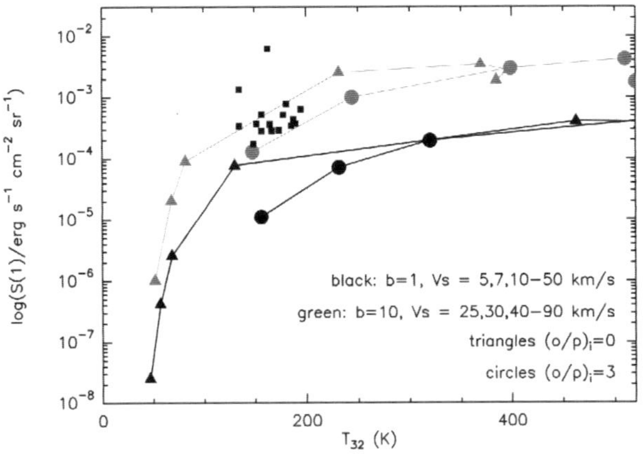

Figure 3. C-shocks model predictions for the intensity of the H_2 S(1) line as a function of the excitation temperature derived from the S(1) and the S(0) lines for different shock velocities (v_s), $b = B(\mu G)/\sqrt{n(\text{cm}^{-3})}$ (black lines: 1, light lines: 10), and initial ortho/para ratio (triangles: 0, circles: 3). The solid squares represent the intensity measured for the GC clouds [28]

The GCR clouds also present widespread emission and high abundances of molecules like SiO, SO, CH_3OH, C_2H_5OH and other organic molecules that have been ejected from the dust grains. The abundances of these molecules also imply the presence of shocks in the CMZ.

(i) The quantity of warm NH_3 measured in the GC clouds [11] cannot be explained by thermal evaporation of the grain mantles. First, the dust temperature is lower than 35 K [30]. Second, this molecule is easily dissociated in the presence of UV radiation. Taking into account the photo-dissociation rates and the NH_3 abundances in the CMZ one derives that a few 10^{22} cm^{-2} of H_2 should be heated every $\sim$ 10 yr, which is not realistic [30]. The easiest way to explain the NH_3 abundance in gas phase is grain mantle erosion by mechanical processes.

(ii) The abundance of CH_3OH, C_2H_5OH and other complex organic molecules is high in the CMZ clouds that are not associated with thermal continuum sources [21, 14, 25]. In contrast, their abundances decrease by more than a factor 10 in the clouds located in the vicinity of thermal continuum sources [21, 25].

(iii) The SiO is particularly interesting since the abundance of this molecule shows extreme variations in the Galaxy from $< 5\,10^{-12}$ in dark clouds [43] to $10^{-11} - 10^{-10}$ in diffuse translucent clouds and in spiral arms clouds [9, 15]. The SiO abundance is also very low ($\sim 10^{-11}$) in PDRs [35]). In contrast, the SiO abundance in shocked regions associated to star formation is $\sim 10^{-8}$ [17, 2]. In these regions, the Si and SiO abundances in the gas phase are increased by the processing of the dust grains by shock waves [36]. The SiO abundances in GCR clouds are $\sim 10^{-8}$ [18, 11], as high as those found in shocked regions.

Shocks can explain the molecular chemistry and the high temperatures of the neutral gas in the GCR, and the $l = 1.3°$-complex. However the origin of the shocks is not clear. For instance, it could be related to the interaction of evolved massive stars with the surrounding interstellar medium [20]. Alternatively, cloud-cloud collisions induced by the large scale dynamics of the

Galaxy could also play a role [42, 11, 21, 33, 27, 30]. Perhaps the best evidence of the effect of the Galactic dynamics on the physical conditions of the GC molecular clouds is the high SiO abundance measured in the $l = 1.3°$-complex [11], since this region seems to be the interaction region between gas moving in elongated orbits along the bar and the gas moving in circular orbits in the GCR [8].

2.2. Clouds with non-circular velocities

Different energetic phenomena are taking place in the inner regions of the GC (birth and death of massive stars, energetic radiation, shocks,...) making it very complex to interpret the observations of the GCR clouds. The study of less complex regions can give us important hints on the origin of the physical conditions of the CMZ. The clouds with non-circular velocities do not present signs of star-formation [24, 32]. Furthermore, as discussed in Sect. 1, their physical conditions can be affected by the global gas dynamics. However, in contrast to the GCR cloud complexes, the physical conditions of the gas with non-circular velocities that constitute the characteristic shape of the lv-diagram are not well-known. The few large scale maps that have been used to study the physical conditions of the ensemble of the GC molecular gas show that the CO(2-1) to CO(1-0) ratio is rather constant (~ 1) in the GC clouds (including Clump 2) [24, 34]. A few clouds of the upper-right corner of the lv-diagram have been observed in SiO and NH_3 [12, 11] and in H_2 and fine structure lines [30]. Their densities and temperatures are similar to the GCR clouds.

Recently, it has been presented the results of a survey of CS(2-1) and SiO(2-1) in 161 clouds belonging to all the kinematical components of the Galactic center in the longitude range $3.5° > l > -1.5°$ [32]. The results are displayed graphically in Fig. 4. Blue triangles and bars represent the peak velocity and the full width at half maximum (FWHM) of the sources detected in CS. Black points and bars show the peak velocity and FWHM of the sources detected in CS and SiO.

2.2.1. Density In addition to the GCR and the $l = 1.3°$-complex, CS emission is detected in all the kinematical structures of the GC lv-diagram (Fig 4) like the faint structures that constitute the upper and lower edges of the lv-diagram (K, M, R), including the vertical appendix of the component K at $l \sim 1.7°$, which can be the signature of mass transfer from the elongated orbits to the GCR [8] or the clouds in the right edge of the lv-diagram (L, N, P) and those in the structure J. These observations demonstrate that there are dense CS-emitting clouds ($n > 10^4$ cm^{-3}) in all the kinematical components seen in the CO lv-diagram and not only in the GCR and the $l = 1.3°$-complex clouds detected by previous works [3]. The critical density to withstand the tidal shear at the galactocentic radius of the clouds with non-circular velocities is $\sim 10^3$ cm^{-3} [38]. Therefore, the CS clouds in non-circular orbits are gravitationally bound and the dust lanes do not contain diffuse gas only [32]. However, there are no signs of star formation in the dust lanes. Indeed, the lack of star formation along the dust lanes is common in the observations of other barred galaxies, where there is star formation in the nuclear ring and at the extremities of the bar, but there are no signs of star formation in the dust lanes along the bar [40, 26, 37]. The inhibition of the star formation can be an effect of the high velocities, relative to the gas in the dust lanes, of the dense clouds whose orbits intersect the dust lanes [40]. These velocities are much higher in barred than in normal galaxies. The high velocity can cause a quick compression of the clouds followed by a rapid expansion which disperses the majority of the cloud. This mechanism has also been invoked to explain the observations of NGC 1530 [26] and it could also inhibit the star formation in the gas with non-circular velocities in the GC [32].

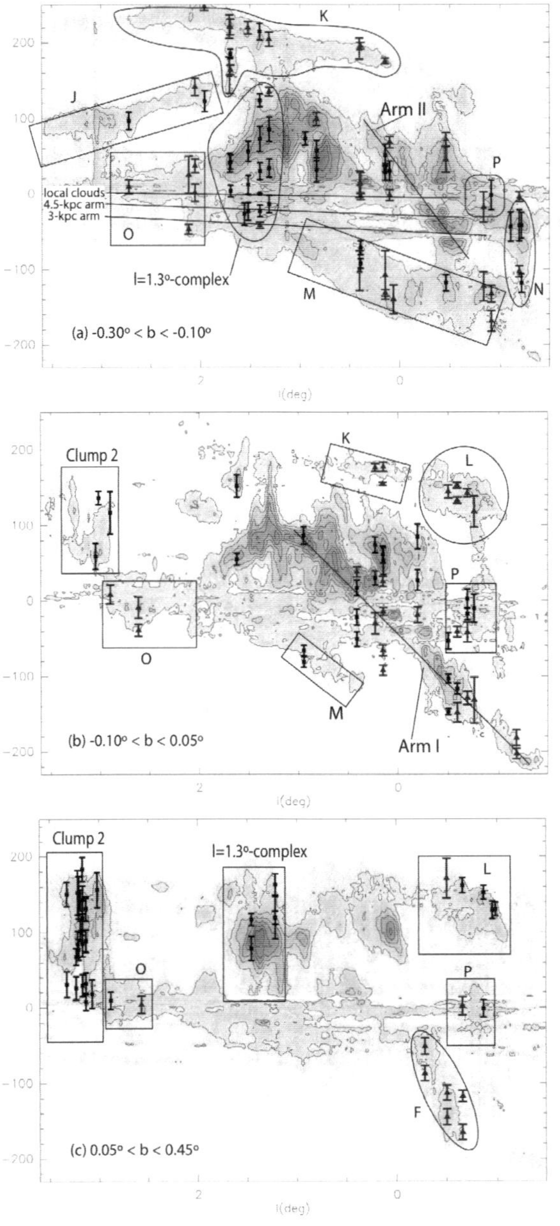

Figure 4. In grey-levels and contours we show the longitude-velocity diagram obtained with the CO(1-0) data by [3] at different latitude ranges. We also indicate the different kinematical features discussed in this paper (note that not all the features that are seen in different panels are labeled in all of them). Triangles and bars represent the peak velocity and the full width at half maximum (FWHM) of the sources detected in CS. Black points and bars show the peak velocity and FWHM of the sources that have been detected in CS and SiO.

2.2.2. SiO abundances SiO(2-1) emission has also been detected in at least a few clouds of all the kinematical components [32]. The SiO(2-1) line has been detected in all the observed clouds of the $l = 1.3°$-complex and Clump 2. The SiO line has also been detected in two clouds of the component K: one cloud near the CO maximum at $l \sim 1.4$ and one cloud in the vertical appendix at $l \sim 1.7°$. SiO emission has also been detected in the lower edge (component M) and the negative longitude extremity of the lv-diagram (components P and N, including a few clouds with different velocities along the structure N) as well as towards several positions in the structures J and L. In most of the clouds where the SiO line has not been detected the CS is indeed rather weak and the upper limits to the SiO/CS line ratio are not significant.

The SiO abundances in the clouds with non-circular velocities with detected SiO are $(0.5-8)\,10^{-9}$ [32], including those in the structures K and M. These abundances are in agreement with those measured in the GCR and the $l = 1.3°$-complex [11, 18, 19]. For one source in the Sgr B2 area ($l = 0.83°$, $b = -0.18°$, $v = 57$ km s^{-1}) and two in the feature P we have even measured a SiO abundance of $\sim 10^{-8}$, which is comparable to the highest abundances derived in the $l = 1.3°$-complex [11].

The high SiO abundances measured in the GC clouds with non-circular velocities should be due to the strong shocks that created the dust lanes themselves. This scenario has also been proposed to interpret the detection of HNCO, CH_3OH and SiO emission along the bar of IC 342 [22, 41].

3. Conclusions

We have compared the properties of the clouds moving in elongated orbits along the Galactic bar with the better known clouds of the Galactic center ring (GCR). The gas of the GCR is dense, warm and exhibits a rich chemistry. The UV radiation can only account for the heating of 10-30 % of the warm gas. The rest of the warm gas can be heated by C-shocks if the magnetic field intensity is higher than 300 μG. The high abundance of molecules linked to the grain chemistry also points to the presence of shock waves. A recent survey has shown that all the kinematical components of the Galactic center contain dense clouds that can withstand the tidal shear. This includes the structures with non-circular velocities like the dust lanes along the bar. However, in the GC there are no signs of star formation except in the nuclear ring (the GCR). The high relative velocities and shear expected in other kinematical structures, in particular in those associated with the dust lanes, could inhibit the star formation. The high measured abundances of one of the most representative molecules of the GC shock chemistry (SiO) in the cloud components associated with the dust lanes can be due to the large scale shocks that created the structures themselves.

References

[1] Athanassoula, E. 1992, MNRAS, 259, 345
[2] Bachiller, R. & Perez Gutierrez, M. 1997, ApJ, 487, L93
[3] Bally, J., Stark, A. A., Wilson, R. W., & Henkel, C. 1987, ApJS, 65, 13
[4] Binney, J., Gerhard, O. E., Stark, A. A., Bally, J., & Uchida, K. I. 1991, MNRAS, 252, 210
[5] Bitran, M., Alvarez, H., Bronfman, L., May, J., & Thaddeus, P. 1997, A&AS, 125, 99
[6] Cabrit, S., Lefloch, B., Cernicharo, J., Pineau Des Forêts, G., Lebourlot, J., & Flower, D. 2004, The Dense Interstellar Medium in Galaxies, 587
[7] Dahmen, G., Huettemeister, S., Wilson, T. L., Mauersberger, R., Linhart, A., Bronfman, L., Tieftrunk, A. R., Meyer, K., Wiedenhoever, W., Dame, T. M., Palmer, E. S., May, J., Aparici, J., & Mac-Auliffe, F. 1997, A&AS, 126, 197
[8] Fux, R. 1999, A&A, 345, 787
[9] Greaves, J. S., Ohishi, M., & Nyman, L.-A. 1996, A&A, 307, 898
[10] Güsten, R. 1989, in IAU Symp. 136: The Center of the Galaxy, ed M. Morris, 89
[11] Hüttemeister, S., Dahmen, G., Mauersberger, R., Henkel, C., Wilson, T. L., & Martin-Pintado, J. 1998, A&A, 334, 646
[12] Hüttemeister, S., Wilson, T. L., Bania, T. M., & Martin-Pintado, J. 1993, A&A, 280, 255
[13] Jackson, J. M., Heyer, M. H., Paglione, T. A. D., & Bolatto, A. D. 1996, ApJ, 456, L91+
[14] Lis, D. C., Serabyn, E., Zylka, R., & Li, Y. 2001, ApJ, 550, 761
[15] Lucas, R. & Liszt, H. S. 2000, A&A, 355, 327
[16] Martin, C. L., Walsh, W. M., Xiao, K., Lane, A. P., Walker, C. K., & Stark, A. A. 2004, ApJS, 150, 239
[17] Martin-Pintado, J., Bachiller, R., & Fuente, A. 1992, A&A, 254, 315
[18] Martín-Pintado, J., de Vicente, P., Fuente, A., & Planesas, P. 1997, ApJ, 482, L45
[19] Martín-Pintado, J., de Vicente, P., Rodríguez-Fernández, N. J., Fuente, A., & Planesas, P. 2000, A&A, 356, L5
[20] Martín-Pintado, J., Gaume, R. A., Rodríguez-Fernández, N., de Vicente, P., & Wilson, T. L. 1999, ApJ, 519, 667
[21] Martín-Pintado, J., Rizzo, J. R., de Vicente, P., Rodríguez-Fernández, N. J., & Fuente, A. 2001, ApJ, 548, L65

[22] Meier, D. S. & Turner, J. L. 2005, ApJ, 618, 259
[23] Morris, M., & Serabyn, E. 1996, ARA&A, 34, 645
[24] Oka, T., Hasegawa, T., Hayashi, M., Handa, T., & Sakamoto, S.1998a, ApJ, 493, 730
[25] Requena-Torres, M. A., Martín-Pintado, J., Rodríguez-Franco, A., Martín, S., Rodríguez-Fernández, N. J., & de Vicente, P. 2006, ArXiv Astrophysics e-prints, arXiv:astro-ph/0605031
[26] Reynaud, D. & Downes, D. 1998, A&A, 337, 671
[27] Rodríguez-Fernández, N. J., Martín-Pintado, J., de Vicente, P., Fuente, A., Hüttemeister, S., Wilson, T. L., & Kunze, D. 2000, A&A, 356, 695
[28] Rodríguez-Fernández, N. J., Martín-Pintado, J., Fuente, A., de Vicente, P., Wilson, T. L., & Hüttemeister, S. 2001, A&A, 365, 174
[29] Rodríguez-Fernández, N. J., Martín-Pintado, J., de Vicente, 2001, A&A, 377, 631
[30] Rodríguez-Fernández, N. J., Martín-Pintado, J., Fuente, A., & Wilson, T. L. 2004, A&A, 427, 217
[31] Rodríguez-Fernández, N. J., Martín-Pintado, J., 2005, A&A, 429, 923
[32] Rodriguez-Fernandez, N., Combes, F., Martin-Pintado, J., Wilson, T. L., & Apponi, A. 2006, ArXiv Astrophysics e-prints, arXiv:astro-ph/0603851
[33] Sato, F., Hasegawa, T., Whiteoak, J. B., & Miyawaki, R. 2000, ApJ, 535, 857
[34] Sawada, T., Hasegawa, T., Handa, T., Morino, J.-I., Oka, T., Booth, R., Bronfman, L., Hayashi, M., Luna Castellanos, A., Nyman, L.-Å., Sakamoto, S., Seta, M., Shaver, P., Sorai, K., & Usuda, K. S. 2001, ApJS, 136, 189
[35] Schilke, P., Pineau des Forêts, G., Walmsley, C. M., & Martín-Pintado, J. 2001, A&A, 372, 291
[36] Schilke, P., Walmsley, C. M., Pineau des Forets, G., & Flower, D. R. 1997, A&A, 321, 293
[37] Sheth, K., Regan, M. W., Vogel, S. N., & Teuben, P. J. 2000, ApJ, 532, 221
[38] Stark, A. A., Martin, C. L., Walsh, W. M., Xiao, K., Lane, A. P., & Walker, C. K. 2004, ApJ, 614, L41
[39] Tsuboi, M., Handa, T., & Ukita, N. 1999, ApJS, 120, 1
[40] Tubbs, A. D. 1982, ApJ, 255, 458
[41] Usero, A., Garcia-Burillo, S., Martin-Pintado, J., Fuente, A., & Neri, R. 2005, ArXiv Astrophysics e-prints
[42] Wilson, T. L., Ruf, K., Walmsley, C. M., Martin, R. N., Batrla, W., & Pauls, T. A. 1982, A&A, 115, 185
[43] Ziurys, L. M., Friberg, P., & Irvine, W. M. 1989, ApJ, 343, 201

Institute of Physics Publishing
doi:10.1088/1742-6596/54/1/007

Journal of Physics: Conference Series **54** (2006) 42–46
Galaxy Center Workshop 2006

NH_3 observations of the Central Molecular Zone with Kagoshima 6 meter radio telescope

Toshihiro Handa[1], Toshihiro Omodaka[2], Takumi Nagayama[2], Hayati Bebe Iahak[2] and Naohito Matsuyama[2]

[1] University of Tokyo, Japan
[2] Kagoshima University, Japan

E-mail: handa@ioa.s.u-tokyo.ac.jp

Abstract. We made a wide survey in two Ammonia lines of the major part of the Central Molecular Zone in the Galactic center region. The ammonia emission can be detected at most of the observed positions, which means ammonia emission is extended over the Central Molecular Zone. We identify 4 clouds, which correspond Sgr A 20-km s^{-1} cloud, Sgr A 50-km s^{-1} cloud, sgr B cloud complex, and the 1.3-degree cloud. Besides them diffuse emission can be detected over the Central Molecular Zone. The intensity ratio of the (2,2) transition to (1,1) transition is uniform over most of the observed region and is about 0.7–0.8, which means that the molecular gas is 24–42 K. It means most dense gas in the Galactic center region is as cool as majority of dust there. This cool temperature is consistent to the values derived previously, although their observed coverage was smaller than ours. On $l - v$ plane the NH_3 emitting region is surrounded by clouds with the high intensity ratio of CO (2–1) to CO (1–0) lines. It means the dense molecular gas is surrounded by hot molecular gas in 10–100 pc scale.

1. Introduction

There is a huge molecular cloud complex in the Galactic center region between $-1^\circ \leq l \leq +1.5\,\mathrm{deg}$ called the Central Molecular Zone (CMZ). The CMZ is rich in dense molecular gas and well traced in dense gas tracer such as CS [1]. Several observations in radio and infrared have been done and gas and dust temperatures are estimated. However, there is a discrepancy between gas and dust temperatures; most observations show gas temperature is hotter than dust temperature. The gas temperature is as hot as 60–70 K from millimeter line observations [2], and the dust temperature is as cool as 15–22 K from submillimeter and infrared observations [3, 2].

Ammonia line observations gave a big hint to solve this discrepancy. Hüttemeister et al. shows that temperature from NH_3 (1,1) and (2,2) transitions is about 25 K and temperature from NH_3 (4,4) and (5,5) transitions si about 200 K [4]. However, there is not NH_3 observations which cover the whole CMZ [5, 6]. Here, we present an ammonia map of the whole CMZ.

2. Observations

We have observed a main part of the CMZ using the Kogoshima 6-m telescope, simultaneously observing at the spectral lines of NH_3 (2,2) line at 23.694 GHz and (1,1) line at 23.722 GHz. The full width at half maximum (FWHM) beam size is $9.5'$. The observed region is a rectangular area of $-1^\circ \leq l \leq +1.625^\circ$, and $-0.315^\circ \leq b \leq +0.250^\circ$, which covers most of the CMZ.

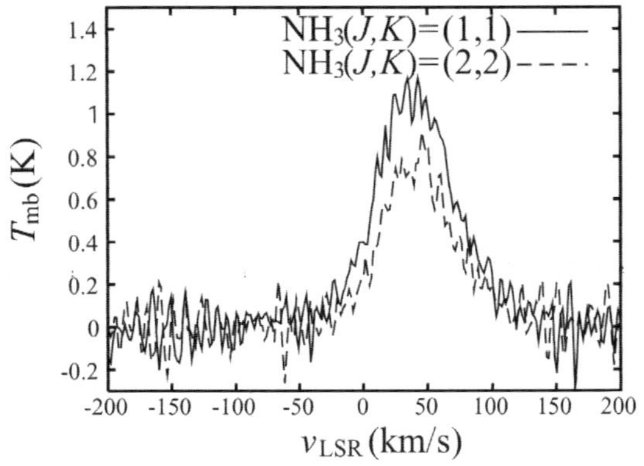

Figure 1. A profiles in NH_3 (1,1) and (2,2) transitions at $(l, b) = (0.750°, -0.125°)$. The velocity resolution is 1.6 km s^{-1}, which is the original resolution of the spectrometer before smoothing process.

The spacing of the observation grid is 0.125°, which corresponds to 18.5 pc at the distance to the galactic center, 8.5 kpc. The main beam efficiency at 23 GHz is 0.59. The velocity resolution at 23 GHz is 1.5 km s^{-1}, although we smoothed all spectra to be with 5 km s^{-1} resolution. The typical rms noise level is 0.047 K at 5 km s^{-1} resolution.

3. Results

3.1. Profiles

An NH_3 inversion transition line has 5 hyperfine components. The outermost hyperfine components are about 20 km s^{-1} separated from the main component and the inter components are about 8 km s^{-1}.

In the Galactic center region molecular gas shows violent motion within a 9′-beam. It makes hyperfine components to blend into a single line. Figure 1 shows profiles in both NH_3 transitions of Sgr B cloud. Apparent linewidths in both transitions show wider than 50 km s^{-1}. Although we have no chance to separate the satellite lines, such wide lines make it simple. In an optically thin case, apparent NH_3 line widthes in (1,1) and (2,2) shows almost the same apparent linewidths, when intrinsic linewidths are wider than 40 km s^{-1}. (figure 4b in [8]). Actually our profile shapes of these two lines are very similar. Therefore, we can use a blended line as a single line and the gas temperature can be derived from the ratio of (2,2) to (1,1) transitions.

3.2. Overall distribution

Figure 2 shows integrated intensity maps in NH_3 (1,1) and (2,2) lines over -200 km s^{-1} to +200km s^{-1}. The overall distribution is similar to that in CO with the same spatial resolution [7]. Three peaks corresponds to Sgr A clouds (+20 km s^{-1} and +50 km s^{-1} clouds), Sgr B cloud and 1.3-degree cloud.

although most flux is concentrated to these clouds, some diffuse emission is observed. It extends over the whole CMZ. The

3.3. Four major cloud complexes

In the $l - b - v$ data cube we identify 4 clouds, which correspond Sgr A 20-km s^{-1} cloud, Sgr A 50-km s^{-1} cloud, sgr B cloud complex, and the 1.3-degree cloud. Due to violent gas motion in the Galactic center region and our large beamsize, the satellite lines of ammonia emission are highly blended to its main line. Therefore, we cannot estimate opacity of the lines. However, we can estimate the intrinsic line width if we assume that the line is optically thin using the method developed by McGray and Ho [8].

Using the estimated line width we derived the virial masses. From integrated intensity we derive the LTE masses with assumption that NH_3 lines are optically thin and $n(NH_3)/n(H_2) =$

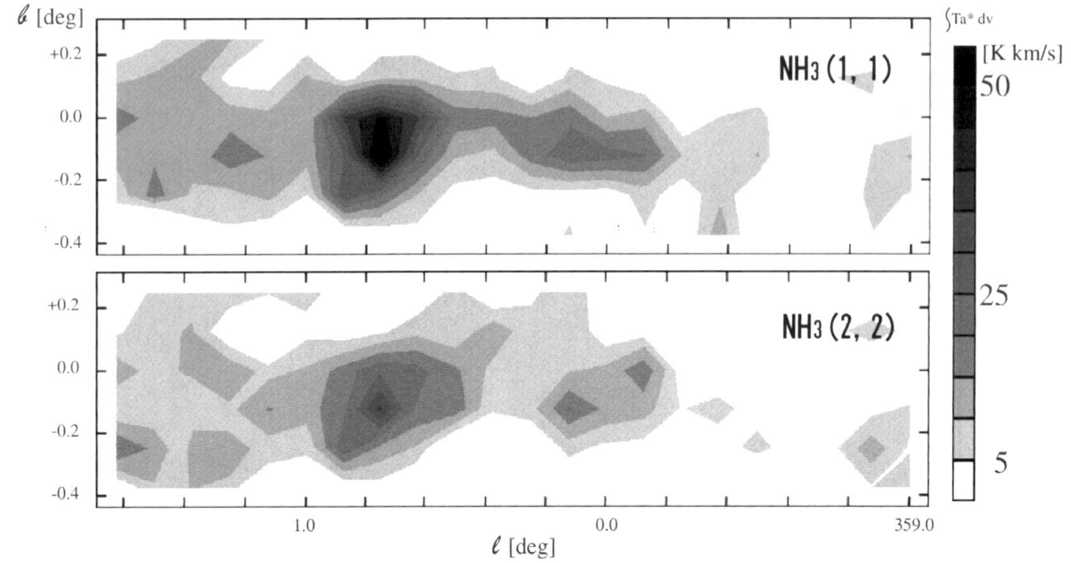

Figure 2. Integrated intensity of NH_3 (1,1) and (2,2) lines over -200 km s^{-1} to +200km s^{-1}. The intensity scale is shown in the figure.

Table 1. Estimated mass of the 4 clouds

name	Sgr A +20km s^{-1}	Sgr A +50km	Sgr B	the 1.3 degree
size	18 pc	33 pc	33 pc	42 pc
Δv(obs)	35 km s^{-1}	45 km s^{-1}	70 km s^{-1}	20 km s^{-1}
Δv(int)	25 km s^{-1}	39 km s^{-1}	70 km s^{-1}	9 km s^{-1}
$N(NH_3)$	6.6×10^{14}cm^{-2}	5.9×10^{14}cm^{-2}	1.7×10^{15}cm^{-2}	3.8×10^{14}cm^{-2}
$M_{\rm virial}$	$2.8 \times 10^6 M_\odot$	$1.2 \times 10^7 M_\odot$	$4.0 \times 10^7 M_\odot$	$8.5 \times 10^5 M_\odot$
$M_{\rm LTE}$	$1.1 \times 10^6 M_\odot$	$3.3 \times 10^6 M_\odot$	$1.0 \times 10^7 M_\odot$	$3.3 \times 10^6 M_\odot$

10^{-8}. The resultant masses are shown in table 1. The virial mass and LTE mass are consistent in the order of magnitude. Therefore, the NH_3 lines are not so opaque for these clouds.

4. Discussion

4.1. Molecular gas temperature derived from two NH_3 lines

The intensity ratio of the (2,2) transition to (1,1) transition is uniform over most of the observed region and is about 0.7–0.8, which means that the molecular gas has uniform in temperature.

Although we cannot estimate the opacity of the line, we can estimate the gas temperature using the method developed by Morris et al. [6]. The derived temperature from the line intensity ratio of NH_3 (2,2) to (1,1) is almost independent of gas density, but depends on opacity of the line. A high spatial resolution observations show that optical depth is about $\tau \sim 3-10$ for dense cloud cores [5], and $\tau = 2.3 \pm 1.0$ for weak sources and $\tau \sim 4$ for intense clouds [4]. Because our observations are more biased to weaker extended sources, optical depth may be much smaller than these results. So we estimated the temperature in both $\tau \ll 1$ and $\tau = 10$ cases. The estimated gas temperature corresponded to the ratio of 0.7–0.8 is 36–42 K in optically thin case and 24–34 K in optically thick case. This is consistent to the previous estimation of 25 K for the selected clouds [4]. Our full survey of CMZ show that the dense molecular gas in CMZ is dominated by gas as cool as majority of dust there.

For 4 major clouds we estimate average intensity ratio of NH_3 (2,2) to (1,1) lines using the

Table 2. Estimated temperature of the 4 clouds

name	Sgr A +20km s^{-1}	Sgr A +50km	Sgr B	the 1.3 degree
$R_{NH_3(2,2)/(1,1)}$	0.72 ± 0.07	0.75 ± 0.06	0.68 ± 0.02	0.82 ± 0.06
$T_K(\tau \ll 1)$	37 ± 3	38 ± 3	35 ± 1	42 ± 3
$T_K(\tau = 10)$	26 ± 4	28 ± 4	23 ± 1	34 ± 6

correlation plot between these two lines. The estimated ratio and derived temperature are shown in table 2.

4.2. A cool dense cloud complex with a hot molecular envelope

Although we cannot estimate the molecular gas temperature where no NH_3 emission is detected, we can make rough evaluation using other spectral lines, such as CO. The intensity ratio of CO ($J = 2 - 1$) over CO ($J = 1 - 0$), $R_{CO(2-1)/(1-0)}$, is a good indicator of gas pressure, $n(H_2\ T_K$, for molecular clouds in the Galactic center region [7].

Our NH_3 observations are made with the same beamsize and the same sampling grid as observations in CO ($J = 1-0$) with the CfA 1.2-m telescope and in CO ($J = 2-1$) with Tokyo-Nobeyama 60-cm telescope [9, 7]. Therefore, we can make direct comparison with $R_{CO(2-1)/(1-0)}$ distribution.

Figures 3 and 4 show the longitude–velocity diagram of our observations overlaid on the distribution of $R_{CO(2-1)/(1-0)}$. The diagram shows that NH_3 emitting regions do not correspond to high $R_{CO(2-1)/(1-0)}$ regions. When the co-existing molecular gas is under pressure equilibrium, NH_3 gas is the same pressure as CO gas whose pressure is indicated by $R_{CO(2-1)/(1-0)}$. The NH_3 emitting region contains a dense molecular gas, because NH_3 is a dense gas tracer. Therefore, the NH_3 emitting region is not so hot. This is consistent to the estimation from the NH_3 line intensity ratio in section 4.1.

Most of the high $R_{CO(2-1)/(1-0)}$ region surrounds the NH_3 emitting region. It seems that only the blue-shifted edge of Sgr B cloud ($l = 0.9°$) has no high $R_{CO(2-1)/(1-0)}$ region. However, at $v = 0$kms^{-1} the CO data are strongly affected by foreground gas in the Galactic disk and the low $R_{CO(2-1)/(1-0)}$ should not be real.

In the high $R_{CO(2-1)/(1-0)}$ region is not dense, because NH_3 emission is absent there. This means gas temperature there should be high.

At $b = 0°$ Sawada et al. gives gas pressure distribution [7]. For the NH_3 emitting region, $nT_K = (5 - 10) \times 10^4$Kcm^{-3} from their LVG result and $T_K = 40$K from our NH_3 ratio. On the other hand, for the high $R_{CO(2-1)/(1-0)}$ region $nT_K > 10^5$Kcm^{-3} from their LVG result and $nH_2 < 10^3$cm^{-3} because of absence of NH_3 emission. These value means $T_K > 100$K for the high $R_{CO(2-1)/(1-0)}$ region. This means that the high $R_{CO(2-1)/(1-0)}$ region is "hot molecular envelope".

This gives us a view that the hot molecular envelope surrounds the cool dense clouds. It suggests that the envelope may be heated by irradiation of intense interstellar radiation field or intense X-ray in the Galactic center.

References

[1] M. Tsuboi, T. Handa and N. Ukita 1999 *ApJS* **120** 1

[2] D.C. Lis, E. Serabyn, R.Zylka and Y. Li 2001 *ApJ* **550** 761

[3] D. Pierce-Price, J.S. Richer, J.S. Greaves, W.S. Holland, T. Jenness, A.N. Lasenby, G.J. White, H.E. Matthews, D. Ward-Thompson, W.R.F. Dent, R. Zylka, P. Mezger, T. Hasegawa, T. Oka, A. Omont and G. Gilmore 2000 *ApJ* **545** L121

[4] S. Hüttemeister, T.L. Wilson, T.M. Bania and

[5] R. Güsten, C.M. Walmsley and T. Pauls 1981 *A&A* **103** 197

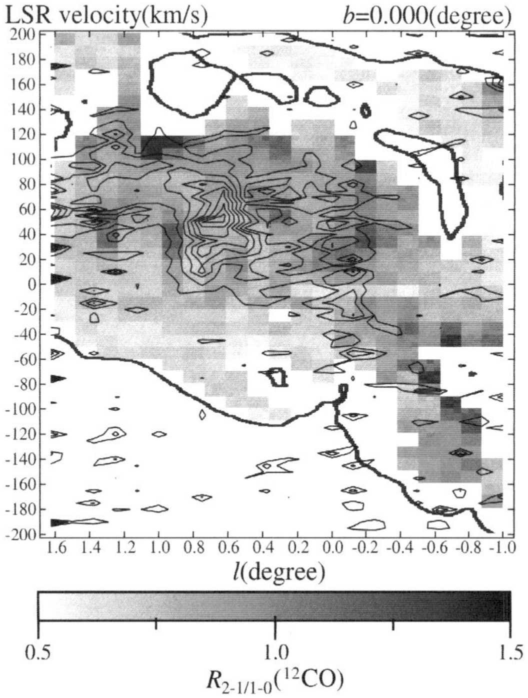

Figure 3. The longitude–velocity diagram of NH_3 (1,1) at $b = 0.000$ deg (contour) superposed on CO intensity ratio (gray-scale) [7]. Both the lowest contour and the contour step are 0.095 K.

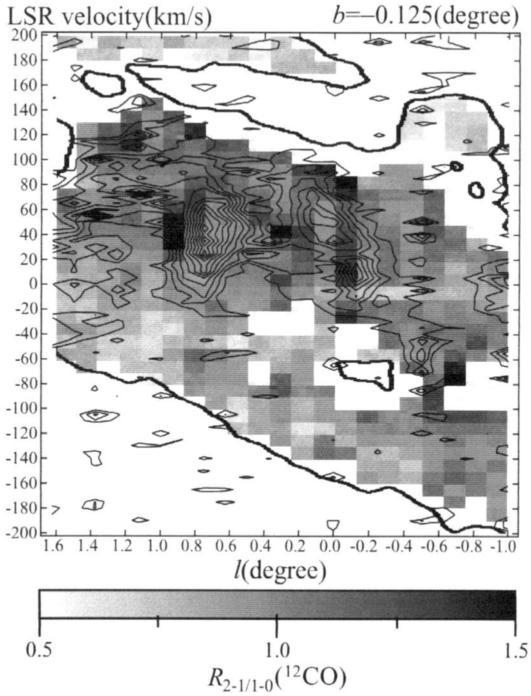

Figure 4. The longitude–velocity diagram of NH_3 (1,1) at $b = -0.125$ deg (contour) superposed on CO intensity ratio (gray-scale) [7]. Both the lowest contour and the contour step are 0.095 K.

[6] M. Morris, N. Polish, B. Zuckerman and N. Kaifu 1983 *AJ* **88** 1228 J. Martín-Pintado 1993 *A&A* **280** 255
[7] T. Sawada, T. Hasegawa, T. Handa, J.-I. Morino, T. Oka, R. Booth, L. Bronfman, M. Hayashi, A. Luna C., L.-A. Nyman, S. Sakamoto, M. Seta, P. Shaver, K. Sorai and K. S. Usuda 2001 *ApJS* **136** 189
[8] R.S. McGray and P.T.P. Ho 2002 *ApJ* **577** 757
[9] M. Bitran, H. Alvarez, L. Bronfman, J. May and P. Thaddeus 1997 *A&AS* **125** 99

Institute of Physics Publishing
doi:10.1088/1742-6596/54/1/008

Journal of Physics: Conference Series **54** (2006) 47–51
Galaxy Center Workshop 2006

An extremely large column density cloud G0.11-0.11 in the Galactic Center Region

Toshihiro Handa[1], Masaaki Sakano[2], Seiichiro Naito[3], Masaaki Hiramatsu[3], and Masato Tsuboi[4,5,3]

[1] Institute of Astronomy, University of Tokyo, Mitaka, Tokyo 181-0015, Japan
[2] University of Leicester, UK
[3] Department of Astronomy, University of Tokyo, Japan
[4] Nobeyama Radio Observatory, Japan
[5] Graduate University for Advanced Studies (Soken-Dai), Japan

E-mail: handa@ioa.s.u-tokyo.ac.jp

Abstract. We obtained the detail map in $H^{13}CO^+$ and in thermal SiO lines of G0.11−0.11, which is a molecular cloud located between the Galactic Center Arc and Sgr A. From line intensity ratios we found $H^{13}CO^+$ line is optically thin, whereas the thermal SiO lines are optically thick for this cloud. The cloud has a large column density up to $N(H_2) = (6-7) \times 10^{23}\mathrm{cm}^{-2}$, which corresponds to about 640 − 740 mag in A_V or 10 − 12 mag in $A_{25\mu m}$. This is the largest known even in the Galactic Center region. The intensity ratio of SiO to CS lines suggests that emitting gas is highly inhomogeneous for SiO abundance on a scale smaller than the beam with about 35′.

1. Introduction

Large scale molecular line surveys have revealed that molecular clouds in the central ∼ 100 pc of the Galaxy are different from those in the Galactic disk. For example, the incidence of relatively dense clouds is higher in the Galactic Center region. The Nobeyama Radio Observatory (NRO) CS survey has shown that density of most molecular clouds there is over 10^4 cm^{-3} [1].

Of dense molecular clouds in the Galactic Center region, the molecular cloud G0.11−0.11 is unique and one of the most interesting objects. It is located between Sgr A and the Galactic Center Arc (GCA). The CS observations reveal that G0.11−0.11 has a large molecular mass and large velocity width at the eastern[1] and western edges of the cloud. The eastern edge appears to have an interaction with the GCA [2]. G0.11−0.11 is bright at the X-ray fluorescent iron line [3], which suggests that dense gas in the cloud is excited by intense X-ray or high energy particles.

The previous molecular line observations of G0.11−0.11 were done in optically thick lines, such as CS and CO. The morphologies of G0.11−0.11 in CS and CO lines are similar. To obtain the physical properties of the cloud, we made observations in $H^{13}CO^+$ ($J = 1-0$) line, which should be optically thin because of its very low abundance. At the same time, we present a high resolution view in thermal SiO lines. The thermal SiO lines are thought to be a good tracer

[1] In this paper, all directions on the sky are in terms of the galactic coordinates.

of hot and shocked regions, because this molecule is in gas phase only under high temperature environment [4].

2. Observations

We have observed G0.11−0.11 using the Nobeyama 45-m telescope, simultaneously observing at the spectral lines of $H^{13}CO^{+}$ $J = 1 - 0$ (86.75 GHz), SiO $J = 1 - 0, v = 0$ (43.42 GHz), and SiO $J = 2 - 1, v = 0$ (86.85 GHz). The full width at half maximum (FWHM) beam sizes at 43 and 86 GHz are $35''$ and $18''$, respectively. The observed region is a rectangular area of $0°\ 4' \leq l \leq 0°10'$, and $-0°10' \leq b \leq -0°4'$, which covers the whole cloud. The spacing of the observation grid is $20''$, which corresponds to 0.82 pc at the distance to the Galactic Center, 8.5 kpc. The main beam efficiencies at 43 and 86 GHz are 0.81 and 0.50, respectively. The velocity resolutions at 43 and 86 GHz are 0.87 km s^{-1} and 0.44 km s^{-1}, respectively, although we smoothed all spectra to be with 5 km s^{-1} resolution.

3. Results

3.1. Features and morphology of the cloud

Figure 1 shows the integrated intensity maps in the three lines in the velocity range of 15 km s^{-1} $\leq v_{\rm LSR} \leq$ 45 km s^{-1}. The spatial resolutions are adjusted to $45''$ by applying gaussian convolution. The appearance in the SiO lines resembles that in the CS $J = 1 - 0$ line (see Fig. 1 of [2]).

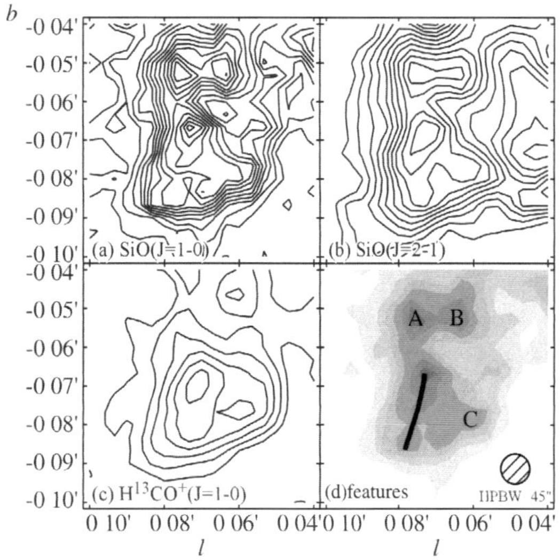

Figure 1. Integrated intensity maps of G0.11−0.11 in (a) SiO $J = 1 - 0, v = 0$, (b) SiO $J = 2 - 1, v = 0$, and (c) $H^{13}CO^{+}$ $J = 1 - 0$, and (d) a schematic chart of main features. The velocity range is $15 \leq V_{\rm LSR} \leq 45$ km s^{-1}. The FWHM beam sizes are adjusted to $45'$, shown in panel (d). The intensity scale is in integrated mainbeam brightness temperature, $\int T_{\rm MB}\, dv$. Both the first contour level and the contour interval are 2 K km s^{-1} for panels (a) to (c). The gray scale in panel (d) is the same map as in panel (a).

However, several differences are apparent in the images in the $H^{13}CO^{+}$ $J = 1 - 0$ and SiO lines. The $H^{13}CO^{+}$ $J = 1 - 0$ intensity is significantly concentrated to the southern half of the cloud, although the cloud seems to extend beyond $b \geq -0°6'$ in SiO and CS images. These discrepancies are presumably due to the difference in optical depths between the SiO and $H^{13}CO^{+}$ lines (see detail in Section 4.1). Namely, the $H^{13}CO^{+}$ $J = 1 - 0$ line intensity traces the column density, but the SiO lines do not. Thus, G0.11−0.11 shows significant difference in column density below and above a front at $b = -0°6'$.

G0.11−0.11 shows four distinctive features in these lines. Along the eastern edge of the cloud, a prominent ridge is seen in all the three lines. We dub it E-ridge hereafter (as indicated with a solid line in Figure 1). At $l = 0°7'$, it extends from $b = -0°6'$ to $b = -0°9'$ perpendicular to the Galactic plane. On the northern end of the E-ridge, a peak is seen in both the SiO lines at

$l = 0°8', b = -0°5'20''$ (peak A, hereafter). At $l = 0°6'20'', b = -0°5'20''$, another peak (peak B) is seen in the SiO lines. The other prominent feature is a peak at $l = 0°6', b = -0°8'$ (peak C). Figure 1 illustrates these features.

The E-ridge extends for $3'$ in the Galactic latitude, equivalent to 7 pc at a distance of 8.5 kpc. A corresponding feature is also seen in the CS line (Fig. 2 in [2]). The E-ridge is extended in the direction parallel to the GCA. This morphology may suggest an interaction of the molecular gas with the GCA. However, this interaction is probably not so strong, if it exists, because the E-ridge is not the most prominent feature in the SiO image. In the $H^{13}CO^+$ $J = 1-0$ map the E-ridge is also distinguishable, but less prominent than in the SiO lines, and very weak in $b \geq -0°6'$.

The peak A is seen in the SiO maps between 10 km s^{-1} $< v_{LSR} <$ 35 km s^{-1}. At the high-red-shift end, the peak A is merged into the ridge along the Galactic plane through the peak B. In the SiO line images, the peak A appears to be somehow connected with the E-ridge. However, the $H^{13}CO^+$ line image shows no feature corresponding to the peak A or B ($v_{LSR} <$ 35 km s^{-1}), whereas it shows the E-ridge apparently. Hence, the peak A is unlikely to be a part of the E-ridge.

The peak B is seen in the SiO maps at $v_{LSR} >$ 30 km s^{-1}. Beyond $v_{LSR} >$ 45 km s^{-1} the position of the peak B is shifted toward the north by $20''$. In the $H^{13}CO^+$ map a clear counterpart is seen only beyond $v_{LSR} >$ 45 km s^{-1}. It suggests that the peak B may be a double source and separable at $v_{LSR} =$ 45 km s^{-1}. In any case, the peak B with $v_{LSR} <$ 45 km s^{-1} is only seen in the SiO lines.

The peak C is seen between 30 km s^{-1} $< v_{LSR} <$ 45 km s^{-1}. The $H^{13}CO^+$ map shows its counterpart clearly. The peak C is morphologically connected to the E-ridge in $l-b-v$ space. The E-ridge and the peak C might be two main parts of G0.11−0.11.

3.2. Intensity ratio

To evaluate the morphological resemblance among the SiO lines and difference between the $H^{13}CO^+$ and SiO lines quantitatively, we estimate intensity ratios of observed lines. They are keys to determine the optical depth and/or physical conditions of the emitting gas in G0.11−0.11.

First, we estimate an intensity ratio of two SiO lines, $R_{SiO(2-1)/SiO(1-0)}$. To calculate an average value, we use an intensity correlation for all the observed points in a box assigned in $l-b-v$ space for each feature. To remove the difference in resolution due to different beam size at the three lines, we reduce the resolution to be $45''$ by appropriate gaussian convolution. We found the ratios of SiO $J = 2-1$ line to SiO $J = 1-0$ line to be 0.9–1.0 for the E-ridge and the three peaks, and also found no significant difference of the ratios among the regions in the cloud.

We also estimated the line intensity ratios of $H^{13}CO^+$ $J = 1-0$ to SiO $J = 1-0$, and found them to be uniform in each feature, although they differ significantly between northern and southern parts of G0.11−0.11. For the E-ridge and the peak C, they are 0.5. For the peaks A and B, they are about 0.2 or smaller, although the S/N is poor.

4. Discussion

4.1. Column density and mass of the cloud

The $H^{13}CO^+$ $J = 1-0$ line is expected to be optically thin, because of its small abundance. We can check it from the $H^{13}CO^+$ $J = 1-0$ intensity compared with the CS $J = 2-1$ intensity. We should note that, since the excitation parameters of both the lines are similar, their intensity ratio ought not to be a strong function of the physical conditions of the gas; accordingly the only causes of variation in this ratio must be variations in either the relative abundances of the species or in their relative optical depth. Using an $H^{13}CO^+$ abundance of 10^{-10} and a CS abundance of 10^{-8} [5, 6] together with excitation parameters of the lines, the expected intensity ratio of

$H^{13}CO^+$ $J = 1-0$ to CS $J = 2-1$ is about 6×10^{-3}, if both lines are optically thin. Using published CS data [2], the line intensity ratio of the $H^{13}CO^+$ $J = 1-0$ to the CS $J = 2-1$ is calculated to be 0.12–0.14 in the southern part of G0.11−0.11. It follows that the CS line in this locality must be optically thick, whereas the optical depth of $H^{13}CO^+$ $J = 1-0$ is about 0.1. In the northern part of the cloud, the ratio is about 0.05 or smaller and the $H^{13}CO^+$ $J = 1-0$ line is therefore optically thin.

Then, we estimate column density of the southern part of G0.11−0.11 and molecular mass of the whole cloud from $H^{13}CO^+$ $J = 1-0$ intensity under the condition of local thermal equilibrium (LTE). When we assume $T_K = 70$ K, the column density of molecular hydrogen at the E-ridge is derived to be $N(H_2) = (6-7) \times 10^{23} cm^{-2}$. This corresponds to about 640–740 mag in A_V (visual extinction) at the typical gas-to-dust ratio expected for dense clouds, and about 80–90 mag and 10–12 mag in A_K and $A_{25\mu m}$ (extinction at $25\mu m$), respectively [7]. Similar values are obtained for peak C. This large $A_{25\mu m}$ is consistent with the fact that G0.11−0.11 is observed as a shadow in an infrared map with MSX [8]. The shadow has the similar spatial extension on the sky to that in $H^{13}CO^+ J = 1-0$.

In submillimeter continuum map we can find the counterpart of G0.11−0.11, although it is less prominent than major submillimeter features [9]. Using the same conversion from gas column density to submillimeter brightness as theirs, $N(H_2) = 6 \times 10^{23} cm^{-2}$ corresponds to 20 Jy $8''$-beam^{-1} at 450 μm and 8 Jy $15''$-beam^{-1} at 850 μm, respectively. The maps with SCUBA [9] show about 15–20 Jy beam^{-1} at 450 μm and 3–4 Jy beam^{-1} at 850 μm, respectively. The estimated values are consistent, because both estimations are based on assumptions with some uncertainty. The gas-to-dust mass ratio may be reduced in the cloud, because strong thermal SiO line of the cloud suggests dust evaporation. The molecular abundance of $H^{13}CO^+$ may be smaller than the value we assumed. Moreover, inhomogeneity in the cloud may affect the conversion factors from the observable values to the true mass of the cloud.

The estimated column density of G0.11−0.11 is one of the largest ever observed even in the Galactic Center region. For several X-ray sources in the Galactic Center region, total hydrogen column densities were estimated to be $N_H \leq (1-3) \times 10^{23} cm^{-2}$ [10, 11]. The cloud G0.11−0.11 shows a larger column density by at least factor of 5 than the ordinary environments in the Galactic Center region. It is extraordinarily large, being comparable only with Sgr B2, which is the most massive cloud in the Galaxy.

Other than G0.11−0.11, many dark features in the MSX map are found in the Galactic Center region. They are called as the MSX dark clouds [8]. Some of them were also observed in the H_2CO line [12]. Typical column densities for MSX dark clouds are estimated to be $N(H_2) = 10^{23-25} cm^{-2}$. Our estimated column density of G0.11−0.11 is as large as the typical MSX dark clouds; n.b., G0.11−0.11 itself was not observed.

4.2. The SiO emitting clump and its structure

The intensity ratio of the two SiO lines, $R_{SiO(2-1)/SiO(1-0)}$ is about 0.9–1.0 for all the features in G0.11−0.11 (Section 3.2). This value implies two possibilities; one is that the both the SiO lines are optically thick, and the other is that the both the lines are optically thin and the density of molecular hydrogen in the SiO-emitting gas is $10^{3.7-3.8} cm^{-3}$. However, the latter case is unlikely for the following two reasons. If the hydrogen density were $10^{3.7-3.8} cm^{-3}$, the CS lines would be optically thin. But our estimation (Section 4.1), as well as the previous estimation [2], shows that the CS line is (at least moderately) optically thick.

Because the observed antenna temperature at the SiO line is much lower than the expected gas kinetic temperature, the beam filling factor must be smaller than unity, i.e. the telescope beam is not filled with the emitting surface. Hence, we should employ a "clumpy model" ([13]) to consider a physical state of G0.11−0.11. Using the "clumpy model", the observed line-intensity ratio depends on the opacity of the emitting clumps, and the observed antenna temperature is reduced by the beam filling factor.

Using "clumpy model" and optically thick line, we can roughly estimate some parameters of a clump. We find that the mainbeam brightness temperatures of most features in G0.11−0.11 are $T_{\rm MB} = 1 - 2$K. When $T_{\rm K} = 70$K, the typical beam filling factor is 0.02. Because G0.11−0.11 does not show a discrete clump with $35''$ beam, there must be 10 or more clumps in a beam. Thus the diameter of a clump is smaller than $1.5''$ or 0.06 pc. The averaged gas density in a clump is then estimated to be higher than $2 \times 10^8 {\rm cm}^{-3}$.

In this case a clump may be unstable, because the free fall time of the clump is much shorter than the sound crossing time. However, it can be stable when the size of a clump is much smaller. Given a beam averaged column density and beam filling factor, the free fall time is proportional to square root of the clump size. On the other hand, given a gas temperature, the sound crossing time is proportional to the size. Therefore, the clump can be stable when the size is smaller than 4×10^{-4} pc for our estimated values. Note that the critical size may be much larger, if the clump is supported by magnetic field.

Large opacity even in dense gas tracers and large extinction even in mid-infrared suggest that the cooling time may be longer than cores in starforming regions in the Galactic disk region. Using virial mass analysis, Sawada et al. show that molecular clouds in the Galactic Center region are under high pressure as $nT_{\rm K} = 10^5 {\rm K\ cm}^{-3}$ [14]. By this external pressure, emitting clumps in G0.11−0.11 may be compressed.

From our observations, G0.11−0.11 is found to be likely composed from many hot and dense clumps, which can be hardly cooled down because of large extinction even in infrared. This condition is greatly different from that of starforming clouds in the Galactic disk region. Under such condition, star formation should be much different. This may be a reason why a dense cluster of massive stars is seen only in the Galactic Center region. Although G0.11−0.11 is a good site to investigate this speculation, high resolution observations in rarer molecules such as CS isotopes are required to unveil optically thick clump.

References

[1] Tsuboi, M., Handa, T. and Ukita, N. 1999 *ApJS* **120** 1

[2] Tsuboi M., Ukita N., Handa T. 1997, *ApJ* **481** 263

[3] Yusef-Zadeh, F., Law, C. and Wardle, M. 2002, *ApJ* **568** L121

[4] Ziurys, L. M., Friberg, P. and Irvin, W. M. 1989, *ApJ* **341** 857

[5] García-Burillo, S., Martín-Pintado, J., Fuente, A. and Neri, R. 2000, *A&A* **355**, 499

[6] Irvine, W. M., Goldsmith, P. F. and Hjalmarson, A. 1987, in Interstellar Processes, ed. D. J. Hollenbach & Hjalmarson (Dordrecht: Reidel), 561

[7] Mathis, J. S. 2000, in "Allen's Astrophysical Quantities, fourth edition", eds by A. N. Cox, 523

[8] Egan, M. P., Shipman, R. F., Price, S. D., Carey, S. J., and Clark, F. O. 1998, *ApJ* **494** L199

[9] Pierce-Price, D., Richer, J. S., Greaves, J. S., Holland, W. S., Jenness, T., Lasenby, A. N., White, G. J., Matthews, H. E., Ward-Thompson, D., Dent, W. R. F., Zylka, R., Mezger, P., Hasegawa, T., Oka, T., Omont A. and Gilmore G. 2000, *ApJ* **545** L121

[10] Sakano, M., Koyama, K., Nishiuchi, M., Yokogawa J., Maeda, Y. 1999, *Adv. Space Res.* **23** 969

[11] Sakano, M., 2000, Ph.D. thesis, Kyoto University

[12] Carey, S. J., Clark, F. O., Egan, M. P., Price, S. D., Shipman, R. F., and Kuchar T. A. *ApJ* **508** 721

[13] Snell, R. L., Mundy, L. G., Goldsmith, P. F., Evans II, N. J. and Erickson N. R. 1984, *ApJ* **276** 625

[14] Sawada, T., Hasegawa, T., Handa, T., Morino, J. -I., Oka, T., Booth, R., Bronfman, L., Hayashi, M., Luna, A. C., Nyman, L. -Å., Sakamoto, S., Seta, M., Shaver, P., Sorai, K. and Usuda K. S. 2001, *ApJS* **136** 189

Institute of Physics Publishing
doi:10.1088/1742-6596/54/1/009
Journal of Physics: Conference Series **54** (2006) 52–56
Galaxy Center Workshop

3D Simulations of the 180pc Molecular Ring

Myoungwon Jeon, Sungsoo S. Kim

Department of Astronomy and Space Science, Kyung Hee University, Yongin-shi, Kyungki-do 446-701, Korea

E-mail: myjeon@ap1.khu.ac.kr

Abstract. Understanding the kinematics and morphology of molecular clouds in the Central Molecular Zone (CMZ) is important for studying the mode of star formation in the central region of the Galaxy. A quasi-continuous ring structure with a radius of 180pc, so-called "180-pc molecular ring", is interpreted as the transition region between the two stable orbits in the bar potential, x1 & x2 [2]. We present, for the first time, 3-D high resolution hydrodynamic simulations of the gas flows in this transition region, and compare the results with previous 2-D studies. We will also see if an m=1 perturbation, which is revealed as a displacement of the CMZ, is responsible for some of the discrepancies between the previous simulations and observations.

1. Introduction

The origin and stability of the "180-pc molecular ring" seen in the center of our galaxy is a long-standing puzzle in galactic center dynamics. In the previous studies, the 180 pc molecular ring feature was considered as an expanding ring [1]. In 1991, Binney et al. [2] suggested that the longitude-velocity (ℓ–v) plot of the 180 pc molecular ring, a parallelogram rather than an ellipse, can be understood as a projection of the innermost stable x1-orbit [3] in a barred potential. They also showed that the locations of the giant molecular clouds at the Galactic Center (GC), such as Sgr B and Sgr C, could be explained by the x2 orbits.

There have been many hydrodynamical simulations to study the structure and kinematics of the GC molecular clouds [4, 5, 6, 7]. These simulations have been successful in reproducing structures of the gas motions on global scales, but they do not resolve some important discrepancies between simulations and observations such as the large asymmetries in positive longitude. According to observations, more molecular clouds are seen at positive longitudes than at negative longitudes in the GC region. Different mechanisms have been proposed to explain the observed lopsidedness of the distribution of molecular gas. One of the possibilities is that an m=1 perturbation may be responsible for the asymmetry.

Here, we try to answer the influence of an m=1 perturbation on the motion of the molecular clouds through smoothed particle hydrodynamics (SPH) simulations. We then make detailed comparisons between the observations and simulations to understand the distribution and kinematics of the GC molecular clouds.

2. Model & methodology

We have studied various types of galactic potential models used in other recent papers and turned several different potentials into numerical codes. Each type of model has some attractive aspects in modeling the potential of our galaxy. These models have been successful in reproducing the structure of the gas motions, but several models were incapable of reproducing x1- and x2-type orbits at the desired locations. In our numerical simulations, we represent the galaxy by a fixed potential with three components: an exponential stellar disk, a power-law ellipsoidal bar, and a dark matter (DM) halo. For the density of the dark halo, we use a Navarro-Frenk-White(NFW) model.[8]:

$$\rho\left(\mathrm{r}\right)=\frac{\rho_{\mathrm{s}}}{\left(\mathrm{r}/\mathrm{r}_{s}\right)\left(1+\mathrm{r}/\mathrm{r}_{s}\right)^{2}}$$

Where r_s and ρ_s are characteristic scale and density of the halo, respectively. For the disk component, we adopt Freeman's [9] exponential disk that has the following form,

$$\Phi\left(\mathrm{R}\right)=-\pi\mathrm{G}\Sigma_0\mathrm{R}\left[\mathrm{I}_0\left(\mathrm{y}\right)\mathrm{K}_1\left(\mathrm{y}\right)-\mathrm{I}_1\left(\mathrm{y}\right)\mathrm{K}_0\left(\mathrm{y}\right)\right] \qquad \text{where} \qquad \mathrm{y}=\frac{\mathrm{R}}{\mathrm{R}_{\mathrm{d}}}$$

Here, R_d is the disk scale length, Σ_0 is the central surface density, and I_n and K_n are modified Bessel functions of the first and second kinds. As for the bar component, we adopt a density/potential pair similar to that of the prolate bar proposed by Binney et al. [2] with a power law logarithmic slope α=1.8 and a normalization constant a_0=1.2 kpc.

$$\rho\left(\mathrm{a}\right)=\rho_0\left(\frac{\mathrm{a}}{\mathrm{a}_0}\right)^{-\alpha} \qquad \mathrm{a}=\sqrt{\mathrm{x}^2+\left(\mathrm{y}^2+\mathrm{z}^2\right)/\mathrm{q}_m^2}$$

We find that a good match to the CO data [10] is obtained when q_m=0.75.

The gas particles of the molecular clouds are placed on x1 orbits as an initial condition and are represented by SPH particles. In our simulations, we used the following system of units in which the gravitational constant $G = 1$, the unit of length is chosen to be $R = 1$ pc. The total mass of the galaxy within 1 kpc was assumed to be M_{10}=7× 10^9 $M_\odot$. We calculated the simulations to 0.2 Gyr to allow the gas flow to form a ring and to 0.3 Gyr to verify that the ring is stabilized.

All of the simulations in this study were performed with GADGET-2 [11], a new massive parallel TreeSPH code, capable of following a collisionless fluid with the N-body method, and a gas by means of SPH method. The SPH method has the advantage of allowing for a spatially adaptive resolution length. Fluid quantities are approximated by averaging over neighbouring particles. Furthermore, the SPH scheme includes an artificial viscosity to allow for shocks in the simulated gas flow. We assumed that the gas is isothermal and that its temperature is 10^4 K since previous numerical simulations of gas motion in disc galaxies [12] show that radiative cooling is very effective for shocked gas and that the gas temperature is maintained at 10^4 K.

3. m=1 instability

One of the puzzles regarding the distribution of molecular gas toward the GC region is its lopsidedness. Different mechanisms have been proposed to explain the observed feature. One possibility is that an m=1 perturbation may be responsible for the asymmetric distribution of the molecular clouds. We have considered two possible causes for the creation of $m = 1$ perturbations. One possibility is that a fly-by encounter with a satellite galaxy may provide a perturbing influence which helps to excite an m=1 instability. The other possibility is that an $m = 1$ excitation is provided by a slight displacement of the center of mass of the dark matter halo with respect to the GC. We tried to answer these effects through N body simulations, First, we have simulated the response of molecular clouds to a perturbation by a fly-by encounter. A Plummer model with $M_{Plummer}$= 10^8 $M_\odot$ was chosen for the density of the perturbing object and a hyperbolic orbit for its orbit. Second, the response of molecular clouds to a DM halo that is off centered with respect to the center of the galaxy was investigated. Levine & Sparke [13] found significant asymmetries in their models to explain lopsidedness of galaxies when the disc lies off centered in the dark matter halo. Viewed in this light, our hypothesis is that an offset between GC and center of mass of the DM halo will be responsible for the $m = 1$ instability.

4. Results

4.1. Simulations without perturbation

Some recent simulations [6, 7] of the molecular clouds in the GC have overlooked some potentially important physical effects such as 3-dimensionality, self-gravity of gas, and so on. We therefore set up 3-dimensional simulations including self-gravity of gas particles. In order to emphasize the important role of an $m = 1$ perturbation concerning gas dynamics, we first present numerical results of a simulation without an $m = 1$ perturbation (Fig. 1). Over the course of a 0.3 Gyr integration, the particles rapidly lose energy and angular momentum, and quickly move onto x2-orbits. Consequently, a small perpendicularly aligned bar of heavily populated x2-orbits develops at the center.

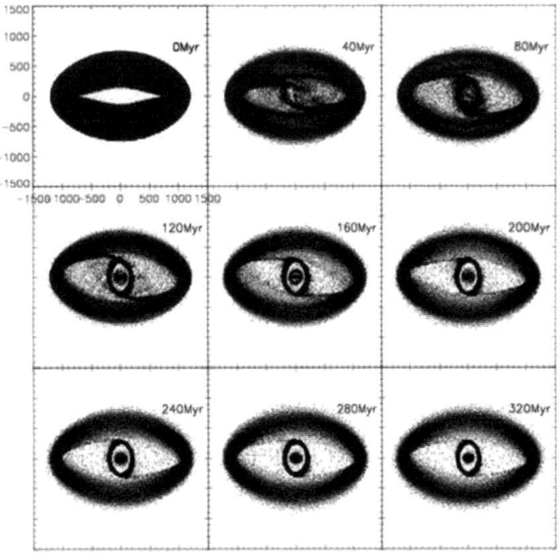

Figure 1. Distribution of the gas particles in the galactic center without perturbation. The length unit is 1 pc.

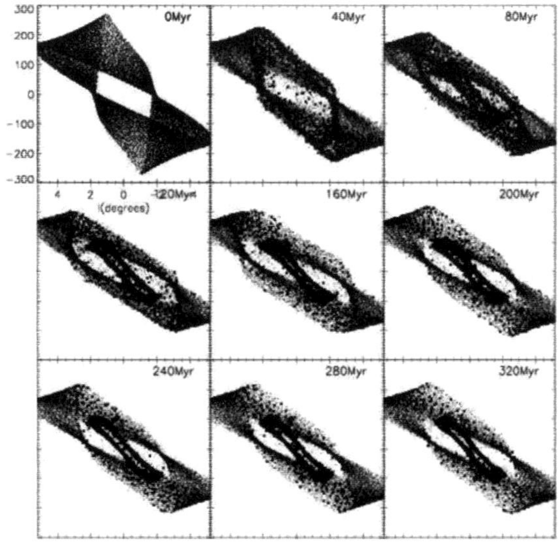

Figure 2. Longitude-velocity plots for the simulation shon in the left panel. The abscissa gives longitude in degrees and the ordinate gives velocity in $\mathrm{km\,s^{-1}}$.

In spite of the fact that our simulations, that comprise the effects of the self-gravity of the gas and of 3-dimensionality, are more realistic than earlier studies, significant difference to previous results were not found. Therefore, the self-gravity of the gas appears to be negligible because it is insufficient to explain the asymmetry of the molecular gas successfully.

4.2. Simulation of fly-by encounter

The effects of a fly-by encounter on a coplanar orbit that varies with various values for the pericenter and for the core radius of the perturbing object were investigated. If the perturber has a small core radius and close pericenter, the gas particles suffer several instabilities and the perturbation produces a lump of ejected particals. Also, gas from the ring is shown to accrete onto the central core. In an attempt to reduce the perturbation by the Plummer sphere, we

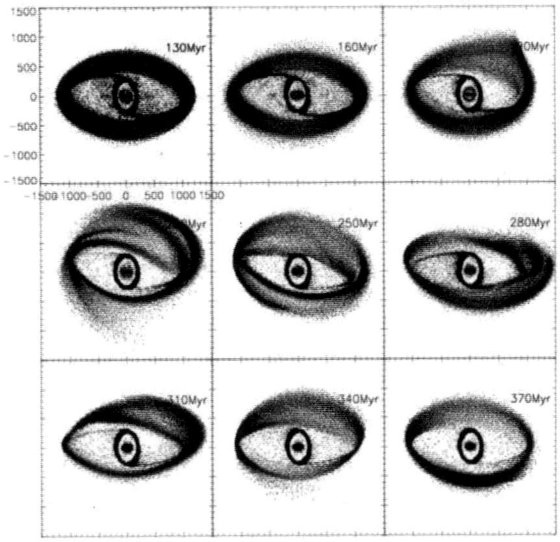

Figure 3. Distribution of the gas particles after a fly-by encounter. The core radius of the perturber is 300 pc and its pericenter 1 kpc. The length unit is 1 pc.

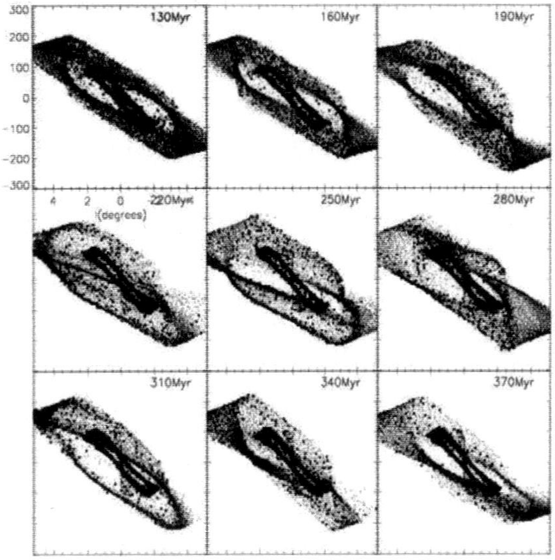

Figure 4. Longitude-velocity plots for the simulation shown in the left panel.

have also tried orbits with large pericenter distances. In this case, there is a smaller deviation in the gas distribution compared to earlier results.

We find that, if the perturber has a core radius of almost 300 pc, and moves on an orbit which has a pericenter of nearly 1 kpc, the result looks similar to the observed lopsidedness in the GC (Fig. 3). Once an $m = 1$ perturbation is excited, these models take many orbits to decay. Therefore, molecular clouds are sensitive to the parameters of encounters such as the core radius and pericenter distance of the perturber. An interesting prospect is that it may be possible to detect indirectly the presence of dwarf galaxies from their effect on the molecular clouds at the GC.

4.3. Simulations with an off centered DM halo

In this approach the lopsidedness of molecular clouds is related to a displacement between the center of mass of the DM halo and the GC. For these simulations, we have calculated the period of the DM halo circling around the GC and have modeled it as a fixed potential. We then examined several simulations with an off-centered DM halo at different offset distances. As the offset increases, the instability of gas particles grows and produces a lump of ejected particles. But if the offset distance was taken to be about 300-400 pc, we were able to reproduce the ℓ–v map of the lopsided molecular clouds.

5. Summary

- We have performed 3-D SPH simulations of the 180 pc molecular ring in the galactic center.
- The self-gravity of the gas is considered to be a negligible factor since it cannot explain the lopsidedness successfully.
- We have studied the response of molecular clouds to perturbations and the subsequent formation of an $m = 1$ instability induced by a fly-by encounter or an off-centered DM halo to try to understand the lopsided distribution of the GC molecular clouds.

- In order to understand these $m = 1$ instability effects, we have made several calculations. Our simulations show that morphological asymmetries occur during the simulations which are able to reproduce the ^{12}CO ℓ–v plot. However, these simulations are in need of tightly constrained parameters, such as, e.g., a sufficiently close pericenter distance and a small core radius of the perturber during a fly-by encounter.

References

[1] Kaife et al. 1972, *Nature Phys. Sci.*, **238**, 105

[2] Binney J., Gerhard O. E., Stark A. A., Bally J., & Uchida K. I. 1991, *MNRAS*, **252**, 210

[3] Contopoulos G., & Mertzanides C. 1977, *A&A*, **61**, 33

[4] Wada K., & Habe A. 1992, ,*MNRAS*, **258**, 82

[5] Friedli D., & Benz W. 1993, *A&A*, **268**, 65

[6] Lee C. W., Lee H. M., Ann H. B., and Kwan K. H. 1999, *ApJ*, **513**, 242

[7] Jenkins A., & Binney J. 1994, *MNRAS*, **270**, 703

[8] Navarro J. F., Frenk C. S., White S. D. M., 1997 *ApJ*, **490**, 493

[9] Freeman K. C., 1970 *ApJ*, **160**, 811

[10] Bally J., Stark A. A., Wilson R.W. & Henkel C. 1987 *ApJS* , **65**, 13

[11] Springel V., 2005 *MNRAS* , **364**, 1105

[12] Hernquist L. E., 1989 *Nature*, **340**, 687

[13] Levine S. E., & Sparke L. S., 1998 *ApJ*, **496**, L13

Institute of Physics Publishing
doi:10.1088/1742-6596/54/1/010

Journal of Physics: Conference Series **54** (2006) 57–61
Galaxy Center Workshop 2006

ISAAC M-band spectroscopy of dust embedded sources at the Galactic Center

Jihane Moultaka[1], Andreas Eckart[1], Thomas Viehmann[1] and Rainer Schödel[1]

[1] Universität zu Köln, I. Physikalisches Institut, Zülpicher Str. 77, D-50937 Köln, Germany

E-mail: `moultaka@ph1.uni-koeln.de`

Abstract. We present our most recent observations in the M-band, ranging from 4.4 to 5.1 micron, of the central parsec of the Galaxy, obtained with the ISAAC spectrograph of the VLT. These observations resulted in unprecedented high quality spectra of the brightest sources of the central parsec which allow us to analyse the spatial distribution, locally and along the line of sight, of the solid and gaseous phase of the CO molecule.

1. Introduction

At a distance of 8 kpc, the Galactic Center (GC) is the closest galactic central region. It can thus be studied in detail by the direct observation of individual stars, gas, and dust. It represents an ideal and unique case allowing us to analyse thoroughly the direct environment of a central supermassive black hole.

The Galactic Center cavity of about 3 pc in diameter, is obscured by extinction from the diffuse interstellar medium (ISM) present along the line of sight and is surrounded by a circumnuclear ring of dense gas and dust showing clumpy extinction [5]. The exact magnitude and spatial distribution of this obscuration are complex but a mean value of the visual extinction towards prominent sources within the central stellar cluster reaches $\sim$ 27 mag [7, 13]. Moreover, the extinction across the central 10" to 20" is shown to be smoothly distributed [13].

Recently, we studied the distribution of the water ice and hydrocarbon features at 3.0 μm and 3.4-3.48 μm, respectively, in the central parsec. Our results show that most of these features are predominantly produced locally in the central parsec and are associated with individual bright sources [10, 11].

In this contribution, we present our most recent observations of the central parsec of the Galaxy, performed in the M-band spectral region using the ISAAC instrument at the ESO VLT UT1 (Antu) telescope. This allows us to analyse the spatial distribution of the CO molecule in its gas- and solid-phase.

2. Observations

We have observed about 20 bright sources of the central parsec using the ISAAC NIR spectrograph of the VLT. We used the long wavelength, low resolution (R=800) spectroscopic mode in the spectral range of the M filter from 4.4 μm to 5.1 μm.

To correct for sky emission, the chopping technique was used with chopper throws of $\sim$ 20" along

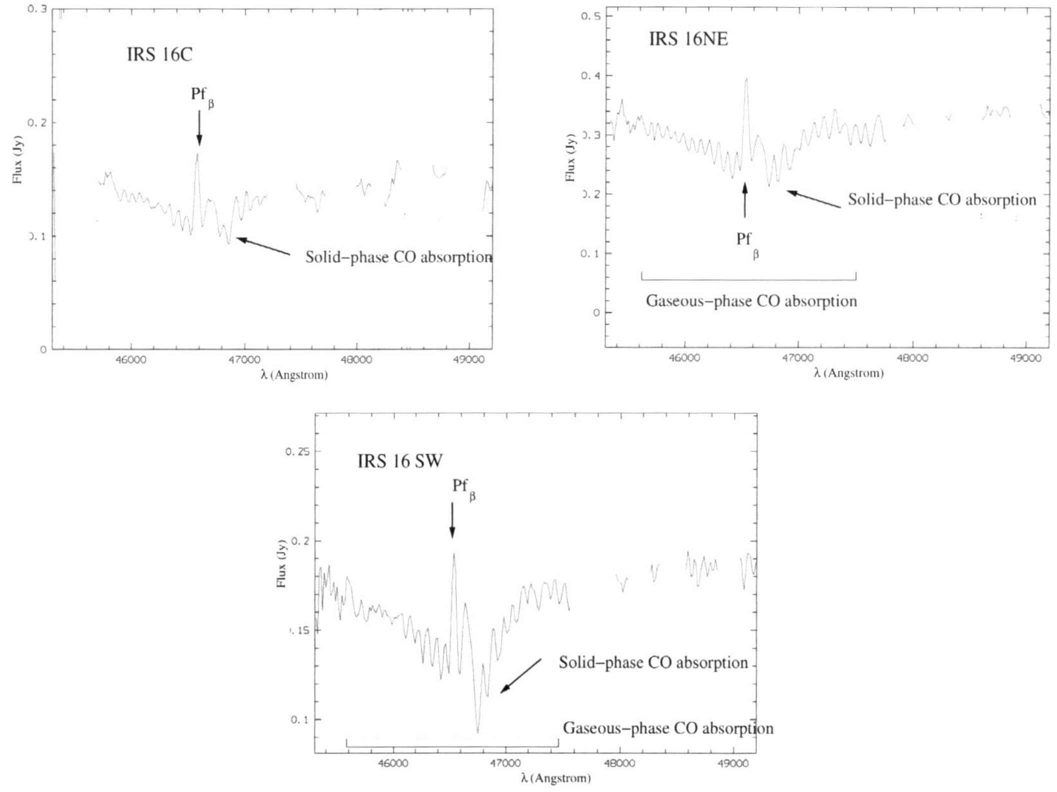

Figure 1. M-band spectra of the He emission line stars of the IRS 16 complex. The CO gaseous and solid phase absorptions are also shown. All blanked spectral regions correspond to residuals from the telluric lines.

the slit. All images were divided by the flatfield, corrected for cosmic rays and for dispersion related distortions. They were also wavelength and flux calibrated and corrected for telluric lines.

This resulted in the first M-band good quality spectra (good resolution and signal to noise ratio) of a large number of GC sources.

An absolute flux calibration was applied to the resulting spectra using the M-band magnitudes of the sources published by [15]. For the case of IRS 2L and IRS 5, for which there are no M-band magnitudes, we calibrated the spectra using the magnitudes of IRS 2S and IRS 10W, respectively.

3. Results

As it can be seen in figures 1, 2, 3 and 4, in addition to the Pf_β Hydrogen line at 4.65μm, two spectral features of the CO molecule are present in this spectral band:

- The absorption band of the ^{13}CO gas-phase with its fundamental vibration-rotation P and R branch lines around 4.666μm.
- The absorption feature at 4.675μm of the solid-phase CO.

The interstellar solid CO at 4.675μm was detected for the first time in the deeply embedded young stellar object W33A by [6]. In the Galactic Center, the solid-phase CO feature was observed in IRS 12 by [8]. The authors also detected the absorption band of the gas-phase CO in

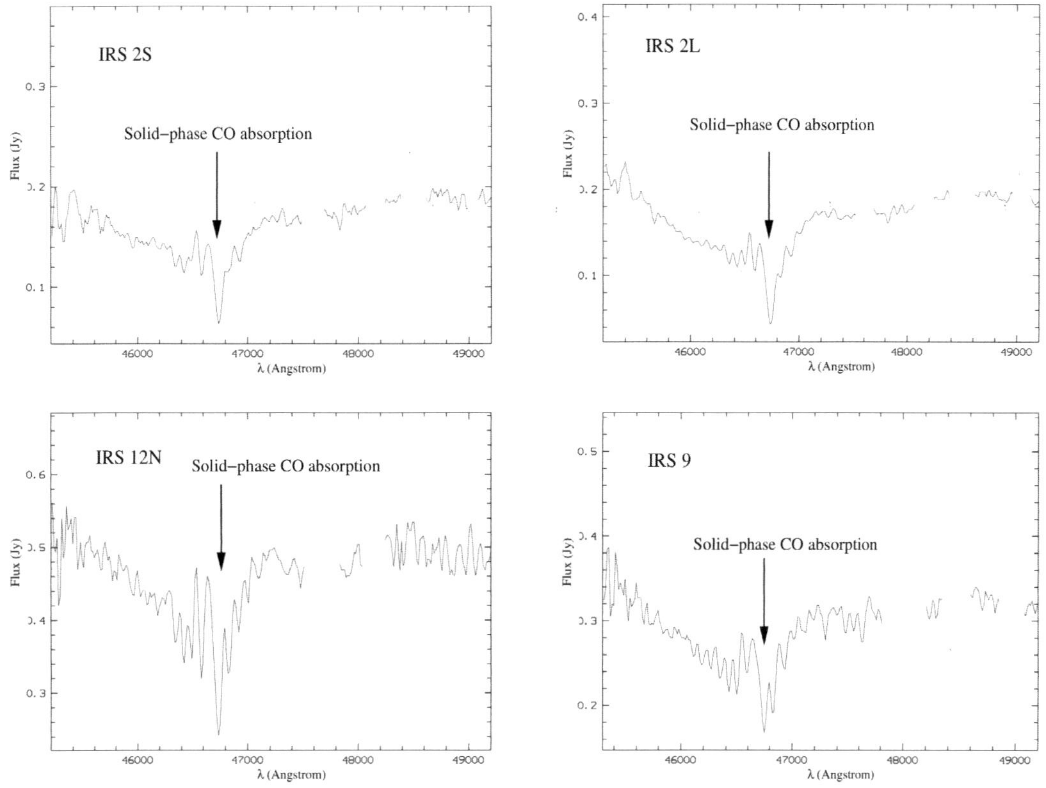

Figure 2. M-band spectra of the cool stars in the Galactic Center. All blanked spectral regions correspond to residuals from the telluric lines.

IRS 3, IRS 7 and IRS 12 but the P and R branches were not resolved. The CO absorption in IRS 3 was also previously observed by [3]. However the Pf_β Hydrogen emission line is absent in all three spectra.
From the solid-phase CO feature detection towards IRS 12 and its absence towards IRS 3 and IRS 7, [8] conclude that the distribution of the cold molecular clouds is not uniform in the central cluster. They also conclude that the molecular clouds cannot be close to the Galactic Center because the CO molecule cannot survive temperatures higher than 20K unless they are located in an environment of water ice and other molecules.

Here we present for the first time, good quality M-band spectra of 15 bright sources of the central parsec including the He emission-line stars of the IRS 16 complex, cool stars and the dust-embedded sources of the northern arm of the minispiral. Unlike [8], in our spectra, the P and R branches of the gas-phase absorption feature are resolved. This will allow us to derive column densities of the CO gas-phase. We also observe a prominent Pf_β emission line in IRS 12 that was not detected previously and a CO solid-phase absorption feature in IRS 7 that was also absent in the IRS 7 spectrum published by [8].
In our spectra, all dust free He-stars of the IRS 16 complex show the solid CO feature that is probably present in the foreground material of the ISM. IRS 16C shows almost no gas-phase absorption, while IRS 16NE and IRS 16SW do. This can be due to the fact that the latter are closer to the minispiral by which the observed gas-phase feature is probably caused. Moreover,

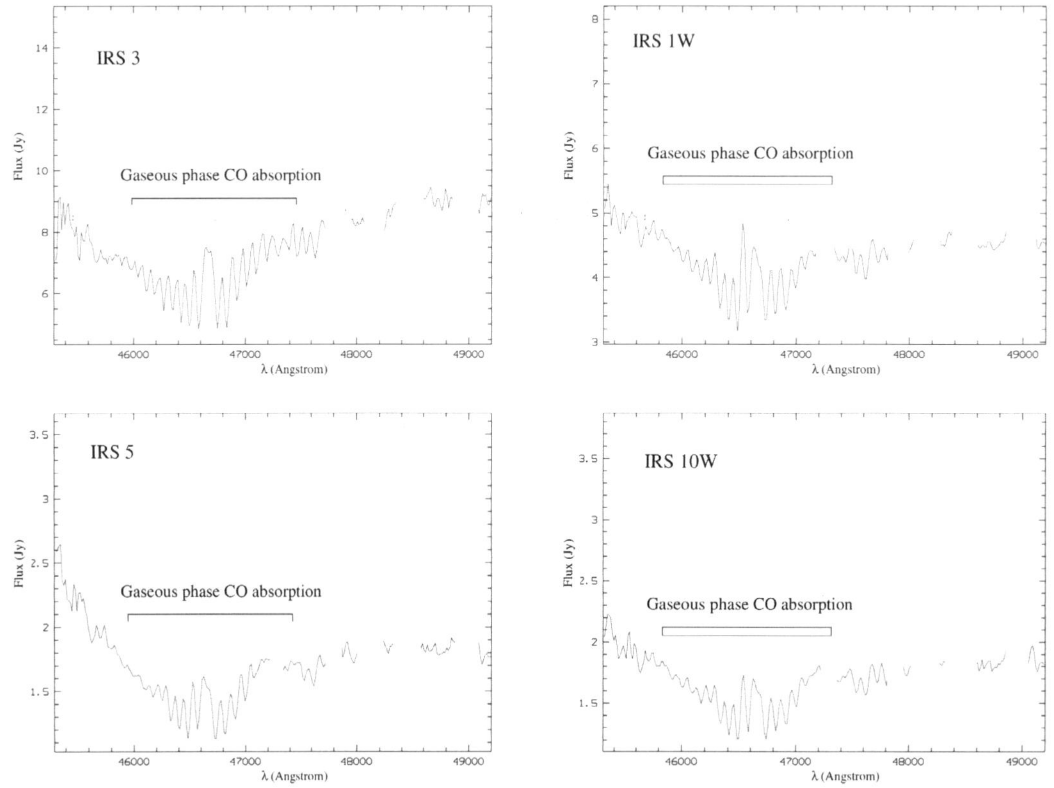

Figure 3. M-band spectra of the dust embedded sources in the northern arm of the minispiral. All blanked spectral regions correspond to residuals from the telluric lines.

all cool stars (IRS2S, 2L, 9, 12N) also show the solid absorption in their spectra which looks deeper than in the spectra of the He-stars. This suggests that part of this absorption is local and associated with these sources, which is an expected result, since the solid-phase CO can survive in the direct environment of these stars. On the other hand, all the dust-embedded sources of the northern arm (IRS1W, 5 and 10W) of the minispiral, except IRS21, as well as IRS3, 29 and 13E (classified as dusty Wolf-Rayet stars), show the gas-phase feature in their spectra. The observed gas-phase CO is most probably located in the minispiral and heated by the hot massive dust-embedded sources.

This work is still in progress and will be pursued by a more detailed and quantitative study in Moultaka et al. (in preparation).

Acknowledgments

This work was supported in part by the Deutsche Forschungsgemeinschaft (DFG) via grant SFB 494.

References

[1] Armstrong J.T. & Barrett A.H. 1985 *ApJSS* **57, 535**
[2] Eyres S.P.S., Geballe T.R. et al. 2004 *MNRAS* **350, L9**
[3] Geballe T.R. 1986 *A&A* **162, 248**
[4] Geballe T.R. Baas F., Wade R. 1989 *A&A* **208, 255**
[5] Güsten R., Genzel R. et al. 1987 *ApJ* **318, 124**

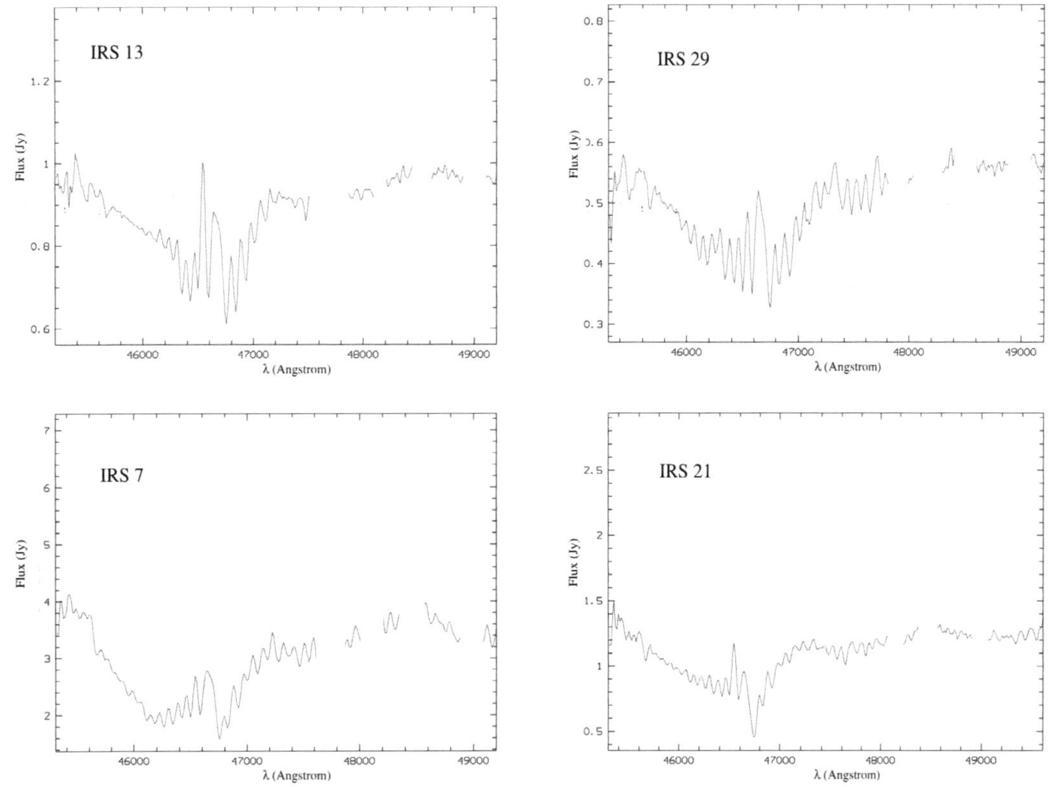

Figure 4. M-band spectra of the IRS 13 complex, IRS 29, the dust embedded source IRS 21 and the cool supergiant IRS 7. All blanked spectral regions correspond to residuals from the telluric lines.

[6] Lacy J.H., Baas F. et al. 1984 *ApJ* **276, 533**
[7] Lebofsky M.J. 1979 *AJ* **84, 324**
[8] Mc Fadzean A.D., Whittet D.C.B. et al. 1989 *MNRAS* **241, 873**
[9] Moneti A., Cernicharo J., Pardo J.R. 2001 *ApJ* **549, L203**
[10] Moultaka J., Eckart A., Viehmann T., et al. 2004 *A&A* **425, 529**
[11] Moultaka J., Eckart A., Schödel R., et al. 2005 *A&A* **443, 163**
[12] Sandford S.A., Allamandola L.J. et al. 1988 *ApJ* **329, 498**
[13] Scoville N.Z., Stolovy S.R. et al. 2003 *ApJ* **594, 294**
[14] Tielens A.G.G.M., Tokunaga A.T. et al. 1991 *ApJ* **381, 181**
[15] Viehmann T., Eckart A., Schödel R., Moultaka J., et al. 2005 *A&A* **443, 117**

Institute of Physics Publishing
doi:10.1088/1742-6596/54/1/011
Journal of Physics: Conference Series **54** (2006) 62–66
Galaxy Center Workshop 2006

Structure of the Galactic Bulge: Is the Milky Way a Double-barred Galaxy?

Shogo Nishiyama[1,2], Tetsuya Nagata[3], and IRSF/SIRIUS team[1,2,3]

[1] National Astronomical Observatory of Japan, Mitaka, Tokyo, Japan
[2] Department of Astrophysics, Nagoya University, Nagoya, Japan
[3] Department of Astronomy, Kyoto University, Kyoto, Japan

E-mail: shogo@optik.mtk.nao.ac.jp

Abstract. Using the data of the IRSF/SIRIUS infrared survey along the Galactic plane ($|\ l\ | \leq 10^{\circ}\!.5$ at $b = +1^{\circ}$), we find a distinct structure, which is probably a *secondary bar*, inside the primary bar of our Galaxy. The apparent magnitude peak of Bulge red clump stars changes continuously from $K_S \sim 13.5$ ($l = -10^{\circ}$) to $K_S \sim 12.3$ ($l = +10^{\circ}$), and this can be explained by the bar structure of the Galactic Bulge. However, the apparent magnitude changes by only ~ 0.1 mag over the central 8°, and this indicates that there is a distinct structure inside the primary bar. In the process of the distance derivation, we have used the infrared extinction law in the J, H, and K_S bands toward the Galactic center newly determined from our survey.

1. IRSF/SIRIUS Galactic Center Survey

The IRSF/SIRIUS Galactic center (GC) survey project is a deep near-infrared (J, H, K_S) survey of the central region of our Galaxy. The survey has been carried out with the 1.4m telescope IRSF and the SIRIUS camera. IRSF/SIRIUS were constructed and have been operated by Nagoya University (Japan), National Astronomical Observatory of Japan, and the South African Astronomical Observatory. The SIRIUS camera [1, 2] can provide three (J, H, K_S) images simultaneously with a field of view of $7'.7 \times 7'.7$ and a pixel scale of $0''.45$. The limiting magnitudes of our survey are 17.0 (J), 16.6 (H), and 15.6 (K_S), which are $\sim 2-3$ mag deeper than 2MASS at the GC. A more detailed description of our survey is given in the PhD thesis by Nishiyama [3]. A three color composite image of the GC, whose field size is $100'$ in R.A. and Dec, is shown in Fig. 1[8707].

2. Bulge Red Clump Stars As a Standard Candle

Red clump (RC) stars are the equivalent of the horizontal-branch stars for metal rich population. They can be used as a good standard candle due to their abundance, brightness ($M_{K_S} = -1.52$ for Bulge population [4, 5, 6]) and narrow distribution in luminosity and color. The *Hipparcos* catalogue allows us exact calibration of absolute magnitudes. Owing to their abundance, RC stars are the only type of standard candles that do not suffer from a large statistical error.

[8707] You can find this image at the web site
http://www.z.phys.nagoya-u.ac.jp/~sirius/press/ASJM04a/gc100min.jpg

Figure 1. Three color (J, H, K_S) composite image of the Galactic Center. The field size is $100'$ in R.A. and Dec., and the Galactic Plane runs from upper left to lower right. The Galactic Center is the bright yellow blob in the center.

3. New Extinction Law in J, H, and K_S bands toward the Galactic Center [7]

With an increasing number of observation of highly obscured objects by interstellar extinction in the GC, a precise determination of the extinction law is important. However, the large difference between the extinction law quite often used ([8] and [9]) and those obtained from new observations has been reported in the optical wavelengths [10, 11]. Hence we have determined the interstellar extinction law toward the GC in the near infrared (J, H, K_S) wavebands from the observation of the region $| \, l \, | \leq 2^{\circ}\!.0$ and $0^{\circ}\!.5 \leq | \, b \, | \leq 1^{\circ}\!.0$ [7]. Using the "RC method", the ratios of total to selective extinction have been *directly* obtained (Table 1). The ratio of absolute extinction $A_J : A_H : A_{K_S} = 1 : (0.573 \pm 0.009) : (0.331 \pm 0.004)$ is also obtained. The power law

Table 1. Interstellar Extinction Laws obtained in our study and previous studies.

	IRSF	van de Hulst[12]	R&L[8]	Cardelli[9]	Indevetouw[13]
A_{K_S}/E_{H-K_S}	1.44 ± 0.01	1.58	–	1.83	1.82
A_K/E_{H-K}	–	1.33	1.78	1.63	–
A_{K_S}/E_{J-K_S}	0.494 ± 0.006	0.55	–	0.73	0.67
A_K/E_{J-K}	–	0.49	0.66	0.68	–
A_H/E_{J-H}	1.42 ± 0.02	1.38	1.64	1.88	1.63
A_H/A_J	0.573 ± 0.009	0.58	0.62	0.65	0.62
A_{K_S}/A_J	0.331 ± 0.004	0.36	–	0.42	0.40
A_K/A_J	–	0.33	0.40	0.40	–

approximation $A_\lambda \propto \lambda^{-1.99\pm0.02}$ is shown to be good enough in the wavelength range $1.2-2.2\mu$m.

The most widely used technique to obtain the interstellar extinction law is the color-difference (CD) method, where the extinction curves are determined in the form of the ratios of color excesses, e.g., $E_{\lambda-V}/E_{B-V}$, as a function of λ^{-1}. The absolute extinction A_λ can be derived from the ratios of color excesses, only if the ratio of total to selective extinction $R_V = A_V/E_{B-V}$ is evaluated $[A_\lambda/E_{B-V} = E_{\lambda-V}/E_{B-V} + R_V)]$. The value of R_V can only be deduced by the extrapolation of the extinction curve to $\lambda^{-1} = 0$, but the extrapolation has uncertainty due to the contamination by dust emission and the neutral extinction by large grains.

In the RC method, we expect the difference between the observed and absolute magnitudes to be the same for every RC star in the Galactic bulge except for differences produced by variations in the amount of interstellar extinction along the line of sight to each star. They are thus located in a straight line in a color-magnitude diagram with the slope of $R_\lambda = A_\lambda/E_{\lambda'-\lambda}$, in accordance to the variable extinction to each of them. We can obtain R_λ *directly* from the observed data, and therefore R_λ can be much more reliable than that obtained by the extrapolation of the extinction curve in the CD method. The reliable R_λ provides the precise absolute extinction value A_λ.

Previous results in the literature are listed for comparison in Table 1 [8, 9, 12, 13]. Our results show a clear difference to those obtained in previous works. The ratios of total to selective extinction determined in our study are smaller. This corresponds to a faster decrease in the absolute extinction A_λ when the wavelength increases. Our results are the closest agreement to the van de Hulst No. 15 curve [12] which was based on the Mie scattering theory and has been used to estimate the wavelength dependence.

Rieke & Lebofsky[8] (R&L) determined the optical-infrared extinction law applying the CD method to their observations of stars near the GC. Their results and that derived by the analytic formula of the average extinction law by fitting the data of R&L [9] have quite often been referred to as the standard extinction law, especially toward the GC. Since their determination requires the ratio of total to selective extinction, R&L set $E_{V-K}/E_{B-V} = 2.744$ and assumed $R_V = A_V/E_{B-V} = 3.09$ for the GC sources to evaluate A_λ/E_{B-V} and A_λ/A_V. Although the lower limit of R_V in R&L was determined by the extinction at L, M, 8, and 13μm, the observations at these wavelengths have relatively large uncertainties, and a small change in R_V results in a large difference in A_λ/E_{B-V} at infrared wavelengths. We also note that the extinction law they derived in the wavelength range L, M, and 8μm is inconsistent with those obtained toward the GC, in particular by spectroscopy with the Infrared Space Observatory (*ISO*) for $\lambda \geq 2.4\mu$m [14]. Therefore, although the inconsistency might be the result of spatial variation of the extinction law [7], it is also quite possible that the inconsistency was caused by

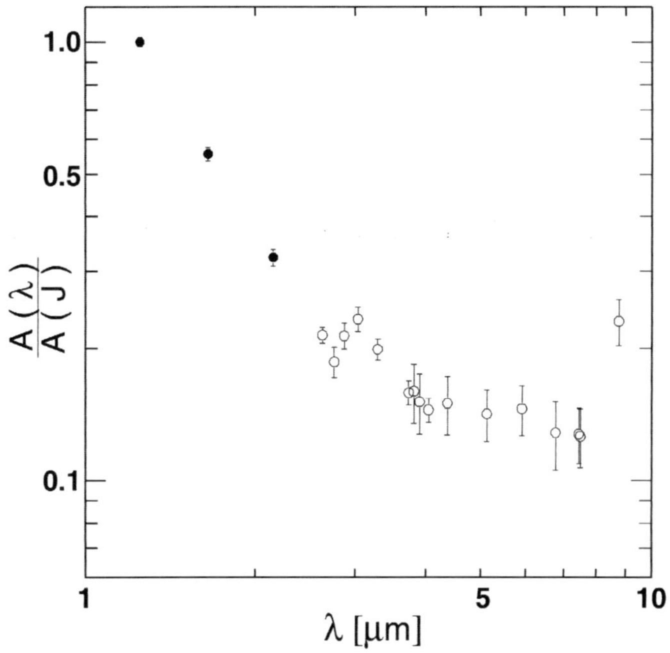

Figure 2. Derived wavelength dependence of the interstellar extinction (A_λ/A_J) in the range of 1 to 10 μm toward the Galactic center. The results of our study [7] and Lutz *et al.* [14] are shown by filled and open circles, respectively. A_λ/A_V by Lutz *et al.* [14] is converted to A_λ/A_J by assuming a $\lambda^{-1.99}$ extinction law for K_S and $\lambda = 2.625\mu$m.

the uncertainty of R_V in the CD method.

Fig. 2 shows A_λ/A_J for $1.2\mu\text{m} \leq \lambda \leq 10\mu$m toward the GC. A_λ/A_J for the J, H, and K_S bands are obtained in our study [7], and those for $\lambda \geq 2.6\mu$m are derived by Lutz *et al.* [14], who observed H recombination lines toward the GC by using the Short Wavelength Spectrometer on board *ISO*. Their A_λ/A_V is converted to A_λ/A_J by assuming a $\lambda^{-1.99}$ extinction law for K_S and $\lambda = 2.625\mu$m because the continuity of the power-law upto $\sim 2.5\mu$m is suggested toward the GC [15]. Also the data of the *Spitzer Space Telescope* have the potential to establish the extinction law for $\lambda \geq 3\mu$m.

4. A Distinct Structure Inside the Galactic Bar [16]

The existence of a bar structure in our Galaxy is by now established. However, its parameters (size, axis ratio, alignment, etc.) have not been well determined yet. Bulge RC stars are good standard candles, and the distribution of their apparent magnitude along the Galactic plane shows an asymmetry which presents clear evidence for the Galactic bar. Large scale surveys are important to determine the detailed structure of the Galactic bar, and deep infrared observations are also required because of the large extinction near the Galactic plane. The limiting magnitude of 2MASS ($m_{K_S} \sim 13$ mag) is too shallow to detect the whole RC stars near the GC.

We surveyed a strip extending over $|\, l \,| \leq 10\overset{\circ}{.}5$ at $b = +1^\circ$, with a latitude width of $\sim 8'$ [16]. To analyze the distribution of RC stars, (1) color-magnitude diagrams were constructed using the H and K_S bands, (2) stars on the red giant branch were extracted, (3) the extinction free magnitude K_{H-K} was defined as $K_{H-K} \equiv K_S - 1.44 \times E_{H-K_S}$, where $A_{K_S}/E_{H-K_S} = 1.44$ and $(H - K_S)_0 = 0.07$ [6], and (4) de-reddened luminosity (K_{H-K}) functions were constructed and fitted with the sum of exponential (red giant branch) plus Gaussian (RC stars) functions.

The dependence of the RC peak magnitude on the Galactic longitude l is shown in Fig. 3. Assuming the absolute K_S magnitude of the Bulge RC stars to be $M_{K_S} = -1.52$ mag, the absolute peak distances are shown using the scale on the right in Fig. 3. The distance modulus to the Bulge RC stars is given by $(m - M)_0 = K_{H-K} - M_{K_S} + \Delta M_K$, where ΔM_K is the population correction calculated from theoretical stellar evolution models. Here we adopt

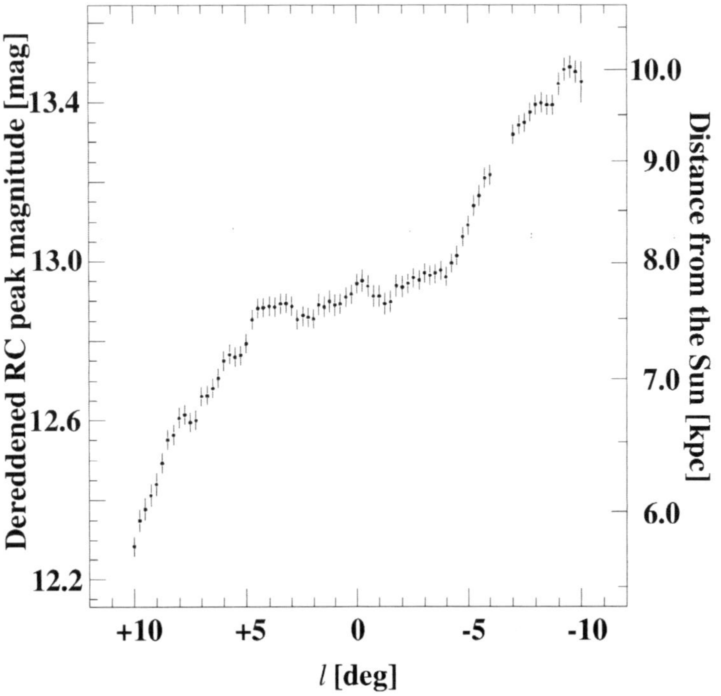

Figure 3. Dependence of the peak of K_{H-K} on the Galactic longitude l. This figure is slightly different from Fig. 4 of Nishiyama *et al.* [16] because K_{H-K} and distance were re-calculated by using a more realistic RC absolute magnitude and color.

$\Delta M_K = -0.07$ (scaled to solar metalicity [5]).

As shown in Fig. 3, the RC peak becomes progressively brighter at greater l from $K_{H-K} \sim 13.5$ ($l = -10°$) to $K_{H-K} \sim 12.3$ ($l = +10°$), indicating the non-axisymmetric Bulge structure in support of previous studies. However, the slope of the data points is not constant, becoming shallower at $\mid l \mid < 5°$. This variation suggests the presence of an inner structure inside the large-scale bar. Since the inner structure seems to be straight and the RC peak magnitude changes only slightly, the inner structure may well be an inner bar.

References

[1] Nagashima C, Nagayama T, Nakajima Y, Tamura M, Sugitani K, Nagata T, Hirao T, Nakaya H, Yanagisawa K and Sato S 1999 *Proceedings of Star Formation 1999, Editor T. Nakamoto, Nobeyama Radio Observatory* 397

[2] Nagayama T, Nagashima C, Nakajima Y, Nagata T, Sato S, Nakaya H, Yamamuro T, Sugitani K and Tamura M 2003 *Proc. SPIE* **4841** 459

[3] Nishiyama S 2006 *PhD thesis, Nagoya University* (http://www.z.phys.nagoya-u.ac.jp/thesis/nishiyamaPhD.pdf)

[4] Alves D R, Rejkuba M, Minniti D and Cook K H 2002 *Astrophys. J* **573** L51

[5] Salaris M and Girardi L 2002 *Mon. Not. R. Astron. Soc* **337** 332

[6] Bonatto C, Bica E and Girardi L 2004 *Astron. Astrophysics* **415** 571

[7] Nishiyama S *et al.* 2006 *Astrophys. J* **638** 839

[8] Rieke G H and Lebofsky M J 1985 *Astrophys. J* **288** 618 (R&L)

[9] Cardelli J A, Clayton G C and Mathis J S 1989 *Astrophys. J* **345** 245

[10] Udalski A 2003 *Astrophys. J* **590** 284

[11] Sumi T 2004 *Mon. Not. R. Astron. Soc* **349** 193

[12] van de Hulst H C 1946 *Rech. Astr. Obs. Utrecht* **11** 1

[13] Indebetouw R *et al.* 2005 *Astrophys. J* **619** 931

[14] Lutz D *et al.* 1996, *Astron. Astrophysics* **315**, L269

[15] Nagata T, Kobayashi N and Sato S 1994 *Astrophys. J* **423** L113

[16] Nishiyama S *et al.* 2005 *Astrophys. J* **621** L105

Institute of Physics Publishing
doi:10.1088/1742-6596/54/1/012

Journal of Physics: Conference Series **54** (2006) 67–71
Galaxy Center Workshop 2006

A Large-scale CO J=3–2 Survey of the Galactic Center

Tomoharu Oka[1], Makoto Nagai[1], Kazuhisa Kamegai[2], and Kunihiko Tanaka[2]

[1] Research Center for the Early Universe and Department of Physics
The University of Tokyo, Bunkyo-ku, Tokyo, Japan
[2] Institute of Astronomy, Faculty of Science, The University of Tokyo, Mitaka, Tokyo, Japan

E-mail: tomo@taurus.phys.s.u-tokyo.ac.jp

Abstract. We have surveyed the central molecular zone (CMZ) of our Galaxy in the CO J=3–2 line with the Atacama Submillimeter-wave Telescope Experiment (ASTE). Molecular gas in the Galactic center shows high J=3–2/J=1–0 intensity ratio ($\sim$ 0.9) while gas in the Galactic disk shows the lower ratio ($\sim$ 0.5). The high-velocity compact cloud CO 0.02–0.02 and the hyperenergetic shell CO 1.27+0.01, as well as gas in the Sgr A region exhibit high J=3–2/J=1–0 intensity ratio exceeding 1.5. We also found a number of small spots of high ratio gas. Some of these high ratio spots have large velocity widths and some seem to associate with nonthermal 'threads' or filaments. These could be spots of hot molecular gas shocked by unidentified supernovae which may be abundant in the CMZ.

1. Introduction

The 'Central Molecular Zone' (CMZ) of the Galaxy, a region of radius $\sim$ 200 pc, contains large amount of dense molecular gas ($n \geq 10^4$ cm^{-3}; e.g., [1]) and exhibits uniformly high gas temperature (T_k =30–60 K; [2]). Molecular gas in the CMZ shows highly complex distribution and kinematics which are characterized by expanding arcs/shells and filaments [3]. A population of high-velocity compact clouds (HVCCs) is also unique in the CMZ [3,4,5]

Large amount of hot plasma [6] and widespread distribution of radioactive nuclei there [7] may suggest that a number of supernova explosion have occurred in the CMZ in the recent past. The widespread SiO emission could be understood as a result of shocks such as fossil superbubbles [8,9]. These suggest that the boisterous molecular gas kinematics in the CMZ may be a result of violent release of kinetic energy by a number of supernova explosions.

To assess the effect of supernova explosions on the kinematics and physical conditions of the CMZ, it is crucial to pick up supernova/molecular cloud interacting zone completely. For this objective, we are performing a large-scale, high resolution CO J=3–2 survey of the CMZ. Here we report results of the survey in 2005 and their implications.

2. Observations

The CO J=3–2 (345.79599 GHz) mapping observations were carried out with the Japanese 10 m submillimeter-wave telescope at Pampa la Bola, Chile (ASTE; [10]). The telescope has a FWHM beam size of 22″ at 345 GHz. The observations were done 2005 July 19–25 in good, stable

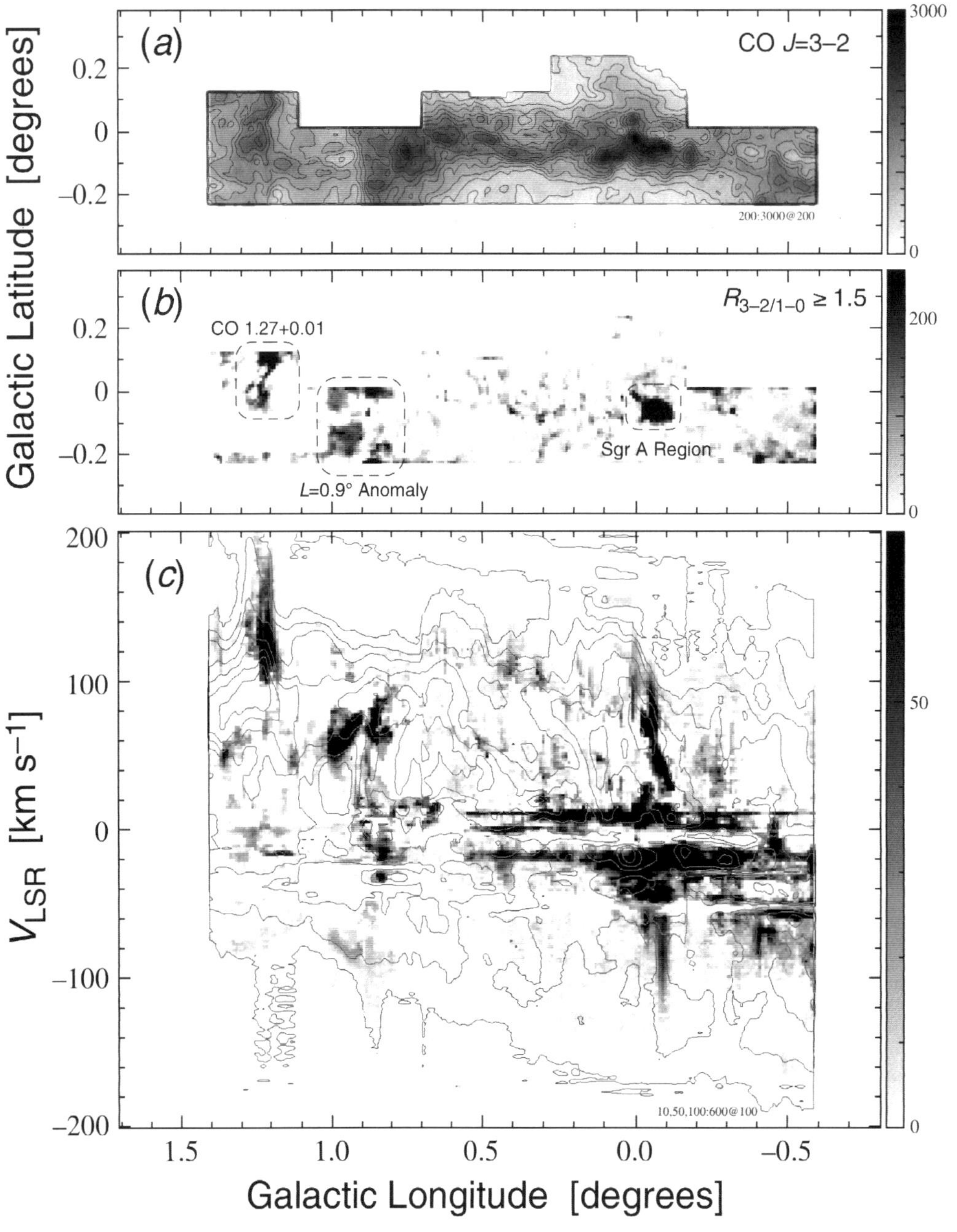

Figure 1. (*a*) The velocity-integrated intensity map of CO *J*=3–2 emission. (*b*) The same map as (*a*) for data with $R_{3-2/1-0} \geq 1.5$. Data severely contaminated by the foreground disk gas have been excluded. (*c*) Longitude-velocity map of CO *J*=3–2 emission integrated over the observed latitudes (contours), superposed on the same map for data with $R_{3-2/1-0} \geq 1.5$ (grays).

weather conditions. All spectra were obtained with an XF-type autocorrelation spectrometer in the 512 MHz bandwidth (1024 channel) mode. We mapped a $\Delta l \times \Delta b = 2° \times 0.5°$ area with a 34″ grid spacing, collecting 6990 spectra in total. The on-source integration time was 10 seconds for each position and rms noise was 0.3 K in T_A^*. The complete description of the data acquisition and reduction procedure will be presented in the forthcoming paper [11].

3. Results

A map of CO J=3–2 emission integrated over velocities between $V_{\rm LSR} = -200$ km s^{-1} and +200 km s^{-1} is presented in Fig.1*a*. CO J=3–2 emission extends over the current spatial coverage of the ASTE survey. Its distribution closely follows characteristics of the CMZ delineated by the CO J=1–0 surveys (e.g., [3]). Four major cloud complexes are seen also in the CO J=3–2 map, from left to right, the $l = 1.3°$ complex, the Sgr B complex near $l = 0.7°$, the Sgr A complex near $l = 0.0°$, and the Sgr C complex near $l = 0.5°$. At the center of the $l = 1.3°$ complex, we see a clear hole of emission, which is the hyperenergetic shell CO 1.27+0.01 [5]

4. CO J=3–2/J=1–0 Intensity Ratio

We took the CO J=3–2/J=1–0 intensity ratio to extract highly excited gas from the CMZ, since the ratio is sensitive to the temperature and density of molecular gas. Fig.2 shows a frequency distribution of CO J=3–2/J=1–0 intensity ratio ($R_{3\text{–}2/1\text{–}0}$) weighted by CO J=1–0 intensity. The $R_{3\text{–}2/1\text{–}0}$ distribution has a prominent peak at 0.85 which is the typical value in the CMZ, and a shoulder at ~ 0.5 which is mostly attributable to the foreground spiral arms. The ratio close to unity indicates that the bulk of molecular gas in the CMZ is thermalized and moderately opaque.

Generally, $R_{3\text{–}2/1\text{–}0}$ can be a measure of temperature and density. High ratios are found in UV-irradiated cloud surfaces near early-type stars and shocked molecular gas adjacent to supernova remnants. We tried to extract highly excited gas from the CO data sets by $R_{3\text{–}2/1\text{–}0} \geq 1.5$. One-zone LVG calculations say that $R_{3\text{–}2/1\text{–}0} \geq 1.5$ corresponds to $n({\rm H_2}) \geq 10^{3.6}$ cm^{-3} and $T_{\rm k} \geq 48$ K when $N_{\rm CO}/dV = 10^{17}$ cm^{-2} (km s^{-1})$^{-1}$. The spatial distribution of high ratio gas in the CMZ (Fig.2*b*,*c*) is characterized by a number of clumps or spots as well as some diffuse components described as follows.

4.1. Sgr A Region

The most prominent one is a high ratio clump at Sgr A, which consists of several high-velocity features. The high ratio clump has a size of $6' \times 4'$. We see a pair of high-velocity

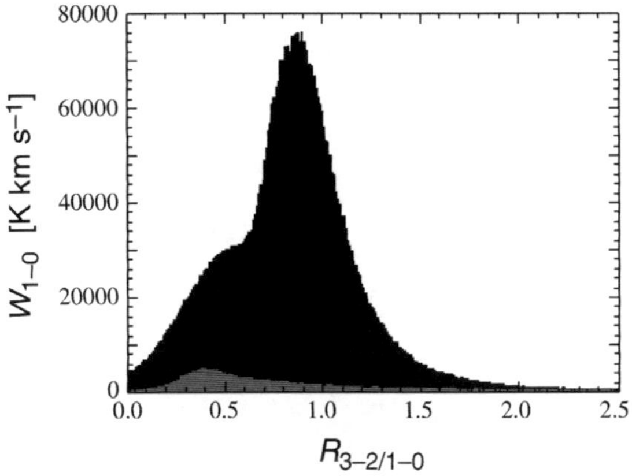

Figure 2. Frequency distribution of $R_{3\text{–}2/1\text{–}0}$ weighted by CO J=1–0 intensity. Gray area shows the contribution of four spiral arms in the Galactic disk.

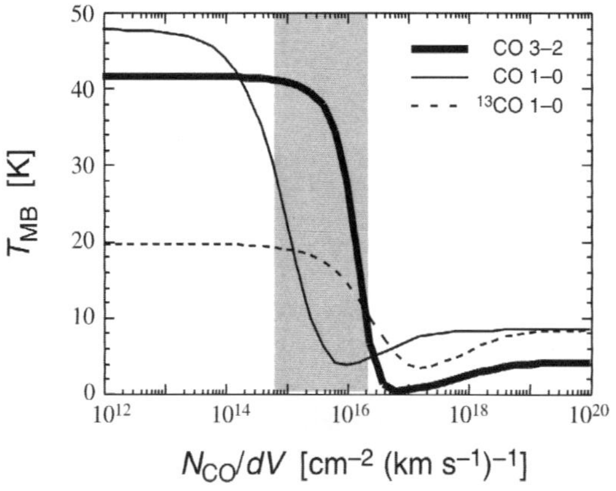

Figure 3. Curves of CO antenna temperatures as a function of $N_{\rm CO}/dV$ of the cool absorber. Shaded area indicates the $N_{\rm CO}/dV$ range where $R_{3\text{–}2/1\text{–}0}$ exceeds 1.5.

emission, which is much larger than the circumnuclear disk (CND; $d \sim 100''$; [13]). We suggest this high ratio clump include the CND itself and its extension. A small red-shifted clump at $(l, b) \simeq (0.01°, -0.02°)$ is a well-known HVCC, CO 0.02–0.02 [4]. High ratio clumps which may be generated by the interaction with Sgr A East [12] appear at $l \simeq -0.03°$, as three velocity components at $V_{\rm LSR} \simeq -70$, +30, +70 km $\rm s^{-1}$. Another component at $(l, V_{\rm LSR}) \simeq (-0.1°, +40\,{\rm km\,s^{-1}})$ with a relatively narrow velocity width may be a cloud surface irradiated by UV photons from the central cluster.

4.2. l=1.3° Complex

The $l = 1.3°$ complex is the large molecular feature having a prominent elongation toward positive latitude [3]. Two expanding shells have been identified at the center of $l = 1.3°$ complex (CO 1.27+0.01; [5]). High ratio gas is associated with both of the expanding shells, and it is abundant especially in the Galactic northwestern rim of the 'minor' shell. The association of high $R_{3\text{–}2/1\text{–}0}$ gas with these expanding shells ensures that they are generated by a series of supernova explosions, suggesting that a microburst of star formation has occurred there in the recent past.

4.3. l=0.9° Anomaly

An anomalous gathering of high ratio gas in the $\Delta l \times \Delta b \sim 0.2° \times 0.2°$ area centerd at $(l, b) \simeq (+0.9°, -0.1°)$. This region corresponds to the edge of the Sgr B complex and a tangent of the star forming ring (Arm I; [14]). The spatial distribution and kinematics of high ratio gas is highly complex. Although the origin of this anomaly is unknown, some compact components in the anomaly can be attributable to the previously identified features. A spot at $(l, b) \simeq (+0.87°, -0.23°)$ has a large velocity width, lying in the rim of an expanding shell-like feature with a diameter of $\sim 0.7°$. This expanding shell touches with the SiO rich core M 0.83–0.18 [9]. Blue-shifted wing emission at $(l, V_{\rm LSR}) \simeq (0.85°, -120\,{\rm kms^{-1}})$ sems to be associated with the radio shell of SNR 0.9+0.1, which is bright in TeV γ-ray [15].

4.4. Absorption by Foreground Gas

We see several narrow velocity-width features in $|l| \leq 0.6$, $-60 \leq V_{\rm LSR} \leq 15$ km $\rm s^{-1}$. They avoid the l-V loci of the foreground spiral arms. These features can be explained by the existence of cool absorber in interarm regions. We made a model with a warm opaque cloud veiled by a layer of cool absorber (Fig.3). Since the J=2 level is subthermally excited in the cool absorber, the J=3–2 line is hardly absorbed unless $N_{\rm CO}/dV \geq 10^{16}$ $\rm cm^{-2}$ (km $\rm s^{-1})^{-1}$ where radiative excitation dominates. This effect raises $R_{3\text{–}2/1\text{–}0}$ without participation of highly excited, less opaque gas. Thus we must see the data in velocities of interam gas with great care.

4.5. High $R_{3\text{–}2/1\text{–}0}$ Spots

We found a number of high ratio spots (Fig.4). Most of these spots have compact appearances and prefer high-velocity ends of giant molecular clouds within the CMZ. Many high ratio spots have compact entities with large velocity width, exhibiting signs of shocked gas. Some of those associate with nonthermal threads or filaments. Four high ratio spots seem to be relevant to the bundle of nonthermal filaments of the Galactic Center Radio Arc.

These facts strongly suggest that supernova/molecular cloud interaction plays an important role in accelerating electrons and forming nonthermal threads or filaments, which are unique and abundant in the CMZ. It has been reported that shocked molecular gas is associated with the nonthermal filament 'Snake' at the intersection with the SNR G359.1–0.5 [16]. The association of shocked gas with nonthermal flaments prefers the hypothesis that localized magnetic tubes with a milligauss field are illuminated by relativistic electrons at these filaments.

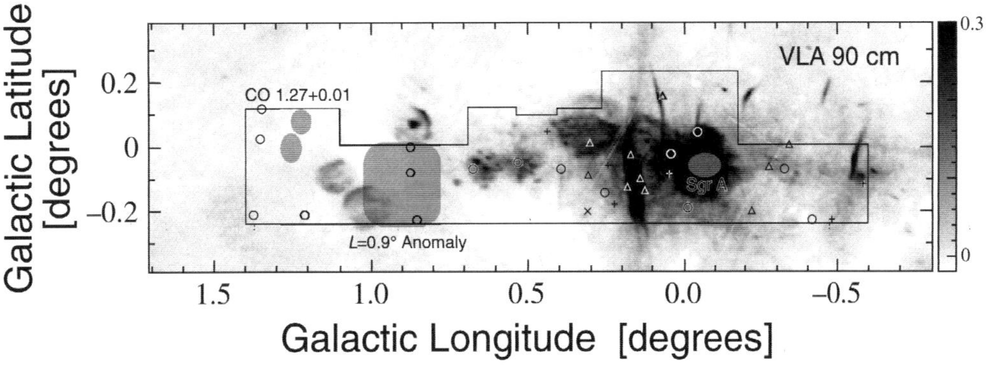

Figure 4. Spatial distribution of high ratio ($R_{3-2/1-0} \geq 1.5$) spots superposed on the VLA 90 cm image [17]. Circles indicate high-velocity compact clouds (HVCCs) or high-velocity wings with high ratio. Triangles are high ratio spots in velocity ends of clouds. Crosses are those in cloud edges. The 'x' denotes a high ratio cloud in the 20 km s^{-1} arm.

5. Summary

We have surveyed a $\Delta l \times \Delta b = 2° \times 0.5°$ area of the CMZ in CO J=3–2 line. The CO J=3–2 distribution closely follows characteristics of the CMZ delineated by the CO J=1–0 surveys. The analysis of the J=3–2/J=1–0 intensity ratio shows that the bulk of molecular gas in the CMZ is thermalized and moderately opaque. We found clumps and small spots of high ratio gas as well as diffuse high ratio features in the CMZ. Most of them are likely shocked gas generated by the interaction with unidentified supernovae, while some could be UV-irradiated surfaces of molecular clouds. Continuation of the CO J=3–2 survey, as well as follow-up studies of high ratio features detected, will reveal the ubiquity and origin of shocked molecular gas in the CMZ.

Acknowledgments

The authors would like to thank all members of the ASTE team for excellent performance of the telescope and every support in observations.

References

[1] Paglione, T. A. D., Jackson, J. M., Bolatto, A. D. & Heyer, M. H. 1998, ApJ, 493, 680
[2] Morris, M., Polish, N., Zuckerman, B., & Kaifu, N. 1983, AJ, 88, 1228
[3] Oka, T., Hasegawa, T., Sato, F., Tsuboi, M., & Miyazaki, A. 1998, ApJS, 118, 455
[4] Oka, T., White, G. J., Hasegawa, T., Sato, F., Tsuboi, M., & Miyazaki, A. 1999, ApJ, 515, 249
[5] Oka, T., Hasegawa, T., Sato, F., Tsuboi, M., & Miyazaki, A. 2001, PASJ, 53, 787
[6] Yamauchi, S., Kawada, M., Koyama, K., Kunieda, H., & Tawara, Y. 1990, ApJ, 365, 532
[7] von Ballmoos, P., Diehl, R., & Schüonfelder, V. 1987, ApJ, 318, 654
[8] Martéin-Pintado, J., de Vicente, P., Fuente, A., & Planesas, P. 1997, ApJ, 482, L45
[9] Hüuttemeister, D., Dahmen, G., Mauersberger, R., Henkel, C., Wilson, T. L., & Martéin-Pintado, J. 1998, A& A, 334, 646
[10] Kohno, K., et al. 2004, in The Dense Interstellar Medium in Galaxies, ed. S. Pfalzner et al. (Berlin: Springer), 349
[11] Oka et al. 2006, PASJ submitted
[12] Christpher, M. H., Scoville, N. Z., Stolovy, S. R., & Yun, M. S. 2005, ApJ, 622, 346
[13] Tsuboi et al. 2006, this issue
[14] Sofue, Y. 1995, PASJ, 47, 527
[15] Aharonian, F., et al. 2005, A& A, 432, L25
[16] Yusef-Zadeh, F., Uchida, K. I., & Roberts, D. 1995, Sci, 270, 1801
[17] LaRosa, T. N., Kassim, N. E., Lazio, T. J. W., & Hyman, S. D. 2000, ApJ, 119, 207

Institute of Physics Publishing
doi:10.1088/1742-6596/54/1/013

Journal of Physics: Conference Series **54** (2006) 72–76
Galaxy Center Workshop 2006

Odin spectral line observations of Sgr A and Sgr B2 at submm wavelengths and in the 118-GHz band

Aa. Sandqvist[1], P. Bergman[2,3], P. Bernath[4], U. Frisk[5], Å. Hjalmarson[2], B. Larsson[1], M. Lindqvist[2], M. Olberg[2], A. O. H. Olofsson[2], L. Pagani[6] and the Odin Team[7]

[1] Stockholm Observatory, AlbaNova University Center, SE-106 91 Stockholm, Sweden
[2] Onsala Space Observatory, SE-439 92 Onsala, Sweden
[3] European Southern Observatory, Alonso de Cordova 3107, Casilla 19001, Santiago, Chile
[4] Department of Chemistry, University of Waterloo, Waterloo, ON N2L 3G1, Canada
[5] Swedish Space Corporation, PO Box 4207, SE-171 04 Solna, Sweden
[6] LERMA & FRE 2460 du CNRS, Observatoire de Paris, FR-75014 Paris, France
[7] http://www.snsb.se/Odin/Odin.html

E-mail: aage@astro.su.se

Abstract. Since its launch in 2001, the Odin satellite has been observing the Galactic Centre Sgr A Complex (CND, +20 and +50 km s^{-1} Clouds) as well as the nearby star formation region, Sgr B2, a number of times. Observations have been made in the 118-119 GHz and 486-581 GHz bands. A limited mapping of the Sgr A Complex in the $H_2^{16}O$ line has been performed and new observations of the $H_2^{18}O$ line took place in 2006. In the 118-119 GHz band, a strong line of HC_3N ($J = 13 - 12$) has been detected at a number of positions - sensitive upper limits have been obtained for the O_2 ($1_1 - 1_0$) and the SiC ($^3\Pi_2, J = 3 - 2$) lines. Towards Sgr B2, submm observations have yielded absorption profiles of $H_2^{16}O$, $H_2^{18}O$, $H_2^{17}O$, NH_3, and $^{15}NH_3$.

1. Introduction

The Odin satellite [1] was launched on 20 February 2001 and is now entering its sixth year of operation (Fig. 1). It is equipped with a 1.1 m offset Gregorian telescope and a cryogenic receiver package of four tunable, SSB, submm Schottky mixers covering the 486-504 and 541-581 GHz frequency range and a fixed-tuned HEMT receiver at a frequency of 118-119 GHz. This is a mission shared between aeronomy and astronomy on a 50/50 basis. The Swedish-led project is funded jointly by the Swedish National Space Board (SNSB), the Canadian Space Agency (CSA), the National Technology Agency of Finland (Tekes) and the Centre National d'Etudes Spatial (CNES, France). The Swedish Space Corporation (SSC) is the prime industrial contractor and is also responsible for the satellite operation.

The Galactic Centre, with its important sources, the Sgr A Complex and Sgr B2, has been of prime importance in Odin's observing program. The first detection of water in the Sgr A Complex with Odin was reported in 2003 [2]. In this paper we concentrate on further observations of the Sgr A Complex and refer to [3] for a preliminary presentation of the Sgr B2 results.

Observations in the Sgr A Complex have been mostly towards three positions, namely Sgr A* with the CircumNuclear Disk (CND), the +50 and the +20 km s^{-1} molecular clouds. Spectral lines probed have been 119- and 487-GHz O_2, 492-GHz C I, 548-GHz $H_2^{18}O$, 557-GHz $H_2^{16}O$, 572-GHz NH_3, and 576-GHz CO ($J = 5 - 4$). We present here only 118-119 GHz, $H_2^{16}O$, and CO ($J = 5 - 4$) results.

Figure 1. The Odin satellite.

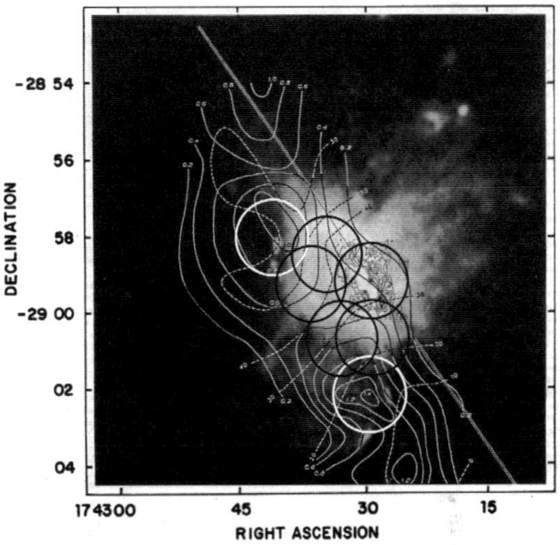

Figure 2. Observed positions in the Sgr A Complex are marked by circles whose 2-arcmin diameter indicates the Odin submm beam. The whole region depicted is slightly larger than the 10-arcmin 119-GHz Odin beam. The Complex consists of the 20-cm continuum radiograph, the CND (thin lines) in HCN, the Molecular Belt and its +20 (SW white circle) and +50 (NE white circle) km s^{-1} Cloud components in 2-mm H_2CO (solid lines) with isovelocity contours (broken lines). The coordinates are epoch 1950.0. (adapted from [4])

2. The submillimetre observations

A limited mapping has now been performed in the Sgr A Complex (Figs. 2 and 3) using the $H_2^{16}O$ line. Water is seen in emission from the CND and the +20 and +50 km s^{-1} Clouds, and in absorption from the Expanding Molecular Ring (EMR) near −130 km s^{-1}, the 3-kpc spiral Arm near −50 km s^{-1}, the −30 km s^{-1} Arm, and the Local Sgr Arm near 0 km s^{-1}. There is also evidence of absorption from a High Positive Velocity Gas (HPVG) component between 100 and 200 km s^{-1} in some positions. No baselines have been subtracted and the effect of the submm dust continuum emission from the Molecular Belt is clearly noticed in the levels of the profiles.

In 2003 [2] we presented the Odin observation of $H_2^{18}O$ towards the Sgr A* CND position. No emission (or absorption) was visible in that profile whose channel resolution was 0.6 km s^{-1} and whose rms noise was 0.02 K. In the late spring of 2006 we scheduled 150 orbits for Odin to observe the +50 km s^{-1} Cloud position. If these observations lead to a positive detection of the $H_2^{18}O$ line, a similar observing period will probably be scheduled for the +20 km s^{-1} Cloud in early 2007.

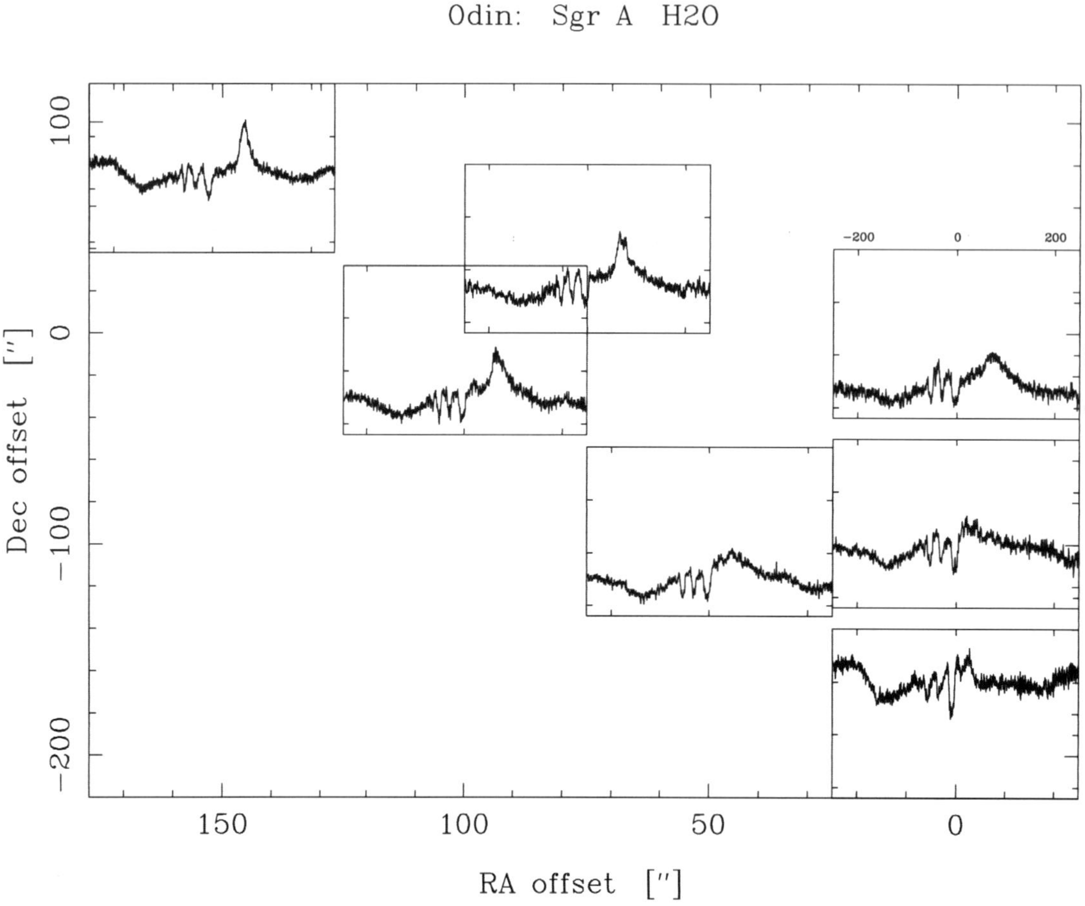

Figure 3. The Odin observations of the 557-GHz $H_2^{16}O$ line in the seven positions mapped in the Sgr A Complex and shown in Fig. 2. The coordinate offsets are relative to the position of Sgr A*. The range of velocity of the profiles is −250 to +250 km s^{-1} and the range of antenna temperature is −0.1 to +1.5 K. These range values are also indicated on the x- and y-axes, respectively, of the profile inset at the Sgr A* position at (0,0).

The Odin observations of the 576-GHz CO ($J = 5 - 4$) line towards the +50 km s^{-1} Cloud, the CND and the +20 km s^{-1} Cloud are shown in Figs. 4-6 where they are compared with ground-based observations of CI, CO ($J = 7 - 6$) and CO ($J = 4 - 3$) lines [5], smoothed to the Odin 2-arcmin angular resolution. Note that the pronounced absorption features seen in the $H_2^{16}O$ line at 0, −30 and −50 km s^{-1} are not seen in the Odin CO ($J = 5 - 4$) line but only in the lower excitation CO ($J = 4 - 3$) line. It is interesting that similar infrared absorption features arising from cold H_3^+ have also been detected [6].

3. The 118-GHz millimetre observations

One of the main motivations for the Odin project is to search for interstellar O_2 and a specially designed radiometer is dedicated to this mission at a frequency around 119 GHz. It has turned out that this is a very elusive molecule and many long-integration-time searches towards a variety of objects have only yielded upper limits, implying very much lower abundances than expected from theoretical calculations [7]. The one important exception to this is the detection of a weak O_2 emission feature in the ρ Oph molecular cloud [8].

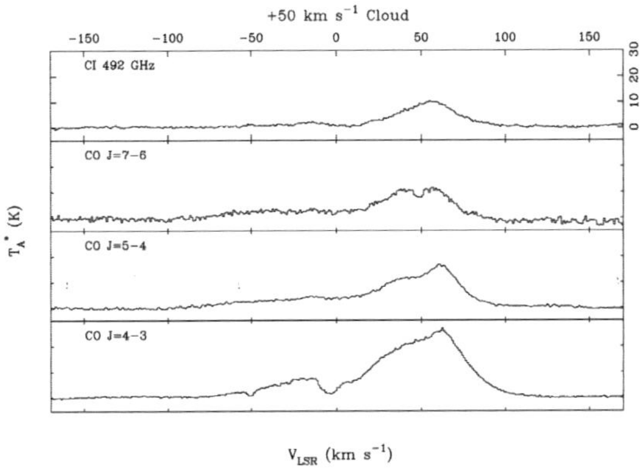

Figure 4. CO (5-4), the +50 km s^{-1} Cloud.

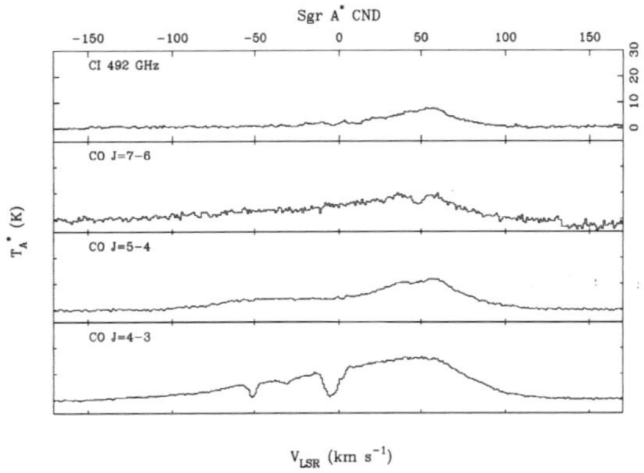

Figure 5. CO (5-4), the CND.

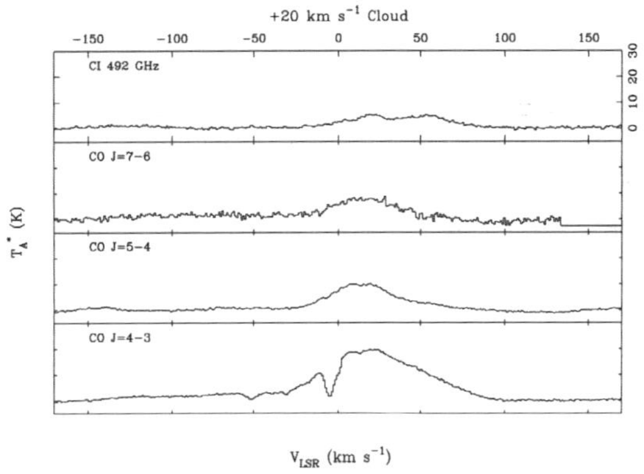

Figure 6. Odin observations of CO ($J = 5 - 4$), the +20 km s^{-1} Cloud. In Figs. 4-6 Odin-beam-smoothed observations of CI, CO (7-6) and CO (4-3) [5] are included for comparison.

The 118 – 119 GHz observations were part of the larger Galactic Centre project, where Sgr A* CND, the +20 and +50 km s^{-1} Clouds were observed at submillimetre wavelengths. These three components of the Sgr A Complex are only separated by a couple of arcmin and all lie within the 10 arcmin Odin 119-GHz beam, irrespective of the position observed. Thus all the mapping positions can be integrated together to obtain sensitive 119-GHz limits. Furthermore, the 119-GHz receiver on board Odin is unfortunately not properly phase-locked but slowly drifts in frequency. However, regularly during each orbit, Odin observes the Earth's atmosphere which allows for monitoring the stability of the receiver and correcting for any frequency drift using the telluric oxygen line [9]. Due to this slow frequency drift, the O_2 frequency has moved to the edge of the band but instead admitted other line frequencies to be observed.

The 118 – 119 GHz observations of the three positions during 2004 have been coadded after careful frequency calibrations and the resulting profile is shown in Fig. 7. The strong line present in the middle of the band we propose is due to the $J = 13 - 12$ transition of the HC_3N (cyanoacetylene) molecule which has a rest frequency of 118.2707322 GHz. The clarity and sharpness of the strong HC_3N feature renders confidence in the reliability of the frequency calibration. Due to an uncertainty of low-level features in earlier 119-GHz observations from 2003, we have indicated in Fig. 7 the expected positions of both the O_2 ($1_1 - 1_0$) line at a rest frequency of 118.7503430 GHz and the SiC ($^3\Pi_2, J = 3 - 2$) line at a rest frequency of 118.1122437/93 GHz [10]. The rms noise is not uniform across the band, which has a channel velocity resolution of 1.6 km s^{-1}, and we therefore have two limits: for the O_2 line the rms value is 4.4 mK, and for the SiC line the rms value is 7.9 mK.

In Fig. 8, we show the different mapping observations which were coadded in Fig. 7. It is

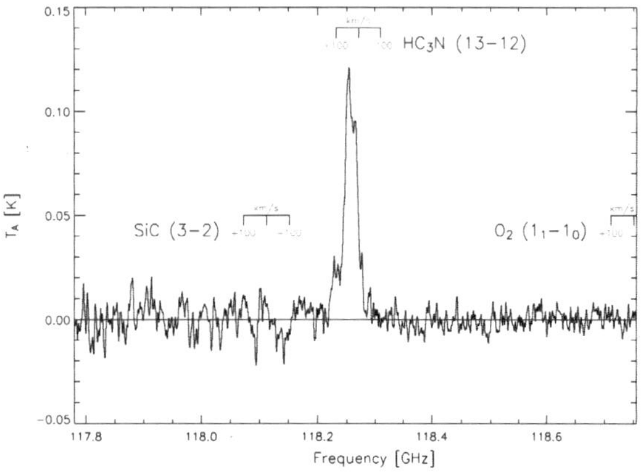

Figure 7. Odin 118-GHz observations of the Sgr A Complex region in the Galactic Centre - total integrated profile, channel resolution: 1.6 km s^{-1}. Small velocity scales are shown above each expected position of line features.

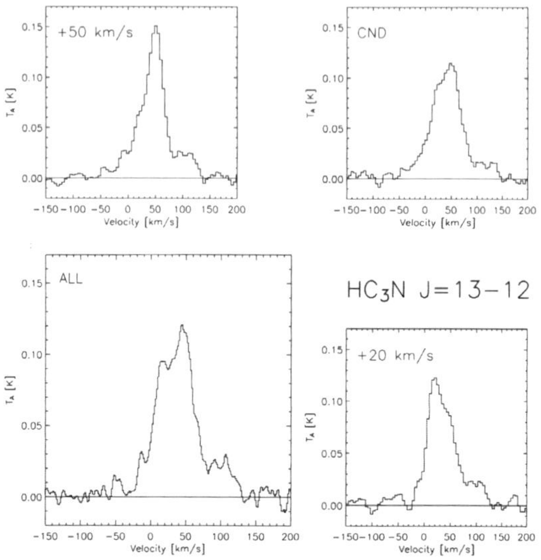

Figure 8. HC_3N "map" showing the three profiles obtained towards the +50 km s^{-1} Cloud (upper left), Sgr A* CND (upper right) and the +20 km s^{-1} Cloud (lower right) - channel resolution is 4.7 km s^{-1}. The total integrated profile with a channel resolution of 1.6 km s^{-1} is shown in the lower left.

clear from both the velocity half-widths (which are of the order of 50 km s^{-1}) and the velocities of the peak temperatures that the HC_3N profiles show behaviour consistent with their dominant sources being the +20 and +50 km s^{-1} Clouds in the Molecular Belt. These are relatively warm ($\approx$ 100 K), high density ($\approx 10^{4.5}$) Giant Molecular Clouds, with moderate velocity dispersion which, however, have only few signs of star formation.

References

[1] Frisk U., Hagström M., Ala-Laurinaho J., et al. 2003 *A&A* **402** L27

[2] Sandqvist Aa., Bergman P., Black J.H., et al. 2003 *A&A* **402** L63

[3] Hjalmarson Å., Bergman P., Biver N., et al. 2005 *Advances in Space Research* **36** 1031

[4] Sandqvist Aa. 1989 *A&A* **223** 293

[5] Martin C.L., Walsh W.M., Xiao K., et al. 2004 *ApJS* **150** 239

[6] Oka T., Geballe T.R., Goto M., et al. 2005 *ApJ* **632** 882

[7] Pagani L., Olofsson A.O.H., Bergman P., et al. 2003 *A&A* **402** L77

[8] Larsson B., Liseau R., Pagani L., et al. 2006 *A&A* (submitted)

[9] Larsson B., Liseau R., Bergman P., et al. 2003 *A&A* **402** L49

[10] Pickett H.M., Poynter R.L., Cohen E.A., et al. 1998 *J. Quant. Spectrosc. & Rad. Transfer* **60** 883 - http://spec.jpl.nasa.gov/

Institute of Physics Publishing
doi:10.1088/1742-6596/54/1/014

Journal of Physics: Conference Series **54** (2006) 77–80
Galaxy Center Workshop 2006

OH 1720 MHz Masers in the Galactic Center: Sagittarius A East and the Circumnuclear Disk

Y.M. Pihlström[1] and L.O. Sjouwerman[2]

[1] University of New Mexico, Albuquerque NM, USA
[2] National Radio Astronomy Observatory, Socorro NM, USA

E-mail: ylva@unm.edu, lsjouwerman@aoc.nrao.edu

Abstract. We present new and archival Very Large Array (VLA) radio interferometry observations of the 1720 MHz OH masers in the Galactic Center (GC). Most 1720 MHz OH masers arise in regions where the supernova remnant Sgr A East is interacting with the interstellar medium. However, here we also present some previously detected positive velocity masers together with some newly detected negative velocity masers. In combination, those masers appear to bracket Sgr A* in position and velocity. The location and velocities of those masers are consistent with their being in the circumnuclear disk (CND).

1. Results from VLA observations

A large fraction of the Galactic Center 1720 MHz OH masers arise in regions where the supernova remnant Sgr A East is interacting with the interstellar medium (ISM). Their velocities are positive with values of about 40–60 $\mathrm{kms^{-1}}$ consistent with the expansion velocity of the supernova remnant [2, 4, 5, 8]. However, a few masers have been detected that appear to have large velocity offsets as compared to the Sgr A East masers. Those include a bright maser detected multiple times at $+132$ $\mathrm{kms^{-1}}$ [8], and a weaker maser detected at a single occasion at -132 $\mathrm{kms^{-1}}$ [4].

In 2005 we conducted new, higher sensitive VLA observations to search for more masers with offset velocity in order to understand their nature. In addition, three VLA archival data sets had a velocity coverage, spatial resolution and sensitivity that were comparable to the new data. These three data sets were reduced and re-analyzed and used in combination with the new data. We detected 13 new and 13 previously published masers brighter than ten times the channel rms; two previous detections were not re-detected (see Fig. 1). Three new high-negative velocity masers confirm the existence of such masers after a single, never confirmed re-detection of the -132 $\mathrm{kms^{-1}}$ maser. They are conjugate to the $+132$ $\mathrm{kms^{-1}}$ masers with respect to Sgr A* and thus must originate in the CND. They are probably unsaturated, and therefore physically different from the SNR OH masers (in green in Fig. 1).

2. Nature of the offset velocity masers

The same VLA data also provides us with OH absorption data which shows that absorption around 60 $\mathrm{kms^{-1}}$ occurs in the eastern part of the SNR shell, while absorption at lower velocities ($\sim$ 20–40 $\mathrm{kms^{-1}}$) also covers the western part of the shell (Fig. 2). This is consistent with the distribution of the $+50$ and $+20$ $\mathrm{kms^{-1}}$ clouds that compose the major parts of the molecular belt extending over a large part of the Galactic Center. The most striking feature of the absorption data is that, at any velocity, little or no absorption can be seen against the minispiral and Sgr A*. While no absorption is seen towards the minispiral and Sgr A*, there appears to be weaker absorption along the western edge of the CND (seen for velocities between +100 and −115 kms^{-1}). The velocity structure of this absorption is in remarkable agreement with the HCN (J=1–0) emission which probes the CND via a clumpy and irregular ring structure [1, 3, 7]. The OH gas seen in absorption along the CND is therefore almost certainly part of the same circumnuclear ring as is the HCN emission.

The location of the OH gas does not give support for a direct interaction between the CND material and Sgr A East, ruling out the supernova remnant shock as the pump source for the high velocity masers. Instead the CND masers may be pumped by dissipation between clumps in the CND. We note that NICMOS imaging of the H_2 line in the GC has revealed emission associated with the CND [9]. It is argued that dissipation of random motion of molecular clumps is the most likely cause of the excitation of the H_2 molecules via shocks, implying the presence of C-type shocks within the CND. Similar to the C-shock chemistry predicted for SNR/ISM masers, the post-shock regions in the CND should produce suitable conditions for regions of enhanced OH abundances [6].

3. Conclusions

These results imply that 1720 MHz OH masers do not occur only in star-forming regions and in SNR/ISM interactions, but can also arise in clumpy and disturbed circumnuclear disks at small radii. If similar conditions occur in circumnuclear disks of extragalactic objects, sufficiently bright 1720 MHz OH maser lines could tentatively serve as an additional tool to study circumnuclear gas dynamics.

References

[1] Christopher, M.H., Scoville, N.Z., Stolovy, S.R., & Yun, M.S., 2005, ApJ, 622, 346
[2] Ekers, R.D., van Gorkom, J.H., Schwarz, U.J., & Goss, W.M., 1983, A&A, 122, 143
[3] Jackson, J.M. et al. 1993, ApJ, 402, 173
[4] Karlsson, R., Sjouwerman, L.O., Sandqvist, Aa., & Whiteoak, J.B. 2003, A&A, 403, 1011
[5] Nord, M.E., Lazio, T.J.W., Kassim, N.E., Goss, W.M., & Duric, N. 2004, ApJL, 601, L51
[6] Wardle, M. 1999, ApJL, 525, L101
[7] Wright, M.C.H., Coil, A.L., McGary, R.S., Ho, P.T.P. & Harris, A.I. 2001, ApJ, 551, 254
[8] Yusef-Zadeh, F., Roberts, D.A., Goss, W.M., Frail, D.A., & Green, A.J. 1996, ApJL, 466, L25
[9] Yusef-Zadeh, F., Stolovy, S.R., Burton, M., Wardle, M., & Ashley, M.C.B. 2001, ApJ, 560, 749

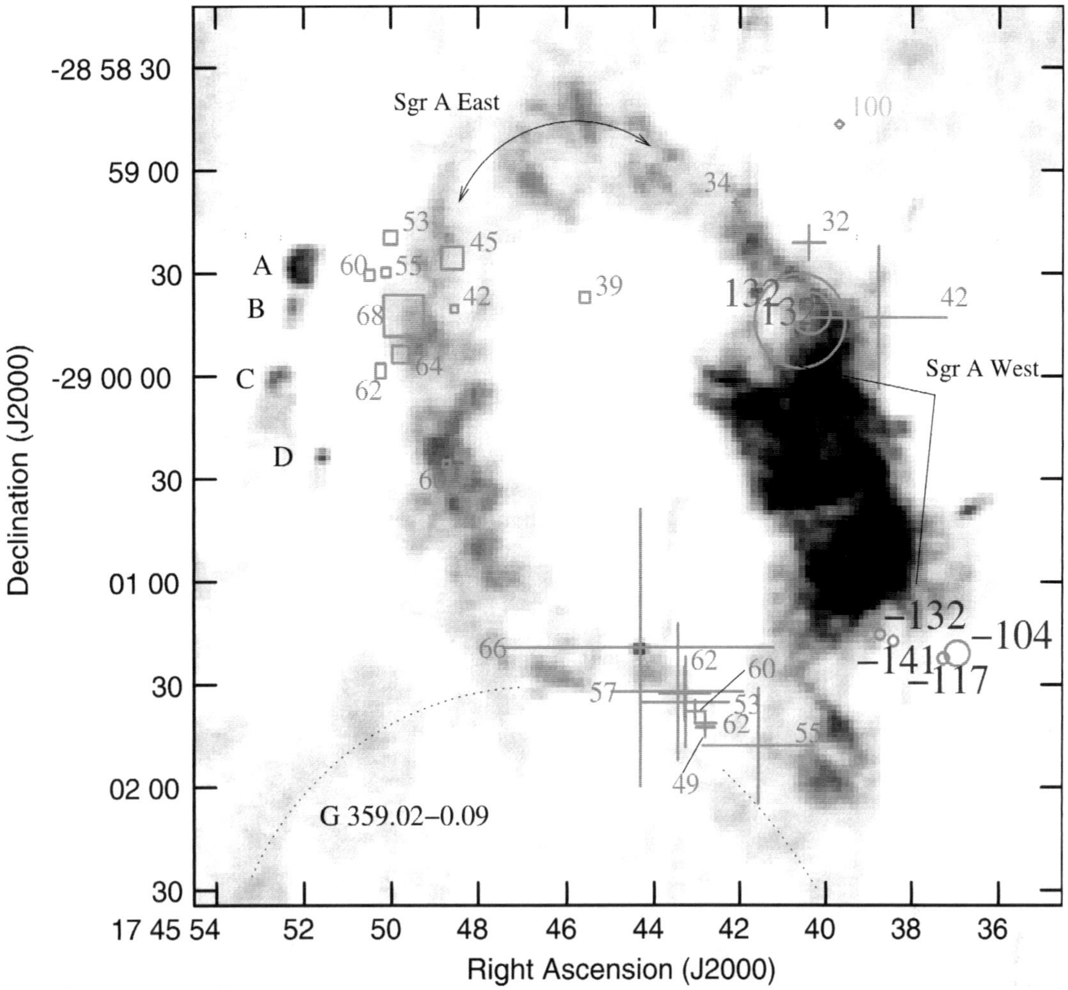

Figure 1. OH 1720 MHz masers superimposed on our 1.7 GHz radio continuum. Numbers associated with symbols are the maser LSR velocities. Pluses and squares denote masers associated with the SNR shell (green), circles are masers associated with the CND (red and blue), and the diamond (pink) represents a newly detected though unrelated maser. The size of the symbol scales with the brightness temperature, except for the brightest maser at +66 km/s, for which the symbol size is divided by a factor of five in order to fit the symbol in the figure. The dotted semi-circle outlines the SNR G359.02-0.09 and labels A-D identify HII regions. As indicated, the large shell-like feature is the supernova remnant Sgr A East, and the bright spiral structure is known as Sgr A West.

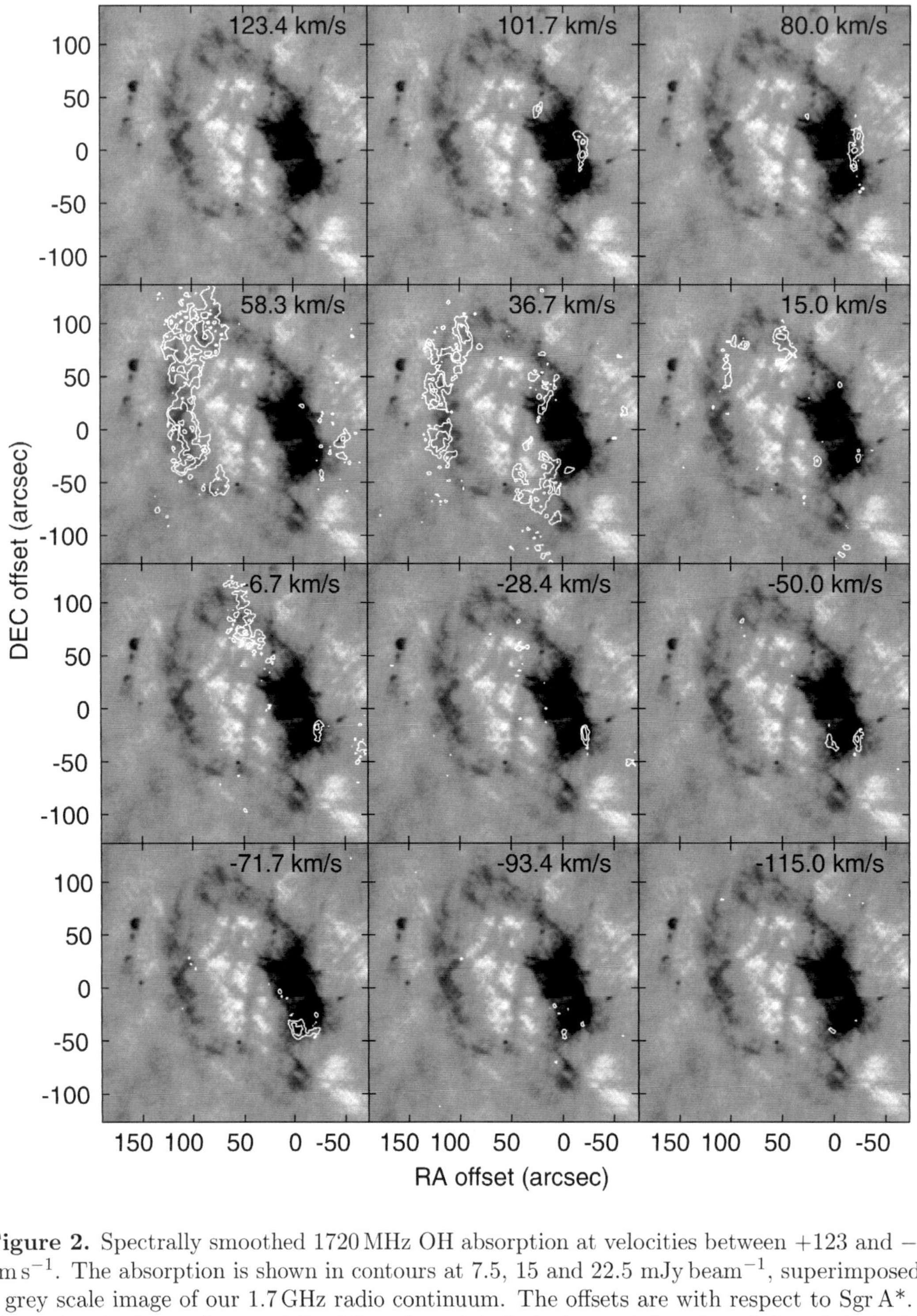

Figure 2. Spectrally smoothed 1720 MHz OH absorption at velocities between +123 and −115 km s^{-1}. The absorption is shown in contours at 7.5, 15 and 22.5 mJy beam^{-1}, superimposed on a grey scale image of our 1.7 GHz radio continuum. The offsets are with respect to Sgr A*. No absorption is observed toward Sgr A* or the minispiral. Around the edge of the minispiral, the CND can been observed in absorption. Absorption at +58 and +37 km s^{-1} mainly illustrates the extent of the molecular belt.

Institute of Physics Publishing
doi:10.1088/1742-6596/54/1/015

CN in the Sgr A* environment - first results

M Stanković[1], E R Seaquist[1] and S Mühle[1]

[1] Department of Astronomy and Astrophysics, University of Toronto, 60 St. George St, Toronto, Ontario, M5S 3H8

E-mail: stankovic@astro.utoronto.ca

Abstract.
The radical CN, through the CN/HCN ratio, has proven to be an exceptionally good tool to study the photodissociation regions (PDRs) dominated by FUV radiation. Our region of interest, the circumnuclear disk (CND) surrounding Sgr A*, is exposed to the strong UV radiation field originating from the central stellar cluster. It contains potentially star forming cores recently elucidated by interferometric HCN observations with OVRO by Christopher et al. (2005). We present the results of preliminary observations of CN (N=2–1) transitions with the IRAM 30-m telescope and report, for the first time, detection of CN emission in the GC region indicative of PDR activity. This work motivates the further investigation of the role of PDRs in the clumpy medium of the Galactic center (GC) environment in global photodissociation and in promoting or inhibiting star formation in the CND.

1. Introduction

Within the inner 10 pc of our Galaxy lie several fascinating structures not seen elsewhere in the Galaxy. The most prominent molecular gas feature surrounding Sgr A* is an asymmetric, clumpy ring of warm and dense molecular gas commonly referred to as the circumnuclear disk (CND), with outer radius 3–4 pc and inner radius 1.5–2 pc (e.g. [2, 4, 7]). It is immersed in a strong UV radiation field from massive stars in the central cluster responsible for ionizing gas inside the ring. The origin of these massive stars is not yet understood ([1] and references therein). This ionized gas component forms a remarkable system of streamers (the so-called minispiral) associated with gas accreting onto the central supermassive black hole. The non-ionizing (FUV) radiation, responsible for photodissociation of molecular species in the central cavity, diffuses farther out into the clumpy molecular ring, heating it and promoting gas phase chemical reactions, thereby producing photodissociation regions (PDRs). FUV photons not only govern in part the physical properties of associated PDRs, but they may also play an important role in regulating the star formation processes [9].

Recent interferometric studies of the CND in HCN and HCO^+ at $\sim 4''$ resolution by [4] reveal the presence of exceedingly dense cloud cores ($10^7\,cm^{-3}$) with an average diameter of $7''$ (0.3 pc). These core densities are high enough to withstand tidal shear from Sgr A* and the central stellar concentration, thereby representing possible star forming zones that could provide a source for the young stars in the central parsec. Mid-IR observations of the CND by [15] show unexpectedly high dust temperatures in the regions of high optical depth, mostly in the southwestern lobe of the CND which contains some of the most massive cores discovered by [4]. Similarly, in their $30\mu m$ observations, [11] identify at least 16 color temperature peaks well within the CND. These peaks may be young stellar sources embedded in the ring which locally heat the gas.

The PDR models predict ([6]) that FUV heating of molecular gas clumps generates photoevaporative outflows and drives strong shocks, compressing the gas to higher densities. These predictions indicate that investigation of the PDR structure in the GC region will significantly contribute to resolving persisting questions about the evolution of the CND and star formation in the hostile Sgr A* environment.

1.1. Diagnostic Properties of CN

The radical CN has proven to be an exceptionally good tool to study dense regions illuminated by a strong UV radiation field when supplemented by data from other selected molecular tracers [3, 5], and yet there are no published observations of CN in the GC region. CN is a by-product of HCN photodissociation and the photons of very high energy (>12.4 eV) are required to destroy this molecule. In particular, the CN/HCN intensity and abundance ratios have been identified as particularly good diagnostic probes of FUV irradiated molecular gas [3, 5, 13]: the CN molecules are more abundant in zones closer to the PDR surface where the UV radiation field is only partly attenuated, whereas HCN becomes abundant at larger (and more shielded) cloud depths closer to the cloud core, implying that large CN/HCN ratios arise naturally in the surface layers of dense PDRs [5, 13].

Observing the CN emission is a further benefit in such studies because the structure in the CN spectra shows fine and hyperfine structure components (Figure 1) that provide information on the physical parameters of the PDR zone. In particular, resolvable fine and hyperfine structure components are important to the interpretation of molecular observations because the optical depth of the observed transition, and hence the CN column density in PDR zones, can be obtained by comparing their relative intensities within a single rotational transition. For optically thin lines the intensity ratio is proportional to the ratio of the corresponding transition strengths; the ratio approaches unity for optically thick lines (assuming that all hyperfine levels have the same excitation temperature).

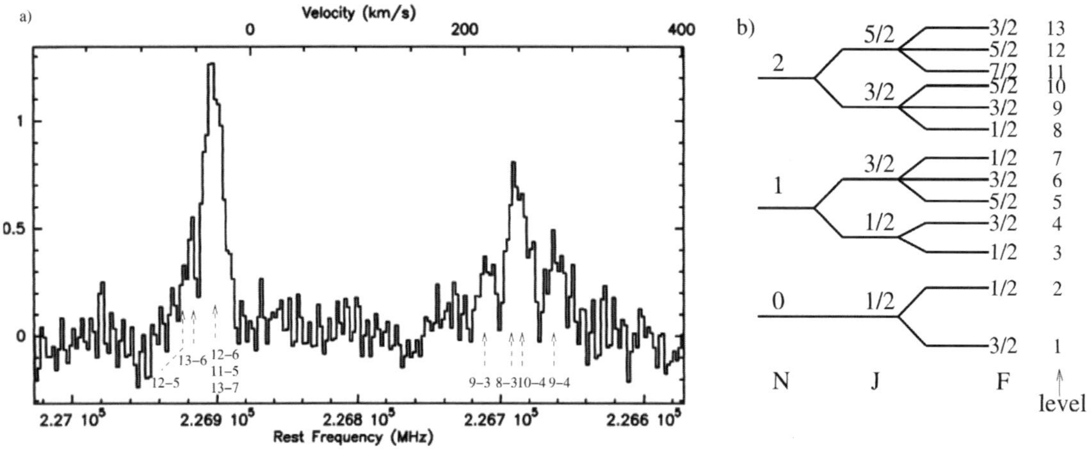

Figure 1. a) An illustration of resolved fine and hyperfine structure components in the CN spectra: the observed CN(N=2–1) transition toward CRL2688 (post AGB star) used as a calibrator source for our Galactic center observations. b) Energy level diagram for CN.

2. Observations and First Results

The observations were made with the 9 pixel dual-polarization heterodyne receiver array, HERA, on the IRAM 30-m telescope in the CN (N=2–1) line (226.875 GHz (2–1, J=5/2–3/2), 226.660 GHz (2–1, J=3/2–1/2)) in October 2005. The WILMA backend was operated in wideband mode offering a total bandwidth of 1 GHz (~1300 km s^{-1} at 230 GHz) sufficient to accommodate the velocity range of the GC structures, and channel separation of 2 MHz (~2.5 km s^{-1}). The beamwidth (FWHM) of the IRAM 30-m telescope is 11$''$ at 230 GHz and the main beam efficiency is 0.51. We position-switched to an emission-free off position at $\Delta\alpha$=120$''$, $\Delta\delta$=-800$''$ relative to Sgr A*, integrating for a total of 4 minutes per point. This reference position was checked to be free of emission by comparison with other positions away from the GC.

The array receiver was centered on Sgr A* and four other pointing positions at offsets of $\Delta\alpha = \pm 2'$ and $\Delta\delta = \pm 2'$ with respect to Sgr A* (Figure 2a). Our observations show that the CN emission decreases away from the CND (Figure 2b). The strongest emission corresponds to the so called southwestern lobe of the CND, containing the most massive cores and showing enhanced HCN emission from the CND. This is also a region of a good spatial agreement between the HCN and H_2 emission at 2.12μm [16], possibly excited either by UV fluorescence or shocks. We detect CN (N=2–1) emission towards Sgr A* as well, whereas other molecular species are photodissociated by the strong UV radiation originating from the central stellar cluster. Our spectra show resolvable fine structure components which provide an excellent diagnostic of the CN optical depth (Figure 2b). We are currently developing a method to deconvolve complex velocity structure of the GC region allowing us to distinguish between fine and possibly hyperfine structure components.

3. Future Work - Investigation of PDR structure in the GC region

In order to understand the fundamental problems of star formation in the GC region and the CND evolution, we intend to investigate the structure of the PDR in the CND and surrounding molecular clouds. PDRs unambiguously indicate the presence of young, massive stars in the central parsec, whose FUV radiation significantly influences the thermal balance and affects the chemistry and physical state of the gas [9]. The incident FUV heats up and pressurizes the surface layers, inducing the mass loss which causes photoevaporation of dense clumps, and drives shocks into the clumps, compressing them to higher densities and making them more stable against the strong tidal stresses from Sgr A* [6]. The compression of non-collapsing clumps driven by shock waves could possibly induce inner core instability and gravitational collapse, thereby triggering star formation. The clump parameters and state of the interclump medium will regulate the dynamical evolution and determine the timescale of photoevaporation of these clumps (and potentially of the entire CND), thereby influencing star formation, either by inhibiting or inducing it. The presence of atomic neutral gas in the central cavity (e.g. [10]), devoid of molecular gas, is already a signpost of ongoing photodissociation at the inner edge of the CND.

A detailed investigation of the PDR structure of the entire CND and its environs could readily lead to an estimate of the lifetime of the CND against complete dissociation of its individual clumps and whether some clumps may survive to form stars. An assessment of pressure confinement of the dense clumps by their surrounding diffuse gas will be necessary to test whether the interior conditions of the PDR associated clumps are favorable to collapse. The investigation of the PDR structure at the inner rim of the CND would help us to understand whether this zone represents an advancing front of dissociation into the CND. Comparisons with other data such as the H_2 fluorescence [16] in this region will be fruitful. It may be possible, from the estimated clump dissociation rates, to date the onset of the FUV continuum for comparison with the age of its source - the young massive stellar population in the central parsec.

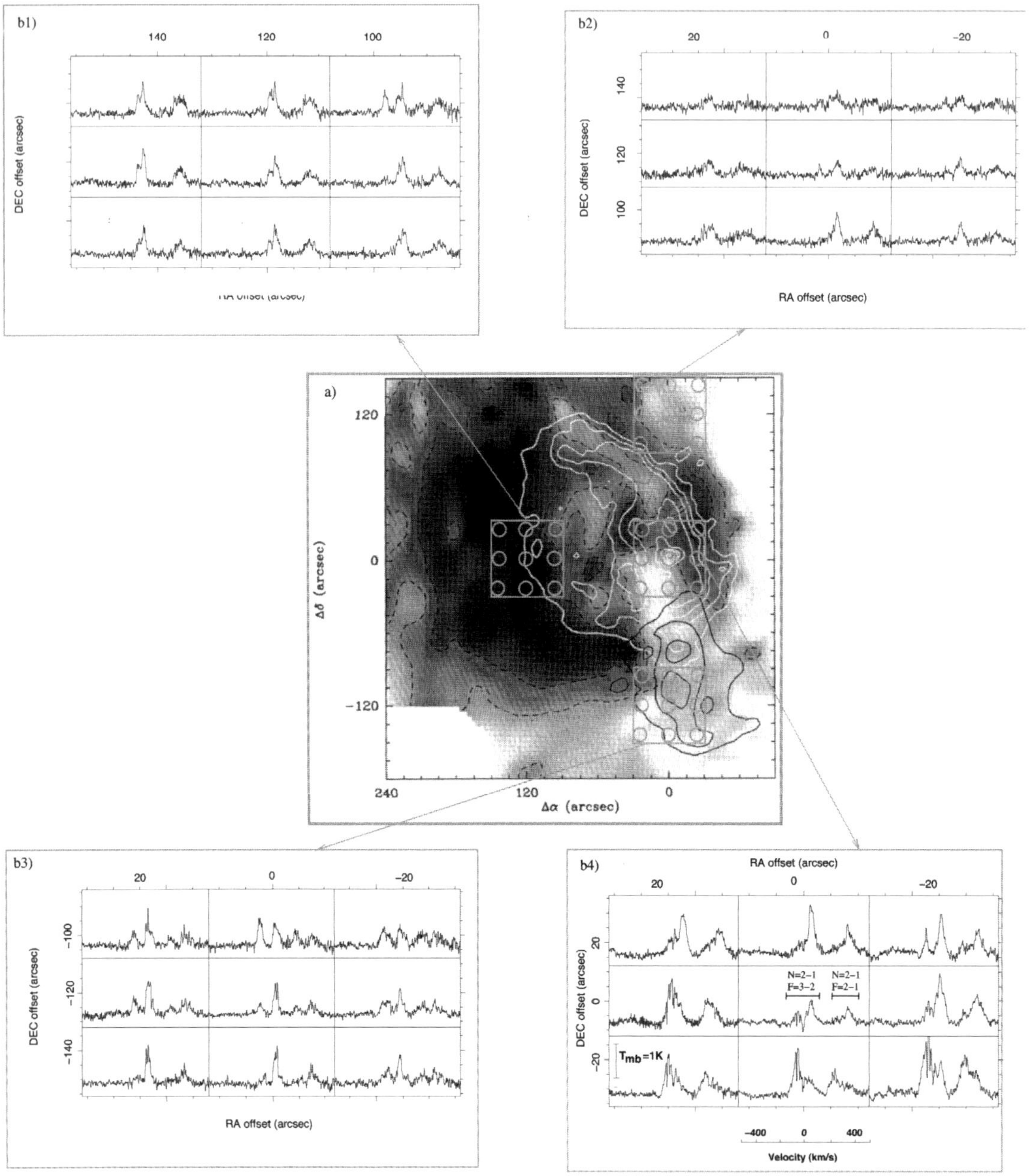

Figure 2. a) Locations of the pointing positions observed with HERA at the IRAM 30-m telescope, superposed on a map of the Sgr A* region from [12]. The grey scale image shows the distribution of CI(1–0) 492 GHz emission, and the white contours the distribution of nonthermal continuum emission from Sgr A East. The western pointing position at $\Delta\alpha = -2'$ and $\Delta\delta = 0'$ is not shown in this figure. Open circles show the positions and beamwidth of the pixels of the HERA array for our CN(2–1) observations. **b)** CN(2-1) spectra shown in b1-b4, observed with HERA at the IRAM 30-m telescope at the positions indicated under a). Our spectra show resolvable fine structure components (as indicated for the spectrum at the central position under b4), leading to direct estimates of the line optical depth from measurements of a single rotational transition. We are currently developing a method to deconvolve the complex velocity profiles in the GC from these components.

The results reported here are based on test observations of CN and they motivate the further exploration of the PDR structure in the CND and surrounding molecular clouds. We intend to conduct a multi-transition survey of a $5' \times 5'$ region centered on Sgr A* that contains the CND and its environs, using a set of strategically selected lines at three different frequencies at a similar resolution ($12''$-$15''$). Such projects are now feasible because of the availability of new heterodyne array receivers at the JCMT, the IRAM 30-m and the NRO 45-m telescopes, and new PDR and radiative transfer codes permitting an increasingly sophisticated analysis of such data. This is ongoing work and we intend to start a larger survey in the coming fall semester of 2006 at the JCMT.

To investigate the PDR structure, densities and temperatures in different physical regimes need to be related by a constraining physical theory and we intend to use sophisticated PDR and radiative transfer models that can relate these nonuniform conditions by photochemical reactions in a PDR zone [8, 14]. In addition, to model the evolution of CND clumps we intend to fit the theory of clump evolution in [6] to the derived PDR clumps parameters, based on a geometric model involving simple spheres with a radial density gradient. For example, the basic clump photoevaporation timescale is set by the radius, the core density, and the density and sound speed in the outer clump PDR zone, all of which can be estimated from the PDR model. The theory also allows an estimate of whether part of the clump will survive, and the virial theorem with external pressure and tidal stresses included can be used to determine whether further contraction of the cold remnant to more extreme densities is likely.

Acknowledgments
This research is being supported by a grant to E.R.S. from the Natural Sciences and Engineering Research Council of Canada and by fellowship support from the University of Toronto.

References

[1] Alexander T 2005 *Phys. Rev.* **419** 65
[2] Becklin E E, Gatley I and Werner M W 1982 *ApJ* **258** 134
[3] Boger G T and Sternberg A 2005 *ApJ* **632** 302
[4] Christopher M H, Scoville N Z, Stolovy S R, Yun M S 2005 *ApJ* **622** 346
[5] Fuente A, García-Burillo S, Gerin M, Teyssier D, Usero A, Rizzo J R and de Vicente P 2005 *ApJL* **619** 155
[6] Gorti U and Hollenbach D 2002, *ApJ* **573** 215
[7] Herrnstein R M and Ho P T P 2005 *ApJ* **620** 287
[8] Hogerheijde M and van der Tak F 2000 *A&A* **362** 697
[9] Hollenbach D J and Tielens A G G M 1999 *RvMP* **71** 173
[10] Jacskon J M, Geis N, Genzel R, Harris A I, Madden S, Poglitsch A, Stacey G J and Townes C H 1993 *ApJ* **402** 173
[11] Latvakoski H M, Stacey G J, Gull G E and Hayward T L 1999 *ApJ* **511** 761
[12] Morris M and Serabyn E 1996 *ARAA* **34** 645
[13] Sternberg A and Dalgarno A 1995 *ApJS* **99** 565
[14] Störzer H, Stutzki J, Sternberg A, 1996, *A&A* **310** 592
[15] Telesco C M, Davidson J A, Werner M W 1996 *ApJ* **456** 541
[16] Yusef-Zadeh F, Stolovy S R, Burton M, Wardle M and Ashley M C B 2001 *ApJ* **560** 749

Institute of Physics Publishing
doi:10.1088/1742-6596/54/1/016

Journal of Physics: Conference Series **54** (2006) 86–94
Galaxy Center Workshop 2006

High Energy Emission from the Galactic Center

Andrea Goldwurm[1,2]

[1] Astroparticule et Cosmologie, Paris, France
[2] Service d'Astrophysique/DSM/DAPNIA/CEA-Saclay, 91191 Gif sur Yvette, France

E-mail: `agoldwurm@cea.fr`

Abstract. The Center of our Galaxy has been observed with high energy instruments since the very beginning of the X-ray and gamma-ray astronomy space programs to search for the high energy radiation from the nearest super-massive black hole, now associated to the peculiar radio source Sgr A*. In the recent years, with the launch of Chandra, XMM-Newton and INTEGRAL and with the start of the operations of the ground-based VHE observatory HESS, this quest has led to some fundamental discoveries with the detection of flaring activity from Sgr A* and of several other high energy point-like and diffuse sources in the region. I will review, after an introductory summary of historical developments, the recent results in the field with particular attention to the findings related to Sgr A* and the observations of the INTEGRAL observatory.

1. Introduction

The Galactic Center (GC) is a very dense and complex sky region of approximately 600 pc size ($\sim 4°$ in projection) where a number of interesting high-energy phenomena take place. Located at about 8 kpc distance it hosts the nearest super massive black hole (SMBH) to the sun, surrounded by a variety of objects that interact with each other. The high energy processes generated in this extreme environment are possibly common to other galactic nuclei, and therefore, given its proximity, it represents a unique laboratory for modern astronomy. Totally obscured in the optical wavelengths by the Galactic plane dust, the GC is mainly observed from radio to infrared (IR) and again at high energies. Some important results have been recently obtained with the new generation of X-ray and gamma-ray observatories: Chandra, XMM-Newton, INTEGRAL and HESS. After a short introduction on the characteristics of the Galactic Center (see also [42, 43, 27, 41]) I will review the historical developments of high energy observations of this peculiar region and will summarize some of the recent results.

The radio images of the GC obtained with the VLA at 90 cm [32] show all the complexity of the central few square degrees of the Milky Way (Fig. 1). This region indeed contains $\sim 10\%$ of the galactic interstellar molecular gas, concentrated in the dense giant molecular clouds (MC) like Sgr B, Sgr C and those of the Sgr A radio and molecular complex. Shell-like supernova remnants (SNR) (e.g. G 359.1-005) heat the interstellar matter with their expanding shells which in some cases appear to be interacting with the molecular clouds. While several other non-thermal filaments demonstrate the presence of accelerated particles spiraling the strong (~ 1 mG) magnetic fields of the region (like the large structure known as the *radio arc*), other structures have thermal radio spectra and are in fact HII regions ionized by close-by, hot, and young star clusters. The central 30 pc are dominated by the Sgr A complex, formed by a few MCs (M-0.02-0.07, M-0.13-0.06) an expanding SNR, Sgr A East, and a central HII region surrounding the bright compact radio source Sgr A*, the radio manifestation of the central SMBH. Sgr A East is a non-thermal radio source composed of diffuse emission of triangular shape and an inner oval shell ($3' \times 4'$, or 7 pc $\times$ 9 pc) centered about $50''$ ($\approx$ 2 pc) west of Sgr A*. The shell appears to be expanding, compressing the molecular cloud M-0.02-0.07 and probably creating the string of 4 HII regions and the OH masers observed around the shell. The first estimates of shell energy were well above the typical release of a SN, and it was proposed that Sgr A East was the result of 40 SN or of the explosive tidal disruption of a star by the SMBH. In the inner regions, a molecular ring rotates rapidly around the compact source Sgr A*, and appears to join onto a thermal diffuse nebula, Sgr A West, with the characteristic shape of a minispiral that is also centered on Sgr A*. Sgr A West is ionized by a cluster of hot young and massive stars, centered at about $2''$ from Sgr A* and known as the IRS 16 cluster. Some of these stars emit powerful stellar winds which interact with the surrounding medium and must feed the SMBH.

While the dynamics of matter at radial distances > 2 pc are dominated by the central core of the Galactic Bulge star cluster, the large velocities of gas and stars observed in the innermost regions imply the presence of a massive black hole. Adaptive optics NIR measurements of the velocities and proper motions of the brightest and closest stars to Sgr A* (the so called "central star cluster"), made with the NTT, the VLT and the Keck over the last 10-15 yr, have provided precise orbital parameters for several of them [57, 20, 24]). The derived parameters imply the presence of a dark mass of 3-4 10^6 $M_\odot$ enclosed within a radius < 100 AU. Only a SMBH can explain such densities. The dynamical center of the central star cluster is coincident (within 10 mas) with the bright ($\approx$ 1 Jy), compact, variable, synchrotron (flat power law spectrum) radio source Sgr A*. Since its discovery 30 years ago [7], it has been considered the counterpart of the massive black hole of the Galaxy. The source is linearly polarized at sub-mm frequencies, where the spectrum also presents a bump indicating that the emission becomes optically thin. The proper motion of Sgr A* is < 20 km s^{-1} and its size, measured at frequencies of 3 mm where the interstellar scattering is small, is of the order of 0.1-0.3 mas, about 15-20 R_S, where $R_S = \frac{2GM}{c^2} \sim 10^{12}$ cm = 0.06 AU is the Schwarzschild radius for a 3.5 10^6 $M_\odot$ BH.

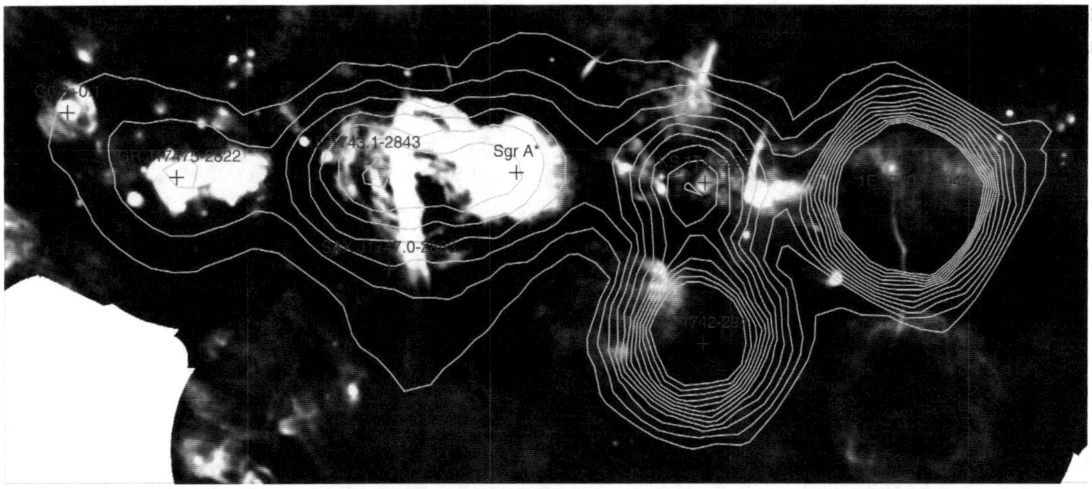

Figure 1. The VLA radio image (2.5° $\times$ 1.5°) at 90 cm of the Galactic Center region, in galactic coordinates with grid lines separarted by 0.5° and the 20-40 keV INTEGRAL/IBIS contours (see Fig. 4) [10].

2. The quest for the high energy emission from Sgr A*

The first detections of high energy emission from the GC direction obtained in the 1970-1980 seemed to support the presence of a SMBH at the center of the Milky Way, predicted first by Lynden Bell and Rees in 1971 [36] and suggested by the discovery of Sgr A* of Balick and Brown in 1974 [7]. However as the resolution and sensitivity of the high energy telescopes improved it was realized that the Galactic SMBH itself was actually very weak. Proctor et al. (1978) [53] presented the picture of the known X-ray emission from the GC at the eve of the launch of the HEAO II satellite, the Einstein Observatory. The Galactic Center X-ray source (GCX-1) detected by Uhuru in 1970 was already attributed to a composition of several sources, some of which variables, but whose positions were known with error boxes of typical sizes of 0.1° - 0.5°.

The first soft X-ray (< 4 keV) images with arcmin resolution were obtained only with the Einstein observatory in 1979 [60] and showed that the central $20'$ of the Galaxy were dominated by diffuse emission and several weak point-like sources one of which associated to the Sgr A* complex. The latter was resolved in 3 weaker sources with Rosat [52] and this set an upper limit of only 10^{34} erg s^{-1} to the soft X-ray luminosity of the nucleus. On the other hand the BH could still shine in hard X-rays or even at 511 keV, the positron-electron annihilation line, since important fluxes were observed from the general direction of the GC since the early 70s at these energies. After all, BH binaries like Cyg X-1 often emit the bulk of their accretion luminosity at $>$ 100 keV. However in the 1990s observations in hard X-rays (3-30 keV) with XRT/SL2 [58] and with ART-P/GRANAT [49] and then in soft gamma rays (30-1000 keV) with SIGMA/GRANAT [26, 27] showed that the GC hard X-rays were rather due to the peculiar hard source 1E 1740.7–2942, a BH X-ray binary and microquasar located at $>$ 100 pc away from the center, and that Sgr A* also was under-luminous at these energies. The SIGMA telescope also set upper limits on the presence of a point-like source of 511 keV line [38] while in those same years OSSE/CGRO showed that the bulk of the 511 keV line emission was not variable but rather diffuse and extended over the whole Galactic bulge [54]. The pre-Chandra/XMM and pre-INTEGRAL/HESS era ended with the detection of a GC gamma-ray source at energies $>$ 100 MeV with EGRET/CGRO [40]. This source is now positioned slightly away from the nucleus ($\sim 0.2^\circ$) [30] but it could still be linked to Sgr A*.

These high energy results set the total Sgr A* bolometric luminosity to less than (0.5 - 1) 10^{37} erg s^{-1} [27, 47] and prompted the development of new accretion models where the radiative efficiency is very low. The most popular were those named Advection Dominated Accretion Flows (ADAF) which assume that the energy exchange between protons and electrons is inefficient and electron temperature remains well below the virial temperature reached by the protons. Energy is therefore advected into the BH before being released as radiation [47]. These models have been used for other BH accreting systems with sub-Eddington accretion rates. They predict very low efficiencies, but very hard X-ray spectra extending to the gamma-ray bands.

A new era of high energy astronomy was opened with the launch of Chandra and XMM-Newton (1999) X-rays observatories, the launch of the gamma-ray mission INTEGRAL (2002) and the start of operations of the HESS Atmospheric Cherenkov ground telescope (2003) for very high energy (VHE) gamma-rays.

3. The Galactic Center in the X-ray band

In the 2-10 keV X-ray band the GC has been deeply monitored by Chandra and XMM-Newton in the recent years. The emission is dominated by few bright, sometimes transient, X-ray binaries probably not associated with the GC (e.g. 1E 1740.7–2942, 1E 1743.1–2843, etc.). However, the region also contains a large population of weak point-like persistent and transient sources, distributed as the old stellar population of the Galactic Bulge cluster. However the visible point sources can account only for 10% of the total emission, the rest being produced by diffuse emission with 3 distinct components, a soft thermal one (kT $\sim$1 keV), probably due to SNR heated gas, a hot (kT ~8 keV) possibly thermal component and a non-thermal one characterized by a strong 6.4 keV line (the fluorescent line of neutral or weakly ionized iron). Several SNR, non-thermal filaments and star clusters are also detected. The central 20 pc emission is dominated by the thermal emission from Sgr A East, while Sgr A* itself appears very weak.

*3.1. Quiescence and flaring activity of Sgr A**

As mentioned above, a SMBH surrounded by dense environment is an ideal system to generate accretion powered X-ray emission, but it was quickly realized that the Galactic SMBH is very weak, even in the hard X-ray domain (where BH binaries in hard state emit the bulk of their accretion luminosity) [26, 27]. The total (from radio to X-rays) Sgr A* luminosity amounts to less than 5 10^{36} erg s^{-1}, i.e. some 10^{-8} times the Eddington luminosity (L_E) of a 3.5 10^6 $M_\odot$ BH. Since the IRS16 stellar winds are supposed to feed the BH at a rate of few 10^{-4} $M_\odot$ yr^{-1} (but see [17] for an update on this issue) which implies accretion luminosities of 0.02 L_E, this led to the development of theories of very inefficient accretion flows (e.g. Bondi models and ADAF models) [41, 47]. The Chandra observatory in 1999, with its unprecedented angular resolution of 0.5$''$ confirmed the very low X-ray luminosity of Sgr A* (2 10^{33} erg s^{-1} in the 2-10 keV band) but measured a steep spectrum (photon index $\alpha \sim 2.5$), not compatible with the ADAF thermal bremsstrahlung models ([6]). One year later Chandra made the dramatic discovery of a powerful X-ray flare from Sgr A*. During this event, of a total duration of 3 hr, the flux increased by a factor 50 to reach luminosities of 10^{35} erg s^{-1} and displayed a hard spectral slope ($\alpha \sim 1.3$) [5]. XMM-Newton confirmed the presence of such bright, hard flares from Sgr A* [28] and discovered the most powerful one with an increase factor of 200 and, this time, a significantly steeper spectrum (α ~2.5) [50] (see Fig. 2). The flare duration (few hours) and the observed short-time-scale variations (200 s) indicate that the X-ray emission is produced within 20 R_S. This cannot be accounted for by the standard ADAF model (for which the bulk of the X-ray emission is produced from the whole accretion flow starting at the accretion radius), so several other models are now considered where non-thermal emission plays a major role. The Liu and Melia picture [33, 34] assumes that after a Bondi-like flow with sub-equipartition magnetic field, the accreted matter circularizes in a small, very hot, highly-magnetized Keplerian disk where quasi-relativistic electrons produce synchrotron radiation in the sub-mm band and, by inverse Compton, the steep X-ray spectrum. Flares can be produced either by sudden increase in accretion rate or rapid release of magnetic energy, and the two different X-ray spectral slopes can be explained. Markoff et al. [39] locate the main energy release at the base of a relativistic jet rather than in the accretion disk. Substantial modification of ADAF models (inclusion of outflows, convection and non-thermal component) were also considered [61] to explain the Sgr A* X-ray flares. The different models can account for the observed spectral shapes, but they predict different correlations between sub-mm, NIR and X-ray fluxes. Simultaneous observations of Sgr A* flares at different frequencies could strongly constrain the models, and therefore several multiwavelength campaigns on Sgr A* have been performed in the recent years (see section 3.4).

3.2. The diffuse X-ray emission

The Chandra and XMM-Newton surveys of the GC have also provided several new results on the diffuse X-ray emission. The first one is the confirmation that the central few hundred parsecs are permeated by a hard diffuse emission peaked towards the center and extending along the plane, as observed by previous instruments and in particular by ASCA [31]. Continuum and line spectra of this emission [48, 44] (and in particular the strong 6.7 and 6.9 keV lines of ionized iron) seem to indicate that it is thermal with a temperature of about 8 keV. Such a hot plasma cannot be confined in the region by the gravitational potential, it would escape in $< 4\ 10^4$ yr and its origin is therefore still unexplained (but see [13]). Chandra detected more than 2000

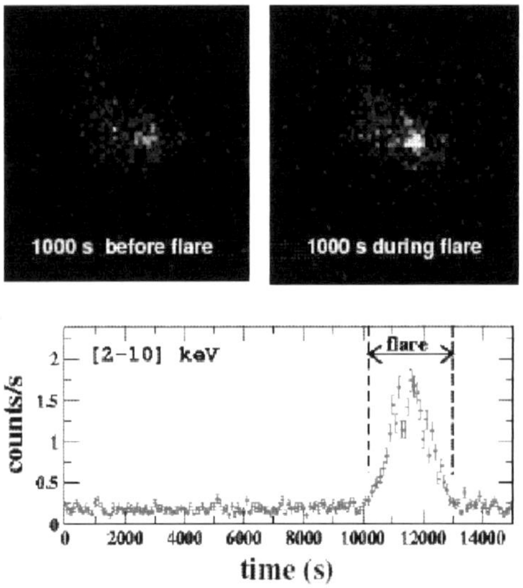

Figure 2. XMM-Newton Sgr A* X-ray flare detections. Top: MOS 2-10 keV images (5′ size) before and during the 2001 Sep flare [28]. Bot: PN light curve of the brightest flare [50].

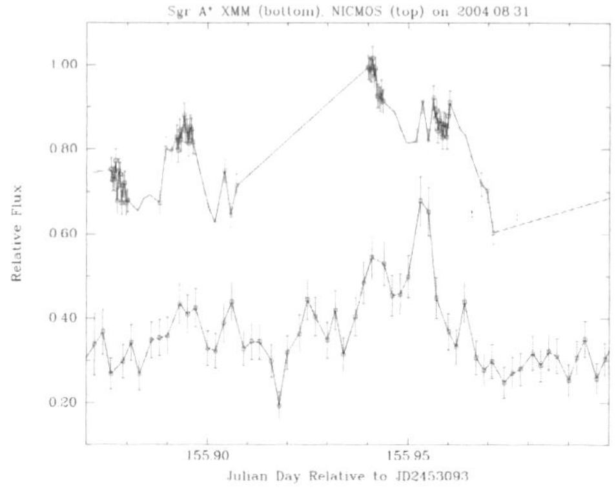

Figure 3. Simultaneous HST near-IR (top: blue for 1.60 μm, green for 1.87 μm and red for 1.90 μm) and XMM X-ray (bottom) light curves during the 2004 Aug 31st Sgr A* flare. Units are flux density (mJy) for NIR and count rates for X-ray curves [63].

point-like weak sources in the central 17′ × 17′, and in spite of the fact that the cumulative spectrum of these sources is very similar to the one of the hot component, this population cannot explain more than 10% of the diffuse emission. Muno and collaborators [44] argued that since the spatial distribution of the diffuse component is not compatible with the one expected from point sources the hot diffuse component cannot be given by the cumulative effect of weak unresolved point-like sources. On the other hand, some features (the continuum is sometimes associated to the 6.4 keV line rather than the 6.7 keV one) are difficult to reconcile with a pure thermal nature, so a few authors have proposed a non-thermal origin for this component, i.e., cosmic ray interaction with the ISM [59, 62], or the effect of SN ejecta in dense gas regions [14]. Indeed the other distinct component of the GC diffuse emission is the 6.4 keV line of neutral weakly-ionized iron which has a different morphology than the 6.7 keV line, a very large equivalent width and is certainly due to non-thermal processes, such as reprocessing of external high energy radiation or cosmic ray interactions within dense molecular clouds.

3.3. Sgr B2 and Sgr A East

The 6.4 keV image of the region shows a very strong peak at the position of the Sgr B2 giant MC. This was interpreted as fluorescent line due to scattering of hard X-ray emission coming from an external source, possibly Sgr A* itself [31, 46]. A strong transient outburst of hard X-rays from the SMBH occurred some 300 yr ago could have traveled the distance to Sgr B2, illuminating the dense cloud and generating the Fe line and the hard X-ray emission. Indeed hard (> 10 keV) X-ray emission from Sgr B was initially detected with the GRANAT/ART-P, and now clearly observed up to 200 keV with INTEGRAL [55].

The recent X-ray observations of the Sgr A complex have demonstrated that the bright X-ray source Sgr A East is a mixed morphology SNR, where the non-thermal radio shell surrounds a centrally peaked thermal X-ray emission [37]. The X-ray plasma has 2 components, one at 1 keV and the other at 4 keV [56]. High abundances in the center of the source indicate that part of the emission is due to heated SN ejecta. However the X-ray data show now that Sgr A East, apart from the high plasma temperature and from expanding into a very dense medium, is not an exceptional SNR. It appears to be the product of a typical SN II or a SN Ia occurred about 10^4 yr ago. Assuming a certain distance of the SN from Sgr A*, the shell of swept up ISM could have reached the SMBH feeding it and triggering a Sgr A* outburst of hard emission, later reflected by Sgr B2 [37]. As mentioned before, this interpretation remains controversial, the main problem being that other 6.4 keV peaks observed with XMM-Newton and related to MCs cannot be due to the same outburst event unless very special geometries and distance scales between the MCs of the region are assumed.

3.4. Recent results on Sgr A flares: NIR / X-ray detections and modulation*
The years 2003 and 2004 have seen several new developments in the domain. In particular a series of large multiwavelength campaigns have been performed in order to obtain simultaneous broad-band measurements of the variable emission from Sgr A*. One of them was carried out in 2004 and was based on a XMM-Newton approved large project ($\approx$ 550ks) for the X-ray monitoring of Sgr A* with half the observation performed at the end March and the other half at the end of August. This program involved radio (VLA, ATCA), sub-mm (CSO, SMT, NMA, BIMA), NIR (VLT, HST) and gamma-ray (INTEGRAL, HESS) observatories. Although not all planned observations could be realized and simultaneous coverage was limited, the campaign was highly successful with the detection of two new bright (factor 40 increase) X-ray flares with XMM-Newton [9], several NIR flares with the HST (one of which simultaneous to an X-ray one) and radio and sub-mm variability at various levels [63]. One of the 2 X-ray flares (31st August) was particularly long ($\sim$ 10 ks) and some interesting time variability analysis could be performed. A bright X-ray and radio transient at $3''$ from the SMBH was also detected and studied during this program [51]. In this respect it has to be mentioned that several transients in the GC region have been discovered and studied with XMM and Chandra and that these studies have provided evidence of an intriguing overabundance of such sources in the central pc of the Galaxy [45].

These campaigns were motivated by the fact that already in 2003 it was revealed using the VLT [22] and then the Keck [24] observatories that Sgr A* also flares in the NIR band. The NIR flares are more frequent (several per day) than the X-ray ones ($\sim$ 1 per day) and their steep non-thermal spectra extending in the MIR domain [20, 23, 15] confirm that synchrotron is the dominant radiation mechanism. The IR observations also provided the spectacular evidence that the emission appears modulated with a period of 17 min [22]. If such a period is associated to the last stable orbit of an accretion disk then the timescale implies that the SMBH is rotating at 50% of the maximum allowed spin. Revisiting X-ray flare variability Aschebach et al. [4] found a set of possible quasi periodic oscillations (QPO) in the power spectra. These periods appeared to correspond to the characteristic frequencies (Keplerian, vertical and radial epicyclic oscillations and Lense-Thirring precession) of a disk orbiting around a BH with maximum spin. A more detailed timing analysis of the X-ray data, however, has been recently presented by Bélanger et al. (2006b) [11] [12]. These authors found in the Sgr A* 10 ks X-ray flare of 31st August 2004 a very significant QPO at 22.2 min (1330 s). Again this period is shorter than the period at the LSO for a Schwarzschild BH and it implies, if associated to stable orbits, a Kerr BH with spin parameter a $\sim$ 0.2 or higher. These authors also re-analyzed the previous XMM data and found no evidence for the other modulations reported previously. Although these results are extremely exciting they need to be confirmed by other significant measurements. Firm detections of QPO from Sgr A* with periods in the range 15-30 min would certainly favor the accretion disk models (vs. jet models) and would provide strong constraints on the mass and spin of the GC SMBH.

The first flare simultaneously observed in IR and X-rays was detected using Chandra and the VLT [18]. The flare was however very weak and no clear light curve and spectral shape could be derived from the X-ray data. A much brighter event was observed simultaneously with XMM and the HST during our large 2004 multiwavelength campaign on Sgr A* mentioned above [63]. The 31st August X-ray flare could indeed be observed with the NICMOS camera of the HST (Fig. 3) at frequencies between 1.6 and 1.9 μm (Fig. 3). The simultaneous NIR and X-ray flares were very similar in shape and the lack of significant time lags indicate that the same population of particles must produce the 2 components. The synchrotron cooling time for producing X-rays in the typical 10 G magnetic field of the accretion flow is much shorter (< 1 min) than the duration of flares, so the X-rays are most probably produced by inverse Compton of sub-mm photons off transiently-accelerated electrons that generate the NIR emission via the synchrotron mechanism [63]. The different spectral slopes of flares could be caused by a varying electron distribution or magnetic field [63, 35]. Similar results were derived from other simultaneous detections of X-ray and NIR flares with Chandra and the VLT [19]. The main efforts are now directed to the simultaneous measurement of spectra over a broad band (from sub-mm to X-rays), since such data would provide information about the particle distribution, magnetic fields, and sizes of the emitting region [35]. Indeed, as for X-rays, NIR flares have been observed with both blue and red spectral slopes and the presence of a flux-slope correlation (the more intense the harder) even within a single event was recently reported [25], although this result has not been firmly confirmed. The detection of a sub-mm flare coincident with a NIR flare, also obtained during our 2004 campaign, could imply that emission decay is not due to synchrotron cooling (since for sub-mm photons cooling time is longer than for NIR) but rather to adiabatic expansion, which may indicate plasma outflows [63]. The inverse Compton emission responsible for the X-rays could also extend to the gamma-ray regime depending on the electron distribution and magnetic field. We therefore searched for simultaneous flares using the INTEGRAL observatory.

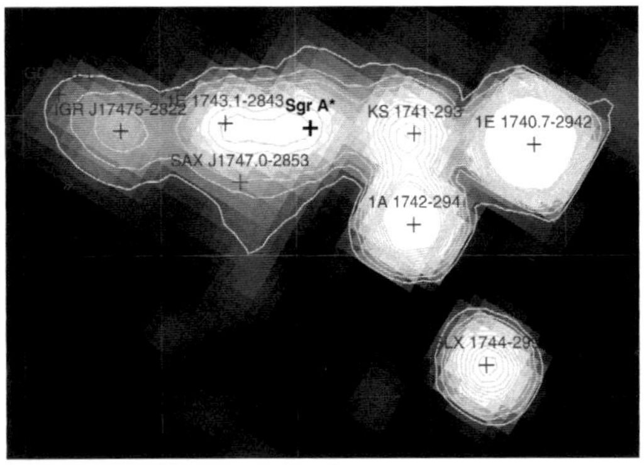

Figure 4. INTEGRAL/IBIS 20-30 keV image of the Galactic Center from the 2003-2004 survey, in galactic coordinates with grid lines separated by 0.5°. Contours mark iso-significance linearly-spaced levels from 9.5 to 75 σ [10].

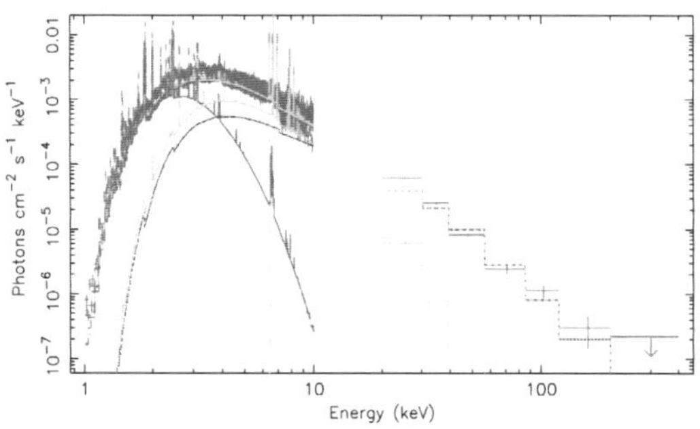

Figure 5. Broad band XMM/INTEGRAL spectrum of IGR J17456–2901 where the 1-10 keV spectrum was obtained from XMM data of the region within 8′ from Sgr A*. The best fit model is an absorbed 2 temperature plasma with kT $\approx$ 1.0 (red) and 6.6 keV (green) plus a broken power-law of indexes 1.5 and 3.2 with break energy at 27 keV (blue) [10].

4. The Galactic Center in gamma-rays

The INTEGRAL observatory monitored, with the IBIS/ISGRI telescope, the GC region for more than 4.7 Ms of effective exposure between 2003 and 2004, obtaining the most precise images of the GC ever collected in the 20-600 keV band [8, 29, 10]. In addition to the bright X-ray binaries, INTEGRAL detected a faint and persistent high energy emission coming from the very center of the Galaxy (Fig. 4), compatible with a pointlike source located (within the 1′ error radius) at the Sgr A* position. Due to the IBIS angular resolution ($\sim 13'$ FWHM) this source (IGR J17456–2901) cannot be clearly associated to the SMBH or to other objects of the dense central region. The lack of variability and of a bright discrete X-ray counterpart suggests that it is rather a diffuse emission concetrated in the inner central 10-20 pc of the Galaxy. The INTEGRAL spectrum was compared to the 1-10 keV one obtained from XMM-Newton (partly simultaneous) data integrating over the region of the IBIS point spread function [10]. The normalizations of the spectra match well (Fig. 5), but the thermal plasma with kT of 6 keV used to model the bulk of the X-ray diffuse emission cannot explain the data at > 20 keV, and neither can the contributions of the transient point sources seen by Chandra or XMM in 2003 and 2004. A non-thermal component extending up to 120-200 keV with spectral slope of photon index 3 is clearly present and its origin is still unexplained. Simultaneous XMM-INTEGRAL observations performed during the 2004 campaign are not conclusive on the possible detection with INTEGRAL of the Sgr A* X-ray flares since the 2 events were observed with XMM during the INTEGRAL passage into the radiation belts, when the instruments did not recorded data. However, even if the Sgr A* X-ray flares extend to > 20 keV with their hard slope, they are too infrequent to fully account for the central gamma-ray source.

In addition, INTEGRAL observed constant hard emission, from Sgr B2 (IGR J17475-2822 in Fig. 4). This supports the hypothesis that the iron emission from Sgr B2 is a reflection nebula generated by a large eruption of Sgr A* that occurred 300 years ago [55].

IGR J17456–2901 on the other hand could be linked to the VHE gamma-ray emission observed by several Atmospheric Cherenkov Detectors. HESS, the most sensitive and precise of them, reported the presence of a TeV source centered within $1'$ from Sgr A* [1]. The source is constant and displays power-law spectrum extending from 300 GeV up to 10 TeV. This emission cannot be explained by heavy dark matter particle annihilation and is probably due to interactions of particles accelerated at very high energies. Recently [2] the HESS collaboration also reported TeV diffuse emission closely correlated to the giant molecular clouds of the Galactic Center region. The spectrum and distribution of this emission are consistent with the idea that the central source accelerated in a recent past the hadrons now interacting with the molecular gas. However the mechanism and site of acceleration, the expanding shell of the Sgr A East SNR [16] or the regions close to the SMBH horizon [3], have not yet been identified. The EGRET source observed between 50 MeV and 10 GeV (3EG J1746–2852) located at $0.2°$ from Sgr A* [40, 30] seems too far to be the 1 GeV counterpart for the INTEGRAL and the HESS sources, but in this complex region the EGRET data are not conclusive. Gamma-ray emission from the GC has now been clearly detected but its origin and nature are not yet fully understood.

5. Conclusions

The Chandra, XMM-Newton, INTEGRAL and HESS monitoring of the GC will continue in the coming months and years, hopefully coupled to NIR, sub-mm and radio observing programs. These programs will possibly settle the issue of periodicities in the X-ray flares and will provide measurements of the broad band spectra of the Sgr A* flares. In the near future GLAST will probably unveil the mystery of the EGRET source at the GC and the next generation of ACD detectors will map the region at TeV energies with increased precision. To solve the puzzle of the hard X-ray emission and to fully understand the several items still open will however necessitate focusing instruments in this energy domain such as Simbol–X, which is expected to fly at the beginning of the next decade [21].

References

[1] Aharonian, F., et al., 2004, *A&A*, 425, L13
[2] Aharonian, F., et al., 2006, *Nat*, 436, 695
[3] Aharonian, F. & Neronov, A., 2005, *ApJ*, 619, 306
[4] Aschenbach, B., et al., 2004, *A&A*, 417, 71
[5] Baganoff, F. K., et al., 2001, *Nat*, 413, 45

[6] Baganoff, F. K., et al., 2003, *ApJ*, 591, 891
[7] Balick, B. & Brown, R. L., 1974, *ApJ*, 194, 265
[8] Bélanger, G., et al., 2004, *ApJL*, 601, L163
[9] Bélanger, G., et al., 2005, *ApJ*, 635, 1095
[10] Bélanger, G., et al., 2006, *ApJ*, 636, 275
[11] Bélanger, G., et al., 2006b, *ApJ*, subm., astro-ph/0604337
[12] Bélanger, G., et al., 2006c, *these proceedings*
[13] Belmont, R., et al., 2005, *ApJL*, 631, L53
[14] Bykov, A. M., 2003, *A&A*, 410, L5
[15] Clenet, Y., et al., 2005, *A&A*, 439, L9
[16] Crocker, R. M., et al., 2005, *ApJ*, 622, 892
[17] Cuadra, J., et al., 2006, *MNRAS*, 366, 358
[18] Eckart, A., et al., 2004, *A&A*, 427, 1
[19] Eckart, A., et al., 2006, *A&A*, in press, astroph/0512440
[20] Eisenhauer, F., et al., 2005, *ApJ*, 628, 246
[21] Ferrando, P., et al., 2005, *SPIE Proc.*, 5900, 195
[22] Genzel, R., et al., 2003, *Nat*, 425, 934
[23] Ghez, A. M., et al., 2004, *ApJL*, 601, L159
[24] Ghez, A. M., et al., 2005, *ApJ*, 620, 744
[25] Gillessen, S., et al., 2006, *ApJL*, 640, L163
[26] Goldwurm, A., et al., 1994, *Nat*, 371, 589
[27] Goldwurm, A., 2001, *ESA-SP*, 459, 455
[28] Goldwurm, A., et al., 2003, *ApJ*, 584, 751
[29] Goldwurm, A., et al., 2004, *ESA-SP*, 552, 237
[30] Hartman, R. C., et al., 1999, *ApJS*, 123, 79
[31] Koyama, K., et al., 1996, *PASJ*, 48, 249
[32] La Rosa, T. N., et al., 2000, *AJ*, 119, 207
[33] Liu, S. & Melia, F., 2002, *ApJL*, 566, L77
[34] Liu, S., et al., 2004, *ApJL*, 611, L101
[35] Liu, S., et al., 2006, *ApJ*, 636, 798
[36] Lynden-Bell, D. & Rees, M., 1971, *MNRAS*, 152, 461
[37] Maeda, Y., et al., 2002, *ApJ*, 570, 671
[38] Malet, I., et al., 1995, *ApJ*, 444, 222
[39] Markoff, S., et al., 2001, *A&A*, 379, L13
[40] Mayer-Hasselwander, H., et al., 1998, *A&A*, 335, 161
[41] Melia, F. & Falcke, H., 2001, *ARAA*, 39, 309
[42] Mezger, P. G., et al., 1996, *A&A Rev.*, 7, 289
[43] Morris, M. & Serabyn, E., 1996, *ARA&A* , 34, 645
[44] Muno, M. P., et al., 2004, *ApJ*, 613, 326
[45] Muno, M. P. et al., 2005, *ApJL*, 622, L113
[46] Murakami, H., et al., 2000, *ApJ*, 534, 283.
[47] Narayan, R., et al., 1998, *ApJ*, 492, 554
[48] Park, S., et al., 2004, *ApJ*, 603, 548
[49] Pavlinsky, M. N., et al., 1994, *ApJ*, 425, 110
[50] Porquet, D., et al., 2003, *A&A*, 407, L17
[51] Porquet, D., et al., 2005, *A&A*, 443, 571
[52] Predehl, P. & Truemper, J., 1994, *A&A*, 290, L29
[53] Proctor, R. J., et al., 1978, *MNRAS*, 185, 745
[54] Purcell, W. R., et al., 1997, *ApJ*, 491, 725
[55] Revnivtsev, M. G., et al., 2004, *A&A*, 425, L49
[56] Sakano, M., et al., 2004, *MNRAS*, 350, 129
[57] Schödel, R., et al., 2003, *ApJ*, 596, 1015
[58] Skinner, G. K., et al., 1987, *Nat*, 330, 544
[59] Valinia, A., et al., 2000, *ApJ*, 543, 733
[60] Watson, M. G., et al., 1981, *ApJ*, 250, 142
[61] Yuan, F., et al., 2003, *ApJ*, 598, 301
[62] Yusef-Zadeh, F., et al., 2002, *ApJ*, 568, L121
[63] Yusef-Zadeh, F., et al., 2006, *ApJ*, 644, 198

Institute of Physics Publishing
doi:10.1088/1742-6596/54/1/017
Journal of Physics: Conference Series **54** (2006) 95–102
Galaxy Center Workshop 2006

New X-ray views of the Galactic Center observed with *Suzaku*

Katsuji Koyama, Yoshiaki Hyodo and Tatsuya Inui
Department of Physics, Graduate school of Science, Kyoto University, Sakyo-ku, Kyoto 606-8502
E-mail: koyama@cr.scphys.kyoto-u.ac.jp

Abstract.
We report the diffuse X-ray emission from the Sgr A and B regions observed with *Suzaku*. From the Sgr A region, we found many K-shell transition lines of iron and nickel. The brightest are Kα lines from FeI, FeXXV and FeXXVI at 6.4 keV, 6.7 keV and 6.9 keV. In addition, Kα lines of NiI and NiXXVII, Kβ of FeI, FeXXV and FeXXVI, and Kγ of FeXXV and FeXXVI are detected for the first time. The center energy of Kα of FeXXV favors collisional excitation as the origin for this line emission. The ionization temperature determined from the flux ratio of Kα of FeXXV and FeXXVI is similar to the electron temperature determined from the flux ratio of Kα and Kβ of FeXXV, which are in the range of 5−7 keV. Consequently, the Galactic Center diffuse X-rays (GCDX) are consistent with emission from a plasma nearly in ionization equilibrium. The radio complex Sgr B region also exhibits Kα lines of FeI, FeXXV and FeXXVI. The 6.7 keV line (FeXXV) map exhibits a local excess at $(l,\ b) = (0^\circ.612,\ 0^\circ.01)$, and could be a new young SNR. The 6.4 keV image is clumpy with local excesses near Sgr B2 and at $(l,\ b) = (0^\circ.74,\ -0^\circ.09)$. Like Sgr B2, this latter excess may be another X-ray reflection Nebulae (XRN).

1. Introduction

The *AstroE*2 satellite was launched on 10th July in 2005 and was named *Suzaku* [1] after an imaginary red bird in the ancient oriental mythology. The X-ray Imaging Spectrometer (XIS) [2] is one of the major instruments on *Suzaku*. The XIS consists of four sets of X-ray CCD cameras, placed on the focal planes of the X-ray Telescope (XRT) [3]. Three XIS sensors contain front-illuminated (FI) CCDs, and the other has a back-illuminated (BI) CCD [2]. The other focal plane instrument is the micro-calorimeter (XRS) [4], which had extremely high energy resolution. The XRS was cooled down to 60 mK, and its overall energy resolution was originally 7 eV at 5.9 keV. These temperature and energy resolution were on-orbit world records. The XRS was able to resolve, e. g., the iron He-like triplets, and hence should have opened a new window of X-ray astronomy. However, regretfully, all the liquid He evaporated very quickly. We nevertheless still have the XIS, which have the best performances among space CCD cameras. The background level for diffuse sources, in particular at the energy of 5–10 keV range is nearly one order of magnitude lower than *Chandra* and *XMM*. The effective area and energy resolution are also comparable to or better than *XMM*. Therefore the XIS on *Suzaku* is best suited for observations of the diffuse X-rays from the Galactic Center (GC). Early *Suzaku* results will appear soon in a refereed journal ([5], [6]). This paper reports the summary and essence of these papers.

2. The Sgr A Region

We have made narrow band GC maps in the 6.34–6.46 keV and 6.62–6.74 keV bands, which are shown in figures 1 and 2. Although these line energies are very close to each other, the X-ray maps are entirely different. The 6.4 keV line is the Kα of FeI, hence figure 1 traces molecular clouds. While the 6.7 keV line is Kα of FeXXV, hence figure 2 traces very hot plasma.

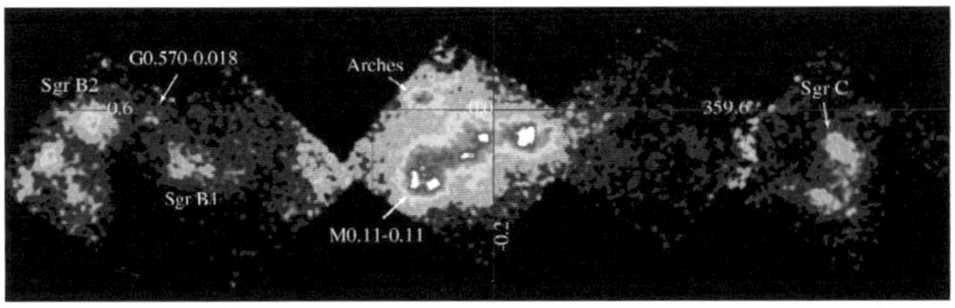

Figure 1. The narrow band map at the 6.4 keV line (the 6.34–6.46 keV band). Coordinates are galactic l and b in degrees.

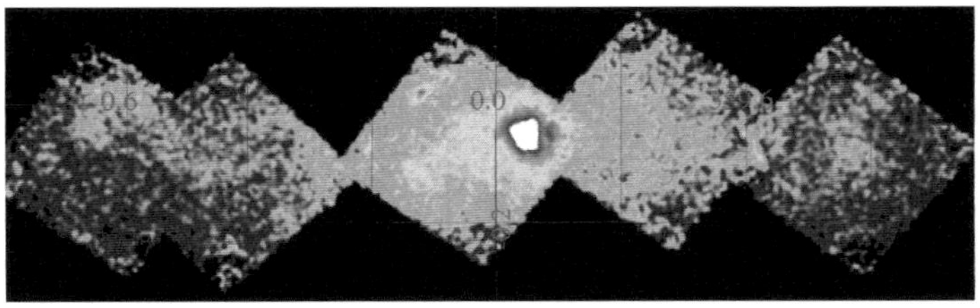

Figure 2. Same as figure 1, but for the 6.7 keV line (the 6.62–6.74 keV band).

Table 1. The detected lines from the Sgr A region.

Emission lines	
Observed Center Energy* (eV)	Identification Line
6409	FeI Kα
6680	FeXXV Kα
6969	FeXXVI Lyα
7069	FeI Kβ
7490	NiI Kα
7781	NiXXVII Kα
7891	FeXXV Kβ
8220	FeXXVI Lyβ
8264	FeXXV Kγ
8681	FeXXVI Lyγ

$*$ systematic line error is $^{+2}_{-7}$ eV.

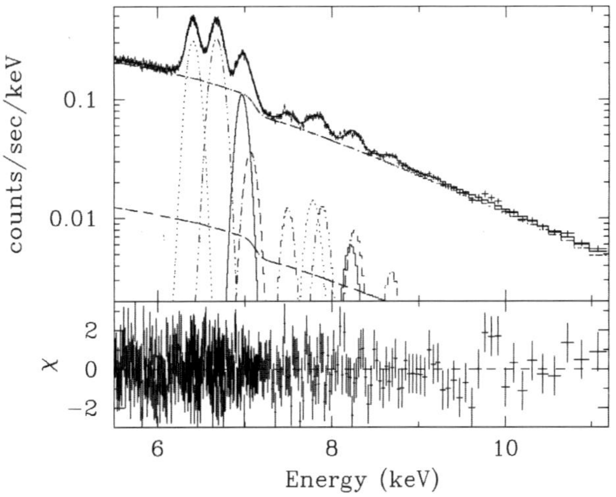

Figure 3. The averaged spectrum of the 3 FIs in the 5.5–11.5 keV band. The data are taken from the full FOV excluding the calibration source regions and Sgr A East. The best-fit result is shown with the dot-dashed line. The long-dashed line is the model of the cosmic X-ray background (CXB) (after Koyama et al. 2006b, *PASJ*, submitted).

2.1. The center energy and width of Kα line of FeXXV

The spectrum of the GCDX from the three FIs is given in figure 3, where the non X-ray background (NXBG) made from the night earth data is subtracted. We fit the spectrum with a phenomenological model (thermal bremsstrahlung + Gaussians + absorption edge). The best-fit line center energies and the best line candidates are given in Table 1, where the possible systematic errors of the line center energies are $^{+2}_{-7}$ eV. The center energy of Kα of FeXXV is 6680±1 eV. This line is a blend of resonance, inter-combination and forbidden lines, hence the energy depends on the flux ratio of these 3 lines.

For the origin of Kα lines of FeXXV and FeXXVI, we discuss two possible cases: (1) collisional ionization and (2) recombining plasma, while the latter case (case 2) includes a photo-ionized plasma and charge exchange processes. The most distinct difference between the two cases is the flux ratio of resonance and forbidden lines, such that collisional ionizing plasma (1) emits the stronger resonance line than the forbidden line by factor of 2–3, and vice versa for the recombining plasma (2). In fact, the *Chandra* and $XMM-Newton$ grating data of many stellar coronae (case 1) have confirmed the relative dominance of the resonance line in comparison to the forbidden line. On the other hand, the case 2, the recombining of a low temperature electron is seen in a photo-ionized plasma, with the typical case being NGC 1068 [7], shows a larger flux from the forbidden line than from the resonance line. Thus the center energy of Kα of FeXXV should be different for the two cases.

The center energy of the collisional ionization plasma (1) and that of the charge exchange process (2) were measured from iron ion beam experiments in the laboratory [8]. The center energy depends on the energy of iron beam and target materials. Within a reasonable range of conditions, the collisional ionization plasma (1) and the charge exchange process (2) give center energies of 6680−6685 eV and 6666 eV, respectively. We also simulated the collisional-ionization case using the plasma codes APEC and MEKAL, and also find central energies of 6680 eV (MEKAL) and 6685 eV (APEC), respectively. Our observed energy of 6680 eV agrees well with collisional ionization plasma of the laboratory data and the MEKAL model, or at most, differ only 5 eV from the APEC model. The line width (1 σ) of Kα of FeXXV is determined

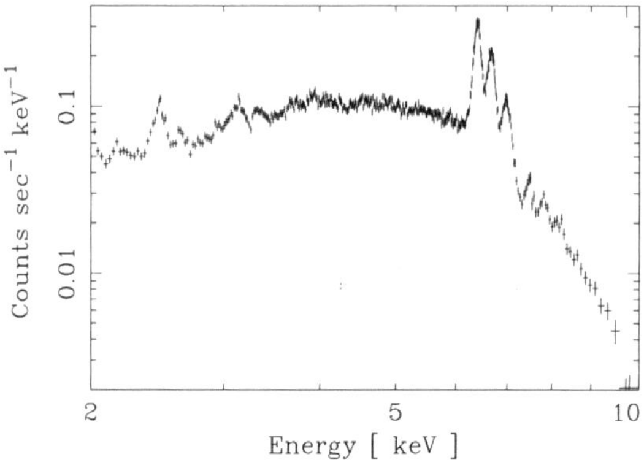

Figure 4. The X-ray spectra from the full FOV of the XIS, but the CCD corners irradiated by the build-in calibration sources are excluded. All the four XIS data are co-added

to be 38 eV, exceeding the systematic effect of ~30 eV (the width of the calibration line). This large line broadening would be due mainly to the triplet lines and satellite lines.

2.2. Constraints on the cosmic ray iron velocity in the charge exchange scenario
The other constraint on the charge exchange process is the Lyman-limit hump. If an electron is captured to a level of a large principal quantum number of an iron nucleus, the subsequent decay to the ground state produces a hump at Lyman series limit (~ 9.2 keV). However we see no significant hump with the upper limit of 9×10^{-6} photons s^{-1} cm^{-2} (we assumed a Gaussian line with width 30 eV), which is about 6% of the intensity of Lyα. This upper limit sets the lower limit of the collision energy to > 100 eV amu^{-1} [9], [10]. Thus the velocity of bare iron nuclei must be larger than ~150 km s^{-1}, or the line width of Kα of FeXXVI (Lyα) must be larger than 3 eV. Since the observed line width is smaller than that of the calibration line, the real line broadening should be nearly zero. This is inconsistent with no hump at 9.2 keV. However, we note that the accurate measurement of line broadening of this level (~ 3 eV) is very difficult with the XIS, so further confirmation with better resolution instruments is very important.

Together with the discussion of the previous section, we conclude that the 6.7 keV and 6.9 keV lines in the GCDX are caused by collisional ionization and do not originate from the charge exchange process.

2.3. The flux ratio of K-shell transition lines of FeXXV, XXVI and NiXXVII
The ionization temperature determined from the best-fit Kα flux ratio between FeXXVI and FeXXV is similar to the electron temperature determined from the best-fit flux ratio of Kβ and Kα of FeXXV, within the error range of 5–7 keV. The observed flux ratios of Kβ and Kα of FeXXVI, and Kγ and Kα of FeXXV, XXVI are all consistent with an electron temperature of $\sim 5 - 7$ keV. With the assumption that the relative abundances of iron and nickel are proportional to the solar value, the Kα ratio between NiXXVII and FeXXV is consistent with an ionization temperature of ~5 keV. We therefore conclude that the the plasma in the GC is in collisional-ionization equilibrium with a temperature of $\sim 5 - 7$ keV .

3. The Sgr B Region

The X-ray spectrum of all the Sgr B region is given in figure 4. The spectra of the four XISs (XIS0-XIS3) are co-added and the NXBG is subtracted. With the superior energy resolution of the XIS for diffuse sources, we can clearly resolve the 6.4 keV, 6.7 keV and 6.9 keV lines.

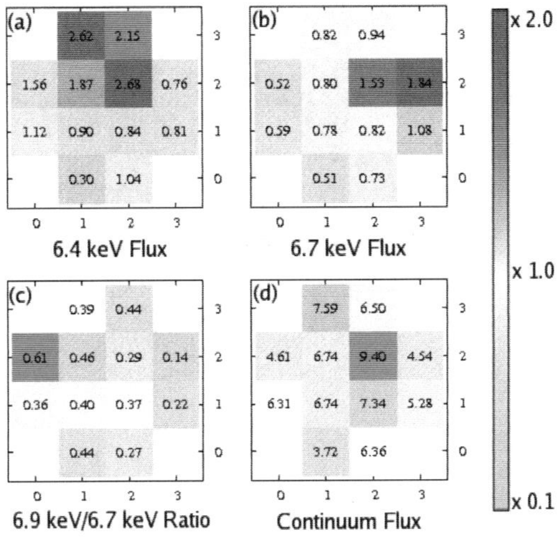

Figure 5. The 6.4 keV (a), 6.7 keV (b), the flux ratio of 6.9 keV/6.7 keV (c), and the continuum flux map (d). The embedded numbers are fluxes in units of 10^{-5} photons cm^{-2} s^{-1} $arcmin^{-2}$ or flux ratios. The color means that the average value is white, and the increasing (decreasing) excess above (below) the average is given with increasing deepness of red (green), where the maximum (minimum) is 2.0 (0.1) times of the average value.

In order to study this region more quantitatively, we have divided the XIS FOV into 16 (4×4) areas as is shown in figure 5 by the solid grids. For brevity, we assign coordinates for each square position such as (0, 0) and (3, 3), where the former and latter are lower-left and upper-right positions, respectively. Since the satellite roll-angle of the observation was nearly 0°, the north of this coordinate is also the north of the sky coordinate. The four corners, (0, 0), (0, 3), (3, 0) and (3, 3) contain the MnI Kα and Kβ line from the built-in calibration sources, so we exclude these 4 areas in the following analysis.

We then made X-ray spectra from the 12 squares. The NXBG spectra are constructed from night earth data in the same detector area of the 12-square regions and subtracted from the spectra of the 12 Sgr B regions. The NXBG-subtracted spectra are fitted with a model of thermal bremsstrahlung plus Fe absorption edge at the 7.1 keV and four emission lines, which are the Kα lines from FeI (6.4 keV), FeXXV (6.7 keV) and FeXXVI (6.9 keV) and the Kβ line from FeI (7.06 keV). Using the best-fit fluxes, we made color-coded maps (figure 5), which are the fluxes of the 6.4 keV (figure 5a) 6.7 keV lines (5b), the flux ratio of the 6.9 keV/6.7 keV lines (5c), and the continuum flux (5d).

Both the 6.4 and 6.7 keV line maps show a large contrast in red and green colors, which means that the fluxes have large variations from position to position. On the other hand, the 6.9 keV/6.7 keV ratio is rather smooth (the contrast of colors is weak) except the positions of (3, 2) and (0, 2). If we exclude these positions, the mean value of the other 10 squares is 0.36. For a plasma in collisional ionization equilibrium (CIE), this value corresponds to a temperature of 6–7 keV. Koyama et al. (2006b) reported that the mean temperature of the Galactic Center diffuse X-ray emission (GCDX) is 5–7 keV. We thus conclude that the origin of the diffuse emission is the same. At the Sgr B region, the surface brightness of the GCDX is about a half of that in Sgr A.

The significant excess of the 6.7 keV flux and deficit of the 6.9 keV/6.7 keV flux ratio at the position of (3, 2) indicates that an extra component with strong 6.7 keV line but no strong 6.9 keV line is present at this position. The rather smooth distribution of the continuum X-ray

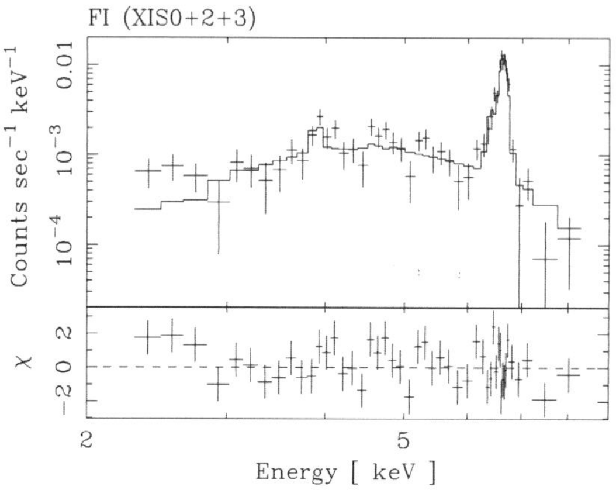

Figure 6. The X-ray spectrum of the sum of 3 FI CCDs for a new SNR (G0.61+0.01) with the best-fit VNEI model (after Koyama et al. 2006c, *PASJ*, submitted).

fluxes (see figure 5d) also suggests that the excess emission is mainly due to the 6.7 keV line. Since the 6.7 keV line is most likely the Kα of FeXXV, a probable origin of the local 6.7 keV excess is a hot plasma with a temperature lower than the large scale GCDX. The 6.9 keV/6.7 keV flux ratio at the position (3, 2) is 0.14, which can be converted to an ionization temperature of ~4–5 keV. This temperature is typical for a young SNR like Sgr A East. The 6.4 keV flux (figure 5a) shows local excesses at the position of Sgr B2 (2, 2) and the east of Sgr B2 (1, 3). The presence of the 6.4 keV line is strong evidence for dense and cool molecular clouds.

3.1. Discovery of a new young SNR

Since the 6.7 keV excess at the position of (3, 2) is likely to be a new SNR (see discussion below), we designate this source as Suzaku J1747.0-2824.5 (G0.61+0.01) from its center position. The background-subtracted spectrum of G0.61+0.01 is given in figure 6, for the 3 FIs case, where the background was taken from the southeast region. We see pronounced peak at 6.7 keV, but the line shape is asymmetric with a tail at lower energy. This tail would be due to the presence of a faint Kα line of FeI at 6.4 keV.

The FI and BI spectra are simultaneously fitted with a thin thermal plasma in non-equilibrium ionization (VNEI) [11], adding two Gaussian lines at 6.4 keV and 7.06 keV. These two lines represent the Kα and Kβ lines of FeI, where the flux of latter line is fixed to 12.5% of the former. The best-fit result for the FI is shown in figure 6. The best-fit temperature is ~3 keV and the abundances of Ca and Fe are 3-4 times of solar. The $N_{\rm H}$ value is 1.6×10^{23} cm^{-2}, larger than the typical value to the GC (6×10^{22} H cm^{-2}) [12]. Therefore, G0.61+0.01 would be located behind or in the rim of the Sgr B2 cloud. Since G0.61+0.01 is located in the south of an expanding radio shell [13], which is probably interacting with the Sgr B2 cloud rim, we assume that the distance of G0.61+0.01 is the same as that of Sgr B2, 8.5 kpc [14]. Then the size of G0.61+0.01 of about 2.4 arcmin (the radius of the major axis) corresponds a real size of 6 pc. Dividing this size by the sound velocity of the 3 keV plasma, we obtain the dynamical time scale of $\sim 4 \times 10^3$ years. The 2–10 keV band luminosity is estimated to be 1.5×10^{34} ergs s^{-1}. These values are typical for an ejecta-dominanted, young SNR.

Since G0.61+0.01 is found at the edge of the XIS field, some parts of this candidate SNR would be out of the XIS field. One possibility is that G0.61+0.01 is a part of the expanding radio shell discovered by Oka et al. [13]. The kinetic energy of the radio shell is a few of 10^{52} erg s^{-1}, within the range of single or multiple supernova explosions.

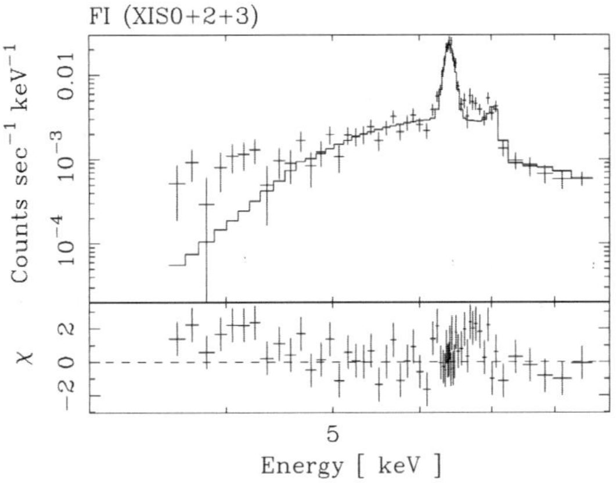

Figure 7. The X-ray spectrum of the sum of the 3 FI CCDs for Sgr B2 with an absorbed power-law model and two Gaussian lines (after Koyama et al 2006c, *PASJ*, submitted).

4. Discovery of a New XRN

The 6.4 keV map shows two bright regions at (2, 2) and (1, 3). One is Sgr B2, which has been already found as a strong 6.4 keV source [15], and the other is a newly discovered source Suzaku J1747.7−2821.2 (M0.74−0.09). We show the background-subtracted spectra of M0.74−0.09 and Sgr B2 in figures 7 and 8. The latter is for comparison to the former new source. We simultaneously fit the FIs and BI spectra with a model of absorbed power-law plus two Gaussians near at 6.4 and 7.06 keV, which are for the Kα and Kβ lines of FeI. This model nicely fits the data except for an excess near the 6.7 keV line in the Sgr B2 spectra. One possibility is that the 6.7 keV enhancement is a part of the new SNR candidate G0.61+0.01, because it is located in the close vicinity of Sgr B2. The other possibility is that the 6.7 keV enhancement is from high-mass young stellar objects (YSO) embedded in the center of Sgr B2, because some YSOs show a hint of 6.7 keV line emission [19].

The Sgr B2 cloud has been studied extensively with *ASCA* and *Chandra* [15],[17]. The authors concluded that the 6.4 keV emission is fluorescence induced by strong X-rays coming from Sgr A*, hence the feature is called an X-ray Reflection Nebula (XRN). *Suzaku* found a clear Kβ line at 7.06 keV of the ~10% of Kα and deep Fe edge at 7.1 keV. These discoveries provide additional support for the XRN scenario of Sgr B2.

The spectrum of M0.74-0.09 exhibits a strong 6.4 keV line with an equivalent width of 1.6 keV, a 7.06 keV line, and an edge structure at 7.1 keV. All these lines and edge energies are consistent with being from Kα, Kβ and K$_{\rm edge}$ of FeI. The flux of the 7.06 keV line is about 10% of that of the 6.4 keV line, which is also consistent with the fluorescent X-rays. Unlike Sgr B2, no hint of YSOs is found so far. No bright point source is found in the *Chandra* image. Therefore, the X-rays cannot be the scattering and fluorescence by embedded YSOs or any internal source. If the X-rays from Sgr B2 and M0.74−0.09 are caused by the Thomson scattering and fluorescence of the same irradiating external source like Sgr A*, then the $N_{\rm H}$ ratio between these sources should be same as the 6.4 keV line flux ratio. The observed $N_{\rm H}$ ratio and the 6.4 keV line flux are in good agreement with each other, hence supports the fluorescent scenario. Therefore the XRN scenario successfully applied to Sgr B2 can be also applied to M0.74−0.09.

The counter-scenario against the XRN is that the 6.4 keV line emission is produced by the collision of electrons with neutral iron. Since the cross section of iron K-shell ionization is maximum at the electron energy of a few 10 keV [18], the most probable source is low energy electrons (LEE) as proposed for the origin of the Galactic Ridge iron K-shell emission [19]. Since

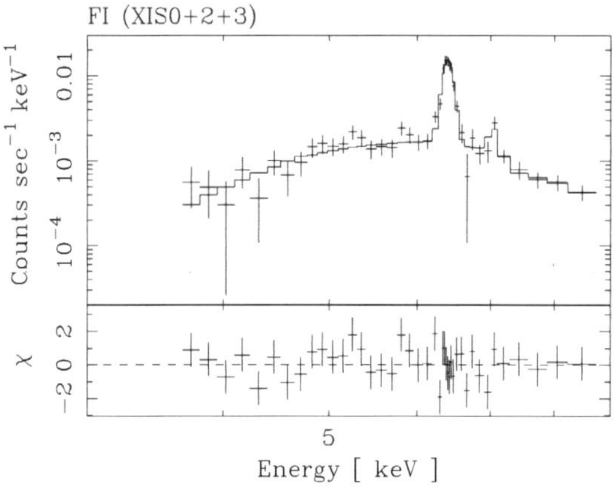

Figure 8. Same as figure 7, but for a new source M0.74-0.09 (after Koyama et al. 2006c, *PASJ*, submitted).

a few 10 keV electrons is absorbed in less than 10^{22}H cm^{-2} of depth [18], the produced X-ray spectrum has no large absorption edge. Our observation, however, shows clear absorption of (4–10) $\times 10^{23}$H cm^{-2}, in far excess to the Galactic interstellar absorption [12]. Thus the LEE origin is unlikely, unless we assume a special geometry such that the 6.4keV source is deep in or behind the dense cloud.

Acknowledgments
The authors express sincere thanks to all the *Suzaku* team members and our Laboratory colleagues. Y.H. and H.N. are supported by JSPS Research Fellowship for Young Scientists. This work is supported by the Grant-in-Aid for the 21st Century COE "Center for Diversity and Universality in Physics" from the Ministry of Education, Culture, Sports, Science and Technology (MEXT) of Japan.

References

[1] Mitsuda, K., et al. 2006, *PASJ,*, submitted
[2] Koyama, K., et al. 2006a, *PASJ*, submitted
[3] Serlemitsos, P. et al. 2006, *PASJ*, submitted
[4] Kelly et al. 2006, *PASJ*, submitted
[5] Koyama, K., et al. 2006b, *PASJ*, submitted
[6] Koyama, K., et al. 2006c, *PASJ*, submitted
[7] Kinkhabwala, A., et al. 2002, *ApJ*, **575**, 732
[8] Beiersdorfer, P. 2003, *ARA&A*, **41**, 343
[9] Perez, J. A., Olson, R. E., & Beiersdorfer, P. 2001, J. *Phys. B*, **34**, 3063
[10] Wargelin, B. J., Beiersdorfer, P., Neill, P. A., Olson, R. E., & Scofield, J. H. 2005, *ApJ* , **634**, 687
[11] Borkowski, K. J., Lyerly, W. J., & Reynolds, S. P. 2001, *ApJ*, **548**, 820
[12] Sakano, M., Koyama, K., Murakami, H., Maeda, Y., & Yamauchi, S. 2002, *ApJ*,**138**, 19
[13] Oka, T., Hasegawa, T., Sato, F., Tsuboi, M., & Miyazaki, A. 1998, *ApJS*, **118**, 455
[14] Reid, M. J., Schneps, M. H., Moran, J. M., Gwinn, C. R.,Genzel, R., Downes, D., & Roennaeng, B. 1988, *ApJ*, **330**, 809
[15] Murakami, H., Koyama, K., & Maeda, Y. 2001, *ApJ*, **558**,687
[16] Takagi, S., Murakami, H., & Koyama, K. 2002, *ApJ*,**573**, 275
[17] Koyama, K., Maeda, Y., Sonobe, T., Takeshima, T., Tanaka, Y., & Yamauchi, S. 1996, *PASJ*, **48**, 249
[18] Tatischeff, V. 2002, *astro-ph/0208397*
[19] Valinia, A.,Tatischeff, V., Arnaud, K., Ebisawa, K., & Ramaty, R. 2000, *ApJ*, **543**, 733

Institute of Physics Publishing
doi:10.1088/1742-6596/54/1/018
Journal of Physics: Conference Series **54** (2006) 103–109
Galaxy Center Workshop 2006

The 6.7-keV iron-line emission in the Galactic Centre

Robert Warwick[1], Masaaki Sakano[1] and Anne Decourchelle[2]

[1] Department of Physics and Astronomy, University of Leicester, Leicester LE1 7RH, UK
[2] CEA/DSM/DAPNIA, Service d'Astrophysique, C.E. Saclay, 91191 Gif-sur-Yvette Cedex, France

E-mail: rsw@star.le.ac.uk

Abstract. We use recent *XMM-Newton* observations to study the "diffuse" X-ray emission seen in the Galactic Centre Region. Spectrally, the emission can be separated into three major components, each characterised by a prominent spectral line. Using these lines as tracers, we investigate the underlying spatial distribution of the various components. Specifically, we find the 6.7-keV line of helium-like iron, has a relatively smooth, circularly symmetric distribution centred on Sgr A* and a surface brightness which falls off with radius as $r^{-0.87\pm0.06}$ over the range $r = 3' - 12'$. This mirrors the distribution of the underlying stellar population and adds strong support to the hypothesis that the 6.7-keV line and the associated hard thermal continuum (with $kT \approx 8$ keV) originates in the summed emission of faint point sources.

1. Introduction

Diffuse X-ray emission pervades both the Galactic Plane and the Galactic Centre Region. Prominent features of this emission are a hard continuum and an associated 6.7-keV emission line from helium-like iron, which, if thermal in origin, implies the presence of a plasma at a temperature of $5 - 10$ keV [1]. This hard "Galactic Ridge X-ray Emission" (GRXE), has until recently, remained a major puzzle in the Galactic X-ray Astronomy [2]. However, since the 6.7-keV line is particularly bright in the Galactic Centre, one might hope that studies of this region would provide the best opportunity for identifying the processes giving rise to the phenomenon [3].

Taking account of its apparently diffuse nature, the most straightforward conjecture is that the GRXE is energised largely by supernova explosions. However, individual thermal supernova remnants are generally characterised by temperatures much lower than the 5–10 keV required to collisionally excite the observed 6.7 and 6.9-keV lines. A further difficulty is that if this hot emission is truly diffuse, the total amount of energy residing in the plasma is tremendous; for example, the Galactic Centre component alone requires an energy of $\sim 10^{54}$ erg, leading to a major problem of energetics. In fact, the temperature and pressure is so large that such a hot plasma cannot be confined by the Galaxy's gravitational potential or bound by its magnetic field, implying a very high rate of energy loss ($\sim 10^{43}$ erg s^{-1} across the whole galaxy).

Chandra observations have recently led to important advances in this field. With its fine angular resolution, *Chandra* can resolve point sources down to an X-ray flux limit of $\sim 3 \times 10^{-15}$ erg s^{-1} cm^{-2} (2–10 keV) in deep exposures. Even at this level, the bulk of the hard component remains unresolved both in the Galactic plane [4] and in the Galactic Centre [5]. Although this finding would appear to favour the diffuse hypothesis, recent studies of the large-scale distribution of the GRXE demonstrate that it closely follows the near-infrared background light of the Galaxy, which in turn is a good tracer of the stellar mass density [6]. This reinforces the alternative view [7], that we are in fact measuring the integrated emission of faint X-ray sources, most of which are likely to be cataclysmic variables (CVs) [6, 8]. It would appear that Chandra surveys reaching fluxes well below $\sim 10^{-16}$ erg s^{-1} cm^{-2} (2–10 keV) will be required to resolve a significant fraction of the hard thermal emission into sources. Based on the *Chandra* source counts, the number of sources required in the Galactic Centre in order to fully account for the hard emission far exceeds the predicted number of CVs [6, 21]. However, since the nature of Galactic X-ray source population detected at faint fluxes is poorly understood, the uncertainties remain large. In fact, we have demonstrated with *XMM-Newton* that a population of weak point sources showing strong 6.7-keV line emission does exist in the Galactic Centre [9] and the *ASCA* and *Chandra* teams report similar findings [10, 21]. Judging from their strong 6.7-keV lines, these sources are most likely CVs, possibly intermediate polars, although some of them are unusually luminous and may represent the brightest members of a different class of object.

Historically the tracing of the GRXE into the Galactic Centre region has been complicated by both the confusion of luminous point sources and also the presence in the Galactic Centre of bright diffuse components of different origins. For example, much cooler (kT typically ~ 1 keV) optically thin thermal emission is apparent throughout the region, presumably as a result of the material and energy released into the relatively dense interstellar environment by supernovae and the stellar winds of high-mass stars. A further diffuse X-ray component which, to date, has only been observed in Galactic Centre Region, is fluorescent X-ray emission characterised by the 6.4-keV K_α line of neutral or near-neutral iron. This fluorescence has been observed from several dense molecular clouds, most noteably the most massive cloud in the Galaxy, Sgr B2 [3, 11, 12, 13]. There remains some uncertainty as to the excitation process. One of the scenarios proposed involves the X-ray illumination of the clouds by external (or perhaps internal) sources, one possibility being that Sgr A* was much brighter in the past and hence provided the required illumination. An alternative model involves the excitation of the surfaces of the molecular clouds by an intense incident flux of low-energy cosmic-ray electrons (LECRe) [14, 15].

2. Observations and preliminary data reduction

In this paper we use recent *XMM-Newton* observations to explore the spectral and spatial distribution of the various components of the diffuse X-ray emission seen in the Galactic Centre Region. Specifically we use data accumulated from the series of *XMM-Newton* observations targeted at Sgr A*, which are available from the *XMM* public archive. For the present work we focus on images and spectra obtained from the EPIC pn camera, which has a 30′ diameter field of view. Both single and double pixel events (pattern 0–4) were selected and all the datasets were screened for periods of high background due to the incidence of soft protons. For each observation flat-fielded images in various wavebands were created by subtracting a constant (non-vignetted) background particle rate (estimated from the corners of the CCD detector not exposed to the sky) and dividing by the appropriate exposure map. The images from the different observations were then mosaiced and spatially smoothed prior to further analysis. For the present work, a composite Galactic Centre spectrum was accumulated from an annular region of inner radius 4′ and outer radius 10′ centred on Sgr A*. A background spectrum was similarly taken between 10′ and 12′. In both cases bright point sources and extended sources of particularly high surface brightness, such as the Arches cluster and the G0.1-0.1 molecular cloud, were excluded.

3. Results

3.1. The X-ray spectrum of the Galactic Centre Region

Fig. 1 shows the count-rate spectrum measured in the EPIC pn camera for the Galactic Centre Region. Between 2–4.5 keV the continuum plus prominent helium-like lines due to S, Ar and Ca imply thermal emission at a temperature of ~ 1 keV. Above 6 keV, the underlying hard continuum and the prominent helium-like (6.7 keV) and hydrogen-like (6.96 keV) iron K_α lines are consistent with an 8-keV thermal component. The fluorescent iron K_α line at 6.4 keV identifies a third spectral component. Here we model the first two spectral components as thin thermal emission from a plasma in collisional equilibrium (via the Mekal code). We model the third component as the combination of fluorescent line emission and non-thermal bremsstrahlung produced at the surfaces of dense clouds by an incident flux of LECRe [14]. As is evident from Fig. 1, the helium-like sulphur line at 2.46 keV, the helium-like iron line at 6.7 keV and the fluorescent iron line at 6.4 keV can serve as tracers of these three distinct components.

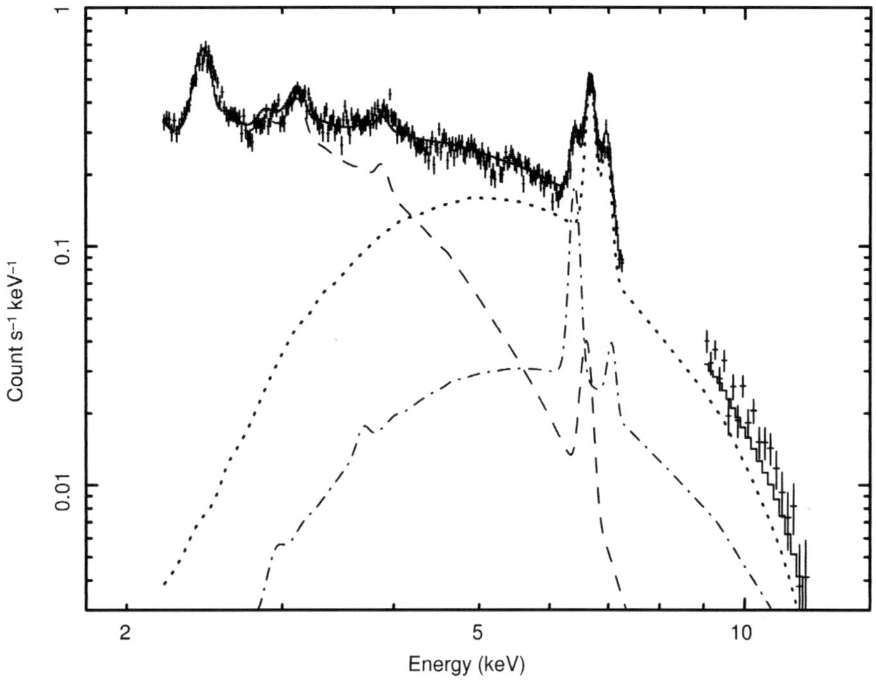

Figure 1. *XMM-Newton* pn spectrum of the Galactic Centre Region. The full spectrum can be modelled (solid line) as the sum of three separate components. These comprise a 1-keV thermal component (dashed line), an 8-keV thermal component (dotted line) and fluorescent line emission plus an associated non-thermal continuum (dot-dashed line).

3.2. Imaging the Galactic Centre Region in the continuum and lines

Fig. 2 (top left panel) shows an image of the Galactic Centre Region produced in the 4.5–6 keV band, which is relatively clear of line emission. All three spectral components identified earlier will contribute to this band. The other panels in Fig. 2 show narrow-band continuum-subtracted images centred on each of the spectral lines identified earlier. Clearly the spatial distributions of the three emission lines are very different. The sulphur emission is concentrated in an extended region to the north-east of the Galactic Centre. The 6.4-keV fluorescent iron emission has a similarly asymmetric distribution with very bright clumps associated with known molecular complexes (*e.g.*, the G0.1-0.1 cloud). On the other hand the 6.7-keV iron line exhibits a much smoother distribution which, to a first approximation, is circularly symmetric about the

Galactic Centre.

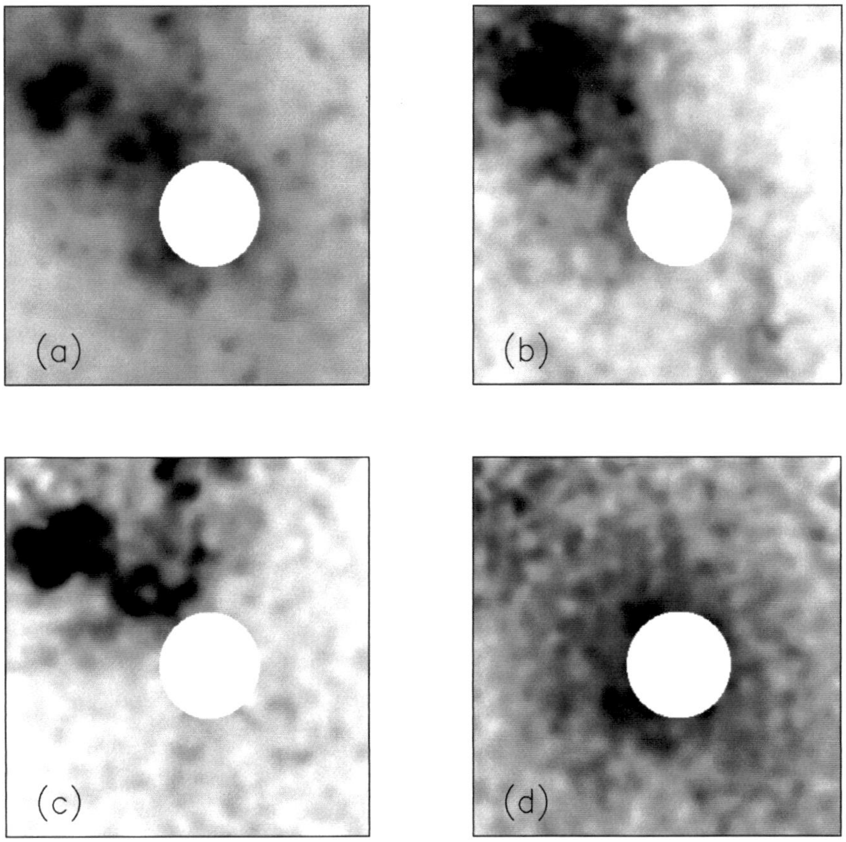

Figure 2. *XMM-Newton* EPIC pn images of Galactic Centre Region in various wavebands: (a) the 4.5–6 keV X-ray continuum band; (b) a narrow band centred on the 2.46 keV sulphur line; (c) a narrow band centred on the 6.4-keV iron line; (d) a narrow band centred on 6.7-keV helium-like iron line. The three line images are all continuum subtracted. The grey scale is inverted such that the X-ray bright regions appear black. Bright point sources have been excluded, as has the region within $3'$ of Sgr A* (the latter region is blanked out). These images have been spatially smoothed with a circular Gaussian mask with $\sigma = 20''$. Each image covers a square region of dimension $21.3'$ on a side.

As a check of the hypothesis that the bulk of the diffuse emission seen in the Galactic Centre Region can be associated with one of the three spectral components identified earlier[1], we have compared the observed 4.5–6 keV continuum image to one constructed by applying a scale factor to each of the line images and then simply summing together the scaled versions of the images. In this process, best-fit values for the three scale factors were determined by minimising a χ^2 statistic calculated on the basis of the pixel-by-pixel differences between the constructed and observed images. Using the best-fitting scale factors, we obtained a very good match to the observed continuum image, as is illustrated by the correlation in Fig. 3. These best-fitting

[1] Clearly this is only approximation, which excludes other known contributors to the Galactic Centre X-ray emission. Nevertheless, it does appear that the *bulk* of the emission resides in just three spectral components.

scale factors can be interpreted, in terms of the assumed spectral models, as measures of the abundance, Z, relative to solar. We obtained $Z = 0.5 - 1$ for the sulphur line (in the 1-keV thermal plasma), $Z \sim 1$ for the 6.7-keV iron line (in the 8-keV thermal plasma) and $Z \sim 2.5$ for the 6.4 keV fluorescent iron line (in the LECRe model), all of which are consistent with the spectral modelling constraints.

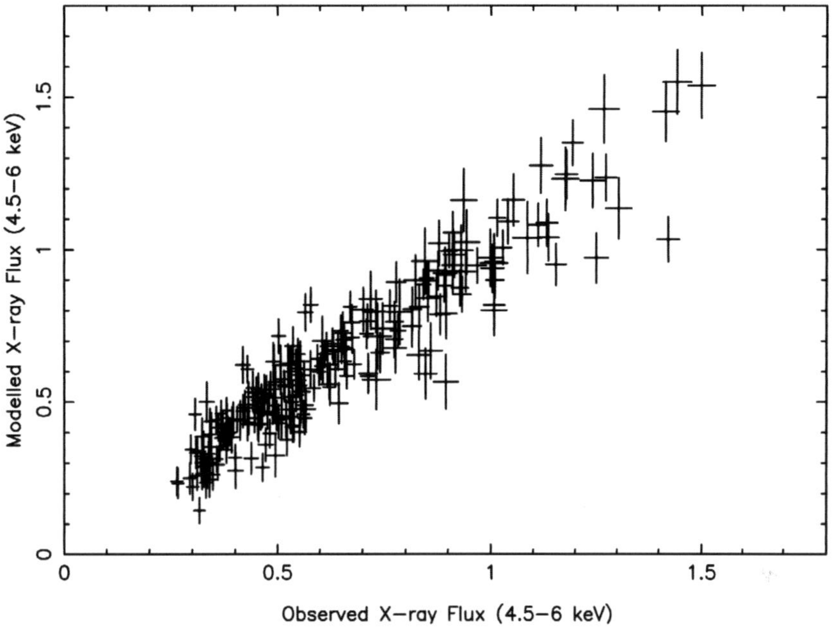

Figure 3. Comparison of the fluxes in the 4.5–6 keV band image with those obtained by appropriately scaling and summing the sulphur, 6.4-keV iron and 6.7-keV iron line images. To reduce the size of the error bars, the original images were compressed to give a correlation based on pixels of dimension $72'' \times 72''$.

3.3. The radial distribution of the 6.7-keV component

As noted above, the 6.7-keV image shows a relatively smooth, circularly symmetric distribution. However, when we fit the observed surface brightness distribution with a simple $r^{-\alpha}$ function, where r is the offset from Sgr A*, we find evidence in the residuals for excess brightness in the north-east of the image. This excess correlates reasonable well with the sulphur distribution and, in fact, we might expect some contamination of the 6.7-keV image by iron-line emission associated with the 1-keV thermal component (see Fig. 1). The approach we have adopted, therefore, is to fit the measured 6.7 keV distribution with a two-component spatial model consisting of a power-law radial dependence plus a scaled version of the sulphur image. The measured radial profile of the 6.7-keV emission (after subtracting the sulphur correction) is shown in Fig. 4. This profile follows a power-law form $r^{-0.87\pm0.06}$ over the range $r = 3' - 12'$.

The residuals to the two-component fit are shown in Fig. 5. Although regions around bright continuum sources were excluded when fitting the two-component model to the data, we have specifically not excluded these regions in Fig. 5, so as to illustrate the 6.7 keV emission in the Galactic Centre Region over and above the modelled distribution. Although a few bright 6.7

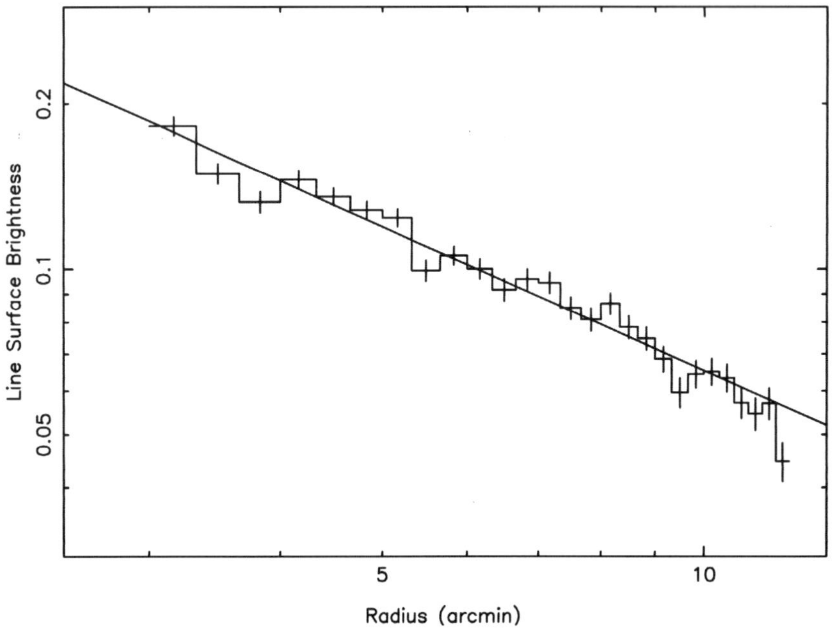

Figure 4. The radial surface brightness distribution of the 6.7-keV iron line. The solid line has a power-law slope of -0.87.

keV sources stand out in these plots, *e.g.*, the Arches cluster $\sim 11'$ north of Sgr A* and the Sgr A East SNR within the central $3'$ field, there is no evidence for any additional large-scale components. Furthermore the fitted radial profile appears to extrapolate reasonably well to within $\lesssim 1'$ of Sgr A*.

4. Discussion

The measured fall-off in the surface brightness of the 6.7 keV emission in Fig. 4 is rather similar to the $r^{-1.0\pm0.1}$ decline in the surface density of the X-ray point sources resolved by *Chandra* within $9'$ of Sgr A* [5]. Within this central region, resolved point sources with fluxes $\sim 3 \times 10^{-15}$ erg s^{-1} cm^{-2} contribute about 10% of the hard diffuse flux [16], with a factor of 40-100 downward extrapolation of the measured source counts required to fully account for the bulk of the diffuse flux. The 6.7-keV surface brightness distribution also mirrors that of the underlying stellar distribution as deduced from infra-red observations of the Galactic Centre [17, 18, 19, 20]. This demonstrates that the correlation of the GRXE intensity with infrared-light observed in the Galactic plane and bulge/bar regions [6], also extends all the way into the central stellar cusp of the Galaxy. As there is no compelling reason to believe that a truly diffuse plasma would show the same spatial profile as that of the underlying stellar distribution, on scales ranging from arcmins to tens of degrees, this is strong evidence that the 6.7 keV represents the integrated emission of a population of discrete X-ray sources embedded within the general stellar population. Recent work suggests that the bulk of the GRXE is due to CVs and coronally active binaries and that the spectral characteristics of these classes of sources are fully consistent with the observed GRXE energy spectrum [6, 8]. In the context of the Galactic Centre a challenge for the future is to confirm that the 2×10^5 sources with X-ray luminosities in

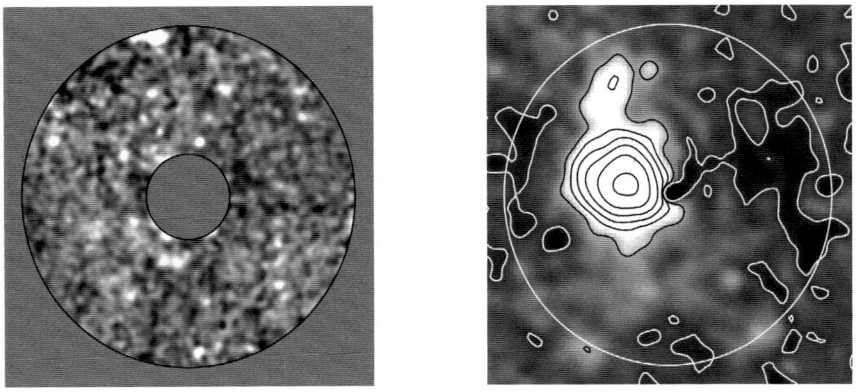

Figure 5. 6.7-keV band images of the Galactic Centre Region after subtracting the model radial distribution and a sulphur-related component. *Left panel:* The annular region between a radius of 3′ and 12′ centred on the position of Sgr A*. In producing this image the regions around bright point and extended sources were specifically *included*; as a consequence several bright 6.7 keV sources, *e.g.*, the Arches cluster at the northern rim, are apparent. *Right panel:* The central 3′-radius region. The black contours delineate the Sgr A East SNR. The white contours trace low-level negative residuals, which peak on Sgr A*.

the range $2 \times 10^{29} - 2 \times 10^{31}$ erg s^{-1}, which are required within the central 40 pc diameter region [16], really do exist and to define the nature and properties of this extremely faint Galactic X-ray source population.

References

[1] Koyama K, Makishima K, Tanaka, Y and Tsunemi H 1986 *PASJ* **38** 121

[2] Tanaka Y 2002, *A&A* **382** 1052

[3] Koyama K, Maeda Y, Sonobe T, Takeshima T, Tanaka Y and Yamauchi S 1996 *PASJ* **48** 249

[4] Ebisawa K *et al* 2005 *ApJ* **635** 214

[5] Muno M P, Baganoff F K, Bautz M W, Brandt W N, Broos P S, Feigelson E D, Garmire G P, Morris M R, Ricker G R and Townsley L K 2003 *ApJ* **589** 225

[6] Revnivtsev M, Sazonov S, Gilfanov M, Churazov E and Sunyaev R 2006 *A&A* **452** 169

[7] Worrall D M and Marshall F E 1983 *ApJ* **267** 691

[8] Sazonov S, Revnivtsev M, Gilfanov M, Churazov E and Sunyaev R 2006 *A&A* **450** 117

[9] Sakano M, Warwick R S, Decourchelle A and Wang Q D 2005 *MNRAS* **357**

[10] Terada Y, Kaneda H, Makishima K, Ishida M, Matsuzaki K, Nagase F and Kotani T 1999 *PASJ* **51** 39

[11] Murakami H, Koyama K, Sakano M, Tsujimoto M and Maeda Y 2000 *ApJ* **534** 283

[12] Murakami H, Koyama K and Maeda Y 2001 *ApJ* 558 687

[13] Revnivtsev M G, Churazov E M, Sazonov S Yu, Sunyaev R A, Lutovinov A A, Gilfanov M R, Vikhlinin A A, Shtykovsky P E and Pavlinsky M N 2004 *A&A* **425** L49

[14] Valinia A, Tatischeff V, Arnaud K, Ebisawa K and Ramaty R 2000 *ApJ* **543** 733

[15] Yusef-Zadeh F, Law C and Wardle M 2002 *ApJ* **568** L121

[16] Muno M P, Baganoff F K, Bautz M W, Feigelson E D, Garmire G P, Morris M R, Park S, Ricker G R and Townsley L K 2004 *ApJ* **613** 326

[17] Serabyn E and Morris M 1996 *Nature* **382** 602

[18] Haller J W, Rieke M J, Rieke G H, Tamblyn P, Close L and Melia F 1996 *ApJ* **456** 194

[19] Haller J W, Rieke M J, Rieke G H, Tamblyn P, Close L and Melia F 1996 *ApJ* **468** 955 *Erratum*

[20] Launhardt R, Zylka R and Mezger P G 2002 *A&A* **384** 112

[21] Muno, M P, Arabadjis J S, Baganoff F K, Bautz M W, Brandt W N, Broos P S, Feigelson E D, Garmire G P, Morris M R and Ricker G R 2004 *ApJ* **613** 1179

Radio pulsars and transients in the Galactic center

Joseph Lazio[1], J. S. Deneva[2], Geoffrey C. Bower[3], J. M. Cordes[2], Scott D. Hyman[4], D. C. Backer[3], R. Bhat[5], S. Chatterjee[6,7], P. Demorest[3], S. M. Ransom[7] and W. Vlemmings[8]

[1] Naval Research Laboratory, 4555 Overlook Ave. SW, Washington, DC 20375-5351, USA
[2] Astronomy Department and NAIC, Cornell University, Ithaca, NY 14853, USA
[3] Astronomy Department and Radio Astronomy Laboratory, University of California, Berkeley, CA 94720, USA
[4] Department of Physics and Engineering, Sweet Briar College, Sweet Briar, VA 24595, USA
[5] Centre for Astrophysics and Supercomputing, Swinburne University, Hawthorn, Victoria 3122 Australia
[6] National Radio Astronomy Observatory, 520 Edgemont Road, Charlottesville, VA 22903 USA
[7] Harvard-Smithsonian Center for Astrophysics, 60 Garden Street, Cambridge, MA 02138 USA
[8] Jodrell Bank Observatory, University of Manchester, Macclesfield, Cheshire, SK11 9DL, UK

E-mail: `Joseph.Lazio@nrl.navy.mil`

Abstract. Radio pulsars and transients provide powerful probes of the star formation history, interstellar medium, and gravitational potential of the Galactic center. Historical radio observations of the Galactic center have not emphasized the *time domain* aspect of observing this region. We summarize a series of recent searches for and observations of radio transients and pulsars that make use of two advances in technology. The first is the formation of large fields of view ($\gtrsim 1^\circ$) at relatively longer wavelengths ($\lambda > 1$ m), and the second is the construction of receivers and instruments capable of collecting data on microsecond time scales at relatively short wavelengths (≈ 3 cm).

1. Introduction

Transient emission—bursts, flares, and pulses—marks compact sources or the locations of explosive or dynamic events. The discovery of transient sources, most notably radio pulsars, can be used to probe a number of aspects of the Galactic center (GC):

- The spatial distribution of objects, as well as period and age distributions of pulsars, in the region, along with a careful tallying of selection effects, would help describe the stellar population in the region and discriminate between hypotheses attempting to explain the presence of the cluster of massive stars: stellar collisions and mergers, migration, and a past episode of intensive star formation.
- The density of the ionized medium can be inferred from its imprint on the radio signals, and potentially its structure, any turbulence within it, and any magnetic field threading it can also be determined.
- Pfahl & Loeb [1] argue that a population of stellar-mass black holes (BHs) exists in the GC due to tidally induced migration, e.g., 100 BHs within the central $1''$ alone. Pulsars discovered within this central field would be experiencing measurable gravitational

perturbations from the BHs and timing them would allow indirect study of this exotic population. The stellar density is probably sufficiently high that, by analogy with globular clusters, the region should be rich in pulsar binaries and millisecond pulsars.

- Orbits of pulsars close to Sgr A* will show many relativistic effects. Careful timing of pulsars would allow us to measure their Einstein delay, first and second order Shapiro delays, and frame-dragging effects. This in turn would provide methods for better constraining the mass of the central BH, and even estimating its spin. Combined with the effects of the proposed population of stellar-mass BHs, these measurements also would allow mapping out the shape of the central gravitational potential.

Generally, historical radio observations of the GC have not emphasized the *time domain* aspect of observing this region or, if so, have done so with only very infrequent observations (e.g., once per month). More recently, a variety of technological improvements and new recognition of the importance of these objects in the GC have spurred new observations. Though there can be common threads to the study of radio transients and pulsars (e.g., some classes of radio transients may be powered by neutron stars), the observational details are sufficiently different that we discuss the two kinds of objects separately.

2. Radio Transients

There have been a few historical radio transients detected toward the GC, though these typically have been discovered serendipitously in the process of other observations, typically Sgr A* itself. Examples of these historical transients include A1742−28 [2] and the Galactic Center Transient [3].

Effective searches for radio transients requires that the quantity $A\Omega(T/\Delta t)$ be "large." Here A is the collecting area or sensitivity of the telescope, Ω is the solid angle coverage, and $T/\Delta t$ is a measure of the cadence or time resolution achieved in the search. High-energy telescopes have been able to find many transients with relatively modest sensitivities by having a large field of view. Recent advances in both computing and imaging algorithms now make it practical to image relatively wide fields of view, of order 1°, on a regular basis at wavelengths $\lambda \approx 1$ m [4]. Motivated by those successes, specific radio transient search campaigns have been conducted toward the GC with the Very Large Array, with recent observations also acquired at the Giant Metrewave Radio Telescope.

This transient campaign suggests that the duty cycle of *radio-selected* radio transients at λ1 m with flux densities above approximately 0.1 Jy is about 10%. That is, in approximately 10% of the observing time, a radio transient has been detected (with one of the transients detected multiple times). Two specific transients that have been found include GCRT J1746−2757 [5] and GCRT J1745−3009 [6, 7]. Even these two objects hint at the potential diversity of radio transients. The former, GCRT J1746−2757, brightened and faded on a time scale longer than a day and potentially as long as a few months. In contrast to many radio transients that are found via their X-ray emission (e.g., microquasars and X-ray binaries), GCRT J1746−2757 did not appear to emit in the X-ray. Conversely, GCRT J1745−3009 showed rapid variability, bursting several times during the course of a 6-hr observation (Figure 1). If located in the GC, the brightness of this transient requires that it emit coherently, and it may represent the first in a new class of radio objects. Like GCRT J1746−2757, it was not detected in the X-ray.

Finally, another recent transient detected is CXOGC J174540.0−290031 [8]. It had the morphology of a classical Faranhoff-Riley Type II radio galaxy, namely two expanding lobes of radio-emitting plasma. Unlike a radio galaxy, however, the proper motion of one of the lobes could be measured. Midway between the two lobes was a central X-ray source, prompting the identification of this object as the result of a bipolar jet of relativistic plasma from a central engine powered by a low-mass X-ray binary. Intriguingly, this transient, or one like it, may be able to create a cavity of ionized gas, potentially explaing the nearby "mini-cavity" [9].

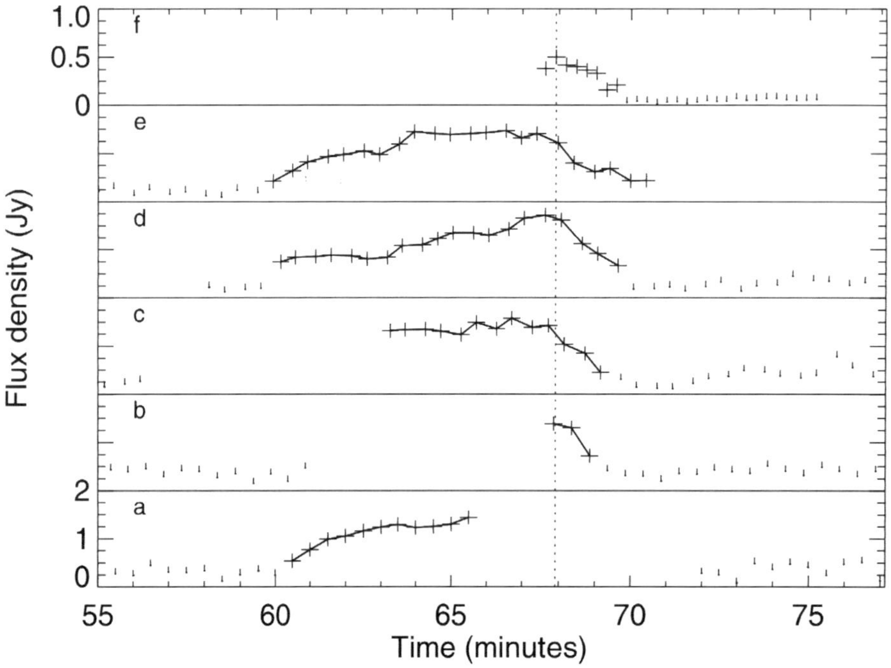

Figure 1. Light curves of GCRT J1745−3009 [7]. The top panel shows the single detected burst from 2003 September 28; the remaining panels show the bursts from 2002 September 30, arranged with the first burst shown in the bottom panel to the fifth burst shown in the second from the top panel. The ordinate ranges from 0 to 1 Jy in the top panel and 0 to 2 Jy in the other panels. For the 2002 September 30 bursts, the light curve has been folded at the apparent 77.1 minute periodicity. For the 2003 September 28 burst, the light curve has been aligned in time to be consistent with the decay portions of the 2002 bursts. In many cases, because the existence of GCRT J1745−3009 was not known at the time of the observation, the full burst is not captured as the observations were interrupted for calibration observations. The arrows represent 3σ upper limits for nondetections. The vertical dotted line is placed at the fitted position of the fourth 2004 burst for reference.

3. Radio Pulsars

Of the 1700 pulsars presently known, fewer than 1% are within 2° of the GC. This deficit of pulsars has long been attributed to the effects of radio wave propagation and scattering through the ionized interstellar medium. As scattering effects typically scale strongly with wavelength, "short wavelength" surveys (e.g., 20 cm) were initiated to mitigate the scattering effects [10].

Only more recently has the extent to which scattering vitiates pulsar searches toward the GC become apparent. Lazio & Cordes [11] (see also [12]) have shown, via observations of background sources and OH/IR stars, that the GC is either immersed in or surrounded by a region of intense scattering. Extrapolating their results to the impact on pulsar signals, Cordes & Lazio [13] estimated that the level of pulse broadening could be $2\,\mathrm{milliseconds}\lambda_{\mathrm{cm}}^4$ for the wavelength measured in centimeters. At the typical wavelength used for pulsar searches ($\lambda \gtrsim 20$ cm), this level of pulse broadening can exceed *2000 seconds*, rendering pulsars undetectable as pulsed objects. The recent discovery of PSR J1745−2912 and PSR J1746−2856 from observations at 9.7 cm, and their non-detection at 3.6 cm, lead Johnston et al. [14] to suggest that the scattering may be even more intense than these models predict. Both of these pulsars are within 0.3° of the GC and are likely located between 150 and 500 pc, but on the near side of the GC.

The typical pulsar search methodology, a periodicity search, is to survey the desired region, collecting long time series which are then Fourier transformed and searched for periodic signals (or multiple harmonics from periodic signals). This methodology fails completely if pulsars do not pulse, as they would not at typical search wavelengths. Searching at wavelengths short enough to mitigate the pulse broadening is difficult, both because of the resulting relatively small telescope beams as well as the typically steep pulsar spectra. Cordes & Lazio [13] and Lazio & Cordes [11] outlined an alternate search methodology in which pulsars might be detected as angularly broadened compact sources, analogous to how Sgr A* and the OH masers associated with various OH/IR stars have been detected. Angularly broadened compact objects with pulsar-like characteristics could then be targetted for more intensive at wavelengths short enough to mitigate the pulse broadening.

Two surveys of the GC using the Very Large Array have been carried out for compact radio sources. Lazio & Cordes [15] surveyed the central 1° at a wavelength of 20 cm and found 175 sources. Bower (2006, unpublished) surveyed the central 17′ at 3.6 cm, in an effort to match a field also surveyed with *Chandra*, and found 85 sources. Candidate lists from these surveys are now the subject of periodicity searches with the 100-m Green Bank Telescope at a wavelength of 3.3 cm. A blind survey of the central few arcminutes also has been conducted as part of these observations. Although a blind survey of the entire GC region is too expensive in terms of telescope time, a blind survey of the central 1′ or so is justified given the presence of the central stellar cluster. In addition, a survey at 6 cm has been conducted with the 100-m Effelsberg telescope [16].

4. Future Observations

There are a number of radio telescopes, either under construction or in the design and development phase, for which searches for radio transients, radio pulsars in the GC, or both form an important part of the science case. We describe these briefly.

The Expanded VLA[1] (EVLA) is a project designed to replace the (original) electronics in the VLA with more modern versions, including replacing the original analog correlator with a modern digital version. The result will be larger bandwidths and a greater access to the spectrum. The increased spectral access will be particularly important at wavelengths around 3 cm, at which it result in a considerable improvement in sensitivity, and the new correlator will enable efficient radio pulsar searches for the first. If the GC population is similar to the disk population of radio pulsars, the EVLA is likely to discover several (or more!) pulsars in the GC.

The Allen Telescope Array[2] (ATA) is a telescope under construction with the aim of providing several new capabilities for observing in the 3–60 cm wavelength range. Composed of 6-m diameter parabolic antennas, the ATA will have a relatively large field of view ($\approx 1°$ deg.^{2} at 30 cm) and the ability to access multiple regions within this field of view simultaneously. In addition, it is expected to be able to access its entire wavelength range instantaneously. Radio transients (potentially including transmissions from extraterrestrial transmitters) are among its key science goals.

The Long Wavelength Array[3] (LWA) is a telescope designed to open for the first time long-wavelength regime of the spectrum ($\lambda > 3$ meters) to high resolution imaging. Radio transients form an explicit part of its science case. It will be constructed of "stations" of dual-polarization dipoles, each station performing equivalent to a more traditional parabolic antenna, and it

[1] http://www.nrao.edu/
[2] http://www.seti.org/
[3] http://lwa.unm.edu/

will be located in the southwestern U.S. state of New Mexico. Pointing will be accomplished electronically. The LWA will be able to access the GC, though because of the GC's elevation as seen from New Mexico, there will be some loss of sensitivity. Nonetheless, if there are other objects such as GCRT J1745−3009, and they emit at long wavelengths, the LWA should still have the sensitivity to detect them. Importantly, the LWA will also have multiple, independent fields of view, potentially allowing for considerably more search time, one of the major limitations with existing instrumentation.

The Low Frequency Demonstrator[4] of the Mileura Wide-field Array (MWA-LFD) is a telescope in the design and development phase. Like the LWA it will be composed of dual-polarization dipoles phased into stations. Its design wavelength range range is approximately 1.25 m to 4 m. Located in Western Australia, where there are few anthropogenic emissions, it will be equisitely sensitive in the FM radio band ($\nu \approx 100$ MHz). Radio transients form a key part of its science case, and the GC will pass directly overhead.

The Square Kilometre Array[5] (SKA) is a telescope in the design and development phase. Intended to cover wavelengths from 1.25 cm to 3 m, the SKA will have a sensitivity some 50 times greater than that of the EVLA. Radio pulsars in the GC form an explicit part of its science case, and it will have the capability to search both in an imaging mode (as done with current VLA searches) as well as in a periodicity mode at high sensitivity and at short wavelength.

With this suite of telescopes becoming available over the next decade, we expect a considerable increase in the number of both radio transients and radio pulsars known toward the Galactic center.

Acknowledgments
SDH is supported by the Research Corporation. Basic research in radio astronomy at the NRL is supported by the Office of Naval Research.

References

[1] Pfahl, E and Loeb, A 2004, *Astrophys. J.*, **615**, 253–258

[2] Davies, R D, Walsh, D, Browne, I W A, Edwards, M R and Noble, R G 1976, *Nature*, **261**, 476–478

[3] Zhao, J, et al. 1992, *Science*, **255**, 1538

[4] LaRosa, T N, Kassim, N E, Lazio, T J W and Hyman, S D 2000, *Astron. J.*, **119**, 207–240

[5] Hyman, S D, Lazio, T J W, Kassim, N E and Bartleson, A L 2002, *Astron. J.*, **123**, 1497–1501

[6] Hyman, S D, Lazio, T J W, Kassim, N E, Ray, P S, Markwardt, C B and Yusef-Zadeh, F 2005, *Nature*, **434**, 50–52

[7] Hyman, S D, Lazio, T J W, Roy, S, Ray, P S, Kassim, N E and Neureuther, J L 2006, *Astrophys. J.*, **639**, 348–353

[8] Bower, G C, Roberts, D A, Yusef-Zadeh, F, Backer, D C, Cotton, W D, Goss, W M, Lang, C C and Lithwick, Y 2005, *Astrophys. J.*, **633**, 218–227

[9] Wardle, M and Yusef-Zadeh, F 1992, *Nature*, **357**, 308–310

[10] Johnston, S, Walker, M A, van Kerkwijk, M H, Lyne, A G and D'Amico, N 1995, *Mon. Not. R. Astron. Soc.*, **274**, L43–L45

[11] Lazio, T J W and Cordes, J M 1998, *Astrophs. J.*, **505**, 715–731

[12] Bower, G C, Backer, D C and Sramek, R A 2001, *Astrophys. J.*, **558**, 127–132

[13] Cordes, J M and Lazio, T J W 1997, *Astrophys. J.*, **475**, 557–564

[14] Johnston, S, Kramer, M, Lorimer, D R, Lyne, A G, McLaughlin, M, Klein, B and Manchester, R N 2006, *Mon. Not. R. Astron. Soc.*, submitted; astro-ph/0606465

[15] Lazio, T J W and Cordes, J M 2006, *Astrophys. J. Suppl.*, submitted

[16] Klein, B, Kramer, M, Müller, P and Wielebinski, R. 2004, *Young Neutron Stars and Their Environments* ed F Camilo and B M Gaensler (San Francisco: Astronomical Society of the Pacific) p 133

[4] http://haystack.mit.edu/

[5] http://www.skatelescope.org/

Institute of Physics Publishing
doi:10.1088/1742-6596/54/1/020

Journal of Physics: Conference Series **54** (2006) 115–125
Galaxy Center Workshop 2006

Chandra Observations of the Galactic Center: High Energy Processes at Arcsecond Resolution

Q. Daniel Wang[1]

[1] Department of Astronomy, University of Massachusetts, Amherst, MA 01003

E-mail: wqd@astro.umass.edu

Abstract. About 2 million seconds of *Chandra* observing time have been devoted to the Galactic Center (GC), including large-scale surveys and deep pointings. These observations have led to the detection of about 4000 discrete X-ray sources and the mapping of diffuse X-ray emission in various energy bands. In this review, I first summarize general results from recent studies and then present close-up views of the three massive star clusters (Arches, Quintuplet, and GC) and their interplay with the Galactic nuclear environment.

1. Introduction

While Sgr A* itself is only weakly active at present, much of the high-energy activity in the GC is initiated apparently by the three young massive stellar clusters located within a 50 pc radius of the super-massive black hole Sgr A* (Fig. 1): Arches [with an age of $(2-3)\times 10^6$ yrs], Quintuplet [$(3-6)\times 10^6$ yrs], and GC [$(3-7)\times 10^6$ yrs] (e.g., Figer et al. 2004; Stolte et al. 2002, 2005; Genzel et al. 2003). Massive stars themselves can be moderately bright X-ray sources (e.g., colliding stellar wind binaries). Such stars also release large amounts of mechanical energy in form of fast stellar winds and supernovae, heating and shaping the surrounding interstellar medium (ISM) and affecting the accretion onto the black hole. Furthermore, the stellar end-products of massive stars (neutron stars and black holes) can also be strong X-ray sources. X-ray observations are thus a powerful tool for probing such high-energy phenomena and processes.

The GC is a region where the spatial resolution of X-ray observations matters. This is why *Chandra* has invested heavily in the GC, conducting repeated large-scale raster surveys [existing 360 ks (Wang et al. 2002a) and upcoming 600 ks (PI: Muno)] and numerous deep pointed observations: about 1 Ms on Sgr A (Baganoff et al. 2003), 100 ks each on Sgr B (Takagi et al. 2002), Sgr C (PI: Murakami), and Arches (Wang et al. 2006a), as well as 50 ks on the Radio Arc (Yusef-Zadeh et al. 2002a). The main instrument used in these *Chandra* observations is the ACIS-I, which covers a field of $17'\times 17'$ and an energy range of 0.5-10 keV (although below ~ 2 keV, X-rays from the GC are heavily absorbed by the ISM). The spatial resolution ranges from $\sim 1''$ on-axis to $\sim 10''$ at the outer boundaries of a field.

While much of the data analysis is still ongoing, here I will first summarize some of the general results on detected discrete sources and diffuse X-ray emission and will then focus on the three massive star clusters, highlighting various high-energy phenomena and processes involved.

2. General results on point-like X-ray sources

The *Chandra* observations have led to the detection of about 4000 discrete sources; about 2400 of them are from the deep Sgr A observations (Wang et al. 2002a; Muno et al. 2003, 2006; Wang et al. 2006a). Most of these sources are located physically in the vicinity of the GC, judging from their X-ray spectral characteristics and number statistics. The number-flux relation (or the so-called log$N$-log$S$ relation) of the GC sources can be approximated as a power law with an index of 1.5 ± 0.1. However, the relation is flatter (over the luminosity range of 5×10^{31} ergs s$^{-1} \lesssim L_x \lesssim 10^{34}$ ergs s^{-1}) in the vicinity of the Arches and Quintuplet clusters (Muno et al. 2006; Wang et al. 2006a), apparently due to the concentrations of massive stars and possibly their end-products.

The GC X-ray sources represent a heterogeneous population of high-energy objects. Bright sources (1E 1740.7-2942 and 1E 1743.1-2843) with 2-10 keV luminosities $L_x \gtrsim 10^{36}$ ergs s^{-1} are low-mass X-ray binaries (LMXBs). There is a clear dearth of X-ray sources with 10^{34} ergs s$^{-1} \lesssim L_x \lesssim 10^{36}$ ergs s^{-1}, although a few transients are detected in this intermediate luminosity range (e.g., Muno et al. 2005; Sakano et al. 2005; Wijnands et al. 2006; Wang et al. 2006a). Sources within the range of 10^{33} ergs s$^{-1} \lesssim L_x \lesssim 10^{34}$ ergs s^{-1} tend to be colliding wind massive star binaries (Wang et al. 2006a) and possibly young pulsars (Wang et al. 2002b; Wang et al. 2006b). The natures of fainter sources (10^{31} ergs s$^{-1} \lesssim L_x \lesssim 10^{33}$ ergs s^{-1}) are less certain; many of them are likely to be CVs, consistent with the lack of bright stellar counterparts in near-IR and radio surveys (e.g., Laycock et al. 2005; Bandyopadhyay 2005). A fraction of these sources have very hard X-ray spectra (e.g., with power law photon indices less than 0; Muno et al. 2004a). Such sources are most likely intermediate polars, in which X-ray-emitting regions are at least partially obscured and the observed X-rays represent the reflected emission (Ruiter, Belczynski, & Harrison 2006).

3. General results on diffuse X-ray emission

Despite their large number, the detected discrete X-ray sources typically account for only about 10% of the total observed X-ray emission from the GC. The spectrum of the remaining "diffuse" emission and the accumulated spectrum of the sources are similar, in terms of both the continuum shape and the presence of prominent emission lines such as the He-like S and Fe Kα transitions, which indicate a broad temperature range for the thermal plasma of $\sim 1 - 10$ keV (Wang et al. 2002a; Muno et al. 2004b). The spatial distribution of the diffuse emission appears to follow closely the K-band stellar light. Therefore, the bulk of the emission likely originates in the old stellar population (coronally active binaries and CVs), as has been proposed for the Galactic Ridge hard X-ray emission (Revnivtsev et al. 2005).

There is no clear evidence for a significant presence of truly diffuse gas with $T \gtrsim 8$ keV, as was believed for many years. However, a non-thermal diffuse hard X-ray component does seem to be important (Wang et al. 2002a; see also the presentation by R. Warwick). The strongest evidence for this component is the ubiquitous presence of the 6.4-keV line emission, globally correlated with the molecular gas in the GC region. The line emission arises from the filling of inner shell vacancies of neutral or weakly ionized irons, which can be effectively produced by ionizing radiation with energies $\gtrsim 7.1$ keV. Because of the high equivalent width ($\gtrsim 1$ keV) of the observed line, the source of the ionizing radiation must be currently obscured or absent. This reflection interpretation gives the most convincing interpretation for the 6.4-keV line emission associated with the giant molecular clouds, Sgr B2 and Sgr C. The required ionizing source is believed to be Sgr A*, which might have been substantially brighter ($\gtrsim 10^{39}$ ergs s^{-1}) several hundred years ago. Indeed, the continuum emission from Sgr B, presumably due to the reflection through electron Thompson scattering, is shown to have a hard X-ray spectrum (power law photon index $\Gamma \approx 1.8$) in the 2-200 keV range, consistent with what is expected from a typical AGN (Revnivtsev et al. 2004). However, the reflection interpretation is less successful in explaining the 6.4-keV line emission observed in fields closer to Sgr A* (e.g., Wang et al. 2002a). While the line intensity is globally correlated with the emission from trace molecules, a peak-to-peak correlation is often absent. The lack of such a correlation is not expected, because molecular clouds should be optically thin to the ionizing radiation. However, it is possible that their density peaks may be optically thick to molecular emission, such as the CS (J=2-1) line.

An alternative mechanism to produce the K-shell vacancies is the collision of iron with low-energy cosmic-ray electrons (LECRe; Valinia et al. 2000). In this case, one also expects hard X-ray emission from the bremsstrahlung process. The problems with this mechanism are 1) the requirement for an unusually high iron abundance ($\gtrsim 4\times$ solar) of the gas and 2) the low efficiency of the emission (typically $10^{-4} - 10^{-5}$; e.g., Tatischeff 2003); most of the energy is consumed by ionization losses. It is not clear how the required energy density of the LECRe could be maintained *globally*, even in the GC region.

4. *Chandra* view of the massive star clusters in the GC region

Fig. 1 presents the global perspective of the Arches and Quintuplet clusters in the GC environment. This most active region also includes various prominent thermal and nonthermal radio filaments, dense molecular clouds, and strong diffuse X-ray emission, all of which are concentrated on the positive Galactic longitude side of Sgr A* (Wang et al. 2006a and references therein). The lopsided distribution of these features may partly be a chance coincidence. However, some of the features are likely to be related, although they seem to have very different line-of-sight velocities. The thermal mid-IR-emitting filaments are clearly produced by the radiative heating of the Arches and Quintuplet clusters, while the strong 6.4-keV line emission is associated with the dense molecular gas (Wang et al. 2002a; Yusef-Zadeh et al. 2002b). Fig. 2 illustrates a scenario that provides a unified interpretation of various distinct interstellar features observed in the region. The densest cloud in the region, G0.13-0.13 (Handa et al. 2006), is projected inside the west part of the distinct mid-IR cavity (Fig. 1; Price et al. 2001) and is probably located at the far side of a tunnel dug out by the cloud (Fig. 2). This tunnel in projection is seen as the cavity and is apparently filled with hot plasma, which is responsible for the enhanced diffuse soft X-ray emission (Fig. 1; Wang et al. 2006a). This scenario is consistent with the lack of enhanced near-IR extinction toward G0.13-0.13 against GC stars and with the high excitation of molecular gas associated with the cloud, which is apparently due to shock-heating (Handa et al. 2006). The compression of the inter-cloud medium (the cavity wall in Fig. 2), and hence the attached magnetic field, may even explain the nonthermal radio filaments. The diffuse X-ray emission, both thermal and nonthermal, are likely products of the mechanical energy input from the massive star clusters. Much work is still required to clarify what is actually going on in this unusually dynamic region of the GC.

4.1. The Arches and Quintuplet Clusters

Both the Arches and Quintuplet clusters are associated with local enhancements of X-ray emission (e.g., Fig. 3), which was discovered serendipitously at large off-axis angles in early *Chandra* observations (Yusef-Zadeh et al. 2002a; Wang et al. 2002a; Law & Yusef-Zadeh 2004). A recent 100 ks observation allows for an in-depth study of both discrete and diffuse components of the X-ray emission from these two clusters (Wang et al. 2006a).

The three bright X-ray sources in the Arches core region all have near-IR counterparts, classified as WN stars (Fig. 3b). These sources have remarkably similar spectra, in terms of both the continuum shape and the strong presence of the 6.7-keV line (Fig. 4). Their spectra can be characterized by optically-thin thermal plasma with temperatures of ~ 2 keV and metal (chiefly iron) abundances of $1.8\times$ solar. The 0.3-8 keV luminosity of each source is $\sim 1 \times 10^{34}$ ergs s^{-1}. With these properties, the sources are most likely colliding wind massive star binaries, with luminosities comparable to those of the most extreme such objects (e.g., η Carina; Seward et al.

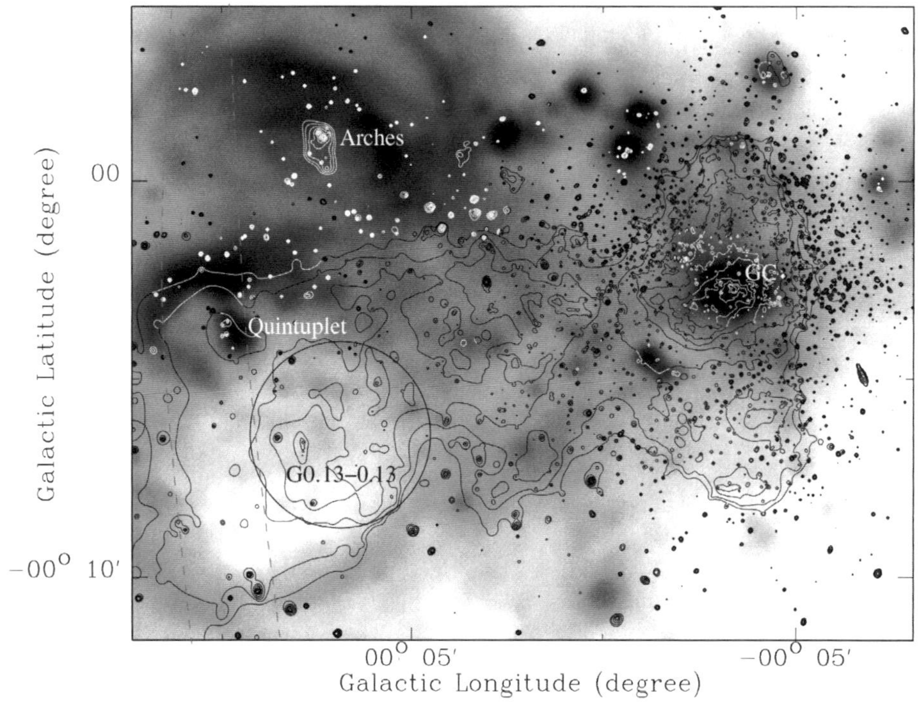

Figure 1. Panoramic views of the GC environment of the three massive star clusters: MSX 24 μm intensity distributions (gray-scale with the white representing the faintest emission; Price et al. 2001) and *Chandra* ACIS-I 1-9 keV intensity contours. The region with strong nonthermal radio filaments are outlined by the dashed lines, while the cloud G0.13-0.13 is circled.

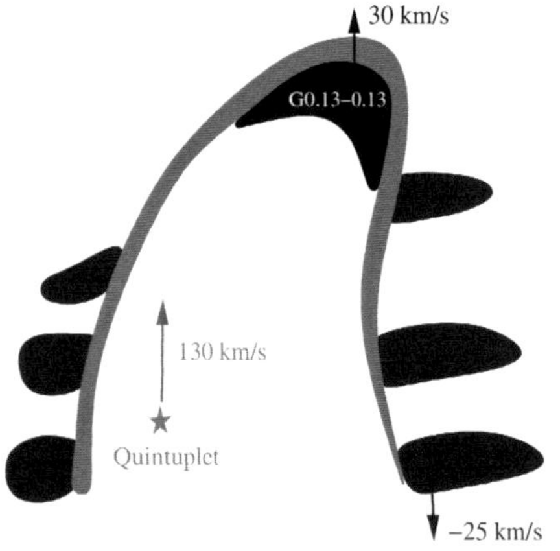

Figure 2. A plausible configuration of several major ISM components in the vicinity of the Quintuplet clusters, as viewed from the Galactic north pole. The observed line-of-sight velocities of the components are marked.

2001). The X-ray metal abundance measurement, based primarily on the He-like Fe Kα line, is also interesting, which is insensitive to the exact temperature of the plasma. The optically-thin thermal emission process is also quite simple, astrophysically. Furthermore, the iron abundances

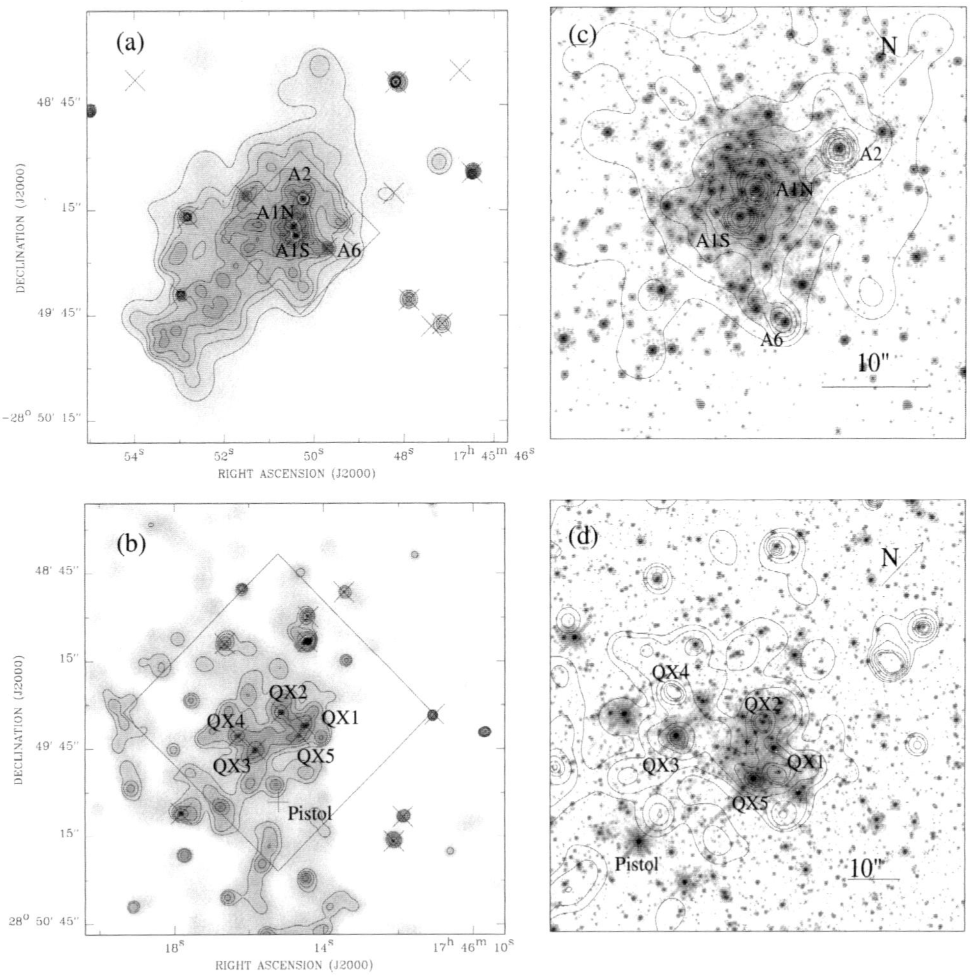

Figure 3. Adaptively smoothed ACIS-I 1-9 keV band images of the Arches (a) and Quintuplet (b) clusters. The intensity contour levels are at 20, 23, 27, 33, 57, 57, 80, 114, 180, 314, 482, 682, 1351, and 3358 (above a local background of 13.4) for (a), and at 17, 29, 33, 42, 54, and 72 (above 17) for (b); all in units of 10^{-3} counts s^{-1} $arcmin^{-2}$. The two large squares in (a) and (b) outline the fields covered by the *HST* NICMOS near-IR images of the Arches (c) and Quintuplet (d), respectively (Figer et al. 2004). These two NICMOS images (the north to the upper right) are rotated with respect to the panels (a) and (b). The contours are the same as in (a) and (b), except for excluding the first four levels in (c) for clarity. The detected sources are marked with *crosses* in (a) and (b). The location of the Pistol star is marked as a *plus*. Several bright X-ray sources named previously are labeled (Yusef-Zadeh et al. 2002a; Law & Yusef-Zadeh 2004).

in the stellar winds should not be contaminated by the nuclear synthesis of these stars and thus reflect the value in the general ISM of the GC.

The nature of the source-removed "diffuse" emission is more complicated. The overall spectrum of the emission shows both 6.4-keV and 6.7-keV emission lines. The 6.7-keV line arises predominately in the region close to the core of the Arches cluster. The region has an

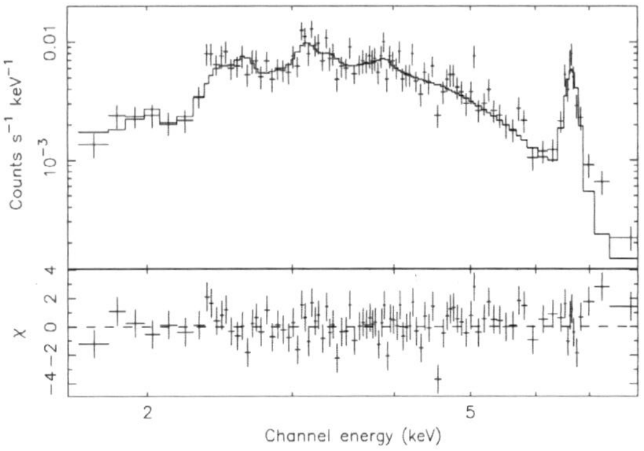

Figure 4. Accumulated ACIS-I spectrum of the three brightest X-ray sources (A1N + A1S + A2) in the Arches cluster and the best-fit thermal plasma models.

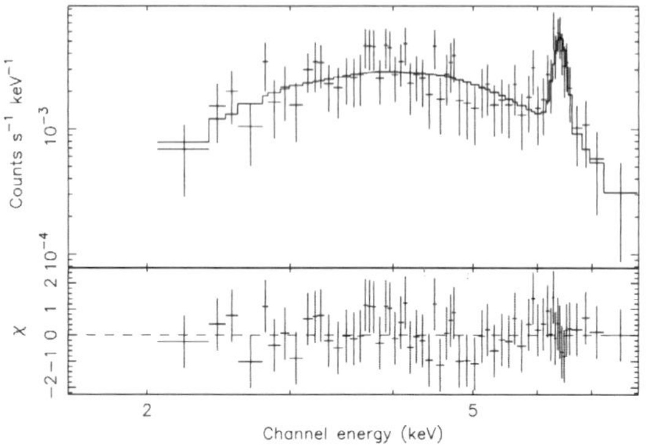

Figure 5. ACIS-I spectrum of the diffuse X-ray emission southeast of the Arches cluster (Fig. 6) and the best-fit power law plus 6.4-keV Gaussian line model.

extent of $\sim 30''$ and appears to be elongated towards the east, morphologically matching an extinction deficit around the cluster (Stolte et al. 2002). The 6.4-keV line emission is more widely distributed, but is particularly enhanced in the southeast extension of the diffuse X-ray emission (Fig. 3a and Fig. 5). All these can be interpreted as the collision between the cluster wind of the Arches cluster and a dense gas cloud (Fig. 7). The steep decline of the observed surface intensity of the diffuse X-ray emission with the off-cluster radius within $\sim 10''$ is consistent with the prediction of emission from the expanding cluster wind (Fig. 8a), whereas the flattening of the intensity distribution at larger radii (but $\lesssim 15''$) likely reflects the reverse shock heating and confinement of the wind.

The presence of a cloud in the vicinity of the Arches cluster is known. In the CS map made by Serabyn & Gusten (1987) with the 30 *IRAM* telescope, the cloud is labeled as "Peak 2". Fig. 6 shows a recent interferometry map made with *OVRO* (Wang et al. 2006a). The cloud has a line-of-sight velocity of -25 km s^{-1} (i.e., moving toward us), which is similar to many other clouds or filaments in the region. In comparison, the cluster is moving at ~ 95 km s^{-1} away from us. Therefore, the relative velocity between the two is at least about 120 km s^{-1}. Because the filling factor of molecular clouds in the region is estimated to be about 0.3 (Serabyn & Gusten 1987), a chance collision between the cluster and such a cloud is not rare. The far-IR spectroscopy further shows the presence of a component of dusty gas at a velocity of -70 km s^{-1}, unique at the location of the Arches cluster (Cotera et al. 2005). This component may represent shocked cloud gas, deflected toward us.

The collision between the cluster and the cloud (Fig. 7) also provides a plausible explanation of the large-scale diffuse X-ray emission (Fig. 6). Indeed, the overall morphology of the diffuse X-ray emission resembles a bow shock. Presumably, the motion of the cloud relative to the cluster is roughly toward the east in the sky. The diffuse X-ray emission, except in the core region around the Arches cluster, is reasonably well correlated with the 6.4-keV line intensity, but shows little peak-to-peak correlation with the CS line intensity. The diffuse X-ray emission is enhanced particularly in the region southeast of the cluster, probably tracing the eastern edge of the cloud. But the northern part of the cloud, where the CS line emission is the strongest, shows no enhancement in the 6.4-keV line emission. The cloud does not seem to be dense enough to be optically thick to the CS line emission. Therefore, the radiation reflection interpretation does not work in this case. Certainly, the source of the radiation cannot be the Arches cluster itself. The luminosity of the cluster falls several orders of magnitude short of that required. The probable origin of the diffuse X-ray emission is then LECRe that may be produced locally in

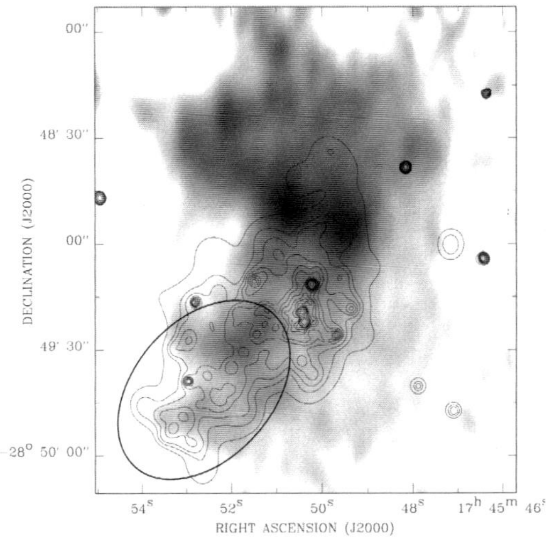

Figure 6. ACIS 1-9 keV diffuse emission intensity contours overlaid on the CS (J=2-1) line emission image. The X-ray contours are the same as in Fig. 3a. The ellipse outlines the region from which the spectrum in Fig. 5 is extracted.

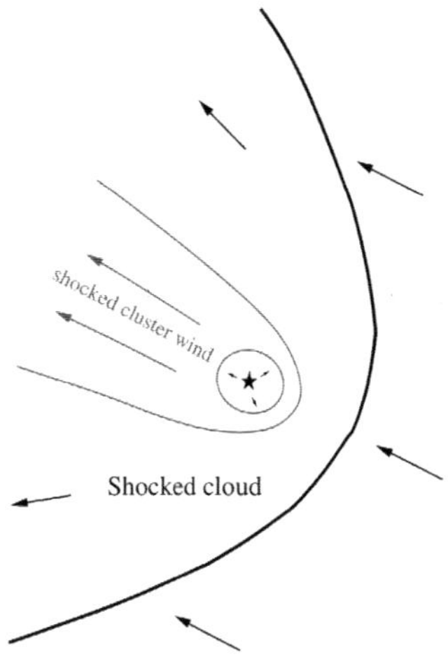

Figure 7. An illustration of the proposed cluster-cloud collision scenario. The shocked cloud gas is partly traced by the CS and 6.4-keV lines (Fig. 6), whereas the shocked cluster wind plasma near the cluster is by the 6.7-keV line.

and around the clusters (Wang et al. 2006a).

The older and looser Quintuplet cluster is weak in X-ray emission, both point-like and diffuse (Fig. 3b and d, Fig. 8b). The X-ray sources in the core of the cluster also show more diverse spectral characteristics, with substantially different intrinsic spectral shapes and absorptions. Some of these sources are probably colliding dusty winds in massive star binaries. The Pistol star, despite of its enormous bolometric luminosity, is not detected. But the 3σ upper limit to the 0.3-8 keV luminosity, 3×10^{33} ergs s^{-1}, is consistent with the nominal relation $L_x/L_{bol} \sim 10^{-7}$ for massive stars. The diffuse X-ray emission from the Quintuplet cluster ($L_X \sim 2 \times 10^{33}$ ergs s^{-1}) is about a factor of 10 lower than that from the Arches cluster and can be naturally explained by the cluster wind and a limited number of low-mass pre-main sequence young stellar objects (YSOs).

There appears to be a general deficiency of YSOs in the two clusters, relative to the prediction from the standard Miller & Scalo initial mass function (IMF). Compared with the X-ray emission from YSOs in the Orion nebula (Feigelson et al. 2005), the observed total diffuse X-ray luminosities from the Arches and Quintuplet clusters suggest that they contain no more than 2×10^4 and 3×10^3 YSOs. These numbers are a factor of 10 and 5 smaller than what would be expected from the IMF and the massive star populations observed in the cores of the two clusters. One possibility is that the IMF flattens at intermediate masses, as indicated in a near-IR study of the inner regions of the Arches cluster (Stolte et al. 2005). The top heavy IMF may be a result of the star formation in the extreme environment of the GC. In particular, collisions between dense molecular clouds such as G0.13-0.13 may be responsible for the formation of massive clusters.

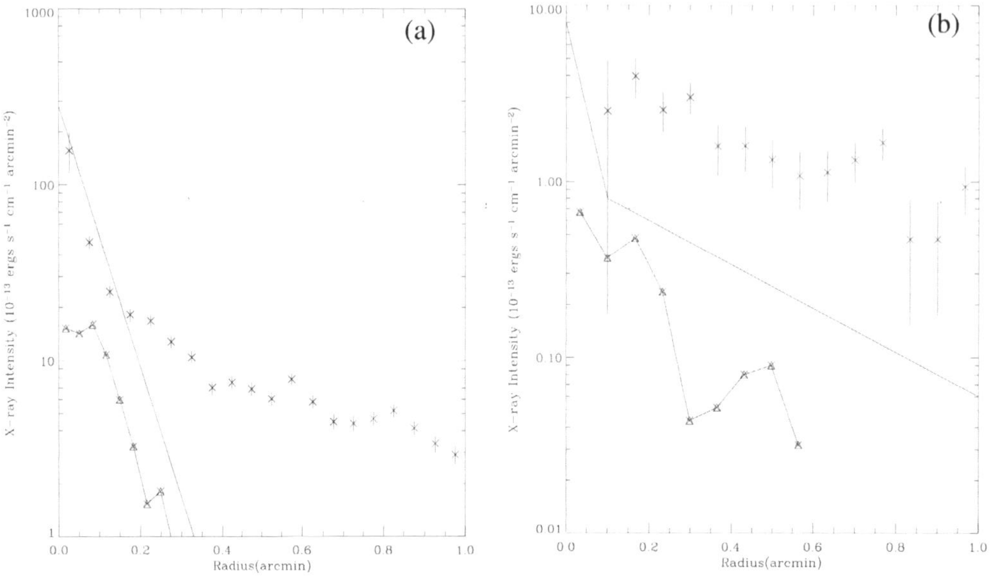

Figure 8. Radial ACIS-I 1-9 keV intensity profiles (*crosses* with 1σ error bars) around the Arches (a) and Quintuplet (b) clusters, compared with the respective NICMOS F205W stellar light distributions (connected *triangles*). The cluster wind predictions are shown approximately as the solid line from 3-D simulations for the "standard" stellar wind mass-loss rates of the two clusters (Rockefeller et al. 2005)

4.2. The GC cluster

The deficiency of YSOs in the GC cluster has been proposed by Nayakshin et al. (2005), based on the observed diffuse X-ray intensity of the region (Fig. 9; Baganoff et al. 2003). From the X-ray constraint on the total mass of the cluster, they further conclude that it cannot be a remnant of a massive star cluster that originated at several tens of parsecs away from Sgr A* and then dynamically spiralled in. The GC cluster was thus most likely formed *in situ* in a self-gravitating circum-nuclear disk. These conclusions would be weakened only slightly if the age-dependence of YSO X-ray luminosities (a factor of ~ 2; Preibisch & Feigelson 2005) were accounted for. Therefore, the top heavy IMF appears to be a general characteristic of star formation in the GC region. However, the exact form of the IMF is yet to be determined for the clusters.

The diffuse X-ray emission from the GC cluster region appears significantly different from those from the Arches and Quintuplet clusters. The spectrum of the emission shows a Fe Kα emission line at an intermediate energy ~ 6.55 keV (e.g., Fig. 10b; Baganoff et al. 2003), which is a clear signature for the non-equilibrium ionization (NEI) state of the X-ray-emitting plasma with an ionization timescale $n_e t \sim 3\times10^{10}$ cm^{-3} s (Wang et al. 2006b). The source of the plasma is likely to be the stellar winds from massive stars in the GC cluster. We may assume that the pre-shock stellar winds are cold ($T \lesssim 10^4$ K) and generally have velocities of $\sim 1 \times 10^3$ km s^{-1}. Fig. 9 shows that the distribution of the stars is not very centrally concentrated, except for the compact sub-cluster complex IRS 13. The density of the shocked wind plasma is typically not very high (order ~ 10 cm^{-3}). Thus, much of the shock-heated plasma may be out of the collisional ionization equilibrium, at least in individual wind-wind shock regions.

Further constraints on the properties of the plasma may be obtained from studying its interaction with individual stars or sub-clusters. The X-ray study of IRS 13 provides such a possibility. The X-ray emission from this sub-cluster is clearly resolved and shows an NEI plasma

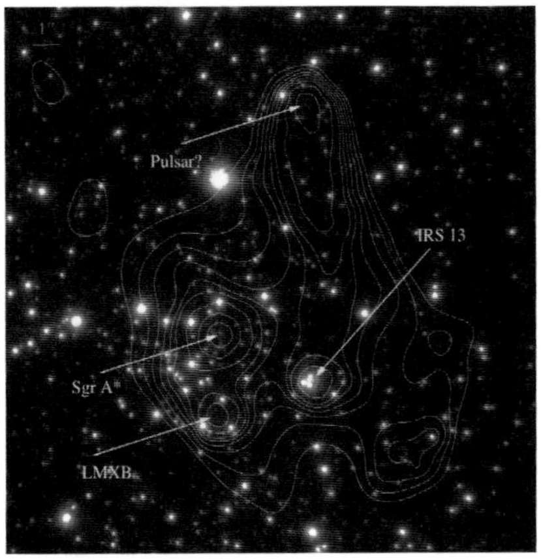

Figure 9. SINFONI near-IR image of the GC cluster (Eisenhauer et al. 2005) with overlaid ACIS-I 1–9 keV intensity contours (Wang et al. 2006b).

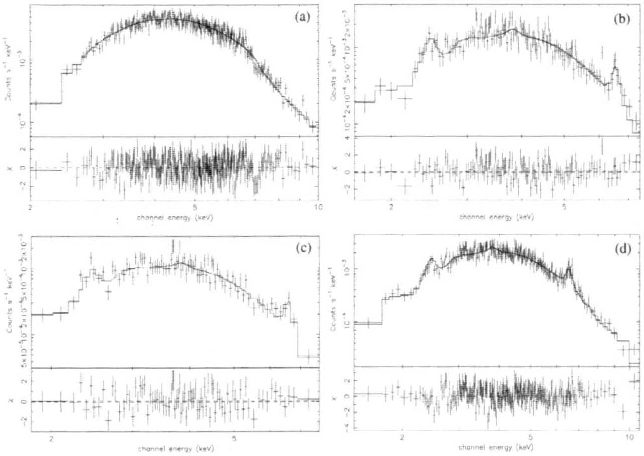

Figure 10. ACIS-I spectra of Arches (a), Diffuse (b), IRA 13 (c), and Sgr A* (d). The relative deviations from the best-fit models (Wang et al. 2006b) are shown in the respective bottom panels.

spectrum (Fig. 10c). Therefore, the source most likely represents stellar wind-wind collisions. The X-ray emission may be enhanced due to a strong ram-pressure confinement by the overall GC cluster wind, plus a potential outflow from Sgr A*, as predicted in the radiatively-inefficient accretion flow models (Yuan, Quataert, & Narayan 2003). The confinement is indicated by the X-ray morphology that is slightly offset from the centroid position of IRS 13 and elongated toward to the west (Wang et al. 2006b).

The properties of the plasma around Sgr A* should also affect its accretion. The quiescent X-ray emission from Sgr A* is resolved to have a size of about a couple of arc-seconds. This size corresponds to the Bondi accretion radius of the super-massive black hole. The accretion flow is believed to be responsible for the X-ray emission. Interestingly, the observed spectrum of the emission resembles that of the surrounding diffuse plasma and shows the Fe Kα line at $\sim$ 6.6 keV (Fig. 10d; Wang et al. 2006b; Xu et al. 2006). Therefore, the flow is also in an NEI state. A detailed modeling of the flow, confronted with the observed spatial and spectral distributions of the X-ray emission, will help to constrain the accretion dynamics.

The X-ray emission from the GC region is further complicated by the presence of a comet-like pulsar wind nebula (PWN), a few arc-seconds northwest of Sgr A* (Fig. 9; Wang et al. 2006b). The suspected pulsar, presumably moving at a high speed relative to the ambient medium, is located at the northern head of the nebula. This PWN interpretation naturally explains not only the morphology, but also the nonthermal spectrum (Fig. 10a) and the spectral steepening with the off-source distance due to the synchrotron cooling of ultra-relativistic electrons (and positrons). Furthermore, the inverse-Compton scattering of the intense ambient infrared photon field by the electrons provides a ready explanation for the TeV emission from the GC (Aharonian et al. 2004). The lack of a synchrotron-emitting radio or infrared counterpart of the PWN is at least partly due to the fast Compton cooling (hence little accumulation) of relatively low-energy electrons, for which the Klein-Nishina suppression of the inverse-Compton scattering efficiency is not important. The presence of this young pulsar raises a number of questions: where does it originate? and where is the supernova remnant? The most logical origin of the pulsar is the GC cluster itself. However, the remnant is more difficult to isolate, the properties of which depend sensitively on the density and temperature structure of the medium (e.g., Tang & Wang

2005). Considering the orbital motion around Sgr A*, it is even possible that the pulsar may be produced together with the well-known SNR Sgr A East, although an alternative pulsar candidate has already been proposed for that remnant (see the contribution by S. Park).

Clearly, the above is a very incomplete and subjective review of recent *Chandra* results on the GC. We will certainly learn a great deal more from the existing and upcoming *Chandra* observations before GC-2009 Workshop in China.

Acknowledgments

Some of the ideas presented above are developed during or after many stimulating discussions with workshop participants, to whom I am grateful. I thank my collaborators for their contributions to the various research projects mentioned above, the organizers of the workshop for the invitation and the hospitality, and M. Muno for a careful editing and suggestions that led to the improvement of this presentation. The financial support by NASA through the grant SAO/CXC GO4-5010X is acknowledged.

References

Aharonian, F., et al. 2004, A&AL, 425, 13

Baganoff, F. K., et al. 2001, Nature, 413, 45

Baganoff, F. K., et al. 2003, ApJ, 591, 891

Bandyopadhyay, R. M, et al. 2005, MNRAS, 364, 1195

Cotera, A. S., Colgan, S. W. J., Simpson, J. P., & Rubin, R. H. 2005, ApJ, 622, 333

Eisenhauer, F., et al. 2005, ApJ, 628, 246

Feigelson, E. D., et al. 2005, ApJS, 160, 379

Figer, D. F., Rich, R. M., Kim, S. S., Morris, M., & Serabyn, E. 2004, ApJ, 601, 319

Genzel, R., et al. 2003, ApJ, 594, 812

Handa, T., et al. 2006, ApJ, 636, 261

Law, C., & Yusef-Zadeh, F. 2004, ApJ, 611, 858

Laycock, S., et al. 2005, ApJL, 634, 53

Muno, M. P., et al. 2003, ApJ, 589, 225

Muno, M. P., et al. 2004a, ApJ, 613, 1179

Muno, M. P., et al. 2004b, ApJ, 613, 326

Muno, M. P., et al. 2005, ApJL, 622, 113

Muno, M., Bauer, F. E., Bandyopadhyay, R. M., & Wang, Q. D. 2006, ApJS, 165, 173

Nayakshin, S., & Sunyaev, R. 2005, MNRAS, 364, 23

Preibisch, T., & Feigelson, E. D. 2005, APJS, 160, 390

Price, S. D., et al. 2001, ApJ, 121, 2819

Revnivtsev, M. G., et al. 2004, A&AL, 425, 49

Revnivtsev, M. G., et al. 2005, A&A, submitted (astro-ph/0510050)

Rockefeller, G., Fryer, C. L., Melia F., & Wang, Q., D. 2005, ApJ, 623, 171

Ruiter, A. J., Belczynski, K., & Harrison, T. E. 2006, ApJL, 640, 167

Serabyn, E., & Güesten, R. 1987, A&A, 184, 133

Sakano, M., Warwick, R. S., Decourchelle, A., & Wang, Q. D. 2005, MNRAS, 357, 1211

Seward, F. D., et al. 2001, ApJ, 553, 832

Stolte, A., Grebel, E. K., Brandner, W., & Figer, D. F. 2002, A&A, 394, 45

Stolte, A., Brandner, W., Grebel, E. K., Lenzen, R., & Lagrange, A. 2005, ApJL, 628, 113

Takagi, S., Murakami, H., & Koyama, K. 2002, ApJ, 573, 275

Tang, S. K., & Wang, Q. D. 2005, ApJ, 628, 205

Tatischeff, V. 2003, EAS Publications Series, 7, 79

Valinia, A., Tatischeff, V., Arnaud, K., Ebisawa, K., & Ramaty, R. 2000, ApJ, 543, 733

Wang, Q. D., Gotthelf, E. V., & Lang, C., 2002a, Nature, 415, 148

Wang, Q. D., Lu, F. J., & Lang, C. C. 2002b, ApJ, 581, 1148
Wang, Q. D., Dong, H., & Lang, C. C. 2006a, MNRAS, in press (astro-ph/0606282)
Wang, Q. D., Lu, F. J., & Gotthelf, E. V. 2006b, MNRAS, 367, 937
Wijnands, R., et al. 2006, A&A, 449, 1117
Xu, Y. D., et al. 2006, ApJ, 640, 319
Yuan, F. Quataert, E., & Narayan, R. 2003, ApJ, 598, 301
Yusef-Zadeh, F., et al., 2002a, ApJ, 570, 665
Yusef-Zadeh, F., Law, C., & Wardle, M. 2002b, ApJ, 568, 121

Institute of Physics Publishing Journal of Physics: Conference Series **54** (2006) 126–132
doi:10.1088/1742-6596/54/1/021 Galaxy Center Workshop 2006

A Candidate Neutron Star within the Radio Shell of Sgr A East

Sangwook Park[1], Michael P. Muno[2], Frederick K. Baganoff[3], Yoshitomo Maeda[4], Mark Morris[2], George Chartas[1], Divas Sanwal[1] and Gordon P. Garmire[1]

[1] Department of Astronomy and Astrophysics, Pennsylvania State University, 525 Davey Laboratory, University Park, PA. 16802, USA

[2] Department of Physics and Astronomy, University of California Los Angeles, Los Angeles, CA. 90095, USA

[3] Kavli Institute for Astrophysics and Space Research, Massachusetts Institute of Technology, Cambridge, MA. 02139, USA

[4] Institute of Space and Astronautical Science, 3-1-1 Yoshinodai, Sagamihara, Kanagawa, 229-8510, Japan

E-mail: park@astro.psu.edu

Abstract. We present imaging and spectral analyses of Sgr A East from deep *Chandra* observations. A nearby hard point-like source CXOGC J174545.5-285829 (dubbed the cannonball) appears to be a high-velocity neutron star. The estimated total Fe mass of Sgr A East supports a core-collapse origin from a massive star. Based on these results and the proximity between the features, we propose that the cannonball is a candidate neutron star created by the supernova that also produced the remnant Sgr A East.

1. Introduction

Sagittarius (Sgr) A East is an extended radio source (angular size $\sim 2'.7 \times 3'.6$) which encompasses the Galactic center supermassive black hole candidate Sgr A* [1]. The non-thermal radio spectrum and the shell-like morphology suggested an identification of a supernova remnant (SNR) for Sgr A East [1]. The first detailed X-ray study of Sgr A East was performed with the *Chandra X-Ray Observatory* by Maeda *et al.* [2]. X-ray emission from Sgr A East was centrally-peaked and originated from a hot thermal gas ($kT \sim 2$ keV) showing strong Heα lines from highly-ionized elements of S, Ar, Ca, and Fe. The estimated X-ray emitting mass and thermal energy were low, which indicated a single supernova origin for Sgr A East, rather than multiple SNRs or a hyper-energetic explosion. Maeda *et al.* also suggested a core-collapse origin (the progenitor mass $\sim 13-20$ $M_\odot$) for Sgr A East based on the metal abundance ratios [2]. The results from the observations of Sgr A East with the *XMM-Newton Observatory* by Sakano *et al.* were generally consistent with those from the *Chandra* [3]. *XMM-Newton*, however, detected Fe Lyα line emission ($E \sim 6.96$ keV) as well as the Fe Heα line ($E \sim 6.7$ keV), which required the presence of hotter gas with $kT \sim 4$ keV in addition to a $kT \sim 1$ keV plasma [3]. Sakano *et al.* suggested that the observed metal abundance pattern could indicate either a Type Ia or a core-collapse origin [3].

As of 2002 June, our continuous *Chandra* monitoring observations of the Galactic center accumulated a total exposure of ~600 ks, which is an order of magnitude deeper than that which was available after the initial *Chandra* observation [4]. We here present the results from these deep *Chandra* data. We focus on the spatially-resolved spectroscopy of the Fe-rich stellar ejecta and the study of a nearby candidate neutron star, in the context of the origin of Sgr A East. While we make a brief presentation of the main results in this paper, the detailed description of the data analysis and the extensive discussion of the results can be found in Park *et al.* [5].

2. Sgr A East: X-Ray Images

A "true-color" X-ray image of Sgr A East is presented in Fig. 1. The enhancements of the blue emission around the center of the SNR are remarkable. The X-ray line equivalent width (EW) images show that this enhanced hard X-ray emission is primarily from the Fe Heα line emission (Fig. 2). The high Fe EW at the center of the SNR is most likely caused by the enhanced Fe abundance there, and indicates the distribution of Fe-rich stellar ejecta (§ 3). Red, soft X-ray emission extending toward the northern side of the SNR, the so-called "plume", is also evident (Fig. 1) [2, 4]. With the deep exposure, a hard point-like source (CXOGC J174545.5−285829) is detected at the "tip" of the plume (Fig. 1) [6]. This hard point-like source is a candidate high-velocity neutron star (see §§ 4, 5, & 6 for more discussion), which might have also produced the bow shock-like plume: we name CXOGC J174545.5−285829 "the cannonball".

Figure 1. The exposure-corrected *Chandra* image of Sgr A East. Red: 1.5−4.5 keV. Green: 4.5−6.0 keV. Blue: 6.0−8.0 keV. Each sub-band image has been adaptively smoothed and the detected point sources have not been removed.

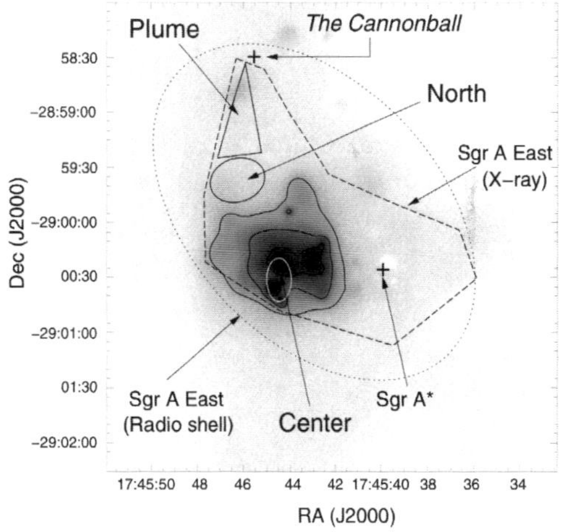

Figure 2. Gray-scale EW image of Fe Heα line ($E \sim 6.6$ keV) from Sgr A East. Linear gray-scales (darker gray-scales are higher EWs) range from 200 eV to 2500 eV with three contours at 800 eV, 1400 eV, and 2000 eV. Some key features within the field of view and regions used for the spectral analysis are marked.

3. Sgr A East: X-Ray Spectra

The "center" region (Fig. 2) clearly shows the strong Fe Heα line at $E \sim 6.6$ keV (Fig. 3). A single temperature thermal plasma model ($kT \sim 2$ keV) cannot adequately describe the observed spectrum, because of the residual feature at $E \sim 7$ keV, which is most likely the weak Fe Lyα line ($E \sim 6.96$ keV). The detection of the Fe Lyα line is statistically significant ($\sim 7\sigma$). A hot plasma component with $kT \sim 5$ keV, in addition to the $kT \sim 1$ keV gas, is required in order to properly fit the Fe Lyα line and the overall spectrum (Fig. 3 & Table 1). Additionally, a strong enhancement of the Fe abundance is evident (Table 2), indicating the presence of the Fe-rich stellar ejecta at the center of the SNR. The best-fit two-temperature thermal plasma model is presented in Fig. 3 and Tables 1 & 2.

The "plume" region (Fig. 2) can be fitted with a single-temperature ($kT \sim 1.3$ keV) thermal plasma model (Fig. 4, Tables 1 & 2). The Fe Heα line is significantly weaker than the center region. The Fe abundance is consistent with the solar value, which is considerably lower than in the central region. These results indicate that the plume is emission from shocked ISM rather than from metal-rich ejecta. The Fe Lyα line feature is not substantial in the plume region, and thus the hot component ($kT \sim 5$ keV) is not necessary to describe the X-ray spectrum.

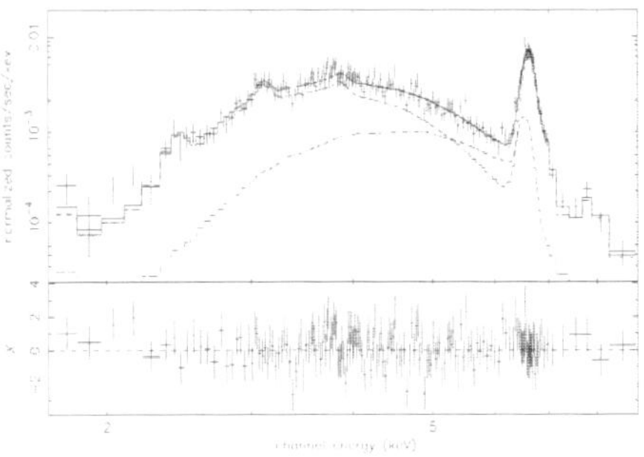

Figure 3. The *Chandra*/ACIS spectrum of the "center" region. The best-fit two temperature thermal plasma model is overlaid.

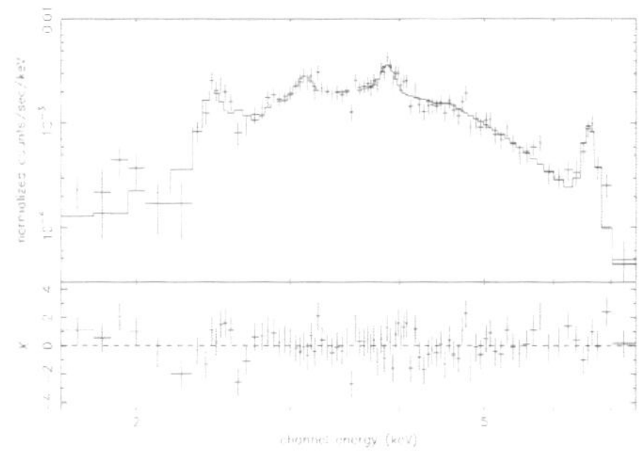

Figure 4. The *Chandra*/ACIS spectrum of the "plume" region. The best-fit single temperature thermal plasma model is overlaid.

Table 1. Results of Spectral Fittings of Sgr A East[a]

Region	N_H (10^{22} cm^{-2})	kT_{soft} (keV)	kT_{hard} (keV)	EM_{soft} (10^{57} cm^{-3})	EM_{hard} (10^{57} cm^{-3})	χ^2/ν
Center	$18.9^{+2.3}_{-1.8}$	$1.05^{+0.19}_{-0.13}$	$5.28^{+0.66}_{-0.57}$	$7.1^{+5.5}_{-2.8}$	$0.3^{+0.1}_{-0.1}$	167.9/163
Plume	$14.7^{+1.5}_{-1.9}$	$1.32^{+0.24}_{-0.11}$	-	$1.8^{+0.6}_{-0.6}$	-	100.5/83

[a]The errors are 2σ uncertainties.

Table 2. Best-Fit Metal Abundances of Sgr A East[a,b]

Region	S	Ar	Ca	Fe
Center	$0.69^{+0.74}_{-0.56}$	$1.76^{+0.80}_{-0.72}$	$1.37^{+0.80}_{-0.76}$	$5.81^{+1.65}_{-1.11}$
Plume	$1.91^{+0.98}_{-0.69}$	$1.20^{+0.65}_{-0.58}$	$2.50^{+0.77}_{-0.71}$	$0.96^{+0.38}_{-0.27}$

[a]Abundances are with respect to solar.
[b]The errors are 2σ uncertainties.

4. The Cannonball: X-Ray Morphology

A hard point-like source is detected at the edge of the plume and is designated as CXOGC J174545.5−285829 by Muno *et al.* [6], the so-called cannonball [5] (There is another point-like source, CXOGC J174545.2−285828, just ∼3″ west to the cannonball. The soft spectrum, the long-term variability, and the detection of an optical counterpart identify this source as a foreground star [7].) The cannonball is apparently extended with a cometary tail directing southward, which roughly traces back to Sgr A East (Figs. 5a & 6a). The deconvolved image and the comparison with the detector PSF show that the extended morphology of the cannonball is intrinsic and is not caused by the PSF (Fig. 6). The angular extent of the tail is ∼3″, which corresponds to a projected physical size of ∼0.1 pc (Hereafter, we assume $d \sim 8$ kpc to the Galactic center [8]).

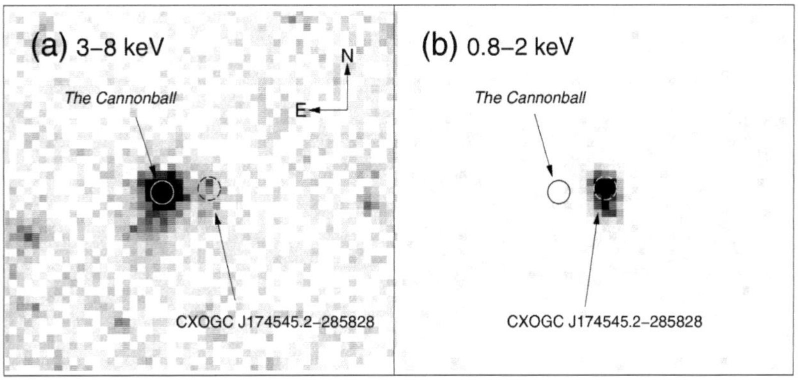

Figure 5. The *Chandra*/ACIS image of the cannonball. (a) The hard band (3 − 8 keV) image. (b) The soft band (0.8 − 2 keV) image. Darker gray-scales are higher intensities.

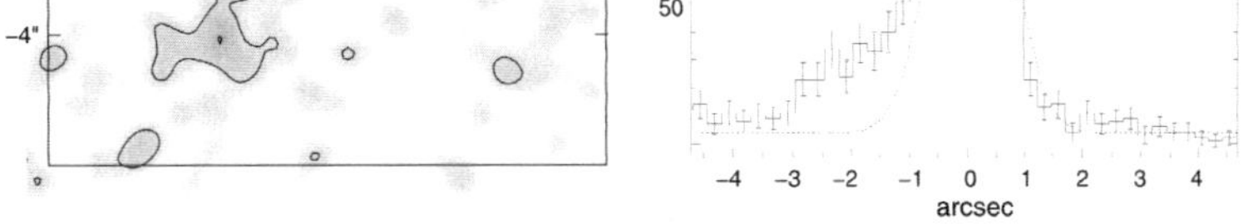

Figure 6. (a) The deconvolved *Chandra*/ACIS images of the cannonball (left source) and CXOGC J174545.2−285828 (right source). The image has been smoothed with a Gaussian of $\sigma = 0''.25$ for the purpose of display. Note that the soft source to the west of the cannonball is resolved into two point-like sources. (b) The source intensity profile projected in the north-south direction taken from the raw ACIS image. The histogram represents the cannonball, and the projected PSF at the source position is overlaid with a dashed curve.

5. The Cannonball: X-Ray Spectrum

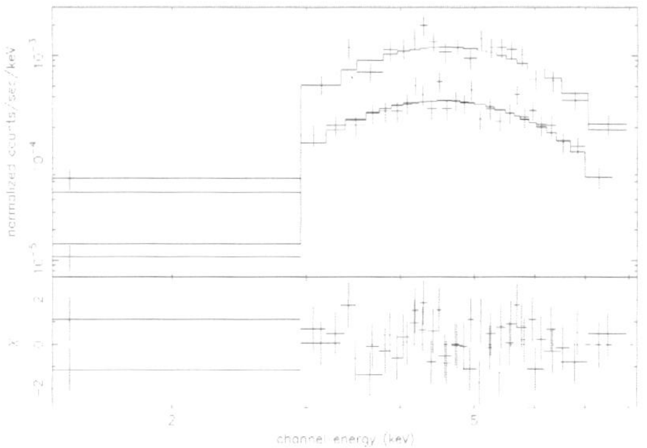

Figure 7. The X-ray spectrum of the cannonball. The upper and lower plots are the spectrum where the source was detected on-chip and on the ACIS-I chip-gap, respectively, due to the various roll-angles. The best-fit power law model is overlaid.

The spectrum of the cannonball can be fitted with a power law (Fig. 7). The best-fit parameters are summarized in Table 3. The high foreground absorption indicates its co-spatial location with Sgr A East in the Galactic center. The best-fit photon index and the derived X-ray luminosity are then consistent with the typical values from Galactic pulsars [9]. Because of the relatively low photon statistics (~1000 counts) and the heavy absorption, the spectrum may also be fitted with thermal plasma models with a very high electron temperature ($kT \sim 25$ keV) but with extremely low metal abundances. Thus, although unlikely, a thermal origin of the X-ray spectrum of the cannonball (e.g., a cataclysmic variable) cannot be completely ruled out.

Table 3. Results of Spectral Fittings of The Cannonball

Γ^a	N_H^a (10^{22} cm^{-2})	$f_{X(2-10keV)}{}^a$ 10^{-13} ergs cm^{-2} s^{-1}	$L_{X(2-10keV)}$ 10^{33} ergs s^{-1}	χ^2/ν
$1.59^{+0.58}_{-0.39}$	$16.5^{+3.3}_{-3.2}$	$1.9^{+3.6}_{-1.2}$	3.1	37.0/46

[a]The errors are 2σ uncertainties.

6. Discussion

The spectral characteristics of the cannonball are consistent with an identification of a pulsar in the Galactic center. The cometary tail (~0.1 pc) extends southward, toward the center of Sgr A East. The projected sky position of the cannonball is just *interior* to the radio shell boundary of Sgr A East while sitting on the tip of the X-ray plume. These spectral and morphological X-ray characteristics suggest that the cannonball may be a high-velocity neutron star, moving toward the north away from the SNR Sgr A East.

We estimate the velocity of the cannonball using three independent methods. Assuming an age of ~5000 – 10000 yr for Sgr A East [2, 3, 10], and taking the angular separation of ~2′ between the cannonball and the center of the SNR, we find $v \sim 455 - 912$ km s^{-1}/sinβ, where β is the angle between the line of sight and the actual traveling direction of the cannonball. Secondly, assuming that the plume is a bow-shock produced by the cannonball, we estimate an opening angle of $\theta \sim 53°$ for the bow-shock and thus the Mach number Ma = $[\sin(\theta/2)]^{-1}$ is ~2.2. The velocity is then $v = c_s$Ma = ~880 km s^{-1} (we here use the sound speed of the ambient gas $c_s = [\gamma kT/\mu m_p]^{\frac{1}{2}} \sim 400$ km s^{-1} where $\gamma = 5/3$ for a monatomic, adiabatic gas, $\mu = 1$ for protons, m_p is the proton mass, and $kT \sim 1$ keV plasma in the plume region). We may also derive the pulsar velocity by assuming pressure equilibrium between the pulsar wind nebula (PWN) and the ambient medium. The best-fit photon index of the cannonball ($\Gamma = 1.6$) implies $\dot{E} \sim 7.6 \times 10^{36}$ ergs s^{-1} based on an empirical relationship between $\dot{E}$ and the power law photon index of the PWN [11]. For the pressure balance among the the PWN, thermal and ram pressures of the ambient ISM, P_{PWN} $[= \dot{E}/(4\pi c R^2)] = P_{th}$ $(= 2n_e kT) = \epsilon_k$ $(= \rho v^2)$, the velocity is derived to be $v \sim 550$ km s^{-1}. In these estimates, n_e is ~ 7.4 cm^{-3} from the best-fit volume emission measure (EM) of the plume, and ρ is the mass density of the ISM derived from the spectral analysis of the plume region. The radius of the PWN, assuming a spherical geometry, is derived to be $R \sim 2.5 \times 10^{16}$ cm which corresponds to $\sim$0$''$.2 at $d \sim 8$ kpc. This small R is consistent with the pointlike detection of the "head" of the cannonball by the ACIS. Considering various assumptions and embedded uncertainties in the above three independent approaches for the velocity estimates, the agreement among all three results are remarkable. Moreover, the estimated velocity range is in good agreement with that of typical high-velocity pulsars in the Galaxy [12], which lends these estimates additional credibility.

No evidence for long-term (observation by observation) variability of the cannonball was detected [7]. The constant light curve with a long time-basis of $\sim$3 yr is remarkably unusual among the Galactic center sources. This non-variability over a few years indicates that the cannonball is unlikely a background AGN, and supports the neutron star identification. The logN-logS relation for the contribution from extragalactic sources in the Galactic center [6] indicates a low probability of $\sim 1.2 \times 10^{-4}$ for the detection of a background extragalactic source at the flux level of the cannonball within the plume region. We then searched for pulsations from the cannonball with a 40-ks archival *XMM-Newton*/EPIC observation pointed at Sgr A*. We do not detect significant evidence of a pulsar. We place an upper limit of $\sim$40% on the pulsed-fraction between 5×10^{-5} and 6.8 Hz.

Assuming that the Fe-rich central region represents the bulk of the total Fe mass ejected from the progenitor star, the total Fe mass estimate can be useful to infer the progenitor mass. We consider some simple geometries in which *all* Fe ejecta material is contained within the central $\sim 40'' - 60''$ diameter region where the Fe EW is the highest (Fig. 2). We, for simplicity, also assume a "pure" Fe-ejecta case for the estimate of the electron density: i.e., *all* electrons originate from the ionized Fe ($n_e \sim 24 n_{Fe}$ for the case that the He-like Fe dominates). The best-fit EMs from the spectral analysis of the center region imply Fe ion density of $n_{Fe} \sim 0.11 f^{-\frac{1}{2}}$ cm^{-3}, where f is the volume filling factor for the X-ray emitting gas. We then estimate the total Fe mass of Sgr A East to be $M_{Fe} \sim 0.15 f^{\frac{1}{2}} - 0.27 f^{\frac{1}{2}}$ $M_\odot$, depending on the assumed geometries (The details of the Fe mass calculation can be found in Park *et al.* [5]). These Fe mass estimates are upper limits because of the pure ejecta assumption. If a significant contribution from H is considered, these Fe mass estimates only become smaller. The estimated Fe mass upper limits are then consistent with a Type II origin from a $13 - 15 M_\odot$ progenitor when compared with standard supernova nucleosynthesis models [13, 14], unless a significant fraction of the Fe-rich ejecta ($\sim$several times larger than the observed Fe ejecta mass) has already been intermixed with the ambient ISM. The suggested Type II core-collapse origin of Sgr A East is self-consistent with the proposed SNR-neutron star association between Sgr A East and the cannonball.

7. Conclusions

The observed X-ray nature of the cannonball suggests an identification as a high-velocity neutron star in the Galactic center. The estimated Fe ejecta mass of Sgr A East supports a core-collapse origin for the SNR. Based on its proximity to Sgr A East, we therefore propose that the cannonball may be a candidate neutron star born from the supernova explosion which also produced the SNR Sgr A East. We note that the discrepancies between the previous results from the early *Chandra* and the *XMM-Newton* observations (e.g., the presence of Fe Lyα line and the extremely hot component ($kT \sim 5$ keV) of the X-ray emitting gas) can be resolved by the high resolution, high photon statistics data from the deep *Chandra* observations.

The non-detection of the pulsation from the cannonball remains an issue for the neutron star identification. However, the limited timing resolution of the archival EPIC/PN data (73.4 ms) did not allow us to search for periods shorter than ~147 ms. The presence of a pulsar for the cannonball thus cannot be ruled out yet. Follow-up X-ray and/or radio observations with high time resolution instruments would therefore be essential for the conclusive identification of the cannonball.

Acknowledgments

This work has been supported in parts by NASA contract NAS8-39073, NAS8-01128 for the *Chandra X-Ray Observatory.* MPM is supported by the Hubble Fellowship program through grant number HS-HF-01164.01-A.

References

[1] Ekers R D, van Gorkom J H, Schwarz U J and Goss W M 1983 *Astron. & Astrophys.* **122** 143-50

[2] Maeda Y, Baganoff F K, Feigelson E D, Morris M, Bautz M W, Brandt W M, Burrows D N, Doty J P, Garmire G P, Pravdo S H, Ricker G R and Townsley L K 2002 *Ap. J.* **570** 671-87

[3] Sakano M, Warwick R S, Decourchelle A and Predehl P 2004 *Mon. Notices Royal Astron. Soc.* **350** 129-39

[4] Baganoff F K, Maeda Y, Morris M, Bautz M W, Brandt W N, Cui W, Doty J P, Feigelson E D, Garmire G P, Pravdo S H, Ricker G R and Townsley L K 2003 *Ap. J.* **591** 891-915

[5] Park S, Muno M P, Baganoff F K, Maeda Y, Morris M, Chartas G, Sanwal D, Burrows D N and Garmire G P 2005 *Ap. J.* **631** 964-75

[6] Muno M P, Baganoff F K, Bautz M W, Brandt W N, Broos P S, Feigelson E D, Garmire G P, Morris M R, Ricker G R and Townsley L K 2003 *Ap. J.* **589** 225-41

[7] Muno M P, Arabadjis J S, Baganoff F K, Bautz M W, Brandt W N, Broos P S, Feigelson E D, Garmire G P, Morris M R and Ricker G R 2004 *Ap. J.* **613** 1179-1201

[8] Reid M J 1993 *Annual Rev. Astron. & Astrophys.* **31** 345-72

[9] Becker A and Aschenbach B 2002 *Proc. 270th WE-Heraeus Seminar on Neutron Stars, Pulsars, and Supernova Remnants* ed W Becker *et al.* (MPE Rep. 278; Garching: MPE) p 64-86

[10] Mezger P G, Zylka R, Salter C J, Wink J E, Chini R, Kreysa E and Tuffs R 1989 *Astron. & Astrophys.* **209** 337-48

[11] Gotthelf E V 2003 *Ap. J.* **591** 361-5

[12] Cordes J M and Chernoff D F 1998 *Ap. J.* **505** 315-38

[13] Nomoto K, Iwamoto K, Nakasato N, Thielemann F -K, Brachwitz F, Tsujimoto T, Kubo Y and Kishimoto N 1997a *Nucl. Phys. A* **621** 467-76

[14] Nomoto K, Hashimoto M, Tsujimoto T, Thielemann F -K, Kishimoto N, Kubo Y and Nakasato N 1997b *Nucl. Phys. A* **616** 79-90

Institute of Physics Publishing
doi:10.1088/1742-6596/54/1/022

Journal of Physics: Conference Series **54** (2006) 133–139
Galaxy Center Workshop 2006

The Arches Cluster and G0.1−0.1 cloud — A view with fluorescent lines

Masaaki Sakano[1], **Robert S. Warwick**[1] **and Anne Decourchelle**[2]

[1] Department of Physics and Astronomy, University of Leicester, Leicester LE1 7RH, UK
[2] CEA/DSM/DAPNIA, Service d'Astrophysique, C.E. Saclay, 91191 Gif-sur-Yvette Cedex, France

E-mail: mas@star.le.ac.uk

Abstract. We present the X-ray results on the Arches Cluster and G0.1−0.1 cloud obtained with *XMM-Newton*. In particular we have found a big loop-like annular structure, adjacent to the Arches Cluster, with the diameter of ∼3 arcmin (∼7 pc) and width of ∼1 arcmin (∼2 pc). The structure is found to show a non-thermal X-ray spectrum with a best-fitting photon index of $\Gamma = 1.4 \pm 0.6$, together with a strong fluorescent Kα line from neutral iron at 6.4-keV with an equivalent width of 1 keV. The G0.1−0.1 cloud shows a reasonable correlation among the 6.4-keV line, very hard continuum and 2.3-keV He-like sulphur line. We discuss the possible origin of these structures and further give a speculation to explain the origin of the fluorescent X-ray lines in the Galactic Centre Region

1. Introduction

Since the hard X-ray imaging era began in the last decade, X-ray observations have given new insights into the Galactic Centre Region (GCR). Arguably the discovery of the diffuse fluorescent iron lines (Koyama *et al* 1996) is the most exciting result among the diffuse structures there. It shows a very patchy distribution, which does not resemble any other distribution, and its origin is yet unsolved.

The fluorescent X-ray Kα-line from neutral or low-ionised iron at 6.4 keV (hereafter, simply a 6.4-keV line) is particularly characteristic of the GCR. The 6.4-keV line is a common feature in the X-ray spectrum of type-II active galactic nuclei (AGNs) and of some X-ray binaries. It is thought to come from neutral materials enveloping or surrounding a central strong X-ray emitter. When the continuum itself is blocked or absorbed by something, possibly the envelope itself, the continuum is suppressed, and as a result the 6.4-keV line appears prominent.

The GCR is full of dense molecular clouds, including the most massive cloud in the Galaxy, Sgr B2. Hence, there are a large number of potential emitters of a 6.4-keV line in principle. However, being different from AGNs or X-ray binaries, the GCR lacks an apparent central X-ray emitter bright enough to produce the observed line. One of the scenarios proposed is that Sgr A* used to be bright and that X-rays from Sgr A* at that time were reflected at the clouds and are now observed as a 6.4-keV line (Koyama *et al* 1996; Murakami *et al* 2000, 2001; Churazov *et al* 2002). Another scenario explains it as the interaction between cosmic rays and molecular clouds (Valinia *et al* 2000; Yusef-Zadeh *et al* 2002; Bykov 2003).

The prototype of these fluorescent clouds in the GCR is Sgr B2. It is barely distinguishable in the continuum band, but is very distinctive in the 6.4-keV line (*e.g.*, Koyama *et al* 1996).

Interestingly the brightest spot in the 6.4-keV line in the GCR is not Sgr B2, but the G0.1−0.1 cloud(*e.g.*, see the X-ray image in Koyama *et al* 1996). More interestingly, G0.1−0.1 apparently has more complex structure than Sgr B2, *i.e.*, it shows thermal and/or continuum components (Koyama *et al* 1996; Wang *et al* 2002; Yusef-Zadeh *et al* 2002; Law & Yusef-Zadeh 2004).

In this paper we present the X-ray results from *XMM-Newton* observations of the Arches Cluster and G0.1−0.1. The former is one of the most massive star clusters in our Galaxy (*e.g.*, Nagata *et al* 1995; Figer *et al* 2005). We have detected a strong iron 6.4-keV line from both the sources. Then we further discuss the possible origin of the 6.4-keV line in the GCR.

2. Observations

We have observed the Arches Cluster and G0.1−0.1 in six occasions with *XMM-Newton* (See Sakano *et al (in prep)* for the observation log). All the data with EPIC MOS and pn cameras on board *XMM-Newton* were analysed. In fitting the spectra, we fitted each spectrum from each camera simultaneously if it had enough statistics, or summed up the whole data set per camera for faint sources to increase the statistics before spectral fitting was carried out.

3. Results on the Arches Cluster

3.1. Images

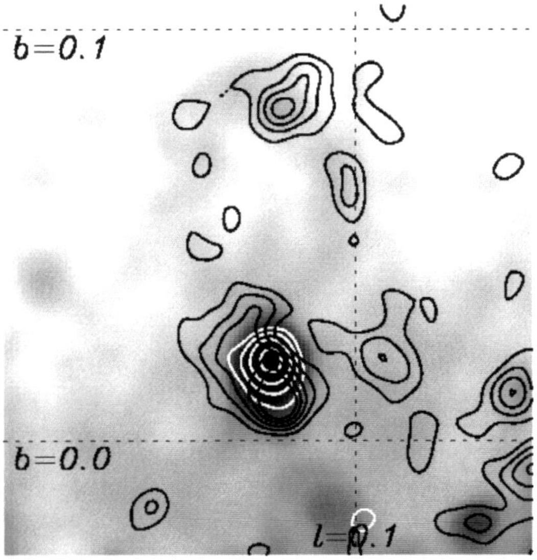

Figure 1. *XMM-Newton* MOS1+2 mosaiced image of the Arches Cluster in the 2.0–9.0 keV band. White and black contours trace the 6.7 and 6.4 keV line emission, respectively, which are Kα-lines from He-like and neutral iron. The underlying continuum is subtracted for both the line emission (see text). The colour level and contours are linearly spaced. The Galactic north is up, where the galactic coordinates are given with dashed lines.

Figure 1 shows the X-ray image of the Arches Cluster taken with *XMM-Newton*/MOS1+2 in the 2–9 keV band, where the different exposures of MOS1 and 2 are summed. The surface brightnesses of two iron K-lines at 6.7 and 6.4 keV are displayed with overlaid contours, where the underlying continuum is subtracted, assuming the same spectral shape for the whole region (See Sakano *et al* 2004a for the detailed technique).

The peak of the whole energy band image coincides with the peak of the 6.7-keV line from helium-like iron. The peak position is also consistent with the *Chandra* one (Law & Yusef-Zadeh 2004), and so with the central area of the Arches Cluster.

The most striking result is the existence of the big loop-like annular structure, extending to the galactic north from the Arches Cluster, with the diameter of ∼3 arcmin (∼7 pc) and width of ∼1 arcmin (∼2 pc). It is brighter in the north and west sides. The south end is smoothly spread towards the diffuse emission surrounding the Arches Cluster; thus it is not clear whether the loop structure is physically interacting with the Arches Cluster or not, within the current data and statistics.

The peak of the 6.4-keV line is found to be shifted by 0.8 arcmin to the galactic south of the peak of the full-energy band image and of the 6.7-keV line (or A3 component in Law & Yusef-Zadeh 2004). Interestingly a 6.4-keV line is emitted from the north end of the loop-like structure, as well as, though less significantly, from the west side.

3.2. Spectra

For both MOS1 and 2, we accumulated source spectra from regions of radii 28 arcsec centred on the positions of the peaks of the core and 6.4-keV emission. Note that pn spectra are not used because the area is located in the chip gap during most of the observations. A background spectrum was collected from the annular region that is centred on the averaged position of both the source regions, and has inner and outer radii of 60 and 88 arcsec. This was subtracted from the source spectra. We then simultaneously fitted the spectra from both detectors (*i.e.*, for each detector for each exposure) for each of the core and 6.4-keV regions with trial models of either a thin thermal plasma or a power-law, modified with interstellar absorption.

We have found that the spectrum of the core region is well described with a thin-thermal plasma model with the best-fitting temperature of 2.7±0.4 keV, absorption of $N_{\rm H}$=$(9.2 \pm 0.9) \times 10^{22}$H cm^{-2}and respective abundances for silicon to calcium and for iron of 2±1 and 0.85 ± 0.15 solar in 90% confidence. The equivalent width of the 6.4-keV line is 90±70 eV.

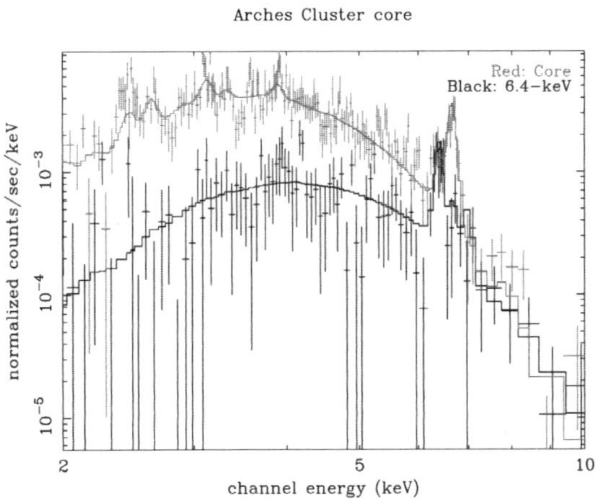

Figure 2. *XMM-Newton* MOS1+2 summed spectra of the core (red) and 6.4-keV peak (black) regions of the Arches Cluster, where all the exposures are summed, and the background is already subtracted appropriately (see text).

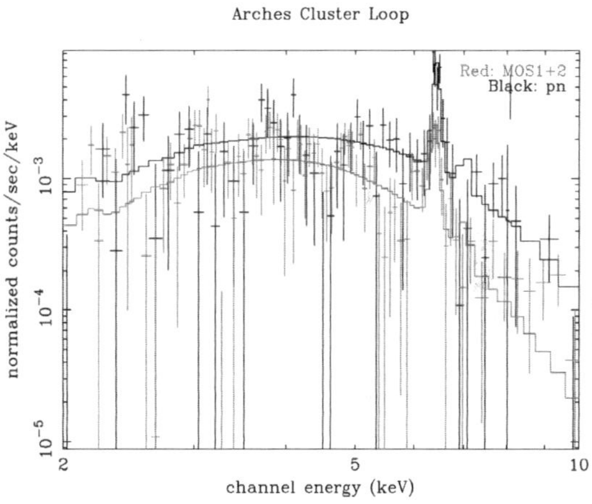

Figure 3. *XMM-Newton* MOS1+2 (red) and pn (black) spectra of the loop region, where the background spectra taken from the adjacent region are subtracted.

On the other hand, the modelling of the spectra from the 6.4-keV-line peak region is more complicated. It must include some of the component seen in the core region due to the mirror response. Thus we fit the spectra with a power-law model, added to the best-fitting thermal model derived from the above with a fixed normalisation calculated from the response, and find that the best-fitting model gives the photon index of $\Gamma = 2.2 \pm 1.3$ and the absorption of $N_{\rm H}$=$(14 \pm 6) \times 10^{22}$H cm^{-2}. The equivalent width of the 6.4-keV line is 600±350 eV, whereas the absolute best-fitting intensity of the line is two times higher than that of the core region.

Figure 2 shows the spectra of those two regions, summed up for MOS1 and 2 and for all the exposures for clarity, together with the best-fitting models. The figure clearly depicts in more quantitative way the strong and weak lines, which have been demonstrated in the image analysis. Moreover the above spectral analysis further shows that the 6.4-keV-peak region has 50% more absorption together with a flatter continuum than the core region.

Next, we accumulated the spectrum from the half-circular annular region that traces the brightest part of the loop region. The background spectra were collected from both the immediate inner and outer vicinities of the loop region, and were subtracted from the above source region. This time we also used the pn spectra. The resultant spectra are found to be even flatter than the 6.4-keV-peak region (Figure 3), with the best-fitting absorption and photon index of $N_{\rm H}$=(7.2±1.4)$\times 10^{22}$H cm^{-2}and Γ = 1.4±0.6, respectively. The 6.4-keV line equivalent width is very large, 1000±250 eV.

4. Results on G0.1−0.1

Figure. 4 shows the summed pn images of the Galactic Centre, where G0.1−0.1 is bright at the 6.4-keV line. The area is also bright in the very high energy band of 9–12 keV. Indeed an overall correlation between 9–12 keV band and 6.4-keV line images are seen. Above all one of the brightest spots at the 6.4-keV line in G0.1−0.1 is also the brightest spot in the 9–12 keV band apart from the Sgr A region, which is at the centre of the image.

Interestingly G0.1−0.1 is also bright at the 2.3-keV line, which is a Kα line from helium-like sulphur. In this case, the correlation between 2.3-keV and 6.4-keV lines are less clear. For example, the above mentioned brightest spot in the 9–12 keV is clearly dim at the 2.3-keV line. The spatial distribution of the 2.3-keV line is more extended than that of the 6.4-keV line, and possibly slightly shifted towards the galactic north. Nevertheless a rough correlation between them is seen; both of them are spread in the east side of the Galactic Centre in rather negative galactic-latitude and are peaked somewhere at G0.1−0.1. Comparing with the circular symmetric distribution of the 6.7-keV line image (see Warwick *et al* 2006 in this issue), there is certainly much resemblance between 2.3 and 6.4-keV line distributions.

5. Discussion

5.1. Origin of the X-ray emission from the Arches Cluster

The thermal emission is consistent with the scenario of heated plasma by the stellar wind collision within the cluster, as is previously suggested (Law & Yusef-Zadeh 2004), because the cluster is very massive and dense.

Both the X-ray emissions from the 6.4-keV-peak region and the loop region are, on the other hand, quite unusual. Being fitted with a power law, both of the regions are clearly non-thermal dominated. To produce the 6.4-keV photons through fluorescence, the existence of photons or particles with an energy higher than 7.1 keV is essential. Any thermal emission is unlikely to produce a large flux of those high energy photons. Hence, providing the strong 6.4-keV line is observed, it is natural that the continuum emission is non-thermal rather than thermal.

At the same time, the detection of a strong 6.4-keV line implies a heavy absorption. A heavy absorption is actually observed from the 6.4-keV-peak region, of which the absorption is 50% larger than the adjacent core region of the Arches Cluster. The absorption from the loop region is only a half of the 6.4-keV-peak one: $N_{\rm H}\sim 7\times 10^{22}$H cm^{-2}. This is probably because the loop is located much farther away from the Galactic Plane, where the averaged absorption in the GCR is a steep function of the angular distance from the Galactic Plane (Sakano *et al* 1999).

Now, although the equivalent widths of the 6.4-keV line in these structures are large and unusual, being 600–1000 eV, they are not extreme. They could be reproduced by external primary sources or even internal sources, where the direct continuum from them can be observed.

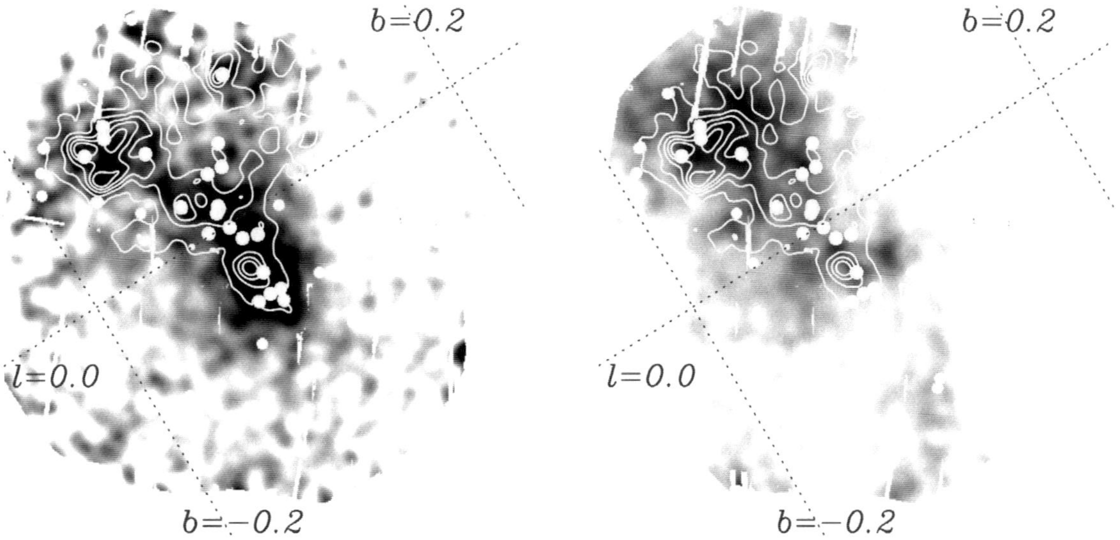

Figure 4. *XMM-Newton* pn summed images in the 9–12 keV band (*Left*) and of a 2.3 keV line from He-like sulphur (*Right*), overlaid with the 6.4-keV line contour. The underlying continua for the line images are already subtracted (see Section 3.1 for the method). Apparent point sources are masked (white filled circles). The celestial north is up, whereas the galactic coordinates are displayed with dashed lines. The brightest spot in the 9–12 keV band at the centre is the Sgr A complex, including Sgr A East and West, the latter of which contains Sgr A*. A bright region in the upper-left side is G0.1−0.1.

Whichever the case is, the observed non-thermal continuum must be the remain of the original primary sources. Then, what and where are the primary sources that finally produce these observed non-thermal X-ray emission, associated with the 6.4-keV line? As is already discussed the above, the thermal emission from the Arches Cluster itself is too soft to be the candidate.

No obvious point source around the Arches Cluster is found. Because there are a couple of 6.4-keV structures near the Arches Cluster, if the primary source is a point source(s), it is likely to be more than one source and to be close enough to each 6.4-keV source. However, it is highly unlikely that ordinary point sources are distributed to form a loop-like structure as is seen, therefore the multiple point source synthesis must be rejected.

A more plausible scenario would be that on-site particle acceleration is occurring at each 6.4-keV cloud or in its close vicinity. Supernova remnants (SNRs) have been confirmed as a site for particle acceleration via their shocks, which would produce X-ray emission (*e.g.*, Koyama *et al* 1995; Hughes *et al* 2000; Bamba *et al* 2005).

Then, one possibility is that the loop structure adjacent to the Arches Cluster is just another shell-type SNR, which must be independent of the Arches Cluster. There is also a fairly smooth bridge-like structure in the 6.4-keV line between the loop-like structure and G0.1−0.1 from more global point of view (see Figure. 4). Then this loop-like structure may just happen to be located adjacent to the Arches Cluster.

Alternatively, considering that a strong shock causes the particle acceleration, and that in this case low-energy electrons with an energy of a mere few tens of keV are all we need, even the starformation activity of the Arches Cluster may drive the shock. If this is the case, stellar winds or possibly a jet from the Arches Cluster are accelerating particles, following the loop-like shape, and then those high-energy particles produce the fluorescent lines as well as the continuum via the interaction with cold matter.

5.2. G0.1−0.1 and further: Origin of the fluorescent line clouds

We have found rough correlations in the spatial distribution of the 6.4-keV line and that of the very high energy-band continuum, as well as that of helium-like sulphur line. The former is a naturally expected one, taking into account the emitting process of the 6.4-keV line. However, the latter is not straightforward to explain, because the plasma and 6.4-keV lines have, generally speaking, no relationship in terms of the emission process.

The most simple solution may be just that the location coincides by chance, since we do not know the actual distance to each structure. However, if this is truly a result of some cause, the following speculation may work. First, a SNR is made. It produces a shock with ambient matter (possibly the molecular clouds), then the matter is heated to become a plasma, emitting soft X-rays characterised by a He-like sulphur line. At the same time, this shock also causes particle acceleration, possibly at the boundary with molecular clouds, since it will enhance the process in general as is suggested (Fukui *et al* 2003). These accelerated particles interact with the ambient clouds, producing a 6.4-keV line as well as a non-thermal continuum.

If this speculation is the case, the sulphur line must be distributed close to the fluorescent clouds, but not exactly. And this indeed is the case for G0.1−0.1. The sulphur line distribution looks slightly shifted towards galactic north. Above all, the brightest spot in the very high energy band and at a 6.4-keV line is dim in the sulphur line. In fact, this situation is similar in the Arches Cluster. The peak position in the 6.4-keV line is shifted from the plasma peak. In addition, Sgr C cloud, the third brightest region with the 6.4-keV line in the GCR also shows a similar situation in a even clearer way (Sakano *et al* 2004b).

Some counter-arguments against this speculation may be possible. Sgr B2 does not show this kind of situation. However, Sgr B2 has the largest column density, which is likely to make everything unobservable in the lower energy band, even if a low-temperature plasma exists, unless the plasma is located in front of the cloud.

Why then do star-forming galaxies do not show a 6.4-keV line, which may have to be seen in this scenario? The reason why we see the 6.4-keV line structures in the GCR is because we can spatially resolve the structure. If we take up all the spectra in the GCR, which would represent what could be seen from star-forming galaxies, the 6.4-keV line gets smeared heavily, and will be hardly observed any more. Moreover, star-forming galaxies show even much larger plasma emission, making it even harder to identify a 6.4-keV line.

Another argument is that galactic SNRs show hardly any 6.4-keV line as far as reported. Therefore in the GCR, there must be either an enhancement mechanism of the particle acceleration, *e.g.*, the strong magnetic field, or exceedingly dense molecular clouds, or possibly both.

Finally we argue that whether this shock-origin speculation is true or not, the high-energy particles are the more likely sources for producing a 6.4-keV line than X-rays from discrete sources. The intensity of the (absorption-corrected) 6.4-keV line is simply proportional to the column density of the cloud and luminosity of the illuminating source(s) and is inversely-proportional to the distance of the source(s). Hence, the most influential factor to determine the intensity of the clouds is the distance from the primary source(s). A number of the 6.4-keV line structures have been reported in the GCR (*e.g.*, Park *et al* 2004), in addition to Sgr B2, G0.1−0.1 and Sgr C. The distribution is quite patchy; for example, most of them are concentrated in the east side of the Galactic Centre, excepting Sgr C. Added to that, some of them lack of molecular cloud counterparts. Although the brightest three have a molecular cloud counterpart, even these three do not have a perfect matching with molecular clouds in a finer scale less than 1 arcmin ($\lesssim$ 2pc).

If the primary source is a single origin, such as Sgr A*, only the parameter left is the column density of the cloud, apart from the luminosity history of the primary source, once the coordinates of each cloud are determined. Thus, the 6.4-keV intensity should basically follow the distribution of radio molecular clouds. However, observations tell it doesn't.

Another possible hypothesis is that some of the X-ray transient sources, which happen to be located close to molecular clouds, may illuminate the clouds (probably by X-rays), producing a 6.4-keV line. It is well known that many transient sources exist in the GCR, which are on and off from time to time (*e.g.,* Wijnands *et al* 2006). If they are responsible for the 6.4-keV line, their spectra must be hard and they should have a fairly high duty ratio as a whole. However their nature is yet unclear. Moreover, no transient source has been identified as a candidate source for any 6.4-keV cloud. Therefore this hypothesis is not very likely, if not excluded.

Alternatively, if the primary source itself is a diffuse source, such as accelerated particles, they may also have a patchy distribution. In this case, wherever the diffuse structure closely meets a dense cloud, fluorescent lines are emitted. Hence the patchy distribution of fluorescent lines is relatively easily reproduced.

Acknowledgments
The authors would like to thank the referee Dr. M. Muno for his careful proof reading and valuable comments.

References

[1] K. Koyama, Y. Maeda, T. Sonobe, T. Takeshima, Y. Tanaka, and S. Yamauchi. Asca view of our galactic center: Remains of past activities in x-rays? *PASJ*, 48:249–255, April 1996.

[2] H. Murakami, K. Koyama, M. Sakano, M. Tsujimoto, and Y. Maeda. ASCA Observations of the Sagittarius B2 Cloud: An X-ray Reflection Nebula. *ApJ*, 534:283–290, May 2000.

[3] H. Murakami, K. Koyama, and Y. Maeda. Chandra Observations of Diffuse X-rays from the Sagittarius B2 Cloud. *ApJ*, 558:687–692, September 2001a.

[4] E. Churazov, R. Sunyaev, and S. Sazonov. Polarization of X-ray emission from the Sgr B2 cloud. *MNRAS*, 330:817–820, March 2002.

[5] A. Valinia, V. Tatischeff, K. Arnaud, K. Ebisawa, and R. Ramaty. On the Origin of the Iron K Line in the Spectrum of The Galactic X-ray Background. *ApJ*, 543:733–739, November 2000.

[6] F. Yusef-Zadeh, C. Law, and M. Wardle. The Origin of X-ray Emission from a Galactic Center Molecular Cloud: Low-Energy Cosmic-Ray Electrons. *ApJL*, 568:L121–L125, April 2002.

[7] A. M. Bykov. Faint hard X-ray sources in the Galactic Center region. *A&A*, 410:L5–L8, October 2003.

[8] Q. D. Wang, F. Lu, and C. C. Lang. X-ray Thread G0.13-0.11. *ApJ*, 581:1148–1153, December 2002.

[9] C. Law and F. Yusef-Zadeh. X-ray obs. of stellar clusters near the GC. *ApJ*, 611:858–870, August 2004.

[10] T. Nagata, C. E. Woodward, M. Shure, and N. Kobayashi. Object 17: Another cluster of emission-line stars near the galactic center. *AJ*, 109:1676–1681, April 1995.

[11] D. F. Figer. An upper limit to the masses of stars. *Nature*, 434:192–194, March 2005.

[12] M. Sakano, R. S. Warwick, A. Decourchelle, and P. Predehl. XMM-Newton observations of Sagittarius A East. *MNRAS*, 350:129–139, May 2004a.

[13] M. Sakano, K. Koyama, M. Nishiuchi, J. Yokogawa, and Y. Maeda. Mass estimation of cold matters in the galactic center region using bright hard x-ray sources. *Adv. in Space Res.*, 23(5/6):969–972, 1999b.

[14] M. Sakano, R. S. Warwick, A. Hands, and A. Decourchelle. Three years progress of the XMM-Newton Galactic Centre/Plane Survey. *Memorie della Societa Astronomica Italiana*, 75:498–499, 2004b.

[15] K. Koyama, R. Petre, E. V. Gotthelf, U. Hwang, M. Matsuura, M. Ozaki, and S. S. Holt. Evidence for shock acceleration of high-energy electrons in the SNR SN1006. *Nature*, 378:255, November 1995.

[16] J. P. Hughes, C. E. Rakowski, D. N. Burrows, and P. O. Slane. Nucleosynthesis and Mixing in Cassiopeia A. *ApJL*, 528:L109–L113, January 2000.

[17] A. Bamba, R. Yamazaki, and J. S. Hiraga. Chandra Observations of Galactic Supernova Remnant Vela Jr.: A New Sample of Thin Filaments Emitting Synchrotron X-Rays. *ApJ*, 632:294–301, October 2005.

[18] Y. Fukui and et al. Discovery of Interacting Molecular Gas toward the TeV Gamma-Ray Peak of the SNR G 347.3–0.5. *PASJ*, 55:L61–L64, October 2003.

[19] S. Park, M. P. Muno, F. K. Baganoff, Y. Maeda, M. Morris, C. Howard, M. W. Bautz, and G. P. Garmire. Characteristics of Diffuse X-ray Line Emission within 20pc of the GC. *ApJ*, 603:548–559, March 2004.

[20] R. Wijnands and et al. The XMM-Newton/Chandra monitoring campaign of the Galactic center region. Description of the program and preliminary results. *A&A*, 449:1117–1127, April 2006.

Institute of Physics Publishing Journal of Physics: Conference Series **54** (2006) 140–146
doi:10.1088/1742-6596/54/1/023 Galaxy Center Workshop 2006

The H.E.S.S. View of the Central 200 Parsecs

Jim Hinton[1,2] **for the H.E.S.S. Collaboration**

[1] Max-Planck-Institut für Kernphysik, P.O. Box 103980, D 69029 Heidelberg, Germany
[2] Landessternwarte, Universität Heidelberg, Königstuhl, D 69117 Heidelberg, Germany

E-mail: `Jim.Hinton@mpi-hd.mpg.de`

Abstract. The inner few hundred parsecs of our galaxy provide a laboratory for the study of the production and propagation of energetic particles. Very-high-energy γ-rays provide an effective probe of these processes and, especially when combined with data from other wavebands, γ-rays observations are a powerful diagnostic tool. Within this central region, data from the H.E.S.S. instrument have revealed three discrete sources of very-high-energy γ-rays and diffuse emission correlated with the distribution of molecular material. Here I provide an overview of these recent results from H.E.S.S..

1. Introduction

The central 200 pc of our Galaxy is a unique region that harbours many remarkable objects — including several potential sites of effective particle acceleration. Non-thermal emission (particularly in the radio, X-ray and γ-ray bands) can be used to trace the energetic particle populations of this region. X-ray and radio observations of synchrotron emission provide information on the product of the local magnetic field energy density and the density of relativistic electrons. The flux of inverse Compton γ-rays, on the other hand, is proportional to the radiation field density and the electron density. The combination of γ-ray and X-ray measurements therefore provides a powerful tool for probing both magnetic field strength and energetic particle content. Moreover, γ-rays provide an effective tracer for hadronic particles: proton-proton interactions in the interstellar medium lead to the production and decay of pions and hence γ-ray production. The combination of γ-ray measurements with tracers of atomic and molecular material may be the *only* way to effectively trace energetic hadrons in our galaxy.

The usefulness of γ-ray observations has traditionally been limited by poor angular resolution and modest sensitivity. A major step forward in the very-high-energy (VHE, $>$ 100 GeV) domain has recently been taken with the commissioning of H.E.S.S. (High Energy Stereoscopic System). H.E.S.S. is an array of four, 13 m diameter, imaging Cherenkov telescopes located in the Khomas highlands of Namibia [1], a southern hemisphere location ideal for observations of the Galactic Centre (GC). H.E.S.S. has an angular resolution of a few arc-minutes and a locational accuracy of $\sim 30''$ for typical point sources. The instrument reaches an energy flux sensitivity of 10^{-12} erg cm^{-2} s^{-1}, an order of magnitude lower than the previous generation of VHE instruments. The wide field of view of H.E.S.S. (5° in diameter) enables us to simultaneously monitor the entire central 200 parsec region. Results from the first two years of H.E.S.S. GC observations are described here.

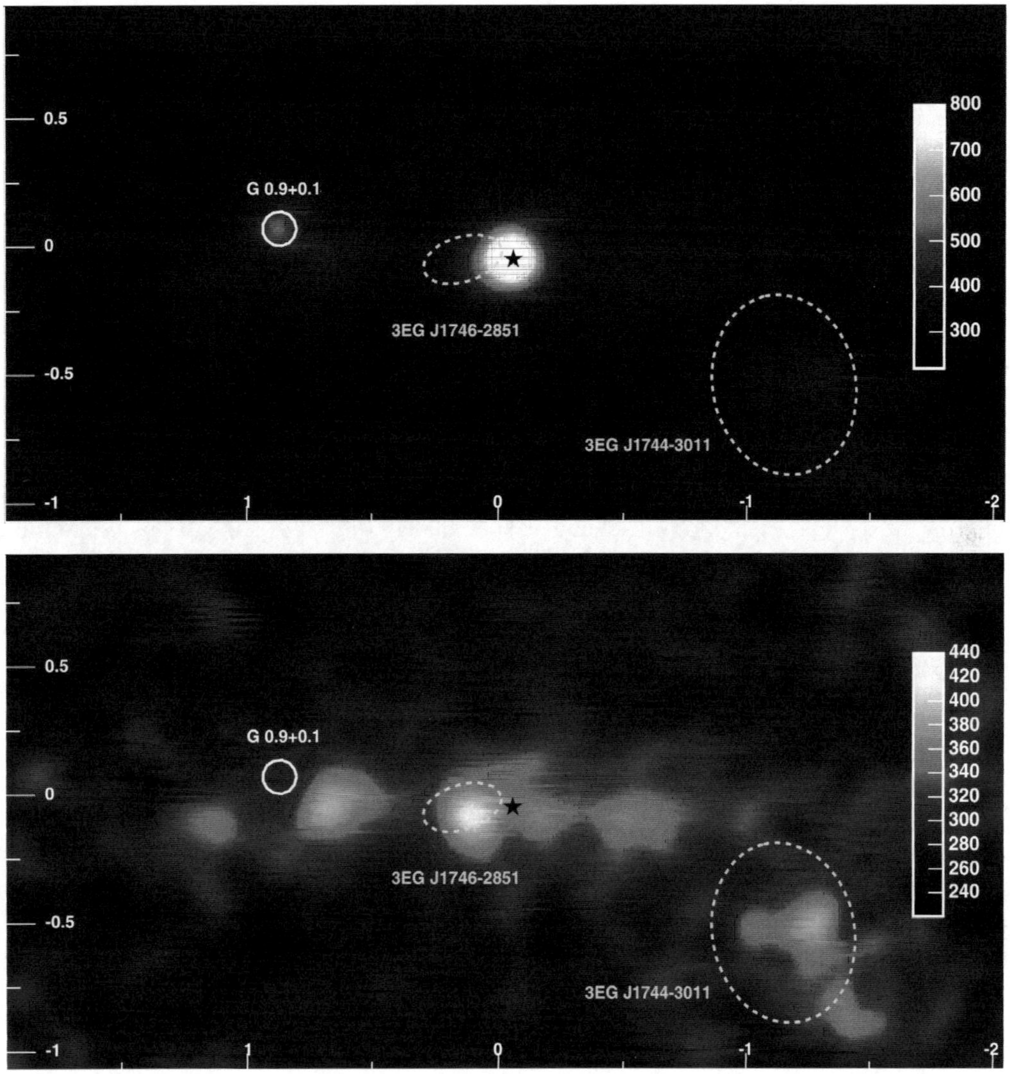

Figure 1. The H.E.S.S. view of the central 200 parsecs (reproduced from [2]). Top panel: smoothed count map (without background subtraction) for data taken with H.E.S.S. in 2004. Bottom panel: the same data after subtraction of point-like excesses at the positions of Sgr A* and G0.9+0.1. The location of Sgr A* is marked with a black star and the G 0.9+0.1 as a yellow circle. 95% confidence regions for the positions of EGRET sources are shown as dashed ellipses [3].

2. The central VHE gamma-ray source: HESS J1745-290

During 2004, detection of a compact source of VHE γ-rays close to the GC was claimed almost simultaneously by groups using three Imaging Air-Cherenkov Telescope (IACT) systems: CANGAROO [4], Whipple [5] and the partially-complete H.E.S.S. array [6]. The best localisation of this source (HESS J1745-290) is that derived from H.E.S.S. data, which places the emission within 1′ of Sgr A*. The initial spectral results from the three groups were originally in serious disagreement. While all groups agreed on a power-law spectrum ($F \propto E^{-\Gamma}$), the photon indices of H.E.S.S. ($\Gamma = 2.21 \pm 0.09$) and CANGAROO ($\Gamma = 4.6 \pm 0.5$) differed substantially, and the single flux point at 2.8 TeV published by the Whipple collaboration was a factor 1.5 above the H.E.S.S. flux. However, further analysis of the CANGAROO data yielded a much larger error on the photon index: $4.6^{-1.2}_{+5.0}$ [7] and a more detailed analysis of the Whipple data yielded a photon index of $2.44^{+0.22}_{-0.48}$ and a flux consistent with the H.E.S.S. value [8]. The final confirmation of the hard energy spectrum seen by H.E.S.S. came with the detection of this source by MAGIC [9]. The MAGIC flux and photon index (2.2 ± 0.2) show good agreement with H.E.S.S. values.

The construction of the H.E.S.S. array was completed in late 2003 and 50 hours of full-sensitivity observations of the GC took place in 2004. The upper panel of Fig. 1 shows a smoothed count map of the central four degrees of our galaxy from these 2004 data [2]. Two VHE γ-ray sources are clearly apparent: one coincident with the SNR G 0.9+0.1 (discussed in section 4) and the unidentified source HESS J1745-290 in the Sgr A region. The higher sensitivity observations in 2004 confirm the spectral shape and source position derived from the H.E.S.S. two-telescope data. The source position derived from the 2004 dataset lies $5'' \pm 10''_{stat} \pm 20''_{sys}$ from Sgr A*. The updated photon index is $\Gamma = 2.29 \pm 0.05_{stat} \pm 0.15_{sys}$ [10]. No significant variability is found on time-scales of days or hours in the 2004 dataset.

Within the error circle of the H.E.S.S measurement lie three compelling candidates for the origin of the VHE emission: the shell-type supernova remnant (SNR) Sgr A East (see e.g. [11], the newly-discovered candidate pulsar wind nebula G 359.95−0.04 [12] and the supermassive black hole Sgr A* itself (see e.g. [13]). Several radiation mechanisms have been discussed for these three objects. Plausible mechanisms include inverse Compton scattering of energetic electrons, the decay of pions produced in the interactions of energetic hadrons with the interstellar medium or dense radiation fields and finally curvature radiation of ultra-high energy protons close to Sgr A*. For some production scenarios, correlated variability is expected in X-rays and γ-rays. Simultaneous observations of the Sgr A region with H.E.S.S. and Chandra took place in July 2005 and may help to clarify this situation.

A widely discussed alternative to these astrophysical origins is the annihilation of dark matter in the central cusp in the density profile of our galaxy. However, the hard power-law spectrum of the central source is hard to reconcile with a dark matter interpretation (see [14] for a detailed discussion).

3. Diffuse emission from the Central Molecular Zone

The dense gas clouds of the central molecular zone provide effective targets for the production of γ-rays via the interactions of hadrons. Indeed, the recent H.E.S.S. observations reveal TeV emission correlated with these clouds (see the lower panel of Fig. 1). This is the first case in which such a correlation has been seen at these energies. Galactic longitude and latitude distributions of the diffuse VHE emission are shown in Fig. 2. In Fig. 2, maps of CS (J=1-0) emission from [15] have been used to estimate the column density of the GC clouds. The ratio of γ-ray emission to column density provides a measure of the density of TeV cosmic rays (CRs) in the Galactic Centre region. Assuming that the locally measured CR spectrum is valid in the GC, we can predict the expected π^0-decay γ-ray spectrum in the Central Molecular Zone. Fig. 3 shows this expectation as a grey shaded box. The measured data show a significantly harder energy spectrum than this expectation, indicating an excess of high energy hadrons in the GC, relative to the solar neighbourhood. Interestingly, the spectrum of the diffuse emission is very similar to that of the central source HESS J1745-290 (dashed line in Fig. 3).

From the lower panel of Fig. 2 it appears that the ratio of γ-ray emission to molecular density varies with galactic longitude, with a pronounced dip at $l \approx 1.3°$. The implied non-uniform CR density can be explained if CRs injected close to Sgr A have not yet had time to diffuse out

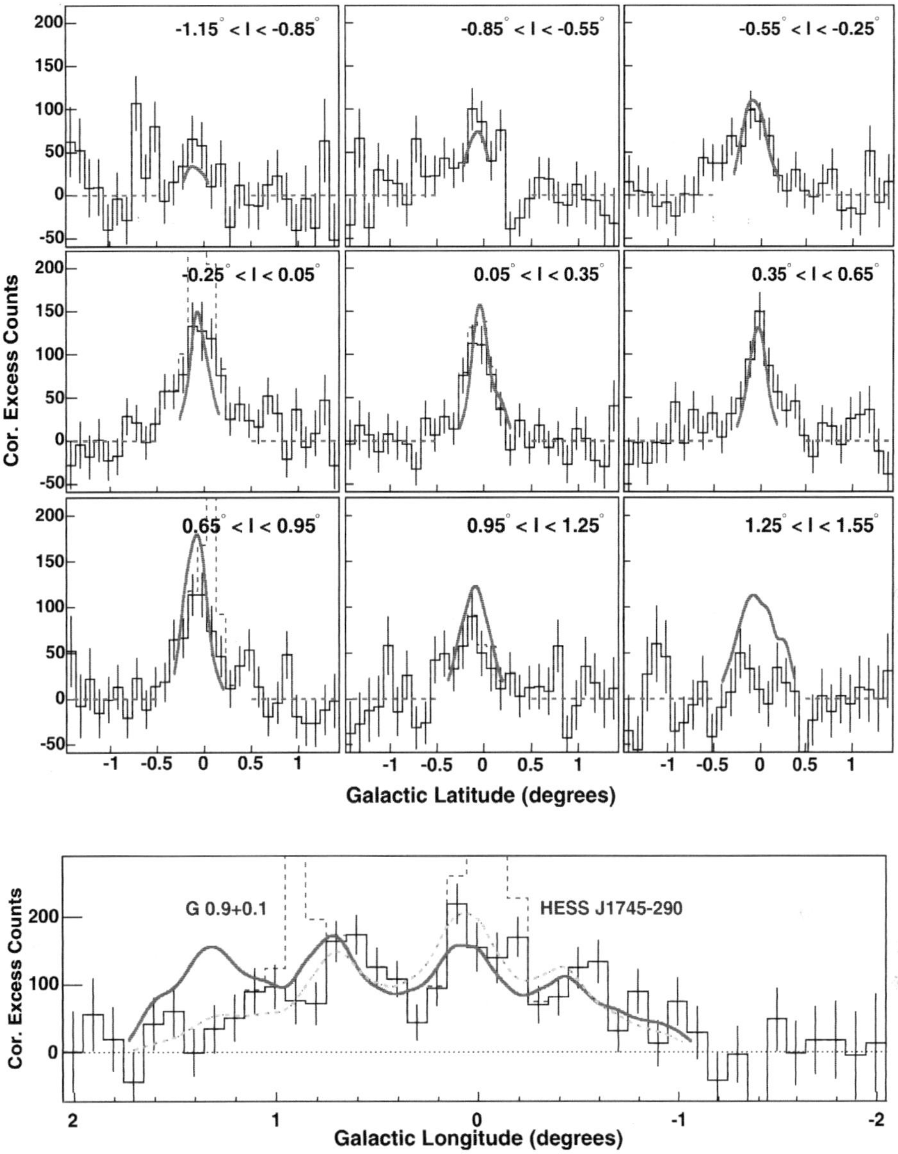

Figure 2. Slices through the VHE γ-ray emission of the galactic centre region. Top panels: latitude slices in different longitude bands. Bottom panel: longitude slice for $\pm 0.2°$ in latitude. The dashed (solid) histograms show the γ-ray emission before (after) the subtraction of point-like sources at the positions of G 0.9+0.1 and Sgr A*. The red curves show slices through smoothed CS emission maps from [15] normalised to the γ-ray data in the region of complete CS coverage. The dashed green curve in the lower panel shows the expected γ-ray emission for a population of CRs injected $\sim 10^4$ years ago, assuming nominal 'disc-like' diffusion (figure reproduced from [2])

to $l \approx 1.3°$. Assuming a diffusion coefficient of $D = 10^{30}$ cm^{2}s^{-1}, the approximate value in the Galactic Disk at TeV energies, an angular displacement of 1° at the distance of the GC

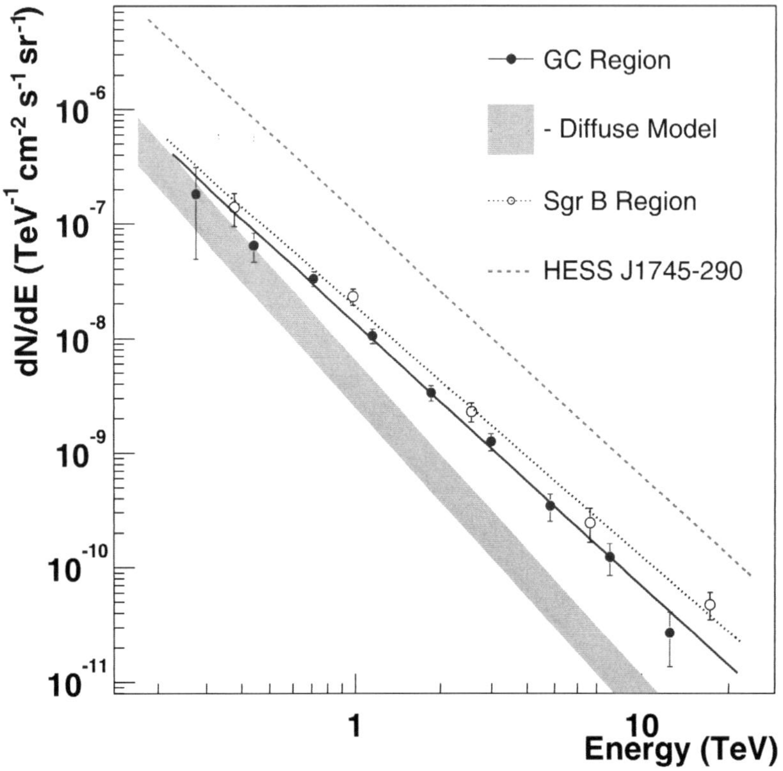

Figure 3. The differential energy spectra of the diffuse GC emission (solid line), the Sgr B molecular complex (dotted line) and the central source HESS J1745-290 (dashed line). The grey shaded box illustrates the expected flux for the GC region from the interactions of CRs with a spectrum identical to that measured at the Earth. The width of the box indicates the uncertainty on the total molecular mass of the CMZ (figure reproduced from [2]).

corresponds to a diffusion time of $\sim 10^4$ years, close to the age of the SNR Sgr A East [16]. The energy required to fill the central 200 parsecs with CRs with the observed density (extrapolating the measured H.E.S.S. spectrum down to 1 GeV) is 10^{50} ergs, remarkably close to the CR energy input required from a typical Galactic SNR in the paradigm of CR acceleration in these objects. A historical burst of activity in Sgr A* also provides a natural explanation for the observed CR excess.

4. The Pulsar Wind Nebula in G0.9+0.1

G 0.9+0.1 is a composite radio SNR with a clear 'core plus shell' morphology (see for example [17]). Observations with Chandra [18] and XMM [19] revealed hard X-ray emission from the core and weak (likely thermal) emission from the SNR shell. The characteristic softening of the X-ray spectrum with distance from the core indicates the likely plerionic nature of this source, although no radio pulsar has been identified (likely due to the large dispersion associated with the GC region). VHE γ-ray emission coincident with G 0.9+0.1 was discovered by the H.E.S.S. collaboration in 2005 [20]. Fig. 4 shows the position of the H.E.S.S. source in comparison to radio measurements. The VHE emission is unresolved by H.E.S.S. and an upper limit of 1.3′ on the rms size of the emission region has been derived. This limit allows us to exclude a γ-ray origin in the shell of the SNR and strongly suggests a common plerionic origin of the γ-ray and X-ray emission, arising respectively from inverse Compton scattering and Synchrotron radiation of energetic electrons within the nebula.

Pulsar wind nebulae (PWN) are prominent members of the VHE γ-ray source catalogue. Recent γ-ray observations with H.E.S.S. of known X-ray PWN such as Vela-X [21] and MSH 15-52 [22] show close correlations between X-ray and VHE γ-ray emission. Combined X-ray/γ-ray observations provide information on the magnetic field in the nebula and on the spatial distribution of electrons. For the distant PWN of G 0.9+0.1 the emission is currently unresolvable in γ-rays but the average magnetic field in the nebula can be estimated at $\sim 6\mu$G [20].

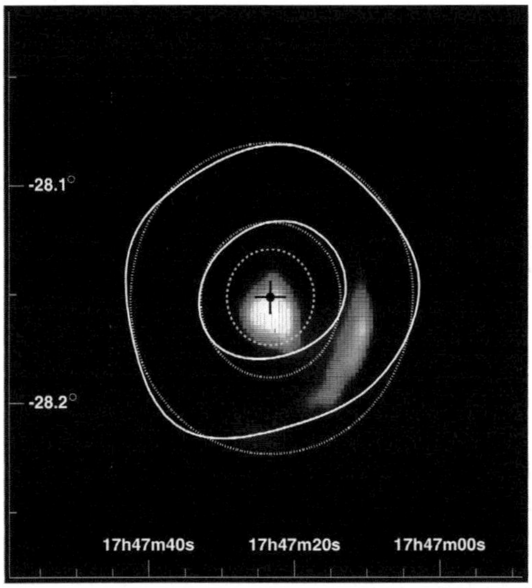

Figure 4. Comparison of the γ-ray source HESS J1747-281 with 90 cm radio emission from the composite SNR G0.9+0.1 (colour scale, [17]). The shape of the signal (solid contours) is compatible with the point-spread-function of the instrument (dotted contours). The limit on the rms size of the VHE emission region is shown as a green dashed circle. The best fit position is marked with a cross (reproduced from [20]).

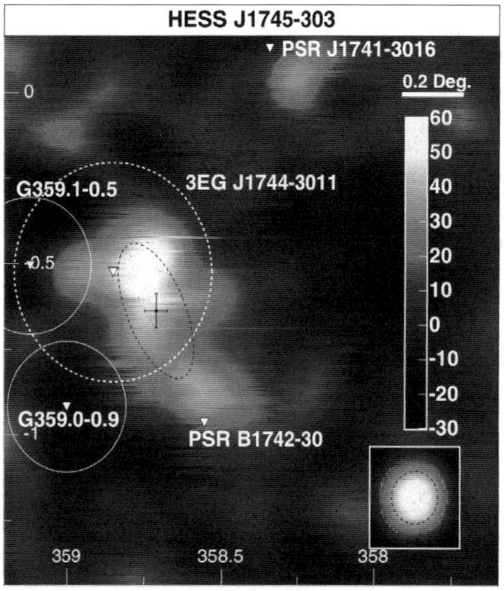

Figure 5. Smoothed and background subtracted γ-ray count map of the region around HESS J1745-303 in galactic coordinates. The best fit position and size of the H.E.S.S. source are marked with a black cross and a dashed ellipse. Nearby SNR and energetic pulsars are marked with triangles (reproduced from [23]). The dashed white circle illustrates the 95% confidence location error for the EGRET source 3EG 1744-3011.

G 0.9+0.1 is not seen in VHE γ-ray observations with MAGIC, but the non-detection is not in conflict with the H.E.S.S. result, given the lower sensitivity of the MAGIC dataset [9].

5. Unidentified Source HESS J1745-303

The H.E.S.S. survey of the inner galaxy [23], combined with deep exposures on targets such as the GC and RX J1713.7-3946 [24] has resulted in the detection of many extended ($\sim 10''$) γ-ray sources without clear counterparts at other wavelengths. One such source is HESS J1745-303, located $\sim 1^\circ$ from Sgr A. Fig. 5 shows a count map of this source, illustrating the positions of nearby supernova remnants and energetic pulsars. The most compelling potential counterpart is the (also unidentified) lower energy γ-ray source 3EG 1744-3011. A fit of a elliptical Gaussian source profile yields a position of $l = 358.71 \pm 0.04^\circ$, $b = -0.64 \pm 0.05^\circ$ and an rms source size of $5' \times 13'$. Deeper observations in all wavebands are clearly desirable to help identify this MeV-TeV γ-ray source.

6. Conclusions

Present and future γ-ray observations will play a key role in our understanding of the physical processes at work in the Galactic Centre. Key outstanding issues, such as the spatial distribution and strength of magnetic fields and the energy density of relativistic particles, can be addressed with the help of such observations. H.E.S.S. has provided the first sensitive view of this region in very-high-energy γ-rays. In a few years time the combination of the second phase of the H.E.S.S. project and the GLAST satellite will provide unbroken sensitive coverage of the 10^8–10^{13} eV γ-ray domain, with important consequences for Galactic Centre research.

References

[1] Hinton, J.A. New Astron. Rev., 48, 331 (2004)
[2] Aharonian, F. A., et al., Nature, 439, 695 (2006)
[3] Mattox, J. R., Hartman, R. C. & Reimer, O., Astrophys. J. Supp. 135, 155 (2001)
[4] Tsuchiya, K., et al., Astrophys. J., 606, L115 (2004)
[5] Kosack, K., et al. Astrophys. J., 608, L97 (2004)
[6] Aharonian, F. A., et al., Astron & Astrophys., 425, L13 (2004)
[7] Katagiri, H., et al. Astrophys. J., 619, L163 (2005)
[8] Kosack, K. P., PhD Thesis, Washington Uni. (2005)
[9] Albert, J., et al. Astrophys. J., 638, L101 (2006)
[10] Rolland, L. & Hinton, J. A. (for the H.E.S.S. collaboration), Proc. 29th ICRC, vol 4, p109 (2005)
[11] Crocker, R. M., et al., Astrophys. J. 622, 892 (2005)
[12] Wang, Q. D., Lu, F. J. & Gotthelf, E. V., MNRAS 367, 937, (2006)
[13] Aharonian, F. & Neronov, A., Astrophys. & Space Sci. 300, 255 (2005)
[14] Profumo, S., Phys. Rev. D, 72, 103521 (2005)
[15] Tsuboi, M., Toshihiro, H & Ukita, N. Astrophys. J. Supp. 120, 1 (1999)
[16] Rockefeller, G. et al., Astrophys. J. 635, L141 (2005)
[17] LaRosa T.N. et al., Astron. J., 119, 207. (2000)
[18] Gaensler, B. M., Pivovaroff, M. J., Garmire, G. P., Astrophys. J., 556, L107 (2001)
[19] Porquet, D., Decourchelle, A., Warwick, R. S., Astron. & Astrophys., 401, 197 (2003)
[20] Aharonian, F. A., et al., Astron & Astrophys., 432, L25 (2005)
[21] Aharonian, F. A., et al., Astron & Astrophys., 448, 43 (2006)
[22] Aharonian, F. A., et al., Astron & Astrophys., 435, L17 (2005)
[23] Aharonian, F. A., et al., Astrophys. J., 636, 777 (2006)
[24] Aharonian, F. A., et al., Astron & Astrophys., 449, 223 (2006)

Institute of Physics Publishing
doi:10.1088/1742-6596/54/1/024
Journal of Physics: Conference Series **54** (2006) 147–151
Galaxy Center Workshop 2006

Viscous Dissipation in the Galactic Center Region

Renaud Belmont[1,2], Michel Tagger[2]

[1] Dept. of Physics and Astronomy, University of California at Los Angeles, Knudsen Hall, Los Angeles, CA 90095-1547, USA

[2] CEA Service d'Astrophysique, UMR "AstroParticules et Cosmologie", Orme des Merisiers, 91191 Gif-sur-Yvette, France

E-mail: belmont@astro.ucla.edu

Abstract. For more than two decades, X-ray observations have revealed strong iron line emission from the central few hundred parsecs of the Galaxy. We have recently suggested that, hydrogen having escaped from the Galactic potential, it might be a helium plasma. But this leaves open the problem of heating the plasma. Here we present a possible heating mechanism in which the gravitational and kinetic energy of cold molecular clouds are dissipated in the surrounding plasma by the strong viscosity acting there. The possible existence of a vertical magnetic field in this region has a strong influence on the viscous properties and is fully taken into account. The MHD wakes of the moving clouds are analyzed in detail and it is shown that its Alfvén component can be responsible for an energy dissipation strong enough to drive efficient heating.

1. Introduction

Whereas the Galactic Center region is strongly obscured at optical wavelengths, it is responsible for intense X-ray emission. This emission is strongly peaked in the 300 first parsecs [1]. Whereas the soft part of the X-ray spectrum (between 1 and 5 keV) has been successfully identified as originating in supernova remnants, the medium responsible for hard emission (between 5 and 10 keV) remains uncertain. The continuum and the two strong K-α lines from H-like and He-like iron at 6.7 and 6.9 keV cannot originate in the $\sim$1 keV plasma of SNRs. Although many hypotheses have been suggested to explain the hard emission, only two of them are still consistent with the most recent observations with Chandra [2], XMM and Suzaku (see the contribution by K. Koyama). The hard emission could originate in hot plasma from numerous discrete point sources that are not resolved by current observations, mostly Cataclysmic Variables [3]. However from what can be inferred from existing observations, it seems there are not enough of them by a factor of a few to account for the X-ray luminosity in the central region. Also, this emission could originate in a truly diffuse plasma whose temperature must be 8 keV to account for the very ionized iron lines. It has long been noted that the temperature is too high for an ordinary interstellar plasma to be bound by the gravitational potential, so the plasma should outflow quickly from the central region, requiring an unreasonable power for its heating. However, we have recently pointed out that this plasma is weakly collisional, the collision time for protons being longer than their escape time. Protons could thus escape without bringing out heavier ions, which are gravitationally bound. This selective evaporation, similar to that in planetary atmospheres, would naturally leave a heavy plasma, mostly composed by helium, confined in the Galactic plane [4]. If the X-ray emission originates in a bound plasma, the energetic requirements are much more reasonable. If there is no other cooling mechanism, the heating only needs to balance the X-ray luminosity in the entire region, e.g. $L \approx 4. \times 10^{37}$ erg s^{-1}. Still no precise mechanism has been proposed for this heating.

Here we present such a mechanism. It uses the gravitational and kinetic energy of the cold molecular clouds flowing in this region. The intense radio and far-infrared emission observed in the Central Molecular Zone (CMZ, [7]), extending out to about 150 pc, reveals that the molecular content is condensed in cold (70K) and dense (10^3-10^4 cm^{-3}) molecular clouds. Numerous surveys have given a knowledge of their statistical properties [8, for instance]: there are at least 100 of them in the CMZ and their typical size is about 10 pc. In a first approximation, they move on very elongated orbits [9, 7] but the velocity dispersion is large and they can have up to 130 km^{-1} forbidden velocities. Whether or not the hot gas rotates at some Keplerian velocity, most of the clouds must flow relative to it, with a typical velocity of 100 km s^{-1} [10]. The orbital and dispersion velocities represent a huge reservoir of kinetic and gravitational energy that can be tapped to heat the surrounding plasma.

The mechanism we have investigated dissipates this energy thanks to the strong viscosity of the hot plasma. However, observations of many magnetized non-thermal filaments [5, 6] seem to trace a pervasive and vertical magnetic field [7]. Its intensity is strongly debated but ranges between 10 μG and a few mG. Such a field modifies dramatically the viscous properties of the plasma, namely it strongly inhibits its efficiency.

In section 2, we briefly review the viscous properties of a 8 keV plasma. Then in section 3, we present the inviscid MHD wake of clouds moving in a surrounding magnetized plasma. The way the viscosity can act on this wake and dissipate energy is detailed in section 4 and the overall efficiency is eventually discussed in section 5.

2. The Braginskii viscosity

The viscous coefficient of a non-magnetized 8 keV plasma is [11]:

$$\eta_0 = 630 \text{ g cm}^{-1}\text{s}^{-1} \left(\frac{k_B T}{8 \text{ keV}}\right)^{5/2} \tag{1}$$

Because of its high temperature, the diffuse plasma in the Galactic Center might be very viscous if not magnetized. For instance, the flow past a cloud would lead to Reynolds numbers of: $\mathcal{R}_e \approx 10^{-2}$.

However, magnetized plasmas have different properties [12]. The general viscosity can be described as the sum of two kinds of viscosity. The off-diagonal terms in the stress tensor are responsible for the usual *shear* viscosity proportional to terms in $\partial_i v_j$, e.g., to the shear of the velocity field. The diagonal terms are responsible for the *bulk* viscosity, proportional to terms in $\partial_i v_i$, e.g., to the compression of the fluid. For subsonic motion, the compression is weak: $\partial_i v_i < \partial_i v_j$, so that bulk viscosity is often neglected.

Any magnetic field makes this stress tensor anisotropic: in magnetized plasmas the diffusive properties are very different in the directions parallel and perpendicular to the local magnetic field. A significant magnetic field reduces the shear viscosity whereas it leaves bulk viscosity unchanged. Namely, off-diagonal terms scale with the diagonal ones as inverse powers of $\Omega_c\tau$, where Ω_c is the cyclotron frequency and τ is the Coulomb collision time. Even for the lowest estimates for the magnetic field intensity (B$\sim$ 1 μG), the largest coefficient for the shear viscosity is divided by more than ten orders of magnitude compared to a non-magnetized situation: $\Omega_c\tau \approx 10^{11}$. As a result, the usual shear viscosity is fully inhibited and only bulk viscosity remains. The local dissipation caused by the bulk contribution is:

$$Q = \frac{\eta_0}{3}\left(\vec{\nabla}.\vec{v} - 3\partial_{\|}v_{\|}\right)^2 \tag{2}$$

where $\|$ refers to the direction parallel to the field lines. The dissipation results from the compression $\vec{\nabla}.\vec{v}$ or from some velocity parallel to the field $v_{\|}$. At this point, it is not easy to estimate the dissipation efficiency. On the one hand, the viscous coefficient is large, because the temperature is high. On the other hand, the compression is weak, because the motion is subsonic. A precise estimate of the compression is required.

3. The wake of molecular clouds

In this section, we describe the inviscid wake of a moving molecular cloud. The compression in the wake is emphasized, since this is what drives the dissipation.

Since molecular clouds are ionized at least on their surface, they can be considered to a good approximation as conductors. The motion of such bodies moving in a magnetized plasma has already been investigated in different astrophysical conditions. From work on the motion of artificial satellites in the Earth's magnetosphere or the motion of Io in the Jovian magnetosphere, it is known that these bodies mostly excite Alfvén waves. These waves propagate along the field lines, carrying a strong energy flux [13, 14]. Other modes are also excited, but are likely to remain weak in amplitude [15, 16] .

In a general manner, the wake of a molecular cloud moving in a magnetized plasma can be considered to be the sum of three components characteristic of the three MHD modes that can propagate in magnetized plasmas: the Alfvén and the slow and fast magnetosonic components. Because these modes have different propagation and polarization properties, the three components are spatially separated and are subject differently to bulk viscosity. The typical MHD wake is presented on figure 1a.

The Alfvén waves propagate along the field lines at the Alfvén velocity v_A. In a picture where the field is vertical and the cloud moves horizontally at v_c, the Alfvén perturbations excited on the cloud surface remain always localized along the Alfvén characteristics. They form a cylinder inclined of an angle $\alpha_A = \tan^{-1}(v_c/v_A)$ from the vertical. As the motion of the clouds in the Galactic Center region is subalfvénic, these *Alfvén wings* are almost vertical. Alfvén waves are not compressible so that there is no density perturbation associated with their propagation. Also, in linear analysis, the velocity perturbation has no component parallel to the field. As a result, they cannot be linearly damped by bulk viscosity.

Most of the velocity perturbation resulting from the flow around the cloud is Alfvénic. However, other modes must also be excited. For instance, the non-linear analysis of the Alfvén wings shows that they drive a small net outflow of matter. To balance this outflow at the cloud surface, sonic perturbations must be associated with some net inflow onto the cloud.

The magnetosonic waves are compressible by nature, so that they are natural candidate for the dissipation. They have complicated propagation properties but for sake of simplicity, we assume that the field is super-equipartition ($\beta > 100\mu G$). In this limit, the slow magnetosonic waves propagate along the field lines at the sound speed c_s, so that they can form a wing, exactly as Alfvén waves do; and fast magnetosonic waves propagate isotropically, so that their amplitude decays with the distance to the cloud because of geometrical effects and they remain localized in the cloud vicinity as usual sonic waves would do. No precise amplitude for the sonic components has been found analytically. However, upper estimates can be found. Based on the mass balance between the Alfvén and the slow components and on the properties of pure sonic waves in non magnetized plasmas, it can be shown that the amplitude of both the slow and fast perturbations remains very weak in comparison to the Alfvén one.

4. Dissipation

In this section, we analyze the effect of viscosity on each component of the wake. The dissipation is then summed over all the clouds in the entire central region and compared to the X-ray luminosity.

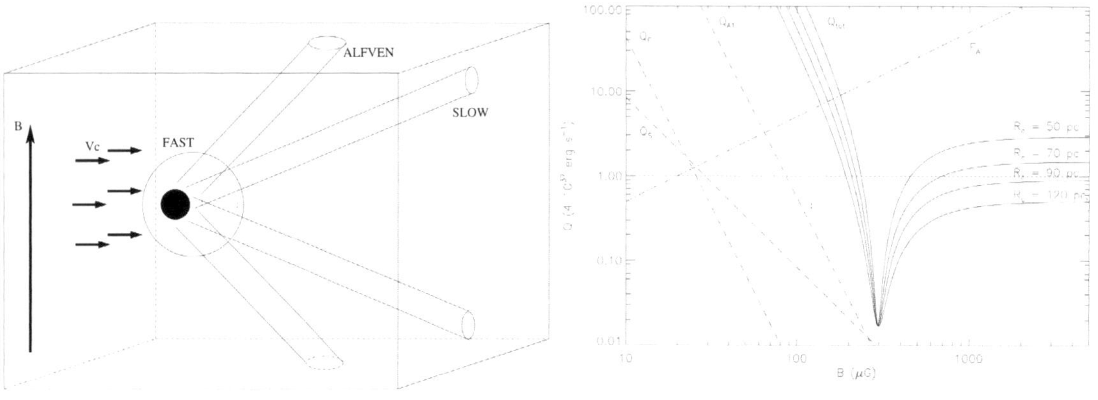

Figure 1. a) *Left panel:* The wake of a conducting molecular cloud moving in a magnetized plasma. b) *Right panel:* Alfvén flux F_A and dissipation rates in the different components of the wake, normalized to the X-ray luminosity. Q_s and Q_f are the rate in the slow and fast perturbation respectively. Q_{A1} is the dissipation rate in the non-linear Alfvén wing. Q_{tot} is the total dissipation dominated by the dissipation of the Alfvén wings in a curved field, for several curvature radii.

4.1. The sonic components

As was explained earlier, the amplitude of the magnetosonic perturbations is weak, so that the overall associated dissipation in weak. On the one hand, the slow wing is conditioned by the non-linear properties of the Alfvén wing; on the other hand, the fast perturbation is governed by the cloud speed. In both cases, the compression goes as some power of $v_c/v_A << 1$. The strong magnetic field also inhibits the dissipation by bulk viscosity. The overall dissipation in the central region is plotted on figure 1b. For any value of the field intensity, the dissipation in the sonic components is insufficient to balance the radiative cooling. Only weak fields seem to permit an effficient heating, but in this limit, the assumption we have used to simplify the study of sonic waves fails.

4.2. The Aflvén component

It we mentioned earlier that linear Alfvén wings are incompressible, so that they cannot be damped by the bulk viscosity. However, the energy flux carried away by these Alfvén waves is large: $F_A = 2. \times 10^{38}\mathrm{erg\ s^{-1}}(B/0.1\ \mathrm{mG})$. For the smallest estimates, this flux is comparable to the power needed to balance the radiative cooling and for for the largest, it is more than 2 orders of magnitude larger! It is thus tempting to create a more realistic model that dissipates this available energy. There are many ways to damp Alfvén waves. Here, we present two of them.

First, as mentioned earlier, non-linear Alfvén wings involve a parallel velocity component, so that although they are incompressible, non-linear Alfvén waves can be dissipated by bulk viscosity (see eq. 2). Again, since the clouds' motions are subalfvénic, the non-linear terms are weak, so that only a small fraction of the Alfvén flux can be dissipated. However, for small fields ($B < 100\mu G$), this is enough to heat the plasma (see figure 1b).

Second, Alfvén waves are truly incompressible only in a perfectly straight magnetic field. Obviously, the field lines in the Galactic Center region are not straight, as can be inferred from the slight bending in most of the non-thermal filaments. In a curved geometry, Alfvén waves couple to slow magnetosonic waves and acquire a compressional component. They can thus be subject to bulk viscosity. Since the compression of these Alfvén waves relies basically on geometrical effects, the compression is $\vec{\nabla}.\vec{v} \approx v_c/R_c$ where R_c is the local curvature radius. The heating efficiency is plotted on figure 1b.

5. Discussion and Conclusions

From the previous results, it can be concluded that the dissipation of the kinetic and gravitational energy of the clouds a) is dominant in the Alfvén wings, b) results mostly from the curvature of the magnetic field lines and c) is most efficient for $B < 150\ \mu G$ and $B > 600\ \mu G$. With the typical estimates for the cloud number and size of $n_c \approx 100$, $r_c \approx 5$ pc, the plasma temperature and density of $k_B T$=8 keV, $n \approx 0.1$ cm^{-3}, the field curvature radius of $R_c \approx 100$ pc, and within the expected range of field intensity, this dissipation is strong enough to balance the radiative cooling, allowing a temperature equilibrium.

One main assumption in this model is the pervasive and vertical nature of the magnetic field. This is most consistent with observational evidence, but strongly debated. A turbulent or horizontal field would probably give very different results and it is difficult to predict its impact on the dissipation rate. However, it is likely that the dissipation could be at least as efficient since the temperature would still guarantee a large coefficient for the bulk viscosity. In addition, as most of the energy leaves as Alfvén waves, and is only partially dissipated by bulk viscosity, the strong coupling between the compressible magnetosonic sonic modes and the Alfvén ones in a tangled field would probably enhance the dissipation rate. Even if the Alfvén flux is weakly dissipated, it cannot leave the Galactic plane as in a straight field geometry and must remain in the plane until it is eventually dissipated.

Also, we have worked with the guideline that the hard emission originates in a hot diffuse plasma. As mentioned in the introduction, it might also come from unresolved discrete point sources. Presently, there is no observational evidence that permits one to discriminate between these two models. On the one hand, it seems that there are not enough point sources to account for the hard emission. On the other hand, until now there was no precise mechanism able to heat the plasma at the inferred temperature. Here we have presented such a mechanism, showing that the idea of a diffuse plasma in the Galactic Center is consistent with our current understanding of this region.

References

[1] Yamauchi, S., Kawada, M., Koyama, K., Kunieda, H., & Tawara, Y. 1990, *ApJ*, **365**, 532

[2] Muno, M. P., et al. 2004, *ApJ*, **613**, 326

[3] Revnivtsev, M., Sazonov, S., Gilfanov, M., Churazov, E., & Sunyaev, R. 2006, *A&A*, **452**, 169

[4] Belmont, R., Tagger, M., Muno, M., Morris, M., & Cowley, S. 2005, *ApJL*, **631**, L53

[5] Yusef-Zadeh, F., & Morris, M. 1987, *ApJ*, **322**, 721

[6] LaRosa, T. N., Lazio, T. J. W., Kassim, N. E., & Lang, C. C. 2000, *Bulletin of the American Astronomical Society*, **197**, 406

[7] Morris, M., & Serabyn, E. 1996, *ARAA*, **34**, 645

[8] Oka, T., Hasegawa, T., Sato, F., Tsuboi, M., Miyazaki, A., & Sugimoto, M. 2001, *ApJ*, **562**, 348

[9] Binney, J., Gerhard, O. E., Stark, A. A., Bally, J., & Uchida, K. I. 1991, *MNRAS*, **252**, 210

[10] Belmont, R., & Tagger, M. 2006, *A&A*, **452**, 15

[11] Spitzer, L. 1962, *Physics of Fully Ionized Gases*, New York: Interscience (2nd edition), 1962

[12] Braginskii, S. I. 1965, *Reviews of Plasma Physics* , **1**, 205

[13] Drell, S. D., Foley, H. M., & Ruderman, M. A. 1965, *JGR*, **70**, 3131

[14] Neubauer, F. M. 1980, *JGR*, **85**, 1171

[15] Wright, A. N., & Schwartz, S. J. **1990**, *JGR*, 95, 4027

[16] Linker, J. A., Kivelson, M. G., & Walker, R. J. 1991, *JGR*, **96**, 21037

Institute of Physics Publishing
doi:10.1088/1742-6596/54/1/025

Journal of Physics: Conference Series **54** (2006) 152–155
Galaxy Center Workshop 2006

Confirmation of New Supernova Remnants near the Galactic Centre

S. Roy[1] and S. Bhatnagar[2]

[1] Astron, Oude Hoogeveensedijk 4, Dwingeloo, The Netherlands
[2] NRAO, Socorro, New Mexico, USA

E-mail: roy@astron.nl,sbhatnag@aoc.nrao.edu

Abstract. We have made deep observations with the GMRT at 330 MHz of seven, yet-unclassified objects in the Molonglo Galactic Centre Survey (MGCS) in order to confirm the nature of these sources. Based on morphology and non-thermal spectral indices, five of them are now confirmed to be SNRs. This makes the number density of SNRs in the region a factor of two higher than expected from a smooth distribution of SNRs in the Galaxy.

1. Introduction

Present catalogues of supernova remnants (SNRs) are thought to be incomplete because of selection effects against detection of old, low-surface-brightness, large SNRs, as well as small-diameter, young SNRs because of the poor resolution/poor sensitivity of previous surveys [1]. This problem is further compounded in the Galactic Centre (GC) region where there is a very complex distribution of sources. Eleven SNRs were identified within $-5^\circ \leq l \leq 5^\circ$, $-2.5^\circ \leq b \leq 2.5^\circ$ region of the Galaxy before the MOST Galactic Centre survey (MGCS). The MGCS [2] at 843 MHz detected 17 candidate SNRs in the region. These detections, if confirmed, indicate that the number density of SNRs in the GC region is twice that in the rest of the Galaxy [3]. Confirming that these candidates from the MGCS are SNRs is important as it suggests a possible correlation between the dense environment in the GC region and a higher SNR density. Out of these seventeen candidates, ten have already been studied and eight have been confirmed as SNRs [4], [5] [6] and [7]. Therefore, we have observed the seven remaining candidates from the above list at 330 MHz with the Giant Metrewave Radio Telescope (GMRT). The GMRT with its high sensitivity at metre wavelength and high angular resolution ($\sim 10''$) is well suited to discovering supernova remnants. Here we present preliminary results of these observations.

2. Observations and data reductions

We have observed 3 fields G355.5+00.0, G358.6+00.9 & G358.8−00.9 at 330 MHz such that all the 7 candidate SNRs G355.4+0.7, G355.6−00.0, G358.1+01.0, G358.7+00.7, G359.1+00.9, G358.5−00.9, and G359.2-01.1 remain within the FWHM of the primary beam of one of these fields. Observations of these 3 fields were carried out on 18th and 19th of March 2004 with the default spectral line mode of the correlator with a total IF bandwidth of 32 MHz. The multiple snapshot mode of observations was used to improve *uv*-coverage and the total time spent on each field was ~3 hours. Absolute flux density calibration was performed using

Table 1. Observed parameters of the SNRs confirmed by these observations

Field name (*Gl±b*)	RA (J2000)	Dec (J2000)	Size (′)	Type	S_{330MHz} (Jy)	S_{843MHz} (Jy)	$S_{8.35GHz}$ (Jy)	$S_{14.35GHz}$ (Jy)
G355.4+00.7	17 31 20	−32 25 40	25	S	8.9±1.3	6?	1.0	1.0
G355.6−00.0	17 35 15	−32 38 05	6×8	S	3.3±0.5	2.6	–	1.0
G358.1+01.0	17 37 00	−29 58 59	20	PS	6.0±2.5	2?	0.9	–
G358.5−00.9	17 46 10	−30 38 56	17	S	8.0±2.5	–	1.0	–
G359.1+00.9	17 39 36	−29 10 50	11	S	4.3±1.0	4.3	0.9	–

3C286. 1822−096 and 1714−252 were used as secondary calibrators. In the absence of system temperature measurements, we have performed gain calibration following [8] with automatic level control (ALC) on. After calibration and editing, a pseudo-continuum data base was made to avoid bandwidth smearing within the primary beam. Images of the fields were formed by Fourier inversion and CLEANing (using the IMAGR task of the AIPS package). The initial images were improved by phase-only self-calibration (self-cal).

3. Results

3.1. Field G355.5+00.0

A low resolution 330 MHz GMRT image of this field is shown in Fig. 1. Two candidate SNRs G355.4+00.7 & G355.6−00.0 are located in this field. Based on MOST maps, G355.4+00.7 is described in [2] as an SNR with a wispy structure, brightened northern cap and a diameter of about 25′. A bridge of emission appear to connect it to a HII region $\sim 30''$ south of it. This candidate SNR as seen in Fig. 1 matches its above description. A higher resolution image of the above field in Fig. 2 also shows a diffuse southern boundary of this object. To determine the spectral index of this and all the other candidate SNRs, we estimate the upper limit on the flux densities of these objects from either 8.35 or 14.35 GHz NRAO single dish images [9] of these fields. Total flux densities of G355.4+00.7 at different frequencies are given in Table 1. In this table, S_{330MHz}, S_{843MHz}, $S_{8.35GHz}$ and $S_{14.35GHz}$ indicates flux densities at frequencies of 330 MHz, 843 MHz, 8.35 GHz and 14.35 GHz respectively. Type code 'S' indicates shell type, and 'PS' indicates partial shell. From Table 1, spectral index α ($S(\nu) \propto \nu^{\alpha}$) for G355.4+00.7 is estimated to be ≤ -0.6. A non-thermal spectral index and a shell-like morphology confirms it as a SNR.

The other candidate SNR G355.6−00.0 is described in MGCS to have a shell structure of size 6′. It appears the same in Fig. 1. Based on its flux densities at different frequencies in Table 1, we determine a spectral index of ≤ -0.3. A non-thermal spectral index and a shell-like morphology confirms it as a SNR.

3.2. Field G358.6+00.9

A 330 MHz GMRT image of this field is shown in Fig. 2. Three candidate SNRs G358.1+01.0, G358.7+00.7, and G359.1+00.9 are located in this field. G358.1+01.0 is a low-surface-brightness partial shell as seen in Fig. 2. The major feature of this object is the vertical feature running from the southern rim to the centre of the shell. The same feature is also seen in the MOST image of this object. Total flux densities of this object at different frequencies are given in Table 1, which shows its spectral index to be ≤ -0.6. A non-thermal spectral index and a shell-like morphology confirms it as a SNR.

In MGCS, G358.7+00.7 is described as a partial shell of very weak emission with a diameter of about 17.5′. The estimated flux density is ~2 Jy for this object at 843 MHz. However, we could not detect this object in our observations (see Fig. 2). Based on the non-detection at 330 MHz, we put a upper limit on the flux density at 330 MHz as 2 Jy, suggesting that it does not

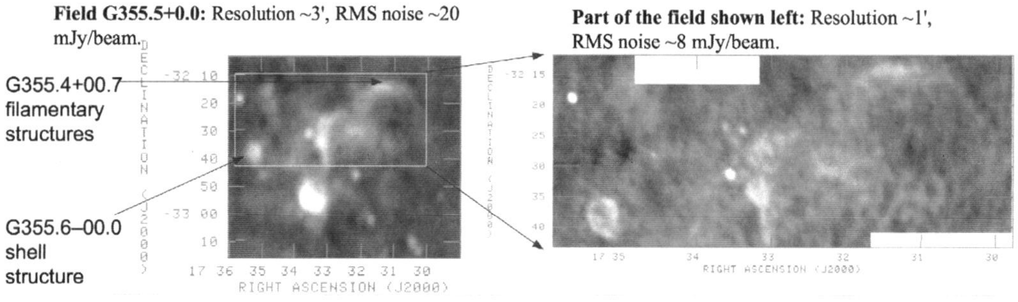

Figure 1. 330 MHz image made with the GMRT of the field containing the source G355.5+00.0. The images in the left and the right panels are at a resolution of 3′ and 1′ respectively.

have a spectral index consistent with non-thermal emission. This source therefore, is unlikely to be a supernova remnant.

G359.1+00.9 appears to be a shell-like object in Fig. 2. Its morphology is the same in MGCS. It has a diameter of about 11′ and the pulsar PSR B1736−29 is located few arc-minutes away. Total flux densities of this object at different frequencies are given in Table 1, which shows its spectral index to be ≤−0.5. A non-thermal spectral index and a shell-like morphology confirms it as a SNR.

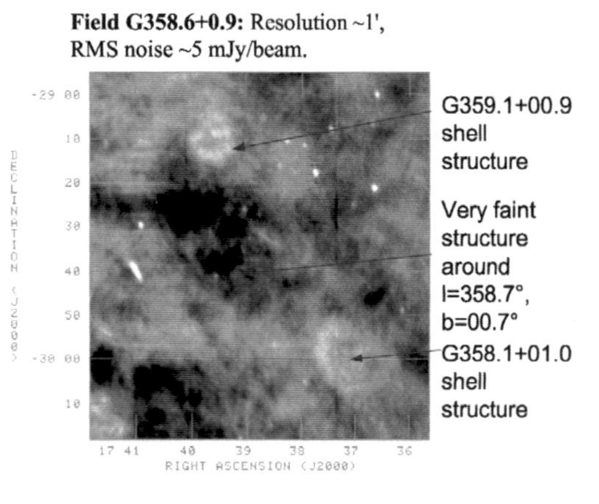

Figure 2. 330 MHz image made with the GMRT of the field containing the SNR G358.6+00.9.

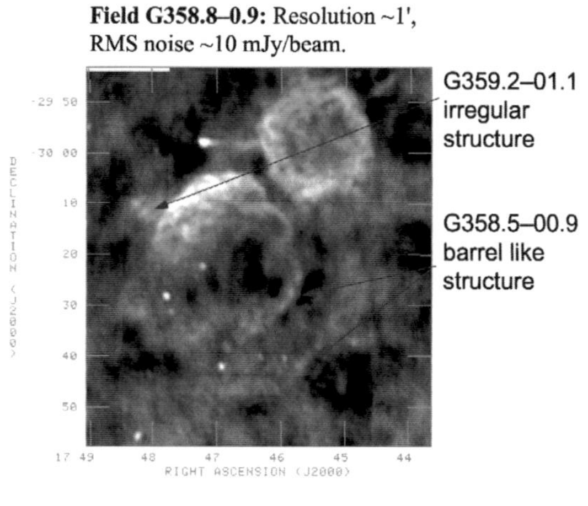

Figure 3. 330 MHz image made with the GMRT of the field field containing the SNR G358.8−00.9.

3.3. Field G358.8−00.9

A 330 MHz GMRT image of this field is shown in Fig. 3. Two candidate SNRs G358.5−00.9 and G359.2−01.1 are located in this field. G358.5−00.9 is described as a faint barrel-like object in MGCS. We barely detect this structure in Fig. 3. It has a size of about 17′. The total flux densities of this object at different frequencies are given in Table 1, which shows its spectral index to be ≤−0.6. A non-thermal spectral index and a shell-like morphology indicates it as a

SNR. However, future observations with better signal-to-noise ratio will be useful to establish its structure.

G359.2−01.1 is located adjacent to G359.0−0.9. In MGCS, it is described as an asymmetric shell-type object with rather diffuse edges, as opposed to the sharp edges normally associated with SNRs. This object shows an irregular structure in the 330 MHz GMRT image. As SNRs do not show this morphology, we rule out that this object is a SNR.

4. Discussions

Out of the 7 candidate SNRs observed, 5 have been confirmed to be SNRs from the present study. This makes the total number of supernova remnants confirmed from MGCS to be 13. With 24 SNRs within $-5° \leq l \leq 5°$, $-2.5° \leq b \leq 2.5°$ of the Galaxy, this makes the number density of SNRs in the region to be almost a factor of 2 higher than expected ($\sim$14) from a smooth distribution of SNRs in the Galaxy [3]. This indicates that these SNRs are the Galactic Centre (GC) component of the SNR distribution in the Galaxy. Massive stars are more likely to form in the GC, which ends up in supernova explosion and our results support such a scenario.

5. Conclusions

We have observed the 7 yet-unclassified candidate SNRs from the MGCS at 330 MHz with the GMRT. Our observations confirm 5 of these to be SNRs. This makes the number density of SNRs in this region almost a factor of 2 higher than expected and suggests a possible correlation of dense and turbulent GC ISM with a higher number density of SNRs.

Acknowledgments

The authors would like to thank the staff of the GMRT who made these observations possible. GMRT is run by the National Centre for Radio Astrophysics of the Tata Institute of Fundamental Research.

References

[1] Green, D. A. 1991 *PASP* **103** 209
[2] Gray, A.D. 1994b *MNRAS* **270** 847
[3] Gray, A.D. 1994a *MNRAS* **270** 861
[4] Bhatnagar, S. 2002 *MNRAS* **332** 1
[5] Roy, S. & Rao, A.P. 2002 *MNRAS* **329** 775
[6] Gaensler, B.M. 1998 *ApJ* **493** 781
[7] Kassim, N.E. & Frail, D.A. 1996 *MNRAS* **283** L51
[8] Roy, S. & Rao, A.P. 2004 *MNRAS* **349** L25
[9] Langston, G., Minter, A., D'Addario, L., et al. 2000 *AJ* **119** 2801

Institute of Physics Publishing
doi:10.1088/1742-6596/54/1/026
Journal of Physics: Conference Series **54** (2006) 156–160
Galaxy Center Workshop 2006

Low frequency studies of the Galactic Centre with GMRT

Subhashis Roy[1], **A. Pramesh Rao**[2]

[1] Astron, Oude Hoogeveensedijk 4, Dwingeloo, The Netherlands
[2] NCRA-TIFR, Pune University Campus, Pune-7, India

E-mail: roy@astron.nl

Abstract. We have observed the GC region at 255 and 154 MHz with the GMRT. From the 255 MHz map we detect 28 compact sources, and an anti-correlation of their deconvolved sizes with angular distance from the GC is observed. This is the first direct indication of scatter broadening for a statistically large sample of extragalactic sources in the region and we discuss its implications in determining the properties of the screen. Our multifrequency observations at 255 and 154 MHz in tandem with previous observations of the GC shows that the $7'$ halo in Sgr A complex is a non-thermal source rather than a mixture of thermal and non-thermal electrons as was proposed before.

1. Introduction

The central region of the Galaxy is characterised by dense and turbulent environment not seen elsewhere in the Galaxy. A large number density of stars causes a good fraction of gas in the region to ionise which scatters any incident electromagnetic wavefront. Consequently the central source Sgr A* appears scatter broadened in radio frequencies to an extent not seen elsewhere for any sources of Galactic origin at the same frequency. Scatter broadening of similar magnitudes are also observed towards maser sources in the region [1]. This raises the question on the location and properties of this screen. Earlier work [1] suggested the screen to be at a distance of ~kpc from the GC, but [2] suggested a 'hyperstrong scattering region' within 0.5° from the GC which is located at a distance of ~133 pc from it. However, there is no direct estimate of its distance from GC and [3] discovered a source G359.87+0.18 seen through the scattering screen of [2], but scattered almost an order of magnitude less than their model. If extragalactic sources are observed through this region then it can be shown that

$$\theta_{xgal} = \frac{D_{GC}}{\Delta_{GC}} \theta_{Gal} \tag{1}$$

[1], where θ_{xgal} is the characteristic diameter of an extragalactic source, D_{GC} is the distance to Sun from the GC, Δ_{GC} is the distance to the scattering screen from GC and θ_{Gal} is the characteristic diameter of a Galactic source. The above equation can be used to determine the distance to the screen from the GC. Scattering is proportional to the square of wavelengths and extrapolation of measured scattering diameter of a few extragalactic sources a few degrees away from GC [4] to lower frequencies indicates a scattering size comparable to the synthesised beam size at 250 MHz of GMRT, which can be measured.

Being located in the central few arc minute of the Galaxy, Sgr A complex has attracted our attention for many decades. This complex from larger to smaller scale is essentially composed of 4 separate objects:
(i) The $7'$ halo, which is modelled as a mixture of thermal and non-thermal electrons [6]. (ii) Sgr A East, which is a supernova remnant (SNR). (iii) Sgr A West, which is a HII region and being located closer than Sgr A East, absorbs its emission through free-free absorption at low radio frequencies [6]. (iv) Sgr A*, which is coincident with the supermassive black hole at the Galactic Centre and is found to be located in front of Sgr A West [5]. The emission from Sgr A East, Sgr A West and Sgr A* is believed to be non-thermal, thermal and non-thermal respectively. The high frequency spectral index of Sgr A East is ~ -1 ($S(\nu) \propto \nu^{\alpha}$), but its flux density at 330 MHz is only about 1.5 higher than what is measured at 1.4 GHz. In order to explain this anomaly, [6] proposed the $7'$ halo to be a mixture of thermal and non-thermal electrons, and estimated its optical depth at 330 MHz to be $\sim$1. However, they suggested it on the basis of observations done at only 2 different frequencies (1.4 GHz and 330 MHz), but free-free absorption has at least 2 free parameters. So, their model of $7'$ halo remains highly questionable and observations at frequencies lower than 330 MHz will better constrain any free-free absorption by $7'$ halo.

In order to measure the properties of ionised gas at the GC through scatter broadening and to determine the nature of $7'$ halo through low frequency observations, we have observed this region with high angular resolution at 255 and 154 MHz with GMRT and our tentative results are presented here.

2. Observations and data reductions

Observation of the Galactic Centre region at 255 MHz was carried out on 13th March 2003, and the 154 MHz observation was done on 20th August 2005. To reduce radio frequency interference (RFI) on the data, a narrow IF bandwidth of 6 MHz having relatively less RFI was chosen for observations at 255 and 154 MHz (other steps like employing 14 db Solar attenuator at the preamplifier stage was also taken to minimise amplifier saturation in presence of strong RFI). Absolute flux density calibration was performed using 3C48. 1830-36 and 1714-252 were used as secondary calibrators. In the absence of System temperature measurements, we have performed gain calibration following [5] with automatic level control (ALC) off. After calibration and editing, a pseudo-continuum data base was made to avoid bandwidth smearing within the primary beam. The initial images were improved by phase-only self-calibration (self-cal).

3. Results & Discussions

3.1. Scatter broadening of extragalactic sources at 255 MHz

To detect small diameter sources in the region, we first made a map of Sgr A complex with all the data and subtracted the Fourier transform of its Clean components from the data and re-imaged with a short UV cutoff to resolve out any extended emission of size larger than $7'$. The synthesised beam size of the resulted map is $17.3'' \times 12.5''$. A total of 28 small diameter sources are detected and Fig. 1 shows a plot of source size as a function of Galactic longitude and latitude. In the plot, X-axis errorbars indicate the average half-width of the synthesised beam, and Y-axis errorbars indicate the deconvolved average (by averaging the major and minor axis sizes) size of the sources. To make the symbols visible, actual angular scales measured in source sizes and synthesised beam have been multiplied by a factor of 30 in these plots. As can be seen from the plot, source sizes appear to increase near the GC. To investigate a possible correlation of a decrease of source size as a function of distance from GC, we plot in Fig. 2 source sizes as a function of angular distance from GC. Since we are measuring scattering diameter of extragalactic sources, we have removed 2 possible Galactic sources from the sample before plotting. A good fraction fraction of the remaining 26 sources are shown to be extragalactic from their spectral index and morphology [7]. Sources which are not observed in [7] have counterparts in 1.4 GHZ VLA Galactic plane survey ([10], [11]) and have a steep spectral index in the range expected for extragalactic sources. From the data plotted, Spearman rank correlation of decreasing source size with angular distance from GC is estimated to be 0.54, which makes the correlation valid at 99% confidence level. Since intrinsic sizes of extragalactic sources are independent of angular separation from GC, this correlation establishes GC contribution in causing the scatter broadening of sources. The scattering diameter of these sources extrapolated close to GC at 255 MHz is about $25''$. This is at least a factor of 6 smaller as compared to what is expected from [2].

The scattering diameter measured for small diameter sources in our map is comparable to what is expected from VLBA observations of extragalactic sources a few degrees away from GC[4]. The scattering screen we detect produce scatter broadening within a factor of two of what is expected for maser sources in the GC region extrapolated to this frequency. Following Equation (1), if this scattering screen is uniformly distributed, it is located $\geq$4 kpc away from the GC. A circularly symmetric screen around the GC at this distance will produce perceptible effects even 10 degrees away from GC along the Galactic plane. However, Fig. 2 shows the extent of this screen from GC to be only $\sim 1°$, which shows the screen is not symmetrically distributed around the rotation axis of the Galaxy and could be just a line of sight ionised cloud.

Other than G359.87+0.18, we have also detected one source G359.6+0.3 (Fig. 3) [7] which is a FRII radio galaxy and is located within the hyperstrong scattering regime [2], but is scatter broadened almost a factor of ten less than the hyperstrong scattering regime. Consequently, if the hyperstrong scattering medium exists, it ought to have a small filling factor, so that all the above extragalactic sources are not seen through it. We note that the maser sources could be preferentially located in the clouds, ionised surfaces of which could be responsible for producing any possible hyperstrong scattering. However, extragalactic sources are seen randomly and unless both Galactic and extragalactic sources are seen through the same cloud, comparison of scattering diameters of Galactic vs. extragalactic sources to find the distance of the screen from GC may not be correct.

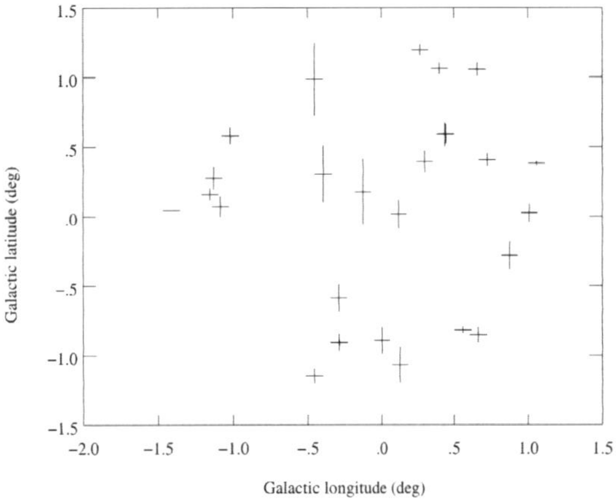

Figure 1. Plot of source size as a function of Galactic longitude and latitude. X-axis errorbars indicate the average half-width of the synthesised beam, and Y-axis errorbars indicate the deconvolved average size of the sources. Source sizes and synthesised beam size have been multiplied by a factor of 30 in this plot

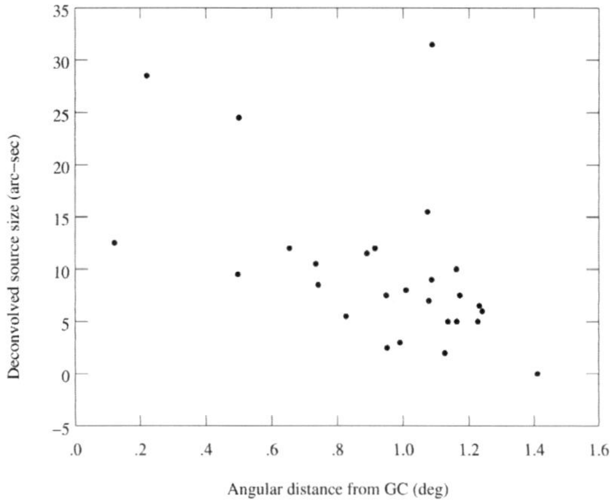

Figure 2. Plot of source size as a function of distance from GC

3.2. 7′ halo and Sgr A East at low radio frequencies

A false colour image of GC at 154 MHz is shown in Fig. 4, which is the highest sensitivity, high resolution image of the region below 330 MHz, and shows the presence of several known sources Sgr A East, Radio arc, Sgr C filament and SNR G0.33+0.0. To model the emission properties of sources in the Sgr A complex, we measure the total flux densities of 7′ halo, Sgr A East, Sgr A West at frequencies of 1.4 GHz, 620, 320, 255 and 154 MHz following the procedure given in [5]. Fig. 5 shows the plot of flux density as a function of frequency for Sgr A East, and Fig. 6 shows the same for 7′ halo. The low frequency turnover of both these objects occur at similar frequencies and they appear absorbed to almost the same extent at 154 MHz suggesting absorption extrinsic to these sources, possibly by a line of sight diffuse ionised gas. To quantitatively verify this, we assume free-free absorption of synchrotron emission emitted from Sgr A East and 7′ halo and plot the fit of the following equation (solid curve) to the data in Figs. 5 & 6 respectively.

$$S = S_0 \times \nu^{\alpha} \times \exp(-\tau) \tag{2}$$

$$\tau = 0.2 \times n_e^2 \times \nu^{-2.1} T_e^{-1.35} \tag{3}$$

where S is the measured flux density at a frequency of ν GHz, τ is the free-free optical depth at that frequency, S_0 is the flux density of the background source at 1 GHz, and α is the synchrotron spectral index. From the fit, we estimate for Sgr A East α=−0.5±0.2, and the absorbing screen has $n_e^2 \times T_e^{-1.35}$=0.23±0.025. For 7′ halo we estimate α=−0.27±0.2, and the absorbing screen has $n_e^2 \times T_e^{-1.35}$=0.18±0.05. It shows the free-free absorption caused by the screen is the same within the measurement errors towards both Sgr A East and 7′ halo at any frequency. Therefore, line of sight absorption by a single foreground screen can adequately explain the observed emission properties of both the sources. 7′ halo can simply be modelled as a synchrotron source, which is contrary to what was proposed by [6]. We note the synchrotron spectral index of both the sources between 1.4 GHz and 154 MHz is significantly flatter than what is estimated from higher frequencies [6], and the change in their spectral index is ≤0.5. This indicate that there is a break in the synchrotron spectrum at frequencies around 1.4 GHz for these objects. If the magnetic field in Sgr A East is taken to be 3.5 mG [8], then following [9], estimated age of Sgr A East is about 10^4 years. If the magnetic field in 7′ halo is assumed to be the same as in Sgr A East, its age will also be the same.

4. Conclusions

We have observed the GC region at 255 and 154 MHz with the GMRT. From the 255 MHz map we detect 28 compact sources, and an anti-correlation of their deconvolved sizes with angular distance from the GC is observed. This shows scatter broadening due to GC ionised medium. If this medium is homogeneous, it is located at least 4 kpc away from the GC. However, it appears more likely that the ionised medium is clumpy and a good fraction of extragalactic sources are not seen through the clumpy screen, where the maser sources are preferentially located. Our multifrequency observations at 255 and 154 MHz in tandem with previous observations of the GC shows that the 7′ halo in Sgr A complex is a non-thermal source rather than a mixture of thermal and non-thermal electrons as was proposed before.

Acknowledgments

The authors would like to thank the staff of the GMRT who made these observations possible. GMRT is run by the National Centre for Radio Astrophysics of the Tata Institute of Fundamental Research.

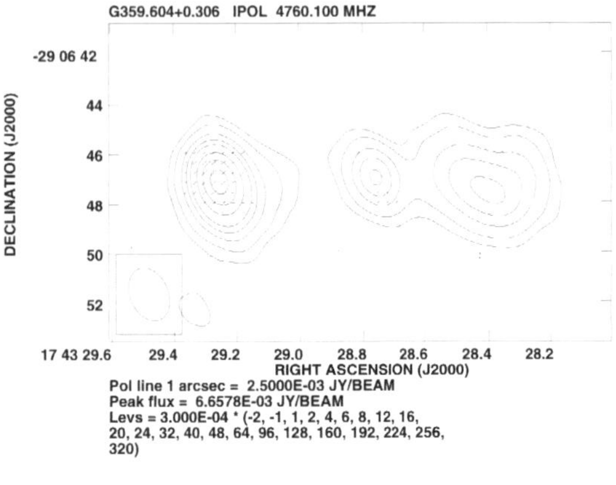

Figure 3. A 5 GHz VLA map of the FRII galaxy G359.6+0.3

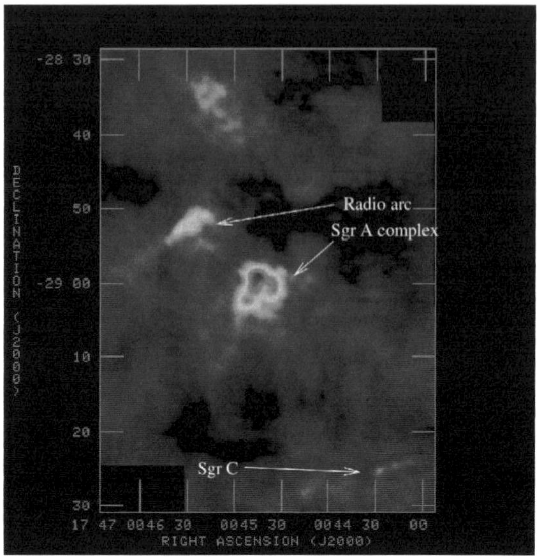

Figure 4. Central 1 degree region of the Galaxy at 154 MHz. Resolution ~45″, rms noise is about 25 mJy/Beam.

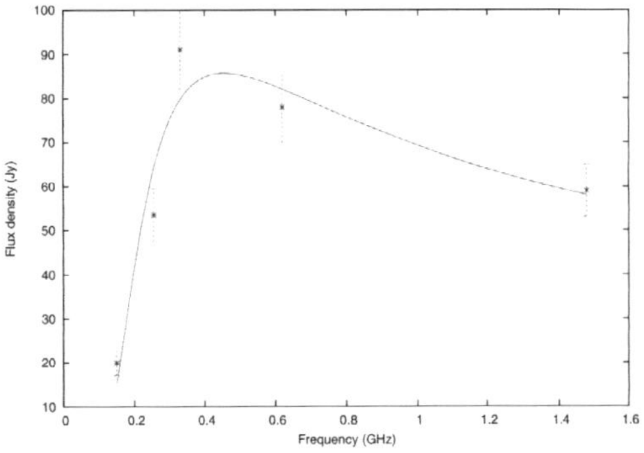

Figure 5. Plot of flux density of Sgr A East between 1.5 and 0.15 GHz

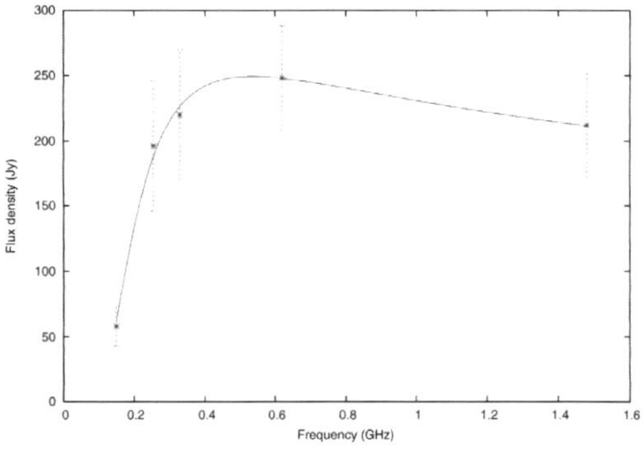

Figure 6. Plot of flux density of 7′ halo between 1.5 and 0.15 GHz

References

[1] van Langevelde, H. J., Frail, D. A., Cordes, J. M., & Diamond, P. J. 1992 *ApJ* **396** 686
[2] Lazio, T. J. W. & Cordes, J. M. 1998 *ApJS* **118** 201
[3] Lazio, T. J. W., Anantharamaiah, K. R., Goss, W. M., Kassim, N. E., & Cordes, J. M. 1999 *ApJ* **515** 196
[4] Bower, G. C., Backer, D. C., & Sramek, R. A. 2001 *ApJ* **558** 127
[5] Roy, S. & Rao, A.P. 2004 *MNRAS* **349** L25
[6] Pedlar, A., Anantharamaiah, K. R., Ekers, R. D., et al. 1989 *ApJ* **342** 769
[7] Roy S., Rao A. P. & Subrahmanyan R. 2005 *MNRAS* **360** 1305
[8] Yusef-Zadeh, F. 1999 *ApJ* **512** 230
[9] Kardashev, N. S. 1962 *Soviet Astronomy* **6** 317
[10] Zoonematkermani, S., Helfand, D. J., Becker, R. H., White, R. L., & Perley, R. A. 1990 *ApJS* **74** 181
[11] Helfand, D. J., Zoonematkermani, S., Becker, R. H., & White, R. L. 1992 *ApJS* **80** 211

Institute of Physics Publishing
Journal of Physics: Conference Series **54** (2006) 161–165
doi:10.1088/1742-6596/54/1/027
Galaxy Center Workshop 2006

Spitzer IRS Observations of the Galactic Center

Janet P Simpson[1,2], Sean W J Colgan[1], Angela S Cotera[2], Edwin F Erickson[1], David J Hollenbach[3], Michael J Kaufman[4] and Robert H Rubin[1,5]

[1] NASA Ames Research Center, Mail Stop 245-6, Moffett Field, CA 94035-1000, USA
[2] SETI Institute, 515 N Whisman Road, Mountain View, CA 94043, USA
[3] NASA Ames Research Center, Mail Stop 245-3, Moffett Field, CA 94035-1000, USA
[4] Dept of Physics, San Jose State University, One Washington Square, San Jose, CA 95192, USA
[5] Orion Enterprises

E-mail: simpson@cygnus.arc.nasa.gov

Abstract. We present Spitzer IRS R$\sim$ 600 spectra (10 – 38 μm) of 38 positions in the Galactic Center (GC), all at the same Galactic longitude of and including the Arches Cluster. Our positions include the Arched Filaments, regions near the Quintuplet Cluster, the "Bubble" lying along the same line-of-sight as the molecular cloud M0.13-0.13, and the diffuse interstellar gas along the line of sight at higher Galactic latitudes. From measurements of the [O IV], [Ne II], [Ne III], [Si II], [S III], [Fe II], [Fe III], and H_2 S(0), S(1), and S(2) lines we determine the gas excitation and ionic abundance ratios. The main source of excitation is photoionization, with the Arches Cluster ionizing the Arched Filaments and the Quintuplet Cluster ionizing the gas nearby and at lower Galactic latitudes including the far side of the Bubble. In addition, strong shocks ionize gas to O^{+3} and destroy dust grains, releasing iron into the interstellar medium (ISM); the shock effects are particularly noticeable in the center of the Bubble but the highly ionized gas is present in all positions. The H_2 lines are formed both in photodissociation regions in the GC and along the line of sight to the GC.

1. Introduction

The Milky Way's Galactic Center (GC) includes a modest-sized black hole, three clusters of young massive stars, and massive molecular clouds. Figure 1 shows a radio image of the region from [17] with the pertinent locations marked. The very center of the Galaxy, Sgr A West, contains a cluster of massive stars and the black hole, which is coincident with radio source Sgr A*. The other two clusters of massive stars, the Arches Cluster and the Quintuplet Cluster, are located about 25 pc in the plane of the sky to higher Galactic longitudes. The Arched Filaments are the lit-up edges of dense molecular clouds [8] and are probably excited by the hot stars in the Arches Cluster [1, 2]. The Sickle region is almost certainly excited by the hot stars of the Quintuplet Cluster [15, 13]. The Radio Arc [18] consists of non-thermally emitting linear filaments perpendicular to the Galactic plane; its relation to the thermal filaments and their exciting star clusters has been and remains controversial. The "Bubble" [9, 13] is a circular open structure south of the Quintuplet Cluster whose rim is faint at radio frequencies (Fig. 1) but prominent in the 21 μm continuum of warm dust [12]. The dense molecular cloud M0.13-0.13 is at the same position on the plane of the sky with the Bubble being foreground to the cloud.

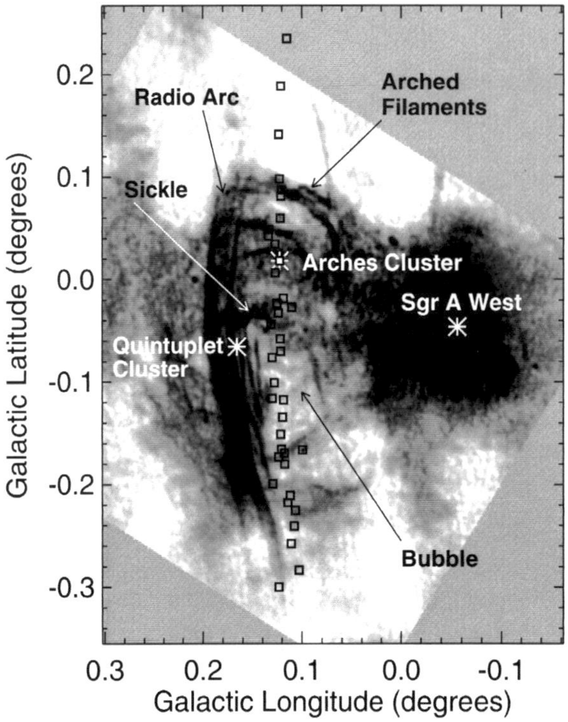

Figure 1. Thermal and non-thermal continuum imaged at 21 cm with 11″ resolution by [17]. The observed positions are indicated by square boxes and other locations of interest are labeled.

In this paper we present Spitzer Space Telescope observations of mid-infrared (MIR) spectra of 38 positions in the GC. Our goals are to clarify the source(s) of excitation of the Arched Filaments and to determine the relationship of the Bubble with the clusters and other filaments.

2. Observations

We observed the Galactic Center region with Spitzer's Infrared Spectrograph [6] with the Short-High and Long-High modules (10 – 38 μm) on 21–23 March 2005. The observed line of positions is shown in Figure 1. The emission from the 2–3 positions at each end of the line probably originate in the diffuse interstellar medium (ISM) all along the line-of-sight to the GC as well as in the GC itself. The data were reduced and calibrated with the S13 pipeline. The basic calibrated data (bcd images) were median-combined and cleaned of rogue pixels and noisy order edges. The spectra were extracted using the Spectroscopic Modeling, Analysis and Reduction Tool [5]. Line and continuum fluxes of the brighter lines were measured by fitting Gaussians plus a sloped continuum to each of the line profiles. Fluxes for the faintest lines were estimated by integrating over the line profile. The extinction correction was estimated by fitting a template of the MIR spectrum of the Orion Nebula [16] to the 10–15 μm spectral region (Simpson et al., in preparation).

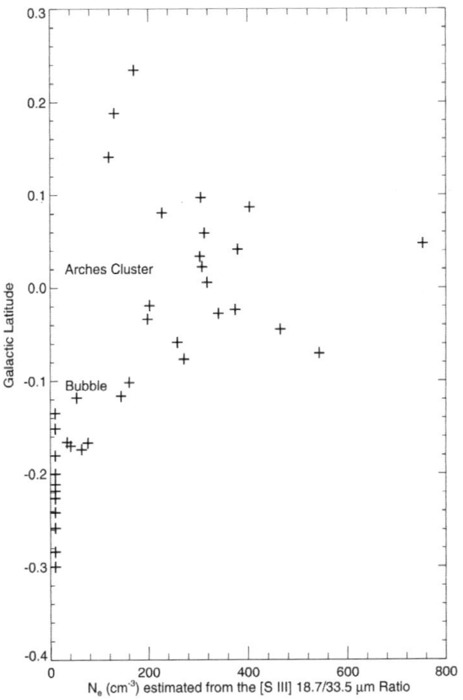

Figure 2. Electron density estimated from the [S III] 18.7/33.5 μm line ratio plotted as a function of Galactic latitude.

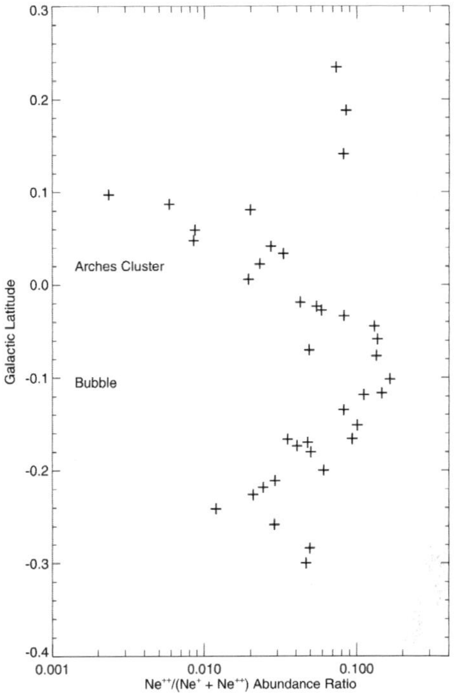

Figure 3. Excitation as measured by the $Ne^{++}/(Ne^{+} + Ne^{++})$ ratio plotted as a function of Galactic latitude.

3. Results

The fluxes of the ionized lines correlate well with the Arches and Bubble structures seen in the radio (e.g., Fig. 1) and MIR continua, from which we conclude that these lines do indeed originate in the GC. However, the H_2 line fluxes peak near the Galactic plane (latitude $\sim -0.05°$), similar to the distribution seen in warm CO J=4–3 [11]. The radial velocities of the H_2 lines also show a distinctly different pattern from those of the ionized lines. The ratios of the three H_2 lines, S(0), S(1), and S(2), show little scatter and indicate that they originate in photo-dissociation regions (PDRs) with densities $\sim 10^4 - 10^5$ and $G_0 < 1000$ cm^{-3}, about what we would expect for a GC dense molecular cloud lying at some distance from the exciting stars [7]. There are probably additional contributions to the H_2 lines from the PDRs along the line of sight to the GC.

Electron densities (N_e) were estimated from the ratios of the density-sensitive [S III] 18.7 and 33.5 μm lines. These densities are plotted in Figure 2. This ratio is also sensitive to extinction and at some positions the 18.7/33.5 μm line ratio is below that produced by even the lowest possible N_e; for these positions the extinction was increased until the [S III] 18.7/33.5 μm line ratio corresponded to a minimum $N_e \sim 10$ cm^{-3}. Since both [S III] lines are very strong, the chief uncertainty is due to uncertainties in the extinction correction. Using these densities and a typical electron temperature of 6000 K for the GC [8], we computed the excitation and abundance ratios plotted in Figures 3–5.

The H II region excitation can be estimated from the ratio of a doubly ionized line with respect to a singly ionized line of the same element if the ionization potentials of both lines are greater than that of hydrogen (13.6 eV). For Spitzer, the best line pair is the [Ne III] 15.6 μm and [Ne II] 12.8 μm lines, since both lines have similar extinction. Figure 3 shows that the highest excitation occurs in the interior of the Bubble region, which may be influenced by the hot stars of the Quintuplet Cluster, and in the diffuse ISM at the ends of the line of positions. Even so, this gas is all "low excitation", since $Ne^{++}/Ne^{+} << 1$. The gas ionized by the Arches Cluster is of lower excitation yet, in spite of the fact that the Arches Cluster is supposedly younger than the Quintuplet Cluster [4]. This difference in excitation was already known from our work on the GC H II regions with the Kuiper Airborne Observatory [3, 1, 15] and ISO [2]. We conclude that the Arched Filaments are ionized by the Arches Cluster since the excitation decreases rapidly as one looks at positions at higher Galactic latitudes from the Cluster. This pattern also holds for the $S^{+3}/(S^{++} + S^{+3})$ ratio and the ionization-parameter-sensitive ratio Si^{+}/S^{++} (Simpson et al. in preparation). Because the gas at the position of the Arches Cluster itself has lower excitation than the two positions at immediately higher Galactic latitudes, we believe that the Arches Cluster lies some distance away from the gas emitting the observed lines and is not directly in contact.

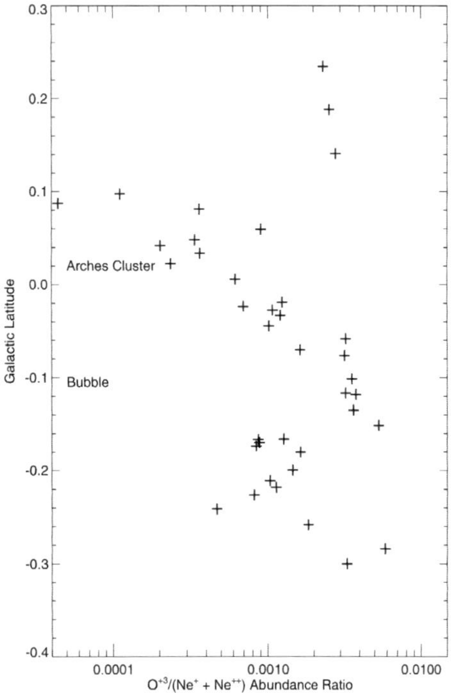

Figure 4. Abundance of shock-ionized gas as measured by the $O^{+3}/(Ne^{+} + Ne^{++})$ ratio plotted as a function of Galactic latitude.

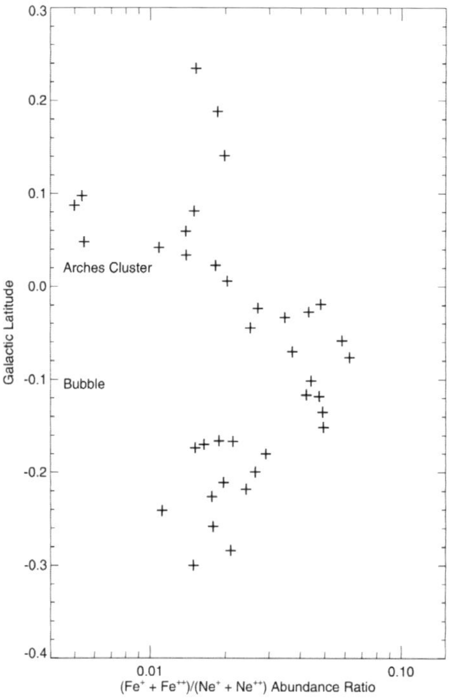

Figure 5. Relative importance of grain-destroying shocks as measured by the $(Fe^{+} + Fe^{++})/(Ne^{+} + Ne^{++})$ ratio plotted as a function of Galactic latitude.

Figure 4 shows the $O^{+3}/(Ne^{+} + Ne^{++})$ ratio (O/Ne $\sim$ 4 in Galactic H II regions and is expected to be approximately constant). We see that O^{+3} is most abundant in the center of the Bubble and in the diffuse ISM at the ends of the line of positions. Triply ionized oxygen is not produced in H II regions, except in circumstances of extremely low metallicity (all extragalactic) where there are exceptionally hot stars that emit photons with energies $>$ 54 eV. However, [O IV] is seen in some galactic nuclei (normally not low metallicity) and is thought to be produced by high velocity shocks (e.g., [10]). In some low metallicity galactic nuclei, the

[O IV] could be excited by hot Wolf-Rayet stars [14]; however, there is no indication that the O^{+3} observed here is produced by any of the WR stars in the Arches Cluster. Most likely, the pervasive O^{+3} seen in the GC is produced by shocks.

For the most part, iron is strongly depleted in the ISM. Figure 5 shows the $(Fe^{+} + Fe^{++})/(Ne^{+} + Ne^{++})$ ratio (Solar Fe/Ne $\sim 0.15 - 0.5$). We see that gas-phase iron is much more abundant in the Bubble than in the Arched Filaments. A likely explanation is that grains in the Bubble region and diffuse ISM have been destroyed by strong shocks, thereby liberating the iron.

4. Conclusions

(i) The highest excitation, as seen by the excitation indicators Ne^{++}/Ne^{+}, Si^{+}/S^{++}, and $O^{+3}/(Ne^{+} + Ne^{++})$, is found in the center of the Bubble and not at the locations of the Arches and Quintuplet Clusters in the plane of the sky. At the present time there is probably a large line-of-sight offset between the clusters and the ionized gas. This is not unreasonable since the clusters, with ages of 2 and 4 My, probably did not form in the same molecular clouds that they are now ionizing.

(ii) The low Galactic latitude rim of the Bubble is ionized by a source at higher Galactic latitude, probably the Quintuplet Cluster. It is not the outer edge of the M0.13-0.13 molecular cloud, which lies along the same line-of-sight as the Bubble.

(iii) The Arched Filaments are ionized by the Arches Cluster.

(iv) There are multiple indicators of strong shocks, peaking in the center of the Bubble. This possibly indicates the location of a violent outflow a few thousand to 10^4 years ago.

Acknowledgments

This work is based on observations made with the Spitzer Space Telescope, which is operated by the Jet Propulsion Laboratory, California Institute of Technology under NASA contract 1407. Support for this work was provided by NASA.

References

[1] Colgan S W J, Erickson E F, Simpson J P, Haas M R and Morris M 1996 *Astrophys. J.* **470** 882
[2] Cotera A S, Colgan S W J., Rubin R H and Simpson J P 2005 *Astrophys. J.* **622** 333
[3] Erickson E F, Colgan S W J, Simpson J P, Rubin R H, Morris M and Haas, M R 1991 *Astrophys. J.* **370** L69
[4] Figer D F, Kim S S, Morris M, Serabyn E, Rich R M and McLean I S 1999 *Astrophys. J.* **525** 750
[5] Higdon S J U *et al* 2004 *Pub. Astr. Soc. Pacific* **116** 975
[6] Houck J R *et al* 2004 *Astrophys. J. Suppl.* **154** 18
[7] Kaufman M J, Wolfire M G and Hollenbach D J 2006 *Astrophys. J.* **644** 283
[8] Lang C C, Goss W M and Morris M 2001 *Astron. J.* **121** 2681
[9] Levine D, Morris M and Figer D 1999 in *The Universe as Seen by ISO*, ed P Cox and M Kessler (ESA SP-427) p 699
[10] Lutz D, Kunze D, Spoon, H W W and Thornley M D 1998 *Astron. Astrophys.* **333** L75
[11] Martin C L, Walsh W M, Xiao K, Lane A P, Walker C K and Stark A A 2004 *Astrophys. J.* **150** 239
[12] Price S D, Egan M P, Carey S J, Mizuno D R and Kuchar T A 2001 *Astron. J.* **121** 2819
[13] Rodríguez-Fernández N J, Martín-Pintado J and de Vicente P 2001 *Astron. Astrophys.* **377** 631
[14] Schaerer D and Stasińska G 1999 *Astron. Astrophys.* **345** L17
[15] Simpson J P, Colgan S W J, Cotera A S, Erickson E F, Haas M R, Morris M and Rubin R H 1997 *Astrophys. J.* **487** 689
[16] Simpson J P, Witteborn F C, Price S D and Cohen M 1998 *Astrophys. J.* **508** 268
[17] Yusef-Zadeh F and Morris M 1987 *Astrophys. J.* **320** 545
[18] Yusef-Zadeh F, Morris M and Chance D 1984 *Nature* **310** 557

Institute of Physics Publishing
doi:10.1088/1742-6596/54/1/028
Journal of Physics: Conference Series **54** (2006) 166–170
Galaxy Center Workshop 2006

Circumstellar CO in OH/IR stars close to the Galactic Centre

A Winnberg[1], S Deguchi[2], H J Habing[3], J Nakashima[4], H Olofsson[5] and M J Reid[6]

[1] Onsala Space Observatory, SE–439 92 Onsala, Sweden
[2] Nobeyama Radio Observatory, Minamisaku, Nagano 384–1305, Japan
[3] Leiden Observatory, P.O. Box 9513, NL–2300 RA Leiden, The Netherlands
[4] Academia Sinica Institute of Astronomy and Astrophysics, P.O. Box 23-141, Taipei 106, Taiwan
[5] Stockholm Observatory, AlbaNova University Center, SE–106 91 Stockholm, Sweden
[6] Harvard-Smithsonian Center for Astrophysics, 60 Garden Street, Cambridge MA 02138, USA

E-mail: anders@oso.chalmers.se

Abstract. A pilot project is carried out to measure circumstellar CO emission from three OH/IR stars close to the GC using the Nobeyama Millimeter Array at 115 GHz and the Sub-Millimeter Array at 230 GHz. An interferometer is necessary as a 'spatial filter' in this region of space because of the confusion with interstellar CO emission. The intention is to find out whether it is possible to later conduct a large-scale survey for mass-loss rates using, for example, ALMA. Thus an important parameter would be added to our understanding of the evolution of the Galactic Bulge. Sources have been detected towards two of the stars with 'correct' positions and radial velocities. However, for one of the stars the line profile is not what one expects for expanding circumstellar envelopes. This surprising result is discussed and our plans for future observations are presented.

1. Introduction

The most rapid stellar mass loss occurs at the so called red-giant (RGB) and asymptotic-giant branches (AGB). Due to the high stellar density and rapid star formation, the Galactic Bulge contains large numbers of RGB and AGB stars. It is important to measure the mass loss rates of these stars in order to get a picture of the recirculation of matter and metal enrichment in this region of the Milky Way.

The most accurate method of estimating the stellar mass-loss rates is based on the CO rotational spectral lines. The first attempt to detect such circumstellar lines in the vicinity of the Galactic Centre (GC) was made by Mauersberger et al. [1] using the 30-m radio telescope in Spain. They detected the $J = 1 - 0$ and $2 - 1$ CO lines from the proto-planetary nebula (PPN) OH0.9+1.3 which has a high radial velocity (~ 100 km s^{-1}) making it likely to be physically close to the GC. They proposed also to look for CO emission from OH/IR stars, since such stars are believed to be the progenitors of PPNe. Winnberg et al. [2] used the same telescope to observe several OH/IR stars close to the GC but detected CO emission only from one star: OH0.3–0.2. This star has a very high radial velocity (–341 km s^{-1}). Observations of all the other candidate stars failed because of confusion with the interstellar background emission.

The present project employs a different observing technique in an attempt to detect the *circum*stellar CO emission in the midst of *inter*stellar CO emission. A radio interferometer with suitable baselines could resolve out the interstellar background but leave the circumstellar emission as unresolved point sources. This use of an interferometer could be called a 'spatial filter'.

In order to find out the optimal baseline lengths and the most favorable CO lines we started a pilot experiment. We chose three OH/IR stars close to the position of the GC with strong IR fluxes and with low-to-moderate radial velocities. In 2003 – 2004 we used the Nobeyama Millimeter Array (NMA) at 115 GHz (CO, $J = 1-0$) and in 2005 we used the Sub-Millimeter Array (SMA; Mauna Kea, Hawaii) at 230 GHz ($J = 2-1$).

This paper presents the main results of these two data sets and outlines our plans for future observations.

Ultimately we hope that our project will lead to larger systematic surveys of mass loss rates for late-type stars in the Galactic Bulge using sub-millimetre arrays on the southern hemisphere such as the Atacama Large Millimeter Array (ALMA).

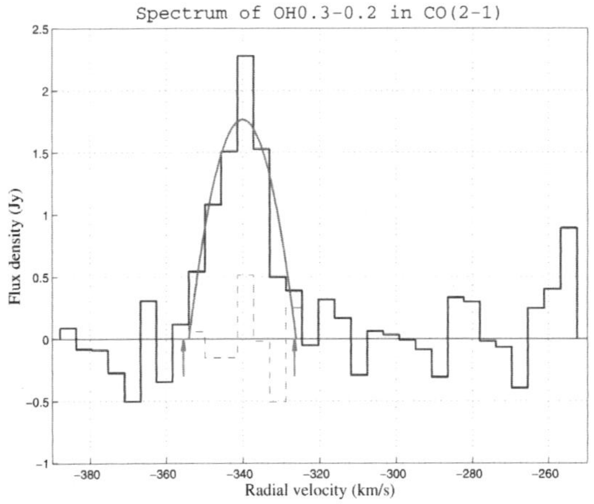

Figure 1. Blue solid line: observed spectrum; red solid line: fitted parabola; blue dashed line: observed spectrum minus parabola; arrows: velocities of OH components

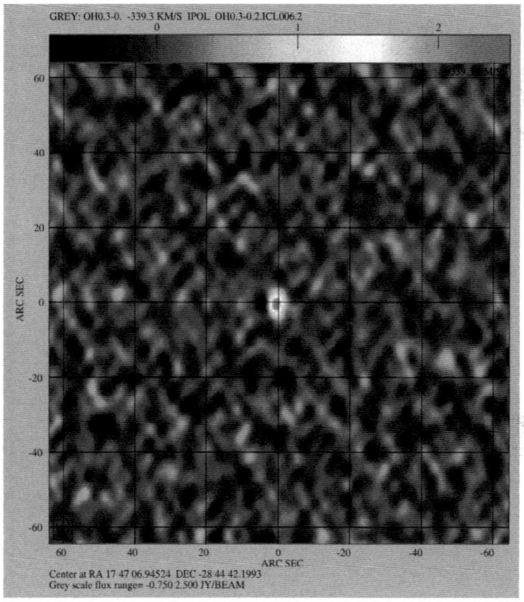

Figure 2. Map of OH0.3–0.2 at the 230-GHz CO line maximum. The zero point is at the position of the OH source [3] that has an uncertainty of about 1″.

2. Observations

The first observations in this project were done with the 6-element array (NMA) at Nobeyama Radio Observatory, Japan, in November 2003 and January 2004 at 115 GHz (CO 1 – 0). Three OH/IR stars were selected for observations on the basis of strong IR fluxes: OH359.117–0.169, OH359.762+0.120 and OH359.971–0.119. A clear signal was obtained from OH359.971–0.119. The position and radial velocity of the CO line agreed with the corresponding data of the 1612-MHz OH line. However, the line width was too narrow for an unresolved circumstellar CO source.

In June 2005 the same three stars plus the star OH0.3–0.2 were observed using the 8-element array (SMA) on Mauna Kea, Hawaii, at 230 GHz (CO 2 – 1). The spectrometer was configured to give a resolution of 3.25 MHz corresponding to 4.2 $\mathrm{km\,s^{-1}}$. The test star OH0.3–0.2 was detected and showing properties in accordance with the single-dish data (Figures 1 and 2).

Typical resolutions were $4'' \times 7''$ with the NMA and $3'' \times 4''$ with the SMA.

3. Data reduction

The *uv* data were investigated for the presence of spatially extended emission by plotting the visibility flux density as a function of projected baseline length. Based on such plots it was decided to exclude baselines shorter than 32.5 m for OH359.762+0.120 and shorter than 39 m for OH359.971–0.119 in order to avoid, as far as reasonable, contamination by residuals of interstellar emission. For OH359.117–0.169 no evidence for significant interstellar emission was found. Unfortunately, no circumstellar emission was found either for this star. No interstellar emission, of course, was found in the IF band of the test source OH0.3–0.2 due to its high radial velocity (–341 $\mathrm{km\,s^{-1}}$).

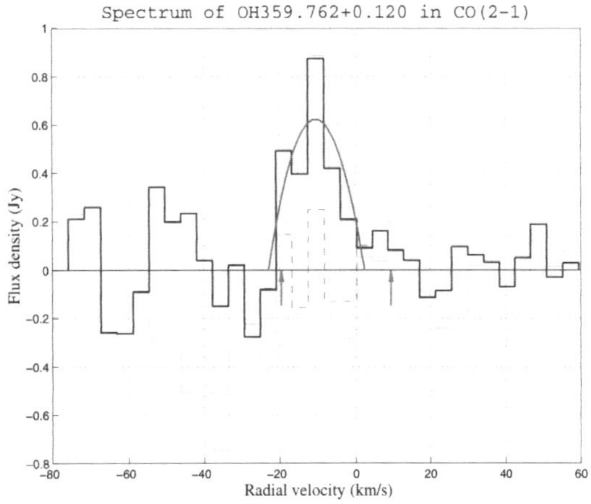

Figure 3. Blue solid line, blue dashed line, red solid line and arrows: as in Figure 1. Green solid line: observed spectrum with all projected baseline lengths included.

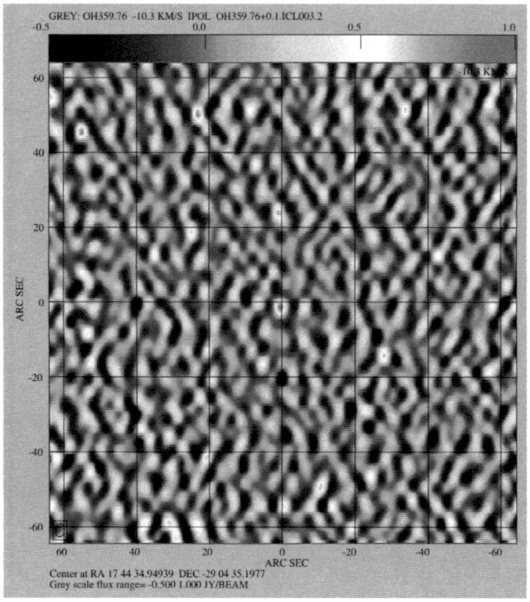

Figure 4. Map of OH359.762+0.120 at the 230-GHz CO line maximum. The zero point is at the position of the OH source [4] that has an uncertainty of about $1''$.

Maps were made with 256×256 pixels of size $0.5''$ with natural weighting. They were cleaned using the standard Högbom/Clark algorithm taking account of the noise fluctuations and the dynamic range of the beam pattern. If a compact source was seen in one of the channels within the OH velocity span at a position close to the phase centre (the position of the OH source) a two-dimensional elliptical Gaussian fit was made of this source and a spectrum was made at the pixel closest to this position.

4. Results

In the SMA maps of OH359.762+0.120 and OH359.971–0.119 there are sources close to the phase centres (Figures 4 and 6). The deviations are well within the limits expected on the grounds of OH position errors, CO phase calibration errors and signal-to-noise ratios, i.e. within $1'' - 2''$. In addition the CO lines have frequencies corresponding to radial velocities (LSR) inside the OH velocity ranges. Finally, the detection of OH0.3–0.2 serves as a guarantee of the fact that the instrument worked correctly.

OH359.762+0.120 was detected with a rather poor S/N. Therefore there are quite large errors associated with the elements of the fitted parabola (Figure 3). Within these errors the mean radial velocity of the line and the line width are compatible with the systemic velocity of the star and its envelope expansion velocity as measured from the OH line profile. The impression of skewness of the line is probably not significant. Notice that the exclusion of short baselines did not improve the detection of the line significantly. It merely improved the spectral baseline.

OH359.971–0.119 was detected with moderate to good S/N. However, the CO line is narrow and close to the 'red-shifted' OH line component, i.e. the backside of the envelope (Figure 5). A similar line was observed with the NMA at 115 GHz. Notice that the line profile obtained when all the *uv* data are included (green line) shows a much stronger and broader 'red-shifted' line and even a 'blue-shifted' counterpart. The 'hole' in between these two line components is probably caused by an unresolved 'dip' in the interstellar background.

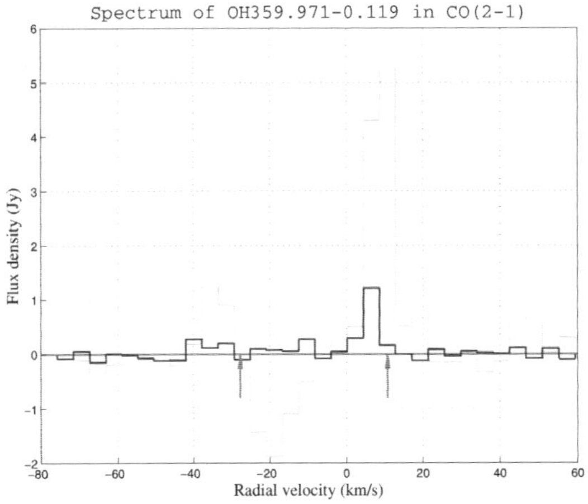

Figure 5. Blue solid line, green solid line and arrows: as in Figure 3.

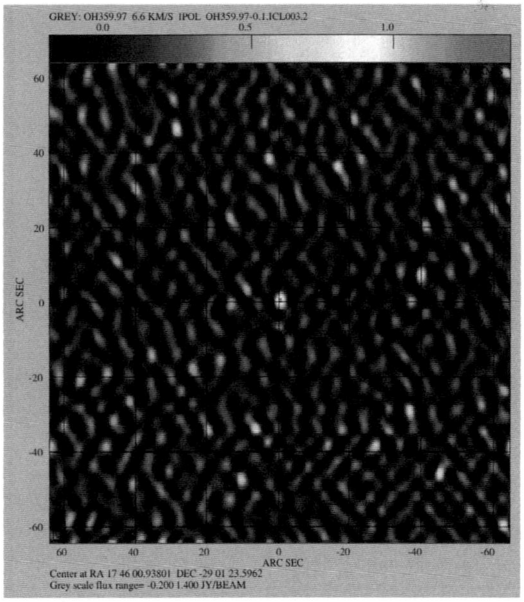

Figure 6. Map of OH359.971-0.119 at the 230-GHz CO line maximum, i.e. the strongest channel in the 'blue profile' in Figure 5. The zero point is at the position of the OH source [4] that has an uncertainty of about $1''$.

5. Discussion

As to OH359.762+0.120 there is little doubt that the detected CO source is a true circumstellar source. The S/N needs improvement, however.

The CO source associated with OH359.971–0.119 is an enigma so far. The line profile is similar to that expected from a resolved stellar envelope where unresolved sources are left at the front and back sides. For such a picture to be true, the distance needs to be so short that the star would not belong to the Galactic Bulge. For example, assuming that the diameter of the CO envelope is 2×10^{17} cm (which might be an overestimate for the $2-1$ transition) and that the angular diameter is $20''$, the distance would be about 700 pc only. The IR properties of this star also are such that it is arguable whether it belongs to the Bulge (cf. [5] and [6]).

Another explanation of the line profile would be that it is the result of strong absorption of interstellar CO in front of the source.

Finally, there is always the (improbable?) explanation that the source is an unresolved remnant of interstellar CO emission that happens to be at the same position and radial velocity as OH359.971–0.119.

6. Plans

We have applied for further observations with the SMA. First of all we want to re-observe all the four stars, including OH0.3–0.2, at 345 GHz ($J = 3 - 2$) with the same compact configuration. Secondly we want to observe the stars OH359.762+0.120 and OH359.971–0.119 plus two new stars (OH359.938–0.077 and OH0.319–0.040) that have somewhat higher radial velocities using the SMA in its extended configuration at 230 GHz.

If the sources are interstellar, or circumstellar and extended, we would expect a weak or even undetected 345-GHz line. If they are circumstellar and absorbed (compact or extended) we would expect a less absorbed 345-GHz line. These prospects are based on the facts that the interstellar CO line at 345 GHz is weaker than those at 230 and 115 GHz whereas its circumstellar counterpart is somewhat stronger or at least as strong as the one at 230 GHz. In addition, of course, the SMA resolution is higher at 345 GHz.

The re-observation of OH359.762+0.120 and OH359.971–0.119 at 230 GHz in the extended configuration is motivated by the fact that less data will be wasted by rejection of short baselines. We also want to explore the dependence of radial velocity by observing the stars OH359.938–0.077 and OH0.319–0.040 that have somewhat higher radial velocities (–83.3 and +74.8 $\mathrm{km\,s^{-1}}$, respectively). We expect the interstellar emission to be less of a concern in connection with these stars.

References

[1] Mauersberger, R., Henkel, C., Wilson, T.L., Olano, C.A., 1988, A&A, 206, L34

[2] Winnberg, A., Lindqvist, M., Olofsson, H., Henkel, C., 1991, A&A, 245, 195

[3] Fix, J.D., Mutel, R.L., 1984, Astron. J., 89, 406

[4] Lindqvist, M., Winnberg, A., Habing, H.J., Matthews, H.E., 1992, A&AS, 92, 43

[5] Blommaert, J.A.D.L., van der Veen, W.E.C.J., van Langevelde, H.J., Habing, H.J., Sjouwerman, L.O., 1998, A&A, 329, 991

[6] Ortiz, R., Blommaert, J.A.D.L., Copet, E., Ganesh, S., Habing, H.J., Messineo, M., Omont, A., Schultheis, M., Schuller, F., 2002, A&A, 388, 279

Institute of Physics Publishing Journal of Physics: Conference Series **54** (2006) 171–175
doi:10.1088/1742-6596/54/1/029 Galaxy Center Workshop 2006

Comparison of Spitzer/IRAC Galactic Center 3.6–8.0 μm survey results with X–ray emission in the central 40×40 parsecs

D. Y. Gezari[1], R. G. Arendt[2], R. Smith[3], S. Stolovy[4], F. Yusef–Zadeh[5], C. Law[5], H. A. Smith[6], H. Moseley[7], S. Ramirez[8], J. Karr[9], A. Cotera[10] and K. Sellgren[11]

[1] NASA/Goddard Space Flight Center, Code 667, Greenbelt, MD 20771
[2] Science Systems & Applications Inc. / GSFC
[3] NASA/Goddard Space Flight Center, Code 662
[4] Spitzer Science Center, Caltech
[5] Northwestern University
[6] Harvard-Smithsonian Center for Astrophysics
[7] NASA/Goddard Space Flight Center, Code 665
[8] Infrared Processing and Analysis Center/Caltech
[9] Academia Sinica
[10] SETI Institute
[11] Ohio State University

E-mail: Daniel.Y.Gezari@nasa.gov

Abstract. We have studied the correlation between 2357 Chandra X–ray point sources identified by Muno et al. (2003) in a 40×40 parsec field and ~20,000 infrared sources we observed in the corresponding subset of our 2×1.5 degree Spitzer/IRAC Galactic Center Survey at 3.6–8.0 μm (see Stolovy et al.; this conference), using various spatial and brightness thresholds. The correlation was determined for source separations of less than 0.5, 1.0 or 2.0 arcsec. No significant correlation was found between the X–ray and infrared point sources on these scales. Only one compact infrared source, IRS 13, can be identified with any of the dozen prominent X–ray emission features in the 3×3 parsec region centered on Sgr A*, and the diffuse X–ray and infrared emission around Sgr A* seems to be anti–correlated on a few–arcsecond scale. We compare our results with previous identifications of near infrared companions to Chandra X–ray sources.

1. Chandra–Spitzer/IRAC point source correlation study

We have calculated the correlation between 2357 hard and soft Chandra X–ray sources identified and catalogued by Muno et al. (2003) in a 40×40 parsec field in the Galactic Center (Figure 1) and the ~20,000 infrared point sources brighter than $L = 12$ that we extracted from the corresponding field in our Spitzer/IRAC Galactic Center Survey. Our survey imaged a 2.0×1.5 degree (265×200 parsec at 7.6 kpc) region of the Galactic Center at 3.6, 4.5, 5.8 and 8.0 μm with 1.2 arcsec pixels and nominal resolution of 2 arcsec for the entire survey (see Stolovy et al.; this conference).

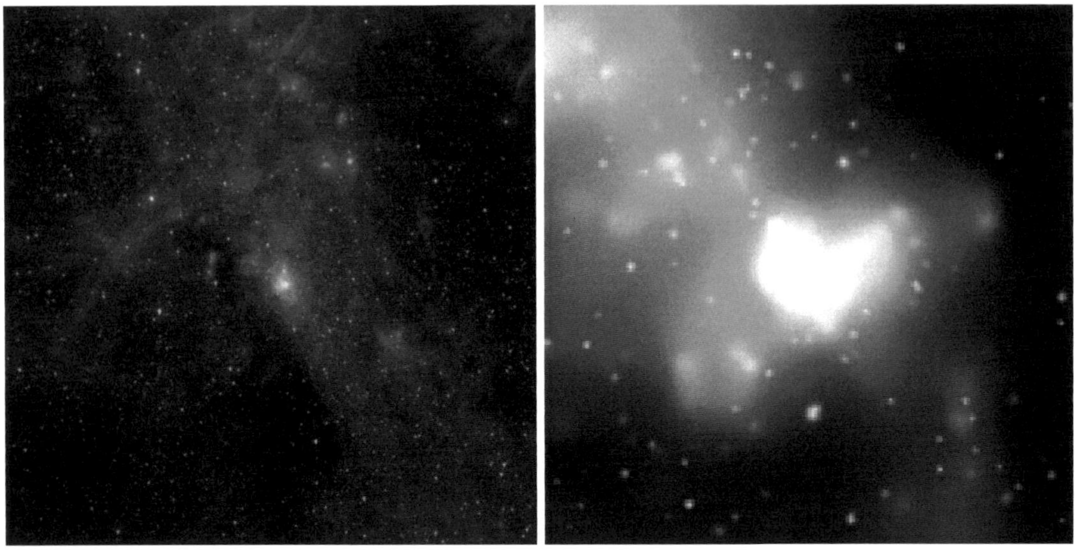

Figure 1. The 40×40 parsec infrared and X–ray field that was used for the X–ray/infrared point source correlation study. *Left:* Composite 3.6, 5.8 and 8.0 μm Spitzer/IRAC image of the 40×40 parsec study field. It yielded ~20,000 infrared point sources. *Right:* Composite Chandra image of the study field (Wang et al. 2003; *red* = 1–3 keV, *green* = 3–5 keV, *blue* = 5–8 keV).

We searched for infrared sources that fell within 0.5, 1.0 and 2.0 arcsec of the Muno et al. (2003) X–ray source positions. The X–ray and infrared sources in the study field show an extremely weak correlation. Fewer than 7% of the 2357 X–ray sources had infrared counterparts in our list of IRAC infrared sources brighter than 12 mag at 4.5 μm in a 1 arcsec sampling radius. However, nearly all of these are likely to be false identifications. We determined the likelihood of false correlations in the crowded infrared field by doing the same correlation analysis with the X–ray source position template offset 5.0 arcsec from the nominal location, in a grid of eight positions N–S and E–W off center. The number of coincidences found in excess of chance is typically 1% of the X–ray sources. For soft X–ray sources there is a modest excess in the number of infrared counterparts above that expected from random chance. There is a 3-4 σ excess in the number of soft X-ray sources with IR counterparts at 3.6 μ compared to a 2 σ excess in the correlations at 8.0 μ. Based on the soft X–rays observed from these sources (and therefore their relatively low column densities), we expect that the soft X–ray sources are foreground objects, and not at the distance of the Galactic Center. The excess soft X-ray/IR correlation is therefore likely due to these sources being nearby, rather than being intrinsically bright in both wavebands. The hard X–ray sources seem to have only as many infrared counterparts as would be expected from random associations.

2. Infrared and X–ray sources are uncorrelated in the central few parsecs

Although casual inspection of the Chandra and Spitzer/IRAC images gives the impression that the bright X–ray central cluster would have many infrared stellar counterparts, close examination shows that only one of the brighter, compact infrared sources is associated with any of the dozen distinct X–ray sources in the central 3×3 parsecs. With the exception of IRS 13 (which notably has been proposed as a candidate host for an intermediate mass black hole; see e.g., Maillard et al. 2004), none of the other infrared point sources coincide with any bright X–ray features. Otherwise the infrared sources are completely uncorrelated with Chandra X–ray sources in the central 3×3 parsec field of view on all but the largest ($>$ 1 parsec) spatial scale.

Figure 2. Near infrared stars (left, Blum et al. 1996), Chandra X–ray sources (center) and Spitzer/IRAC compact 3.6–8.0 μm infrared sources (right) in the central 3×3 parsecs.

None of the X–ray point sources in this field can be identified with any near infrared star ($K < 14$) imaged by Blum et al. (1996), with any of the point sources imaged in the Spitzer/IRAC 3.6–8.0 μm survey, or with the compact 12 μm IRS sources imaged with sub–arcsecond resolution by Gezari et al. (1996). The poor correlation even extends to the bright central knot of extended X–ray emission immediately surrounding Sgr A*. The bright X–ray plume that is elongated N–S adjacent to Sgr A* actually lies in a dark region a few arcseconds north of the bright infrared point sources IRS 7 and IRS 3. The peak of diffuse X–ray emission around SgrA* coincides with the "mini–cavity", a local minimum in infrared and radio–continuum emission.

The radio point source and black hole candidate Sgr A* coincides with one of the brightest Chandra X–ray point sources in the field. Sgr A* itself has no unambiguous infrared counterpart brighter than $K = 14$, although an extremely faint, flaring 2.2 μm point source has recently been detected at the nominal Sgr A* position using advanced imaging techniques (Genzel et al. 2003, Ghez et al. 2005, Eckart et al. 2006, Gillessen et al. 2006). The relationship between SgrA* and the central cluster of X–ray sources has been discussed by Baganoff et al. (2003).

3. Infrared/X–ray source colors

The color–magnitude diagram for the 9542 IRAC sources in the 40×40 parsec study region that were detected in both channels 2 and 3 is shown in Figure 3. The enlarged filled circles indicate 373 stars that are identified with some of the 2357 X–ray sources in the Muno et al. (2003) catalog. The distribution of colors of those infrared stars that have coincident X–ray sources is essentially the same as the distribution of all other infrared stars in our sample. There is also no apparent differentiation between the colors of infrared stars as a function of their distance from the X–ray source, or between the infrared candidates and the larger sample of 9542 infrared sources in the field. The "tail" on the plot at 6 mag is due to saturation of the 4.5 μm IRAC detector.

4. Discussion

Bandyopadhyay et al. (2005), reported that roughly 75% of a sample of 77 hard X–ray sources had faint ($K = 13 - 20$) candidate K–band counterparts within a 1.3 arcsec radius. They interpret these infrared candidates as highly reddened main sequence and supergiant stars at large distances rather than in the foreground. However, they noted that this is exactly the number of random associations that they predict from a Monte Carlo simulation. Thus although Bandyopadhyay et al. identified 58 K–band "candidates" none of these sources are likely to be

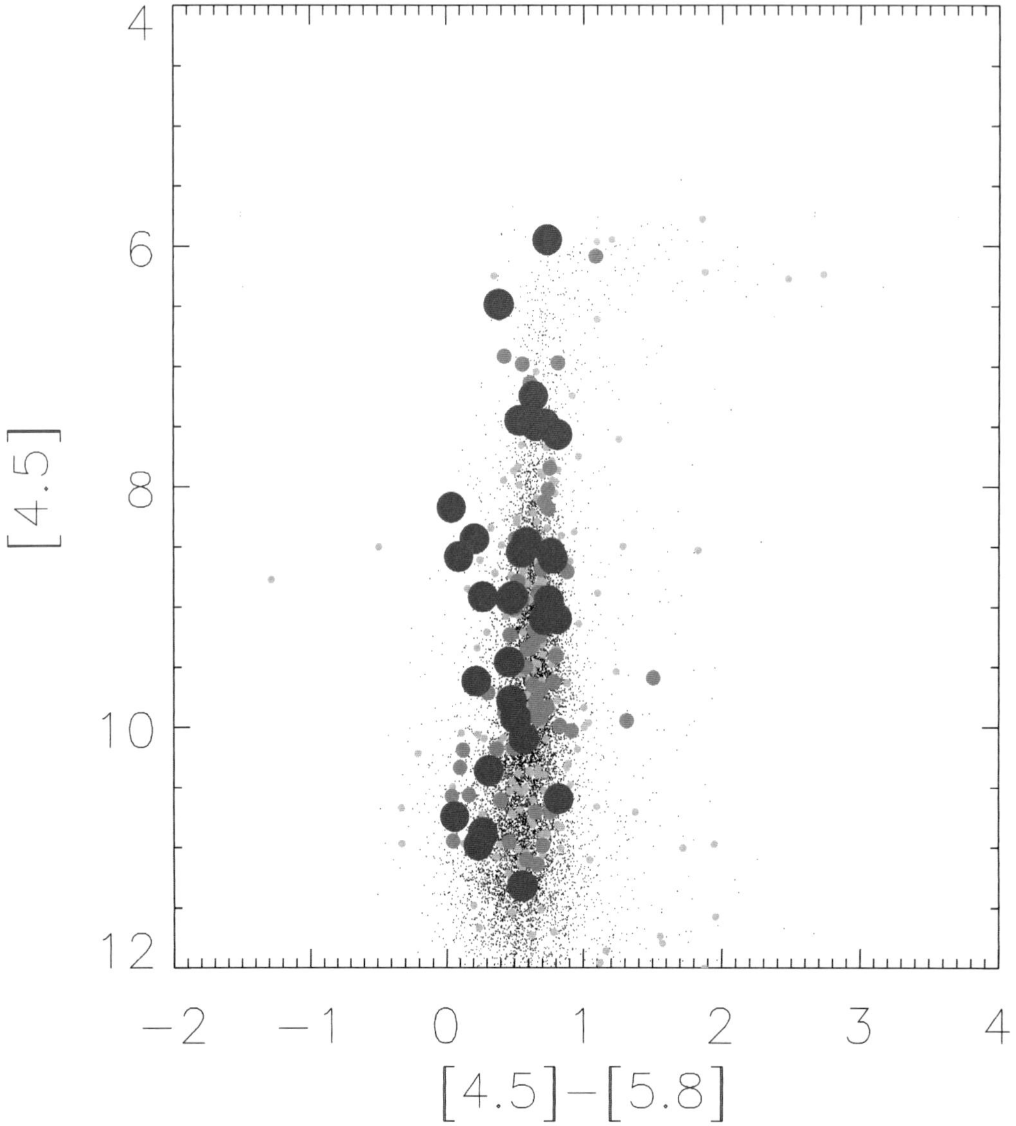

Figure 3. Color–magnitude diagram for 9542 IRAC sources (*black dots*) that were detected in both channels 2 and 3 (4.5 μm and 5.8 μm) in the 40×40 parsec study region. 31 X–ray sources had IRAC sources lying within 0.5 arcsec of the nominal X–ray positions (*large gray circles*), 99 had IRAC sources within 1.0 arcsec (*medium gray circles*), and 331 had IRAC sources within 2.0 arcsec (*small gray circles*).

real infrared counterparts to the hard X–ray sources in their sample. They conclude that most of these hard X–ray sources are likely to be accreting binaries located at the Galactic Center, although the relevance of their correlation analysis to this conclusion is not clear. However, Laycock et al. (2006) compared 1453 hard and 105 soft X–ray sources with near infrared stars from the 2MASS survey and concluded that high–mass X–ray binaries are not the dominant hard X–ray source population.

If the conclusions reached by Bandyopadhyay et al. and by Laycock et al. are both correct we can conclude that the dominant population of hard X–ray sources must be primarily low–

mass X–ray binaries and/or cataclysmic variables, Neither of these would be expected to be infrared–bright, which is consistent with our study result.

5. Summary

We have analyzed the possible correlations between the largest number of candidate sources to date: 2357 X–ray sources (of which 1809 are hard X–ray sources most likely located at the Galactic Center) and ~20,000 infrared point sources that we found with our Spitzer/IRAC survey in the 40×40 parsec comparison field. The lack of any significant correlation between hard X–ray sources and infrared point sources suggests that there is no unique population of sources that are both X–ray and infrared bright. Based on this study, we can set the upper limit on the fraction of all hard X–ray sources that can be both X–ray and infrared bright to be <1.7% (3 sigma).

References

[1] Baganoff, F. K., Maeda, Y., Morris, M., Bautz, M. W. (and 8 others) 2003,Ap. J., 591, 891

[2] Bandyopadhyay, R. M., Miller–Jones, J. C. A., Blundall, K. M., Bauer, F. V. (and 5 others) 2006, M.N.R.A.S. (in press), astro–ph/0509346

[3] Blum, R. D., Sellgren, K., & Depoy, D. L. 1996, ApJ, 470, 864

[4] Eckart, A., Baganoff, F. K., Schoedel, R., Morris, M. (and 16 others) 2006, A&A (in press), astro–ph/0512440

[5] Genzel,R., Schödel,R., Ott, T., Eckart, A., Alexander, T., Lacombe, F., Rouan, D., Aschenbach, B. 2003, Nature, 425, 934

[6] Gezari, D., Dwek, E, & Varosi, F., 1996, Proc. Symposium 169 of the IAU, ed. L Blitz & P. Teuben, 231

[7] Ghez, A., Hornstein, S. D., Lu, J., Bouchez, A. (and 13 0thers) 2005, Ap. J., 635, 1087

[8] Gillessen, S., Eisenhauer, F., Quatert, E., Genzel, R. (and 9 others) 2006, astro–ph/0511302

[9] Laycock, S., Grindlay, J., van den Berg, M., Zhao, P., Hong, J., Koenig, J., Schlegel, E. M., Persson, S. E., 2006 Ap. J. (Letters), in press

[10] Maillard,J.P., Paumard,T., Stolovy,S.R., Rigaut,F. 2004, A&A, 423, 155

[11] Muno, M. P., Baganoff, F. K., Bautz, W. M., Brandt, W. N. (and 6 others) 2003, Ap. J., 589, 255

[12] Schodel, R. Eckart, A., Iserlohe, C., Genzel, R. and Ott, T. 2005, astro–ph/0504474

A mid-infrared survey of the inner 2 x 1.5 degrees of the Galaxy with Spitzer/IRAC

S. Stolovy[1], S. Ramirez[2], R. G. Arendt[3], A. Cotera[4], F. Yusef-Zadeh[5], C. Law[5], D. Gezari[6], K. Sellgren[7], J. Karr[8], H. Moseley[6] and H. A. Smith[9]

[1] Spitzer Science Center, California Institute of Technology, Pasadena, CA 91125
[2] IPAC/Caltech
[3] SSAI/GSFC
[4] SETI Institute
[5] Northwestern Univ.
[6] GSFC
[7] Ohio State Univ.
[8] Academia Sinica
[9] CfA

E-mail: stolovy@ipac.caltech.edu

Abstract. We present a survey of Spitzer Space Telescope/IRAC observations of the central 2 x 1.5 degrees (265x200 pc) of the Galaxy at 3-8μm. These data represent the highest spatial resolution and sensitivity large-scale map made to date of the Galactic Center (GC) at mid-infrared wavelengths. The IRAC data provide a census of the optically obscured stellar sources as well as a detailed map of the highly filamentary structure in the interstellar medium. The diffuse emission is dominated by PAH emission from small grains in star-forming regions. Dark clouds displaying a large variety of sizes and morphologies are imaged, many of which remain opaque at IRAC wavelengths. Using a multiwavelength comparison, we determine which objects are likely to be in the foreground and which are located at the GC. We find no counterparts at IRAC wavelengths to the unique system of linear, nonthermal radio filaments present at the GC.

1. Motivation

The IRAC instrument aboard the Spitzer Space Telescope is extremely well-suited to mapping out large areas of the sky efficiently while penetrating the interstellar dust that makes observing regions such as the GC impossible at optical wavelengths. We chose to map the central 2 x 1.5 degrees of the Galaxy in order to cover much of the Central Molecular Zone of the Galaxy, with a scientific emphasis on studying objects that are located at the GC. At a galactocentric distance of 7.6 kpc ([1]), our survey covers the inner $\sim$ 265 x 200 pc of the Galaxy. IRAC obtains simultaneous 5.2'x5.2' images at wavelengths centered at 3.6, 4.5, 5.8 and 8.0 μm, which we also refer to as Channels (Ch)1-4, respectively. Each of the 4 cameras consists of an array of 256x256 pixels with 1.22"x1.22" pixels. Ch 1, 2, and 3 are undersampled, and Ch 4 is nearly Nyquist sampled (at 8 μm, $\lambda/2D$ is $0.97''$). We consider the spatial resolution of the entire survey to be on order $2''$ (0.07pc at the GC).

IRAC observations provide a stellar census between the Sun and GC as well as a map of the diffuse PAH emission seen near star forming regions. Previous near-IR surveys such as 2MASS and DENIS, although providing a basis for resolving and understanding the stellar population toward the GC, are limited in that the longest wavelength of 2.2 μm is insufficient to observe the most obscured regions, and longer wavelengths are needed to measure IR excesses. The MSX survey of the Galactic Plane imaged the diffuse dust emission from PAH molecules at 8μm and from larger dust grains (out to $\sim$ 21μm) but lacked the spatial resolution to properly resolve individual stars; at 8 μm, the MSX resolution is 18$''$. The infrared dark clouds discovered with MSX ([2]) are now imaged with 10 times better spatial resolution with the IRAC data. The GC was also observed with ISOCAM as part of a galactic plane survey called ISOGAL at 7 and 15μm with 3$''$ pixels ([3]). These data, especially the 15 μm observations, will provide a useful complement to this dataset. However, the ISOGAL survey did not observe the central regions of high interest such as Sgr A and the Arched Filaments due to saturation.

2. Observations and Data Reduction

The entire region was surveyed using what is referred to as the '2 second' exposure time using the full (5.2'x5.2') array, but which represents an actual integration time of only 1.2 second per pixel on the sky. Five 'medium' dithers were taken for each pointing in order to correct for bad pixels, scattered light, and latent images as well as to better sample the point spread function. Therefore the total average integration time on the for the survey is 6 seconds in each channel. At this depth, we are confusion limited everywhere except toward dark clouds that are so opaque that they are dark even at IRAC wavelengths.

We performed additional processing on the pipeline version S13.2 BCD (basic calibrated data) pipeline products in order to correct for various instrumental effects, including: corrections for scattered light, latent images, column pulldown (Ch1 and 2) and banding (Ch 4). The total number of full-array BCD's obtained in each IRAC array was 2895, organized into 12 AORs (Astronomical Observation Requests). The brightest regions of the survey were observed last to minimize the effects of latent images. The full-array map was obtained between March 29 and March 31, 2005. The fields of view projected on the sky are nearly identical for IRAC Channels 1 and 3 and Channels 2 and 4, respectively, but those pairs of channels are offset on the sky by about 400$''$. We mapped a slightly larger region than 2 x 1.5 degrees in order to get full coverage in all 4 channels. In this paper, however, we crop our mosaics slightly to trim the jagged edges; Fig. 1 displays an area covering 2 x 1.4 degrees.

We used mopex (available from ssc.spitzer.caltech.edu/postbcd/download-mopex.html) to create mosaics resampled to 1.0$''$ pixels and to extract point sources. The resultant mosaics represent a significant improvement over available SSC pipeline products. Our mosaics exhibit seamless continuity in the background level. Some detector artifacts remain, however; the most obvious one is the "bandwidth effect" which causes extra 'sources' to appear 4 pixels away (and in some cases 8 pixels away) from very bright sources due to an electronic effect of the 4 readout channels in Ch3 and Ch4.

Although only a tiny fraction of the area covered was saturated, we aimed to produce a high quality mosaic across the entire region. Thus, we obtained data in IRAC's subarray mode for regions that were saturated in the 2 second exposures 6 months after the main survey. In subarray mode, a small section of the array is read out (32x32 pixels = 40"x40"), and we used the shortest exposure time available of 0.02 sec. Sixteen targeted objects (including the brightest regions in the survey, Sgr A West and the 'IR Quintuplet' cluster (see Fig. 2) were observed in subarray mode on Sept. 15, 2005. The entire survey, including both full and subarray exposures, was completed in less than 16 hours of telescope time. The subarray data were substituted for the saturated full array data in the analysis presented here.

These GC observations will be incorporated into the larger GLIMPSEII survey (PI: Churchwell, Program ID= 20201), which maps the Galactic Plane interior to $\pm$ 10 degrees of Galactic latitude. For the longitude range of $\pm$ 2 degrees, GLIMPSEII extends the map out to $\pm$ 2 degrees of latitude to better map the bulge. The GLIMPSEII team will make data products available to the public including BCD's, mosaics and point source catalogs, all of which will be available through the Spitzer Science Center archive. The GLIMPSEII products will differ from our independently derived version. High resolution jpgs (each 7002x5050 pixels) of both the 4-color composite and the 8μm-only image are are available for download from the press release archive at ssc.spitzer.caltech.edu. We encourage readers to access these images; the small format available here cannot convey the rich details that are present. Due to page limits, we show only the Ch1 and Ch4 mosaics; Ch2 looks very much like Ch1 and is dominated by point sources, and Ch3 has both point sources and PAH emission that is weaker by about a factor of 3 compared to the Ch4 mosaic.

3. Properties of Point Sources

Point source extraction and properties of the point sources will be discussed in more detail elsewhere (Ramirez et al. in prep., Cotera et al., these proceedings), but we provide a summary here. We performed both point response function (PRF) fitting and aperture photometry on the BCD's and compared the two methods to make our final catalog. We chose a small aperture to avoid confusion and applied an aperture correction for the aperture flux. We found that the aperture fluxes were systematically lower than the PRF fluxes and we adjusted the aperture fluxes such that the median ratio of the two sets was 1. We produced source-subtracted images to assess the effectiveness of the extraction. We found that the PRF fitting occasionally failed and predicted a flux that was much too high. For the cases where the PRF/aperture flux ratio exceeded 1.3, we adopted the adjusted aperture value. The source-subtracted residual images show even fainter sources but we did not attempt to extract them. Additionally, the brighter sources close to saturation are in the nonlinear regime and therefore do not match the PRF. We have significantly improved our point source extraction techniques since the conference presentation and we present the newest results here.

The total number of sources detected at a level of 3σ or above in each channel was: 893417, 819914, 535809, and 385926 for Channels 1, 2, 3, and 4, resp. The number of sources common within a $1''$ radius all 4 bands was 284,676. The extraction thus far has been done only on the full array data. We intend to update the catalog with extractions from the subarray data in the future. We have bandmerged our IRAC catalog with the 2MASS catalog, which has comparable spatial resolution, and we are in the process of analyzing the SEDs of the sources. The vast majority of point sources are red giants and AGB stars.

Figure 3 shows the flux distribution for each of the channels for the entire survey. We adopt zero magnitude definitions in the Spitzer Observer's Manual of 280.9, 179.7, 115.0, 64.13 Jy for Ch. 1, 2, 3, and 4, resp. In a survey with relatively uniform background, one might be able to use a diagram such as Figure 3 to estimate the confusion limit near the peak of the counts before the dramatic plunge at the faint end. However, there is no single number that characterizes the confusion in this survey due to the large diversity of background levels against which these sources are extracted. To illustrate this point, Fig. 4 shows the flux distributions in Ch. 1 toward diverse regions with equal 4'x4' areas: 1) in the plane in a region of high surface brightness (dash-dot) between Sgr A and the Arches, 2) away from the plane but also away from dark clouds (thick solid line), 3) towards Sgr B2, a GC dark cloud in the plane (dotted line) and 4) towards a dark foreground cloud (thin solid line). The confusion limit in region 1 is about 2 magnitudes brighter than the others (roughly 11.5 vs 13.5), and there are also significantly more bright sources detected in region 1. It's also clear that the foreground dark cloud blocks many of the faint sources behind it that are detected in 2) and 3). Regions 2 and 3 have nearly identical distributions; implications for Galactic structure are that we may be detecting faint stars beyond the distance of the GC (or at least beyond the distance to Sgr B2) when observing away from the plane. Galactic coordinates are given in the caption.

4. Properties of the Diffuse Emission

PAH Emission The diffuse, highly filamentary structure seen in the IRAC mosaics are dominated by emission from very small grains known as polycyclic aromatic hydrocarbons (PAH's). The PAH's are excited by UV photons and emit in infrared bands centered on 3.3, 6.2, 7.7, 8.6, and 11.3 μm. PAH's are observed in Ch 1, 3, and 4, with the 7.7 μm line in Ch 4 being the dominant band. The PAH emission emanates from photodissociation regions (PDRs) and is generally indicative of star forming regions. For GC objects, the PDR and ionized HII regions often appear in projection to be coincident. The PAH's are transiently heated; since they do not achieve thermal equilibrium, dust temperatures cannot be derived by comparing the IRAC bands. In fact, the ratios of the mosaics (for instance, Ch4/Ch3) show remarkable uniformity across the entire region. This supports the idea, without having to do extensive spectroscopy, that the diffuse emission in the IRAC data is dominated by PAH emission. The line strengths of the PAH bands are nearly constant over several orders of magnitude of radiation field strength ([4]). Very weak extended emission is seen in some regions in Ch1 (PAH 3.3 μm band) and Ch2 (Brα , H_2, and CO lines are in this band). A few compact star forming regions, such as the 'H1-8' regions near the base of the thermal Arched Filaments have an excess in Ch4/Ch3 . We interpret the excess as arising from thermal emission from larger dust grains in exceptionally high radiation fields near regions of massive star formation.

Dark Clouds and Extinction Dark clouds are imaged with exquisite detail, and display a large variety of morphologies. Long, stringy, filamentary structures are often seen. The darkest clouds in our images must be located in the foreground, because they block the light from the background red giants and AGBs seen elsewhere. In addition, most of the dark clouds are physically associated with (adjacent to) clouds with bright PAH emission. The presumption is that star formation is progressing through a cloud. Sgr B2 (center left in Fig. 1) is known to be located near the GC; although we do detect several compact diffuse objects coincident with radio emission, it is remains highly obscured. A comparison with images from the SHASSA survey ([5]) reveals that many of the diffuse objects displaced a few tenths of a degree from the Galactic plane have Hα emission associated with them, and in many cases the PDR and ionized regions are displace from each other rather than superposed. Wherever Hα is detected, the extinction must be less than a few visual magnitudes, which places them in the foreground. In contrast to filamentary dark clouds, very compact dark globules are observed in many places in absorption against the bright emission in the plane; some of these globules are only $5''$ in size (less than 0.2 pc if located close to the GC); the size scales are consistent with what has been reported from near-IR observations ([6]). An extinction law was derived by [7] towards diverse regions in the Galactic plane that is nearly constant in the IRAC bands. We estimate that we are sensitive to sources seen through visual extinctions of Av$\sim$ 40-50, and we confirm that there is little "reddening" in the IRAC bands.

The Sickle, Pistol, and Quintuplet As an example of the detail and dynamic range present in the larger mosaics, Figure 2 shows 3.6'x3.6' subimages of Ch1 and Ch4 images of the objects known as the Sickle, Pistol and Quintuplet. With the incorporation of subarray data, the IR Quintuplet is unsaturated. Our data reveal that the Sickle is resolved into finger-like structures, likely photoevaporated by the hot stars in the Quintuplet (see Cotera et al., these proceedings.) The Pistol star is clearly observed and is in the center of a bubble-like nebula. The Pistol nebula is observed *only* in Ch4; the lack of Ch3 emission is consistent with the diffuse emission being dominated by [ArII] emission, which was observed with ISO spectroscopy (Moneti et al. in prep.).

Large-scale filamentary structures The GC Lobe is a loop-like structure that spans the central degree of the Galaxy in the radio continuum as well as at mid-infrared wavelengths. It was first observed in the infrared by MSX ([8]) and often is interpreted as a mass outflow from the GC region. The remarkable twisted pair of filaments seen on the eastern side of the GC lobe was recently observed with Spitzer's MIPS-24 μm instrument and was dubbed the Double Helix Nebula (DHN) by [9]. The DHN is also observed in our Ch3 and Ch4 images, as is the western and brighter side of the GC lobe. The Ch4 image of the DHN is shown in Fig. 5; we note that the southern end is the most collimated part but it abruptly dissipates southward and does not connect in any obvious way to IR-emitting objects located at the GC. The western lobe (to the upper right of the Ch4 mosaic in Fig. 1) shows a more extensive "feathery" morphology but shares similarities with the DHN in that both have undulations with similar spatial frequencies. However, the well-defined helical morphology is seen only in the DHN. The Ch4/Ch3 color ratio implies that both sides of the GC Lobe are dominated by PAH emission in the IRAC data. The UV sources for the PAH emission may simply be the local radiation field, as the PAH emission can be excited by relatively cool stars.

Law (PhD thesis) obtained observations of the Lobe region with the GBT and VLA and found evidence in that the GC Lobe has a single origin and is located in the GC region. In radio continuum emission, the two sides of the Lobe are parallel and have a similar, distinctive curve. A comparison of radio and IR data imply that the GC Lobe exhibits concentric shells of gas observed (from interior to exterior) in: radio recombination lines (at zero velocity), radio continuum, and at the outer edge, the PAH emission. The east and west sides are not symmetrically displaced in longitude from Sgr A*, but they do appear to coincide with the longitudinal positions of the brightest nonthermal filaments; near the Arches to the east and the Sgr C filament to the west. Thus, they may be related to interactions of the relativistic electrons and the ISM rather than a process originating in the central few parsecs.

Another large-scale arc-like structure that is about 0.4 degree long is seen in the southwestern part of the image and can be seen to the lower right in the Ch4 image in Fig 1. It has a more diffuse nature than the GC Lobe and has no radio emission associated with it to date. It appears to originate, at least in projection, from a little-studied HII region near (l,b)=(359.59, -0.30) that also has a dark cloud associated with it and is likely a foreground object. It is possible that this large feature is also located in the foreground. A star forming region to the northeast (upper left) near (l,b)=(0.67,0.66) is clearly a foreground object with Hα filling in the PAH hourglass structure; the HII region is likely ionized by a known foreground B star.

Comparison with Radio Continuum Data The GC is a strong source of both thermal and nonthermal radio emission, which are distinguished by the spectral index at radio wavelengths. The thermal emission arises from the ionized gas as typically seen in star forming regions. The nonthermal emission is observed towards a number of sources, including Sgr A*, supernova remnants (SNRs) and, most remarkably, the complex and mysterious system of long, linear, filaments spanning in some cases tens of parsecs, the brightest of which are perpendicular to the Galactic plane ([10]).

The Ch4 data have a much higher signal/noise in the diffuse emission than the existing radio data. Nearly every known *thermal* region has a Ch 4 counterpart with the exception of Sgr B2, which is so heavily extincted that very little emission is seen even at 8 μm. Filamentary structures without clear radio counterparts are seen in Ch4 (Fig. 6) arching nearly perpendicular to the Galactic plane between Sgr A and the Arched Filaments. In contrast, there are NO IRAC diffuse structures that are spatially coincident with *nonthermal* radio emission. This is not surprising, as the PAH emission and the nonthermal radio emission arise from different environments. We constrain the spectral index in the linear filaments to be steeper than -0.6. Fig. 6 illustrates that the curved thermal arched filaments roughly parallel to the plane do show a close correlation between radio continuum emission and Ch4 emission. Known SNRs from radio data (e.g. Sgr A East, Sgr D SNR) are not detected in our survey; shocked H_2 emission (in Ch2) is likely too weak relative to the high background to be detected.

The Central Few Parsecs Sgr A* itself is too faint to detect in the IRAC data, and the central region is highly confused due to the large pixel size. The central "minispiral" and inner edge of the circumnuclear disk are clearly imaged in Ch4 and we note that IRS8 is exceptionally bright. Thermal HII regions around the edge of the Sgr A East remnant are observed, along with embedded point sources. The 50km/s and 20km/s clouds are seen in extinction.

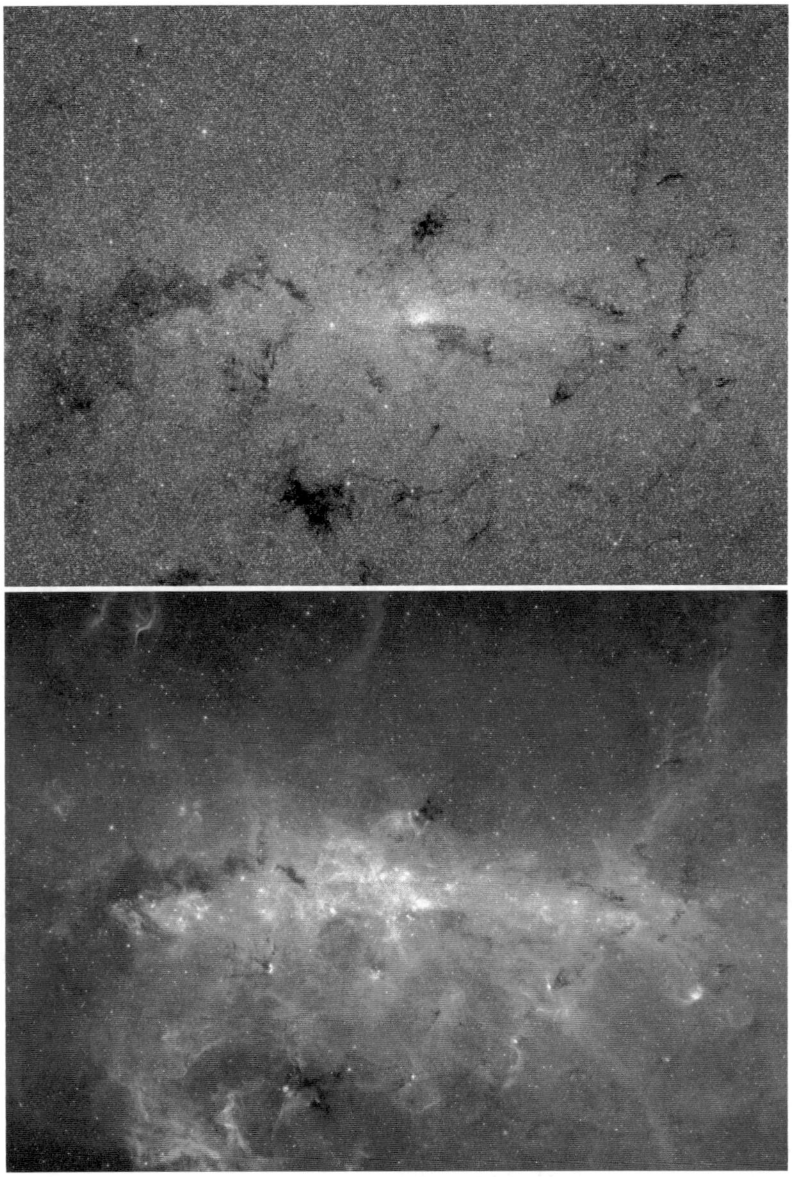

Figure 1. Mosaics of Ch1 (3.6μm, top) and Ch4 (8.0μm, bottom), in logarithmic grayscale. Galactic east is to the left and these images span l=$\pm$ 1 degree and b=$\pm$ 0.7 degree. The dynamic range in each image spans $\sim$ 5 orders of magnitude.

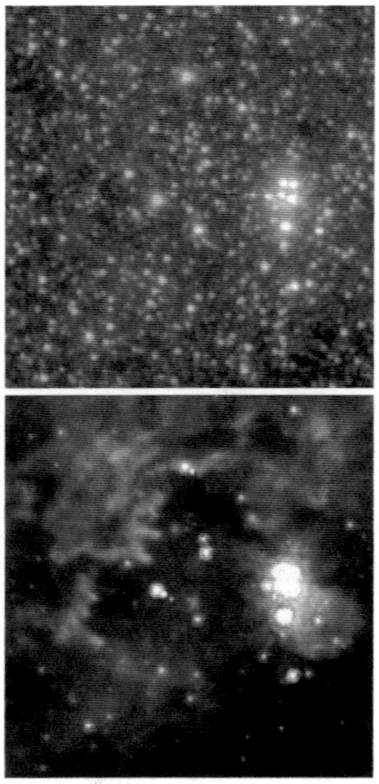

Figure 2. Quintuplet, Pistol, and Sickle region in Ch1 (top) and Ch4 (bottom), combining the full array data with subarray data taken 6 months later. The area shown is 3.6' on a side with Galactic east to the left.

References

[1] Eisenhauer, F. et al. 2005 *ApJ* **628**, 246

[2] Egan, M.P. et al. 1998, *ApJ* **494** 199

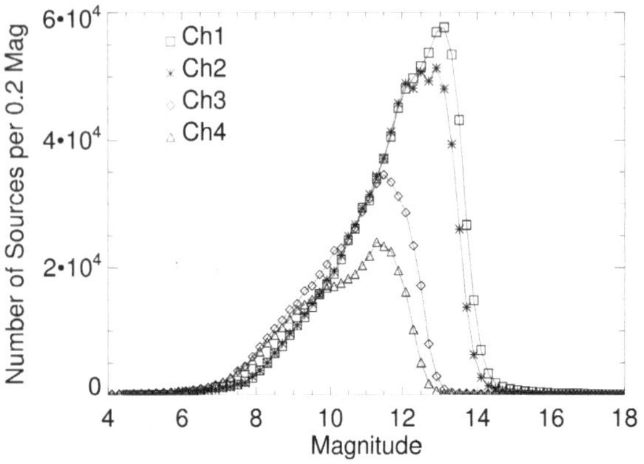

Figure 3. Histogram of point source detections for Ch1 (3.6μm, squares), Ch2 (4.5μm, asterisks), Ch3 (5.8μm, diamonds) and Ch4 (8.0μm, triangles) for the entire survey.

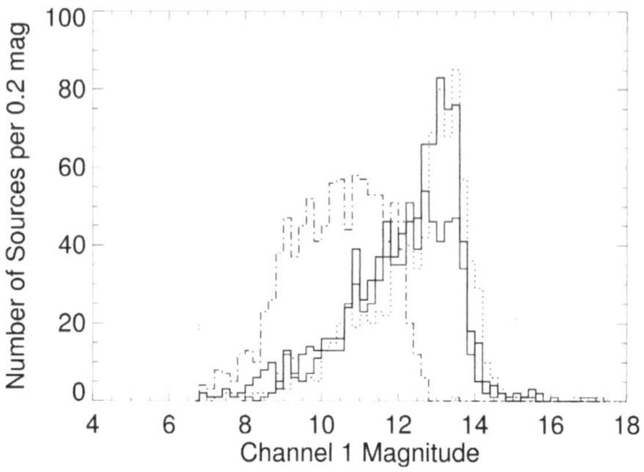

Figure 4. Histogram of Ch1 fluxes in a 4'x4' aperture towards the following regions in (l,b): 1) dash-dot: (0.06, -0.05) (2) thick solid: (359.93,0.64) 3) dotted: (0.63,0.00); 4) thin solid: (359.93,0.17). The total number of sources detected in each region are: 967, 966, 884, and 733 for regions 1-4, resp.

Figure 5. Double Helix Nebula at 8μm, with the fainter point sources removed for clarity. This is the eastern part of the Galactic Center Lobe.

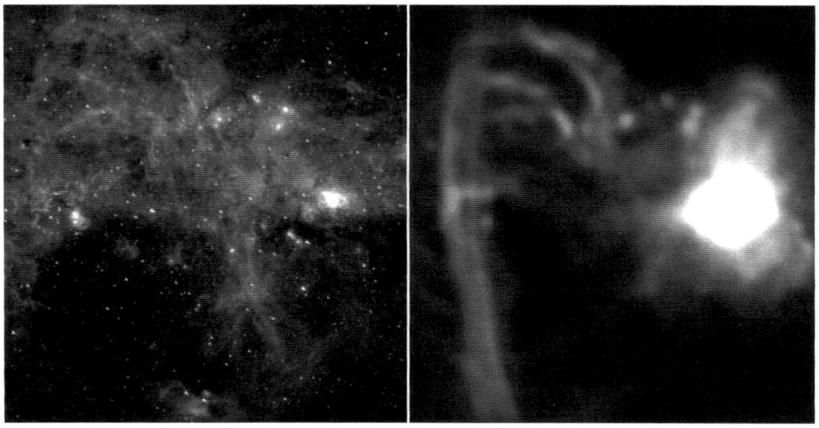

Figure 6. Sgr A/Arched Filaments region in Ch4 (left) and 20cm radio emission. The linear, nonthermal filaments are not observed with IRAC, but Ch4 counterparts to thermal objects are seen. The images are 20.5′ (45.3 pc) on a side with Galactic east to the left and centered at (l,b)=(0.057,-0.060). The radio image (from [10]) has 30″ resolution.

[3] Omont, A. et al. 2003 *A&A* **403** 975
[4] Li, A. & Draine, B. 2001*ApJ* **554** 778
[5] Guastaud, J. et al. 2001 *PASP* **113** 1326
[6] Gosling, A., Blundell, K. and Bandyopadhyay, R. 2006 *ApJ* **640**, 171
[7] Indebetouw, R. et al. 2005 *ApJ* **619** 931
[8] Bland-Hawthorn & Cohen 2003 *ApJ* **582** 246
[9] Morris, M., Uchida, K. and Tuan, D. 2006 *Nature* **440** 308
[10] Yusef-Zadeh, F., Hewitt, J. W., & Cotton, W., 2004 *ApJS* **155** 421

Institute of Physics Publishing Journal of Physics: Conference Series **54** (2006) 183–189
doi:10.1088/1742-6596/54/1/031 Galaxy Center Workshop 2006

The Stellar Population in the Galactic Center: Insights from the *Spitzer Space Telescope*

A. Cotera[1], S. Stolovy[2], S. Ramirez[3], R. Arendt[4], C. Law[5], K. Sellgren[6], D. Gezari[4], F. Yusef-Zadeh[5], H. Smith[7], B. Whitney[8]
[1] SETI Institute, 515 N. Whisman Rd., Mountain View, CA
[2] Spitzer Science Center, California Institute of Technology
[3] IPAC, California Institute of Technology
[4] NASA Goddard Space Flight Center
[5] Northwestern University
[6] Ohio State University
[7] Center for Astrophysics, Harvard University
[8] Space Science Institute

E-mail: acotera@seti.org

Abstract. The *Spitzer* IRAC observations (Stolovy et al., these proceedings) of the central 265 pc ×210 pc provide an opportunity to study the relationships between massive stars, gas, and dust in the Galactic Center at unprecedented resolution. The observations are inclusive of the three known extremely dense clusters of massive hot young stars which ionize the unusual thermal filaments seen at both radio wavelengths and in PAH emission in the IRAC images. Here we explore the effects of the massive stars, particularly in regions including the Quintuplet and Arches clusters, on the nearby diffuse ISM emission (see also Simpson et al. these proceedings). We discuss the diagnostics available using the IRAC colors, and what these show us in regards to known massive stars. Finally, we discuss the 1-8 μm SEDs which are currently under construction, and the preliminary results using a novel SED fitting program to determine the stellar type and extinction to all available point sources in the survey region.

1. Introduction

The formation of stars in the Galactic center has posed a problem since the presence of young stars in the central parsec was first suggested [1, 2]. The canonical view has been that as a result of the high gas temperatures and pressures, large internal cloud velocity dispersions, and strong pervasive magnetic fields, star formation by the standard method of self-gravitational collapse is significantly suppressed. This is thought to be particularly true within a radius of ~100 pc, because of the additional complication of strong tidal forces from the strongly peaked mass within that radius. The identification of very young stars (<2 Myrs) scattered throughout the central 100 parsecs [3, 4, 5, 6, 7], however, contradicts this canonical view. The existence of a truncated initial mass function (IMF) has been suggested based on observations of the Arches Cluster [8, 9], at approximately 6-7 $M_\odot$. A truncated IMF is, however, highly unusual, and often later shown to be the result of resolution or background limitations.

Recently, two more challenges to the canonical idea of the suppression of typical star formation in the GC have arisen: (a) the magnetic fields may not be that strong [10], (b) slow self-gravitational collapse as the preferred mode of star formation, exemplified by low mass stars in the Taurus-Auriga cloud, may not be correct. Recently Lada & Lada [11] have suggested that 70-90% of star formation occurs in rich embedded clusters, most of which contain massive stars. Star formation which is triggered by massive stars, is a viable [12] mode of star formation and may predominate.

Our *Spitzer Space Telescope* IRAC observations of the GC (see Stolovy et al., these proceedings), in addition to providing a detailed map of the filamentary structure of the interstellar medium, provide an amazing opportunity to explore the stellar population throughout an approximate 140 pc radius of the Central Molecular Zone. When combined with observations at NIR wavelengths, the IRAC point source catalog will provide a much needed census of stellar sources throughout the entire region. The identification and distribution of stellar types will provide us with a better understanding of the star formation mechanisms throughout the GC.

2. Regions of known young massive stars

As a first step in enhancing our understanding of the younger stellar population throughout the GC, we turn to the regions of known young massive stars. The four IRAC bands, referred to as Channels 1-4 in Stolovy et al. (these proceedings), provide observations at 3.6, 4.5, 5.8 and 8.0 μm respectively. At 3.6 μm, the observed flux is dominated by emission from stellar atmospheres, whereas at 8.0 μm, photospheric emission is weaker and diffuse ISM emission from polycyclic aromatic hydrocarbons (PAHs), tracing photodissociation regions, becomes an important component of the emission. The combination of the four bands therefore provides an excellent map of the relationship between the stars and the ISM.

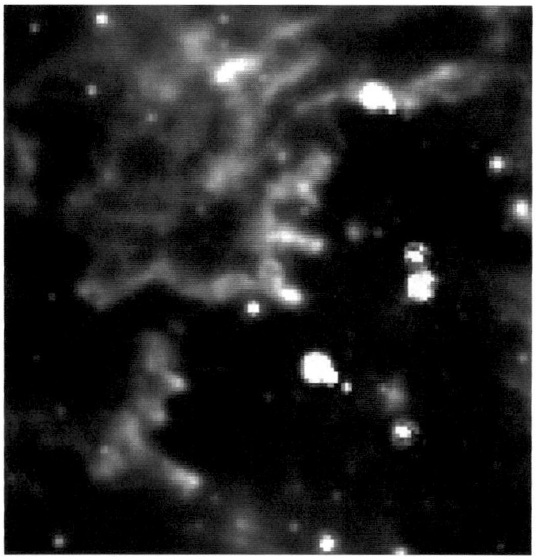

Figure 1. IRAC Ch 4 (8.0 μm) image of the Sickle HII region. The image is in galactic coordinates with a FOV of $4' \times 4'$. For R_{GC} of 8.0 kpc, $4'$ corresponds to $\sim$9.3 pc. At this distance, the fingers are approximately $1-1.5$ pc long and 0.5 pc wide. The ionizing source for both Figs. 1 and 2 are located to the lower right of the image.

Figure 2. IRAC Ch 4 image of the Pillars in M16. The image has been rotated to approximate the orientation of the Sickle. The FOV is $16' \times 16'$, which at the distance of M16, $\sim$2 kpc, corresponds to $\sim$9.3 pc. Thus, Figs. 1 and 2 have the same physical size. Overall, the features in the Sickle are remarkably similar to those in M16.

2.1. The Quintuplet, Sickle and Pistol region

One of the most intriguing early results of this survey is the possibility that we are seeing a photoevaporative region similar to those seen in the HST images of the "elephant trunks" (also known as the "Pillars of Creation") in M16 [12]. In the GC, these features, which we call the "fingers", are associated with the Sickle HII region and are presented in Fig. 1. IRAC observations were also made of M16 as part of the GLIMPSE Legacy program [13], and are shown for comparison in Fig. 2. In M16, the columns are thought to be regions of locally higher density than their surroundings, capable of retaining their natal molecular material despite the strong disruptive influences of the nearby OB association. Simpson et al. [14] demonstrated that the hot stars located within the Quintuplet cluster (e.g. [5]) are the most likely ionizing source for the Sickle. Our *Spitzer* images strongly suggest that the emission from the Sickle comes from fingers analogous to the columns in M16, and are the result of photoionization in a photoevaporative flow from the hot massive stars known to exist in the Quintuplet cluster. Analysis of the GLIMPSE IRAC data of M16 [15], clearly indicates the presence of low mass star formation throughout the M16 region, including the Pillars, despite the presence of the more evolved nearby OB stellar cluster. A detailed comparison of this region and the Sickle is in progress, and will address the uniqueness of the Sickle as a site of current star formation.

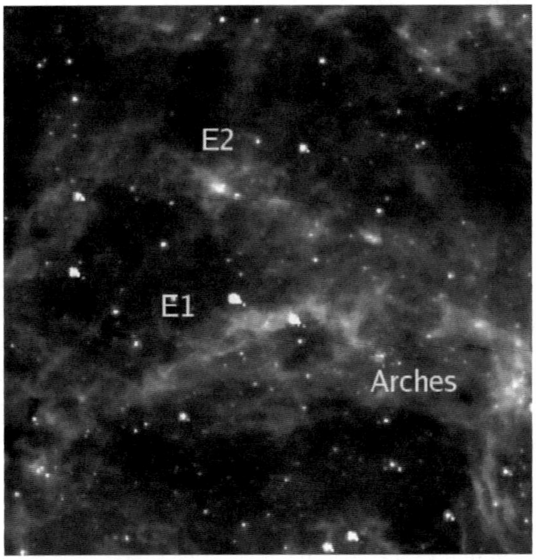

Figure 3. IRAC Ch 4 image of the region inclusive of the Arched Filaments. The image is in galactic coordinates, with a $4' \times 4'$ FOV. The Arches Cluster is very faint at this wavelength, and is located just above the "Arches" label.

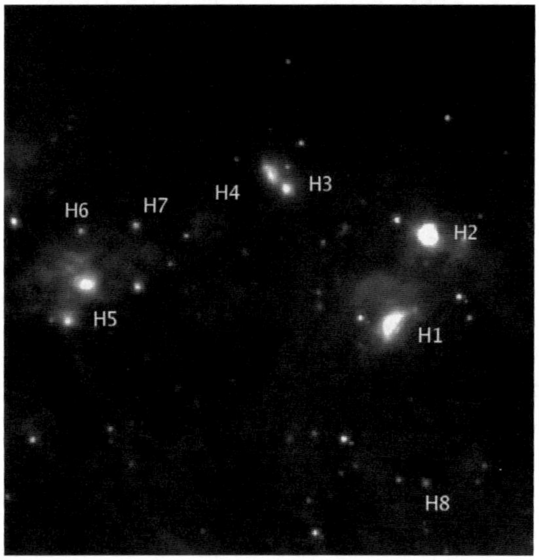

Figure 4. IRAC Ch 4 image of the H1-H8 compact HII regions. The image is in galactic coordinates, with a $4' \times 4'$ FOV. Besides the previously identified radio regions, we see strong 8.0 μm emission from two compact regions near H5, similar to H6-H7.

2.2. The Arched Filaments

The region of the Arched Filaments contains the Arches Cluster, which represents one of the densest known clusters of young massive stars, including ~15 WN stars [16]. The Arches Cluster has been shown to ionize the nearby thermal filaments [17, 18]. The IRAC Ch 4 image of the region surrounding the Arches Cluster is presented in Fig. 3. As discussed in Stolovy et al. (these proceedings) the diffuse emission seen in IRAC Ch 3 and 4 images is predominately from PAHs, and the ratio of the two bands is remarkably uniform throughout the entire region, with

Ch4/Ch3∼3 (Arendt et al. in preparation). Using this global ratio to normalize the Ch 3 to the Ch 4 data, and subtracting off the adjusted Ch 3 image, we are able to identify regions of 8 μm excess. The largest region displaying this excess consists of an elongated bubble centered on the Arches Cluster. The bubble extends to the west, ending parallel to the E2 filament, is bounded on the south by the G0.95+0.07 region, but crosses perpendicularly to the E1 filament. This elongated bubble has not been previously identified, but the underlying mechanism has not yet been determined.

2.3. The H1-H8 HII Regions

The H1-H8 regions are a group of compact HII regions located between the areas containing the Quintuplet and Arches Clusters, and the central few parsecs. At least one young massive star has been identified in this region, associated with H2 [6, 7]. Each of the regions also has strong, compact, 8 μm excess emission. Unlike the known massive stars in the Quintuplet and Arches, however, the stars located throughout this region appear to be forming singly. As such, they perhaps provide a template for a less spectacular, but perhaps more common region of ongoing star formation within the central few hundred parsecs.

3. Photometry

The morphology of the ISM and its juxtaposition with the stellar population, only scratches the surface of the ability of the GC IRAC data set to explore the stellar population throughout the entire 265 pc×200 pc of the galaxy. One of the goals of the project is to create a point source catalog of the entire region, and combine this with 2MASS observations at shorter wavelengths. In addition, we have recently obtained J band images of the central ∼100 pc to significantly fainter magnitudes, and will use this data to enable inclusion of NIR stars to fainter magnitudes. The region is extremely crowded (see Stolovy et al.), and reliable point source extraction of the IRAC fluxes requires combining the results of aperture and PRF extraction which is discussed in more detail in Stolovy et al., and will be discussed fully by Ramirez et al. (in preparation).

One of the challenges in utilizing the IRAC observations to explore the stellar populations is the uniqueness of the filter set. A second consideration, particularly when combining the IRAC observations with both 2MASS and in the future with *Spitzer* MIPS 24 μm data (PI: Yusef-Zadeh), is the number of possible combinations of color-color (CCD) and color-magnitude diagrams (CMD). Fortunately, considerable work has already been done on the parameter space explored by the IRAC filters (e.g. [19, 20]). In Figs. 5 and 6, we present the [4.5] vs. [3.6]−[4.5] CMD, and the [3.6]−[4.5] vs. [5.8]−[8.0] CCD plots typically utilized to begin exploring the stellar populations with the IRAC observations. In Fig. 5 we have convolved Kurucz stellar atmosphere models [21] with the Li & Draine extinction curve [22], and the IRAC filter function available from the SSC website, to produce theoretical magnitudes for normal stars assuming (a) a distance of 2 kpc with no extinction, and (b) 8 kpc with A_V=25. Using the approximate turnover in the number vs. magnitude plots (see Stolovy et al. these proceedings) to set a confidence limit, we find that we can probe the stellar population to approximately BV stars, including M-K giants and all supergiants. In the CCD, we find stars for which we are observing the stellar atmospheres are located at [3.6]−[4.5]= 0.39 ± 0.07, while [5.8]−[8.0]= -0.10 ± 0.06, which is displaced from the origin primarily by extinction. We note that in the [3.6]−[4.5] vs. [5.8]−[8.0] CCD plot, this can be seen as an extinction vector which goes up and to the left.

In Fig. 6, we reproduce Figs. 8 and 9 from Whitney et al. (2004). These figures were produced from theoretical models of YSOs (young stellar objects) of varying temperatures and circumstellar disk orientations, located at a distance of 2 kpc. As can be seen, unlike the case of the stellar atmosphere models, the objects are typically located at [3.6]−[4.5]> 0.5 and [5.8]−[8.0]> 1.0. In the GC, we would expect these numbers to be [3.6]−[4.5]> 0.9 and [5.8]−[8.0]> 0.9. At 2 kpc, our confidence limit translate to [4.5]∼ 8 ± 0.5. Based on the theoretical models, we should be able to observe YSOs with a temperature of >8000 K for much of the survey. In the region of the CCD occupied by massive YSOs, however, we expect

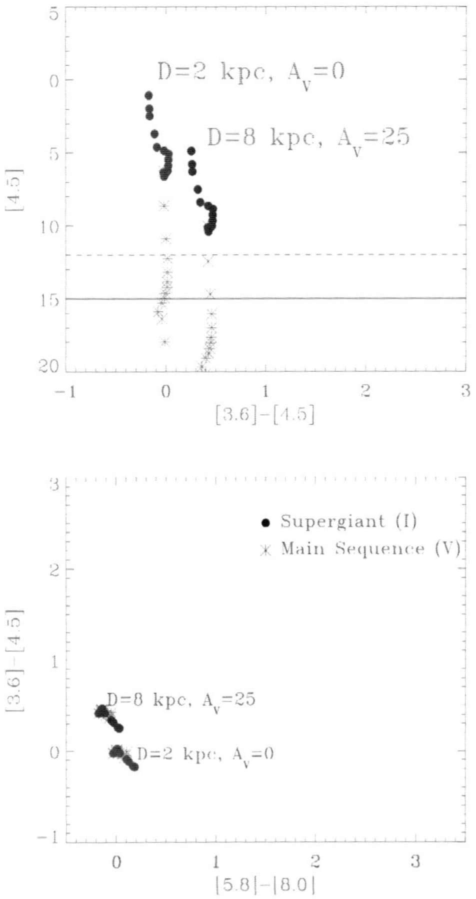

Figure 5. Color-Magnitude (CMD) and Color-Color (CCD) diagrams for Kurucz [19] stellar atmosphere models convolved with the IRAC filter functions. The solid line in the upper figure indicates the approximate confidence limit of the Ch. 2 sources, the dashed line is limit for the combined four band point source catalog.

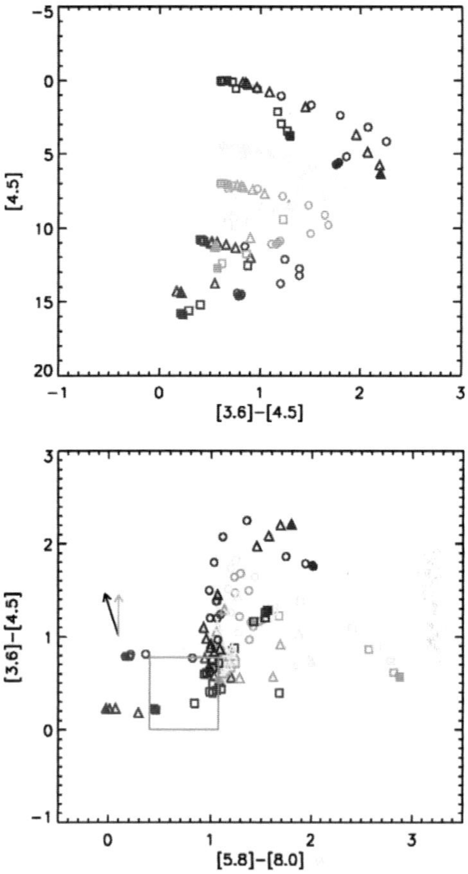

Figure 6. Figs. 9 (upper) and 8 (lower) from Whitney et al. (2004) for sources at a distance of 2 kpc and no line-of-sight extinction. The blue (upper) data points present derived IRAC magnitudes and color for a T=31,500 K YSO at different circumstellar disk inclination angles. Yellow symbols are for T=16,000 K, orange for T=8000 K, and red for T=4000 K. The light grey extinction vector is from [23].

significant contribution from AGB stars, and future work will address differentiating YSOs from AGB stars in our point source catalog.

In Figs. 7-8, we present the same CMD and CCD for the entire IRAC GC survey, and for an $8' \times 8'$ FOV centered on the Quintuplet Cluster; previously known young massive stars are indicated by diamonds. Although the plots for the entire region apparently contains an abundance of sources, they in fact represent only the 284,676 sources which are common to all four IRAC bands. As such, the [4.5] vs. [3.6]−[4.5] CMD presented only represents ~35% of the sources identified in the [4.5] band alone. The CMD clearly shows that the brightest sources are saturated at the shorter wavelength, therefore their colors are not always reliable. The known young massive stars which are not in this saturation region are located in the region expected for supergiant stars in Fig. 5. In the CCD, again the vast majority of the stars are located in the region expected for supergiant stars in the GC, however, the known massive stars have slightly larger [5.8]−[8.0] colors than theoretical predictions for supergiant stars. In addition, we find that only one of the known stars resides in the region with [3.6]−[4.5]> 0.9 and [5.8]−[8.0]> 0.9; this is in fact the Pistol Star, an LBV star. There are numerous other stars in this region of the CCD that have not been spectroscopically classified as yet, and theoretically may be either YSOs or AGB stars. The diagram provides a place to begin our search for additional young massive stars in the GC.

4. Future Work: SED fitting

The *Spitzer Space Telescope* legacy project GLIMPSE, utilized the IRAC camera to image much of the Galactic plane, producing a point source catalog which contains millions of sources. To help analyze this data, Robitalle et al. [25] have produced a grid of YSO models and a tool to fit these models rapidly. The grid also makes extensive use of stellar atmosphere models [21, 25], and will include AGB stars at a future date. The models utilize absolute luminosity, and therefore can fit for distance within a specified range. In addition, since they use at least four data points, typically from combined 2MASS and IRAC catalogs, they can simultaneously fit each source for extinction.

We have used this SED fitting program with a preliminary catalog of our IRAC data. We combined our data with the 2MASS catalog for the region. The version of the 2MASS catalog that we used had detections in all three NIR bands, and for the input to the SED fitting routine we required detections in at least two of the four IRAC bands. Of the 352,235 sources in our combined catalog, 204,035 were fit with stellar atmospheres. There were an additional 147,912 sources that were not well fit by the available stellar atmospheres, but are likely to be stars and their classifications will be improved in future refinements of the fits. There were an additional 1291 sources remaining, which are best fit by the YSO models. Planned future work on the SED fitting program are to include spectra from AGB, Carbon and OH/IR stars which will greatly improve our fits to the GC IRAC data set. The results of the final fitted data will enable us to create a present day mass function over the entire region of the survey, identify possible massive stars and YSO candidates, and create a detailed, extensive extinction map.

References

[1] Rieke, G. H., & Rieke, M. J. 1989, ApJL, 344, 5

[2] Allen, D. A., Hyland, A. R., & Hillier, D. J. 1990, MNRAS, 244, 706

[3] Blum, R. D., Depoy, D. L., Sellgren, K. 1995, ApJ, 441, 603

[4] Cotera, A. S., Erickson, E. F., Colgan, S. W. J., Simpson, J. P., Allen, D. A., & Burton, M. G. 1996, ApJ, 461, 750

[5] Figer, D. F., McLean, I. S., & Morris, M. 1999, ApJ, 514, 202

[6] Cotera, A. S., Simpson, J. P., Erickson, E. F., Colgan, S. W. J.,; Burton, M. G., & Allen, D. A. 1999, ApJ, 510, 747

[7] Muno, M. P., Bower, G. C., Burgasser, A. J., Baganoff, F. K., Morris, M. R., Brandt, W. N. 2006, ApJ, 638, 183

[8] Figer, D. F., Kim, S. S., Morris, M., Serabyn, E., Rich, R. M., McLean, I. S. 1999, ApJ, 525, 750

[9] Stolte, A., Brandner, W., Grebel, E. K., Lenzen, R., & Lagrange, A.-M. 2005, ApJL, 628, 113

[10] LaRosa, T. N., Brogan, C. L., Shore, S. N., Lazio, T. J., Kassim, N. E., & Nord, M. E. 2005, ApJL, 626, 23

[11] Lada, C. J, & Lada, E. A. 2003, ARA&A, 41, 57

[12] Hester, J. J. et al, 1996, AJ, 111, 2349

[13] Benjamin, R. A., et al. 2003, PASP,115, 953

[14] Simpson, J. P., Colgan, S. W. J., Coter a, A. S., Erickson, E. F., Haas, M. R., Morris, M., & Rubin, R. H. 1997, ApJ, 487, 689

[15] Indebetow, R., Robitaille, T., Whitney, B. A., et al. 2006, submitted

[16] Blum, R. D., Schaerer, D., Pasquali, A., Heydari-Malayeri, M., Conti, P. S., & Schmutz, W. 2001, AJ, 122, 1875

[17] Rodríquez-Fernández, N. J., Martín-Pintado, J., & de Vicente, P. 2001, A&A, 377, 631

[18] Cotera, A. S., Colgan, S. W. J., Simpson, J. P., & Rubin, R. H. 2005, ApJ, 622, 333

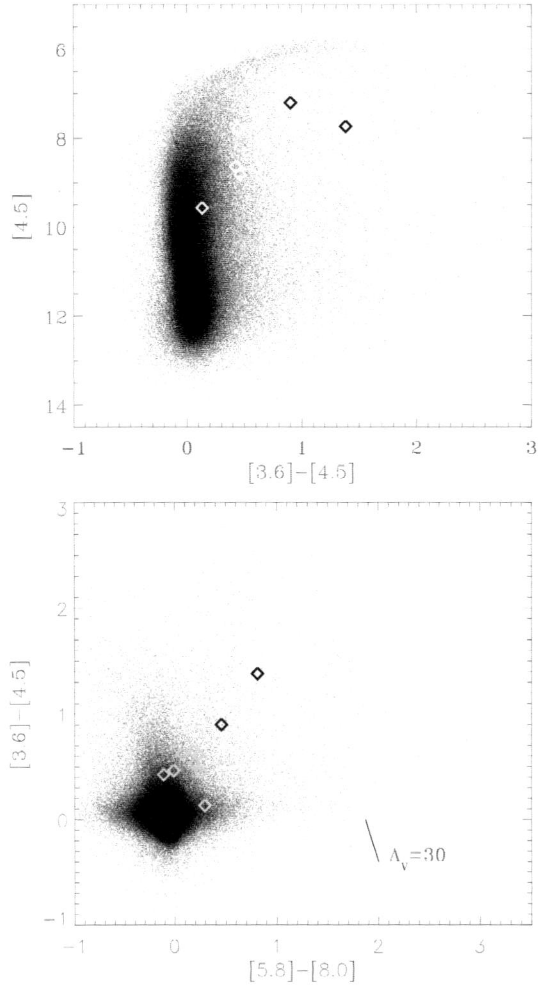

Figure 7. CMD and CCD for the full band merged IRAC catalog. The diamonds indicate previously identified young massive stars. The vast majority of the stars are normal Class I-V stars.

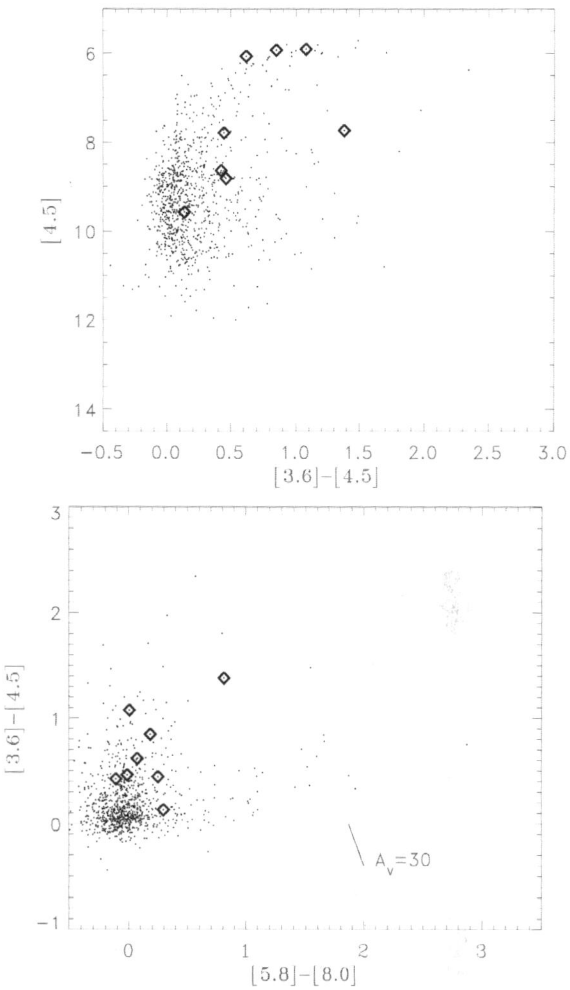

Figure 8. Similar plots as Figs. 5-7 for a $8' \times 8'$ FOV centered on the Quintuplet cluster. The known star at [3.6]−[4.5]$\sim$ 1.4 and [5.8]−[8.0]$\sim$ 0.8 is the Pistol Star

[19] Allen, L., et al. 2004, ApJS, 154, 363

[20] Whitney et al. 2004, ApJ, 617, 1177

[21] Kurucz, R. L. 1992, in The Stellar Populations of Galaxies, ed. B. Barbara & A. Renzini (Dodrecht: Kluwer), P. 225

[22] Li, A. & Draine, B. T. 2001, ApJ, 554, 778

[23] Indebetouw, R. et al. 2005, ApJ, 619, 931

[24] Robatille, T., Whitney, B., et al. 2006, in preparation

[25] Brott, I. & Hauschildt, P. H. 2005 Proceedings of "The Three-Dimensional Universe with Gaia" (ESA SP-576), 4-7 October 2004. Eds: C. Turon, K.S. O'Flaherty, M.A.C. Perryman. P. 565

Institute of Physics Publishing
doi:10.1088/1742-6596/54/1/032
Journal of Physics: Conference Series **54** (2006) 190–198
Galaxy Center Workshop 2006

The formation of high mass stars

Cathie Clarke[1]
[1] Institute of Astronomy, Cambridge, U.K.
E-mail: cclarke@ast.cam.ac.uk

Abstract. I review current controversies and uncertainties surrounding the formation of massive stars in conventional star forming regions. In contrast to the situation for low mass stars, massive star formation is currently not well understood, a difference that may be readily traced to the importance of a variety of feedback processes in the case of massive stars. I shall show that, notwithstanding considerable progress in numerical simulations over the last decade, there are still many shortcomings in such simulations, owing to the difficulty of resolving and following the impact of feedback on a range of spatial scales. Currently, therefore, there is not even any consensus as to whether massive star formation is just a scaled up version of low mass star formation or whether different mechanisms are required. Meanwhile, observations are clarifying a number of necessary properties for successful models. Here we discuss evidence that the star formation process yields a maximum stellar mass, as well as recent data on circumstellar discs around OB stars.

1. Introduction

The possibility that massive stars have been formed in situ in the 'exotic' environment of the Galactic Centre prompts obvious questions about the conditions required for massive star formation. In an attempt to answer these questions, it is natural to look at nearby regions of massive star formation and to the theories that have been developed to explain these. Can such theories be extrapolated to 'extreme environments'? In this contribution I will attempt to answer this question: the short answer, however, is that we are still far short of a compelling theory of massive star formation, even in well studied regions. I will make the case that this shortcoming mainly derives from ignorance of the way that feedback of energy from massive stars operates in star forming environments that are as complex and clumpy as those encountered in reality.

I will start by asking four questions about the upper end of the IMF and by describing the observational and theoretical insights that can be brought to bear on these issues. I then discuss an untraditional model of massive star formation in which the most massive stars form via stellar mergers, touching on both theoretical work in this area and on observational constraints that exist. I conclude with a list of 'burning issues' and a wish-list of studies that could advance our understanding in these areas. It is left to the reader to decide whether any of the ideas discussed in the present paper can offer insights into the conundrum of massive star formation in the Galactic Center.

2. Four issues

Figure 1 contains a schematic depiction of the IMF in the solar neighbourhood. Although we cannot say with any certainty that this functional form is universal (and indeed there is intriguing evidence for a top heavy IMF in the Galactic Center, [1]), it would appear to be a reasonable representation for the IMF in most regions studied to date (see [2] and references therein).

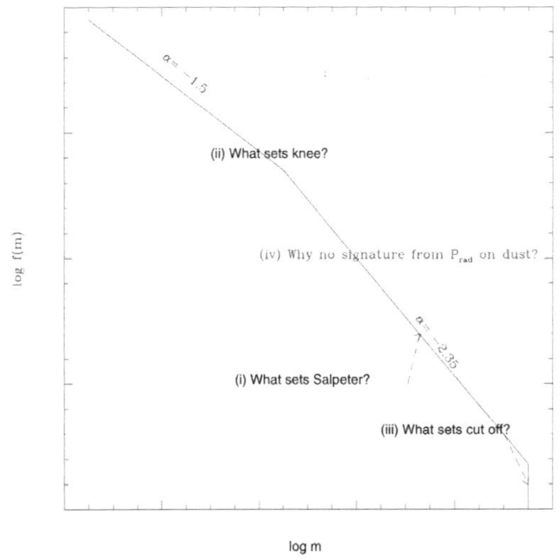

Figure 1. Schematic form of IMF, annotated with key features and critical questions.

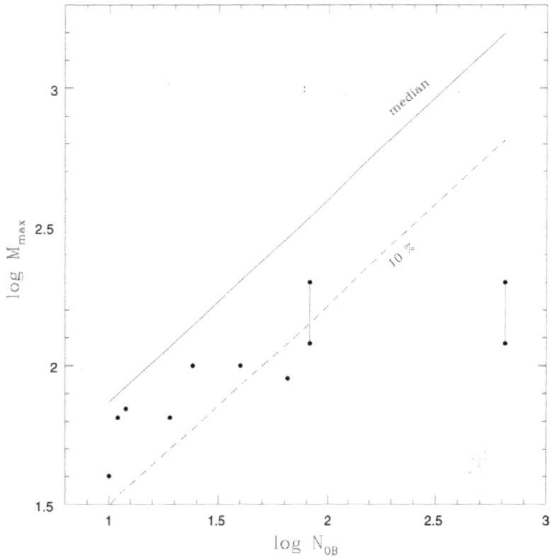

Figure 2. Centiles (random drawing model) of maximum stellar mass as a function of cluster richness compared with cluster data.

The form of this IMF raises the following four questions:

(i) What sets the slope of the power law at high masses (i.e. a Salpeter distribution ([3]) of the form $f(m) \propto m^{-2.35}$, where $f(m)$ is the fraction of stars with masses in the range m to $m + dm$)?

(ii) What sets the value of the 'knee' of the IMF, i.e. the flattening of $f(m)$ at masses below about a solar mass? In general terms, whereas the answer to (i) relates to process, the answer to (ii) imposes a physical scale on the problem. Whether this scale is expected to vary with epoch or environment must depend on whether the scale derives from conditions in the star forming clouds or with some physics associated with this mass scale.

(iii) What sets the maximum mass of stars? In the absence of any physical process that limits star formation at the upper end, one would expect an IMF of the form shown in Figure 1 to be populated up to a mass that depends on the mass of the parent gas reservoir. Since this appears *not* to be the case (i.e. stars more massive than $200M_\odot$ have not been identified even in highly populous regions containing many hundreds of OB stars), we must conclude that there is some physical reason why more massive stars either do not form or do not survive.

(iv) Why is the IMF featureless in the region of $10M_\odot$? This is the mass scale at which radiative feedback of ultraviolet radiation on dust becomes important and, in simulations including this effect, such feedback can prevent some stars from growing beyond $\sim 10M_\odot$. This would be expected to produce a feature in the IMF which is not observed. We thus conclude that the effect of radiation pressure on dust is unlikely to be important in massive star formation and that this offers possible clues about the geometry/mode of massive star formation.

2.1. *What sets the slope of the 'Salpeter' power law at high masses?*

The balance of current evidence is in favour of the form of the upper IMF being remarkably invariant, with the power law slope of the probability density function at high masses being a little steeper than -2 (see e.g. [4] and references therein). Such evidence derives not only from relatively nearby systems where one can directly constrain the IMF through star counts, but is also consistent with population synthesis studies of Lyman break Galaxies at a redshift of ~ 3 ([5]). There are however a few regions where it appears that the upper IMF is discrepant with a Salpeter power law, notably for field OB stars in the Magellanic Clouds, where [6] has claimed a slope of the upper IMF that is as steep as -4.

Over the years, a number of models have been proposed to explain the slope of the upper IMF, which have invoked scale free physical effects as diverse as coalescence, fragmentation and radiative feedback ([7]-[10]). Recent star formation simulations (e.g. [11]) , which however omit most of these effects, also replicate the observed Salpeter slope, even though they include the most minimal set of physical effects (i.e. gravity, thermal pressure and an initial supersonic turbulent velocity field in the star forming gas). Such a result can be simply understood in terms of 'competitive accretion', whereby stars acquire mass in proportion to their accretion cross-section, which is set, in these simulations, by the critical impact parameter for Bondi Hoyle accretion. This implies (in the case of a constant density medium) that the accretion rate scales as the square of the stellar mass. If a star of mass M_i attains a final mass M_f after fixed time t, then it follows that the dependence of M_f on M_i can be written:

$$\frac{dM_f}{dM_i} = \left(\frac{M_f}{M_i}\right)^2$$

Thus a small range of initial masses maps onto a spread of final masses $dM_f \propto M_f^2$. Therefore the number of stars per unit mass interval scales as M_f^{-2} and one recovers an IMF slope of -2. Although the situation in the simulations is considerably more complicated than the above 'toy' model (i.e. there are considerable variations in density, the gas is self-gravitating and stars are subject to a complex history of stellar dynamical interactions: see discussion in [12]), this simple argument (originally due to [13]) is responsible for the fact that the simulations produce a roughly 'Salpeter' slope with remarkable robustness. Although other physical processes may contribute to the upper IMF, it is at least reassuring that this quasi-universal outcome of star formation should be so readily replicated by the simplest simulations (i.e. those that include no feedback, no magnetic fields and a very simple equation of state).

2.2. *What sets the knee of the IMF?*

For an IMF consisting of disjoint power laws, slope $-\alpha$, the mean stellar mass in a given power law section is dominated by the mass of the upper (lower) end of the section if α is respectively less (greater) than 2. It therefore follows that for the IMF shown schematically in Figure 1, the mass scale at which the slope steepens from $\alpha < 2$ to $\alpha > 2$ imprints on the IMF a characteristic stellar mass close to this value. As has often been stressed by Richard Larson (e.g. [14] and references therein), the fact that the observed IMF has this form (regardless of whether we find it convenient to parameterise the observed distribution by disjoint power laws or by a log normal representation) endows the star formation process with a characteristic stellar mass of around $1M_\odot$. Star formation is therefore not an entirely scale free process. The profound consequence of this for the Universe in which we live is that the characteristic mass scale of self-gravitating gaseous objects is roughly the mass of the Sun.

It is obviously important to understand whether this mass scale is a consequence of the typical properties of star forming gas clouds (e.g. pressures or densities) or whether it relates to some piece of physics that might lead to a similar outcome in very different environments. An obvious candidate for this mass scale is the Jeans mass, M_J, which depends on thermodynamic variables according to $M_J \propto T^{1.5}/\rho^{1/2}$ or equivalently $\propto T^2/P^{1/2}$. One then immediately encounters a difficulty, even in the case of isothermal clouds: given the strong density and thermal pressure variations in both observed molecular clouds and in numerical simulations of turbulent fragmentation, what is the appropriate ρ (or P) to employ here? Two extreme possibilities would be the average cloud density and the density of gas in shocked layers. Since these two quantities differ by a factor roughly equal to the square of the Mach number (for isothermal shocks), this introduces an uncertainty in the corresponding Jeans mass of a factor equal to the Mach number (typically ~ 10 in turbulent molecular clouds).

If one simply analyses the density distribution in turbulent cloud simulations and identifies apparently Jeans unstable regions within the simulation volume, one obtains an IMF with a characteristic mass corresponding to the Jeans mass in the shocked gas (e.g.[15]). The thermal pressure in the shocked gas is of necessity of order the turbulent ram pressure in the cloud, while the latter is comparable to (but somewhat more than) the mean pressure in the galactic environment of a star forming cloud. This led [16] to make an estimate of how the characteristic stellar mass varied as a function of galactic halo mass and redshift of formation. Such simple considerations gave the right answer for the characteristic stellar mass at current epochs as well as replicating the high characteristic mass commonly attributed to Population III star formation in low mass haloes virialising at high redshifts. It is however hard to find good observational constraints on the predicted variation of characteristic stellar mass in intermediate situations.

Despite the attractiveness of such a simple framework, the work of [17] has cast doubt on whether it is indeed the turbulent pressure in the parent clouds that sets the characteristic stellar mass (i.e. if it is indeed the thermodynamic state of the shocked gas that is relevant). Their simulations differ from those of [16] in that they follow the collapse of gravitationally unstable regions rather than identifying notionally Jeans unstable clumps from the density distribution. Although there is plenty of material in their simulations that is shock compressed to high densities, this high density does not apparently imprint an associated (low) mass scale on the IMF. Instead they find that the IMF resulting from these simulations has a knee that is close to the Jeans mass corresponding to the *mean* density in the cloud. Although this result is not fully understood, it seems likely that the reason this mass scale is 'remembered' is connected with the two dimensional nature of shocked gas layers. Such compression changes neither the timescale for lateral collapse nor for the lateral passage of sound waves in the layer. Consequently, the actual mass scale for collapse is unaffected.

The conclusion that the IMF knee is simply related to the Jeans mass at the mean cloud density is an uncomfortable one however. It implies, for example, that for a cloud of fixed mass and temperature, the characteristic stellar mass scales as the cloud radius to the power of 1.5. Such a strong dependence on cloud parameters is however hard to reconcile with the fact that the observed 'knee' in the IMF is close to a solar mass in all well observed systems in the Galaxy.

A more promising outcome is however obtained in simulations which abandon the strictly isothermal equation of state and which, following the suggestion of [14], allow for mild cooling by line radiation as the gas is compressed at densities below some critical value and mild heating with compression, as cooling becomes dominated by dust, above this value. Pilot simulations of [18] suggest that in this case the IMF is imprinted with a knee corresponding to the Jeans mass at the density and temperature at which compressive heating sets in, and - in contrast to the strictly isothermal case - is then much less sensitive to initial conditions (i.e. cloud mass and radius). If this result is corroborated through further simulations, it would then firmly link the form of the IMF to radiative physics in the star forming gas (specifically, in this case, the details of cooling via line and dust emission). Any argument for a different characteristic mass scale in the Galactic Center would then need to invoke different cooling physics rather than different cloud parameters.

2.3. What sets the maximum stellar mass?

In the absence of any physical limitations, the upper IMF should be populated up to a mass limit that is simply set by the mass of the star forming gas reservoir. A simple way to compute how the expectation value of the maximum stellar mass should scale with total mass, M, is to calculate the stellar mass such that the probability of a star exceeding this mass is $\sim 1/N$ (where N is the number of stars in the cluster). For a Salpeter IMF with probability density function $\propto m^{-2.35}$, and assuming also that $M \propto N$, one then obtains the scaling $m_{max} \propto M^{\frac{1}{1.35}}$ (e.g. [19, 20]). In order to assess, whether an observed maximum stellar mass in a cluster of mass M is compatible with the hypothesis of random drawing from a Salpeter IMF, we need to compute the theoretical probability density function for maximum stellar mass after N drawings from such an IMF and see where the observed value lies within the theoretical cumulative distribution ([21]). The results of such an exercise are shown in Figure 2, where the observed maximum stellar masses for clusters of various membership number (parameterised according to the total population of OB stars in the cluster) are compared with the 50th and 10th centiles of the theoretical distribution. It is clear that the observed values are systematically too low compared with the theoretical distribution for the more populous clusters: recalling that Figure 2 is plotted on a logarithmic scale, the discrepancy for large N clusters is very marked (in fact, the probability that the right hand datapoint - corresponding to R136 in the core of 30 Doradus in the Large Magellanic Cloud - is taken from the theoretical distribution is 10^{-15}!) Apparently, then, the Salpeter IMF does not continue in a scale free (i.e. power law) manner to arbitrarily high mass and some unidentified physical effect limits stars to masses of less than $200M_\odot$, even in highly populous regions.

This important conclusion requires some qualification. In principle it could be circumvented in a range of circumstances (for example, if more massive stars were too deeply embedded or otherwise not recognised as such, if they formed first (and had hence expired already) or if we had overestimated the lifetimes for such stars: see discussion in [21, 22]). It is also possible that such massive stars are unstable for reasons that are currently not well understood. If none of these considerations apply, we are left only with the possibility that some effect - presumably associated with feedback from the star formation process - prevents stars growing to greater than $\sim 200M_\odot$.

It is not however an easy matter to identify the physical process that would operate selectively at such a mass scale. Indeed, as we shall see below, the more obvious mass scale at which we expect feedback to be operative (in this case due to radiation pressure on dust) is around $10M_\odot$. Stars exceeding $\sim 100M_\odot$ are close to being Eddington limited but there are no obvious reasons why this situation should become increasingly untenable at the highest stellar masses. Both ionising radiation and stellar winds may play an important role in truncating the upper end of the IMF. However, pending the development of numerical algorithms that can model these effects in realistically clumpy gaseous environments (e.g. [23]) it is unclear how the efficacy of feedback depends on the properties of the parent cluster. We must therefore remain agnostic at the present time as to whether the apparent $200M_\odot$ limit is just a property of conventional star forming regions or whether it is expected to hold also in a range of exotic environments.

2.4. Why is there no signature of radiation pressure on dust in the observed IMF?

Since the early studies of [24], it has become a commonplace to assert that the momentum transfer between stellar ultraviolet photons and the dust in accreting gas is a potential problem for forming massive stars by accretion. The radiative acceleration on the dust just before it sublimes (and thus at the point where it experiences the first strike of radiation from the star) depends both on the radiative flux and on the flux weighted opacity of the dust. Both of these increase strongly when the star is around $10M_\odot$: the former on account of the fact that the star is on the steep ($L \propto M^3$) portion of the mass luminosity relation and the latter because the flux weighted opacity for dust increases strongly as the incident spectrum shifts into the ultraviolet.

It is now known that there are several factors that would make accretion onto OB stars much easier than envisaged by [24]. For instance, [25] showed that accretion is favoured even by the simple expedient of relaxing the assumption of steady state flow: in a time dependent calculation, material that is accelerated by gravity prior to the switch-on of the central star can accrete successfully even under the action of a net outward force at later times. Probably the most important effect, however, is that of relaxing the assumption of spherical symmetry. In a situation of low density along the stellar poles (whether generated in a centrifugally supported flow as in the simulations of [26] or through a bipolar outflow as in the study of [27]) , the radiation field is beamed along the poles and thus permits accretion in the equatorial plane.

The calculation of [26] (which includes full frequency dependent radiative transfer coupled to the hydrodynamics) remains the state of the art in this field, and has, in relatively few simulations, certainly demonstrated that it is possible to form stars that are much more massive than $10 M_{\odot}$ in realistic accretion disc geometry. What is currently lacking, however, is the demonstration that stars of this mass are *routinely* formed from a range of initial conditions and thus that radiation pressure essentially presents *no barrier* to OB star formation. Certainly in spherical symmetry, where the relative computational simplicity means that it is possible to explore a range of initial conditions, [26] found that although it is certainly *possible* to form stars of arbitrarily high mass (given sufficiently cold and massive progenitor cores) there are also plenty of instances where the process stalls at around $10 M_{\odot}$. The observed IMF is however featureless in this mass range, so evidently the actual process of massive star formation is much less sensitive to radiation pressure than is the case in these simulations.

3. An alternative scenario: massive star formation via stellar mergers

Although further simulations may demonstrate that in disc geometry the accretion process is essentially oblivious to radiation pressure (as is required by the featureless form of the observed IMF), [28] (see also [29]) have in the meantime explored a more radical way in which massive star formation might be rendered insensitive to radiation pressure effects. In this alternative scenario, OB stars form through the merger of lower mass stars. In its original incarnation, the theory envisaged that stellar collisions occur in an ultradense cluster core, with the requisite high densities being produced by adiabatic contraction of the core in response to mass loading by accreting gas. The stellar densities required for mergers ($\sim 10^8$ stars pc^{-3}) are however far higher than any observed in practice and would need to be explained away in terms of a short lived, heavily embedded phase that would be observationally inaccessible. The re-inflation of such cluster cores to match observed stellar densities requires the operation of vigorous feedback (e.g. through ionising radiation or stellar winds), although obviously this feedback must not be too effective or else the core would not reach the necessary high densities in the first place.

A more recent variant of the model relieves it of some of these difficulties. [30] showed that stellar mergers are important at much lower stellar densities (around 10^6 stars pc^{-3}) if one takes into account the role of hard binaries in the cluster core. Binaries obviously offer a much larger cross section than single stars: the merger does not in this case occur via a single hit but through the successive hardening of binaries in three body encounters until the components are forced to merge at pericenter. The model of course requires a high incidence of close binaries among massive stars, a property that is certainly consistent with current observations (e.g. [31]) .

However, although massive stars undoubtedly undergo merger events in these simulations, careful analysis reveals that such stars actually acquire the bulk of their mass through accretion. Thus although mergers may play a role in the creation of high mass stars in cluster environments, the process cannot be considered as an *alternative* to accretion and thus cannot relieve the potential problem of radiation pressure on dust. (Indeed, in the idealised simulations of [32], which model stars that acquire their mass by Bondi Hoyle accretion due to supersonic passage through the ambient medium, it was found that the effect of radiation pressure for this accretion geometry is actually more problematical than for the case of spherical accretion. This effect has not however been investigated in the case of realistic cluster simulations).

What are the observational constraints that can be brought to bear on the mechanism for massive star formation? Given that it is argued that the stellar merger phase is brief and subject to high extinction, one would not expect it to be directly observable. Alternatively, the possession or otherwise of circumstellar discs by young OB stars is often proposed as a possible observational discriminant. Conventional scenarios would, by analogy with the situation for low mass stars, be expected to yield stars with discs, whereas circumstellar material may be expected to be stripped during the process of stellar mergers. What, then, is the observational situation regarding disc possession in young OB stars?

For stars more massive than $7M_\odot$, circumstellar discs are not detected by the stage that stars are optically visible ([33]), in contrast to the situation for lower mass stars. However, for stars in the mass range $7 - 20M_\odot$ there is good evidence for the existence of discs at an earlier evolutionary stage, i.e. when the star is still embedded in its natal material (see [34] and references therein). These discs are massive (comparable to the central star) and, according to kinematic studies, may be regarded as 'true discs', in the sense of being centrifugally supported. The typical size scale of such 'true discs' is $\sim 100 - 1000$ A.U.. For stars more massive than $20M_\odot$, however, there is no evidence for true (centrifugally supported) discs ([34]). Instead, giant toroids are observed, with masses much larger than that of the central star and size scales in the range $10^3 - 10^4$ A.U. (e.g. [35]).

It is currently unclear how these recent findings should be interpreted in terms of mechanisms for massive star formation. A pro-merger interpretation might be that the disruption of small scale ('true') discs at a mass scale of $20M_\odot$ is due to mergers becoming important for stars at this mass scale. This interpretation however ignores a number of factors: the fact that small scale discs might be present in more massive stars (but obscured by the surrounding circumstellar toroids) as well as the possibility that the disruption of small scale discs might result from radiation pressure (rather than dynamical disturbance during mergers). Finally, we note that the specific angular momentum of material in the 'true discs' ($\sim 10^{-2}$ km s^{-1} pc) may be higher than that found in the gas in the cores of clusters in simulations of stellar mergers ([30]). It is unclear whether this is a numerical issue or whether it could represent a more fundamental problem in such cluster simulations.

4. The Burning Issues

In the final section I will summarise what can be considered the main outstanding questions for theorists of massive star formation, together with an outline of some suggested ways forward:

(i) What sets the maximum stellar mass?
Progress in this area requires better calculations of feedback from ionisation and stellar winds in realistic star forming environments (As an example, see Figure 3 which depicts the results of a hydrodynamic simulation of a star forming cluster where the gas is subject to the effect of ionising radiation from a central star. The inhomogeneous (filamentary) nature of the gas has important consequences for the efficacy of feedback , the generation of outflows and the escape of ionsing radiation from the cluster).

(ii) What determines the mean stellar mass?
Following the recent studies of [14] and [18], more effort is required to couple a proper treatment of the thermal properties of star forming gas to hydrodynamic simulations of star cluster formation. In particular, it is necessary to check whether the approach that is currently in use (i.e. the compressing of information about the thermal properties of the gas into a global barotropic equation of state) is valid in the complex density fields in star forming regions.

Figure 3. Hydrodynamic simulation of star formation in cluster gas irradiated by OB star. Note the complex filamentary structure produced by the interaction of ionising photons with the accretion flow onto the cluster core. From [23]

(iii) If mergers don't obviate the need for accretion onto OB stars, what accretion geometry is allowed or required?
The evidence based on a relatively small number of simulations is that if stars acquire their mass from a core of bound material containing significant angular momentum, then radiation pressure on dust is probably not important in halting accretion onto the star ([26]). If on the other hand, stars acquire their mass from material that is initially unbound, this gas becomes bound by flowing tangentially past the star before dissipating energy in an accretion shock. This almost tangential portion of the flow is highly susceptible to disruption by radiaton pressure ([32]) Further simulations are required which combine realistic initial conditions and the high resolution necessary to study the impact of radiation at the dust sublimation radius.

(iv) Do turbulent fragmentation calculations produce stars whose associated circumstellar material has the high specific angular momentum implied by recent observations of discs around OB stars?
Currently the answer to this is probably 'no', but this may result either from (numerical) angular momentum transport in the simulations or else from suppression of vorticity in the initial conditions due to finite resolution. Further high resolution simulations are required in order to address this issue.

On an observational front, the best way to constrain viable mechanisms for OB star formation probably relates to the circumstellar environments of OB stars. Continued efforts are required to identify possible 'true' (centrifugally supported) discs around the more massive O stars. Another piece of fossil evidence about OB formation mechanisms is contained in the binary statistics for such systems, since this is an area where cluster simulations will be able to yield detailed predictions about the properties of the binaries that survive the putative merger phase.

Acknowledgments
I am grateful to Ian Bonnell for helpful comments on this review.

References

[1] Nayakshin, S., Sunyaev, R., 2005. MNRAS 364, L23.

[2] Corbelli, E., Palla, F., Zinnecker, H., 2005. "The Initial Mass Function 50 Years Later", Astrophysics and Space Science Library, Vol. 237.

[3] Salpeter, E., 1955. ApJ 121,161

[4] Kroupa, P., 2002. Science 295, 5552, p. 82

[5] Pettini, M., Rix, S., Steidel, C., Shapley, A., Adelberger, K., 2003. in Proc. IAU Symposium 212, ed. van der Hucht, K, Herrero, A., Esteban, C., San Francisco: Astronomical Soc. Pacific, p.671

[6] Massey, P., 2002, ApJS 141,81

[7] Silk, J., Takahasi, T., 1979. ApJ 229,242

[8] Murray, D., Lin, D., 1996. ApJ 467,728

[9] Silk, J., 1995. ApJ 438,41.

[10] Adams, F., Fatuzzo, M., 1996. ApJ 464,256

[11] Bonnell, I., Clarke, C., Bate, M., Pringle, J., 2001. MNRAS 324,573.

[12] Bonnell, I., Bate, M., 2006. MNRAS 362,915

[13] Zinnecker, H., 1989. in proc. IAU Colloq. 120, eds. Tenorio-Tagle, G., Moles, M., Melnick, J., Springer-Verlag Berlin, Heidelberg, New York.

[14] Larson, R., 2005. MNRAS 359,211.

[15] Padoan, P., Nordlund, A., Jones, B., 1997. MNRAS 288,145

[16] Clarke, C., Bromm, V., 2003. MNRAS 343,1244

[17] Clark, P., Bonnell, I., 2005. MNRAS 361,2

[18] Bonnell, I., Clarke, C., Bate, M.R., 2006. MNRAS 368,1296

[19] Elmegreen, B., 2000. ApJ, 539, 342

[20] Weidner, C., Kroupa, P., 2006. MNRAS 365,1333

[21] Oey, S., Clarke, C., 2005. ApJ 620,430.

[22] Weidner, C., Kroupa, P., 2004. MNRAS 348,187

[23] Dale, J., Bonnell, I., Clarke, C., Bate, M., 2005, MNRAS 358,291.

[24] Wolfire, M., Cassinelli, J., 1987. ApJ 319,850

[25] Edgar, R., Clarke, C., 2003, MNRAS 338,962

[26] Yorke, H., Sonnhalter, C., 2002. ApJ 569,846

[27] Krumholz, M., McKee, C., Klein, R., 2005, ApJ 630,250

[28] Bonnell, I., Bate, M., Zinnecker, H., 1998, MNRAS 298,93

[29] Bonnell, I., Bate, M., 2002. MNRAS 336,705

[30] Bonnell, I., Bate, M., 2006. MNRAS 362,915

[31] Garcia, B., Mermilliod, J., 2005. A & A 368,122

[32] Edgar, R., Clarke, C., 2004. MNRAS 349,678

[33] Fuente, A., Rodriguez-Franco, A., Testi, L., Natta, A., Bachiller, R., Neri, R., 2003. ApJ 598,L39

[34] Cesaroni, R., Galli, D., Lodato, G., Walmsley, C., Zhang, Q., 2006P rotostars and Planets V, in press.

[35] Beltran, M., Cesaroni, R., Neri, R., Codella, C., Furuya, R., Testi, L., Olmi, L., 2005. A & A 435,901

Institute of Physics Publishing
doi:10.1088/1742-6596/54/1/033
Journal of Physics: Conference Series **54** (2006) 199–207
Galaxy Center Workshop

The two young star disks in the central parsec of the Galaxy: properties, dynamics, and formation

T Paumard[1], R Genzel[1,2], F Martins[1], S Nayakshin[3,4], A M Beloborodov[5,6], Y Levin[7,8], S Trippe[1], F Eisenhauer[1], T Ott[1], S Gillessen[1], R Abuter[1], J Cuadra[3], T Alexander[9,10], A Sternberg[11]

[1] Max-Planck Institut für extraterrestrische Physik (MPE), Giessenbachstraße, 85748 Garching, Germany
[2] Department of Physics, University of California, 366 LeConte Hall, Berkeley, CA 94720, USA
[3] Max-Planck Institut für Astrophysik (MPA), Karl-Schwarzschild-Str. 1, 85741 Garching, Germany
[4] Theoretical Astrophysics Group, Department of Physics & Astronomy, University of Leicester, Leicester, LE1 7RH, United Kingdom
[5] Physics Department and Columbia Astrophysics Laboratory, Columbia University, New York, USA
[6] Astro-Space Center of Lebedev Physical Institute, 84/32 Profsoyuznaya st., Moscow, 117997, Russia
[7] Canadian Institute for Theoretical Astrophysics, University of Toronto, 60 St. George Street, Toronto, Ontario, M5S 3H8, Canada
[8] Sterrewacht Leiden, Leiden University, P.O. Box 9513,NL-2300 RA Leiden, The Netherlands
[9] Faculty of Physics, Weizmann Institute of Science, Rehovot 76100, Israel
[10] William Z. and Eda Bess Novick career development chair
[11] School of Physics & Astronomy, Tel Aviv University, P.O. Box 39040, Tel Aviv 69978, Israel

E-mail: paumard@mpe.mpg.de

Abstract. We report the definite spectroscopic identification of $\simeq 40$ OB supergiants, giants and main sequence stars in the central parsec of the Galaxy. Detection of their absorption lines have become possible with the high spatial and spectral resolution and sensitivity of the adaptive optics integral field spectrometer SPIFFI/SINFONI on the ESO VLT. Several of these OB stars appear to be helium and nitrogen rich. Almost all of the $\simeq 80$ massive stars now known in the central parsec (central arcsecond excluded) reside in one of two somewhat thick ($\langle |h|/R \rangle \simeq 0.14$) rotating disks. These stellar disks have fairly sharp inner edges ($R \simeq 1''$) and surface density profiles that scale as R^{-2}. We do not detect any OB stars outside the central 0.5 pc. The majority of the stars in the clockwise system appear to be on almost circular orbits, whereas most of those in the 'counter-clockwise' disk appear to be on eccentric orbits. Based on its stellar surface density distribution and dynamics we propose that IRS 13E is an extremely dense cluster ($\rho_{\rm core} \gtrsim 3 \times 10^8 M_\odot$ pc^{-3}), which has formed in the counter-clockwise disk. The stellar contents of both systems are remarkably similar, indicating a common age of $\simeq 6 \pm 2$ Myr. The K-band luminosity function of the massive stars suggests a top-heavy mass function and limits the total stellar mass contained in both disks to $\simeq 1.5 \times 10^4 M_\odot$. Our data strongly favor in situ star formation from dense gas accretion disks for the two stellar disks. This conclusion is very clear for the clockwise disk and highly plausible for the counter-clockwise system.

1. Introduction

The Galactic Center (GC) is a unique laboratory for studying galactic nuclei. Given its proximity ($\simeq$ 8 kpc), processes in the GC can be investigated at resolutions and detail that are not accessible in any other galactic nucleus. The GC has many features that are thought to occur in other nuclei (for reviews, see [1–4]). It contains the densest star cluster in the Milky Way intermixed with a bright H II region (Sgr A West or the 'mini-spiral') and hot gas radiating at X-rays. These central components are surrounded by a $\simeq 1.5$ pc ring/torus of dense molecular gas (the 'circum-nuclear disk', CND). At the very center lies a very compact radio source, Sgr A*. The short orbital period of stars (in particular the B star S2) in the central arcsecond around Sgr A* show that the radio source is a 3–4 $\times\ 10^6 M_\odot$ black hole (BH) beyond any reasonable doubt [5, 6]. The larger Galactic Center region contains three remarkably rich clusters of young, high mass stars: the Quintuplet, the Arches, as well as the parsec-scale cluster around Sgr A* itself [7].

Beyond the scope of pure Galactic Center research, the Galactic Center early-type star cluster gives the opportunity to study the wider interest questions of massive star formation at large as well as starburst galaxies. Of course, this cluster is a signature of mass deposition in the central parsec, and is therefore to be put in the context of fueling and growth of supermassive black holes.

The presence of many high-mass, short-lived stars, in the central parsec shows that massive star formation has occurred at or near the Galactic Center within the last few million years. This is surprising. All obvious routes to creating or bringing massive young stars in(to) the central region face major obstacles. In situ star formation, transport of stars from far out, scattering of stars on highly elliptical orbits and rejuvenation of old stars due to stellar collisions and tidal stripping have all been proposed and considered (for a recent review of the rapidly growing body of literature see [4]). No explanation at this point is the obvious winner (or loser). Perhaps the two most prominent and promising scenarios for explaining the young massive stars outside the central cusp, at radii of 3–10″ from Sgr A*, are

(i) the '*in situ, accretion disk*' scenario [8–12]. Here the proposal is that stars have formed near where they are found today, very close to the central black hole. However, in situ star formation is impeded by the tidal shear from the central black hole and surrounding dense star cluster. To overcome this shear, gas clouds have to be much denser ($\simeq 10^{12} R_{1''}^{-3}$ cm^{-3}) than currently observed [13]. The tidal shear can be overcome if the mass accretion was large enough at some point in the past – perhaps as the consequence of the in-fall and cooling of a large interstellar cloud – such that a gravitationally unstable (outside of a critical radius) disk was formed. The stars were formed directly out of the fragmenting disk;

(ii) the '*in-spiraling star cluster*' scenario [14–19]. Here the idea is that young stars were originally formed outside the hostile central parsec and only transported there later on. Individual transport of stars by two body relaxation and mass segregation from further out takes too long a time ($\simeq 10^{7.5}$–10^9 years, [4]). Stars in a bound, massive cluster can sink in much more rapidly owing to dynamical friction [14]. To sink from an initial radius of a few parsec or more to a final radius of $\ll 1$ pc within an O star lifetime (a few Myr) requires a cluster mass $> 10^5 M_\odot$. To prevent the final tidal disruption of such a cluster at too large a radius – resulting in the deposition of its stars there – the core of the original star cluster also has to be much denser ($> 10^7 M_\odot$ pc^{-3}) and more compact ($\ll 1$ pc) than any known cluster. However, as a helpful by-product, dynamical processes in such a hypothetical super-dense star cluster may then lead to the formation of a central, intermediate mass black hole (IMBH; [20, 21]). Such a black hole may help to stabilize the cluster core against tidal disruption and lessen the high density requirement somewhat [22].

In the following, we present the data and results we obtained using SINFONI concerning the parsec-scale early-type star cluster around Sgr A*. These results are presented in greater depth in [23]

2. Observations and data analysis

2.1. Observations

SPIFFI [24, 25] is a near-infrared integral field spectrometer providing a 2048 pixel spectrum simultaneously for a contiguous, 64×32-pixel field. Its salient features include a reflective image slicer and a grating spectrometer with an overall detective throughput (including pre-optics module and telescope) of $\geq 30\%$. Its 2048^2-pixel Hawaii II detector covers the J, H and K (1.1 to 2.45 μm) atmospheric bands. In its 2003 version with a smaller 1024^2-pixel detector the spectrometer provided 1024 spectra for a 32^2-pixel field. Spectral resolving powers range from $R = 1000$ to 4000. Three pixel scales (12.5×25 square milli-arcseconds (mas), 50×100 mas^2 and 125×250 mas^2) can be chosen on the fly. In the SINFONI ESO VLT facility, SPIFFI is mated with the MACAO adaptive optics module [26] employing a 60-element wave-front-curvature sensor with avalanche photodiodes. This mode makes it possible to perform spectroscopy at the smallest (diffraction limited) pixel scale.

We have covered a square field of > 2 pc $\times$ 2 pc centered on Sgr A* at low spatial resolution as well as various subfields at intermediate or high resolution. We looked for early-type stars by visually inspecting the cubes and line maps around features typical of early-type stars. The main signatures we looked for are essentially lines from Hydrogen (Brγ) and He I. The intrinsically most prominent of these lines (Br$\gamma$and He I 2.058 μm) are unfortunately present in the emission from the interstellar medium, which make them hard to use, in particular at lower spatial resolution. There remains however one fairly deep absorption feature in the K-band spectra of OB stars that is not present in the ISM: a compound of 4 He I lines and three N III lines at 2.113 μm. The spectra of the OB candidates so identified were then extracted with the interstellar emission removed by subtracting an off-source spectrum (generally from a ring around each source).

3. Results

3.1. OB stars are finally detected

Our observations have led to the firm detection of 29 OB supergiants, as well as 12 OB stars of luminosity class III and V. In addition we have 18 OB candidates whose identifications we regard as tentative. Those additional stars need to be confirmed. All these detections refer to the region outside the central cusp, with projected radius $p_{\mathrm{SgrA*}} \geq 0.8''$. Our group has already reported 70 mas resolution SINFONI observations of this central cusp [27], with the detection of more than a dozen main sequence B stars, in addition to the late O9/B0 main sequence star S2 (S02) detected earlier by [6].

We also identify 30 post main-sequence blue supergiants and Wolf-Rayet stars, adding several stars to the sample already known from previous work [9, 28–30]. Of these, we classify 17 as Ofpe/WN9 and late nitrogen rich WRs (WNL=WN7–9) stars, and 12 as carbon rich WRs (WC) stars. There is one early WN (WNE, WN5/6) star, IRS 16SE2. The Northern Arm bow-shock star IRS 1W, with Brγ in emission and He I λ 2.058 μm in absorption as only features, is perhaps a Be star (see [31] for spectrum and detailed discussion). There is also one additional tentative WC candidate (IRS 7SE2).

More than 100 early-type stars have now been detected in the nuclear star cluster, and this number is expected to grow in the next years. The 90 sure detections are presented on Fig. 1.

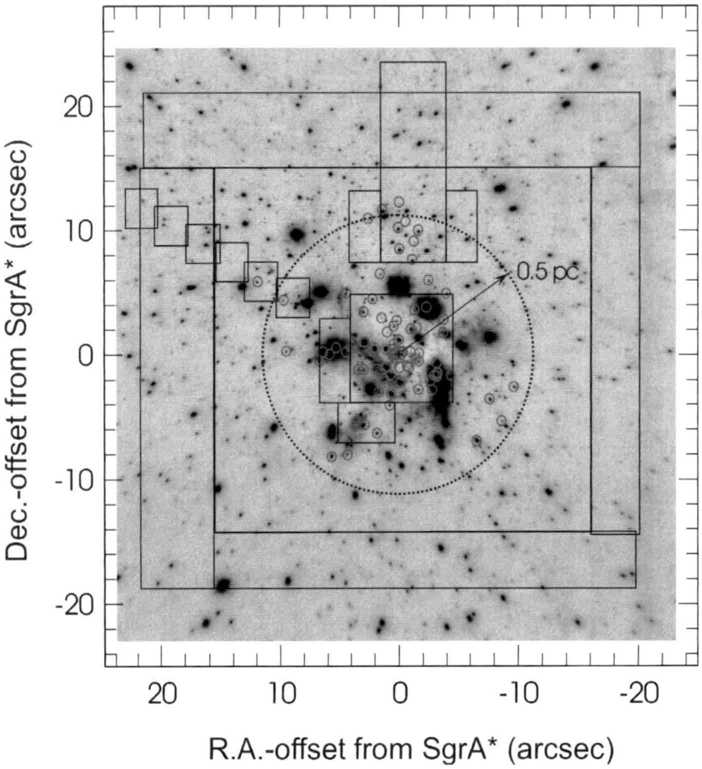

Figure 1. Position of the 90 certain early-type stars detected by SINFONI (*circles*), drawn on an L-band NACO image. The low and medium spatial resolution fields observed with SINFONI are represented by boxes.

3.2. Dynamics

We know the proper motion of most of these stars from our NACO astrometry monitoring program. It is interesting to look at j, the projected angular momentum of each star with respect to Sgr A* (Fig 2). It appears that the early-type stars have typically higher $|j|$ values than late type stars, meaning that they are preferentially on more circular orbits. In that we confirm [9]. This previous work as well as others also claim that the early-type stars belong in one or two stellar disks. To check whether our new data confirm this finding, we introduce a new method for finding disk patterns in kinematic data.

Assume that a set of stars with velocities

$$\vec{v}_k = (v_{x,k}, v_{y,k}, v_{z,k}) = ||\vec{v}_k||(\sin\theta_k \cos\varphi_k, \sin\theta_k \sin\varphi_k, \cos\theta_k)$$

all live in a disk with normal vector

$$\vec{n} = (\sin i \cos\Omega, -\sin i \sin\Omega, -\cos i)\ .$$

Then for each k,

$$\begin{aligned} 0 &= \vec{n}\cdot\vec{v}_k \\ &= \sin i \cos\Omega \sin\theta_k \cos\varphi_k - \sin i \sin\Omega \sin\theta_k \sin\varphi_k - \cos i \cos\theta_k \\ \sin i \sin\theta_k \cos(\Omega+\varphi_k) &= \cos i \cos\theta_k \\ \operatorname{cotan}\theta_k &= \tan i \cos(\Omega+\varphi_k)\ . \end{aligned}$$

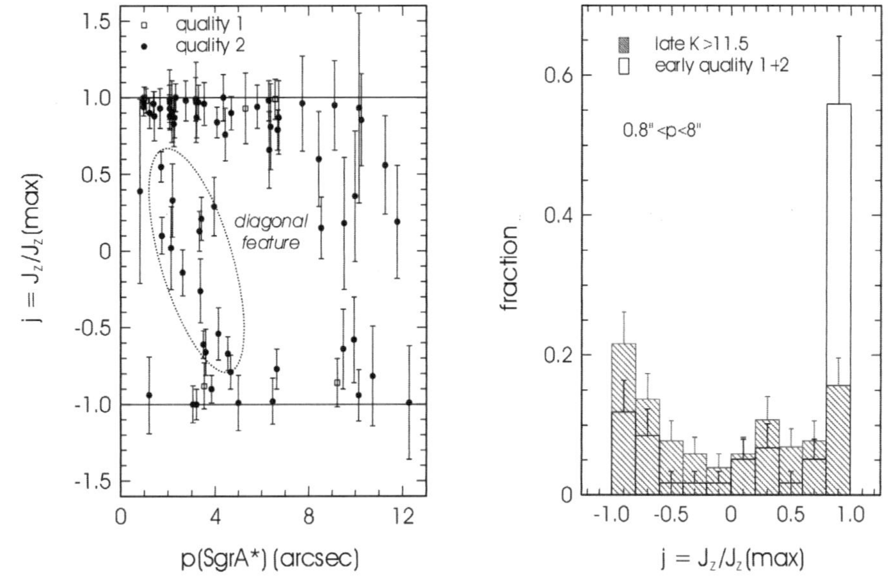

Figure 2. *Left:* j (normalized angular momentum with respect to the line of sight) vs. R, projected distance from Sgr A*. *Right:* j histograms for early and late type stars.

Therefore, if a given set of stars really constitutes a disk, then the θ vs. φ plane must show a telltale cosine pattern. This plane is shown on Fig. 3 for the early-type stars with $j \simeq +1$ (hereafter the *clockwise system* [CWS]) as well as for the early-type stars with $j \simeq -1$ (hereafter the *counter-clockwise system* [CCWS]). Apart from a few stars with large error bars, the two diagrams exhibit a clear cosine pattern. We therefore conclude that the Galactic Center early-type stars indeed belong in two well defined disks. It is interesting to note that these two disks are at large angle from each other and from the Galactic plane, and that the orbital planes of the S stars as determined by [27] seem to be randomly distributed.

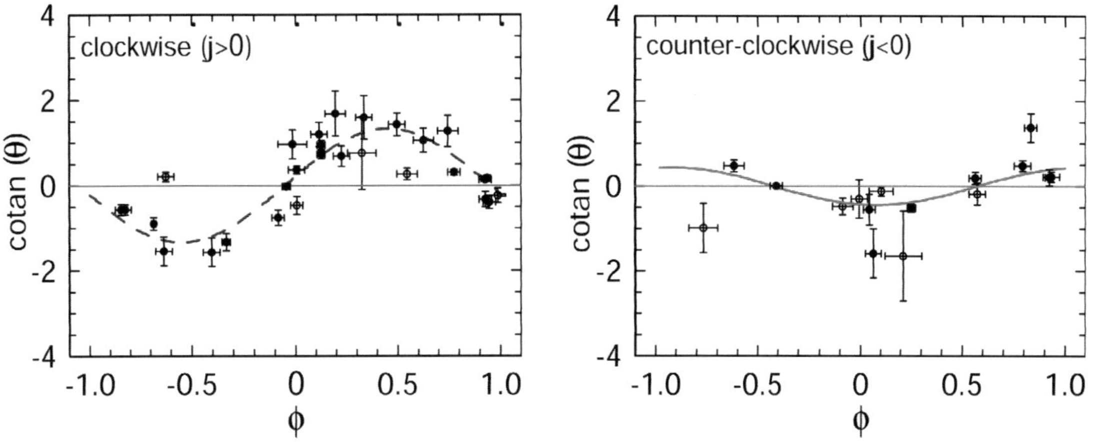

Figure 3. A cosine pattern is very clear in the cotan θ vs. φ diagram of both the CWS (*left*) and the CCWS (*right*). This shows that each system constitutes a disk (see text).

3.3. Physical properties

The two disk systems contain similar fractions of the various subtypes of early-type stars, which is indicative of a common age, within $\simeq 1$ Myr. Both the Hertzsprung–Russel diagram and population synthesis show that the two systems were born 6 ± 2 Myr ago.

They also have a very similar stellar density profile with $\Sigma \propto R^{-2}$ down to an inner radius of $\simeq 1''$ for the CWS and $\simeq$ 2–4$''$ for the CCWS. Outside these inner edges, the two profiles are essentially identical not only in slope but also in absolute value. Inside the innermost edge ($1''$) is located the S star cluster, which does not show any preferential orbital plane as already stated above.

Between $1''$ and $4''$ there is a fourth group of early-type stars that shows up on the j vs. R diagram (Fig. 2) as a diagonal feature (DF) running approximately from $(R = 1, j = 0.8)$ to $(R = 4, j = -0.8)$. At this point, we have no satisfying interpretation for these stars. The "diagonal feature" itself may be an artifact made of randomly distributed stars. During the conference, Jessica Lu has shown Keck data for the inner few arcseconds with which she claimed to demonstrate that the CCWS did not exist (these proceedings). It is important to note that this claim is based on data covering only the inner $4''$ radius, and that therefore these observations do not cover the bulk of the CCWS stars, but mostly the DF group.

For the first time we have developed a method for determining the eccentricity distribution in each disk. Contrary to the other physical properties discussed above, which are similar in the two disks, the eccentricities are not (Fig. 4). Eccentricities in the CWS are fairly small. The distribution peaks at $e \simeq 0.2$ (although it extends up to fairly high eccentricities). On the contrary, the CCWS is dominated by a high eccentricity ($e \simeq 0.8$) population.

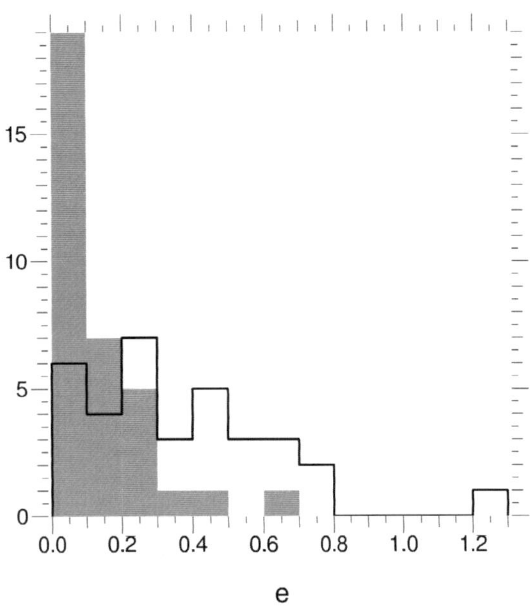

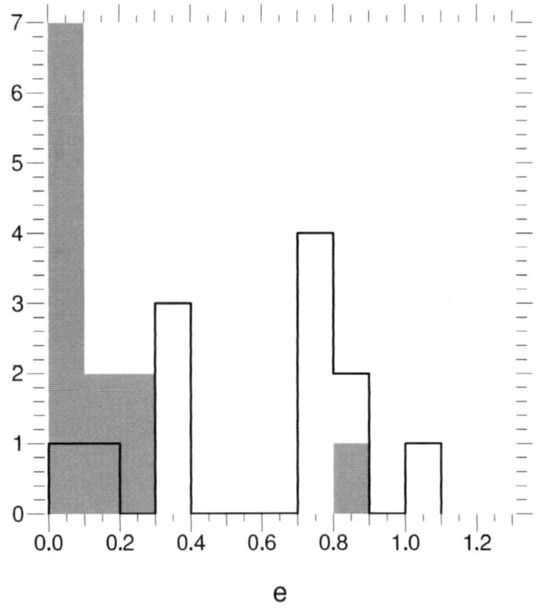

Figure 4. Eccentricity distribution in the CWS (*left*) and the CCWS (*right*). The lines represent the true distributions; the filled histograms are Monte-Carlo simulations showing what our method would give if all stars were on circular orbits, and are representative of the typical error on e.

4. IRS 13E: the cluster within

[32] have attracted a lot of attention by showing that the "Helium star" IRS 13E was indeed

a group of at least seven stars, three of which shared a common velocity. They have suggested that an intermediate mass black hole (IMBH) might be necessary to tie this group together. They believed this group was a strong argument in favor of the inspiraling star cluster scenario. On the other hand, [33] have claimed that this group could not be a stable cluster.

In order to assess whether the object was a significant overdensity rather than a chance alignment, we have analyzed a very deep H-band image of the region surrounding IRS 13E (Fig. 5). We find several dozen faint stars in the group. Using careful statistical arguments, we reject by more than 3σ the possibility of a chance alignment. For this reason we believe this cluster has to be somewhat stable, or it would be very unlikely to observe it. However, with the observations at hand, a conservative lower limit for the total cluster mas is only $\simeq 10^3 M_\odot$. The mass in detected stars amounts to roughly half of that, and is only a lower limit to the total stellar mass. We must conclude that, at this point, the case for an IMBH is not strong. More proper motion or radial velocity measurements for individual stars in the cluster are warranted to bring better constraints on the total mass of the group. It is important to note that this cluster seems to belong in the CCWS.

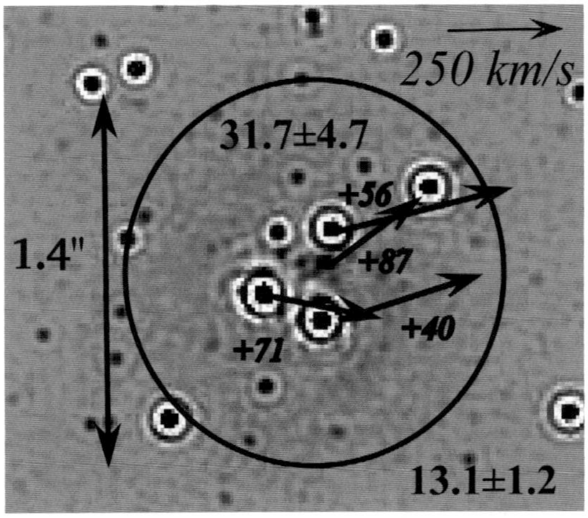

Figure 5. Very deep H-band image of the IRS 13E complex. The arrows and small numbers close to 4 stars give their proper motion and radial velocities. The numbers inside and outside the $0.68''$ circle give the surface density in the these two areas at a limiting magnitude $m_H = 20.4$ (19.4).

A very interesting side-result of this study is the determination of a lower limit to the core density of IRS 13E: $3 \times 10^8 M_\odot\,\mathrm{pc}^{-3}$. To our knowledge, this is one of the highest cluster core densities known to date, second only to the cusp around Sgr A* itself. Such a high density makes direct star–star collisions likely, which may favor the runaway birth of an IMBH.

5. Summary

From the radial density profile, the mass function and stellar counts, we can safely infer that the total mass contained in each disk $M_\star < 10^4 M_\odot$. This is one or two orders smaller than the mass required by inspiraling cluster scenarios. For this reason, we strongly feel that our new data finally rule them out. On the other hand, the facts we uncovered, in particular the density profile and inner radii, are well explained by in situ star formation inside a massive accretion disk.

This scenario has two major apparent difficulties, which are quite possible to overcome:

(i) the presence of a massive core (IRS 13E) and high eccentricities in the CCWS: these two properties seem explainable in the context of rapid star formation in a "dispersion ring" rather than a fully circularized accretion disk (note that the timescale for star formation is roughly the same as that needed for circularization);

(ii) the presence of two co-eval disks at large angle: the timespan of $\simeq 1$ Myr allowed between the two events is sufficient for star formation to remove all gas from the first disk and for a second one to form. The infall of two massive molecular clouds within such a short timescale might be perceived as unlikely. However, they may have been triggered by a single event, such as an interaction with another galaxy.

Finally, we have doubled the number of early-type stars known in the GC, and proved that they are concentrated in two well defined disks. The two systems have probably been born in situ, from massive gaseous accretion disks. However, this explanation fails so far to produce the S star cluster in the innermost arcsecond, for which the "paradox of youth"remains.

References

[1] Genzel R and Townes C H 1987 *ARAA* **25** 377

[2] Morris M and Serabyn E 1996 *ARAA* **34** 645

[3] Mezger P G, Duschl W and Zylka R 1996 *A&A Rev.* **7** 289

[4] Alexander T and Livio M 2004 *ApJ* **606** L21

[5] Schödel R et al. 2002 *Nature* **419** 694

[6] Ghez A M et al. 2003 *ApJ* **586** L127

[7] Figer D 2003 *Astr. Nachr. Suppl.* **324** 255

[8] Levin Y and Beloborodov A M 2003 *ApJ* **590** L33

[9] Genzel R et al. 2003 *ApJ* **594** 812

[10] Goodman J 2003 *MNRAS* **339** 937

[11] Milosavljevic M and Loeb A 2004 *ApJ* **604** L45

[12] Nayaksin S and Cuadra J J 2005 *A&A* **437** 437

[13] Morris M 1993 *ApJ* **408** 496

[14] Gerhard O 2001 *ApJ* **546** L39

[15] McMillan S L W and Portegies Zwart S F 2003 *ApJ* **596** 314

[16] Portegies Zwart S F, McMillan S L W and Gerhard O 2003 *ApJ* **593** 352

[17] Kim S S and Morris M 2003 *ApJ* **597** 312

[18] Kim S S, Figer D F and Morris M. 2004 *ApJ* **617** L123

[19] Gürkan M A and Rasio F A 2005 *ApJ* **628** 236

[20] Portegies Zwart S F and McMillan S L W 2002 *ApJ* **576** 899

[21] Gürkan M A, Freitag M and Rasio F A 2004 *ApJ* **604** 632

[22] Hansen B M S and Milosavljevic M 2003 *ApJ* **593** L77

[23] Paumard T et al. 2006 *ApJ* **643** 1011–35

[24] Eisenhauer F et al. 2003 *The ESO Messenger* **113** 17

[25] Eisenhauer F et al. 2003 *Proc. SPIE* **4814** 1548

[26] Bonnet H et al. 2003 *Proc. SPIE* **4839** 329

[27] Eisenhauer F et al. 2005 *ApJ* **628** 246

[28] Genzel R, Thatte N, Krabbe A, Kroker H and Tacconi-Garman L E 1996 *ApJ* **472** 153

[29] Genzel R, Pichon C, Eckart A, Gerhard O and Ott T 2000 *MNRAS* **317** 348

[30] Paumard T, Maillard J-P, Morris M and Rigaut F 2001 *A&A* **366** 466

[31] Paumard T, Genzel R, Maillard J-P, Ott T, Morris M R, Eisenhauer F and Abuter R 2004 *Young Local Universe, Proceedings of the XXXIXth Rencontres de Moriond, ed. Chalabaev, A., Fukui, T., Montmerle, T., & Tran-Thanh-Van, J. (Paris: Éditions Frontières)* 377, http://www-laog.obs.ujf-grenoble.fr/ylu/ylu_vgr/vgr_index.html

[32] Maillard J-P, Paumard T, Stolovy S R and Rigaut F 2004 *A&A* **423** 155

[33] Schödel R, Eckart A, Iserlohe C, Genzel R and Ott T 2005 *ApJ* **625** 111

Institute of Physics Publishing
doi:10.1088/1742-6596/54/1/034
Journal of Physics: Conference Series **54** (2006) 208–216
Galaxy Center Workshop 2006

Star formation in the GC: observational constraints and theory

Sergei Nayakshin
Dept. of Physics & Astronomy, University of Leicester, LE1 7RH, UK

Abstract. We review some observational aspects and theoretical models for the formation of the young massive stars in the central parsec of our Galaxy. It is concluded that an in-situ star formation model provides the best description of the data. One possible guess for the initial gas configuration is one circular and another eccentric gas discs, not necessarily coeval during their short lifetimes. The circular disc probably resulted from a mass deposition event long compared with the local orbital time, whereas the eccentric disc should have formed more dynamically. Based on X-ray observations of the GC, we argue that the initial mass function (IMF) of stars formed in these discs is top-heavy. Such top-heavy IMF is most likely a result of inefficiency with which these optically thick discs fragment, and thus may be a generic result for star-forming AGN discs. We also present preliminary results of numerical simulations of star forming discs, showing that stars with a top-heavy IMF can be formed in both circular and eccentric discs. However, the inner arcsecond "S-stars", and the massive star cluster IRS13E, do not naturally form in these star-forming discs, and still lack explanation.

1. Introduction

Young massive "He-I" stars dominate the power output of the central parsec [13] of our Galaxy. Many of these stars are now resolved to be located in a very well defined and rather thin stellar disc that rotates clockwise as seen on the sky [15, 5, 27]. The rest of the stars can be arguably classified as a second more diffuse disc or a feature that rotates counter clock-wise [5, 27]. Stars appear to be on more eccentric orbits in that disc [27, 17]. Such bright massive stars seem to be excluded from the central arcsecond ($1'' \approx 0.04$ pc), which instead contains a dozen less massive but still quite young B-type stars [7, 8, 3]. In this paper we will not discuss these latter stars except to say they are clearly important but we do not yet have a good model for these.

"Standard" models of star formation are not easily applicable here due to a huge tidal field of the central object at $R = 0.1$ pc distances from Sgr A*. The required gas density is $n_H > 10^{11}\text{cm}^{-3}(R/0.1\text{pc})^{-3}$.

Two models seem most promising. [6] suggested that the stars may have been formed at a distance of tens of parsecs, thus avoiding the need for the excessive gas density prior to star formation, in a massive star cluster. The cluster's orbit would then decay through dynamical friction with the background stars. Further detailed models showed that the cluster needs to be very massive ($M \sim 10^6\,\text{M}_\odot$), very compact, and also contain an intermediate mass black hole [11, 12, 10]. In the other model, stars are formed in situ out of a massive cold and self-gravitating disc [15, 21]. This process has been predicted by theorists [26, 9] but has not yet been observationally verified.

Below we review some of the evidence for in-situ formation of the young stars near Sgr A*, and top-heavy nature of their IMF. We also discuss models of a gaseous star-forming disc.

2. Evidence for in-situ star formation

[27] find a well defined outer edge of about $\simeq 0.5$ parsec to the observed distribution of young massive stars in the GC. This is difficult to understand if stars migrated into the inner parsec from outside. In the massive young cluster disruption model [6], the cluster is disrupted gradually as it sinks closer in, first loosing stars most loosely bound to it. Detailed models of this process [10] show that stars are distributed anywhere between the cluster's place of birth to the disruption radius. If the "pre-maturely" lost stars were massive stars, then the NIR observations of [27] should have detected them.

In principle, there remains a possibility that the cluster was very strongly mass-segregated, i.e., that massive stars were present only in its core, and the rest of it was made up of low mass stars, $M \sim \mathrm{M}_\odot$. In that case, perhaps, all high mass stars could be transported within the inner parsec. This model seems to be challenged by X-ray data. Recent calibration of X-ray properties of T Tauri stars by [28] revealed that they are very bright in X-rays, i.e. some three orders of magnitude brighter than they are on the main sequence. Their spectra have both a soft and a hard component. The latter is observable even at high extinction regions such as the GC. These young stars also display giant X-ray flares, with some becoming as bright as $\sim 10^{33}$ erg s^{-1}. Importantly, the bulk of X-ray emission clearly comes from magnetic flares on the stellar surface, proving that the emission is due to internal relaxation processes of young stars and is thus independent of the large scale star formation environment.

[24] used these new X-ray results for young stars, and observations of the GC reported in [33], to show that the inner ~ 10 parsec could not be hiding more than 10^4 or so young low mass stars. This is insufficient: some $10^5 - 10^6\,\mathrm{M}_\odot$ of young stellar mass is needed to make the cluster heavy enough for it to sink in during the short lifetime of the young massive stars.

3. Top-heavy (or bottom-light) IMF of stellar discs

Because Sgr A* is so dim in X-rays, it turned out possible to push the X-ray constraints further. [1] described the properties of the unresolved X-ray emission near Sgr A* in detail. [24] used these results to deduce that the area most densely populated by massive young stars, i.e., the inner $R \leq 0.2$ parsec could contain no more than $\sim 10^3\,\mathrm{M}_\odot$ of low-mass YSO ($M < 3\,\mathrm{M}_\odot$). This is interesting since a factor of 10 or so more would be expected if the IMF were the "normal" galactic one, such as [18]. [31] reported adaptive optics NIR observations of the Arches cluster that show a "bottom-light" IMF for the inner parts of the cluster. [32], using *Chandra* X-ray data, also finds a deficiency of low mass stars for both Arches and the Quintuplet clusters.

While this similarity in deficiency of low mass stars between the three young star clusters is suggestive and intriguing, the interpretation is not yet clear. For the central star cluster, due to high velocity dispersions there, the relaxation time is around $\sim 10^9$ years. Hence any low mass stars, if born there on orbits similar to those of "He-I" stars, would still be present now. The present day MF for these low mass stars should thus be the same as their IMF. On the other hand, for Arches and Quintuplet clusters the relaxation times are comparable to their age estimates. It is possible that these clusters had a more normal IMF as low mass stars were evaporated off the clusters.

4. Constraints on total stellar masses from orbital modelling

Suppose that the two stellar systems were created infinitely thin and flat, as they would be in the simplest self-gravitating disc scenario. With time, an isolated stellar disc will thicken due to internal N-body heating, and two stellar discs will warp each other due to their non-spherical gravitational potentials [22, 21]. Both of these effects are stronger the more massive the stellar discs are. Now, if the initial systems were not thin and flat, then the thickening and warping will be even quicker (see [25]). Therefore, demanding that the model orbital configuration fits two stellar planes no worse than the real stars do at the present moment, we can arrive at a constraint on the total mass of these discs. This is useful as it is independent of X-ray and NIR constraints, and could potentially uncover "dark matter" in these discs.

The simplest constraint comes from the small height to radius ratio of the innermost clockwise system [27], $H/R \sim 0.1$. This by itself requires the disc mass to be less than $\sim 10^5\,M_\odot$ [21]. To test disc warping constraints, [25] set up the initial stellar discs as two infinitely thin systems of $N-$body particles on circular orbits and then evolved them for 3 million years. The disc angular momentum vectors were assumed to be given by the best fits to the observed stellar discs [27].

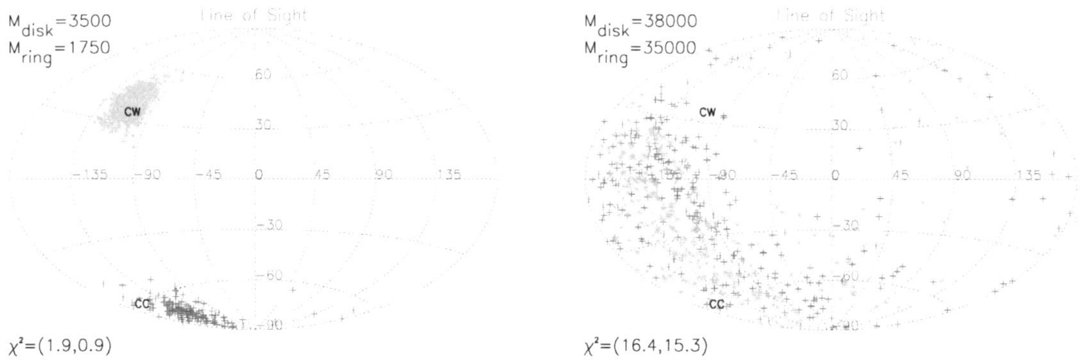

Figure 1. The projections of the angular momentum vectors of individual stars for two different tests from [25]. The positions of the clockwise and counter-clockwise systems from [27] are shown with symbols CW and CC, respectively. Total disc and ring masses are labelled at the upper left corner. The values of the minimum reduced χ^2 fits to the (ring, disc) systems are displayed in the lower left corner of the Figure. The test on left easily passes comparison with the data, whereas the one on the right is strongly ruled out.

Figure 1 shows two such calculations, one with low discs masses, and another with high (the values of disc masses are shown in the upper left corner of each panel). In the low mass case, hardly any orbital precession took place, and hence the systems remain to be very well defined nearly flat stellar discs. The χ^2 fit to these configurations yields small values of χ^2 (see the lower left corner of the panels). The right hand figure shows a very strongly disturbed system. The values of χ^2 from the observations [5, 27] yield limits of around $10^4\,M_\odot$ for the total masses of each of the stellar discs. We believe these are robust lower limits, however one calls or models the second stellar disc/feature, as eccentric or thickened stellar systems are even more vulnerable to warping [25].

5. A basic self-gravitating gaseous disc model

From a theoretical perspective, the most conveniently studied disc model is one which is just on the brink of gravitational collapse. For this to be the case, the disc has to be both massive enough and be able to cool fast enough: Toomre parameter $Q = 1$, and $t_{\rm cool} \lesssim 3\Omega^{-1}$ [26, 4], where $\Omega = \sqrt{GM_{\rm BH}/R^3}$. This theoretical construction, which most properly called a marginally star-forming disc, may be quite relevant to real AGN discs. Indeed, if the disc came into existence through a gradual mass accumulation from larger scales (see §8), it is quite likely that the disc mass was increasing very slowly compared to the local dynamical time, $1/\Omega$. In that case, the disc will always end up in the marginally star-forming state when enough mass is deposited. The subsequent disc evolution will of course depend on whether mass deposition stops or continues after star formation sets in.

Vertically integrated marginally star-forming disc models have been considered recently by [16] and [23] with applications to Sgr A* in mind. Such a model yields the minimum disk surface density, Σ_d, as a function of radius. The upper panel of Figure 2 shows the resulting minimum disc "mass" as a function of radius, $M_d = \pi \Sigma_d R^2$, and the disc midplane temperature (multiplied by 10^3 for clarity).

Note that if we decided to make stars out of a given amount of gas, then the most economical place to form them would be around 0.1 pc, since this would require the minimum amount of gas, i.e. a fraction of $10^4\,M_\odot$ (see Figure 2). Farther out and further in more gas is needed. In fact, detailed models that bridge the gap between the self-gravitating and "standard" disc regimes predict that star formation does not take place within about $R \sim 0.01 - 0.03$ parsec (depending on opacity and other model details) for Sgr A* mass [2]. Essentially, gas flows into Sgr A* far too fast and is too hot to form stars inside that radius.

Summarising, this basic model makes two predictions that are consistent with the observations: (i) the mass budget should be around $10^4\,M_\odot$, and (ii) there should be an inner edge to stellar disc of about an arcsecond.

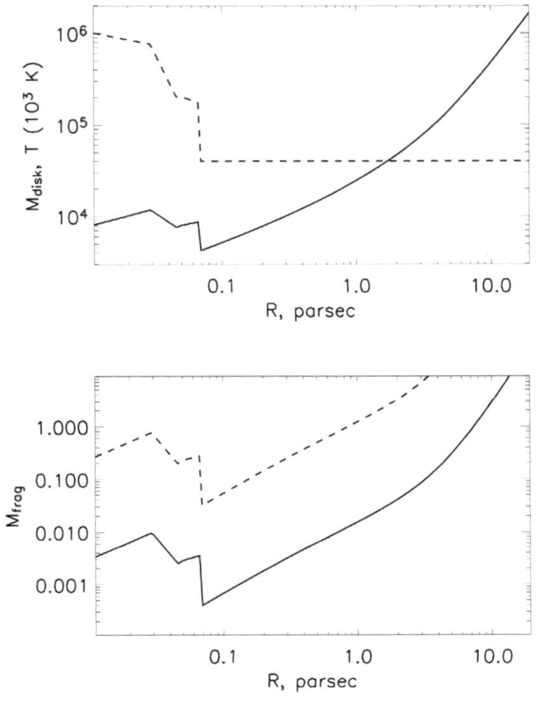

Figure 2. Upper panel: Disk mass (solid curve), $M_d = \pi \Sigma R^2$, in units of Solar masses, and midplane temperature (dashed, in units of $10^3 K$) as a function of distance from the SMBH are shown in the upper panel. The SMBH mass is that of Sgr A*. Lower panel: two estimates of the mass of the first fragments forming in the disk. The realistic value of the fragments mass is likely to be in between these two curves. For more details refer to [23].

6. The IMF of star-forming discs: theory

The estimates for the fragmentation mass, $M_{\rm frag}$, i.e. the mass of the first bound fragments, range between different authors due to an uncertain geometrical coefficient; but they are within the two estimates shown in the lower panel of Figure 2. From the Figure, $M_{\rm frag}$ is always sub-solar in the observationally interesting range of radii, i.e. $R \sim 0.1 - 1$ pc. Hence, if disk were to collapse into clumps of mass of this order, one would expect low-mass stars to dominate the mass spectrum of collapsed objects.

However, in reality, collapse of gas clumps may not be dynamical (as assumed by the simple model), but gradual, regulated by cooling and clump rotation. As gas clump collapse occurs in optically thick conditions, one can show that the minimum mass of a fragment will be $\sim 0.1\,\mathrm{M}_\odot$ (Nayakshin 2006, in preparation). Further, the clumps will have sizes comparable to the dimensions of the "first cores" [14], which are non-negligible. The clumps are thus likely to merge and grow by agglomeration [16]. Therefore one expects that $M_{\rm frag}$ is a strong underestimate of the actual stellar mass. Further, the proto-stars may grow to larger masses by accretion. [23] suggested that luminosity of the young stars accreting gas from within a star-forming disc is sufficient to heat the disc up, increasing the Toomre Q-parameter of the disc above unity and hence making the disc stable to further fragmentation. The proto-stars then continue to accrete mass by accretion, becoming very massive.

It may be that processes discussed by [16] and [23] that lead to a top-heavy IMF both operate, and hence such an IMF may be a general outcome for fragmentation in optically thick self-gravitating AGN discs. Finally, here we neglected possible magnetic fields. If strong enough, these could increase the Jeans mass of the clumps to tens of Solar masses [20].

7. Numerical models of star-forming discs

We shall now present some of the results of numerical simulations of star forming discs that will be published in full elsewhere. We use the SPH/N-body code GADGET-2 [30] to simulate the dynamics of stars and gas in the (Newtonian) gravitational field of Sgr A*. Star formation is treated on an individual star basis, i.e., a "proto-star" is introduced when gas density exceeds a critical density. Some adjustments had to be introduced to model the conditions unchartered by "normal" star formation simulations. Dynamical time scales in our discs are actually shorter than collapse time scale of "first" proto-stellar cores [14]. Therefore, in our approach, the newly born proto-stars have finite sizes, consistent with the size of the first cores (~ 5 AU), and can thus merge with other star particles. Once the proto-stars exceed the mass of $\sim 0.1\,\mathrm{M}_\odot$, the cores should collapse dynamically to much smaller linear scales. Hence stars more massive than $0.1\,\mathrm{M}_\odot$ accrete gas but are not allowed to merge with one another. Radiative cooling of the disc is replaced with a simple locally constant cooling time prescription, $t_{\rm cool} = \beta/\Omega$, where β is a constant of order unity. Value of $\beta = 3$ amounts to assuming that radiative cooling time is *fixed* at the value from the analytical model of marginally star-forming disc.

Circular gas discs without feedback. Following this approach, we ran several models, some designed to test our numerical methods against known results [29]. No star formation feedback was included, assuming that all the radiation generated by star formation leaves the disc freely. The initial condition is a gas disc of mass $2 \times 10^4\,\mathrm{M}_\odot$ in Keplerian circular rotation around Sgr A*, extending from 1" to 4" (1"≈ 0.04 pc). Figure 3 shows a snapshot of a run with $\beta = 3$ well into the non-linear stage, at time $t \sim 10^4$ years, when more than half of the gas were already turned into stars. The innermost disc is almost all turned into stars by that moment.

Surprisingly, gravitational heating generated by stars (scattering off each other and interacting with gas) is sufficient to heat the disc up above its pre-star-formation value, slowing down and even shutting off fragmentation at later times. Clearly, the effect is more pronounced the longer the cooling time, as the disc heats up to higher temperatures during star formation. It should then come as no surprise that the IMF formed in our simulations is a strong function of cooling time (i.e. β), as is evident from Figure 4. The IMF becomes top heavy for $\beta \gtrsim 1$. A cautionary note here is that an accurate modelling of radiative cooling of the disc is necessary for future detailed numerical simulations, as in reality $t_{\rm cool}$ will be a function of time and position. This may be a serious computational challenge for the future, as these discs are often optically thick and nearly razor-thin.

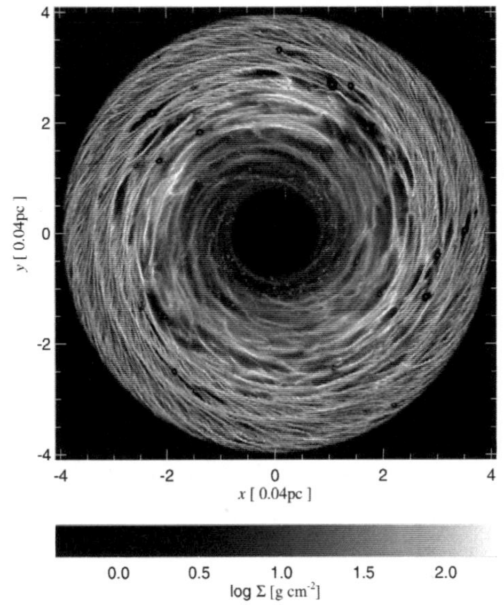

Figure 3. Snapshot of a star-forming disc of initial mass $2 \times 10^4\,M_\odot$ at time $t = 10,000$ years from the beginning of the simulation. Cooling parameter $\beta = 3$. Stars are shown as asterisks. Star formation is fastest in the innermost region, where most of the gas is already depleted. At the end of the simulation essentially all the gas is turned into stars. [Note that, unlike "normal" star formation, feedback from massive stars will not be very effective in blowing the gas away via radiation pressure or winds as the stars are within the deep potential well due to Sgr A*.] The stars thus steal the majority of Sgr A*'s dinner.

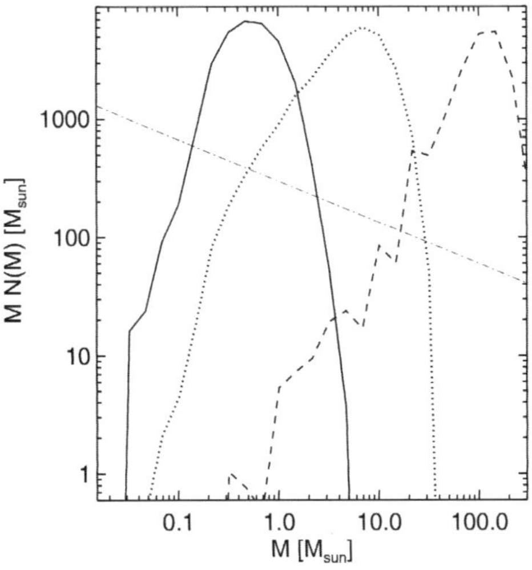

Figure 4. The IMF of three tests with three different values of $\beta = 0.3, 2$, and 3, shown with the solid, dotted and dashed curves, respectively. The Salpeter-mass function power law is shown with the dot-dashed curve for comparison. The main result is: the longer the cooling time, the more massive stars become, as clumps merging and accretion onto stars overtake disc fragmentation.

Simulations with feedback. We have also ran several tests where *thermal* feedback from gas accretion onto stars was added. These did confirm the suggestion made by [23] that feedback will stifle further disc fragmentation. Unfortunately for the simulations, this reduces not only fragmentation but also accretion onto stars, as gas is hotter. As a result, only $\sim 2\%$ of gas was turned into stars in the simulations with feedback, while in the simulations without thermal feedback, most of gas was reprocessed into stars by the same physical time. We could not run the feedback simulations long enough to obtain the resulting IMF, unfortunately, although the average mass of the proto-stars was increasing during the runs, as predicted.

Eccentric gas discs. To the best of our knowledge, all theoretical literature on star-forming discs was devoted to circular gas discs. Perhaps for this reason, suggestions that star formation cannot occur in eccentric discs were made. This would be a serious flaw of disc models for the

GC young stars, as many stars are on significantly eccentric orbits [17, 27].

However, there is no fundamental law of Nature against star formation in eccentric gas discs. Such orbits could even favour fragmentation because the range of radii, and hence physical conditions sampled by the orbit, is larger than for a circular orbit of same semi-major axis. To test this issue numerically, we ran a test completely analogous to those performed for circular discs but with gas placed on a non-circular orbit (see Figure 5). In particlar, we placed the gas in a small segment of a disc bounded by $4'' \leq R \leq 7''$ and azimuthal angle $0 \leq \phi \leq \pi/4$, and assumed the gas velocity to be 0.7 times the local Keplerian value. The cooling parameter $\beta = 3$. We also included the potential of the older stellar cusp as in [5]. For eccentric orbits, the resulting orbital precession leads to mixing of the orbits and shocks.

We found that eccentric discs can fragment and form stars just like circular discs can. One surprise was that the IMF formed in this simulation was even more top-heavy than in the circular orbit test with the same value of β. It is possible that stronger shock heating helps in a further reduction of disc fragmentation in favour of accretion onto existing stars. Note that stellar orbits differ from that of the gas at latter times because they are not affected by the shocks developing in the gas.

8. Discussion and Conclusions

Key conclusions on the current state of research in the field (in my perhaps biased view) are,

(i) **Origin of young stars:** star formation in situ, inside massive gaseous discs. If, instead, stars originated from outside the inner parsec, a trail of young stars would be seen – in the NIR for bright massive ones [27], and in the X-rays for low mass stars [24].

(ii) **IMF** of young stars in the central parsec is top-heavy, as is evident from both the X-ray [24] and NIR observations [27]. Analytical and numerical models of star forming discs suggests that such a top-heavy IMF can form due to inefficiency with which bound gas clumps collapse. Either these clumps are long-lived and merge quickly with other clumps [16], or (and) that the disc fragmentation as a whole is suppressed by star formation feedback [23]. Alternatively, if magnetic fields are important, the initial Jeans mass can be high [20].

(iii) **Inner edge of stellar discs** is predicted to be around 1" [21] for the star-forming disc model, which agrees well with the observed stellar distribution [27].

(iv) Theoretical prediction for the **stellar mass budget** is at least a fraction of $10^4\,M_\odot$ (see Fig. 2), in agreement with the observations.

(v) **Eccentric stellar orbits.** Numerical simulations, while still in an early stage, show that eccentric gas discs can also form stars with a top-heavy IMF. Orbital precession of a stellar disc filled with eccentric orbits is quicker than that of a circular disc [25], and hence such an eccentric geometrically thin disc would evolve into quite a diffuse feature by now, not unlike the observed counter clock-wise one.

(vi) **Origin of gaseous discs:** infall of one or two large molecular clouds. The observed stellar features form large angles with respect to each other, and the Galaxy plane [27]. The stars in the discs are co-eval within about a million years, which is much shorter than the estimated viscous time of discs [21].This implies that the discs could not have been assembled from larger scale discs via viscous angular momentum transport. The same conclusion follows from the fact that one of the discs should have had eccentric orbits. These orbits would be circularised if viscous angular momentum transfer were important, or if the disc formation time were long, as the orbits would have precessed and destroyed non-circular motions by then.

Amongst most interesting outstanding problems are:

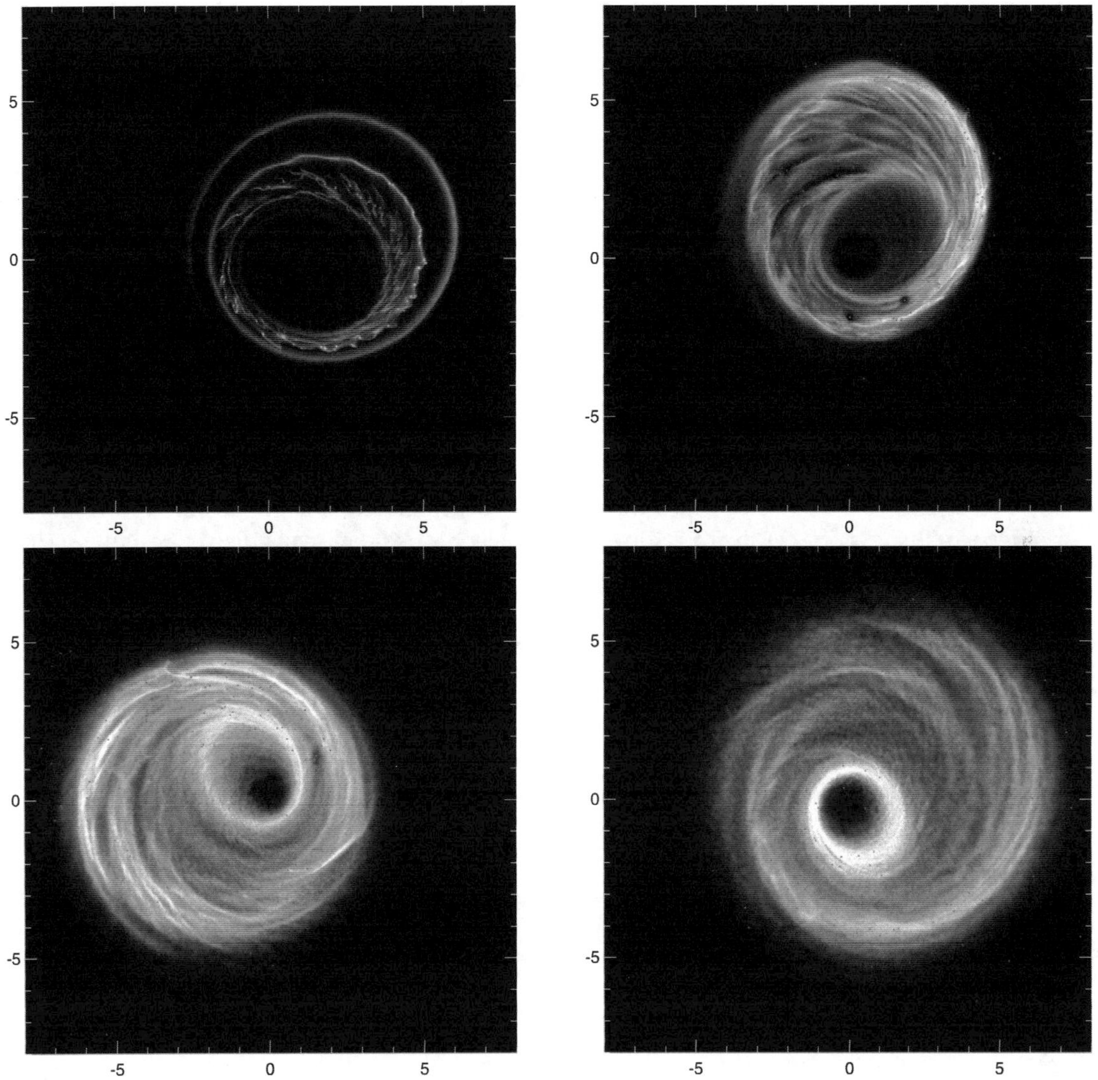

Figure 5. Snapshots of the eccentric disc segment evolution, corresponding to $t = 9, 27, 60$, and 120 thousand years, from left to right and top to bottom. Initially the disc segment is tidally sheared in an eccentric spiral. Precessing, the spiral forms an eccentric disc. Outer and inner edges of the disc precess at different rates. The outer edge of the disc makes slightly more than one revolution during the entire simulation.

(i) **Origin of gas** in the central parsec needs to be quantitatively modelled. Hopefully such a modelling could indicate properties of the gas cloud(s) that formed the discs, e.g., the orbit of the cloud, whether it was clumpy, bound or unbound, etc.

(ii) **IRS13E**: this enigmatic star cluster presents an unsolved problem. In principle, such a cluster could be formed in situ, in an accretion disc, as argued by [19, 21]. In numerical simulations to date, however, gaseous disc fragments on too many independent proto-stars which prevents the growth of a single dominant feature by dynamical interactions between each other. In other words, gaseous mass tends to be spread too evenly in the simulations.

(iii) **S-stars** The early type stars of the inner arcsecond, most of which are on eccentric orbits, remain unexplained in the context of an star-forming disc model (see point *iii* above).

(iv) **Cluster and disc model combined?** In connection with the points (i-iii) just made, we believe it may be premature to rule out the role of a star cluster/IMBH in the formation of the young stars. In particular, a molecular cloud infalling directly into the inner parsec could already have high density regions since "normal" GMC are not homogeneous at all. Some of these regions could start forming stars "on the fly", whereas the lower density gas would have to first get shocked and compressed into the circular clock-wise disc before starting to form stars.

(v) **Implications for AGN and quasars in general.** Young stars near Sgr A^* confirm that even moderate amounts of gas, i.e. $\sim 10^4\,M_\odot$, can become gravitationally unstable and form stars in the inner parsecs of galaxies. This is a fascinating process long predicted by theorists [26]. The problem is that the current models predict too high an efficiency with which gas is turned into stars (e.g., none of our self-gravitating disc models survived long enough to see inside Sgr A^*.) Generic theoretical arguments [9] suggest that the same must be true for massive quasar discs, even if star formation feedback is accounted for. Thus, understanding how Sgr A^* young stars were formed out of a gaseous disc is a small victory in the face of a bigger problem: how do more massive discs feed AGN and quasars instead of forming stars like Sgr A^* did?

References

[1] Baganoff, F. K., et al. 2003, ApJ, 591, 891
[2] Collin, S., & Zahn, J.-P. 1999, A&A, 344, 433
[3] Eisenhauer, F., et al. 2005, ApJ, 628, 246
[4] Gammie, C. F. 2001, ApJ, 553, 174
[5] Genzel, R., et al. 2003, ApJ, 594, 812
[6] Gerhard, O. 2001, ApJL, 546, L39
[7] Ghez, A. M., et al. 2003, ApJL, 586, L127
[8] Ghez, A. M., Salim, S., Hornstein, S. D., Tanner, A., Lu, J. R., Morris, M., Becklin, E. E., & Duchêne, G. 2005, ApJ, 620, 744
[9] Goodman, J. 2003, MNRAS, 339, 937
[10] Gürkan M. A., Rasio F. A., 2005, ApJ, 628, 236
[11] Hansen, B. M. S., & Milosavljević, M. 2003, ApJL, 593, L77
[12] Kim, S. S., Figer, D. F., & Morris, M. 2004, ApJL, 607, L123
[13] Krabbe, A., et al. 1995, ApJL, 447, L95
[14] Larson, R. B. 1969, MNRAS, 145, 271
[15] Levin, Y., & Beloborodov, A. M. 2003, ApJL, 590, L33
[16] Levin, Y. 2006, ArXiv Astrophysics e-prints, arXiv:astro-ph/0603583
[17] Lu, J., Ghez, A., et al. 2006, these proceedings.
[18] Miller, G. E., & Scalo, J. M. 1979, ApJS, 41, 513
[19] Milosavljević, M., & Loeb, A. 2004, ApJL, 604, L45
[20] Morris, M. 1993, ApJ, 408, 496
[21] Nayakshin S., Cuadra J., 2005, A&A, 437, 437
[22] Nayakshin, S. 2005, MNRAS, 359, 545
[23] Nayakshin, S. 2006, to appear in MNRAS (astro-ph/0512255)
[24] Nayakshin S., Sunyaev R., 2005, MNRAS, 364, L23
[25] Nayakshin, S., Dehnen, W., Cuadra, J., & Genzel, R. 2006, MNRAS, 366, 1410
[26] Paczyński B., 1978, Acta Astron., 28, 91
[27] Paumard, T., et al. 2006, ApJ, 643, 1011
[28] Preibisch, T., et al. 2005, ApJS, 160, 401
[29] Rice, W. K. M., Lodato, G., & Armitage, P. J. 2005, MNRAS, 364, L56
[30] Springel, V. 2005, MNRAS, 364, 1105
[31] Stolte, A., Brandner, W., Grebel, E. K., Lenzen, R., & Lagrange, A.-M. 2005, ApJL, 628, L113
[32] Wang, Q. D. 2006, ArXiv Astrophysics e-prints, arXiv:astro-ph/0606414
[33] Muno, M. P., et al. 2004, ApJ, 613, 326

Institute of Physics Publishing Journal of Physics: Conference Series **54** (2006) 217–223
doi:10.1088/1742-6596/54/1/035 Galaxy Center Workshop 2006

The Arches cluster – evidence for a truncated MF in the GC?

Andrea Stolte[1]

[1] Department of Astronomy, UCLA, 475 Portola Plaze, Los Angeles, CA 90095, USA

E-mail: astolte@astro.ucla.edu

Abstract. The Arches cluster serves as an indicator for star formation and cluster evolution in the dense GC environment. A high-resolution adaptive optics study of the starburst suggests a present-day mass function (MF) that deviates significantly from the standard IMF in the intermediate-mass regime. A turn-over is observed around $6 - 7\,M_\odot$, below which the MF appears to decline. Above $\sim 10\,M_\odot$, the integrated MF in the cluster center, $r < 0.4$ pc, displays a moderately flattened slope of $\Gamma = -0.9 \pm 0.15$, close to the Orion IMF with $\Gamma = -1.1$ [8]. In particular in the cluster core, $r < 0.2$ pc, a pronounced peak is visible at $7\,M_\odot$, while a weaker change in the MF slope is found in the annulus, $0.2 < r < 0.4$ pc. We compare the Arches cluster to the starburst cluster NGC 3603 in the Carina spiral arm, which displays a moderately flattened present-day MF with $\Gamma = -0.9$ all the way down to subsolar masses. The deviation in the Arches MF can be caused by accelerated dynamical evolution due to tidal disruption of the cluster in the GC potential, or can be the cause of a deviating intial mass function in the Galactic Center. Theoretical arguments for a potentially deviating IMF in a dense environment are recalled.

1. Introduction

The Arches cluster is with an estimated total mass of $M \sim 10^4\,M_\odot$ and a core density of $3 \times 10^5\,M_\odot\mathrm{pc}^{-3}$ one of the most massive starburst clusters in the Milky Way [3, 4]. The cluster is very compact with a core radius of 0.2 pc [15], but due to it's projected distance of only 25 pc from the GC, it is expected to be rapidly disrupted in the GC tidal field. Due to its young age of only 2 Myr, the cluster has preserved its central compactness, and thus provides us with a unique testbed for a single-age starburst population evolving close to the GC, and thus for the stellar mass function in the GC environment. This is particularly important as the nuclear cluster itself is comprised of mixed age populations likely formed during different cycles of star formation and possibly at different galactocentric radii, complicating the derivation of the stellar mass function close to the GC and especially conclusions on the initial MF.

The sample of resolved starburst clusters in the Milky Way consists of only a few clusters, such that conclusions on the differences in the starburst MF close to the GC and in the outer Galaxy suffer from statistics. A starburst showing very similar characteristics to the Arches is NGC 3603 in the Carina spiral arm. NGC 3603 has an estimated total mass of $7000\,M_\odot$, a core density of $\sim 10^5\,M_\odot\mathrm{pc}^{-3}$, and is with an age of 1 Myr in a comparable stage of stellar and dynamical evolution. We therefore chose NGC 3603 for a direct comparison with Arches.

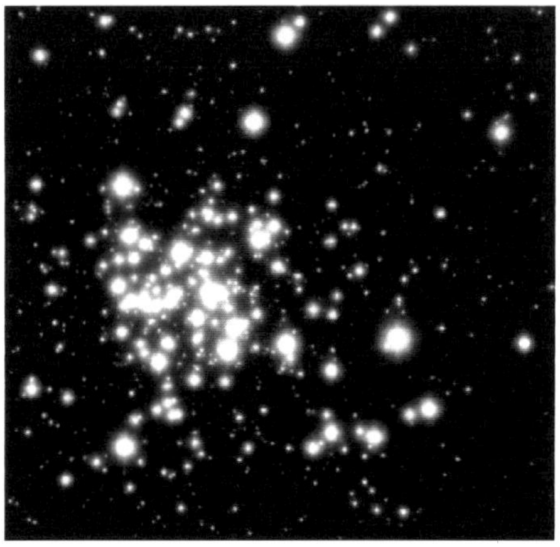

Figure 1. High-angular resolution VLT/NAOS-CONICA *HK* composite image of the Arches cluster at a resulotion of 84 milliarcseconds. The full extent of the image is approximately 1 pc. Note the large number of bright, high-mass cluster members.

Figure 2. Seeing-limited VLT/ISAAC *JHK* composite image of the starburst NGC 3603 in the Carina arm at a resolution of 0.″4. This ISAAC excerpt covers approximately 2 pc, about twice the radial extent of the Arches image.

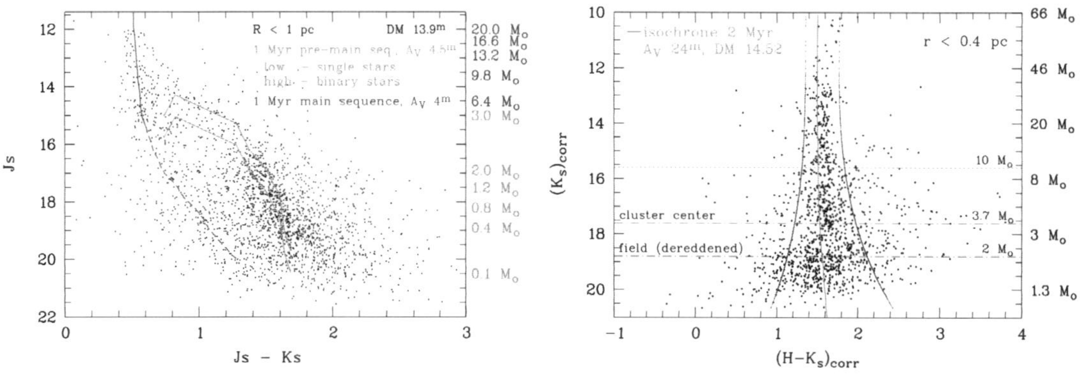

Figure 3. NGC 3603 (left, adapted from Stolte et al. 2006 [14]) and Arches (right, adapted from Stolte et al. 2005 [13]) colour-magnitude diagrams. The brightness is shown on the left axis, while the right axis displays mass estimates derived from the isochrones used to calculate mass functions (see text for details). Magnitudes and colours in the Arches CMD are corrected for a systematic radial extinction variation observed in the cluster field [16]. Main sequence and pre-main sequence isochrones are overlaid for NGC 3603, while for Arches only the main sequence enters the MF. The dark grey lines in the Arches CMD envelope the colour selection of likely cluster members. Note that below $K = 18$ mag, bulge contamination is expected to dominate the stellar population in the direction towards Arches.

2. Starburst Cluster Mass Functions

The MFs were created from the colour-magnitude diagrams (Fig. 3) using Geneva main sequence models to transform magnitudes into stellar masses [6]. In the case of the Arches cluster, a systematic radial increase in visual extinction was corrected, and a red and blue colour cut was

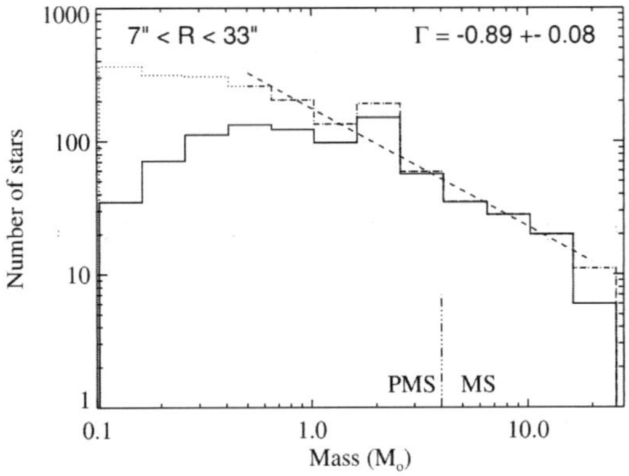

Figure 4. Present-day mass function of NGC 3603. The dashed line shows a linear least-squares fit to the incompleteness corrected MF (dash-dotted line). The MF is consistent with a single slope between 20 and $0.4\,M_\odot$, the 50% completeness limit. The irregularity around $M \sim 2-3M_\odot$ is due to the transition between main sequence and pre-main sequence, where the CMD is very scattered and the evolutionary status of stars is ambiguous.

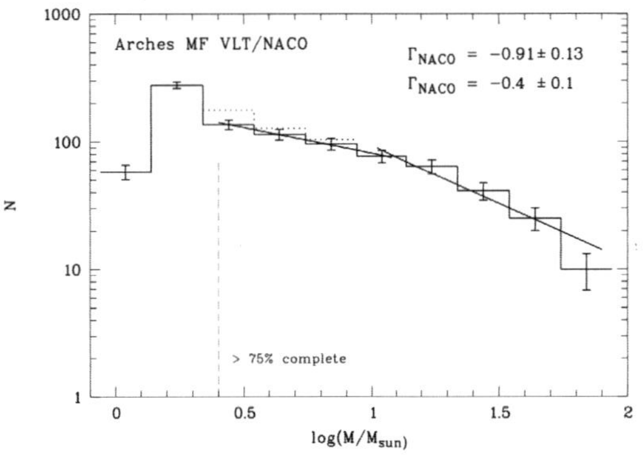

Figure 5. Arches present-day mass function. The slope of $\Gamma = -0.4$ between 4 and $12\,M_\odot$ appears shallower than the slope at the high-mass end, where $\Gamma = -0.9$. The flattened slopes with respect to a Salpeter IMF indicate an overdensity of intermediate- to high-mass stars, which is particularly pronounced in the range $4 < M/M_\odot < 12$. Due to the onset of severe field contamination below $4\,M_\odot$, the low-mass MF does not represent the cluster anymore.

applied to reject background and foreground stars to avoid field contamination. The colour selection of likely cluster members includes sources along the line of sight which by chance have colours similar to Arches main sequence stars. As the field star contribution increases with decreasing luminosity and mass, the remaining field contamination is expected to steepen the present-day MF. The flat MF derived for Arches is therefore a conservative estimate providing an upper limit to the slope of the MF in the cluster center.

In NGC 3603, stars were individually dereddened, and the field star contamination was subtracted statistically. The hydrogen burning turn-on in NGC 3603 occurs at $4\,M_\odot$, and for lower masses a Palla & Stahler 1 Myr pre-main sequence isochrone was emploid [10]. A secondary sequence observed at the high-mass end in NGC 3603 was modelled as equal-mass binaries, and a comparable statistical binary contribution of 30% equal-mass systems was assumed to be hidden in the low-mass PMS population.

The core of both clusters shows a heavy bias to high-mass stars (see Fig. 1), while the outer cluster regions appear closer to a standard Salpeter (1955) IMF. The integrated MFs for NGC 3603 and Arches are shown in Figs. 4 and 5, respectively. These MFs are radially limited to 1 pc in NGC 3603 and 0.4 pc in Arches both due to increasing field contamination, which is more severe in the GC environment. The present-day MF in NGC 3603 displays a slightly flattened, remarkably continuous power-law from the highest to the lowest masses (note that saturation limits us to $M < 20\,M_\odot$) with a slope of $\Gamma = -0.9 \pm 0.1$ for $0.4 < M < 20\,M_\odot$, influenced by the high-mass core causing the integrated slope to flatten.

The Arches MF displays the same slope of -0.9 ± 0.15 at the high-mass end, $10 < M < 65\,M_\odot$. This consistency is surprising given the different derivation methods used for both clusters and the different mass regimes covered. However, below $10\,M_\odot$, the Arches MF flattens to $\Gamma = -0.4$, indicating a lack of intermediate-mass stars. Below $\log M/M_\odot = 0.4$ or $2.5\,M_\odot$, the binned MF rises due to increasing field contamination by bulge stars, and does not represent the cluster population any more.

To investigate the behaviour further and reduce the effect of binning on the derived MF slope, we have created bin-scanned MFs. Bin-scanned MFs were calculated by shifting the starting point of the first mass bin by one-tenth of the bin width or $\log M/M_\odot = 0.02$ while keeping the bin width fixed to $\log M/M_\odot = 0.2$. This scanning corresponds to a small-step move down the colour-magnitude diagram, thus providing a better idea of the changes with decreasing stellar mass. However, note that this also implies that individual points are not statistically independent anymore, and the formal fitting error underestimates the true uncertainty in the MF slope (see [13] for a more detailed discussion of the bin-scan method). The bin-scanned integrated MF of the Arches cluster within a radius of 0.4 pc is shown in Fig. 6. In the mass range $6 < M < 60\,M_\odot$, the MF slope of -0.86 ± 0.1 is in very good agreement with the fit derived from the single-bin MF in Fig. 5. At the high-mass end, the steepening in the slope is most likely caused by statistical truncation when binning the CMD along the few very brightest stars. This is consistent with random sampling of a standard MF at the high-mass end. At the low-mass end, the onset of field contamination now becomes much more obvious at $4\,M_\odot$ ($\log M/M_\odot = 0.6$). Note that the details in the entire regime between 2 and $10\,M_\odot$ were completely wiped out when only one set of mass bins was used. In the bin-scanned MF, a peak appears at $6\,M_\odot$ ($\log M/M_\odot = 0.8$), and the MF appears to *decline* between 4 and $6\,M_\odot$, before field contamination dilutes the cluster population.

Unfortunately, no field data are yet available at comparable resolution, such that the cluster MF below $4\,M_\odot$ is presently unknown. An upper limit to the field was estimated from the very edges of our NACO field of view, which however is still as close as 0.5 pc to the cluster center and may contain a large fraction of low-mass cluster members in the case of dynamical mass segregation. This field estimate is included in Fig. 6, and suggests that as much as 50 % of the stars below $4\,M_\odot$ may be bulge stars.

In Fig. 7, we show the bin-scanned core MF vs. the annulus MF within $0.2 < r < 0.4$ pc. At the high-mass end, $12 < M < 60\,M_\odot$, the slope of $\Gamma = -0.3 \pm 0.15$ suggests a strong bias to high-mass stars in the cluster core, and thus strong mass segregation, which can be either a primordial or dynamical concentration of high-mass stars in the cluster core. Towards lower masses, the MF displays a pronounced peak at $7\,M_\odot$. In the core, the stellar population below $4\,M_\odot$ is intriguingly similar to the outer field at radii beyond 0.5 pc, which can be comprised of both contaminating background stars in the dense GC field as well as dynamically stripped cluster members. This either implies that the core lacks a large number of low-mass stars, or that the low-mass cluster population does not change with radius, while the intermediate- and high-mass stars are added to the core MF. In any event, the large number of high-mass stars in the cluster core, followed by a strong peak at $7\,M_\odot$ and a decline towards lower masses, does not resemble a standard MF power law.

In the subsequent annulus, $0.2 < r < 0.4$ pc, the deviation in the MF is not as extreme as in the core. The low number of high-mass stars in this bin causes irregularities at the high-mass end, such that the fit becomes more uncertain. The turn-over at $6\,M_\odot$ is still present, but the increased contamination of "field" stars below $4\,M_\odot$ renders the interpretation of the MF between 4 and $6\,M_\odot$ very difficult. The flattening in this regime to a slope of essentially $\Gamma = 0$ suggests a lack of intermediate-mass stars comparable to, but not as severe, as in the cluster core. It is clear that a reliable field estimate and subtraction is required at this point before a final conclusion on the cluster *initial* mass function can be made.

3. Observational evidence for a low-mass depleted MF in the GC

Why should the MF in the GC environment be biased to high-mass stars? The effect of the extreme conditions close to the GC can be estimated from the physical properties of

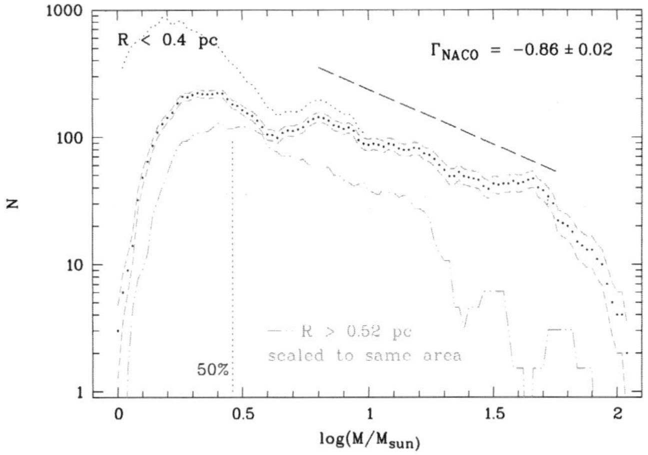

Figure 6. Bin-scanned, integrated MF in the inner 0.4 pc of the Arches cluster. The upper, dotted line is corrected for incompleteness, while the lower, dash-dotted line represents an upper limit to the field contribution derived from the edge of the NACO field of view, $R > 0.52$ pc, scaled to the same area (see Fig. 1). The MF slope, $\Gamma_{\rm NACO}$, is derived from a linear least-squares fit between 6 and 60 $M_\odot$.

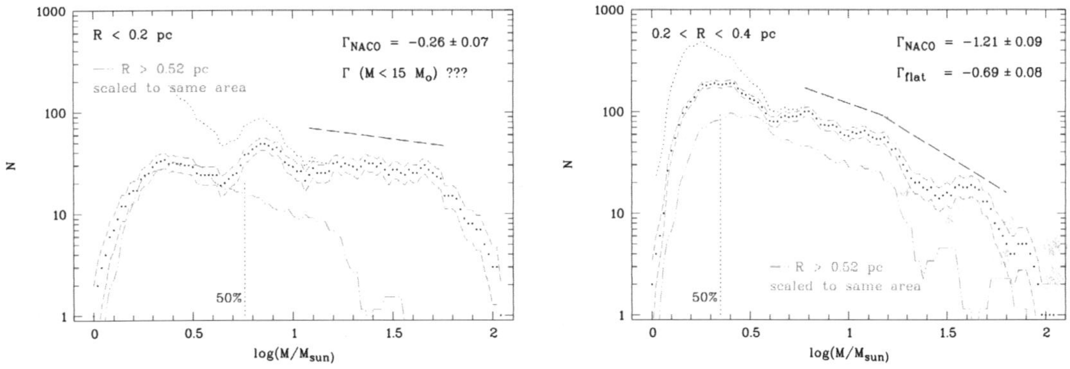

Figure 7. Radial MFs in the cluster center. The dash-dotted line is an upper limit to the field estimate scaled to the respective areas.
Left: Core MF inside 0.2 pc ($R < 5''$). A least-squares fit yields a flat core MF between 10 and 60 $M_\odot$. Right: Annulus MF between 0.2 and 0.4 pc ($5'' < R < 10''$). The MF was fitted in two sections, both of which indicate a steeper slope than the core. Note that the high-mass end of the MF becomes irregular due to the lack of high-mass stars in the annulus.

molecular clouds in the central molecular zone. Enhanced temperatures of 70 K in GC cores vs. 10 K observed in local star-forming cores [7], as well as increased hydrogen densities causing higher thermal pressure could explain a higher characteristic mass than in the solar environment. The characteristic mass can be approximated as the Bonnor-Ebert mass, $m_{BE} = 0.35(T_{core}/10K)^2(n_{H_2}T_{cloud}/10^6\,\mathrm{erg\,cm^{-3}})^{-1/2}$ [1]. In the solar neighbourhood, with $T_{core} = 10$ K, $T_{cloud} = 100$ K, $n_{H_2} = 10^{-4}\mathrm{cm}^{-3}$ this yields the well-known characteristic mass of $0.3\,M_\odot$, consistent with the turn-over in a standard IMF [5], e.g. as observed in the Trapezium cluster [8]. With $T_{core} = 70$ K and a thermal pressure elevated by a factor of 10 this mass increases to

$5\,M_{\odot}$. A similarly high characteristic mass is also suggested with enhanced GC cloud magnetic fields of 100μG [7] following the formulation of Shu [12]. Thus, theoretical arguments support an increased characteristic stellar mass emerging from the star formation process in density and temperature enhanced environments.

Very recently, observational support to a low-mass depleted IMF was also found in the central parsec. From the amount of X-ray sources and integrated light observed in the central parsec, Nayakshin & Sunyaev suggest a lack of low-mass YSOs in the GC [9] (see also Nayakshin, this volume). This depletion is concluded from the fact that young, low-mass stars with $M < 3\,M_{\odot}$ are strong X-ray emitters (as known e.g. from the COUP survey [2]). Nayakshin & Sunyaev investigate the possibility that the massive YSOs in the GC are the remnants of an inspiraling cluster, but conclude that this is unlikely, as such a cluster requires a total mass of $M_{initial} > 10^6\,M_{\odot}$ to survive long enough to channel stars all the way into the center of the Galaxy. The authors arrive at the conclusion that in-situ star formation with a top-heavy IMF that concentrates 99% of its total mass in stars with $M > 10\,M_{\odot}$ can explain the observed YSO population and X-ray constraints. This estimate is surprisingly consistent with a turn-over at $\sim 7\,M_{\odot}$ as observed in the Arches MF.

Nevertheless, one should keep in mind that the Arches MF is a present-day mass function, and no claim can be made from these data alone to reconstruct the *initial* MF. The two largest uncertainties in the MF are first, the unknown contribution of field stars towards lower masses, and second, the rapid dynamical evolution of the cluster in the GC tidal field, which has likely stripped a significant amount of stars from the cluster potential, which are missing in the present-day MF when compared to the IMF.

4. Conclusions

The present-day MF in the Arches cluster displays an irregular behaviour for masses below $\sim 10\,M_{\odot}$, with a turn-over at $6 - 7\,M_{\odot}$ particularly pronounced in the cluster core, but also evident in an annulus outside the core. A similar behaviour is not observed in the starburst cluster NGC 3603 in the Carina spiral arm, suggesting that star formation and cluster evolution of the Arches cluster are influenced by the GC environment. From the present data, we cannot distinguish dynamical evolution from deviations in the initial MF, and no meaningful field subtraction is available at the low-mass end. Although there is observational evidence for missing low-mass stars in the central parsec in addition to the apparent lack of low-mass stars in the Arches MF, a larger area around the Arches cluster has to be investigated before conclusions on the initial MF in the GC starburst and on the outcome of the star formation process in the GC in general can be drawn.

Acknowledgments

A special thanks to Rainer Schödel, who did an excellent job in organising this conference.

References

[1] Elmegreen, B. G. 2000, MNRAS, 311, L5
[2] Feigelson, E. D., Getman, K., Townsley, L. et al. 2005, ApJS, 160, 379
[3] Figer, D. F., Kim, S. S., Morris, M., et al. 1999, ApJ, 525, 750
[4] Figer, D. F., Najarro, F., Gilmore, Diane, et al. 2002, ApJ, 581, 258
[5] Kroupa, P. 2001, MNRAS, 322, 231
[6] Lejeune, T., Schaerer, D. 2001, A&A, 366, 538
[7] Morris, M., Serabyn, E. 1996, ARAA, 34, 645
[8] Muench, A. A., Lada, E. A., Lada, C. J., Alves, J. 2002, ApJ, 573, 366
[9] Nayakshin, S., Sunyaev, R. 2005, MNRAS Letters, 364, 23
[10] Palla, F., & Stahler, F. W. 1999, ApJ, 525, 772
[11] Salpeter, E. E. 1955, ApJ, 121, 161

[12] Shu, F. H., Zhi-Yun, L., Allen, A. 2004, ApJ, 601, 930
[13] Stolte, A., Brandner, W., Grebel, E. K., Lenzen, R., Lagrange, A.-M. 2005, ApJL, 628, 113
[14] Stolte, A., Brandner, W., Brandl, B., Zinnecker, H. 2006, AJ, 132, 253
[15] Stolte, A. 2003, PhD thesis, *Mass functions and mass segregation in young starburst clusters*, University of Heidelberg (http://www.uni-heidelberg.de/archiv/3611)
[16] Stolte, A., Grebel, E. K., Brandner, W., Figer, D. F. 2002, A&A, 394, 459

Institute of Physics Publishing
doi:10.1088/1742-6596/54/1/036
Journal of Physics: Conference Series **54** (2006) 224–232
Galaxy Center Workshop 2006

Metallicity in the GC

Francisco Najarro[1]

[1] Instituto de Estructura de la Materia, CSIC, Serrano 121, 29006 Madrid, Spain

E-mail: `najarro@damir.iem.csic.es`

Abstract. We review quantitative spectroscopic studies of massive stars in the three Galactic Center clusters: Quintuplet, Arches and Central cluster. Thanks to the impressive evolution of IR detectors and the new generation of line blanketed models for the extended atmospheres of hot stars we are able to accurately derive the physical properties and metallicity estimates of the massive stars in these clusters. For the Quintuplet cluster our analysis of the LBVs provides a direct estimate of α-elements and Fe chemical abundances in these objects. For the Arches cluster, we introduce a method based on the N abundance of WNL stars and the theory of evolution of massive stars. For the Central cluster, new observations reveal IRS 8 as an outsider with respect to the rest of massive stars in the cluster both in terms of age and location. Using the derived properties of IRS 8, a new method is presented to derive metallicity from the OIII feature at 2.115μm. Our results indicate solar metallicity in the three clusters.

1. Introduction

The detection of a He I emission line cluster (Krabbe et al. [26]) in the central parsec raised the question about the physical nature of its members and their role on the energetics of this region and triggered a substantial improvement in the atmospheric models for hot stars in the near-infrared. Thus, reliable values for luminosities, temperatures, mass-loss rates, helium abundances and ionizing photons were obtained for the brightest members of the He I cluster (Najarro et al. [34], [35]), which solved the energy puzzle of the Central cluster and placed important constraints on the theory of the evolution of massive stars.

Today, we know that the Galactic Center is clearly a unique environment in the Galaxy concerning massive star formation as it represents 10% of the present Galactic star formation activity. Hence, this region is ideal to address crucial questions such as the universality of the initial mass function (IMF) or the maximum mass a star can possess (Figer [19]). In fact, the Galactic center hosts three dense and massive star clusters that have formed in the inner 30 pc within the past 5 Myr. The Central cluster and the Quintuplet and Arches clusters host more stars with initial masses above 100$M_\odot$ than anywhere else in the Galaxy. Each cluster has a mass around $\sim 10^4 M_\odot$. The extreme youth (2-2.5Myr, Figer et. al [18], Najarro et al. [36]) of the Arches cluster allows to address the above questions using photometry alone (Figer [19]), while the Central and Quintuplet clusters being twice older may have lost their most massive members into the supernova stage.

On the other hand, a fundamental aspect such as metallicity, remains to be addressed. Indeed, the metallicity issue in the galactic center is still a source of controversy. Based on measurements of the gas-phase, Shields & Ferland ([42]) obtained twice the solar metallicity from Argon and Nitrogen emission lines while a solar abundance was derived for Neon. For the cool stars, and based on LTE-differential analysis with other cool supergiants, Carr, Sellgren, & Balachandran ([4]) and Ramirez et al. ([38]) have obtained strong indications for a solar Fe abundance. Further, Maeda et al. ([29]) obtained four times the solar abundance by fitting the X-ray local emission around Sgr. A East, while Koyama et al. ([28]) have recently derived 3.5 times solar abundance from the diffuse GC X-ray emission. It is therefore crucial to obtain metallicity estimates from direct analyses of hot stars and confront them with those from the cool-star and gas-phase analyses. Spectroscopic studies of photospheres and winds of massive hot stars are ideal tracers of metal abundances because they provide the most recent information about the natal clouds and environments where these objects formed.

In this paper, we review progress in both infrared observations and quantitative infrared spectroscopy of massive stars in the Quintuplet, Arches and Central clusters, which allow us to obtain direct abundance estimates of N, C, O, Si, Mg and Fe in the Galactic Center.

2. Improved observations and models

In the field of hot stars, high quality IR-spectra have been obtained during the last decade (Morris et al. [33]; Hanson, Conti & Rieke [21]; Figer, McLean, & Najarro [14]; Blum et al. [3]). Most of these spectra were obtained with low-mid (R$\sim$500-2000) resolution, which is enough to classify the stars but insufficient in most of the cases to perform accurate quantitative spectroscopic studies. An example of mid-high resolution observations of early type stars with strong winds is shown in Fig. 1 which displays the spectra at several IR bands of the "Pistol Star", a LBV in the Quintuplet cluster, which were obtained with UKIRT-CGS4 with a resolution of R$\sim$5000. The number of new observational constraints provided by the new spectroscopic data is striking when compared to previous low-mid observations from which we could gather information only from some H and He lines. Inspecting Fig. 1 we immediately note that the key diagnostic lines would be fully blurred at R$\sim$1000. The availability of high-quality IR spectroscopic data has been substantially improved with a new generation of IR-spectrographs on 8-m class telescopes (ISAAC, NIRSPEC, SINFONI, etc).

The new model (Hillier & Miller [23]) is a line blanketing method based on the standard iterative, non-LTE method to solve the radiative transfer equation for the expanding atmospheres of early-type stars. New species O, Mg, Ca, Si, Na, Al, Fe, etc are included and the blanketing ensures that the effect of continua on lines and lines on the continua as well as overlapping lines are automatically handled (see Hillier & Miller [23], [25] for a detailed discussion of the method). The new model is then prescribed by the stellar radius, R_*, the stellar luminosity, L_*, the mass-loss rate $\dot{M}$, the velocity field, $v(r)$, the volume filling factor f and the abundances of the element considered.

3. The Quintuplet Cluster

The Quintuplet Cluster (Glass, Catchpole, & Whitelock [8], Figer et. al [16], [17]) contains a variety of massive stars , including WN, WC, WN9/Ofpe, LBV and less evolved blue-supergiants. The presence of such stars constrains the cluster age to be about 4Myr old, assuming coeval formation. The cluster provides enough ionizing flux ($\sim 10^{51}$ photons s^{-1}) to ionize the nearby "Sickle" H II region and enough luminosity ($\sim 10^{7.5} L_\odot$) to heat the nearby molecular cloud, M0.20-0.033. Its total mass is estimated to be $\sim 10^4 M_\odot$. The presence of two LBVs ("Pistol Star" & #362, Geballe et al. [9]), with IR-spectra rich in metal lines (see Fig. 1), allows to obtain a direct estimate of the metallicity of the objects and hence constrain the metal enrichment history of the region. Furthermore, using the the Fe II, Si II & Mg II lines (Najarro et. al, in prep.) we can measure the α-elements vs. Fe ratio and infer whether the initial mass function (IMF) is dominated by massive stars or is like in other clusters in the Galaxy with a steeper slope. If the IMF is dominated by massive stars, we should expect enhanced yields of α-elements compared to Fe through a higher than average SNII vs SNIa events.

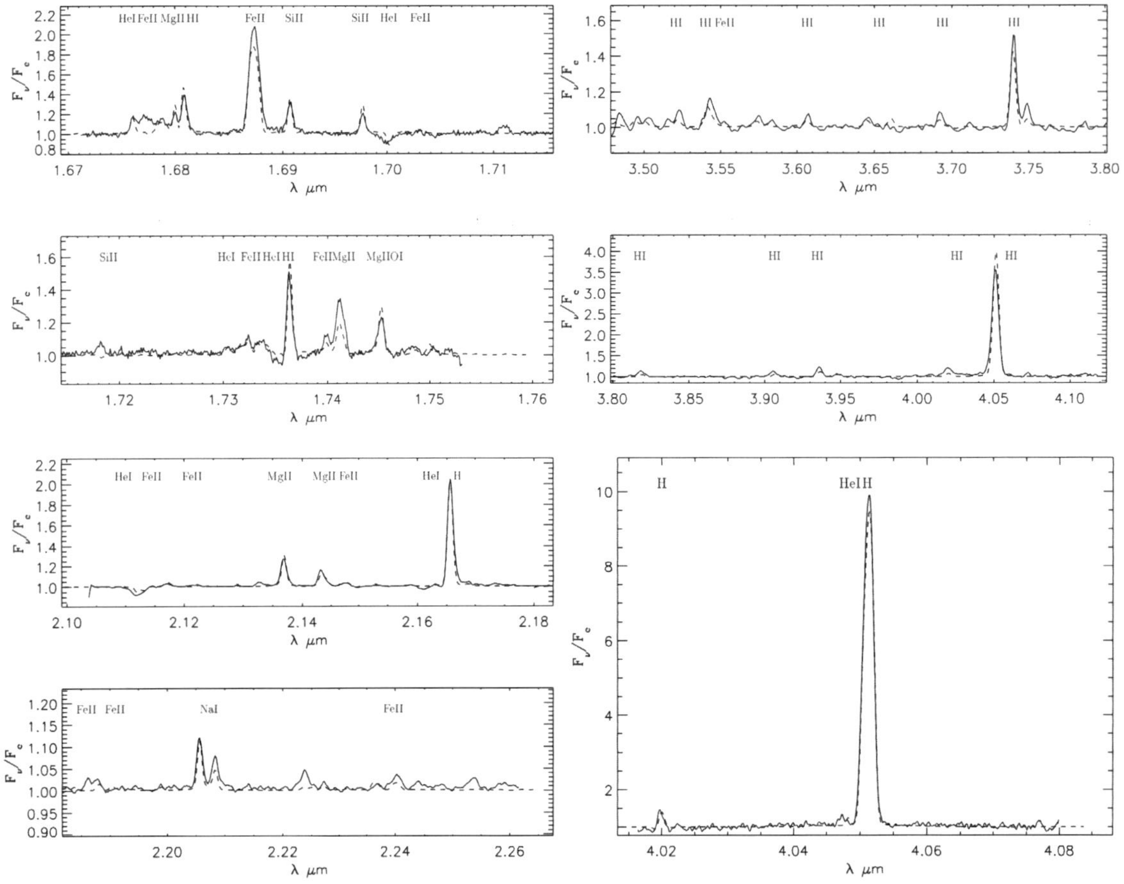

Figure 1. Impact of high S/N and mid-high spectral resolution and new blanketed models for early type stars in the infrared. UKIRT-CGS4 (R∼5000) H-, K-, L-Band and Brα observations of the "Pistol Star" in the Quintuplet cluster and new model fits.

3.1. The "Pistol Star" and Star #362

To model the "Pistol Star" and star #362 we have assumed the atmosphere to be composed of H, He, C, N, O, Si and Fe. The new blanketed models provide a significant improvement in our knowledge of the physical properties of the "Pistol Star" compared to the results we obtained in Figer et al. ([15]) using non-blanketed models. The new model solves the dichotomy between the "high" and "low" Luminosity (Teff) solutions in Figer et al. ([15]) through the analysis of the metal lines (see the excellent fits in Fig. 1). The Si II, Mg II and Fe II lines "choose" the low luminosity model. We find a luminosity around $1.75 \times 10^6 L_\odot$ and a effective temperature of $T_{\rm eff}$∼11000K. This result, which reduces the previous estimate of the star luminosity by a factor of two shows the importance of the new generation of models. Given the $T_{\rm eff}$ and high wind density of the object we do find a degeneracy in the H/He ratio (see also Hillier et al. [24]). In principle, fits of virtually equal quality may be obtained with H/He ratios varying from 10 to 0.05 by number. The only line that may help to break this degeneracy is the He I 2.112μm absorption line, which seems to favor H/He ratios between 3 and 0.05 by number. An important consequence of this degeneracy are the mass fractions derived for Fe, Si and Mg. Our models show that once the H/He ratio falls below one, the resulting metal abundances have to be scaled down. In other words, if H/He≤1 then we will obtain only an upper limit on the metal

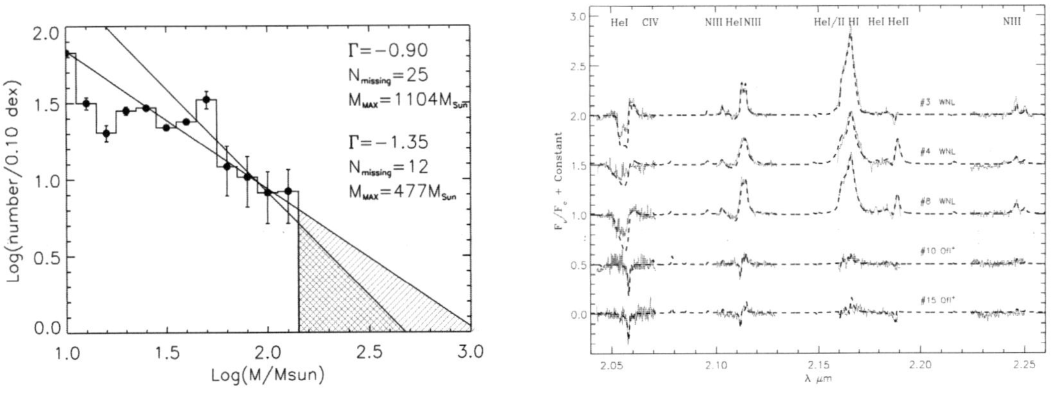

Figure 2. Left. Mass function of the Arches Cluster ([19]). The Salpeter mass function is overploted for comparison purposes. The hatched regions demonstrate that one would expect a significant number of massive stars exceeding $120M_{\odot}$. **Right.** Observed spectra (solid) and model fits (dashed) for the three WNL and two Ofl^{+} Arches stars.

abundances. We obtain solar iron abundance as upper limit for the "Pistol Star" (see line fits in Fig. 1). This estimate is rather robust as there is a large number of Fe II diagnostic lines. Further, if we assume that the object displays He enrichment consistent with an LBV evolutionary phase, H/He≥1, we may conclude that the "Pistol Star" shows **solar** Fe abundance, in agreement with previous estimates from differential analysis of cool stars in the GC (Carr et al. [4], Ramirez et al. [38]). The silicon abundance is obtained from the two Si II lines in the H band. These lines are extremely sensitive to the effective temperature of the star as well as to the transition zone between the star's photosphere and wind. Therefore, our $Si{\approx}1.4Si_{\odot}$ result should be regarded with some caution. Magnesium, on the other hand, provides more diagnostic lines through Mg II both in the H and K-Band. We regard $Mg{\approx}1.6Mg_{\odot}$ as our current best estimate. This slightly α-elements vs Fe enrichment seems to favor the situation of a IMF dominated by massive stars in the GC.

For star #362 (Geballe et al. [9]), a nearly twin of the "Pistol Star" we also obtain a luminosity around $1.710^{6}{\times}L_{\odot}$ and a effective temperature of $T_{eff}{\sim}10500K$. We derive a **solar** iron abundance as upper limit as well and obtain $Si{\approx}1.8Si_{\odot}$ and $Mg{\approx}2.2Mg_{\odot}$ in agreement with the slightly α-elements vs Fe enrichment found for the "Pistol Star".

4. The Arches Cluster

The Arches Cluster (Figer et. al [17], [18]) is the youngest and densest cluster at the Galactic center containing thousands of stars, including at least 160 O stars and around 10 WNLs (WN stars still showing H at their surface, Chiosi & Maeder [6]). The cluster is very young (≤2.5Myr), and the only emission line stars present are WNLs and OIf^{+} having infrared spectra dominated by H, He I, He II, N III lines. Some have weak C III/IV lines. The cluster gathers all requirements to study the high mass IMF slope and to estimate an upper mass cutoff: individual members can be resolved, large amount of mass in stars, young enough so that most massive members are not pre-supernovae and old enough for its stars to have emerged from their natal cocoons. Figure 2 (from Figer [19]) shows the mass function in the Arches cluster extended to very high masses as measured by Figer et al. ([16]). Its slope appears to be shallow with respect to the Salpeter value favoring the formation of very massive stars. Further, we see that one might expect massive stars up to 500-1000$M_{\odot}$, yet none are seen beyond ${\sim}120M_{\odot}$. Figure 2, thus clearly suggests the existence of an upper mass cutoff at ${\sim}150M_{\odot}$.

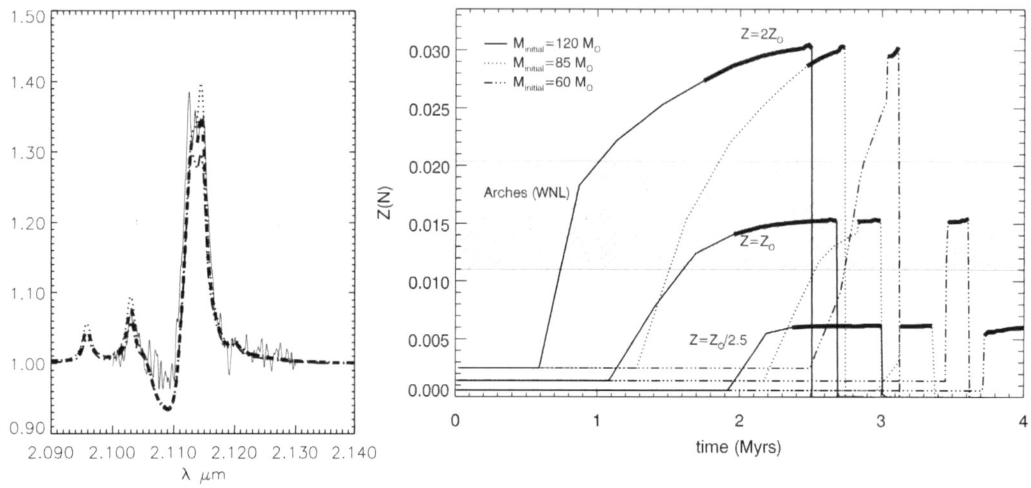

Figure 3. Left. Leverage of error estimates on N abundance for the Arches cluster. Shown is observed 2.10-2.13 μm region (solid) together with current best fit (dashed) and 30% enhanced (dotted) and 30% depleted (dashed-dotted) nitrogen mass fractions. **Right**. Nitrogen mass abundance in the Arches cluster versus time using Geneva models. The measurements require solar metallicity and an age of 2-2.5 Myr.

4.1. Metallicity studies

The absence of late B-supergiants and LBVs prevents one from obtaining direct estimates of the important α-elements vs. Fe metallicity ratio as in the Quintuplet cluster or the Central cluster. Being the youngest cluster at the Galactic center, any hint about its metallicity would constitute our "last-minute" picture of chemical enrichment of the central region in the Milky Way. To analyze the stars in the Arches cluster we have assumed the atmosphere to be composed of H, He, C, N, O, Si and Fe. Observational constraints are provided by the K-Band spectra of the stars (see Figure 2) and the narrow band HST/NICMOS photometry (filters $F_{\rm F110W}$, $F_{\rm F160W}$ & $F_{\rm F205W}$) and Pα equivalent width (filters $F_{\rm F187N}$ & $F_{\rm F190N}$). Object identifications are given according to Figer et al. ([18]). Below we present the results of our analysis (Najarro et al. [36]).

The reduced spectra and model fits are shown in Figure 2. The top three spectra correspond to some of the most luminous stars in the cluster. As described in Figer et al. ([18]), these are nitrogen-rich Wolf-Rayet stars with thick and fast winds. The bottom two spectra in the figure correspond to slightly less evolved stars with the characteristic morphology of OIf$^+$ stars. Of concern are the N III 8-7 lines at 2.103 μm and 2.115 μm as well as the N III 5p ^{2}P–5s ^{2}S doublet at 2.247 μm and 2.251 μm. Figer et al. (1997) showed that these N III lines appear only for a narrow range of temperatures and wind densities, which occur in the WN9h (WNL) stage. The fairly distinct nature and energies of the multiplets involved in each of both N III line sets provide strong constraints for the determination of the nitrogen abundance. Thus, at the S/N of our spectra, our models show that the WNLs N III lines can easily track relative changes as low as 20% in the nitrogen abundance, and a 30% error should be regarded as a safe estimate, as shown in Figure 3-*left*.

Of particular importance is the roughly same surface abundance fraction of N, Z(N), obtained in our analysis for all three WNL objects ($\sim$1.6%) well above the upper limit found for the OIf$^+$ stars ($\sim$0.6%). WNL stars do not exhibit any primary diagnostic line in their K-Band spectra to estimate metallicity. However, the crucial rôle of Z(N) in determining metallicity from WNL stars can be immediately anticipated if we make use of the stellar evolution models for massive stars.

According to the evolutionary models by Schaller et al. ([41]) and Charbonnel et al. ([5]), a star entering the WNL phase still shows H at its surface together with strong enhancement of helium and nitrogen and strong depletion of carbon and oxygen as expected from processed CNO material. During this phase, the star maintains a nearly constant Z(N) value, and the amount achieved essentially depends only linearly on the original metallicity (see Figure 3-*right*), being basically unaffected by the mass-loss rate assumed and the presence of stellar rotation during evolution (Meynet & Maeder [32]). Since we expect the CNO abundance in the natal cloud to scale as the rest of metals, the nitrogen surface abundance must trace the metallicity of the cluster. The parameters derived for these stars (Najarro et al. [36]) indicate that this is indeed the case for objects #3, #4 & #8. The derived Z(N) ($\sim$1.6%) is the one expected for **solar** metallicity from the evolutionary models. The reliability of our method is demonstrated in Figure 3-*right*, where we display the nitrogen mass fraction as a function of time for stars with initial masses of 60, 85, and 120 $M_{\odot}$, and metallicities equivalent to 2, 1, and 0.4 times solar, assuming the canonical mass-loss rates (Schaller et al. 1992). Our results for the WNL and O stars (cross-hatched region) require solar metallicity and an age of 2-2.5 *Myr* (see also Najarro et al. [36]).

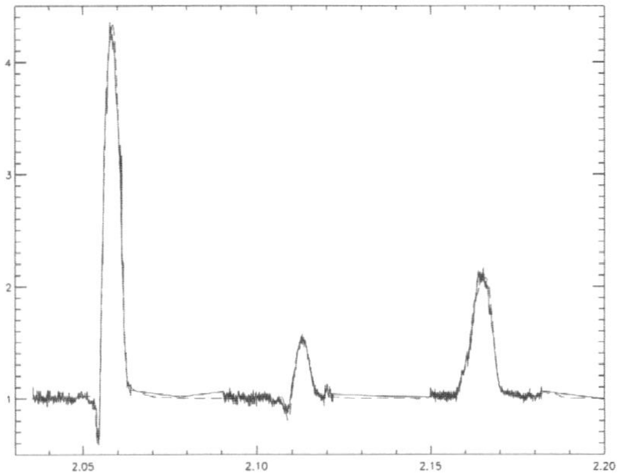

Figure 4. New K-Band high resolution observations and model fits to the AF star. Our models indicate a slightly higher but still consistent temperature and luminosity with the values obtained by Najarro et al. [34], [35].

5. The Central Cluster Revisited

The Central cluster hosts a large number of massive stars that have formed in the past 10 Myr (Becklin et al. [2], Krabbe et al. [26], [27], Najarro et al. [34], [35]).The latest census by Eisenhauer et al. [7] and Paumard et al. [37] (see also these proceedings) includes at least 80 massive stars, with $\sim$50 OB stars close to the main sequence or in their early supergiant phase and 30 more evolved massive stars which appear to be confined to two disks (Paumard et al. [37]). There is also a group of about a dozen of B stars within the central arcsecond (the "S"-stars, see these proceedings for thorough discussions on this topic). Interestingly Paumard et al. [37] do not detect any OB star outside the central 0.5pc and find that the stellar contents of both disks indicate a common age of 6$\pm$2*Myr*, with O8-9I as earliest spectral type detected.

To test whether the new generation of models and observations could alter considerably our current picture of the evolved massive stars in the Central cluster (Najarro et al. [34], [35]) as it has been the case for the "Pistol Star" (see above), we have started a re-analysis campaign using high resolution observations obtained with NIRSPEC (Keck) and our up-todate line blanketed code. Figure 4 displays K-Band obs and new model fits to the AF star. Our new fits indicate a slightly higher temperature and luminosity for this object but still consistent with the values and error estimates obtained by Najarro et al. [34], [35].

5.1. IRS 8: an outsider in the GC

The nature of the Galactic center source IRS 8 (Becklin & Neugebauer,[1]), one of the brightest compact mid-infrared sources in the central infrared cluster, was unknown until adaptive optics H- and K-band imaging revealed that the bulk of its infrared emission originates in a classic bowshock (Rigaut et al. [39], Geballe et al. [10]). Geballe et al. [10] showed that the IRS 8 bowshock is a straightforward consequence of the interaction of a dense and high velocity wind from a hot star (hereafter IRS 8*) that is traversing moderately dense interstellar gas. To investigate the nature of the central source we obtained mid resolution (R~900) K-Band spectra using the Gemini adaptive optics module ALTAIR to feed the near-infrared spectrograph NIRI (see also Geballe et al. [12]).

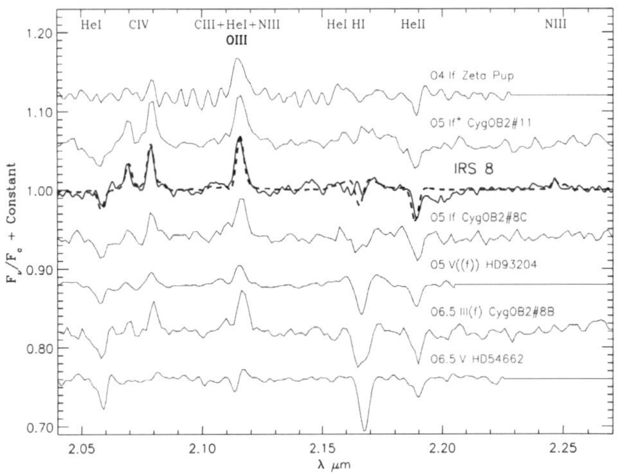

Figure 5. Spectral type determination of IRS 8*. Comparison of the resulting normalized spectrum with K-band spectra from Hanson et al.([21]) degraded to a resolution of R=800. Also displayed (dashed) is a model fit with stellar parameters corresponding to an O5.5If star (see text).

Figure 5 shows the resulting normalized K-band spectrum of IRS 8* compared with online-available K-band spectra from the Hanson et al. [21] catalog for O stars ranging from O4 to O6.5 and different luminosity classes. The resolving powers for all template spectra have been degraded to 800 for direct comparison with the observed spectrum. From Fig. 5 we judge that IRS 8* falls within the O5-O6.5 and III-If ranges, with likely O5-O6 If spectral type and luminosity class. Given the strong spectral similarities of IRS 8* with the O5-6 supergiants in Cyg OB2 (see Fig. 5), we computed model fits covering that parameter domain, drawing from our analysis of the Cyg OB2 stars for which UV, optical and IR spectra are available (Najarro et al. in prep.). Figure 5 displays our best-fitting model (dashed line) using the line blanketing method presented in previous sections (see Geballe et al. [12] for a thorough discussion of the analysis).

Of concern is the re-identification of the strong emission feature at 2.116 μm in IRS 8* which has been attributed in the past to C III and N III n=8–7 transitions and is present over a very wide range of O spectral types and luminosities (Hanson et al. [21]). Our investigation (Geballe et al. [12]) indicates that the 2.116 μm feature in IRS 8* is dominated by O III n=8–7 transitions. Further, the O III component of the 2.116 μm feature largely depends on the oxygen abundance and only slightly on gravity, effective temperature, wind density, and velocity field. Thus, this feature may be a powerful diagnostic of oxygen abundance, and therefore an important metal abundance determiner, over a wide range of O spectral types (Najarro et al. in preparation). Using it we obtain an oxygen abundance of 0.8 to 1.1 times solar in IRS 8*, which indicates **solar** metallicity for the cloud in which IRS 8* formed.

Our analysis suggests that IRS 8*, although only 1 pc from the center, does not fit into the current picture of the central cluster of hot stars. It is of much earlier spectral type than any

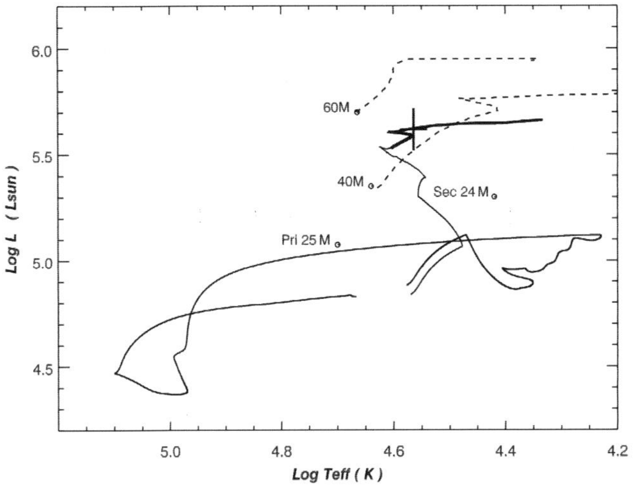

Figure 6. Position of IRS 8* (solid cross) in the HR diagram as estimated from model fits, compared with different evolutionary scenarios. The current location of IRS 8* is reached after 3.6 and 7.1 million years for the single star (dashed) and massive close binary (solid) evolutionary cases respectively (see Geballe et al. [12] for a thorough discussion).

of the stars classified by Paumard et al. [37]. Currently it is the only known OB star outside the central 0.5 pc region of the cluster. Figure 6 shows the position of IRS 8* (solid cross) as estimated from our model fits in the HR diagram compared with different evolutionary scenarios. The age of $2.8 Myr$ and no surface enrichment obtained with tracks of stars without rotation as used by Paumard et al. [37] (Maeder & Meynet [30]) is clearly at odds both with the current estimate for the age of the Galactic center cluster and with the abundance pattern derived from our models. The situation improves when evolutionary models accounting for rotation (dashed-lines in Fig. 6) are considered (3.5 Myr and CNO-processed material on the stellar surface). Except for the age, still well below the estimate obtained by Paumard et al. [37], using rotating models for a single burst scenario we obtain stellar parameters fully consistent with those derived from our modelling.

The crucial question thus is whether this star is really much younger than the cluster and probes the existence of ongoing (or at least much more recent) star formation, or if on the contrary the star is either an impostor or a cluster member that underwent a rejuvenation cure. A possible way out is provided if the star originally was a member of a massive close binary system. In such a case, we could be looking now at the secondary star, with the primary either exploded as supernova or in an evolutionary phase when it is much dimmer at K than the secondary. Using the models by Wellstein & Langer [44] we have found that for a massive close binary system with initial masses of 25 $M_\odot$ and 24 $M_\odot$ (their model 10a) the current position of IRS 8* may be elegantly explained (see Geballe et al. [12] for a thorough discussion) without violating the age of the Galactic center cluster (solid lines in Fig. 6). Similar scenarios are a possible explanation for some of the overluminous He I objects in the central parsec.

6. Conclusions

Our result of solar metallicity for the Central Cluster, the Arches Cluster and the Quintuplet Cluster runs counter to the trend in the disk (Rolleston et al. [40], Smartt et al. [43]) but is consistent with the findings from cool star studies (Carr et al. [4], Ramirez et al. [38]). This may imply that the ISM in the disk does not extend inward to the GC, or that the GC stars are forming out of an ISM that has an enrichment history that is distinctly different from that in the disk. Our result is more consistent with the values found for the bulge (Frogel et al. [20], Felzing & Gilmore [13]).

Acknowledgments

I would like to thank Don Figer, John Hillier, Rolf Kudritzki and Tom Geballe for invaluable discussions. F. N. acknowledges PNAYA-2003-02785-E and AYA2004-08271-C02-02 grants. Thanks Rainer for your patience with the manuscript.

References

[1] Becklin, E. E. & Neugebauer, G. 1975, ApJ, 200, L71
[2] Becklin, E. E., Matthews, K., Neugebauer, G. & Willner, S. P., 1978, ApJ, 219, 121
[3] Blum, R.D, Ramond, T.M., Conti, P.S., Figer, D.F., & Sellgren, K., 1997, AJ, 113, 1855
[4] Carr, J.S., Sellgren, K., & Balachandran, S.C., 2000, ApJ, 530, 307
[5] Charbonnel, C., Meynet, G., Maeder, A., Schaller, G., & Schaerer, D. 1993, A&As, 101, 415
[6] Chiosi, C. & Maeder, A. 1986, Annual Review of A&A, 24, 329
[7] Eisenhauer, F., et al., 2005, ApJ, 628, 246
[8] Glass, I. S., Catchpole, R. M., & Whitelock, P. A. 1987, MNRAS, 227, 373
[9] Geballe, T.R., Najarro, F., & Figer, D.F., 2000, ApJ, 530, L97
[10] Geballe, T. R., Rigaut, F., Roy, J.-R., & Draine, B. T. 2004, ApJ, 602, 770
[11] Feltzing, S., & Gilmore, G. 2000, A&A, 355, 949
[12] Geballe, T. R., Najarro, F., Rigaut, F., & Roy, J. -. 2006, arXiv:astro-ph/0607550
[13] Felzing, S. & Gilmore, G., 2000, A&A, 355, 949
[14] Figer, D.F., McLean, I.S., & Najarro, F., 1997, ApJ, 486, 420
[15] Figer, D. F., Najarro, F., Morris, M., McLean, I. S., Geballe, T. R., Ghez, A. M., & Langer, N. 1998, ApJ, 506, 384
[16] Figer, D.F., McLean, I.S., & Morris, M., 1999a, ApJ, 514, 202
[17] Figer, D.F., Kim, S.S., Morris, M., Serabyn, E., Rich, R.M., & McLean, I.S., 1999b, ApJ, 525, 750
[18] Figer, D.F., Najarro, F., Gilmore, D., et al., 2002, Morris, M., Kim, S.S., ApJ, 581, 258
[19] Figer, D.F., 2004, in proceedins of "IMF50"
[20] Frogel, J.A., Tiede, G.P., & Kuchinski, L.E. 1999, AJ, 117, 2296
[21] Hanson, M. M., Conti, P. S., & Rieke, M. J. 1996, ApJS, 107, 281
[22] Hanson, M. M., Kudritzki, R.-P., Kenworthy, M. A., Puls, J., & Tokunaga, A. T. 2005, ApJS, 161, 154
[23] Hillier, D.J., & Miller, D.L., 1998a, ApJ, 496, 407
[24] Hillier, D.J., Crowther, P.A., Najarro, F., Fullerton, A.W., 1998b, A&A, 340, 483
[25] Hillier, D.J. & Miller, D.L. 1999, ApJ, 519, 354
[26] Krabbe, A., Genzel, R., Drapatz, S., & Rotaciuc, V., 1991, ApJ, 382, L19
[27] Krabbe, A., et al., ApJ, 447, L95
[28] Koyama, L., et. al., 2006, PASJ, arXiv:astro-ph/0609215
[29] Maeda, Y., et. al., 2002, ApJ, 570, 671
[30] Maeder, A. & Meynet, G. 2003, A&A, 404, 975
[31] Meynet, G., Maeder, A., Schaller, G., Schaerer, D., & Charbonnel, C. 1994, A&A Supp., 103, 97
[32] Meynet, G.,& Maeder, A., 2004, A&A, 404, 975
[33] Morris, P.W., Eenens, P.R.J., Hanson, M.M., Conti, P.S., & Blum, R.D, 1996, ApJ, 470, 597
[34] Najarro, F., Hillier, D.J., Kudritzki, R.P., Krabbe, A., Genzel, R., Lutz, D., Drapatz, S., & Geballe, T.R., 1994, A&A, 285, 573
[35] Najarro, F., Krabbe, A., Genzel, R., Lutz, D., Kudritzki, R.P., & Hillier, D.J., 1997, A&A, 325, 700
[36] Najarro, F., Figer, D.F., Hillier, D.J., Kudritzki, R.P., 2004, ApJ, 611, L108
[37] Paumard, T., et al. 2006, ApJ, in press (astro-ph 0601268)
[38] Ramirez, S.V., Sellgren, K., Carr, J.S., Balachandran, S.C., Blum, R., Terndrup, D.M., & Steed, A., 2000, ApJ, 537, 205
[39] Rigaut, F., Geballe, T. R., Roy, J.-R., & Draine, B. T., 2003, in Galactic Center Workshop 2002: The Central 300 Parsecs of the Milky Way, ed. A. Cotera et al., Astron. Nachr., 324, S1, 551
[40] Rolleston, W.R., Smartt, S.J., Dufton, P.L., & Ryans, R.S.I., 2000, A&A, 363, 537
[41] Schaller, G., Schaerer, D., Meynet, G., & Maeder, A., 1992, A&A, 96, 269
[42] Shields, J.C., & Ferland, G.J., 1994, ApJ, 430, 236
[43] Smartt, S.J., Venn, K.A., Dufton, P.L., Lennon, D.J., Rolleston, W.R., & Keenan, F.P., 2001, A&A, 367, 86
[44] Wellstein, S. & Langer, N., 1999, A&A, 350, 148

Institute of Physics Publishing
doi:10.1088/1742-6596/54/1/037

Estimation of the Low-End Mass Function of the Arches Cluster

Jihye Shin, Sungsoo S. Kim

Department of Astronomy and Space Science, Kyung Hee University, Yongin-shi, Kyungki-do 449-701, South Korea

E-mail: jhshin@ap1.khu.ac.kr

Abstract. It has been suggested that the nonstandard star formation environment near the Galactic Center (GC) may lead to an initial mass function (IMF) skewed toward relatively massive stars and having an elevated mass cutoff. Two GC star clusters, the Arches and the Quintuplet, may give an answer to this hypothesis, but the most recent high-resolution IR observations for these clusters can only resolve stars with masses down to $\sim$5M$\odot$. Here, we present a new method for estimating the low-end mass function (MF) from a background-limited observation of a star cluster. We empirically find the mass function that gives the best match between the pixel intensity histograms of the observed image and the model image that is constructed for an assumed mass function. This method also employs Fokker-Planck (FP) models to reproduce the spatial distribution of stars as a function of mass.

1. Introduction

To make up for the limitation of conventional photometric methods, we introduce a novel method that utilizes the pixel intensity histogram of the cluster image. We apply this method to the HST/NICMOS image for the Arches cluster, which was analyzed by Figer et al.[1] with a conventional PSF (point-spread function) fitting technique. Our new method has the advantage of utilizing the unresolved, faint pixels, that are discarded in the conventional PSF fitting technique. Our goal is to search the best-fit initial mass function (IMF) by comparing the intensity histograms of the observed image and the aritificial images created with an assumed IMF. We perform FP simulations to obtain the density profile as a function of stellar mass at the observed age of the cluster for each comparison.

2. The Control Fields

In order to find how the background/foreground sources affect the intensity histogram of the cluster, we analyze the control field image first. The locations of the control fields are close enough to the cluster to be assumed as background/foreground sources of the Arches, and far enough from the cluster to be not affected by outer stars of the cluster. A conventional PSF-fitting photometry was performed to obtain the luminosity function (LF) of bright, resolved stars. We find that our photometry is at least 80% complete down to $\sim$8M$_\odot$ (K $\sim$15mag). For stars fainter than this mass, we adopt a galactic structure model by Wainscoat et al.[2].

The combined LF is shown in Fiqure 1. Artificial images are made by randomly positioning artificial stars with PSFs that are obtained from observations. To make the artificial image

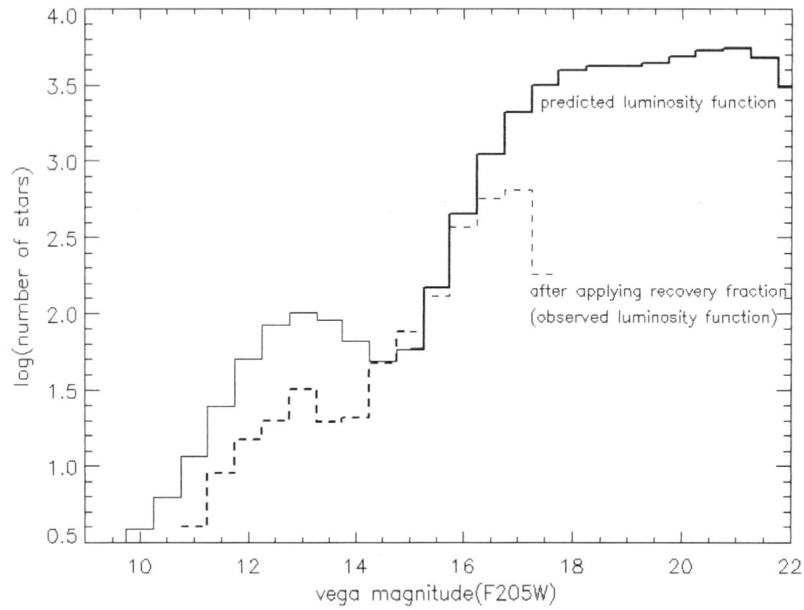

Figure 1. The solid line shows the model LF, and the dashed line shows the observed LF. We connect two LFs at 15th mag where the recovery rate is 80%. The thick line shows the combined LF.

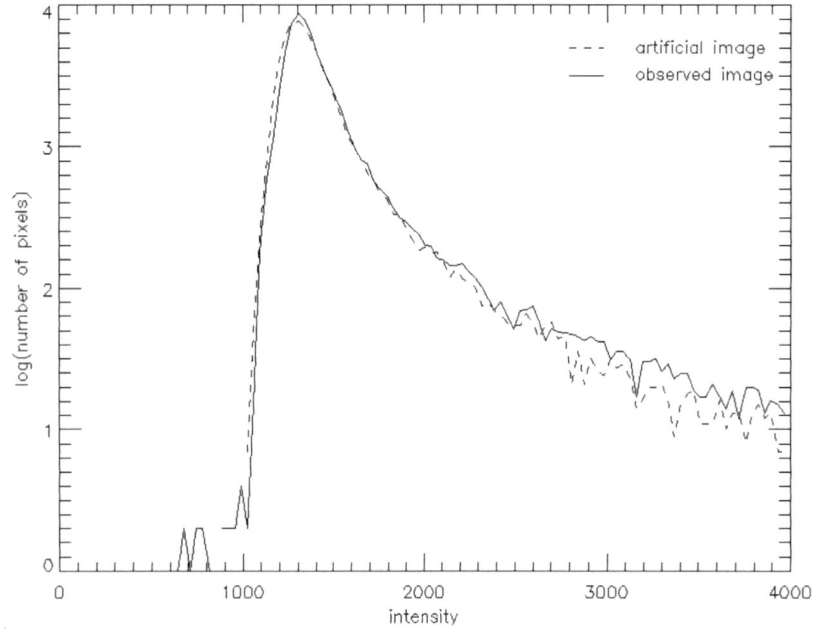

Figure 2. Comparison between intensity histograms of the observed control field image and the artificial image.

resemble the real control field image, several noises are added; dark noise, readout noise and telescope thermal noise. The pixel intensity histograms of the artificial images obtained by this way are in good agreement with that of the observed control field images (Fig.2). It means that we understand the phenomenon that constitutes the control field images and the background/foreground sources of the Arches.

3. The cluster

To relate MF to LF, we adopted the Geneva isochrone and assumed the age of the Arches cluster to be 2Myr [1]. The central region of the cluster is dominated by a number of very bright stars, while the outer part is dominated by background/foreground stars. Thus for the comparison the observation and our model, we consider two radius bins: 6"<r<9" and 9"<r<12" (1"≈0.04pc). Two variables, the peak intensity (hereafter, P) and the height of the peak (hereafter, H) of the

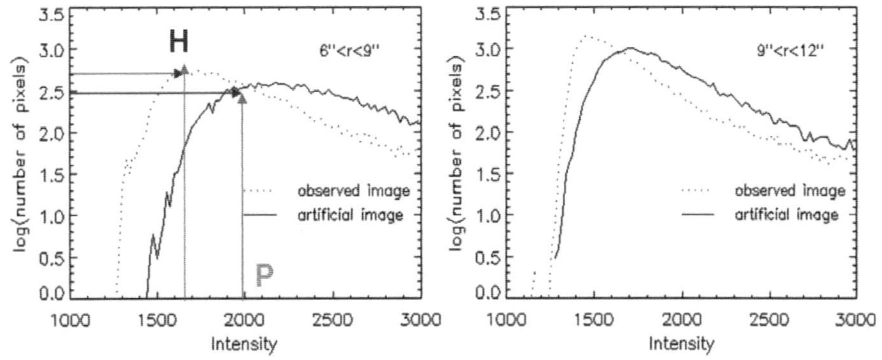

Figure 3. Comparing the intensity histogram of artificial cluster (α=2.3, M_l=0.3M$_\odot$, R_g=30pc) image and the Arches cluster image for each radius bin. Two parameters, P and H determines the shape of the intensity histogram.

histogram, were chosen to compare intensity histograms quantitatively. These two parameters determine the shapes of the intensity histograms (Fig. 3). The intensity histograms are found to be nearly log-normal, and we fit a lognormal function to the intensity histogram to obtain P and H.

3.1. Survey in the Initial [α, M_l] Space

FP simulations were performed to obtain the spatial distribution of stars as a function of mass at the age of the cluster, 2 Myr, under four initial condition: the power-law slope of the IMF (α, α=2.35 being the salpeter IMF), total cluster mass (M_t), tidal radius (R_t), and lower mass cutoff (M_l). Since M_t can be derived from α, R_t, and M_l and R_t can be derived from galcatocentric radius (R_g) and M_t, the number of paramters for initial conditions becomes 3: α=(1.4, 1.7, 2.0, 2.3, 2.6), M_l=(0.1, 0.3, 1.0, 3.0M$_\odot$), and R_g=(30, 60, 90pc); a total of 60 models.

For R_g=30pc, we find that models 9 and 15 show a good agreement to the observation. These models have α=2.3, M_l=0.3 M$_\odot$ and α=2.6, M$_l$=1 M$_\odot$, respectively (Fig. 4). When Rg=60 or 100 pc, no models give a satisfactory agreement between our calculation and the observation.

3.2. Survey in the Present-Day [α, M_l] Space

To see the reliability of the above procedure, we perform the same analysis again without the information on the spatial distribution of stars from the FP simulations (Fig. 5). As for the control fields, the bright-end LF is from the observation and the faint-end LF is modeled as a power-law with various α and M_l values. Since here we are comparing the assumed LF directly

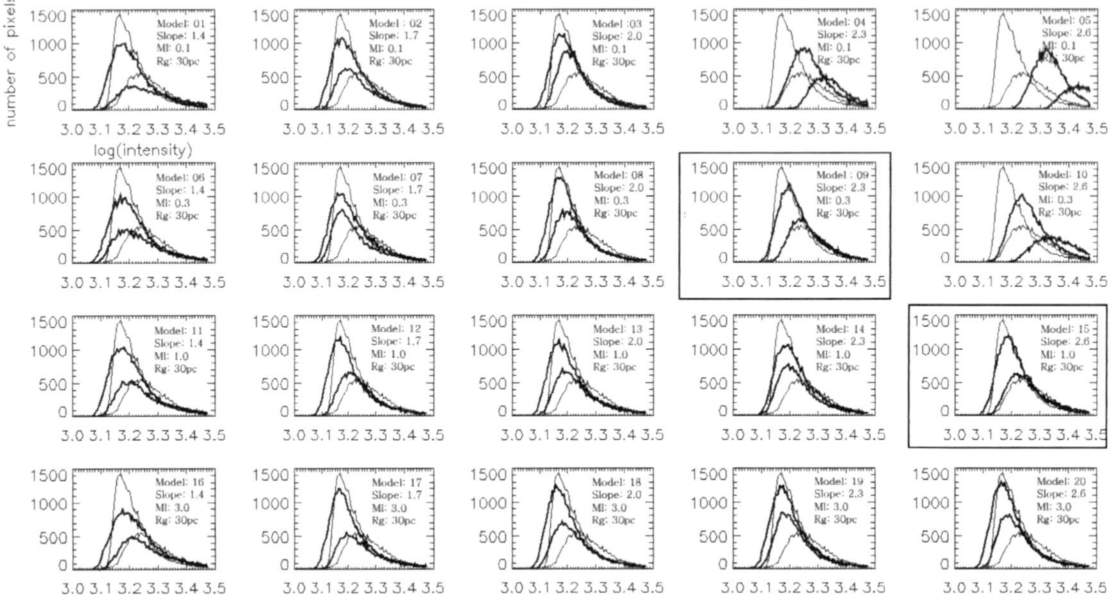

Figure 4. Comparison of the intensity histograms between the artificial images and the observed one (Rg=30pc). Thick lines: Artificial cluster image (upper lines: 6"<r<9", lower lines: 9"<r<12"), Thin lines: Observed cluster image (upper lines: 6"<r<9", lower lines: 9"<r<12"). Figures outlined are assumed to be the most appropriate models.

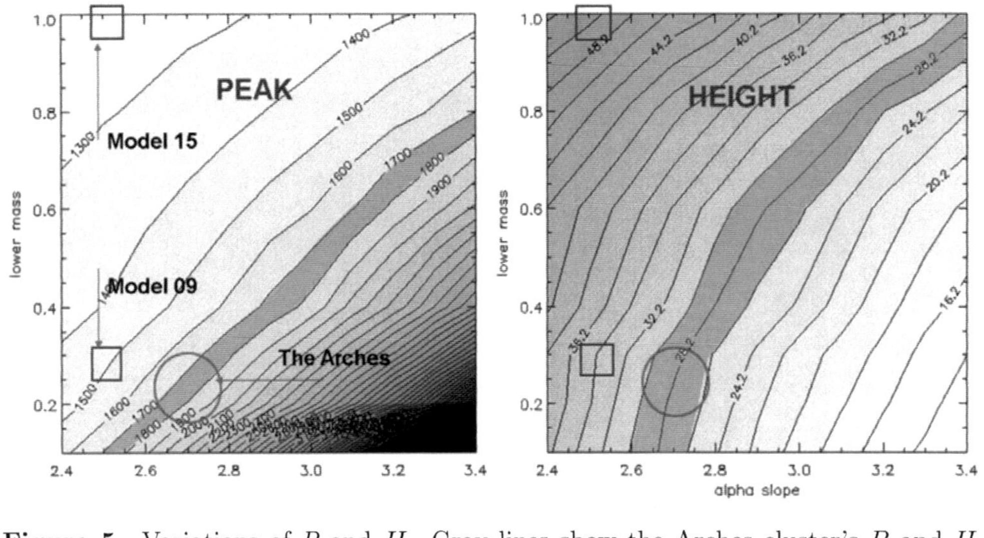

Figure 5. Variations of P and H. Gray lines show the Arches cluster's P and H. The circles show reliable MF crossed by two gay lines. The square show the models of 9 and 15 of Figure 4, the most appropriate FP models for the Arches.

with the observation (i.e., without dynamically evolving the cluster), what we find here is the present-day MF or LF. The region in α and M_l space that results in P and H that are consistent with the observation is shown with the gray lines in Figure 5. The overlapping region between the P and H maps has $\approx\sim$2.7 and $M_l \approx 0.25 M_\odot$. This is very close to the present-day α and M_l

values of Model 09 (2.5 and $0.3M_\odot$, respectively), implying that Model 09 probably represents the observed cluster better.

4. Conclusions

We have analyzed HST/NICMOS data of the Arches cluster near the GC with the implementation of the pixel intensity histogram method. We find that our new method can be used to estimate the faint-end LF that can not be recovered by the conventional photometric techniques. By comparing pixel intensity histograms between the artificial and observed images, we find that the best-fit power-law IMF parameters are α=2.3≈2.6, M_l=0.3≈1.0$M_\odot$. α and M_l values larger or smaller than these values give much less satisfactory fits to observations. None of the Kroupa IMFs with various M_l values give acceptable agreement with the observations. Thus our analyses imply that the IMF of the Arches cluster has a similar power-law slope to the disk population, but a significantly elevated lower-mass cutoff at around 0.3 to 1.0 $M_\odot$.

References

[1] Figer, D. F., Kim, S. S., Morris, M., Serabyn, E., Rich, R. M., McLean, I. S. 1999, *ApJ,* **525,** 750.
[2] Wainscoat, R. J., Cohen, M., Volk, K., Walker, H. J., Schwartz, D. E. 1992, *ApJ,* **83,** 111.

Institute of Physics Publishing
doi:10.1088/1742-6596/54/1/038
Journal of Physics: Conference Series **54** (2006) 238–242
Galaxy Center Workshop

Galactic structure and stellar clusters in obscured HII regions.

Maria Messineo[1], Monika Petr-Gotzens[1], Karl M. Menten[2], Frédéric Schuller[2] and Harm J. Habing[3]

[1] ESO, Garching, Germany
[2] MPIfR, Bonn, Germany
[3] Leiden Observatory, The Netherlands

E-mail: mmessine@eso.org

Abstract. Aiming to improve our understanding of the spatial distribution of Galactic HII regions we have started a program with the integral field spectrometer SINFONI on the ESO-VLT (Very Large Telescope) Yepun to observe obscured stellar clusters in HII regions. Through the detection and spectral classification of individual cluster stars we are able to obtain a complete census of the nature of the brightest stars and to obtain a spectrophotometric distance to the cluster. Such distance estimates are independent and complementary to kinematic distances and free from the ambiguity inherent to the latters' determination. Most importantly, our method is based on near-infrared observations, which are much less limited by interstellar extinction than previous optical programmes, allowing studies along lines of sight where optical measurements are impossible.

1. Introduction

Regions of star formation are good tracers of Galactic structure, in particular of the spiral arms. Unfortunately, determinations of their distances are extremely problematic and often rely only on velocities derived from radio observations. The kinematic distances of inner Galactic objects have an intrinsic ambiguity: the HII region may be "near" or "far". Furthermore it is based on an assumed model of Galactic rotation.

The first depiction of Galactic spiral arms [10] was derived from kinematic distances and spectro-photometric distances, the latter from visual observations of OB stars. Indeed spectrophotometric distances of OB stars may solve the ambiguity. The Georgelin's draw lacked information about the most inner regions of the Galaxy and the far side, directions which are typically not observable in the optical because of high interstellar extinction.

Nowadays the actual number and location of spiral arms is still uncertain [21]. Furthermore, only a few young and massive clusters (the Quintuplet, Arches, and Sgr A clusters) are known to reside inside the central 3 kpc, all of them concentrated within the central 50 parsecs. In contrast to the distribution of late-type stars ([18], [12]), which shows strong signatures of a Galactic bar extending to 3.5 kpc and of a nuclear stellar disk with a radius of about 800 pc, almost all known young stellar clusters are located outside of the co-rotation radius (3.5 kpc) on quasi-circular orbits. This puzzling lack of HII regions within 3 kpc of the Galactic centre suggests a "hole" in the gaseous distribution of the inner Galaxy. However, it could be partly due to the incompleteness in detecting HII regions along lines of sight with high interstellar extinction.

The advent of new and powerful infrared detectors opened a new era of Galactic structure studies. It is now possible to observe highly extincted HII regions at large distances from the Sun. The recent availability of large mid- and near-infrared surveys of the Galactic plane, such as 2MASS, ISOGAL, MSX and GLIMPSE has led to the discovery of several hundreds of new Galactic HII regions, e.g. [11], and candidate infrared stellar clusters. We expect to find many new candidate inner Galactic clusters with the coming UKIDSS data, a deeper near-infrared survey of the Galactic plane, and with future VISTA surveys.

Infrared follow-up studies of these stellar clusters are required in order to resolve each cluster into individual stars and to obtain stellar spectra that provide a complete census of the brightest stars and, thus, allow the determination of the spectro-photometric distance to the cluster.

The study of the clusters' spatial distribution by combining kinematic and infrared spectrophotometric information will greatly improve our knowledge on the structure, formation, and evolution of the Milky Way.

1.1. Spectrophotometric distances of stellar clusters in HII *regions.*

Determination of the spectro-photometric distances to the stars exciting an HII region using infrared data appears to be a very promising way to map the Galactic structure, independent and complementary to the kinematic information, and less limited by interstellar extinction than observations at optical wavelengths.

We have compiled a list of 647 Galactic young clusters which were mostly selected by inspecting 2MASS images at the position of known HII regions by [2] and [1], [7], [8], [15], [5]. Their youth is confirmed by associated continuum emission at 1.4 GHz (NRAO VLA Sky Survey) – for 85% of them. Only for some of these clusters infrared follow-up studies already exist.

New atlases of spectra and line modeling of nearby early type stars have been recently published ([13],[20]) showing that medium resolution (R≈2000–5000) spectra in H- and K-band with a signal-to-noise ratio above 100 yield a classification of early type stars consistent with that derived in the visual within 2 spectral subclasses. The near-infrared spectral classification is based on photospheric hydrogen, H, and helium, He, lines and other infrared lines such as those from carbon and nitrogen – e.g. the He I(4-3) line at 1.700 μm, the Brackett H I(11-4) transition (1.681μm), the Br_γ line at 2.16μm, the NIII(2.1155μm) and CIV(2.078μm) lines, and the He I atomic lines at 2.058μm and 2.1126μm.

Luminosity classes are more uncertain, but they can be estimated from comparison with the spectra in the infrared atlases by considering the fact that the line-widths of the Brackett (Br) lines are broader in dwarf stars than in supergiants. Furthermore, supergiants have He I absorption lines at 2.161μm and at 2.1621μm in the blue wing of the Br_γ ([13], [19]).

Photometric calibration of optical and infrared magnitudes of early type stars are also available, e.g. from [16] and [4] and a new consistent set has recently been published by [17].

The classification of the spectral type of a star and the measurement of its apparent magnitude together with the assumption of a photospheric infrared colour and an absolute magnitude yield simultaneously an estimate of the interstellar extinction and the stellar *distance*. Although the uncertainty of the spectrophotometric distance to each star can be up to 20 – 30%, the accuracy of the distance to the cluster can be improved by averaging the distances to several cluster stars.

1.2. Observations with SINFONI and first results.

We have obtained observations with the Integral Field Spectrometer SINFONI [9] on the ESO-VLT (Very Large Telescope) of eight obscured stellar clusters in HII regions – in W31, W42, W43, W51 and the cluster [DBS2003]8 – in order to determine their spectrophotometric distance.

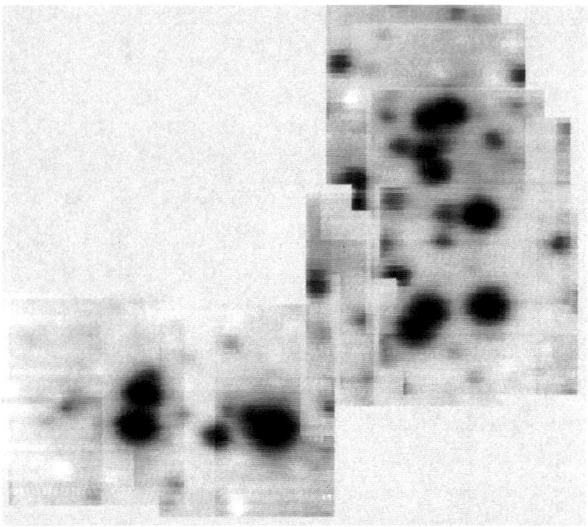

Figure 1. SINFONI K-band data of the stellar cluster W31. The mosaic was obtained with one single observing block (about 1 hour).

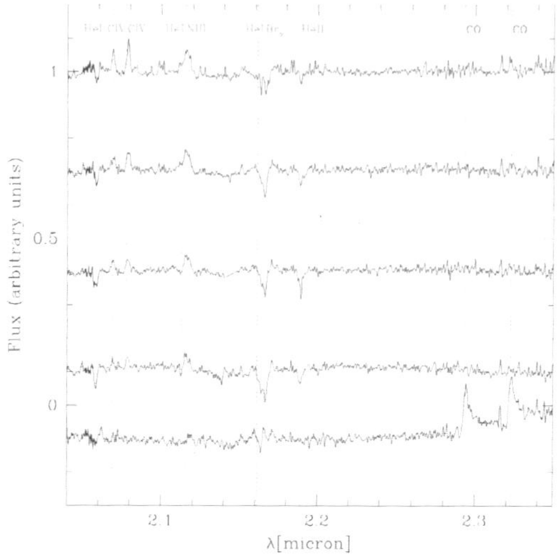

Figure 2. K-band spectra of 5 of the brightest stars in W31 are shown.

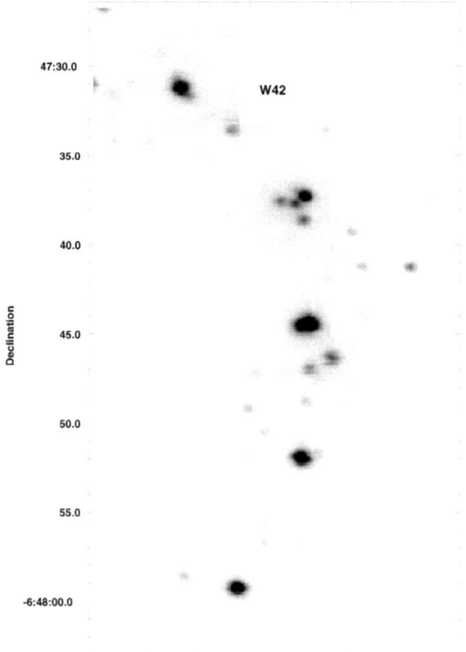

Figure 3. SINFONI K-band data of the stellar cluster W42. The mosaic was obtained with one single observing block (about 1 hour).

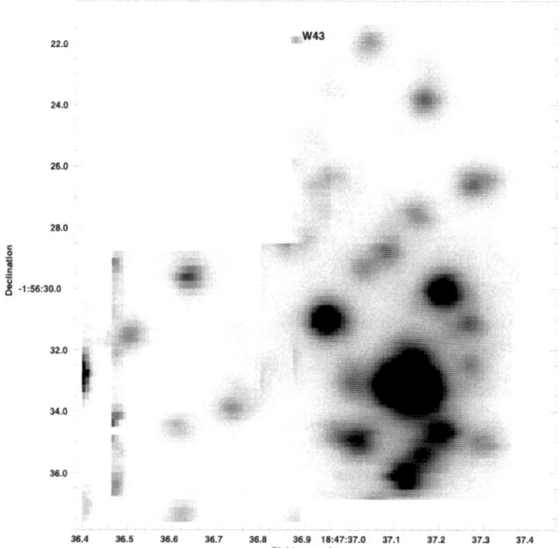

Figure 4. H-band image of the stellar cluster W43 obtained with 1 hour of observation.

The program includes observations of candidate OB stars in both the H (1.45 – 1.85μm) and K (1.95 – 2.45) bands at the highest SINFONI resolution (R $\sim$ 3000 and 4000, respectively). For each cluster, when resolved in the 2MASS K-band image, candidate stars were selected having 2MASS K $<$ 12 mag and (H–K) $>$ 0.6 mag – the latter constraint to exclude foreground objects. Furthermore, the targets are located near the cluster's centre. Since high-mass stars are known to form at such locations, the selected stars are thus highly probable cluster members. Typically, 4 – 6 stars per cluster were chosen.

Why is SINFONI a suited instrument for our project? By performing integral field spectroscopy with SINFONI we simultaneously image and resolve the core of the stellar clusters and obtain spectra. Since often in one field more than one star can be measured, it is typically possible to obtain a complete census of the cluster's brightest stars by employing a few telescope offsets in a single 1 hr observation sequence (e.g. Figs. 1,3,4). Furthermore, SINFONI enables us to detect simultaneously nebular emission (Fig. 6) (such as molecular hydrogen lines, HI recombination lines and ionized iron) and the spectrum of the candidate ionizing star, allowing us to distinguish between these contributions. Finally, the detection of Br_γ emission from the HII region provides an estimate of the extinction to the HII region when compared with radio continuum flux measurements. We have successfully detected photospheric lines in several stars of each cluster considered.

W31

W31 is a large HII complex, with three main HII regions. From ^{12}CO and ^{13}CO emission lines and NH3 absorption features, it was concluded [6] that the complex can be decomposed in two main components: one at a distance of a few kpc and the other at about 15 kpc. A spectro-photometric distance of 3.4 kpc was derived only for one cluster in the complex ([3]). We have obtained SINFONI observations for two different fields in W31. In both cases our preliminary results seem to confirm the near kinematic distance. Five of the brightest stars detected in K-band are shown in Figs. 1 and 2. Four of them have NIII and CIV lines in emission, Br_γ, HeI and HeII lines in absorption. Such lines are typical for stars of spectral type O5 $\pm$1. The profile of the Br_γ suggests that these are dwarfs stars. A young stellar object with the CO band in emission was also discovered.

W42

There is only one published spectro-photometric distance (2.2 kpc, [4]) for the cluster W42 which relies only on one out of the three taken K-band spectra. We have extracted 7 stars from the SINFONI datacubes detected in both H and K-bands (Fig. 3).

W43

In the cluster W43 (Fig. 4) we detected photospheric lines in several of the 13 brightest stars – 9 stars in K and 4 in H band. The census includes 1 WR star, 3 early O-type supergiants, 1 late O-type supergiants, 4 YSOs.

[DBS2003]8

Figure 5 shows the Galactic stellar cluster [DBS2003]8, successfully resolved with SINFONI. The spectrum of the central star shows absorption features at the position of the Br-gamma, Br10, Br11, Br12, Br13, and of the HeI lines at 1.70 and 2.11 μm. which suggest the star is a dwarf O9-B1 star. The spectral type and the determination of apparent magnitudes yield an estimate of the interstellar extinction ($A_K = 3.1$ mag) and a spectrophotometric distance of 2 kpc $\pm$ 0.8 kpc. The derived distance agrees well with one of the two gas complex distances detected along the line of sight ([22]).

These are preliminary considerations, because we are currently completing the data reduction of the full dataset. A detailed study of the spectro-photometric distances will follow soon thereafter.

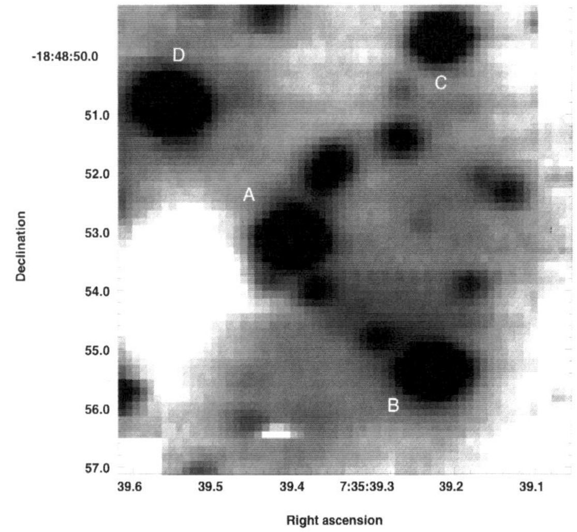

Figure 5. H-band image of the Galactic stellar cluster, [DBS2003]8. The cluster center is resolved and several stars are detected in both H and K band.

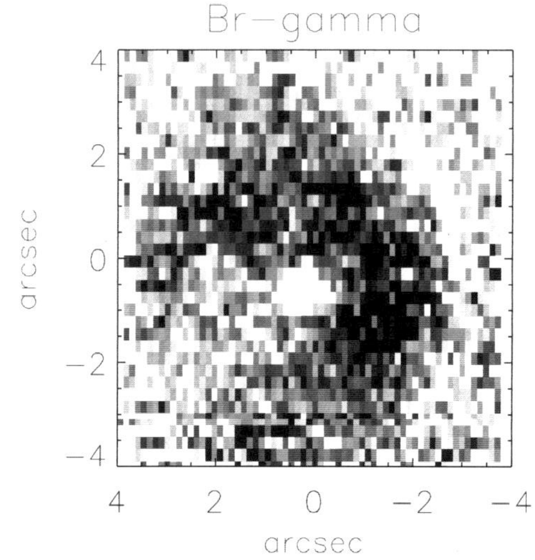

Figure 6. Image of the Br_γ emission detected around the central star of the cluster [DBS2003]8. We also detected H_2 emission, and [FeII] emission at 1.644 μm.

[1] Bica E., Dutra C. M. and Barbuy B. 2003a **A&A** **397** 177
[2] Bica E., Dutra C. M., Soares J. and Barbuy B. 2003b **A&A** **404** 223
[3] Blum R. D., Damineli A. and Conti P. S. 2001 **AJ** **121** 3149
[4] Blum R. D., Conti P. S.and Damineli A. 2000 **AJ** **119** 1860
[5] Borissova J., Pessev P., Ivanov V. D. and et al. 2003 **A&A** **411** 83
[6] Corbel S. and Eikenberry S.S. **A&A** **419** 191
[7] Dutra C. M. and Bica E. 2000 **A&A** **359** L9
[8] Dutra C. M., Bica E., Soares J. and Barbuy B. 2003 **A&A** **400** 533
[9] Eisenhauer F., Abuter R., Bickert K. and et al. 2003 **SPIE** **4841** 1548
[10] Georgelin Y. M. and Georgelin Y. P. 1976 **A&A** **49** 57
[11] Giveon U., Becker R. H., Helfand D. J. and White R. L. 2005 **AJ** **129** 348
[12] Habing H. J., Sevenster M., Messineo M. and et al. 2005 submitted to **A&A**
[13] Hanson M. M., Kudritzki R.-P. and Kenworthy M. 2005 **ApJS** **161** 154
[14] Ivanov V. D., Borissova J., Bresolin F. and Pessev P. 2005 **A&A** **435** 107
[15] Ivanov V. D., Borissova J., Pessev P. and et al. 2002 **A&A** **394** L1
[16] Koornneef J. 1983 **A&A** **128** 84
[17] Martins F. and Plez B. 2006 accepted by **A&A**, astro-ph/0606587
[18] Messineo M., Habing H. J., Sjouwerman L. and et al. 2002 **A&A** **393** 115
[19] Paumard T., Genzel R., Martins F. and et al. 2006 **ApJ** **643** 1011
[20] Repolust T., Puls J., Hanson M. and et al. 2005 **A&A** **440** 261
[21] Russeil D. 2003 **A&A** **397** 133
[22] Russeil D., Georgelin Y. M., Georgelin Y. P. and et al. 1995, **A&AS** **114** 557

Institute of Physics Publishing
doi:10.1088/1742-6596/54/1/039
Journal of Physics: Conference Series **54** (2006) 243–251
Galaxy Center Workshop

Stellar Processes near Sgr A*

Tal Alexander[1,2]

[1] Weizmann Institute of Science, POB 26, Rehovot 76100, Israel
[2]The William Z. & Eda Bess Novick career development chair

E-mail: `tal.alexander@weizmann.ac.il`

1. Introduction

The relative proximity of the Galactic massive black hole (MBH) offers a unique opportunity to study from up close the various modes in which stars in the high density cusp around the MBH interact with it and with each other. Extensive, well documented observational efforts are directed at using stellar dynamics to determine the primary parameters of the central dark object, its mass, size, location, motion and nature (see e.g. [16, 22] for recent reports). In this brief overview I will focus on other aspects of stellar processes near SgrA*: (1) The extreme conditions in the inner Galactic center (GC): high density, strong tidal fields and high velocities, and their implications for the stars there, specifically star-star interactions and mass segregation. (2) The prospects of probing post-Newtonian (PN) physics by high precision orbital solutions and by gravitational lensing effects. (3) The different modes of strong interactions of stars with the MBH and their observable consequences, and the important role of accelerated relaxation mechanisms. A comprehensive treatment of these topics can be found in [3].

2. Stellar dynamics in a high density cusp

The existence of a high density stellar cusp around the Galactic MBH was long predicted theoretically for a wide range of formation scenarios (e.g. [11, 52]), but was only recently established empirically with high confidence ([47] and Schödel et al, this volume). The density in the inner ~ 0.01 pc rises to $\sim 10^8\,M_\odot\,\mathrm{pc}^{-3}$, or $\times 10^9$ higher than in the Solar neighborhood. In such a high density environment, where the typical velocities are $\sim 1000\,\mathrm{km\,s^{-1}}$, collisions are frequent (mean time between collisions shorter than the main sequence (MS) lifetime of a Sun-like star) and energetic (relative velocities above the typical escape velocity from the surface of a MS star). These conditions are quite different from those in a high density core of a globular cluster, where collisions are slow.

2.1. Star-star collisions

High velocity collisions between giants and dwarfs can lead to the disruption of the giant envelope if the collision is close enough to head-on. This leaves a low-luminosity hot core that is invisible in the infrared (IR), where the stars in GC are observed. There are indeed indications of a gradual disappearance of luminous giants toward the center, in the inner $1.5''$. This trend is well modeled by the collisional destruction of giant envelopes [1, 3] (Fig. 1).

In an environment dense enough for effective collisional destruction, near misses, where the stars interact tidally but are not destroyed, are frequent. In such hyperbolic collisions, energy and angular momentum are transferred from the orbit to the stars, but because the orbit is highly unbound, the stars do not form a binary but continue on their separate way. The tidal energy

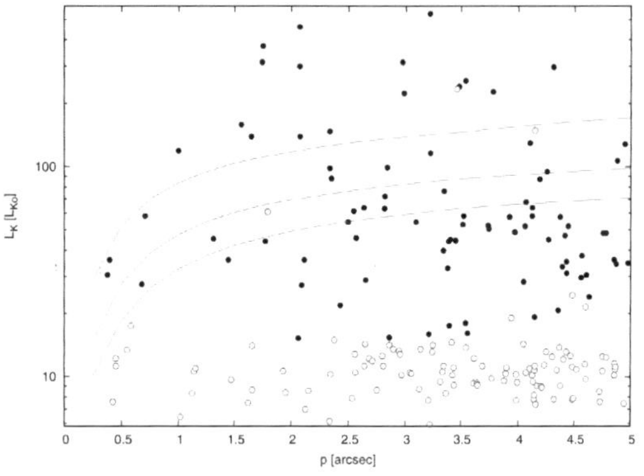

Figure 1. Stellar L-band luminosity as function of projected distance from SgrA*. Filled circles are spectrally identified late-type stars, open circles are spectrally unidentified stars (S. Trippe & T. Ott, priv. comm.). The contours represent results from detailed modeling of collisional destruction in a steep density cusp [1]: from top to bottom, contours correspond to less than 0.5, 1, and 1.5 brighter stars per 0.25″. The model is statistically consistent with the data.

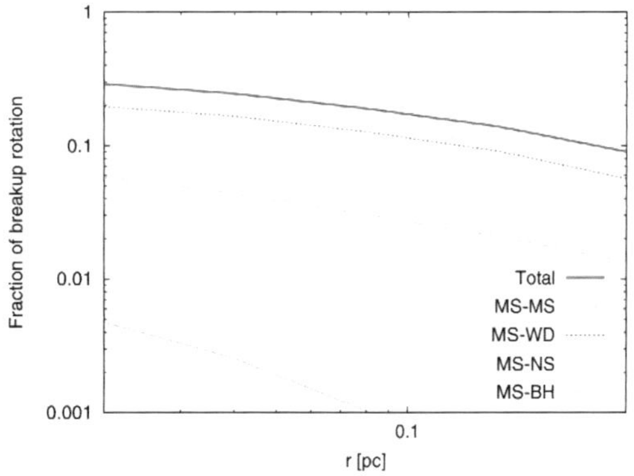

Figure 2. The mean accumulated stochastic spin-up over 10 Gyr of $1\,M_{\odot}$ stars in the GC, due to close tidal star-star encounters, as function of the distance from the MBH [7].

is dissipated relatively quickly, but the excess angular momentum is not so easily discarded. Stochastic tidal spin-up by multiple near-collisions can lead to a high mean stellar spin for long-lived stars if magnetic breaking is inefficient [7]. The mean spin remains high (0.1–0.3 of the centrifugal break-up rotation) over a large volume, $r \lesssim 0.3$ pc, because the decrease in the number of collisions farther away from the MBH is compensated by the higher tidal efficiency in slower collisions (Fig. 2). High spin changes the evolution and appearance of stars [32]. The MBH potential thus indirectly affects the internal properties of stars around it in a substantial fraction of its dynamical radius of influence.

2.2. Extreme mass segregation

The deep potential of the Galactic MBH leads to the accumulation of a very large number of massive objects around it over the lifetime of the Galaxy. Mass segregation particularly affects stellar black holes (SBHs), which are both much more massive than the mean star and essentially eternal [37, 36]. Such a dense cluster of SBHs may contain up to $\sim$0.01 of all Galactic SBHs in a mere 10^{-10} of the Galactic volume. It is a curious twist of scientific progress that at present more is known about the mass function of MBHs than of SBHs. For that reason, and because SBHs may be implicated in a variety of interesting phenomena (gravitational wave (GW) emission, X-ray binaries, collisions with stars and the formation of "exotic objects", accelerated relaxation) that it is of great interest to detect their presence and study their properties. This has turned out to be a very challenging task. There are broadly three detection strategies:

Direct detection Dynamical detection of the diffuse dark mass around the MBH by deviations of stellar orbits from Keplerian motion requires very high precision observations. At present only upper limits are available on the enclosed mass [46, 38], and these are still more than an order of magnitude higher than the drain limit [6] (the maximal number of SBHs that can be packed around a MBH before they start rapidly scattering each other into the MBH). The X-ray emission from accretion of the ISM by isolated SBHs is too weak for detection [42]. The spatial distribution of X-ray binaries suggests a central concentration consistent with mass segregation, but this is still of low statistical significance [40]. SBHs will modify the gravitational lensing properties of the MBH [13], however the lensing event probability is very small [7]. SBHs may form exotic stellar objects by collisions and mergers [50], but the overall appearance of such objects is predicted to be quite similar to normal giants (e.g. [12]).

Environmental effects A high central concentration of massive objects will increase the number and efficiency of destructive stellar collisions and decrease the relaxation time. However, a collisionally depleted region is expected in any case in the innermost parts of the high density cusp, while the determination of the relaxation time is highly uncertain.

Light test particles Mass segregation implies the central depletion of light objects. Several candidate light test particles were considered. Near the MBH, where SBHs ($\sim 10\,M_\odot$) dominate the mass density, neutron stars ($\sim 1.5\,M_\odot$) are light objects. Radio pulsars, if observed, could trace mass segregation [14, 43]. However, electron scattering by the Galactic ISM severely hinders their detection. The distribution of white dwarf cataclysmic variables appears to be centrally depleted [39], but this occurs on the $\sim$ 1 pc scale, where mass segregation is unlikely to be significant. A promising class of light test particle, as discussed further below, are the horizontal branch "red clump" (RC) giants.

RC giants have light (0.5–2 $M_\odot$) and long-lived ($t > 10^9$ yr) progenitors, and are easily observable with an IR magnitude of $K \sim 15.5$ (Fig. 3). There are indications that the RC giants are suppressed in the center compared to other stars ([21] and Schödel et al, this volume). If mass segregation is the underlying cause, then it must differentiate much more sharply between high and low mass stars than is predicted by the Bahcall-Wolf solution ($r^{-7/4}$ for the massive objects, $r^{-3/2}$ for the light ones) [11] (Fig. 4; M. Levi, MSc thesis). There are indeed indications that in a population typical of the GC, which has a much wider range of masses than the globular cluster population considered by Bahcall & Wolf and where the SBHs are a small fraction of the population by number, the slope of the SBHs is as steep or even steeper than r^{-2} [28].

3. Post-Newtonian physics

General Relativity (GR) is the least tested of the theories of the four fundamental forces of nature. It is therefore of interest to test it near the Galactic MBH. GR predicts that physics depends only on the relativistic parameter $\Upsilon \equiv r_g/r$ ($r_g \equiv GM/c^2$) and not on the mass, composition or compactness ($C \equiv r_g/R$) of the massive object. Alternative theories of gravity could behave differently. Figures (5, 6) show that stellar orbits near SgrA* can potentially probe a region in (M, Υ, C) space that has never been probed before, and in terms of M, a region that is inaccessible to compact pulsar binaries, albeit at a lower accuracy than pulsar-based tests of GR [30].

3.1. Post-Newtonian effects in the Doppler shift measurements

Recent developments in integral field IR spectroscopy (e.g. [16]) now allow measurement of the stellar Doppler shift, z, with high accuracies of $\Delta z \sim 10^{-4}$. Some of the stars orbiting the MBH attain velocities at periapse of $\beta_\mathrm{p} \sim$ few $\times$ 0.01. It is therefore possible to detect with high confidence β^2 PN effects (the transverse Doppler shift and gravitational redshift, with $z_\mathrm{PN} \sim \beta_p^2 > \Delta z$) in astrometric and spectroscopic orbital data monitored with existing instruments over a decade [54] (Figs. 7, 6). Neglecting these PN effects when modelling the orbits would bias the measured MBH mass and distance at a level of a few percent. This could interfere with the

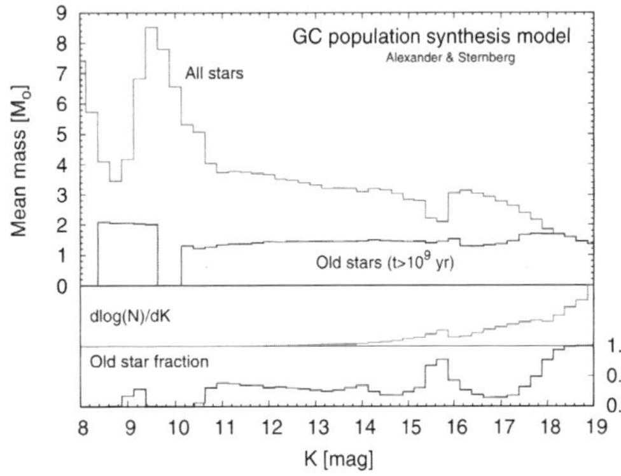

Figure 3. The K-band luminosity function (KLF), mean mass and fraction of old stars in the population according to a continuous star formation population synthesis model for the GC [10]. The RC giants are prominent around K ~ 15.5 mag.

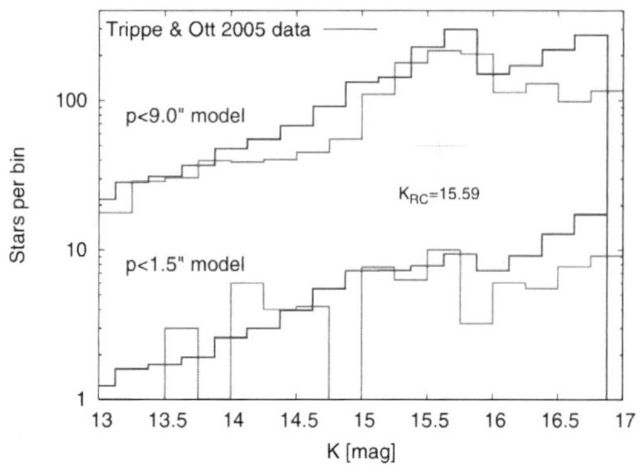

Figure 4. Comparison between the observed KLF inside 1.5″ and 9″ [21] and a strong mass-segregation model (M. Levi, MSc thesis 2006). Note the absence of the RC feature in the smaller field. The expected K magnitude of the RC feature is marked by a cross.

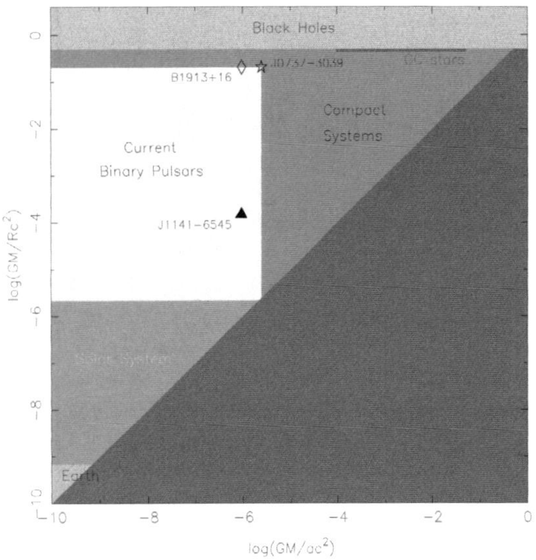

Figure 5. Post-Newtonian exploration space (Υ, C). GC orbital tests (upper right) are compared to pulsar tests exiting instruments and with SKA (kindly modified from [30] by M. Kramer).

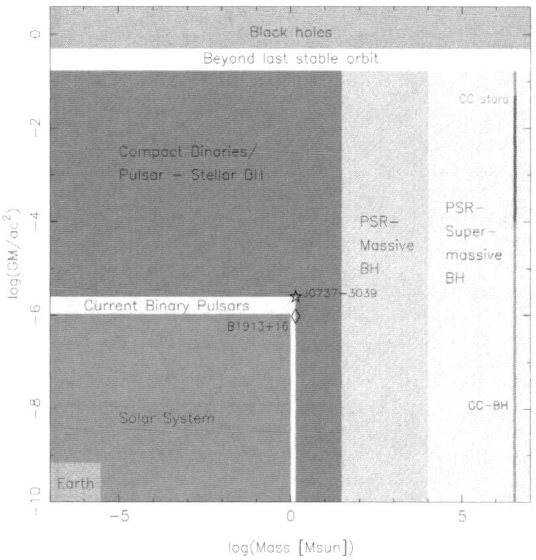

Figure 6. Post-Newtonian exploration space (M, Υ). GC orbital tests (upper right) are compared to pulsar tests with existing instruments and with SKA (kindly modified from [30] by M. Kramer).

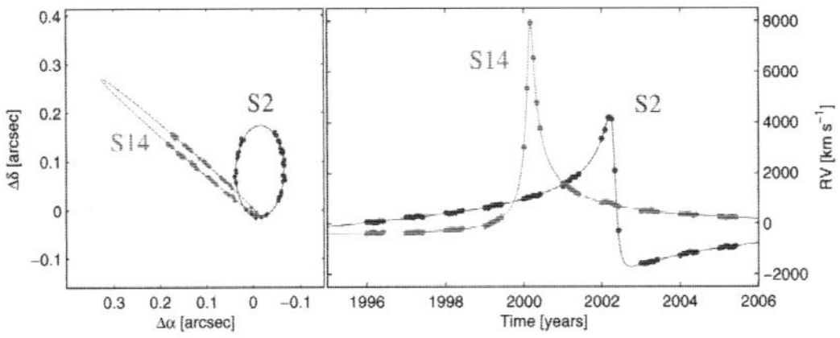

Figure 7. 10 years of simulated orbits and observations, based on the actual orbits of S2 and S14 [54]. (reproduced with permission from the Astrophysical Journal)

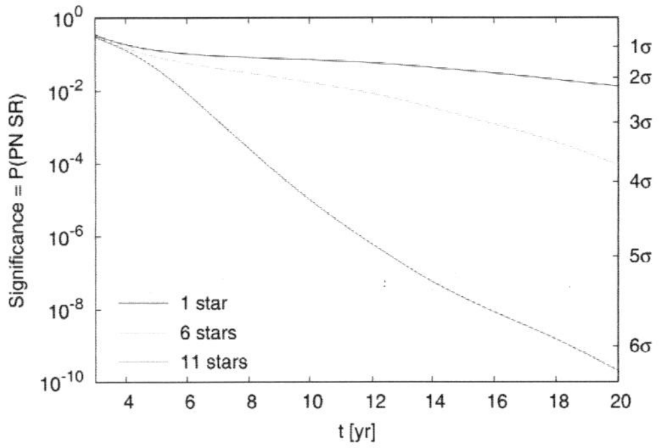

Figure 8. The simulated detection significance of gravitational redshift in stellar orbits for 1, 6 and 11 stars, as function of monitoring time [54]. (reproduced with permission from the Astrophysical Journal)

detection of the diffuse dark mass around the MBH, which is thought to consist a few percent of the enclosed mass on the $\sim$0.1 pc scale.

3.2. Gravitational lensing

Another class of PN phenomena that could be detected in principle in the GC is gravitational lensing (GL) by the MBH. GL effects could be observed as a change in the stellar surface density around the MBH [51], as variable macrolensed (resolved multiple images) far-background stars [2], as microlensed (unresolved multiple images) near-background stars [10], or as microlensing by a composite MBH–SBH lens [4, 13]. However, the prospects of GL by the MBH are not encouraging. Estimates of the optical depth of the far background sources, while highly uncertain, indicate that there may not be enough such sources. The near background sources, while numerous, are not enough to exhibit detectable surface density variations, and their GL cross-sections are too small to be likely sources of microlensing.

4. Strong interactions with the MBH

Some of the most spectacular stellar processes possible in the GC involve strong, close interactions between a star and the MBH. The term "strong" in this context is defined as any interaction that cannot be described by Newtonian gravity operating on point masses. This could be because the stars' internal degrees of freedom can no longer be ignored, or because dissipative interactions have to be included, or because GR effects must be taken into account.

4.1. Modes of interaction

It is instructive to classify strong stellar interactions with the MBH by analogy to atomic processes (Fig. 9). In spite of the fundamental differences between a macroscopic classical system and a microscopic quantum one, this analogy suggests a classification scheme for the various modes of strong star–MBH interactions, which not only provides the "bookkeeping" needed to keep track of all the permutations, but also provides some analytical tools for calculating cross-sections, rates and branching ratios (e.g. [6]). There are three possible end results when an object (star, binary) is deflected into a nearly radial orbit that approaches the MBH: (1) it can be destroyed immediately ("infall"), (2) it can be destroyed after many orbital periods, if some form of orbital dissipation gradually shrinks the orbit ("inspiral") [8, 9, 26], and (3)

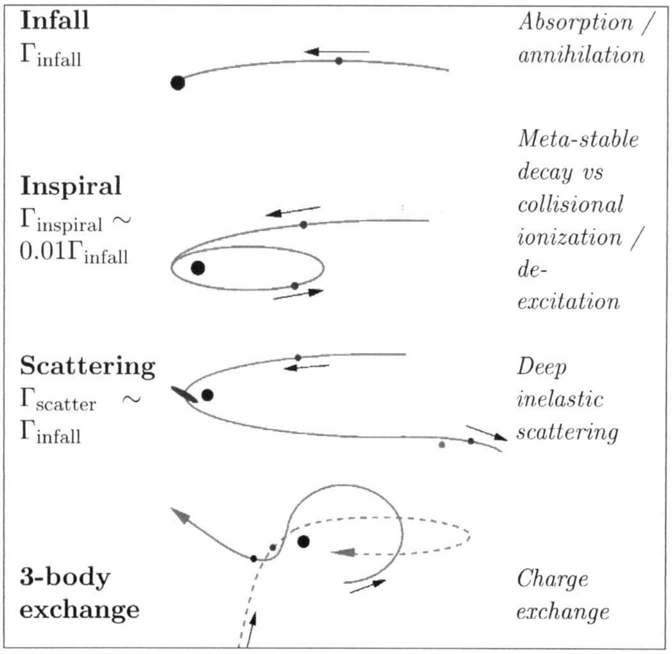

Figure 9. Modes of close stellar interactions with the MBH. Destruction and consumption of stars by direct infall plays a role in the feeding of the MBH and in tidal disruption and detonation followed by tidal disruption flares. Dissipative processes (tidal heating or GW emission) lead to gradual inspiral, if the star can avoid re-scattering. This strongly suppresses inspiral compared to infall. Most stars that approach the MBH scatter off it after undergoing strong tidal deformation or emitting of a short burst of GW. 3-body exchanges between an incoming binary and the MBH, or between an incoming star and a MBH-star "binary" can lead to orbital capture and the ejection of hyper-velocity stars.

it can be re-scattered after one or more cycles to a wide orbit that is safe from the MBH ("scattering") [5]. Infall processes (e.g. tidal disruption) have been studied extensively by many authors [24, 18, 29, 15, 49], and estimates of the infall event rate, Γ_{infall} can be obtained. Inspiral (e.g. by GW emission from compact remnants orbiting the MBH, or by tidal heating) has by comparison a much smaller rate, $\Gamma_{\text{inspiral}} \sim 0.01\Gamma_{\text{infall}}$, since the inspiraling star must avoid re-scattering, and therefore must start its journey much closer to the MBH where the orbital period is short. Scattering (e.g. tidal scattering) has a rate that is is almost equal to the infall rate, $\Gamma_{\text{scatter}} \sim \Gamma_{\text{infall}}$, because stars are easily deflected from a nearly radial orbit. In addition, interactions with binaries involve 3-body exchange scenarios, when an incoming single star interacts with a MBH-star "binary" [6], or when an incoming binary is tidally disrupted by the field of the MBH ([23, 41] and Perets et al, this volume). Such exchange interactions involve the capture of one star and the energetic ejection of the other.

4.2. Loss-cone theory

The rate of the various close interaction processes is determined by the efficiency at which objects are scattered to orbits that take them near the MBH. Phase space diffusion in angular momentum J to low-J orbits ("loss-cone" orbits) is much faster than diffusion in energy to tightly bound orbits. The refilling of the loss-cone proceeds in two regimes (Fig. 10). Close to the MBH, where the loss-cone is geometrically large and the orbital times are short, the change in J per orbit is small relative to the size of the loss-cone J_{lc}, and once a star is inside the loss-cone it is promptly destroyed so the loss-cone is empty. The loss-cone refilling process can therefore be viewed as diffusion across the loss-cone boundary, and it proceeds at a rate $\mathrm{d}\Gamma/\mathrm{d}\log r \sim N_\star(< r)\,/\log[J_c(r)/J_{lc}]t_r(r)$, where J_c is the maximal (circular) angular momentum and t_r is the relaxation time. Far from the MBH, where the loss-cone is geometrically small and

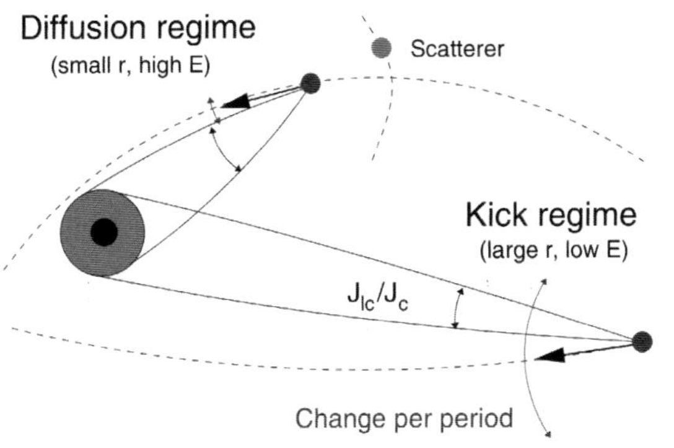

Figure 10. Schematic depiction of loss-cone refilling by 2-body scattering. Stars on orbits that intersect a small volume around the MBH (loss-cone orbits, $J < J_{lc}$) are destroyed. Tight orbits near the MBH have a large loss-cone and short orbital periods, so that the perturbation by scatterers lead to gradual diffusion into the loss-cone and the loss-cone is always empty. In contrast, wide orbits away from the MBH have a small loss-cone and long orbital periods, so that stars scatter across the loss-cone, which is always full.

the orbital times are long, the change in J per orbit is larger than J_{lc} and the loss-cone is always full, and the rate is $\mathrm{d}\Gamma/\mathrm{d}\log r \sim \{M_\bullet/[M_\bullet + M_\star N_\star(<r)]\}\,[J_{lc}/J_c(r)]^2 N_\star(<r)\,/P(r)$, where P is the orbital time [41]. The maximal local contribution to the total rate is attained when the loss-cone is filled. However, the scale that contributes most for the total loss-cone refilling rate depends on the run of $N_\star$, J_c, t_r, P with r and the size of J_{lc}. For 2-body stellar relaxation, which is the minimal assured relaxation process in any stellar system, and for the process of tidal disruption of single stars by the MBH, most of the contribution comes from roughly the radius of dynamical influence of the MBH, $r_h \sim 2$–$4\,\mathrm{pc}$. For the Galactic MBH, the tidal disruption rate is 10^{-5}–$10^{-4}\,\mathrm{yr}^{-1}$ [9]. In contrast, inspiral processes such as GW by compact remnants involve only objects much closer to the MBH, inside few $\times\, 0.01\,\mathrm{pc}$ [26] and therefore much lower rates, 10^{-9}–$10^{-7}\,\mathrm{yr}^{-1}$ per galaxy [26, 28], depending on the degree of mass segregation.

4.3. Efficient loss-cone refilling

Two-body relaxation is a slow process, and it constitutes the bottle-neck for the rates of close interactions with the MBH. Because of the interest in such processes, and their implications for a variety of open issues such as the feeding mechanisms of MBHs and their relation to the M/σ relation [17, 20], or the stalling problem of MBH binaries (e.g. [35]), there have been various attempts to identify more efficient loss-cone refilling mechanisms. These broadly fall into three categories.

Non-spherical potentials The phase space structure of orbits in non-spherical potentials can repopulate the loss-cone even without relaxation. Triaxial potentials lead to chaotic orbits that go through the entire available phase space. Chaotic orbits dramatically enhance the close-interaction rates [34]. However, there are doubts whether triaxiality can be maintained very close to the MBH (e.g. [25]). Axisymmetric potentials lead to centrophilic orbits that preferentially populate the loss cone and can lead to a moderate increase in the tidal disruption rate [33].

Resonant relaxation Approximate symmetries in the potential (in particular the almost point mass potential very near the MBH) significantly slow down the evolution of the scatterers and introduce coherent torques on the scattered stars, which lead to J changes that grow rapidly as $\Delta J \propto t$ rather than in the slow random walk fashion $\Delta J \propto \sqrt{t}$ of normal relaxation [44, 45]. Resonant relaxation may have implications for the observed dynamical structures in the GC [27, 31], as well as for the rates of GW emission events from compact remnants inspiraling into MBHs (see Hopman & Alexander, this volume).

Massive perturbers Galaxies contain sub-structures well above the stellar mass scale, such as giant molecular clouds, stellar clusters, and perhaps also intermediate mass BHs. Massive perturbers (MPs) can significantly shorten the relaxation time [53]. The observed MPs in the GC are indeed numerous enough to play a major role in accelerating loss-cone refilling, in particular for 3-body exchange processes, whose large loss-cone cannot be refilled by regular stellar 2-body relaxation. Relaxation by MPs may thereby explain the presence of the mysterious S-stars very near the MBH and the origin of the observed hyper-velocity stars (see Perets, Hopman & Alexander, this volume). MPs may also play an important role in accelerating the coalescence of merging MBH binaries [41].

5. Conclusion

The many different ways in which stars interact with a MBH and react to its presence reveal a wide and rich range of physical processes in the realms of stellar dynamics, General Relativity and gravitational wave physics, stellar structure, stellar evolution and the physics of compact remnants. The proximity of the Galactic MBH offers us the opportunity to study these processes in greater detail than will ever be possible in other galaxies. These processes have local, testable implications for the GC, as well implications on cosmological scales and for fundamental physics.

Acknowledgments

Support by ISF grant 295/02-1, Minerva grant 8563, and a New Faculty grant by Sir H. Djangoly, CBE, of London, UK, is gratefully acknowledged.

References

[1] Alexander, T. 1999, ApJ, 527, 835
[2] Alexander, T., 2001, ApJ, 553, 149
[3] Alexander, T. 2005, Physics reports, 419, 65
[4] Alexander T. & Loeb, A., 2001, ApJ, 551, 223
[5] Alexander T. & Livio, M., ApJ, 2001, 560, L143
[6] Alexander T. & Livio, M., ApJ, 2004, 606, L21
[7] Alexander, T. & Kumar, P., 2001, ApJ, 549, 948
[8] Alexander, T. & Morris, M., 2003, ApJ, 590, L25
[9] Alexander, T. & Hopman, C., 2003, ApJ, 590, L29
[10] Alexander, T. & Sternberg, A., 1999, ApJ, 520, 137
[11] Bahcall, J. N., & Wolf, R. A., 1977, ApJ, 216, 883
[12] Biehle, G. T., 1991, ApJ, 380, 167
[13] Chanamé, J., Gould, A. & Miralda-Escudé, J., 2001, ApJ, 563, 793
[14] Chanamé, J. & Gould, A., 2002, ApJ, 571, 320
[15] Cohn, H. & Kulsrud, R. M., 1978, ApJ, 226, 1087
[16] Eisenhauer et al., 2005, ApJ, 628, 246
[17] Ferrarese, L. & Merritt, M., 2000, ApJ, 539, L9
[18] Frank, J. & Rees, M. J., 1976, MNRAS, 176, 633
[19] Freitag, M., Amaro-Aeoane, P., Kalogera, V., 2006, ApJ in press (astro-ph/0603280)
[20] Gebhardt, K. et al, 2003, ApJ, 583, 92
[21] Genzel, R. et al, 2003, ApJ, 594, 812
[22] Ghez, A., et al, 2005, ApJ, 620, 744
[23] Gould, A. & Quillen, A. C., 2003, ApJ, 592, 935
[24] Hills, J. G., Nature, 254, 295
[25] Holley-Bockelman, K., Mihos, J. C., Sigurdsson, S., Hernquist, L. & Norman, C., 2002, ApJ, 567, 817
[26] Hopman, C. & Alexander, T., 2005, ApJ, 629, 362
[27] Hopman, C. & Alexander, T., 2006, ApJ, in press (astro-ph/0601161)
[28] Hopman, C. & Alexander, T., 2006, ApJL, in press (astro-ph/0603324)
[29] Lightman, A. P. & Shapiro, S. L., 1977, ApJ, 211, 244
[30] Kramer, M., Backer, D. C., Cordes, J. M., Lazio, T. J. W., Stappers, B. W. & Johnston, S. 2004, NewAR, 48, 993
[31] Levin, Y., 2006, MNRAS submitted (astro-ph/0603583)
[32] Maeder, A. & Meynet, G., 2000, ARA&A, 38, 143
[33] Magorrian, J. & Tremaine, S., 1999, MNRAS, 309, 447
[34] Merritt, D. & Poon, M. Y., 2004, ApJ, 606, 788

[35] Merritt, D. & Milosavljević, M., 2005, Living Reviews in Relativity, 8, 8
[36] Miralda-Escudé, J. & Gould, A., 2000, ApJ, 545, 847
[37] Morris, M., 1993, ApJ, 408, 496
[38] Mouawad, N., Eckart, A., Pfalzner, S., Schödel, R., Moultaka, J., Spurzem, R., 2005, AN, 326, 83
[39] Muno, M. P., et al, 2003, ApJ, 589, 225
[40] Muno, M. P., et al, 2005, ApJ, 622, 113
[41] Perets, H. B., Hopman, C. & Alexander, T., 2006, ApJ submitted (astro-ph/0606443)
[42] Pessah, M. & Melia, F., 2003, ApJ, 585, 29
[43] Pfahl, E. & Loeb, A., 2004, ApJ, 615, 253
[44] Rauch, K. P. & Tremaine, S., NewA, 1996, 1, 149
[45] Rauch, K. P. & Ingalls, B. 1998, MNRAS, 299, 1231
[46] Rubilar, G. F. & Eckart, A., 2001, A&A, 374, 95
[47] Schödel et al, 2005, ApJ, 596, 1015
[48] Schödel et al, 2006, A&A submitted
[49] Shapiro, S. L. & Marchant, A. B., 1978, ApJ, 225, 603
[50] Thorne, K. S. & Żytkow, A. N., 1975, ApJ, 199, L19
[51] Wardle, M. & Yusef-Zadeh, F., 1992, ApJ, 387, 65
[52] Young, P. J., 1980, ApJ, 242, 1232
[53] Zhao, H., Haehnelt, M. G. & Rees, M. J., 2002. NewA, 7, 385
[54] Zucker, S., Alexander, T., Gillessen, S., Eisenhauer, F. & Genzel, R., ApJ, 639, L21

Institute of Physics Publishing
doi:10.1088/1742-6596/54/1/040

Models of mass segregation at the Galactic Centre

Marc Freitag[1,2], Pau Amaro-Seoane[3] and Vassiliki Kalogera[1]

[1]Department of Physics and Astronomy, Northwestern University, Evanston, IL 60208, USA
[2]Institute of Astronomy, University of Cambridge, Madingley Road, CB3 0HA Cambridge, UK
[3]Max Planck Intitut für Gravitationsphysik, D-14476 Potsdam, Germany

E-mail: `freitag@ast.cam.ac.uk`

Abstract.
We study the process of mass segregation through 2-body relaxation in galactic nuclei with a central massive black hole (MBH). This study has bearing on a variety of astrophysical questions, from the distribution of X-ray binaries at the Galactic centre, to tidal disruptions of main-sequence and giant stars, to inspirals of compact objects into the MBH, an important category of events for the future space borne gravitational wave interferometer LISA. In relatively small galactic nuclei, typical hosts of MBHs with masses in the range $10^4 - 10^7\,\mathrm{M}_\odot$, the relaxation induces the formation of a steep density cusp around the MBH and strong mass segregation. Using a spherical stellar dynamical Monte-Carlo code, we simulate the long-term relaxational evolution of galactic nucleus models with a spectrum of stellar masses. Our focus is the concentration of stellar black holes to the immediate vicinity of the MBH. Special attention is given to models developed to match the conditions in the Milky Way nucleus.

1. Method, physics and initial conditions

This contribution summarises our recent series of simulations of galactic nuclei aimed at the investigation of mass segregation [1].

We used a stellar dynamics code based on the Monte Carlo (MC) scheme pioneered by Hénon [2, 3]. This method offers a unique compromise between physical realism and computational efficiency. Like in direct $N-$body codes, the stellar system is represented as a set of particles; this allows the implementation of a rich physics: self-gravity, general stellar mass spectrum and velocity distribution, 2-body relaxation, collisions between stars, stellar evolution, binaries (not included in the present code), tidal disruptions and captures by the central MBH ([4, 5, 6, 7, 8, 9] for a description of the code and previous applications). Unlike $N-$body, the MC method assumes spherical symmetry and dynamical equilibrium. This makes it much faster; while the CPU time per relaxation time increases like N^3 for $N-$body simulations, the MC code has a $N \ln N$ scaling.

We performed about 90 different simulations, to investigate the effects of various physical ingredients, assumptions about their treatment, of the initial nucleus structure and to perform some limited parameter-space exploration. For most models, 4×10^6 particles were used, requiring a few days of computing time on a single-CPU PC. A few cases were computed with 10^6 or 8×10^6 particles to establish that our results are not strongly affected by the limitations in numerical resolution.

All runs included the effects of the gravity of a central MBH, of the self gravity of the stars, of 2-body relaxation, treated in the Chandrasekhar (diffusive) approximation, and of the tidal disruption of main-sequence (MS) stars at the Roche limit around the MBH as well as direct coalescence with the MBH for stars too compact to be tidally disrupted. In most cases, stellar evolution was not included explicitly; instead the stellar population consists, from the beginning of the simulation, of a mixture of MS and compact remnants corresponding to a single star formation episode that took place 10 Gyr ago. In a few models, explicit stellar evolution was included with all stars starting on the MS and turning into compact remnants at the end of their MS life time. For simplicity, giants were not considered because, as far as mass segregation is concerned, only the mass of the star matters and the evolution of the stellar distribution, being a relaxational process requires timescales much longer than the duration of the giant phase. Stellar collisions and large-angle gravitational deflections (not accounted for in the diffusive treatment of relaxation) were considered in a small number of models. We made no attempt to determine whether a given star-MBH coalescence would occur as a gradual "extreme-mass ratio inspiral" (EMRI) detectable by LISA or a direct plunge through the horizon of the MBH (see [10, 11] for background information and references on EMRIs and [12, 13] for the importance of the distinction between EMRIs and plunges).

In all our runs the galactic nucleus is started as an η-model [14, 15], with a central power-law density cusp, $\rho \propto R^{\eta-3}$ and steeper "cut-off" at large radii, $\rho \propto R^{-4}$. In most cases, we used parameters (mass of the MBH, stellar density around it, etc) corresponding to the stellar cluster around Sgr A* at the centre of our Galaxy [16, 17]. We did not try to reproduce the very peculiar spatial and age distribution of the bright IR stars observed within 1 pc of Sgr A*. In this work, we adopt the position that these stars, useful as they are as probes of the gravitational potential, are not representative of the overall stellar population at the Galactic centre, assumed to be much older and therefore amenable to our treatment. This defines a well-posed problem which constitutes an interesting limiting case. Clearly other situations have to be considered in future studies.

In addition to the Sgr A* models, we followed the evolution of galactic nuclei hosting an MBH with a mass $M_\bullet$ in the range $10^4 - 10^7\,\mathrm{M}_\odot$. Based on a somewhat naive application of the $M-\sigma$ relation [18], we scaled the size of the stellar cluster according to $R_{\mathrm{nucl}} \propto M_\bullet^{0.5}$ where R_{nucl} is any characteristic length of the stellar distribution. We have considered two families of models; one with $\eta = 2.0$, the other with $\eta = 1.5$. The interval in $M_\bullet$ was chosen mostly to cover the values that should yield gravitational wave signals in LISA band when a compact star inspiral into the MBH. The present study is a first step towards more robust determinations of the rate and characteristic of such EMRIs. This range of models also covers systems that are both large enough (in terms of the number of stars) to be amenable to treatment with the MC method and small enough for relaxational effects to play a significant role over some 10 Gyr.

To ensure that the MC code, based as it is on a number of simplifying assumptions, yield correct results, we carried out a number of comparisons with simulations performed with the highly accurate (but much more computationally demanding) direct-summation NBODY4 code [19, 20]. In particular, we compared with results obtained by Baumgardt et al. [21] and performed a new NBODY4 simulation of a two-component model with a central massive object using 64 000 particles. On the GRAPE hardware at disposal not more than $\sim 10^5$ particles can be used; hence it is not yet possible to simulate a system with a realistic mass function using direct $N-$body but this 2-component toy model demonstrated for the first time in a direct fashion that the MC code treats mass segregation around a MBH very satisfactorily.

2. Results

For the Sgr A* models, our main results are the following. In all cases, the stellar BHs, being the most massive objects (with a fixed mass of $10\,\mathrm{M}_\odot$ or a range of masses, depending on the model), segregate to the central regions. This segregation takes about 5 Gyr to complete. The nucleus then enters a second evolutionary phase which is characterised by the overall expansion

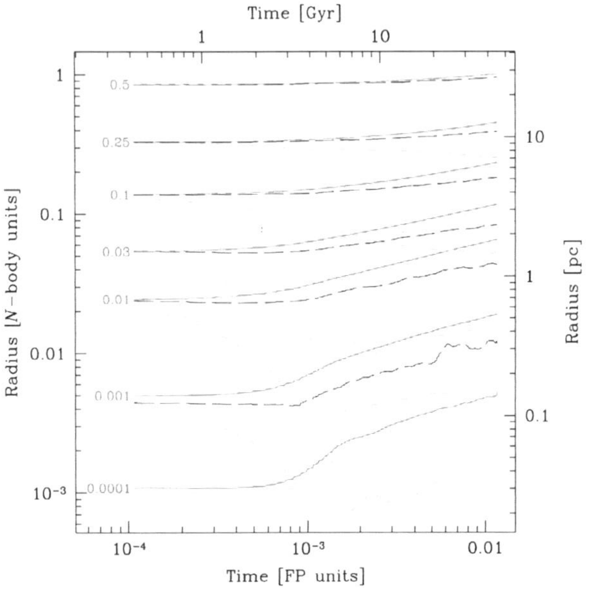

Figure 1. Evolution of Lagrange radii for a "standard" Sgr A* nucleus model. We plot the evolution of the radii of spheres that enclose the indicated fractions of the mass of various stellar species. Solid lines are for MS stars, short-dashed lines for white dwarfs, long-dashed lines for neutron stars and dash-dotted lines for stellar BHs. For this model, a 10 Gyr-old, non-evolving stellar population was used and the central MBH has an initial mass of $M_\bullet = 3.6 \times 10^6\,\mathrm{M}_\odot$. It grows only very little through tidal disruptions and stellar coalescences.

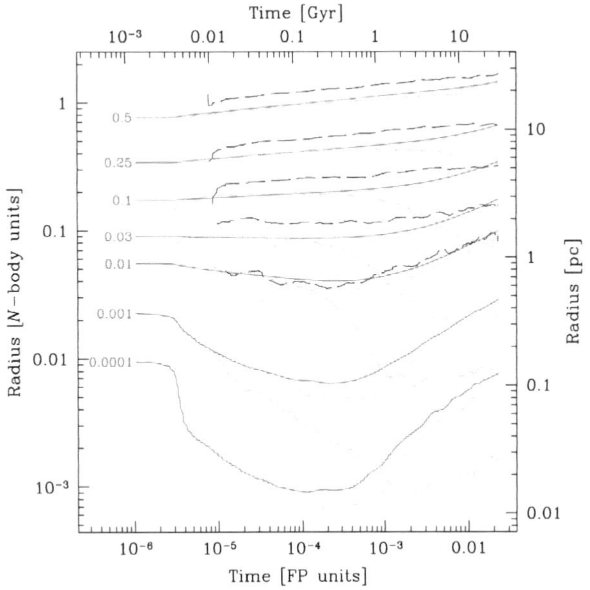

Figure 2. Same as Fig. 1 but for a Sgr A* model with all stars on the MS and $M_\bullet = 1000\,\mathrm{M}_\odot$ at start. The MBH grows by accreting an *ad hoc* fraction of the gas emitted by stellar evolution to reach $M_\bullet \simeq 3.9 \times 10^6\,\mathrm{M}_\odot$ after 10 Gyr. The structure of the nucleus after $\sim$ 5 Gyr is similar to that of the case without stellar evolution. The initial size of the cluster was adjusted to obtain a reasonably good fit to the observed stellar mass profile around Sgr A* at $t = 10$ Gyr.

of the central regions, powered by the accretion of stellar mass (of very negative energy) on to the MBH. Although all species participate in the expansion, mass segregation continues in a relative fashion, as the system of BHs expands slower than the other components. The structure of the nucleus at distances from the centre larger than ~ 10 pc is unaffected by relaxation over a Hubble time. The evolution of two Sgr A* models is shown in Figures 1 and 2.

BHs dominate the mass density within ~ 0.2 pc of the MBH but we do not find them to be more numerous than MS stars in any region we can resolve (down to a few mpc, at $t = 10$ Gyr). Estimating the exponent for the density cusp the BHs form, $\rho \propto R^{-\gamma}$, is difficult because of numerical noise but, in most cases, γ is compatible with the Bahcall-Wolf value $\gamma = 1.75$ [22, 23]. In contrast, the less massive objects, such as MS stars, form a cusp with γ generally in the range $1.3 - 1.4$ which is significantly lower than the value of 1.5 predicted by [23]. This is also found in the 2-component $N-$body simulation. After 5-10 Gyr of evolution, we find of order $2-3\times10^3$, $6-8\times10^3$ and 2×10^4 stellar BHs within 0.1, 0.3 and 1 pc of the centre, respectively. About 10^4 BHs coalesce with the MBH during a Hubble time. Using the formalism of the dynamical friction for objects on circular orbits in a fixed stellar background is an easy alternative for estimating the concentration of massive objects in the central regions. However, for stellar BHs, although this approach offers a qualitatively correct picture, it overpredicts the effectiveness of mass segregation. In the case of a model with $\eta=1.5$ (whose relaxation time does not increase towards the centre), this yields too large a number of BHs accreted by the MBH and *too few* (by a factor 3–5) being present within the inner 1 pc after some 10 Gyr.

All types of objects lighter than the BHs, including the neutron stars, are pushed away from the central regions. Using the observed distribution of these objects to infer the presence of segregated BHs [24] does not seem to be possible, though because, in the absence of BHs, it would take the neutron stars more than 10 Gyr to form a Bahcall-Wolf Cusp of their own, even without natal kick.

These results are not significantly affected by stellar collisions, large-angle scatterings or the initial η value. We also considered three different prescriptions for the masses (and types) of compact remnants and found no strong variations in the simulation outcomes. Most interestingly, an alternative model in which stellar evolution was included and the central MBH was grown from an IMBH seed (by accretion of an *ad hoc* fraction of the mass lost by stars when they turn into compact objects) yields basically the same structure of mass segregation (and same rates of coalescences and tidal disruptions) at $t\simeq10$ Gyr (see Fig. 2). These findings suggest that our main results are not very sensitive to the special "initial conditions" used, as long as they are fine-tuned to produce at $t=10\,Gyr$ a given MBH mass and stellar mass within ~1 pc of the MBH. However, it would be instructive to consider a larger variety of models in future work, including some with extended period of stellar formation. Our present assumption of a single burst of stellar formation maximises the number fraction of stellar BHs and the time available for mass segregation.

We find rates of tidal disruptions and coalescences with the MBH of $\sim4\times10^{-5}\mathrm{yr}^{-1}$ and $\sim3\times10^{-5}\mathrm{yr}^{-1}$, respectively, at $t=10$ Gyr. When stellar collisions are allowed, they occur at a rate about 10 times lower. On average, each MS-MS collision releases about $0.01\,\mathrm{M}_\odot$ of gas; if collisions between MS stars and compact remnants lead to complete disruption of the MS star, collisions (of all types) yield an average of $0.05-0.06\,\mathrm{M}_\odot$ per event.

When large-angle scatterings are explicitly included (essentially as a special case of collisions), they are found to have only little impact on the rate of tidal disruptions or coalescences with the MBH. A stellar BH is about 10 times more likely to be swallowed by the MBH than to be ejected from the nucleus. In contrast with this, in their multi-mass $N-$body simulations, Baumgardt et al. [25] find that all stellar BHs except one are ejected from the cluster and ascribe this result to strong interactions with objects (generally another stellar BH) deeply bound to the IMBH. These interactions are likely to be "resonant", i.e., the three objects (including the IMBH) form a strongly interacting, chaotic configuration for many orbital times until one of the lighter objects is ejected (Baumgardt, personal communication; see e.g., [26]). In principle, this mechanism can be included into the MC code by extending the loss-cone treatment used for tidal disruptions and coalescences to interactions with the binary consisting of the MBH and the most bound stellar object and resorting to explicit integration of 3-body motion for close interactions between the binary and a third object. However a simple analysis based on extrapolation of the cross sections for single-binary encounters [27] suggests that, in galactic nuclei, such ejections are considerably less likely to occur than a direct plunges through the MBH's horizon.

To first order, the evolution of galactic nuclei of different masses (but same initial $\mu=M_\bullet/M_{\rm stars}$ and η values, where $M_{\rm stars}$ is the total stellar mass), can be obtained by rescaling the mass, the size and the time units if stellar evolution is negligible. The evolution timescale is set by the relaxation time which increases approximately like $t_{\rm rlx} \propto M_\bullet^{5/4}$ for our $R_{\rm nucl} \propto M_\bullet^{0.5}$ assumption. This means that small nuclei are affected by faster segregation and MBH-driven expansion and may have been significantly more compact in the past than observed in the local universe. On the other hand, only very little relaxational mass-segregation can have occurred in nuclei with MBHs more massive than $\sim 10^7\,{\rm M}_\odot$ over a Hubble time. The evolution of the region very close to the MBH cannot be scaled from one nucleus to another with different mass and size, however because the processes of tidal disruptions and coalescences with the MBH introduce physical scales. In particular, we find that stellar BHs in more massive nuclei experience less segregation than in smaller nuclei (at a given value of $t/t_{\rm rlx}$). Also the rate of coalescences (in units of $N_*/t_{\rm rlx}$, where N_* is the total number of stars) is higher in more massive nuclei. This is probably an effect of a larger critical radius[1], requiring a larger accretion rate to yield the same amount of energy to power the expansion of the nucleus.

3. Observational consequences and future developments

Although we have not attempted a realistic modelling of the Galactic centre, it is tempting to apply our results to one specific observation of the Sgr A* region. Using the *Chandra* X-ray space-born observatory, Muno et al. [28] have detected 7 transient sources which appear to be much more concentrated around Sgr A* than the overall stellar population. Here we examine whether this may be a direct consequence of mass segregation, if these sources are all stellar BHs accreting from a lower-mass companion. We make the strong assumption that these binaries are not formed or affected by interactions with other stars such as 3-body binary formation, partner exchange, ionisation, etc. Instead we consider that they just react to 2-body relaxation as point objects with a total mass approximated by the mass of the stellar BH. If we pick up at random 7 sources with projected distance from the centre smaller or equal to 23 pc according to the profile for stellar BHs found in our simulations, we find that their distribution is at least as concentrated as the observed one in 15 % of the cases. It is therefore at this point not possible to exclude that the transients owe their peaked profile purely to mass segregation but this seems somewhat unlikely.

As pointed out by Muno et al., the rate of binary interactions should also increase steeply towards the centre and this probably combines with mass segregation to produce the observed distribution. The problem of binary dynamics in the vicinity of a MBH is complicated by the fact that there is no clear-cut definition of the hard-soft transition [29]. The Keplerian velocity dispersion increases virtually without bound when one approaches the centre. This may affect a binary on an orbit of relatively large semimajor axis a around the MBH because 2-body relaxation will cause the orbit to reach down to a value $R_{\rm peri} = (1-e)a \ll a$ over a timescale of order $t_{\rm rlx} \ln(1/(1-e))$ [30].

The most extreme type of dynamical interaction a binary can experience is the tidal separation of its members if its orbit brings it within $\sim a_{\rm bin}(M_\bullet/m_{\rm bin})^{1/3}$ of the MBH. Here $a_{\rm bin}$ is the semimajor axis of the binary itself and $m_{\rm bin}$ its mass. This process is of great interest by itself both as a way to create "hypervelocity stars" and to deposit a star on a tight orbit around the MBH [31, 32, 33, 34, 35].

To our knowledge, the complex question of binary dynamics in a galactic centre has not been investigated yet. This is an ideal subject for self-consistent stellar dynamical simulations of the sort presented here but including binary processes. MC codes are particularly well suited for following the evolution of large systems with a significant fraction of binaries whose interaction can be computed accurately by direct 3− and 4−body integrations [36, 37, 38].

The stellar BHs at the Galactic centre may have detectable consequences even if they are single. If the motions of the S-stars can be tracked with high enough a precision, an extended

[1] The critical radius is essentially the typical value of the semimajor axis for stars swallowed by the MBH.

distribution of non-luminous matter around the Galactic MBH should signal itself through its effect on their orbits. Present-day observations are insufficient to detect the slight Newtonian retrograde precession induced by an extended "dark cusp" [39], but future $30-100$ m telescopes may allow to realise such a measurement and to witness ~ 3 trajectory deflections caused by gravitational encounters per year between any of ~ 100 monitored S-stars and a stellar BH [40]. The compact stars can also collide and merge with MS stars and giants, hence creating unusual objects, once suggested to be the S-stars themselves [41], or increase the rotation rate of extended stars through multiple tidal interactions [42]. In this context, we note that our models predict about $2-3 \times 10^4$ collisions between MS stars and stellar BHs and a similar number of MS–white-dwarf events to occur over a Hubble time in a Sgr A*-type nucleus. Collisions with neutron stars are at least 10 times rarer. Finally, it seems relatively probable that at any given time some stellar BHs in the Galactic centre are bright X-ray sources as they accrete from the clumpy interstellar gas (poster by Patrick Deegan in this conference; Nayakshin & Sunyaev, in preparation).

As for EMRI rates and properties, the determination of how 2-body relaxation shapes the stellar distribution around the MBH is only a first –crucial– step. A robust estimate of the fraction of stars that eventually inspiral into LISA band, rather than plunge directly through the horizon while still on a wide orbit [12, 13] will probably require the development of a specific code. For stars on very eccentric orbits, one needs to follow the combined effects of GW emission and relaxation on a timescale significantly shorter than allowed by the present MC code. Recently Hopman & Alexander ([43] and these proceedings) have considered, for the first time in the study of EMRIs, the role of "resonant relaxation", i.e., of the random changes in eccentricity and orientation of the orbital planes due to the non-vanishing but fluctuating torque exerted on an orbit by the other orbits, each considered as an elliptical mass wire [44]. These authors find that resonant relaxation can increase the EMRI rate by of order a factor 10, an exciting result which is calling for confirmation by other computation techniques. Resonant relaxation can be included in the MC algorithm using the same approximate formulae as Hopman & Alexander. Unfortunately, this treatment relies on the calibration realised through a few very low-N $N-$body simulations [44] and more accurate numerical studies of this potentially important effect are called for (see Touma, these proceedings for a possible high-efficiency method).

Acknowledgments

PAS is indebted to Holger Baumgardt for invaluable help with the modification and use of NBODY4. We thank him for discussions and making available to us the data from his $N-$body simulations. We are grateful to Tal Alexander, John Fregeau, Atakan Gürkan and Frederic Rasio for enlightening discussions. MF received invaluable explanations about statistical tests for the distribution of X-ray sources around Sgr A* from Laurent Eyer.

The $N-$body simulations of 2-component models were performed on the *GRAPE* computers at the Astronomisches Rechen Institut in Heidelberg, thanks to the support of Rainer Spurzem, Peter Berczik and Peter Schwekendiek.

This research was supported by NASA ATP Grant NAG 5-13229. MF and VK acknowledge the Aspen Center for Physics where work on this project has been partially carried out, during the workshop "LISA Data: Analysis, Source and Science". The participation of MF in this workshop was supported in part by NASA grant NNG05G106G. The work of MF in Cambridge is funded through the PPARC rolling grant to the theory group at the IoA. The work of PAS has been supported in the framework of the Third Level Agreement between the DFG (Deutsche Forschungsgemeinschaft) and the IAC (Instituto de Astrofísica de Canarias).

[1] M. Freitag, P. Amaro-Seoane, and V. Kalogera. ApJ accepted, astro-ph/0603280, March 2006.
[2] M. Hénon. *Ap&SS*, 13:284–299, 1971.
[3] M. Hénon. *Ap&SS*, 14:151–167, 1971.
[4] M. Freitag and W. Benz. *A&A*, 375:711–738, August 2001.
[5] M. Freitag and W. Benz. *A&A*, 394:345–374, October 2002.
[6] M. Freitag. *Classical and Quantum Gravity*, 18:4033–4038, October 2001.
[7] M. Freitag. *ApJ Lett.*, 583:L21–L24, January 2003.
[8] M. Freitag, F. A. Rasio, and H. Baumgardt. *MNRAS*, 368:121–140, May 2006.
[9] M. Freitag, M. A. Gürkan, and F. A. Rasio. *MNRAS*, 368:141–161, May 2006.
[10] S. Sigurdsson. *Classical and Quantum Gravity*, 20:45, May 2003.
[11] K. Glampedakis. *Classical and Quantum Gravity*, 22:605, August 2005.
[12] D. Hils and P. L. Bender. *ApJ Lett.*, 445:L7–L10, May 1995.
[13] C. Hopman and T. Alexander. *ApJ*, 629:362–372, August 2005.
[14] W. Dehnen. *MNRAS*, 265:250, November 1993.
[15] S. Tremaine, D. O. Richstone, Y. Byun, A. Dressler, S. M. Faber, C. Grillmair, J. Kormendy, and T. R. Lauer. *AJ*, 107:634–644, February 1994.
[16] R. Schödel, T. Ott, R. Genzel, A. Eckart, N. Mouawad, and T. Alexander. *ApJ*, 596:1015–1034, October 2003.
[17] A. M. Ghez, S. Salim, S. D. Hornstein, A. Tanner, J. R. Lu, M. Morris, E. E. Becklin, and G. Duchêne. *ApJ*, 620:744–757, February 2005.
[18] S. Tremaine et al. *ApJ*, 574:740–753, August 2002.
[19] S. J. Aarseth. *PASP*, 111:1333–1346, November 1999.
[20] S. J. Aarseth. *Gravitational N-body Simulations. Tools and Algorithms.* Cambridge University Press, 2003.
[21] H. Baumgardt, J. Makino, and T. Ebisuzaki. *ApJ*, 613:1133–1142, October 2004.
[22] J. N. Bahcall and R. A. Wolf. *ApJ*, 209:214–232, October 1976.
[23] J. N. Bahcall and R. A. Wolf. *ApJ*, 216:883–907, September 1977.
[24] J. Chanamé and A. Gould. *ApJ*, 571:320–325, May 2002.
[25] H. Baumgardt, J. Makino, and T. Ebisuzaki. *ApJ*, 613:1143–1156, October 2004.
[26] P. Hut. *ApJ*, 403:256–270, January 1993.
[27] D. C. Heggie, P. Hut, and S. L. W. McMillan. *ApJ*, 467:359, August 1996.
[28] M. P. Muno, E. Pfahl, F. K. Baganoff, W. N. Brandt, A. Ghez, J. Lu, and M. R. Morris. *ApJ Lett.*, 622:L113–L116, April 2005.
[29] D. Heggie and P. Hut. *The Gravitational Million-Body Problem: A Multidisciplinary Approach to Star Cluster Dynamics.* Cambridge University Press, 2003.
[30] J. Frank and M. J. Rees. *MNRAS*, 176:633–647, September 1976.
[31] J. G. Hills. *Nature*, 331:687–689, February 1988.
[32] Q. Yu and S. Tremaine. *ApJ*, 599:1129–1138, December 2003.
[33] A. Gualandris, S. Portegies Zwart, and M. S. Sipior. *MNRAS*, 363:223–228, October 2005.
[34] M. C. Miller, M. Freitag, D. P. Hamilton, and V. M. Lauburg. *ApJ Lett.*, 631:L117–L120, October 2005.
[35] E. Pfahl. *ApJ*, 626:849–852, June 2005.
[36] M. Giersz and R. Spurzem. *MNRAS*, 343:781–795, August 2003.
[37] J. M. Fregeau, M. A. Gürkan, and F. A. Rasio. *ApJ Lett.*, 640:L39–L42, March 2006.
[38] M. A. Gürkan, J. M. Fregeau, and F. A. Rasio. ApJ. Lett. in press, preprint astro-ph/0512642, December 2005.
[39] N. Mouawad, A. Eckart, S. Pfalzner, R. Schödel, J. Moultaka, and R. Spurzem. *Astronomische Nachrichten*, 326:83–95, January 2005.
[40] N. N. Weinberg, M. Milosavljević, and A. M. Ghez. *ApJ*, 622:878–891, April 2005.
[41] M. Morris. *ApJ*, 408:496–506, May 1993.
[42] T. Alexander and P. Kumar. *ApJ*, 549:948–958, March 2001.
[43] C. Hopman and T. Alexander. Resonant relaxation near a massive black hole: the stellar distribution and gravitational wave sources. preprint, astro-ph/0601161, January 2006.
[44] K. P. Rauch and S. Tremaine. *New Astronomy*, 1:149–170, October 1996.

Institute of Physics Publishing
doi:10.1088/1742-6596/54/1/041
Journal of Physics: Conference Series **54** (2006) 259–265
Galaxy Center Workshop

The structure of the nuclear stellar cluster of the Milky Way

Rainer Schödel[1] **and Andreas Eckart**[1,2]

[1] I.Physikalisches Institut, Universität zu Köln, Zülpicher Str. 77, 50937 Köln, Germany
[2] Max-Planck-Institut für Radioastronomie, Auf dem Hügel 69, 53121 Bonn, Germany

E-mail: rainer@ph1.uni-koeln.de

Abstract. The structure of the nuclear stellar cluster of the Milky Way is of particular interest because it is the densest stellar cluster in our Galaxy, where the theoretical prediction of the formation of a stellar cusp around the central supermassive black hole, Sagittarius A* (Sgr A*) can be examined. We present high-resolution adaptive optics observations with multiple intermediate band filters of the inner $\sim 20''$ around Sgr A*. From the images, stellar number counts and a detailed map of the interstellar extinction toward the central 0.5 pc of the Milky Way were determined. The extinction map is consistent with a putative southwest-northeast aligned outflow from the central arcseconds. An azimuthally averaged, crowding and extinction corrected stellar density profile presents clear evidence for the existence of a stellar cusp around Sgr A*. We show that the profile of the surface brightness density is dominated by the brightest stars in the central arcseconds and is different from the shape of the stellar cluster as inferred from the number counts. Several density peaks found in the cluster may indicate clumping, possibly related to the last epoch of star formation in the Galactic Center. There is evidence for a common proper motion of the stars in one of these clumps.

1. Introduction

The center of our Milky Way is located at a distance of 7.6 ± 0.3 kpc and contains a $3.6 \pm 0.3 \times 10^6\,M_\odot$ black hole (BH), Sagittarius A* (Sgr A*) [16, 9, 7]. Theoretical work on stellar dynamics in a dense cluster surrounding a black hole that dominates the gravitational potential predicts the formation of a density excess, the so-called *cusp* around the BH. The density of a cusp that forms via two-body relaxation of the cluster is expected to be characterized by a power-law of the form $\rho \propto r^{-1/2 \ldots -7/4}$, where r is the distance to the BH [2, 3, 11, 14]. First evidence for a cusp in the GC was found with NIR speckle imaging observations [6, 1]. The first clear identification of the cusp around Sgr A* was achieved with adaptive optics (AO) imaging observations of the newly commissioned adaptive optics system/near infrared camera NAOS/CONICA ("NACO") at the ESO VLT. It was found that the GC stellar cluster can be described by a broken power-law with $\rho \propto r^{-2.0 \pm 0.1}$ outside of and $\rho \propto r^{-1.4 \pm 0.1}$ inside of $10''$ projected distance from Sgr A* [8]. Here, we present an analysis of the GC stellar cluster that is based on new AO observations that cover a larger field-of-view (FOV) than in the previous work. Further improvements include improved source detection methods, taking extinction explicitly into account, analysing the two-dimensional structure of the cluster, and comparing the methods of star counts vs. surface brightness measurements.

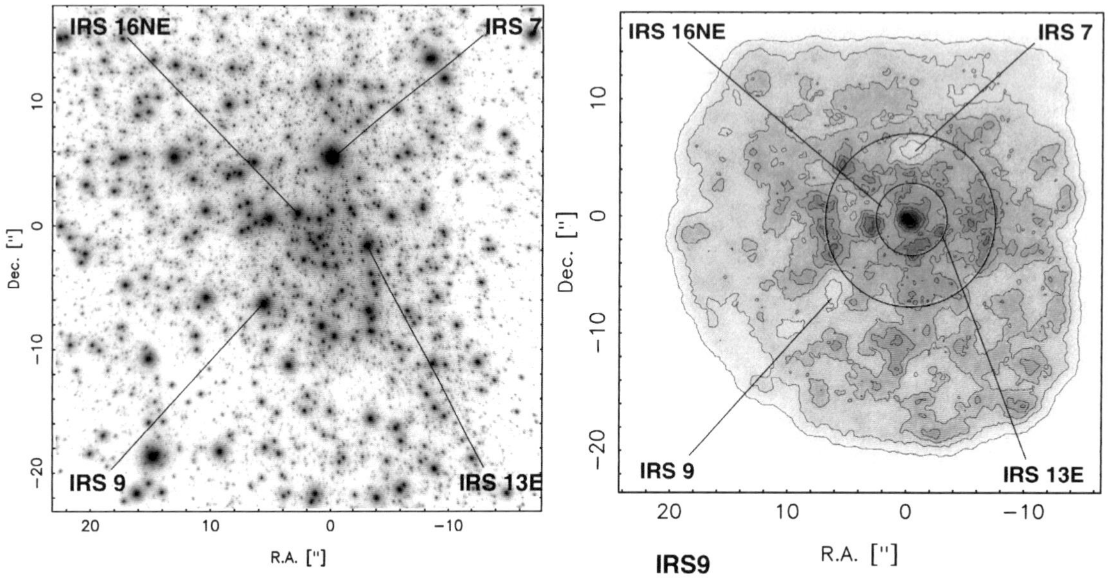

Figure 1. Left: Sum of the 2.27 + 2.30 μm IB filter NACO images. The offsets from Sgr A* in right ascension and declination that are labelled on the coordinate axes serve as an orientation, but do not represent high accuracy astrometry because there may be a slight ($< 1^{\circ}$) remnant rotation present in the images. North is up and east is to the left. Right: Adaptively smoothed, crowding and extinction corrected map of the stellar surface density. The smoothing was performed such that 40 stars contributed to the density at each point in the map. The smoothing radius varies between $0.6''$ near Sgr A* and $2''$ near the edge of the field. IRS 7, IRS 9 and IRS 16NE are particularly bright stars. The detection of faint stars is suppressed in their environment. The low density near these stars is therefore an artefact. IRS 13E is a compact group of co-moving stars, It coincides with a density peak. The circles around Sgr A* indicate radial distances of $3''$ and $7''$. At these distances several clumps appear to be concentrated. Note that bumps in the azimuthally averaged surface density can be seen at these distances in Fig. 3.

2. Observations

AO imaging observations were acquired with NACO/VLT in June and July 2004, using intermediate band (IB) filters of 0.06 μm width, centered at 2.00, 2.06, 2.24, 2.27, and 2.30 μm. More details on the data and their processing can be found in Schödel et al. (2006, submitted to A&A). The program package *StarFinder* [4] was used for point source detection via PSF fitting. The left panel in Fig. 1 shows the sum of the IB 2.27 and 2.30 images, which was used for the extraction of the star counts.

3. Results

3.1. Extinction

The IB 2.00 to 2.27 μm images were used to construct a map of the interstellar extinction. The near-infrared (NIR) spectra of stars are largely free from strong emission or absorption features at these wavelengths. The slope of the SED at these wavelengths is therefore determined by the stellar temperature and interstellar extinction/reddening. Hot stars can be assumed to have Rayleigh-Jeans like SEDs in the NIR. Known hot stars in the FOV [15] were therefore used for photometric calibration. All stars of unknown type were assumed to have an intrinsic blackbody SED with a temperature of 5000 K. An extinction law $A_{\lambda} \propto \lambda^{-1.75}$ was assumed [5]. Foreground

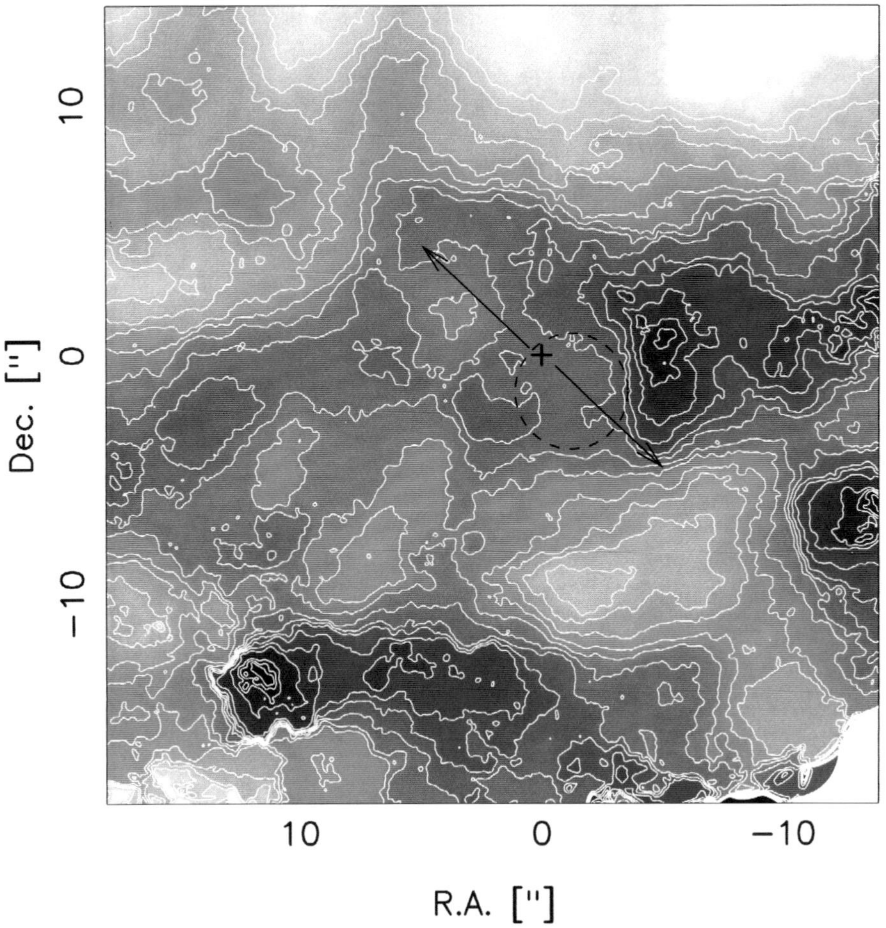

Figure 2. Map of the interstellar extinction at 2.157 μm derived from NACO intermediate band imaging at 2.00, 2.06, 2.24, and 2.27 μm. The contours are spaced from $A_{2.157\mu m} = 2.0 - 3.5$ mag at intervals of 0.1 mag. The cross indicates the position of Sgr A* and the dashed circle the approximate extent of the mini-cavity. The arrows indicate a putative southwest-northeast aligned outflow from the central arcseconds (see also Muzic et al., this volume).

stars were excluded due to their blue SED. The median of all measurements within 2″ of a given point in the FOV was taken.

The resulting extinction map is shown in Fig. 2. The 1σ uncertainty of the map is ≤ 0.1 mag, except in the regions with highest extinction, where it is ≤ 0.2 mag due to the reduced number of detected stars. A comparison with an extinction map derived from recombination lines [18] shows consistent results. The method presented here has the advantages that it is applicable in the absence of line emission, that it is not deteriorated near Sgr A* by the strong radio emission from this source, and that it provides a high spatial resolution. It is intriguing to note that the so called mini-cavity can be distinguished in the extinction map. It has been suggested in previous work that the mini-cavity is caused by fast winds from the Helium stars in the central arcseconds or by an outflow from Sgr A* [10, 13, 19]. The southwest-northeast aligned channel-like feature centered on Sgr A* in the extinction map is consistent with reduced extinction due to such a potential outflow from the central arcseconds along this direction (see also Muzic et al., this volume).

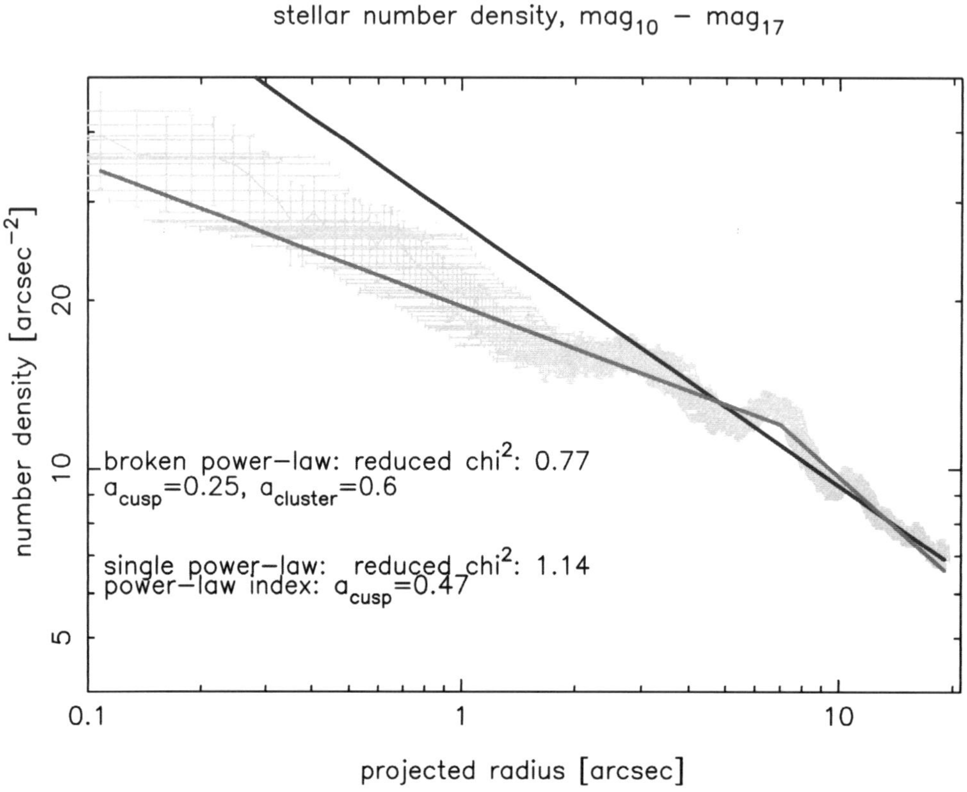

Figure 3. Azimuthally averaged extinction and crowding corrected surface density vs. distance from Sgr A*. Fits to the data with a single (blue) and with a broken (red) power-law are indicated, along with the corresponding power-law indices and reduced χ^2 values. The χ^2-values refer to the *overall* fit. The χ^2-value of the single power law fit is, of course, much worse in the central arcseconds.

3.2. Stellar number density

The stellar number density was determined from the extracted stellar number counts. The counts were corrected for incompleteness due to crowding. The completeness was determined via the standard method of inserting and recovering artificial stars. The right panel of Fig. 1 shows the extinction corrected two-dimensional stellar surface density in the central 0.5 pc of the GC. It can be seen that the GC stellar cluster appears fairly circularly symmetric. A strong density peak is visible centered on Sgr A*. Some regions of reduced density can be attributed to the presence of extremely bright stars, such as IRS 7, that hinder source detection in their immediate environment. It can be also seen that the stellar surface density does not decrease in a homogeneous way from the center, but that there are several density peaks present in the field, preferentially located at radial distances of 3″ and 7″ from Sgr A* (see circles in right panel of Fig. 1 and bumps at 3″ and 7″ in Fig. 3). One of these clumps is coincident with the dense co-moving group of stars in the IRS 13E complex [12, 17].

The azimuthally averaged extinction and crowding corrected surface density is presented in Fig. 3. Masking and a corresponding correction were applied when the completeness of stars of a given magnitude was below 30% in any area of the field. The density was extracted in overlapping rings of increasing width and distance from Sgr A*. The smoothing width was determined from the smoothing radius at a given distance from Sgr A* that was applied when constructing the map of the two-dimensional surface density (see right panel in Fig. 1). Power-

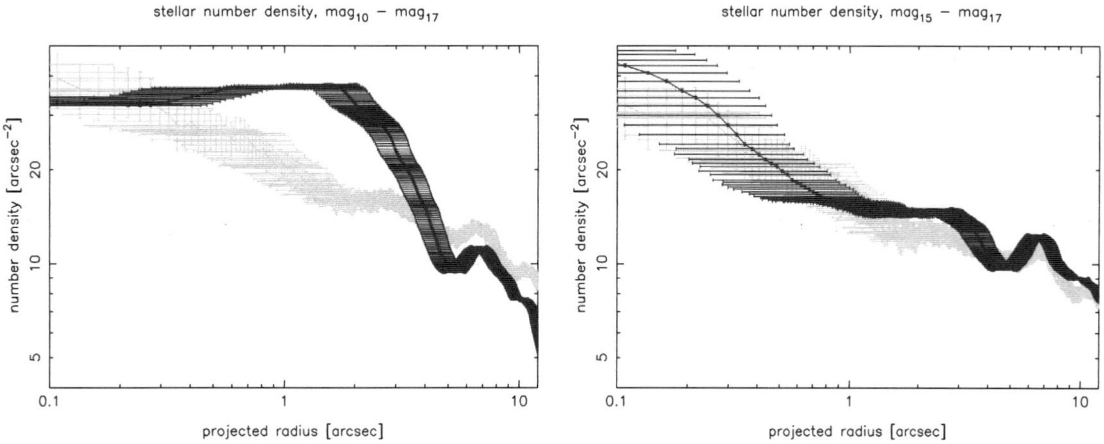

Figure 4. Comparison between the extinction and crowding corrected surface number density (green/grey) of stars in the magnitude ranges 9.75 – 17.75 (left) and 14.75 – 17.75 (right) and the corresponding light density (blue/black). The light density was derived from artificial images, taking into account only the fainter 80% of the pixel brightness distribution. The average light density was scaled to the average number surface density.

laws were fitted to the plot of the stellar surface density. As can be seen in the Fig. 3, a broken power law is necessary in order to provide a reasonable fit in the central few arcseconds. The least-squares fit results in a broken power law with a break radius of $7 \pm 1''$, a power-law index of $\alpha = 0.6 \pm 0.1$ for the outer part, and of $\alpha_{\rm cusp} = 0.25 \pm 0.1$ for the inner part of the cluster.

3.3. Number density vs. light density

In Fig. 4 we show a plot of the stellar surface density and of the stellar light density for stars in the magnitude ranges 9.75 – 17.75 (left) and 14.75 – 17.75 (right). Only the fainter 80% of the pixel brightness distribution was considered in order to dampen the influence of bright stars. While the light density distribution of the faint stars is similar to the stellar number density, there are considerable deviations when including the bright stars. This is the reason for the differences found in the structure of the GC stellar cluster when comparing number counts [8] and surface brightness [18]. The latter approach can lead to incorrect results in the inner arcseconds, that are dominated by bright, young stars. Analyzing the surface brightness of the GC cluster may be misleading due to the presence of bright stars that are not distributed in the same way as the faint population. This fact is further illuminated when analyzing the KLF.

The K-band luminosity function (KLF) of the cluster is shown in the left panel of Fig. 5. The right hand of Fig. 5 illustrates the percentage of the total light of the cluster that is contributed by stars of a given brightness, when assuming a simple power-law form of the cluster (dotted line in the KLF). The light from stars fainter than $\mathrm{mag_K} = 15$ contributes $\leq 1.5\%$ to the total light, i.e. the cluster light is completely dominated by the bright stars, which may not follow the overall distribution of the cluster stars. Via the PSF wings, the bright stars contribute to the diffuse light in AO or even completely diffraction limited images. This may be especially problematic when dealing with observations of lower resolution (e.g. with HST compared to NACO/VLT in the NIR),where crowding of the sources is significantly increased.

3.4. A new co-moving group of stars?

A stellar density peak coincides with IRS 13E (see Fig. 1). This co-moving group of stars may be the remnant core of a stellar cluster [12, 17, 15]. We have identified a potential second dense

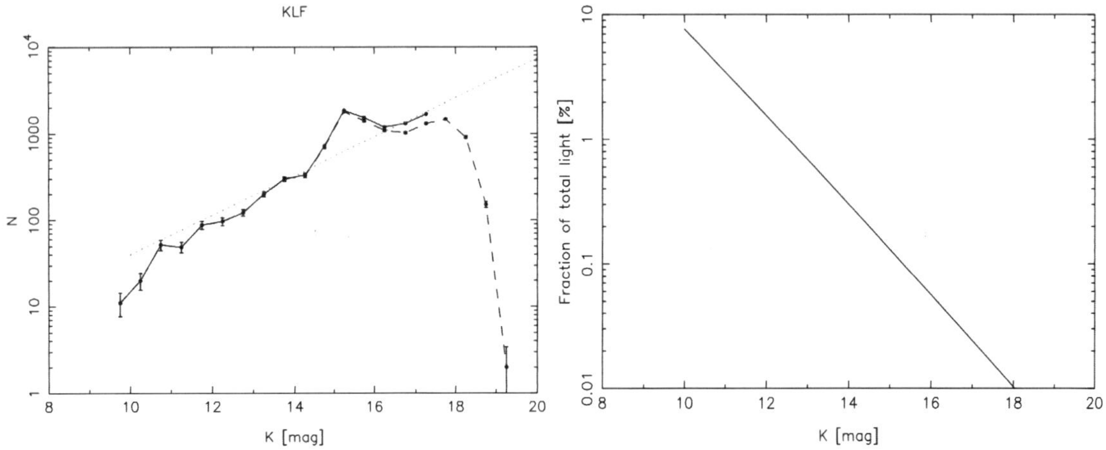

Figure 5. The left panel shows the K-band luminosity function (KLF) of the GC stellar cluster as it was obtained from AO observations with NACO. The dashed line indicates the raw counts, the straight line the completeness corrected counts. The 'bump' at $mag_K = 15.25$ is due to red clump/horizontal branch stars. The dotted line indicates a power law fit to the data – ignoring the red clump bump –with a power law index 0.23 ± 0.02. The right panel illustrates the fraction of the total light of a star cluster that is contributed by stars of a given magnitude, under the assumption that the cluster LF is a power law. Here, the power law derived from the KLF in the left panel above was used.

group of co-moving stars, near IRS15 NE, $\sim 12''$ NNE of Sgr A*. The corresponding field and the proper motions of the stars are shown in Fig. 6. Follow-up observations are needed in order to constrain the proper motions and examine whether this *IRS15E-clump* is indeed a co-moving group of stars. In that case it may be related to clumpy star formation during the most recent epoch of star formation in the central 0.5 pc [15].

4. Summary

Based on new AO observations, we present a detailed analysis of the structure of the stellar cluster in the central 0.5 pc of the Milky Way. The main results are as follows:

- A detailed map of the interstellar extinction in the central 0.5 pc is presented. It is consistent with a hypothetical SW-NE aligned outflow from the central arcseconds.
- A two-dimensional map of the stellar surface density, corrected for extinction and crowding, shows that the GC cluster is largely spherically symmetrical, but not homogeneous. Some density clumps are present, especially at projected distances of $3''$ and $7''$ from Sgr A*.
- The cusp around Sgr A* is detected with high significance. A broken power law with $r_{\rm break} = 7'' \pm 1''$ and power law indices $\alpha_{\rm cusp} = 0.25 \pm 0.1$, for the cusp, and $\alpha = 0.60 \pm 0.1$, for the outer cluster, provides a good fit to the data.
- Stellar light, even the diffuse one, is *not* a good tracer of the GC cluster structure because it is dominated by a few bright stars that are not distributed in the same way as the faint stars.
- We report the detection of a potential new co-moving dense group of stars, labelled the IRS15E-clump, located $\sim 12''$ NNE of Sgr A*.

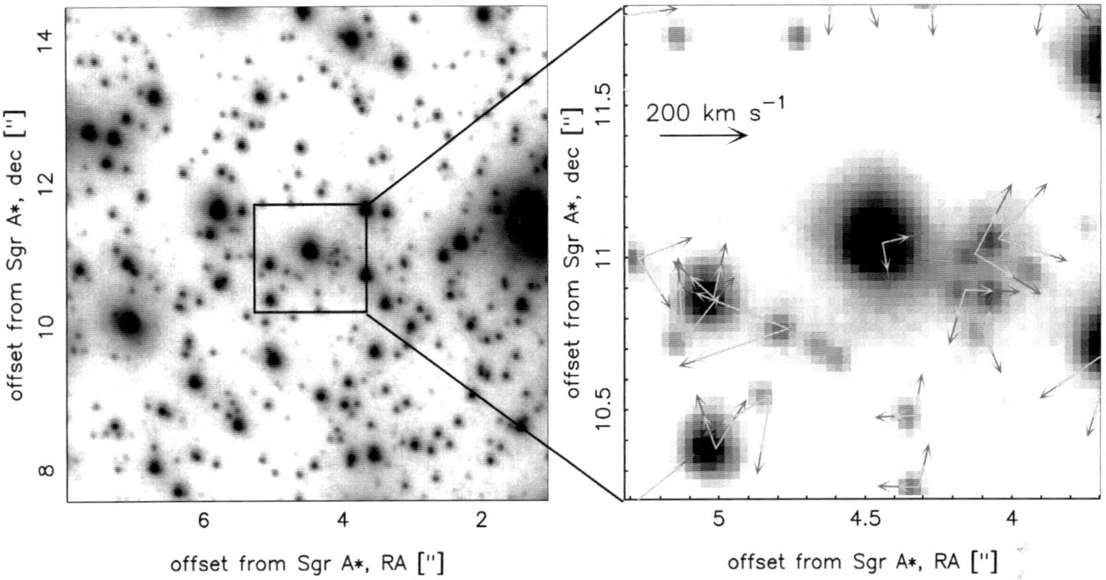

Figure 6. The left panel shows a zoom into the GC stellar cluster east of IRS15 NE. The right panel shows the potential co-moving group of stars that we call the *IRS15E-clump*. The arrows indicate the $\pm 3\sigma$ uncertainties of the magnitude and direction of the proper motions of the stars, as measured from 4 epochs of AO images (2002 to 2004).

Acknowledgments

The authors would like to thank all people that helped in organizing the Galactic Center Workshop 2006.

References

[1] Alexander, T. 1999, *Astrophys. J.*, **527**, 835
[2] Bahcall, J.N. & R.A. Wolf 1976, *Astrophys. J.*, **209**, 214
[3] Bahcall, J.N. & R.A. Wolf 1977, *Astrophys. J.*, **216**, 883
[4] Diolaiti, E., O. Bendinelli, D. Bonaccini, et al. 2000, *Astron. Astrophys. Supplement*, **147**, 48
[5] Draine, B.T. 1989, *Proceedings of the 22nd Eslab Symposium held in Salamanca, Spain, 7-9 December, 1988, ed. B.H. Kaldeich*, 93
[6] Eckart, A., R. Genzel, R. Hofmann, B.J. Sams, & L.E. Tacconi-Garman 1995, *Astrophys. J.*, **445**, L23
[7] Eisenhauer, F., R. Genzel, T. Alexander, et al. 2005, *Astrophys. J.*, **628**, 246
[8] Genzel, R., R. Schödel, T. Ott, et al. 2003, *Astrophys. J.*, **594**, 812
[9] Ghez A.M., G. Duchêne, K. Matthews, et al. 2003, *Astrophys. J.*, **586**, L127
[10] Lutz, D. , A. Krabbe, & R. Genzel 1993, *Astrophys. J.*, **418**, 244
[11] Lightman, A.P. & S.L. Shapiro 1977, *Astrophys. J.*, **211**, 244
[12] Maillard, J.P., T. Paumard, S.R. Stolovy, & F. Rigaut 2004, *Astron. Astrophys.*, **423**, 155
[13] Melia, F., R.F. Coker, & F. Yusef-Zadeh 1996, *Astrophys. J.*, **460**, L33
[14] Murphy, B.W., H.N. Cohn, & R.H. Durisen 1991, *Astrophys. J.*, **370**, 60
[15] Paumard, T., R. Genzel, F. Martins, et al. 2006, *Astrophys. J.*, **643**, 1011
[16] Schödel, R., T. Ott, R. Genzel, et al. 2003, *Astrophys. J.*, **596**, 1015
[17] Schödel, R., A. Eckart, C. Iserlohe, R. Genzel, & T. Ott 2005, *Astrophys. J.*, **625**, L111
[18] Scoville, N.Z., S.R. Stolovy, M. Rieke, et al. 2003, *Astrophys. J.*, **594**, 294
[19] Yusef-Zadeh, F., D.A. Roberts, & J. Biretta 1998, *Astrophys. J.*, **499**, L159

Institute of Physics Publishing
doi:10.1088/1742-6596/54/1/042
Journal of Physics: Conference Series **54** (2006) 266–272
Galaxy Center Workshop

Stellar and wind properties of massive stars in the central parsec of the Galaxy

Fabrice Martins[1], Reinhard Genzel[1,2], Frank Eisenhauer[1], Thibaut Paumard[1], Thomas Ott[1], Stefan Gillessen[1], Sascha Trippe[1]
[1] MPE, Postfach 1312, D-85741 Garching bei München
E-mail: martins@mpe.mpg.de

Abstract. We present a detailed spectroscopic analysis of the massive stars in the central parsec of the Galaxy. Non-LTE atmosphere models including winds and line-blanketing are computed and used to derive stellar and wind properties. The evolutionary status of the stars is discussed, and we show due to the age of the population, the most massive stars (M > 60 $M_{\odot}$) have disappeared. We estimate the total H and He I ionising fluxes emitted by these young stars and show that they are able to account for the nebular observed emission. Finally, we discuss the spectral properties of the S - stars.

1. Introduction

The Galactic Center region contains a large number of massive stars. They are mainly found in three massive clusters: the Arches, the Quintuplet and the central cluster. The latter surrounds the super-massive black hole SgrA* while the two former are a few tens of parsecs away from it. These clusters have slightly different ages: 2.5 Myr for the Arches, 4 Myr for the Quintuplet and 6 Myr for the central cluster. This age spread is responsible for the wide variety of massive stars encountered. This is a very fortunate property which allows a sampling of the various evolutionary states of massive stars, going from main sequence objects to all kinds of Wolf-Rayet stars and to Luminous Blue Variables. Getting access to the physical properties of such a sample of massive stars represents a unique opportunity to constrain quantitatively the evolution of massive stars.

In the present study, we concentrate on the central cluster. Its massive stellar content was first unraveled by [6, 1] and [12] who discovered several luminous objects showing unusually strong H and He emission lines in the K band. At that time, no other star was known to display such a characteristic spectrum, and the objects were nicknamed "Helium stars". The quantitative analysis of these objects by [21, 22] revealed that they were evolved massive stars with relatively cool effective temperatures, placing them close to the so-called Humphreys-Davidson limit above which stars become unstable. The development of powerful near IR observational techniques has recently widened our knowledge of the massive stars GC population. More evolved objects as well as OB stars have been discovered, including the so-called S-stars, a group of stars located within the central arc-second and orbiting the super-massive black hole with periods of a few tens of years.

[2] Also Department of Physics, University of California, Berkely, CA 94720, USA

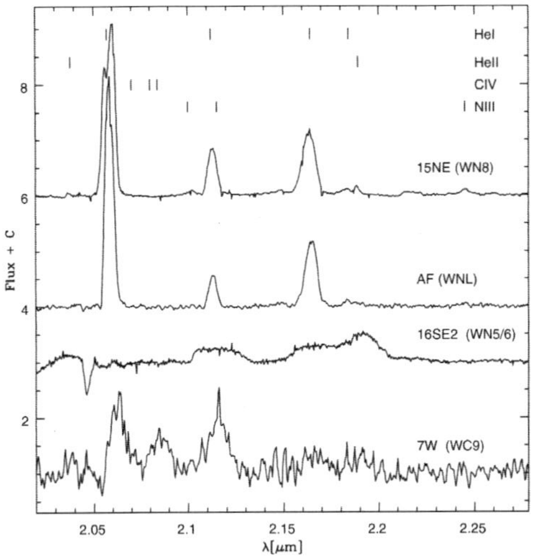

Figure 1. K band spectra of various types of Wolf-Rayet stars observed in the Galactic Center. Important lines are indicated. Compared to the first observations of the GC stars, new spectra display more lines, allowing a better analysis of the stellar and wind properties of these stars.

Here, we present first results of a detailed analysis of SINFONI spectra of the massive stars recently found to gather in two counter-rotating disks ([23] and Paumard et al., these proceedings). The combination of new and better spectra for these stars with the use of state-of-the-art atmosphere models leads to significant improvements in our knowledge of the population of massive stars in the central parsec of the Galaxy.

2. New population of massive stars in the Galactic Center

With the development of powerful observational techniques in the near-IR spectral range, more and more stars can be observed. In that context, 3D spectroscopy is a unique tool since it allows observations of crowded regions such as the Galactic Center central cluster. Not only 3D spectroscopy provides data for tens of stars in one single exposure, but it also allows a much better estimate of nebular contamination than long-slit spectroscopy.

The use of SINFONI in the Galactic Center has proven to be very efficient. During its first year, SINFONI has revealed a new population of massive stars. The spectral classification as well as the dynamical properties of these stars are reported in Paumard et al. (these proceedings and [23]). One of the main discoveries is the presence of numerous OB stars, something which was suspected but not confirmed so far. In addition, several faint Wolf-Rayet stars have also been observed for the first time. Fig. 1 shows a mosaic of spectra of several WR stars observed with SINFONI. It is important to realize that even for stars which have been known for several years, the new spectra are an improvement since their quality allows the detection of weak emission lines such as He II 2.189 μm or N III 2.238 μm. Such lines are extremely useful to derive stellar properties, in particular effective temperatures and abundances. We describe such an analysis and discuss its results in the following sections.

3. New atmosphere models for massive stars

The modelling of atmospheres of massive stars has experienced significant improvements in the last years. Indeed, a realistic modelling requires the inclusion of three main ingredients: first, non-LTE calculations have to be performed since the luminosity of massive stars is so large that radiative processes are dominant in the atmosphere; second, winds must be included because of the large mass loss of these stars creating large expanding envelopes; third, metals have to be included in the models, not only because lines from elements other than H and He are observed, but also because metals strongly modify the atmospheric structure, especially the temperature structure (see [17]). The two first ingredients (non-LTE and winds) are present in most atmosphere models since the early 90's, but the third one, referred to as line-blanketing, was missing until recently. Significant progress in both computers capabilities and numerical techniques now allow a realistic treatment of line-blanketing. Consequently, quantitative analysis of massive stars can be performed and accurate stellar and wind parameters are derived ([2, 18]).

The development of such new atmosphere models for massive stars opens new perspectives for the study of the young stars in the Galactic Center. Indeed, we have seen in the previous section that the improvements in the observational techniques allow the detection of weak lines from various metallic elements and various ionisation states. These diagnostics can now be studied quantitatively and accurate physical parameters can be derived without suffering from approximation such as the lack of line-blanketing as was the case for the pioneering studies of [21, 22]. In the next section we describe the strategy adopted in our quantitative study of SINFONI spectra with these new atmosphere models.

4. Spectroscopic analysis of Wolf-rayet stars

We have computed atmosphere models for the massive stars in the Galactic Center using the code CMFGEN ([10]). CMFGEN makes a non-LTE treatment of the atmosphere in a spherical geometry and includes winds and line-blanketing. In practice, several input parameters are given, the main ones being: effective temperature ($T_{\rm eff}$), luminosity (L), mass loss rate ($\dot{M}$), terminal velocity (v_{∞}), abundances. In addition, a hydrodynamic structure (velocity structure or density structure) has to be provided. Indeed, CMFGEN is not an hydrodynamical code and does not compute the structure itself, but provides the temperature structure, radiation field and emergent spectrum for a given hydrodynamical structure. In practice, the input velocity structure is given by a quasi hydrostatic velocity structure connected to a so-called β velocity law (see [10] for a complete description).

Our general strategy to derive the stellar and wind parameters can be summarised as follows:

- Effective temperature: lines of two successive ionisation states of the same elements are used. He lines were the main indicators, while C lines (C III and C IV) are also used in WC stars.
- Luminosity: the absolute K band magnitude is the main diagnostic. The observed value is estimated from m_K, the distance to the GC (7.62 kpc, see [5]) and the extinction map of [24].
- Mass loss rate: it is derived from the strength of emission lines.
- Terminal velocity: the width of emission lines is the main indicator. When P-Cygni profiles are observed, the blue-ward extension of the absorption dip also serves as an indicator.
- Abundances: they are derived from the intensity of absorption or emission lines.

Two points need to be stressed. First, a huge advantage of the Galactic Center massive stars compared to most other Galactic stars is that their distance is known accurately. Hence, the luminosity suffers from less uncertainty (which is dominated by uncertainties on the extinction). Second, the modelling of near-IR spectra is a highly degenerated problem, and the situation is in fact more complicated than the simple sketch drawn above. Indeed, most indicators depend equally on several parameters. For example, the absolute K band magnitude increases with luminosity, but also with the wind density, and thus the mass loss rate, due to free-free emission in the ionised atmosphere. The determination of stellar and wind parameters is consequently a long iterative process.

Figs. 2 and 3 show the final fits for the stars IRS16SE2 and IRS13E2. We see that in addition to the strong emission lines, weaker lines are also well reproduced. As already mentioned,

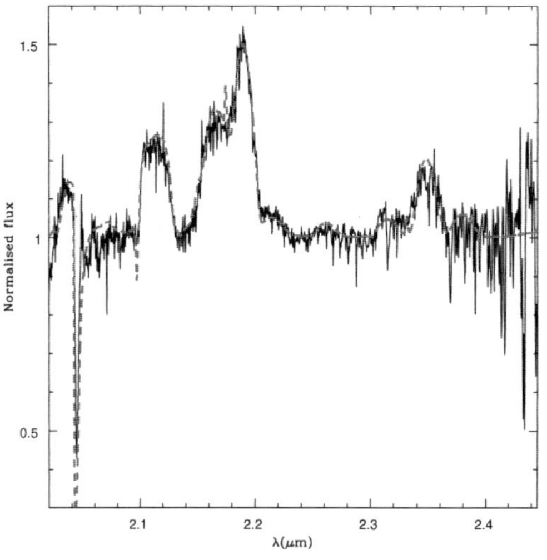

Figure 2. Observed K band spectrum of IRS16SE2 together with the best fit model (bold).

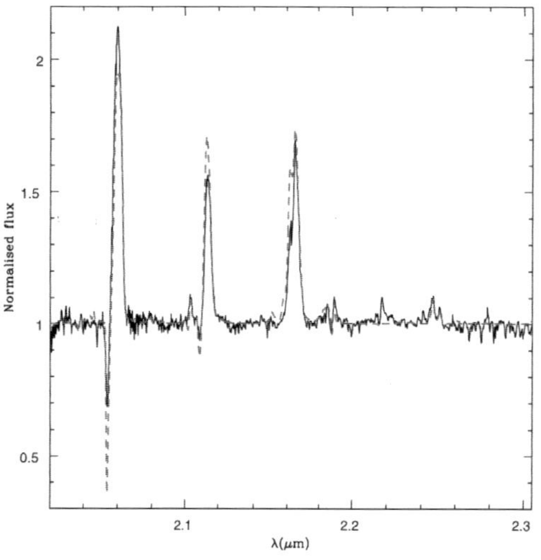

Figure 3. Same as Fig. 2 for IRS13E2.

this leads to a better estimate of the stellar and wind parameters, and enables abundances determinations. We now discuss in more detail some of these parameters.

5. Masses

One of the most important parameters is the initial mass of the stars. In Wolf-Rayet stars and related objects, mass estimates are very difficult since the photosphere is not observable. Hence, standard measurement of gravity from the wing of Hydrogen lines cannot be done. The only solution is to get an estimate of the evolutionary mass, i.e. the mass obtained from evolutionary diagrams. In Fig. 4, we show such a diagram giving the Hydrogen content as a function of luminosity. The comparison of predictions of evolutionary models to the position of the GC massive stars reveals that these stars have initial masses in the range 25-60 $M_\odot$. This means that no very massive stars ($M \sim 100\ M_\odot$) are present. Is it an indication of an upper mass cut-off of the IMF? For a total number of 40 stars in the 25-60 $M_\odot$ mass range (number of Wolf-Rayet stars), one would expect at least 8 stars in the range 60-100 $M_\odot$ for a Salpeter IMF. This is a lower limit since [23] have shown that the mass function in the Galactic Center was top-heavy. Since we do not see any of these very massive stars, one may indeed wonder whether such a cut-off exists since a positive detection would strongly constrain the formation mechanism of the GC young stars. In particular, star formation in a self gravitating disk predicts the presence of such a cut-off ([14, 15]). However, in the present case the most likely explanation is the age of the population. Indeed, [23] showed that the young stars in the GC resulted from a burst of star formation 6$\pm$2 Myr ago. The lifetime of the most massive stars being of only a few Myr (see [19]), we naturally expect the absence of very massive objects in the population of GC massive stars.

6. Ionising radiation and the ISM properties

In the early studies of the Galactic Center massive stars, an important problem appeared: the population of young stars was too cool to account for the ionisation of the interstellar medium.

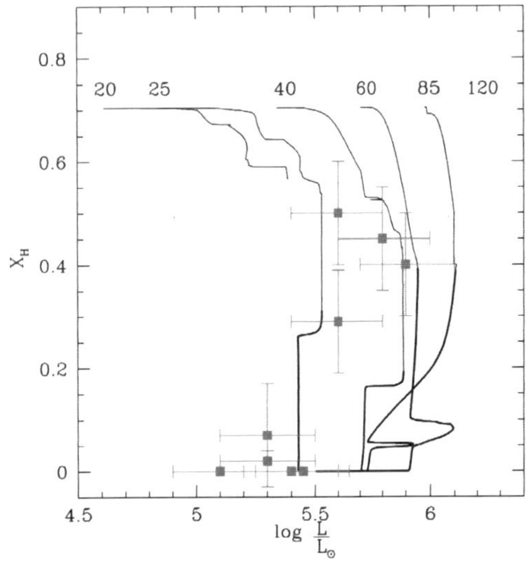

Figure 4. . Hydrogen content as a function of luminosity. Solid lines are predictions of evolutionary models ([19]) and squares are values derived for Wolf-Rayet stars in the Galactic Center. Initial masses for these objects range between 25 and 60 $M_\odot$.

this problem was first highlighted by [22]: their analysis of the brightest "He" stars revealed that they were able to produce enough H ionising photons, but were far too cool to reproduce the nebular He I emission. Indeed, [4] and [13] state that one should expect $\log Q_H = 50.4$ and $\log Q_{HeI} = 48.4$ (Q_H and Q_{HeI} being the H and HeI ionising fluxes), while [22] proved that the He stars provided $\log Q_H = 50.5$ and $\log Q_{HeI} < 46$.

Later, ISO observations of the Galactic Center lead [16] to the conclusion that standard stellar evolution was facing serious problems in this region. Indeed, synthesis population models at an age of 7 Myr produced much more hot stars than was observed in the GC. In addition, the observed Ne excitation was much too low compared to the prediction of photoionisation models. [16] claimed that stellar evolution produced too few cool stars.

Here, we show that this is not true any more. First, a new population of massive stars was discovered. Most of them are either Wolf-Rayet stars fainter than the first "Helium" stars, or OB stars. The latter have effective temperatures of the order 25000-35000 K, much hotter than the values of the He stars derived by [22]. The faint WR stars are also much hotter, with $T_{\rm eff}$ as large as 54000 K. Second, the use of new atmosphere models for the analysis of the He stars lead to an upward revision of their effective temperatures. All in all, this implies that the present population of massive stars in the GC is much hotter than previously estimated, with the consequence of a harder ionising spectrum. For the whole population, we find $\log Q_H = 50.7$ and $\log Q_{HeI} = 49.2$, in very good agreement with the nebular estimates. We also note that the contribution of the He stars to the total H ionising luminosity has dropped to 3 %, in good agreement with the value of the synthesis population models ([16]). From that, we conclude that standard stellar evolution and the present generation of atmosphere models are perfectly adapted to study the Galactic Center massive star population.

7. The S - stars: spectroscopic properties

The so-called S-stars are the stars orbiting the super-massive black hole in the central arc-second of the Galaxy. Since the first spectroscopic identification of one of them – S2 – as a late O – early B star ([7, 8]), there has been a number of observations which have confirmed that the majority of these objects are indeed early to late B stars ([5]). Given that such stars are quite young (no more than ~ 20 Myr for the earliest types), their presence indicates recent star formation. But in this special environment, star formation should not happen ([20]): this is the

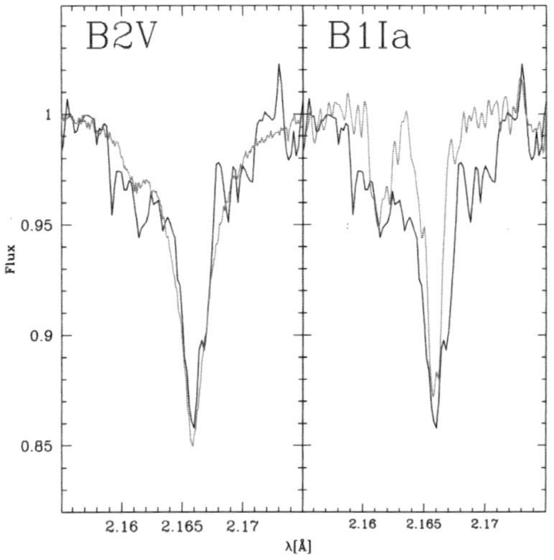

Figure 5. . Comparison between the average Br_γ spectrum of S2 over the last two years to template spectra of a B2V star (left) and a B1Ia star (right) from the atlas of [9]. The width of the observed line is much more compatible with a dwarf than with a supergiant classification.

"paradox of youth" in the Galactic Center ([8]).

With the accumulation of observations, better spectra of the S-stars become available. Consequently, more spectral lines are detected: in addition to the strong Br_γ and He I 2.112 μm absorptions, He I 2.149 μm, He I 2.164 μm and He I 2.184 μm are clearly seen in the spectrum of S2. This confirms the spectral classification as an early B star. With a better signal to noise ratio, the detailed shape of the Br_γ line allows an estimate of the luminosity class of this star. Fig. 5 shows the comparison of this line in S2 (black) to template spectra for an early B dwarf and an early B supergiant (from the atlas of [9]): it is clear that the width of the line is better reproduced by the dwarf template, showing that S2 is not a supergiant. Hence, not only we confirm that spectroscopically S2 is an early B star, but we also argue that it is not a supergiant star. This is another indication of the youth of this star.

In order to solve the paradox of youth, there has been a number of proposal in the last years arguing that the S-stars are not genuine B stars but are older cool giants the envelope of which has experienced extreme changes due to the proximity of the super-massive black hole. Recently, [11] have proposed that the atmosphere of late type giants could be irradiated by some AGN activity of SgrA*. They show that depending on the strength of the irradiation, the resulting spectrum could either still show CO absorption bands, or show no CO but all H and He lines in emission. None of these models correspond to the observed spectrum of the S-stars which display absorption lines (H and He) and no CO. Another scenario proposed by [3] is that the S-stars are the core of AGB stars the envelope of which has been ripped off by tidal forces. Although attractive, this possibility suffers from quantitative problems: the rate at which tidal disruption occurs is too small to account for the number of S-stars; the luminosity of these remaining cores is too low compared to the estimated luminosity of the brightest S-stars; and special abundances are expected in such remnants, while so far the spectra of the S-stars do not differ from normal B dwarfs, i.e. stars with unprocessed He abundance.

In conclusion, the S-stars are spectroscopically B dwarfs and all the scenarios developed to explain them as older stars having experienced strong changes due to the proximity of SgrA* face quantitative problems.

8. Conclusion

We have performed detailed spectroscopic analysis of massive stars in the central parsec of the Galaxy using non-LTE atmosphere models including winds and line-blanketing. The combination of these new models with better spectra displaying more diagnostic lines than in previous study leads to an improved knowledge of the stellar and wind properties of the young stars located in the two counter-rotating disks in the Galactic Center. We show that very massive stars (with M $\sim$ 100 $M_{\odot}$) are absent, mainly due to the age of the population: such stars have already exploded in supernovae after 6 Myr. The average effective temperature of the population is much larger than previously thought, mainly due to both the presence of hot OB and new Wolf-Rayet stars. This has important consequences for the ionisation of the ISM which can now be fully explained by the known population of young stars. This indicates that standard stellar evolution and atmosphere models, once thought to fail to explain the existence of the GC massive stars, now fully account for the observed properties of the population. Finally, we argue that the S-stars are B stars on or near the main sequence.

Acknowledgments

FM acknowledges support from the Alexander von Humboldt foundation. We thank John Hillier for making his atmosphere code CMFGEN available and for interesting discussions.

References

[1] Allen, D.A., Hyland, A.R., Hillier, D.J., 1990, MNRAS, 244, 706
[2] Bouret, J.C., Lanz, T., Hillier, J.C., et al., 2003, ApJ, 595, 1182
[3] Davies, M.B., King, A., 2005, ApJ, 624, 25
[4] Ekers, R.D., van Gorkom, J.H., Schwarz, U.J., et al. 1983, A&A, 122, 143
[5] Eisenhauer, F., Genzel, R., Alexander, T., et al., 2005, ApJ, 628, 246
[6] Forrest, W.J., Shure, M.A., Pipher, J.L., et al., 1987, in "The Galactic Center", ed. D.C. Backer, AIP 155, 153
[7] Genzel, R., Eckart, A., Ott, T., Eisenhauer, F., 1997, MNRAS, 291, 219
[8] Ghez, A.M., Duchêne, G., Matthews, K., et al., 2003, ApJ, 586, 127
[9] Hanson, M.M., Kudritzki, R.-P., Kenworthy, M.A., et al., 2005, ApJS, 161, 154
[10] Hillier, D.J., Miller, D.L., 1998, ApJ, **496**, 407
[11] Jimenez, R., da Silva, J.P., Peng Oh, S., et al., 2006, ApJ, submitted (astro-ph/0601527)
[12] Krabbe, A., Genzel, R., Drapatz, S., Rotaciuc, V., 1991, ApJ, 382, 19
[13] Krabbe, A., Genzel, R., Eckhart, A., et al., 1995, ApJ, 447, 95
[14] Levin, Y, Beloborodov, A.M., 2003, ApJ, 590, 33
[15] Levin, Y., 2006, MNRAS, submitted (astro-ph/0603583)
[16] Lutz, D., 1999, in "The Universe as Seen by ISO", Eds, P. Cox & M.F. Kessler, ESA-SP 427, p. 623
[17] Martins, F., Schaerer, D., Hillier, D.J., 2002, A&A, 382, 999
[18] Martins, F., Schaerer, D., Hillier, D.J., et al., 2005, A&A, 441, 735
[19] Meynet, G., Maeder, A., 2005, A&A, 429, 581
[20] Morris, M., 1993, ApJ, 408, 496
[21] Najarro, F., Hillier, D.J., Kudritzki, R.P., et al., 1994, A&A, 285, 573
[22] Najarro, F., Krabbe, A, Genzel, R., et al., 1997, A&A, 325, 700
[23] Paumard, T., Genzel, R., Martins, F., et al., 2006, ApJ, **643**, 1011
[24] Schödel, R., et al., 2006, ApJ, submitted

Institute of Physics Publishing Journal of Physics: Conference Series **54** (2006) 273–278
doi:10.1088/1742-6596/54/1/043 Galaxy Center Workshop

First infrared VLTI fringes on Galactic Center sources [1]

Jörg-Uwe Pott[1,2], Andreas Eckart[1], Andreas Glindemann[2], and Rainer Schödel[1]

[1] I. Physikalisches Institut, University of Cologne, Zülpicher Str. 77, 50937 Köln, Germany
[2] European Southern Observatory, Karl-Schwarzschildstr.2, 85748 Garching bei München, Germany

E-mail: `pott@ph1.uni-koeln.de`

Abstract. We present the first VLTI observations of targets in the Galactic Center region. The encouraging results are establishing the scientific capabilities of infrared interferometry in the outstanding but – for interferometric means – distant and faint dust obscured Central Parsec. We successfully combined the light of two 8m-class telescopes in K and N band with ESO's AMBER and MIDI beam combiners. The enigmatic dust-forming and spectrally featureless source IRS 3 could be clearly resolved into two components with different physical properties. We estimated lower limits of the 10 μm sizes of the most prominent Northern Arm bow-shocks IRS 1W, 10W, and 8. Investigations on the suitability of IRS 7 as a phase reference for long baseline interferometry at NIR wavelengths are underway.

1. Introduction

The Central Parsec of the Milky Way is due to its proximity a unique place to study star formation and evolution close to a supermassive black hole (Sgr A*: 3.6 $M_{\odot}$, 7.6 kpc; Eisenhauer et al. 2005). The so-called *paradox of youth* in the Galactic Center (GC) describes the existence of numerous massive, young stars in the central stellar cluster, indicating an active star forming history despite the strong gravitational forces, dominated by the black hole (Genzel et al. 2003; Eckart et al. 2004; Moultaka et al. 2004). Observed dust-forming stellar phases are short-lived (e.g., massive Wolf-Rayet stars form dust for not more than 10^{5-6} yr). The presence of many such sources indicates that star formation took place just a few Myr ago. The detailed study of the formation, evolution and kinematics of the circumstellar dust distribution in the GC can reveal and trace mass loss, stellar wind interaction zones, and properties of the underlying enshrouded stellar population.

Since very recently, the VLT Inteferometer (VLTI)[2] offers measurements at an angular resolution on scales of a few to tens of milliarcsec (mas) in several near and mid infrared atmospheric transmission windows. The present sensitivity of the VLTI is sufficient for the investigation of the stellar dust envelopes and bow-shocks of the some sources. In the mid-infrared a two-telescope beam combiner measures the *visibility modulus* of the source with

[1] Based on observations collected at the European Southern Observatory, Paranal, Chile (programs 073.B-0249, 075.B-0113, and 076.B-0863).
[2] http://www.eso.org/projects/vlti/

respected to the used projected baseline length and position angle. The modulus can constrain the one dimensional extension of simple target brightness distributions of circular or elliptical shape with a constant or Gaussian radial intensity profile at a resolution of e.g. 20 mas using a 130 m baseline length and an observing wavelengths of 10 μm. In the near-infrared three VLTI baselines can be combined at the same time, enabling the measurement of the *closure phase* in addition to the visibility moduli. This property gives valuable information about the source asymmetry at high precision, unaltered by atmospheric and instrumental influences. We employed the NIR instrument to analyze the properties of IRS 7. This bright supergiant is to be used as an off-axis fringe tracking reference star for the observations of weak sources in the central arcseconds of the GC, as soon as the star separator facility of the PRIMA instrument[3] will be available.

2. Observations

MIDI, the Mid-Infrared Interferometric Instrument of the VLTI[4], combines the light of two telescopes and measures spectrally dispersed fringes in the 8-13 μm regime. We recorded the correlated and total flux at a spectral resolution of R$\sim$30 using the high-sensitivity mode of MIDI. The observations of the brightest MIR-excess sources in the Central Parsec (IRS 3, 1W, 10W, 2L, 8) were conducted in June 2004 with the shortest baseline UT2-UT3 (46m). The most compact source IRS3 was re-observed in ESO period 75 in 2005 with the UT3-UT4 baseline. Its orientation is nearly perpendicular to UT2-UT3. Thus the new dataset complements the uv-coverage of the previous one and enables the detection of pronounced source asymmetries. The complete N-band flux of IRS 3 was resolved out by observations with the longest baseline UT1-UT4 (130 m).

In March 2006 we could observe the Galactic Center for the first time in the NIR with an optical long baseline interferometer. The AMBER instrument[5] is capable to record spatially dispersed fringes of a telescope- and baseline triple instantaneously. Due to technical problems, we were capable to track fringes continuously only on the intermediate baseline of the used triple (UT1-UT3-UT4).

3. Local dust formation around IRS 3

We presented the first fringes on IRS 3 in Pott et al. (2005). This interesting source eludes any spectral classification so far (Tanner et al. 2006). From the late seventies on it is known to be a hot and compact dusty source. Rieke et al. (1978) estimated a luminosity of $5{\cdot}10^4$ $L_\odot$ and a blackbody temperature of the de-reddened spectrum of about 400 K. Becklin et al. (1978) characterized IRS 3 to be the only dusty source within the Central Parsec with intrinsic silicate absorption of an optical depth of at least 1.3. Viehmann et al. (2006) found four other nearby MIR sources with intrinsic silicate absorption. IRS 3 appears to show similar broadband spectral properties as dust-enshrouded (super-)giants. AGB stars and red supergiants have been shown to have mass loss rates in the range of 10^{-7} to 10^{-3} $M_\odot$ yr^{-1} and stellar wind velocities of 10-20 km s^{-1} (van Loon et al. 1999). These authors have shown, that a few 10-20 carbon-rich super-giants and AGB stars at the very tip of the AGB can dominate the mass loss of an entire sample of thousands of mass losing stars. This underlines the importance that a few individual stars may have for the understanding of mass loss and dust formation at the center of the Galaxy.

Viehmann et al. (2005) have observed a sharp interaction zone in the L- and M-band, where the outer dust shell of IRS 3, resolvable with a beamwidth of a single 8m-class telescope, appears to collide with an outflow from the central arcseconds, possibly due to the fast winds of the He-

[3] http://www.eso.org/projects/vlti/instru/prima/index_prima.html

[4] http://www.eso.org/instruments/midi/; Leinert et al. (2003)

[5] http://www.eso.org/instruments/amber/; Petrov et al. (2003)

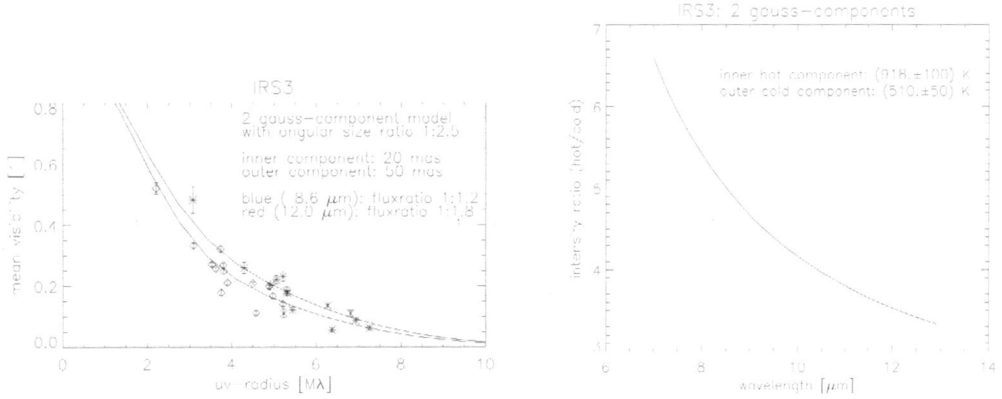

Figure 1. Interpretation of MIDI data of IRS 3. A model of two circular Gaussian brightness distributions fits the data best. The parameters of the model are the absolute size and the flux ratio of both components. Stars and diamonds are indicating 8 and 12 μm data, respectively. Note the relative fluxincrease of the larger component with wavelength. Assuming a blackbody as radiating source, this can be translated into absolute temperatures (right panel).

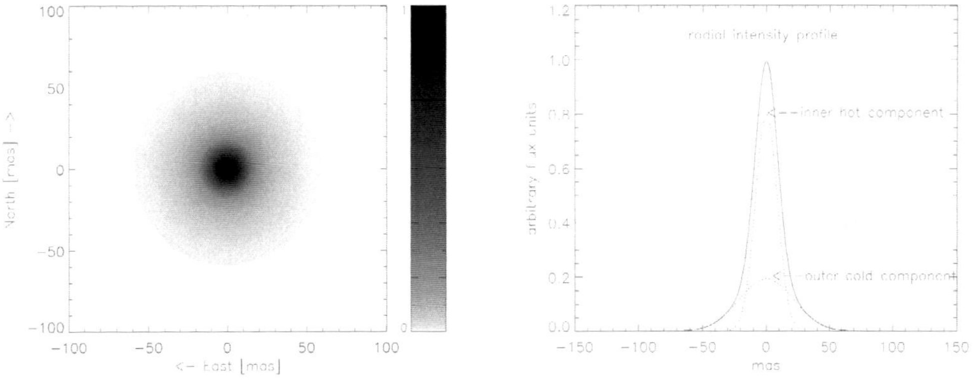

Figure 2. The model of the dust distribution around IRS 3, which fits the data best (Fig. 1). The intensity is plotted in logarithmic greyscale. The model consists of the sum of two circular Gaussian, as shown by the radial profiles in the right panel.

stars. Our MIDI data enable us to constrain the dust distribution around IRS 3 at 10 μm and to investigate deviations from circular symmetry of its innermost region.

An important outcome of the spectrally dispersed visibility data is, that the wavelength dependence of the visibility modulus cannot be fitted by a *single* uniform disc model. The minimum requirement to a successful model are two different components of different angular size. We fitted two blackbodies to the data, representing two dust shells around the embedded stellar source. The flux distribution in each component has a radial decline of Gaussian shape. We cannot derive a relative phase shift between both components from the single-baseline data, due to the lack of most of the phase information. Thus both Gaussian components are centered on top of each other (2). Since we fit a circular-symmetric model, the visibility data can be azimuthally averaged and plotted over the uv-radius (left panel of Fig. 1). This approach is justified by the fact, that we did not observe any significant dependence of the visibility modulus

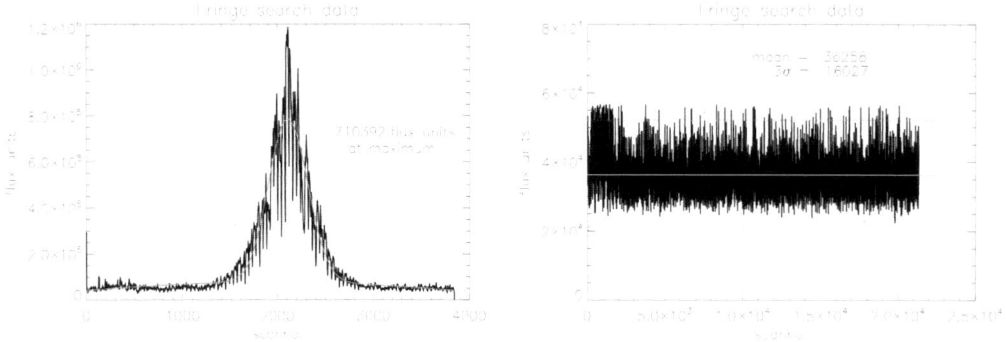

Figure 3. Both figures show the maximum Fourier amplitude of the MIDI fringe search data. A fringe detection is indicated by a significant peak (left: visibility calibrator HD165135). This peak is well above the 3σ level of the search data of IRS 1W as indicated by the two horizontal lines in the right panel. Note the reduced ordinate-scale in the right panel.

Table 1. Upper correlated flux limits give lower limits of the source size at 10 μm.

Sources with existing bow-shock models:		IRS 1W	IRS 10W
Total flux at 8.6micron (VISIR), (extinct. Av=25)	[Jy]	4.6	2.7
Upper limit of the correlated flux density	[Jy]	< 0.3	< 0.25
Visibility limits	[1]	< 0.06	< 0.09
Width of bow-shock models	[mas/AU]	$> 30/240$	$> 20/160$
Sources without existing bow-shock models:		IRS 2	IRS 8
Upper limit of the correlated flux density	[Jy]	< 0.4	< 0.35

on the baseline position angle. We overplotted the model visibility with indicated FWHM sizes of 20 mas (160 AU) for the inner and 50 mas (400 AU) for the outer component. An important fit parameter is the wavelength-dependent flux-ratio of both components. Together with the size-ratio the flux-ratio can be translated into absolute temperatures unambiguously, if we adopt a blackbody SED for each component. The resulting temperatures are (918$\pm$100) K for the hotter inner component and (510$\pm$50) K for the cooler outer component. These simple blackbody components result in bolometric luminosities exceeding $5{\cdot}10^5\,L_\odot$, which suggest that the embedded source is a massive supergiant. The results of detailed radiative transfer calculations leading to a physically self-consistent model of the circumstellar dust distribution and its chemical composition will be published soon.

4. The Northern Arm bow-shocks

A different species of strong MIR emitters is found along the northern mini-spiral. In a few arcsec distance to SgrA*, these deeply embedded sources could indicate young stellar objects and sites of star formation. But recently Tanner et al. (2005) found bow-shock-like morphologies in a NIR AO-added imaging experiment. Such bow-shocks can be created by stellar winds of evolved massive stars ($v_\infty \leq 1000$ km s^{-1}), moving through the ISM of the Minispiral. For the estimation of the underlying stellar parameters the morphology of the bow-shocks has to be known as well as possible.

We investigated the sizes of the bow-shocks in the MIR with MIDI. We did not detect any fringes on the bow-shocks IRS 1W, 8, and 10W or on IRS 2L. Since no technical failure of the

VLTI was apparent and we found fringes on calibrator sources regularly throughout the night, we interpret these non-detections of the correlated flux as low visibilities. That is, the sources are too extended and upper limits on the correlated flux are leading to lower limits on the extension of the bow-shocks at the observing wavelengths.

To derive 3σ upper limits on the correlated flux we investigated and flux-calibrated the fringe-search data of the MIDI experiment. The peak of the Fourier transform along the wavelength dependence of the spectrally dispersed interferometric data indicates the group delay. Its amplitude is proportional to the correlated flux of the source, if the group delay is zero and the amplitude is maximum over the scanned optical path difference range. We used this maximum Fourier amplitude of a calibrator source to flux-calibrate the fringe search data of the bow-shocks (Fig. 3). To transform the thus estimated correlated flux limits into lower limits on the bow-shock size, we added to the NIR bow-shock model of Tanner et al. (2005) a one-dimensional parameter scaling the radial FWHM of the bow-shocks. Comparing the visibility moduli of these scalable models to the measured upper limits results in lower limits of the width of two bow-shocks at 10 μm (Table 1). Furthermore we could derive an overall extension of IRS 1W by deconvolution of the acquisition image. The entire bow-shock structure extends to $\sim$350mas and indicates a significant increase of the MIR emitting bow-shock with respect to the NIR findings.

5. Phase referencing capabilities of IRS 7

The phase and closure phase of the complex visibilities bears important imaging information. A shift in image space transforms into a linear phase slope in the Fourier space of the complex visibilities. Thus in addition to full image reconstruction on the basis of a complete uv-dataset, already a few visibility phases of an unresolved target give a precise distance measurement with respect to a phase reference center. Fringe tracking and phase referencing using off-axis guide stars within an extended 2 arcmin field-of-view will become available with the star separator systems.

Within this distance, the brightest NIR source IRS 7 is the prime target for phase referencing. Its 5.5 arcsec distance to SgrA* minimizes the distance dependent atmospheric deterioration of such interferometric experiments which will possibly allow to study the accretion disc in its flaring phase at unprecedented angular resolution. Blum et al. (1996) estimated a spectral classification of M1I for IRS 7 and gave a $T_{eff} = 3600$ K. This leads to a stellar radius of about 0.5 mas at the distance to the GC, still unresolved for AMBER – a 50m projected baseline length gives an angular resolution of 8mas at 2μm and a linear scale of 8 AU/mas at the GC. Lack of an infrared excess around IRS 7 indicates the lack of a significant extended dust structure in the NIR, which would decrease the signal-to-noise ratio of interferometric measurements. Thus a visibility modulus very close to 1 and a correlated magnitude of more than K=6.6 (Ott et al. 1999) can be expected.

We observed IRS 7 with AMBER and MIDI in March and June 2006 and recorded successfully fringes in both the 2 μm and the 10 μm wavelength regimes. The publication of the flux calibrated data in underway and will clarify the suitability of IRS 7 as phase referencing source in the central parsec.

6. Outlook

The case of IRS 3 presents the capabilities of MIDI in the investigation of dusty sources at the Galactic Center. A full radiative transfer model of IRS 3 could be used to estimate its physical properties. With the upcoming external fringe tracker FINITO, the sensitivity of MIDI will increase significantly and enable us to study in detail several sources in that outstanding region. The interferometric data are constraining the dust distribution and its energetics uniquely. The first detection of fringes with AMBER on IRS 7 without external fringe tracking successfully opens the epoch of NIR-interferometric GC observations. Different interstellar absorption as well as distinct emission properties between the NIR and the MIR domains will lead to complementing data on the stellar and dusty interstellar structures in the center of our Galaxy at highest angular resolutions.

Acknowledgments

The authors would like to thank the ESO VLTI staff for their efficient support of the observations.

References

Becklin, E. E., Matthews, K., Neugebauer, G., & Willner, S. P. 1978, ApJ, 219, 121

Blum, R. D., Sellgren, K., & Depoy, D. L. 1996, AJ, 112, 1988

Eckart, A., Moultaka, J., Viehmann, T., Straubmeier, C., & Mouawad, N. 2004, ApJ, 602, 760

Eisenhauer, F., Genzel, R., Alexander, T., et al. 2005, ApJ, 628, 246

Genzel, R., Schödel, R., Ott, T., et al. 2003, ApJ, 594, 812

Leinert, C., Graser, U., Richichi, A., et al. 2003, The Messenger, 112, 13

Moultaka, J., Eckart, A., Viehmann, T., et al. 2004, A&A, 425, 529

Ott, T., Eckart, A., & Genzel, R. 1999, ApJ, 523, 248

Petrov, R. G., Malbet, F., Weigelt, G., et al. 2003, in Interferometry for Optical Astronomy II. Edited by Wesley A. Traub . Proceedings of the SPIE, Volume 4838, pp. 924-933 (2003)., ed. W. A. Traub, 924–933

Pott, J.-U., Eckart, A., Glindemann, A., et al. 2005, The Messenger, 119, 43

Rieke, G. H., Telesco, C. M., & Harper, D. A. 1978, ApJ, 220, 556

Tanner, A., Figer, D. F., Najarro, F., et al. 2006, ApJ, 641, 891

Tanner, A., Ghez, A. M., Morris, M. R., & Christou, J. C. 2005, ApJ, 624, 742

van Loon, J. T., Groenewegen, M. A. T., de Koter, A., et al. 1999, A&A, 351, 559

Viehmann, T., Eckart, A., Schödel, R., et al. 2005, A&A, 433, 117

Viehmann, T., Eckart, A., Schödel, R., Pott, J.-U., & Moultaka, J. 2006, ApJ, 642, 861

Institute of Physics Publishing
doi:10.1088/1742-6596/54/1/044
Journal of Physics: Conference Series **54** (2006) 279–287
Galaxy Center Workshop

Galactic Center Youth: Orbits and Origins of the Young Stars in the Central Parsec

J. R. Lu[1], A. M. Ghez[1,2], S. D. Hornstein[1], M. Morris[1], K. Matthews[3], D. J. Thompson[4], E. E. Becklin[1]

[1] UCLA Department of Physics and Astronomy, Los Angeles, CA 90095-1562
[2] UCLA Institute of Geophysics and Planetary Physics, Los Angeles, CA 90095-1565
[3] Caltech Optical Observatories, California Institute of Technology, MS 320-47, Pasadena, CA 91125
[4] Large Binocular Telescope Observatory, University of Arizona, Tucson, AZ 85721-0065

E-mail: jlu, ghez, seth, morris, kym@astro.caltech.edu, dthompson@as.arizona.edu, becklin@astro.ucla.edu

Abstract. We present new proper motions for the massive, young stars at the Galactic Center, based on 10 years of diffraction limited data from the Keck telescopes. Our proper motion measurements now have uncertainties of only 1-2 km/s and allow us to explore the origin of the young stars that reside within the sphere of influence of the supermassive black hole whose strong tidal forces make this region inhospitable for star formation. Their presence, however, may be explained either by *in situ* star formation in an accretion disk or as the remnants of a massive stellar cluster which spiraled in via dynamical friction. Earlier stellar velocity vectors were used to postulate that all the young stars resided in two counter-rotating stellar disks, which is consistent with both of the above formation scenarios. Our precise proper motions allow us, for the first time, to determine the orbital parameters of each individual star and thereby to test the hypothesis that the massive stars reside in two stellar disks. Of the 26 young stars in this study that were previously proposed to lie on the inner, clockwise disk, we find that nearly all exhibit orbital constraints consistent with such a disk. On the other hand, of the 7 stars in this study previously proposed to lie in the outer, less well-defined counter-clockwise disk, 6 exhibit inclinations that are inconsistent with such a disk, bringing into question the existence of the outer disk. Furthermore, for stars in the inner disk that have eccentricity constraints, we find several that have lower limits to the eccentricity of more than 0.4, implying highly eccentric orbits. This stands in contrast to simple accretion disk formation scenarios which typically predict predominantly circular orbits.

1. Introduction

The center of our Galaxy has been shown to harbor a supermassive black hole ($M_{SBH} \sim 3-4\times10^6$ $M_\odot$, e.g. [1, 2]) whose close proximity ($R_o \sim 7-8$ kpc, [1, 2, 3]) provides a unique laboratory for studying phenomena in normal galactic nuclei. One peculiar attribute of our Galactic center is the presence of a population of massive ($30-120$ $M_\odot$), young ($\lesssim 10$ Myr) stars within the sphere of influence of the black hole (see e.g. [4, 5]). The origin of such young stars is puzzling given that the present-day gas density is orders of magnitude too low to overcome the extreme tidal forces and collapse to form stars [6, 7]. Proposed resolutions to this "paradox of youth" include scenarios in which these young stars formed in a massive, self-gravitating accretion disk that was once present around the black hole [8] or alternatively, formed far from their current positions, outside the black hole's sphere of influence, as part of a massive star cluster which spiraled in via dynamical friction and deposited the most massive stars where we see them today [9].

Insight into the origins of the massive, young stars may be obtained through understanding the spatial distribution and stellar dynamics of this population. Already, high-resolution infrared imaging of young OB stars located less than $0\rlap{.}''5$ (3680 AU) from the black hole have yielded precise measurements of the stellar orbits which indicate that their orbital planes are randomly oriented [1, 10]. However, the central-arcsecond stars with measured orbits reside close enough to the black hole that their dynamics are likely to have been modified due to various precession mechanisms, which weakens any signature of their origin from their orbits. Orbital analysis of the young stars at larger radii has been limited due to the stars' smaller accelerations and the dominance of optical distortions beyond $0\rlap{.}''5$ [10]. Nonetheless, analyses using the stars' observed three-dimensional velocities have suggested that the stars at larger radii may lie in two perpendicular, counter-rotating disks with inner disk radii of around 1 arcsec [8, 11] and that there may be several co-moving groups or clusters of stars within the area [12, 13]. Both of the above possible formation scenarios predict that the stars should lie in a common orbital plane; however, formation in a self-gravitating accretion disk tends to suggest circular orbits while infalling cluster scenarios could give rise either to circular or eccentric orbits.

In this paper, we present an improved proper motion analysis in which new adaptive optics data allow us to correct for optical distortions and achieve stellar velocities with uncertainties reduced by as much as a factor of 10 (down to $\sim$1 km/s). We also attempt, for the first time, to extend stellar orbit techniques beyond the central arcsecond out to where the young stars within the central parsec may still hold a dynamical signature of their origins.

2. Observations and Analysis

This study uses high-resolution, 2 μm images of the Galaxy's central stellar cluster taken using both speckle and laser guide star adaptive optics (LGSAO) observing techniques at the 10 m Keck telescopes. Two new speckle data sets, taken in 2005, were combined with 25 existing speckle data sets from 1995-2004 [10, 13, 14, 15] and 4 LGSAO data sets from 2004 - 2005 [16]. Prior to LGSAO observations, the NIRC speckle data sets contained a small amount of residual geometric distortion. Although this distortion is negligible near the center of the images, where Sgr A* and the central arcsecond sources are located; the astrometric accuracy of stars at larger radii is dominated by this residual distortion term. Our LGSAO observations offer a substantial improvement in data quality over speckle imaging and, of particular interest to this study, have well characterized distortion. Simultaneous speckle and LGSAO observations have enabled us, for the first time, to fully characterize and correct optical distortion in the speckle data (Figure 1).

While many sources are detected in this rich data set, we focus our attention on a sample of 33 young stars for this study. The sample is selected to be all young stars spectroscopically identified with high confidence ("quality 2") in Paumard et al. [5] that are between $0\rlap{.}''8$ and $3\rlap{.}''2$ from Sgr A*. The inner radius is set by the approximate inner edge of the proposed disks and the outer radius is set by the speckle field of view (6" × 5"). For each star in this sample, the Keck imaging data set provides 15 - 31 astrometric measurements and, additionally, one or more radial velocity measurements are taken from the literature [5, 17, 18, 19, 20]. Velocities in the plane of the sky are derived from the astrometric measurements by fitting first order polynomials ($x = x_o + v_x * \Delta t$) to positions as a function of time, weighted by the positional uncertainties.

For the resulting sample of 33 young stars, we carried out orbit fitting for each star, assuming a Keplerian orbit model in which the gravitational potential arises from a single dominant point mass situated at the photometric centroid of Sgr A*. The mass and the distance to Sgr A* are

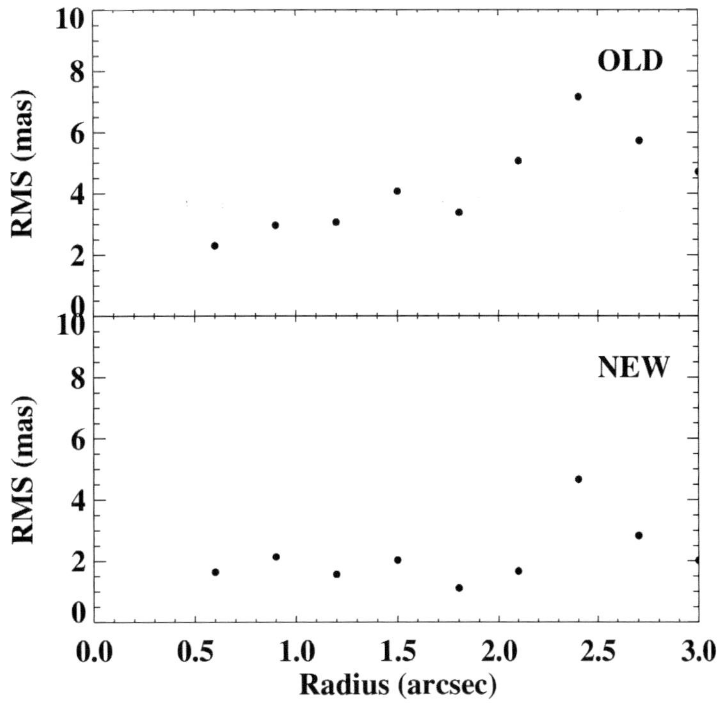

Figure 1. The improvement in positional accuracy at large radii as a result of correcting geometric distortion in speckle data sets. To characterize the systematic positional uncertainty, we take each star at each epoch and calculate the residual positional offset, which is defined as the difference between the measured position and the position as determined by the best fit velocity ($x = x_o + v * \Delta t$). Then the RMS of the residuals is calculated across all epochs for each star. All stars' resulting RMS values are sorted by the distance between the star and Sgr A* (which was at the center of the images) and then averaged over radius bins of 0$''$.3. The radial trend is shown for data prior to the new distortion correction (top) and after the new distortion correction (bottom).

fixed at values set by S0-2's orbit with $M_\bullet$=3.69$\times 10^6$ $M_\odot$ and $R_\circ$=7.36 kpc ([2]). For each star, there are then six free parameters: period (P), eccentricity (e), time of periapse passage ($T_\circ$), inclination (i), position angle of the ascending node (Ω), and the longitude of periapse (ω), not all of which can be well constrained given that the stars only have significant measurements of the 2D positions, 3D velocities, and upper limits to accelerations (see [10] for detailed description of the orbital parameters). Orbit fitting is carried out by minimizing the χ^2 value between the data and the model using the Thiele-Innes method [21]. Uncertainties in the fits are characterized by fully exploring orbital parameter space and empirically determining those orbital solutions with $\chi^2 < \chi^2_{min} + 1$ for 1σ errors.

3. Results

The proper motion measurements, as shown in Figure 2, are as much as ten times better than our most recent work in [13] with typical absolute uncertainties of 1-2 km/s. This substantial improvement is primarily a result of a revised distortion solution for the NIRC-speckle data

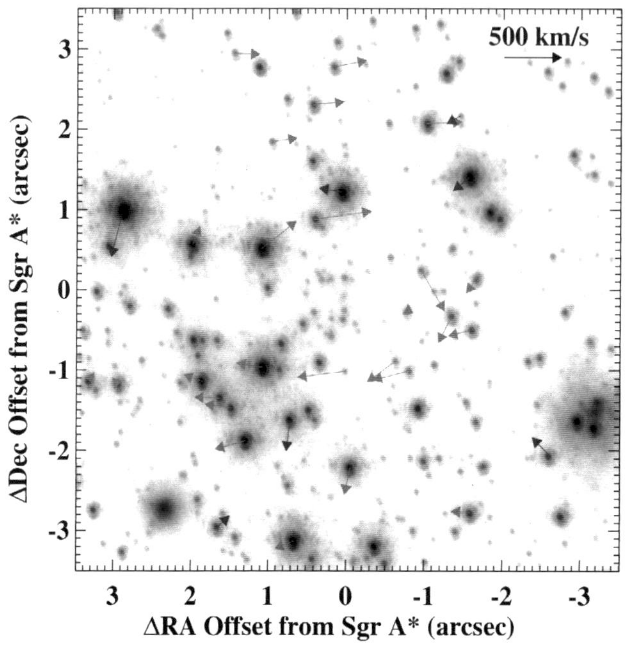

Figure 2. Proper motion vectors of spectroscopically confirmed young stars. Clockwise orbiting stars are drawn in red while counter-clockwise stars are shown in blue. These are overlayed on a K-band LGSAO image shown in grey scale.

sets. Given the high-precision positions and velocities, we attempt, for the first time, to fit orbits to these stars in order to constrain their accelerations and orbital parameters. Figure 3 shows the acceleration limits achieved for all the young stars in the sample as compared to the theoretical curve of acceleration as a function of radius. Assuming that each star is on a bound Keplerian orbit, the measured three-dimensional velocity sets an upper limit to the 3D distance between the star and Sgr A*. The observed 2-dimensional projected distance also sets a lower limit to the total 3D distance (Figure 3, *horizontal arrows*). Additionally, many of the stars show a significant lack of acceleration in the plane of the sky that, when combined with the distance constraints and the 3D velocity information, defines acceptable ranges of orbital parameters for each star. Figure 4 shows some resulting constraints for a sample star, IRS 16CC. These constraints on the orbital parameters can also be expressed as a constraint on the total 3D acceleration for each star (Figure 3, *vertical arrows*). The resulting orbits for the sample of young stars typically have two possible orbital solutions as a result of a degeneracy in the orbital parameter Ω (angle to the ascending node, see Figure 4, *middle*).

Constrained orbital parameters include the angle to the ascending node (Ω) and the inclination (i), which can be combined to determine the orientation of the orbital plane. Figure 5 shows the orientation for all the stars' orbital planes. For each star, the 3σ constraints on Ω and i translate into a cone into which the normal vector to the orbital plane points, and this cone is then projected onto a celestial sphere centered on Sgr A* as an elliptical region. All stars orbiting in a clockwise rotation in the plane of the sky are plotted in red and both of the degenerate solutions are plotted, one as a solid line and one as a dashed line. One of

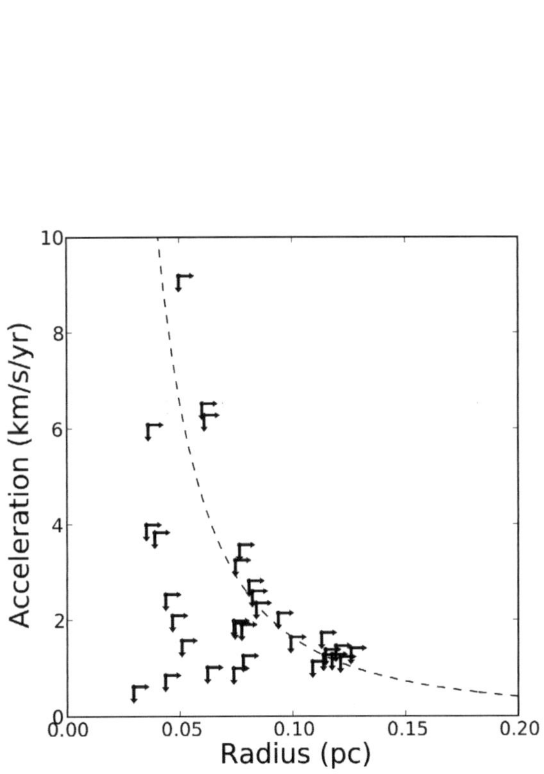

Figure 3. Acceleration limits for the sample of 33 young stars. The theoretical curve for acceleration as a function of total distance is plotted (dashed), as is each star's 3σ acceleration upper limit from orbit fitting (vertical arrow) and radius lower limit from the observed projected distance from the black hole (horizontal arrow).

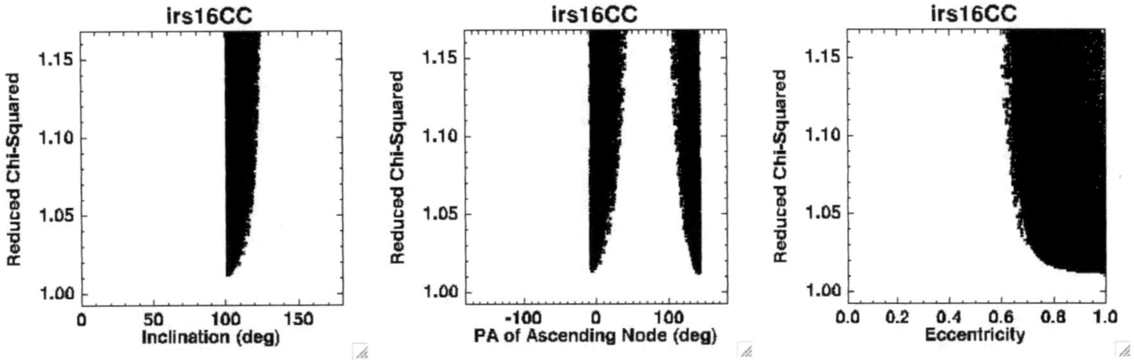

Figure 4. Projected χ^2 parameter space for several orbital elements from a sample star, IRS 16CC. The χ^2 space is mapped by running many trial orbits and those orbits that fall within a 3σ range of χ^2 values are plotted. The inclination (panel 1) and PA to the ascending node, Ω (panel 2), are used to determine the orientation of the orbital plane. This star has an additional constraint on the eccentricity (panel 3).

the solutions for the clockwise stars' normal vectors cluster around a common point indicating that these stars may lie on a common orbital plane. The counter-clockwise stars, plotted in blue, do not appear to cluster around any single point, suggesting that they do not lie on a common orbital plane. Over plotted in black are the candidate orbital planes proposed by [8] and [11] with updated values from [5] for the candidate plane normal vectors (solid black) and disk thickness (dashed black).

Stars in the inner disk also show evidence for eccentric orbits. For more edge-on orbits, a lower limit to the distance from Sgr A* coupled with particular directions of the 3D velocity vector will result in a lower limit to the eccentricity. Of the 26 stars in the clockwise disk, 10 of the stars have 3σ lower limits to their eccentricities of >0.2. Figure 6 shows the 3σ limit on eccentricities for all the stars in the sample. These eccentricity limits are robust as there is a sharp lower limit in chi-squared vs. eccentricity space (see Figure 4, *right* for an example).

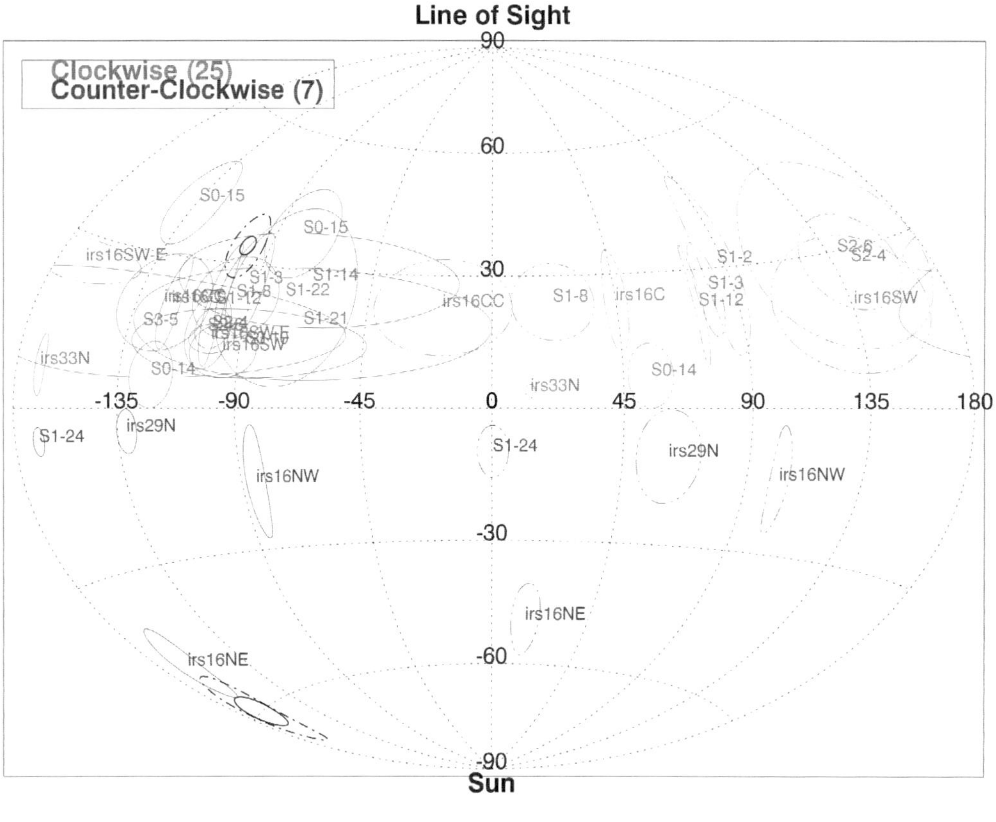

Figure 5. The distribution of orbital planes as determined from individual stellar orbits. The normal vector for each star's orbital plane, plus the associated uncertainty in that vector, describes a cone in 3D space. This cone can be projected as an elliptical region onto a celestial sphere centered on the black hole. The projected 3 sigma region of the normal vectors are plotted on an Aitoff projection of this celestial sphere with clockwise rotators in red and counter-clockwise in blue. For reference, those stellar orbits with normal vectors pointing toward or away from the Sun have face-on orbits as seen from Earth.

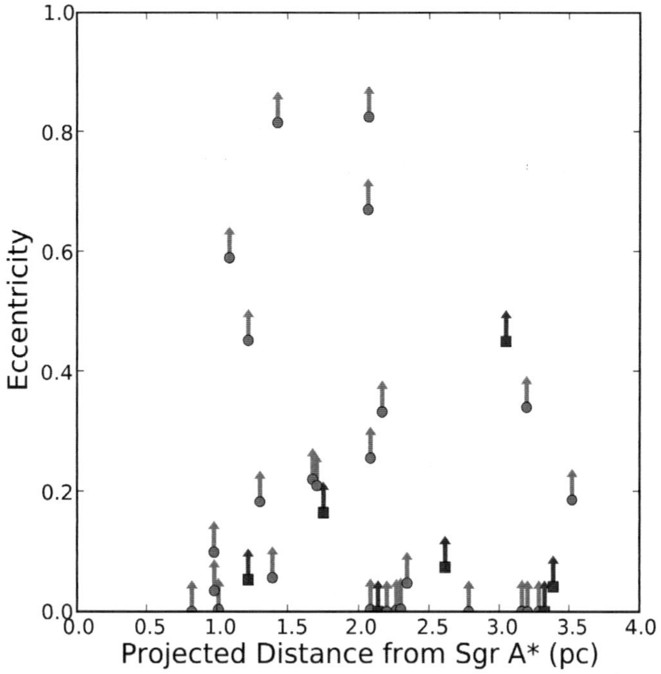

Figure 6. The distribution of eccentricities lower limits as determined from individual stellar orbits. Stars rotating with a clockwise orientation are plotted as red circles and counter-clockwise stars are plotted as blue squares.

4. Discussion

This first attempt at orbital fitting for young stars that reside outside the central arcsecond shows definitive evidence for one disk of clockwise rotating stars. The proposed second disk is not supported by observations of counter-clockwise stars within $\sim 3''$ of Sgr A*; however, our current observations cannot definitively rule out the existence of a 2nd disk until we expand our study to include more counter-clockwise rotating young stars at larger radii from Sgr A*.

Determining the existence of a second disk is critical to understanding the formation scenarios of the young stars within the central parsec. For the self-gravitating accretion disk scenario, it is unlikely that two gas disks could exist simultaneously at the same radii since the disks would collide. Since the clockwise and counter-clockwise populations appear to have similar ages of $\sim$6 million years old to within $\sim$1 million years of each other [5], the existence of two stellar disks at the present epoch would mean that a second gas disk must have formed, become massive enough to collapse under self-gravity, and fragmented to form stars within, at most, 1 million years of the first gas disk. Likewise, for the infalling cluster scenario, two nearly perpendicular disks of stars most likely cannot be formed by the infall of a single star cluster. The existence of two stellar disks would arise from two clusters spiraling in within 6 million years of each other, as set by the clusters' ages, and would thus give an estimate of the frequency of such cluster infall events.

On the other hand, if there is only one disk plus a randomly scattered population, then it may be possible to explain all the stars within the central parsec with a single formation scenario with scattering events giving rise to out-of-the-plane members. This scenario has a difficulty in that IRS 13 appears to be a large cluster of young stars; which either does not lie on the common orbital plane or is counter-rotating within this plane. There is no obvious mechanism for drastically changing the orbit of the highest density portion of the young star disk.

Stars within the well-defined, clockwise disk appear to have high eccentricities ($\gtrsim 0.4$) which may be difficult to explain with a self-gravitating accretion disk formation scenario. Results from Nayakshin & Cuadra [22] indicate that stars born in a circular accretion disk would retain their nearly circular orbits well past the present epoch and currently there is little observational evidence for eccentric accretion disks within the sphere of influence of supermassive black holes in other galaxies (AGN). If the disk mass is built up from many small cloud-infall events, then the disk may circularize prior to becoming massive enough to form stars from self-gravity ($> 10^4$ $M_\odot$); however, for a single massive cloud infall or a cloud-cloud collision event, it remains to be determined if star formation can occur rapidly enough that the cloud's initial eccentricity is imparted to the subsequently formed stars before circularizing.

High eccentricity orbits may be more naturally explained in the infalling star cluster formation scenario. Stars that are stripped from a cluster as it spirals in should have a similar inclination, semi-major axis, and eccentricity as the cluster itself. Therefore, an infalling cluster with an initially eccentric orbit will produce a thin disk of stars with similarly eccentric orbits [23]. However, this formation scenario has its own share of unresolved issues including (1) the required high initial cluster mass ($> 10^5 - 10^6$ $M_\odot$, [24, 25]); (2) the required high central density, possibly even the presence of an intermediate mass black hole [26, 24]; and (3) the lack of observational evidence for tidally stripped stars at larger radii or X-ray emission from pre-main sequence low-mass stars within the central parsec [27].

In summary, the advent of laser guide star adaptive optics has allowed us to retroactively improve our 11 year astrometric data set used for monitoring stars orbiting our Galactic center. This has resulted in as much as a factor of ten improvement in proper motion precisions with resulting uncertainties of ~1-2 km/s. As a result, we have attempted the first orbit fitting of the young stars proposed to lie in stellar disks orbiting the supermassive black hole. The individual orbits for the young stars confirm a disk of young stars at a high inclination rotating in a clockwise sense. The counter-clockwise rotating stars do not appear to be consistent with the previously proposed second disk. Stars within the well-defined, clockwise disk also appear to have high eccentricities, which is challenging to current accretion disk formation scenarios and provides a new observational constraint for models of star formation in the Galactic center.

Acknowledgments

Support for this work was provided by J.R.L.'s NSF Graduate Research Fellowship, NSF grant AST-0406816, and the NSF Science & Technology Center for AO, managed by UCSC (AST-9876783). The W.M. Keck Observatory is operated as a scientific partnership among the California Institute of Technology, the University of California and the National Aeronautics and Space Administration. The Observatory was made possible by the generous financial support of the W.M. Keck Foundation.

[1] F. Eisenhauer, R. Genzel, T. Alexander, R. Abuter, T. Paumard, T. Ott, A. Gilbert, S. Gillessen, M. Horrobin, S. Trippe, H. Bonnet, C. Dumas, N. Hubin, A. Kaufer, M. Kissler-Patig, G. Monnet, S. Ströbele, T. Szeifert, A. Eckart, R. Schödel, and S. Zucker. SINFONI in the Galactic Center: Young Stars and Infrared Flares in the Central Light-Month. *ApJ*, 628:246–259, July 2005.

[2] Ghez et al. Distance to the Galactic Center. *in prep.*

[3] M. J. Reid. The distance to the center of the Galaxy. *ARAA*, 31:345–372, 1993.

[4] D. A. Allen, A. R. Hyland, and D. J. Hillier. The source of luminosity at the Galactic Centre. *MNRAS*, 244:706–713, June 1990.

[5] T. Paumard, R. Genzel, F. Martins, S. Nayakshin, A. M. Beloborodov, Y. Levin, S. Trippe, F. Eisenhauer, T. Ott, S. Gillessen, R. Abuter, J. Cuadra, T. Alexander, and A. Sternberg. The Two Young Star Disks in the Central Parsec of the Galaxy: Properties, Dynamics, and Formation. *ApJ*, 643:1011–1035, June 2006.

[6] R. H. Sanders. The case against a massive black hole at the Galactic Centre. *Nature*, 359:131–+, September 1992.

[7] M. Morris. Massive star formation near the Galactic center and the fate of the stellar remnants. *ApJ*, 408:496–506, May 1993.

[8] Y. Levin and A. M. Beloborodov. Stellar Disk in the Galactic Center: A Remnant of a Dense Accretion Disk? *ApJL*, 590:L33–L36, June 2003.

[9] O. Gerhard. The Galactic Center HE I Stars: Remains of a Dissolved Young Cluster? *ApJL*, 546:L39–L42, January 2001.

[10] A. M. Ghez, S. Salim, S. D. Hornstein, A. Tanner, J. R. Lu, M. Morris, E. E. Becklin, and G. Duchêne. Stellar Orbits around the Galactic Center Black Hole. *ApJ*, 620:744–757, February 2005.

[11] R. Genzel, R. Schödel, T. Ott, F. Eisenhauer, R. Hofmann, M. Lehnert, A. Eckart, T. Alexander, A. Sternberg, R. Lenzen, Y. Clénet, F. Lacombe, D. Rouan, A. Renzini, and L. E. Tacconi-Garman. The Stellar Cusp around the Supermassive Black Hole in the Galactic Center. *ApJ*, 594:812–832, September 2003.

[12] J. P. Maillard, T. Paumard, S. R. Stolovy, and F. Rigaut. The nature of the Galactic Center source IRS 13 revealed by high spatial resolution in the infrared. *AAP*, 423:155–167, August 2004.

[13] J. R. Lu, A. M. Ghez, S. D. Hornstein, M. Morris, and E. E. Becklin. IRS 16 SW: A New Comoving Group of Young Stars in the Central Parsec of the Milky Way. *ApJL*, 625:L51–L54, May 2005.

[14] A. M. Ghez, B. L. Klein, M. Morris, and E. E. Becklin. High Proper-Motion Stars in the Vicinity of Sagittarius A*: Evidence for a Supermassive Black Hole at the Center of Our Galaxy. *ApJ*, 509:678–686, December 1998.

[15] A. M. Ghez, M. Morris, E. E. Becklin, A. Tanner, and T. Kremenek. The accelerations of stars orbiting the Milky Way's central black hole. *Nature*, 407:349–351, September 2000.

[16] A. M. Ghez, S. D. Hornstein, J. R. Lu, A. Bouchez, D. Le Mignant, M. A. van Dam, P. Wizinowich, K. Matthews, M. Morris, E. E. Becklin, R. D. Campbell, J. C. Y. Chin, S. K. Hartman, E. M. Johansson, R. E. Lafon, P. J. Stomski, and D. M. Summers. The First Laser Guide Star Adaptive Optics Observations of the Galactic Center: Sgr A*'s Infrared Color and the Extended Red Emission in its Vicinity. *ApJ*, 635:1087–1094, December 2005.

[17] R. Genzel, C. Pichon, A. Eckart, O. E. Gerhard, and T. Ott. Stellar dynamics in the Galactic Centre: proper motions and anisotropy. *MNRAS*, 317:348–374, September 2000.

[18] T. Paumard, J. P. Maillard, M. Morris, and F. Rigaut. New results on the helium stars in the galactic center using BEAR spectro-imagery. *AAP*, 366:466–480, February 2001.

[19] T. Ott. *Development of the Laser Guide Star System ALFA and Near-Infrared Observations of the Galactic Center.* PhD thesis, Max-Planck-Institut für Extraterrestrische Physik, 2003.

[20] A. Tanner, D. F. Figer, F. Najarro, R. P. Kudritzki, D. Gilmore, M. Morris, E. E. Becklin, I. S. McLean, A. M. Gilbert, J. R. Graham, J. E. Larkin, N. A. Levenson, and H. I. Teplitz. High Spectral Resolution Observations of the Massive Stars in the Galactic Center. *ApJ*, 641:891–904, April 2006.

[21] W. H. van den Bos. *Orbit Determintations of Visual Binaries, Astronomical Techniques.* Univeresity of Chicago Press, 1962.

[22] S. Nayakshin and J. Cuadra. A self-gravitating accretion disk in Sgr A* a few million years ago: Is Sgr A* a failed quasar? *AAP*, 437:437–445, July 2005.

[23] S. J. Berukoff and B. M. S. Hansen. Cluster core dynamics at the Galactic center. *ApJ*, 2006.

[24] S. S. Kim, D. F. Figer, and M. Morris. Dynamical Friction on Galactic Center Star Clusters with an Intermediate-Mass Black Hole. *ApJL*, 607:L123–L126, June 2004.

[25] M. A. Gürkan and F. A. Rasio. The Disruption of Stellar Clusters Containing Massive Black Holes near the Galactic Center. *ApJ*, 628:236–245, July 2005.

[26] B. M. S. Hansen and M. Milosavljević. The Need for a Second Black Hole at the Galactic Center. *ApJL*, 593:L77–L80, August 2003.

[27] S. Nayakshin and R. Sunyaev. The 'missing' young stellar objects in the central parsec of the Galaxy: evidence for star formation in a massive accretion disc and a top-heavy initial mass function. *MNRAS*, 364:L23–L27, November 2005.

Institute of Physics Publishing
doi:10.1088/1742-6596/54/1/045

Journal of Physics: Conference Series **54** (2006) 288–292
Galaxy Center Workshop

Stellar orbits around Sgr A*

S. Trippe[1], S. Gillessen[1], T. Ott[1], F. Eisenhauer[1], T. Paumard[1], F. Martins[1], R. Genzel[1], R. Schödel[2], A. Eckart[2] and T. Alexander[3]

[1] Max-Planck-Institut für extraterrestrische Physik, Postfach 1312, D-85741 Garching, Germany
[2] 1. Physikalisches Institut, Universität zu Köln, Zülpicher Strasse 77, D-50937 Köln, Germany
[3] Faculty of Physics, The Weizmann Institute of Science, PO Box 26, Rehovot 76100, Israel

E-mail: trippe@mpe.mpg.de

Abstract.
In this article we present and discuss the latest results from the observations of stars ("S-stars") orbiting Sgr A*. With improving data quality the number of observed S-stars has increased substantially in the last years. The combination of radial velocity and proper motion information allows an ever more precise determination of orbital parameters and of the mass of and the distance to the supermassive black hole in the centre of the Milky Way. Additionally, the orbital solutions allow us to verify an agreement between the NIR source Sgr A* and the dynamical centre of the stellar orbits to within 2 mas.

1. Introduction

The Galactic Centre (GC) offers due to its proximity the unique opportunity to study the properties of a supermassive black hole (SMBH) in detail. Two parameters of special interest in the last years have been the mass of and the distance to Sgr A*, in direct connection to the question about the nature of the central dark mass. An answer to these questions that was independent of most assumptions became possible, for the first time, in the year 2002, when the star S2 was confirmed to move on a Keplerian orbit around Sgr A* [1]. Follow-up observations covering both radial velocities and proper motions soon made possible the detection of additional stars on orbits. These orbits allow a simultaneous solution for both mass of and distance to the central SMBH.

In the last years those observations have been continued and extended. The rapidly increasing amount of data provides a more and more precise determination of dynamical parameters and thus a more and more precise understanding of the innermost part of our Galaxy ([2], [3], [4], [5], [6]).

2. Instruments and observations

The monitoring of the S-star orbits is necessarily based on diffraction limited imaging and spectroscopy on large telescopes. As the GC is due to strong extinction ($A_V \sim 30$) not visible in the optical wavelength range, we focused on near-infrared (NIR) observations, mainly in the H and K bands (1.5 to 2.5 μm).

Our observations began in 1992 using the speckle imaging camera SHARP I [7] at the 2.2-m-telescope in La Silla. Since 2002, we have used the detector system NAOS/CONICA (NACO for short), consisting of the adaptive optics system NAOS [8] and the NIR camera CONICA [9] at the 8.2-m-UT4 (Yepun) of the ESO-VLT on Cerro Paranal, Chile. NACO produces high quality images with a spatial resolution of ~40 (H band) resp. ~55 mas (K band).

In order to obtain 3-dimensional information on the GC in general and the S-stars in special, we have additionally collected stellar spectra using the integral field spectrometer SPIFFI ([10], [11]) resp.

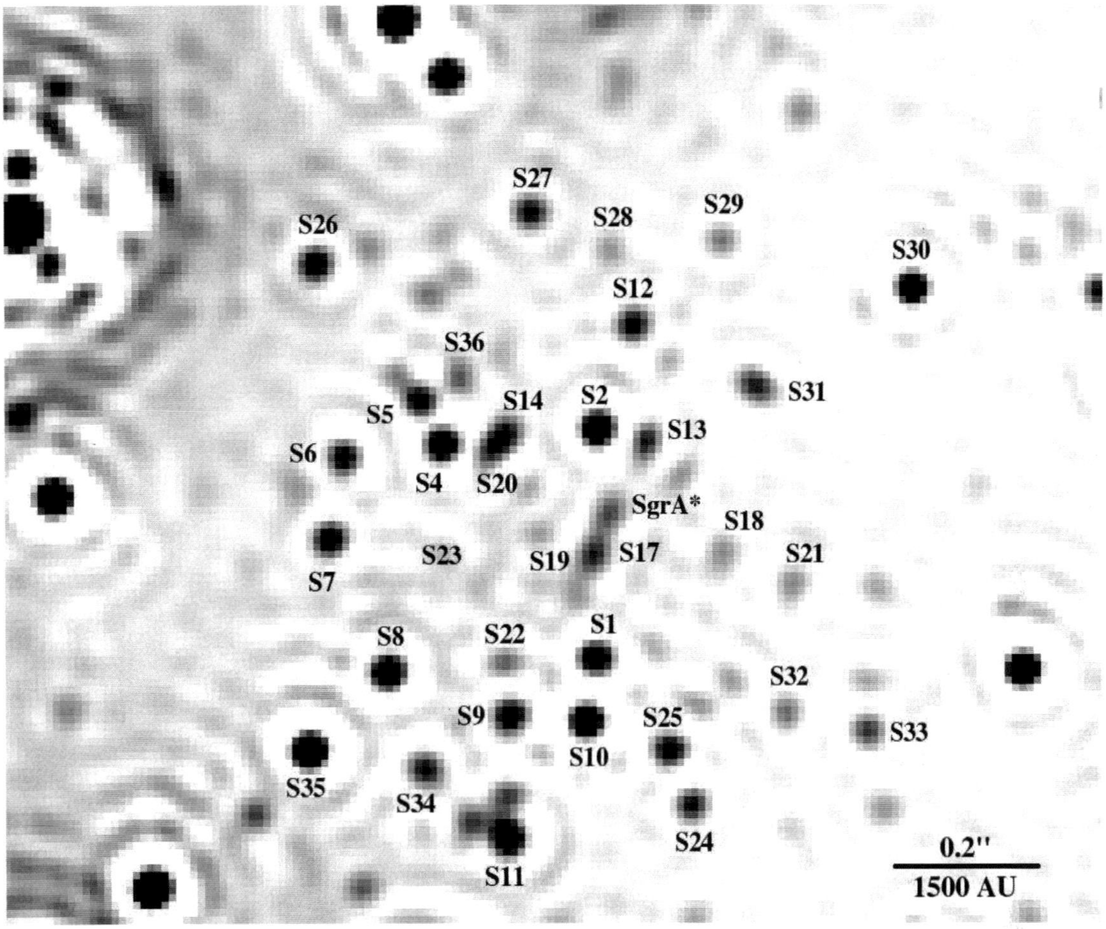

Figure 1. Regularly monitored S-stars. This image is a Lucy-Richardson deconvolved K band NAOS/CONICA image obtained in 2005. At the time of the exposure Sgr A* was active in the NIR and thus can be seen here.

SINFONI, a combination of SPIFFI and the adaptive optics (AO) system MACAO [12, 13]. SINFONI produces diffraction-limited data cubes with 64×32 pixels in the two spatial dimensions and 2048 pixels along the spectral axis, reaching spectral resolutions of ~3500.

3. More and more S-stars

Since the first reports on stellar orbits, the number of S-stars has always been rather limited, the orbital analyses focused typically on ensembles of 6 to 10 stars. As the very special dynamics of those stars is generally of high interest, we investigated the question, if there are more of such objects. Compared to the situation in 2002/2003, where the analysis of orbits was mainly based on speckle data obtained with SHARP, we now have access to a large number of AO-assisted 8-m-class telescope data, both imaging (NACO) and spectroscopic (SINFONI). So we could expect to detect additional stars (see Fig. 1) for the following reasons:

- We collect more accurate data with lower detection thresholds. Thus it is possible to trace fainter sources with accuracies that allow to detect accelerations within shorter time lines (e.g. S18).

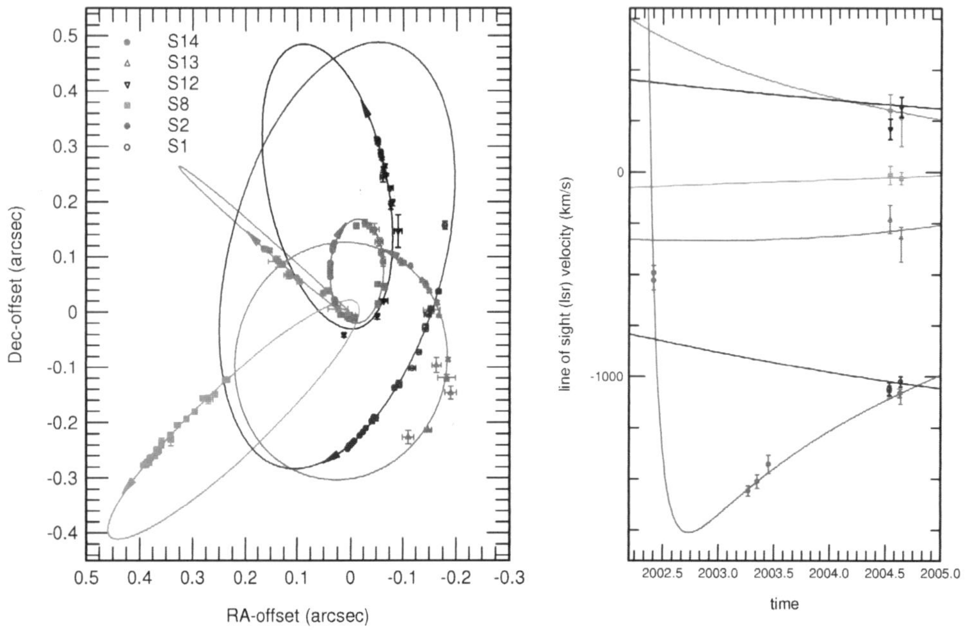

Figure 2. Illustration of stellar motions around Sgr A*. Left: Orbits in the plane of the sky. Sgr A* is located at (0,0), coordinates are given in arcsec. Right: Radial velocities (in km/s) vs. time (in epochs). Figure taken from [6].

- The AO data allow the retrospective detection of sources. Objects, that are too faint or too confused to be identified resp. separated in the elder speckle images alone, can be identified in NACO images and re-found in the SHARP images (e.g. S17).
- High-quality integral field spectroscopy allows the identification of fast sources that appear not to be special in proper motion (e.g. S9).

A good example for this is the identification of the star S18. We noted the fast motion of this object for the first time in a difference map of two NACO images obtained in 2002 and 2005. This quite faint source ($K \simeq 17$) showed a remarkably large proper motion. After identifying S18 in NACO images, we were also able to extract radial velocities out of SINFONI cubes and thus could attach a 3-dimensional orbit to it, using the known Sgr A* potential.

Today our ensemble of S-stars has around 30 members. Orbital solutions exist for about 15 of them.

4. Better and better orbital solutions

The permanently increasing time line of observations, which began in 1992 and are being continued until today, in combination with the much better data quality since 2002, allow a more and more accurate determination of the dynamical parameters of the S-star system, especially mass of and distance to Sgr A*.

In order to compute the dynamical properties of the S-star cluster, we fit all available data simultaneously with a proper potential, which is either a purely Keplerian one or includes general relativistic effects and/or an extended component of the central mass. We simultaneously solve for six orbital parameters per star plus (at most) seven parameters for Sgr A* (mass, 3D position, 3D velocity).

Table 1. Reported uncertainties for mass of and distance to Sgr A*. As one can see easily, the accuracies have increased quite dramatically between 2002 and 2006. Masses given by [3] and [5] are scaled with the distance R to the GC using an assumed distance of 8 kpc, whereas the other mass errors are independent from the distance.

Analysis	mass error [$10^6 M_\odot$]	distance error [kpc]
[1]	1.5	
[4]		0.42
[3]	0.4 $(R/8kpc)^3$	
[5]	0.2 $(R/8kpc)^3$	
[6]	0.32	0.32
MPE, April 2006	0.22	0.22

Technically this procedure is implemented in the high-level tool Mathematica as a χ^2 minimization routine using symplectic integrators. The geometry of the problem is illustrated in Fig. 2.

The mass of and distance to Sgr A*, which have been known to be ~ $3.5 \cdot 10^6 M_\odot$ resp. ~ 7.5 kpc for several years now, have never changed significantly within the reported errors since then. Thus the main effect of the research of the last years was a more and more accurate determination of these values. A major progress in accuracy was made when radial velocities were included into the analysis, thus making assumptions on resp. scaling with the distance to the GC unnecessary. This is illustrated in Table 1, where the reported uncertainties of all main analyses based on the S-star orbits are compared. As one can see, the accuracy in both parameters improved by factors of at least 2 within only three years, which is a quite impressive outcome.

5. Flare position vs. dynamical centre

A welcome side effect of the dynamical analysis described above is the possibility to test whether the dynamical center of the S-star system and the infrared source Sgr A* are in agreement. Sgr A* is generally invisible in the NIR, but shows outbursts ("flares") on time scales of few events per day [14, 15, 16]. So it is possible to extract a position of the NIR source whenever it is active and to compare this position with the foci of the S-star orbits.

We extracted some 10 flare positions from NACO images and compared them to the dynamic centre of the S-star orbits. The results are presented in Table 2. These positions are attached to an astrometric reference frame set by SiO maser radio stars [17] with an accuracy of few mas, thus Sgr A* is not located exactly at (0,0). As one can see easily, the agreement between the two positions is very good: the upper limit for a deviation NIR source – orbit focus is 2 mas.

Table 2. Flare positions compared with the dynamical centre of the S-star orbits. Obviously, the agreement between the two positions is better than 2 mas.

	RA [mas]	DEC [mas]
flares	+2.24±0.49	-4.56±0.72
dyn. centre	+2.29±0.67	-5.52±0.66

6. Summary

We presented recent results from the observations of the S-stars orbiting Sgr A*. These results can be summarized as follows:

- The number of S-stars for which an orbital solution can be found is still increasing and has reached a number of ~15.
- The accuracy of the orbital parameters, especially the mass of and the distance to Sgr A*, has increased substantially within the last three years.
- The position of the NIR source Sgr A* and the dynamical centre of the S-star orbits agree within 2 mas.

References

[1] Schödel, R., et al. 2002a, Nature, 419, 694
[2] Schödel, R., et al. 2002b, ApJ, 596, 1015
[3] Ghez, A.M., et al. 2003, ApJ, 586, L127
[4] Eisenhauer, F., et al. 2003a, ApJ, 597, L121
[5] Ghez, A.M., et al. 2005, ApJ, 620, 774
[6] Eisenhauer, F., et al. 2005, ApJ, 628, 246
[7] Hofmann, R., et al. 1992, Proceedings of the ESO Conference on Progress in Telescope and Instrumentation Technologies, Garching, Germany, 617
[8] Rousset, G., et al. 2003, SPIE, 4839, 140
[9] Hartung, M., et al. 2003, SPIE, 4841, 425
[10] Eisenhauer, F., et al. 2003b, SPIE, 4814, 1548
[11] Eisenhauer, F., et al. 2003c, The ESO Messenger, 113, 17
[12] Bonnet, H., et al. 2003, SPIE, 4839, 329
[13] Bonnet, H., et al. 2004, The ESO Messenger, 117, 17
[14] Genzel, R., et al. 2003, Nature, 425, 934
[15] Ghez, A., et al. 2004, ApJ, 601, L159
[16] Eckart, A., et al. 2006a, A&A, 450, 535
[17] Reid, M.J., et al. 2003, ApJ, 587, 208

Institute of Physics Publishing
Journal of Physics: Conference Series **54** (2006) 293–300
doi:10.1088/1742-6596/54/1/046
Galaxy Center Workshop

Massive perturbers in the galactic center

Hagai B. Perets [1], Clovis Hopman[1,2] and Tal Alexander[1]

Abstract. We investigate the role of massive perturbers, such as stellar clusters or giant molecular clouds, in supplying low-angular momentum stars that pass very close to the central massive black hole (MBH) or fall into it. We show that massive perturbers can play an important role in supplying both binaries and single stars to the vicinity of the MBH. We discuss possible implications for the ejection of high velocity stars; for the capture of stars on tight orbits around the MBH; for the emission of gravitational waves from low-eccentricity inspiraling stars; and for the origin of the young main sequence B stars observed very near the Galactic MBH. Massive perturbers may also enhance the the growth rate of MBHs, and may accelerate the dynamical orbital decay of coalescing binary MBHs.

1. Introduction

There is compelling evidence that massive black holes (MBHs) lie in the centers of all galaxies (e.g. [9]), including the center of our Galaxy [8, 12]. The MBH affects the dynamics and evolution of the galaxy's center as a whole and it also strongly affects individual stars or binaries that approach it. Such close encounters have been the focus of many studies and include a variety of processes such as destruction of stars by the MBH; capture and gradual inspiral of stars into the MBH, accompanied by the emission of gravitational waves (GWs); or dynamical exchange interactions in which incoming stars or binaries energetically eject a star tightly bound to the MBH and are captured in its place very near the MBH [1].

The interest in such processes is driven by their possible implications for the growth of MBHs, for the orbital decay of a MBH binary, for the detection of MBHs, for GW astronomy, as well as by observations of unusual stellar phenomena in our Galaxy, e.g. the puzzling young population of B-star very near the Galactic MBH , or the hyper-velocity B stars at the edge of the Galaxy (e.g. [4]), possibly ejected by 3-body interactions of a binaries with the MBH.

Here we focus on close encounters with the MBH whose ultimate outcome ("event") is the elimination of the incoming object from the system, whether on the short infall (dynamical) time, if the event is prompt (e.g. tidal disruption or 3-body exchange between a binary and the MBH), or on the longer inspiral time, if the event progresses via orbital decay (e.g. through GW emission). Such processes are effective only when the incoming object follows an almost zero angular momentum ("loss-cone") orbit with periapse closer to the MBH than some small distance q. To reach the MBH, or to decay to a short period orbit, both the infall and inspiral times must be much shorter than the system's relaxation time t_r. The fraction of stars initially on loss-cone orbits is very small and they are rapidly eliminated. Subsequently, the close encounter event rate is set by the dynamical processes that refill the loss-cone.

The loss-cone formalism used for estimating the event rate (e.g.[6]) usually assumes that the system is isolated and that the refilling process is 2-body relaxation. This typically leads to a low event rate, set by the long 2-body relaxation time.

[1] Faculty of Physics, Weizmann Institute of Science, POB 26, Rehovot 76100, Israel
[2] Leiden Observatory, P.O. box 9513, NL-2300 RA Leiden, the Netherlands

Two-body relaxation, which is inherent to stellar systems, ensures a minimal loss-cone refilling rate. Other, more efficient but less general refilling mechanisms were also studied with the aim of explaining various open questions, or in the hope that they may lead to significantly higher event rates for close encounter processes [23]. However, most of these mechanisms require special circumstances to work, or are short-lived.

Here we explore another possibility, which is more likely to apply generally: accelerated relaxation and enhanced rates of close encounters driven by massive perturbers (MPs). Efficient relaxation by MPs were first suggested in this context by Zhao, Haehnelt & Rees [37] as a mechanism for establishing the $M_\bullet/\sigma$ relation [9] by fast accretion of stars and dark matter. Zhao et al. also noted the possibility of increased tidal disruption flares and accelerated MBH binary coalescence due to MPs. In this study we investigate in detail the dynamical implications of relaxation by MPs. We evaluate its effects on the different modes of close interactions with the MBH, in particular 3-body exchanges, which were not considered by Zhao et al. and apply our results to the Galactic Center (GC), where observations indicate that dynamical relaxation is very likely dominated by MPs.

2. Loss-cone refilling

In addition to stars, galaxies contain dense massive structures such as molecular clouds, open clusters and globular clusters with masses up to 10^4–$10^7\,M_\odot$. Such structures can perturb stellar orbits around the MBH much faster than 2-body stellar relaxation (hereafter "stellar relaxation"), provided they are numerous enough. The minimal impact parameter still consistent with a small angle deflection in the MP-star scattering is $b_{\rm min} = GM_p/v^2$ (the capture radius), where v is of the order of the local velocity dispersion σ. The total rate of scattering stars into the loss cone, Γ, is obtained by integrating $d\Gamma/dM_p db$ over all MP masses and over all impact parameters between $b_{\rm min}$ and $b_{\rm max} \approx r$. The relaxation rate due to all MPs at all impact parameters is then

$$t_r^{-1} = \int_{b_{\rm min}}^{b_{\rm max}} {\rm d}b \int {\rm d}M_p \left(\frac{{\rm d}^2\Gamma}{{\rm d}M_p db}\right) \sim \log\Lambda \frac{G^2}{v^3} \int {\rm d}M_p \left(\frac{{\rm d}N_p}{{\rm d}M_p}\right) M_p^2 \,, \tag{1}$$

where $\log\Lambda = \log(b_{\rm max}/b_{\rm min})$ is the Coulomb logarithm (here the dependence of $\log\Lambda$ and v on M_p is assumed to be negligible). This formulation of the relaxation time is equivalent to its conventional definition [2] as the time for a change of order unity in v^2 by diffusion in phase space due to scattering, $t_r \sim v^2/D(v^2)$, where $D(v^2)$ is the diffusion coefficient. If the stars and MPs have distinct mass scales with typical number densities $N_\star$ and N_p and rms masses $\langle M_\star^2\rangle^{1/2}$ and $\langle M_p^2\rangle^{1/2}$ ($\langle M^2\rangle \equiv \int M^2({\rm d}N/{\rm d}M){\rm d}M/N$), then MPs dominate if the ratio of the 2nd moments of the MP and star mass distributions, $\mu_2 \equiv N_p\langle M_p^2\rangle / N_\star\langle M_\star^2\rangle$, satisfies $\mu_2 \gg 1$.

The central $\sim 100\,{\rm pc}$ of the GC contain $10^8 - 10^9$ solar masses in stars, and about $10^6 - 10^8$ solar masses in MPs such as open clusters and GMCs of masses $10^3 - 10^7\,M_\odot$ [28, 10, 35, 14, 3]. An order of magnitude estimate indicates that MPs in the GC can reduce the relaxation time by several orders of magnitude,

$$\frac{t_{r,\star}}{t_{r,{\rm MP}}} = \mu_2 \sim \frac{(N_pM_p)M_p}{(N_\star M_\star)M_\star} = 10^3 \left[\frac{(N_\star M_\star/N_pM_p)}{10^2}\right]^{-1} \left[\frac{(M_p/M_\star)}{10^5}\right] \,. \tag{2}$$

This estimate is borne by more detailed calculations (Fig. 1 and table 1), using the formal definition $t_r = v^2/D(v_{||}^2)$ with $M_\star\rho_\star \to \int({\rm d}N_p/{\rm d}M_p)M_p^2{\rm d}M_p$ [2]. In our calculations we follow the Fokker-Planck approach to the loss-cone problem [6], where we recalculate the diffusion coefficients obtained from stellar two body relaxation [2] by taking into account the contribution from the much more massive MPs (for details see [30]).

Although some of the assumptions concerning the loss cone formalism are not necessarily valid in the case of MPs, the loss-cone formalism can be generalized to deal with MPs in an approximate manner with only few modifications [30]. Thus to a good approximation our calculation follows the usual loss cone treatment (e.g. [22]) where the mass of the stars is replaced by that of the MPs, and the integration over the energies of deflected stars (mapped to their distance from the MBH) is done only for regions where MPs exist.

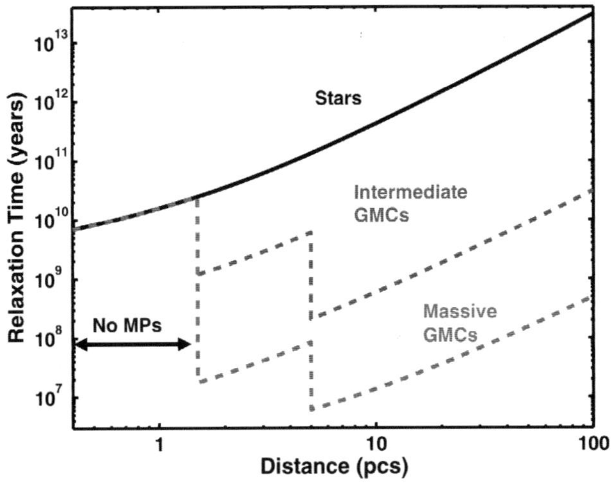

Figure 1. Relaxation time as function of distance from the MBH, for stars (solid line) and for each of the 2 MP models separately, as listed in table 1 and as labelled in the Figure. The discontinuities are artifacts of the assumed sharp spatial cutoffs on the MP distributions. 2-body stellar processes dominate close to the MBH, where no MPs are observed to exist. However, at larger distances massive clumps (at $1.5 < r < 5$ pcs) and GMCs (at $5 < r < 100$) are much more important.

3. Massive perturbers in the Galactic Center

MPs can dominate relaxation only when they are massive enough to compensate for their small space densities. Here we consider only MPs with masses $M_p \geq 10^3 M_\odot$. Such MPs could be open or globular stellar clusters or molecular clouds of different masses, in particular giant molecular clouds (GMCs). Observations of the Galaxy reveal enough MPs to dominate relaxation in the central 100 pc. We adopt here a conservative approach, and include in our modeling only those MPs that are directly observed in the Galaxy. Based on the observations of such MPs we devise several possible models for the MPs in the GC, detailed in table 1. As these observations [28, 10, 35, 14, 3] show that the MPs population is dominated by the GMCs, we consider only GMCs and gaseous clumps in our models. The observed MP species vary in their spatial distributions and mass functions, which are not smooth or regular. For our numeric calculations, we construct several simplified MP models (table 1) that are broadly based on the observed properties of the MPs, and assume that the spatial distribution of MPs follows that of the stars (apart of the inner radius cutoff where no MPs are observed).

The three MP models in table 1: Stars, GMC1 and GMC2, represent respectively the case of relaxation by stars only, by heavy GMCs and light GMCs.

4. Massive perturber-driven interactions with a MBH

The maximal differential loss-cone refilling rate, which is also the close encounters event rate, $\mathrm{d}\Gamma/\mathrm{d}E$, is reached when relaxation is efficient enough to completely refill the loss cone during one orbit. Further decrease in the relaxation time does not affect the event rate at that energy. MPs can therefore increase the differential event rate over that predicted by stellar relaxation, only at high enough energies, $E > E_c$ (equivalently, small enough typical radii, $r < r_c$), the critical energy (radius), separating the full and empty loss cone regimes, where slow stellar relaxation fails to refill the empty loss-cone. The extent of the empty loss-cone region increases with the maximal periapse q, which in turn depends on the close encounter process of interest. For example, the tidal disruption of an object of mass M and size R occurs when $q < r_t$, the tidal disruption radius, $r_t \simeq R\left(M_\bullet/M\right)^{1/3}$. This approximate disruption

Table 1. Massive perturber models

Model	r (pc) [a]	N_p	M_p ($M_\odot$)	β	R_p (pc)	μ_2 [b]
GMC1	5–100	100	10^4-10^7	1.6	5	3×10^5
	1.5–5	30	10^3-10^5	0.9		6000
GMC2	5–100	100	10^3-10^6	1.6	5	3×10^3
	1.5–5	30	10^2-10^4	0.9		60
Stars	5–100	2×10^8	1	—	~ 0	1
Stars	1.5–5	6×10^6	1	—	~ 0	1

[a] $N_p(r)\propto r^{-2}$ assumed.
[b] $\mu_2 \equiv N_p\left\langle M_p^2\right\rangle / N_\star\left\langle M_\star^2\right\rangle$, where $\langle M^2\rangle = \int M^2(\mathrm{d}N/\mathrm{d}M)\mathrm{d}M/N$.

criterion applies both for single stars ($M = M_\star$, $R = R_\star$) and for binaries, where M is the combined mass of the binary components and R is the binary's semi-major axis, a. Stellar radii are usually much smaller than typical binary separations, but stellar masses are only ~ 2 times smaller than binary masses. Binaries are therefore disrupted on larger scales than single stars. In the GC this translates to an empty (stellar relaxation) loss-cone region extending out to $r_c^s \sim 3$ pc for single stars and out to $r_c^b > 100$ pc for binaries. In the GC $r_{\mathrm{MP}} \lesssim r_c^s \ll r_c^b$ (where r_{MP} is the smallest distance from the MBH where MPs are observed) , and so MPs are expected to increase the binary disruption rate by orders of magnitude, but increase the single star disruption rate only by a small factor. Since for most Galactic MP types $r_{\mathrm{MP}} > r_h$ (the radius of influence of the MBH), the disruption rate is dominated by stars near r_{MP}. For example, when the loss-cone is empty, $\sim 50\%$ of the total rate is due to MPs at $r < 2r_{\mathrm{MP}}$; when the loss-cone is full, $\sim 75\%$ of the total rate is due to MPs at $r < 2r_{\mathrm{MP}}$ (see details in [30]).

4.1. Interactions with single stars

Clusters, GMCs and gas clumps in the GC are abundant only beyond the central $r_{\mathrm{MP}} \sim 1.5$ pc, whereas the empty loss-cone regime for tidal disruption of single stars extends only out to $r_c^s \sim 3$ pc. For inspiral processes such as GW emission, r_c is ~ 100 times smaller still [17]. The effect of such MPs on close encounter events involving single stars is thus suppressed (weaker tidal effects by MPs at $r > r_c^s$ are not considered here). This is contrary to the suggestion of Zhao et al. [37], who assumed that the effect of MPs fully extends to the empty loss-cone regime. We find that the enhancement of MPs over stellar relaxation to the single stars disruption rate is small, less than a factor of 3, and is due to stars scattered by gas clumps in the small empty-loss cone region between $r_{\mathrm{MP}} \sim 1.5\,\mathrm{pc}$ and $r_c^s \sim 3\,\mathrm{pc}$. A possible exception to this conclusion is the possible existence of IMBHs population, not modeled here.

4.2. Interactions with stellar binaries

The empty loss cone region for binary-MBH interactions extends out to > 100 pc because of their large tidal radius. On these large scales MPs are abundant enough to dominate the relaxation processes. Here we focus on 3-body exchange interactions [15, 36], which lead to the disruption of the binary, the energetic ejection of one star, and the capture of the other star on a close orbit around the MBH. The event rate is highly dependent on the unknown binary fraction in these regions.

The binary fraction and typical binary semi-major axis depend on the binary mass, and on the rate at which binaries evaporate by encounters with other stars. This depends in turn on the stellar densities and velocities, and therefore on the distance from the MBH. We take these factors into account and estimate in detail the 3-body exchange rate for MP-driven relaxation. The rate is proportional to the binary fraction in the population, which is the product of the poorly-known binary IMF in the GC and the survival probability against binary evaporation.

The capture probability and the semi-major axis distribution of captured stars were estimated by

Table 2. Total binary disruption rate and number of captured young stars

Model	Disruption rate (yr^{-1})		Young Stars[a]	Young Stars[a]
	$r < 0.04\,pc$	$r < 0.4\,pc$	$r < 0.04\,pc$	$0.04 < r < 0.4\,pc$
GMC1	1×10^{-4}	2.8×10^{-4}	33.5	3.8
GMC2	2.1×10^{-5}	6.4×10^{-5}	5	0.24
Stars	3.4×10^{-7}	5.3×10^{-7}	0.15	0.003
Observed	?	?	$10 - 35$[b]	?

[a] Main sequence B stars with lifespan $t < 5 \times 10^7$ yr .

[b] ~ 10 stars with derived $a \lesssim 0.04$ pc. $\gtrsim 30$ stars are observed in the area.

simulations [16, 36]. Numeric experiments indicate that between 0.5–1.0 of the binaries that approach the MBH within the tidal radius $r_t(a)$ are disrupted. Here we adopt a disruption efficiency of 0.75. The harmonic mean semi-major axis for 3-body exchanges with equal mass binaries was found to be [16]

$$\langle a_1 \rangle \simeq 0.56 \left(\frac{M_\bullet}{M_{\rm bin}} \right)^{2/3} a \simeq 0.56 \left(\frac{M_\bullet}{M_{\rm bin}} \right)^{1/3} r_t, \qquad (3)$$

where a is the semi-major axis of the infalling binary and a_1 that of the captured star (the MBH-star "binary"). Most values of a_1 fall within a factor 2 of the mean. This relation maps the semi-major axis distribution of the infalling binaries to that of the captured stars: the harder the binaries, the more tightly bound the captured stars. The periapse of the captured star is at r_t, and therefore its eccentricity is very high [16, 25], $e = 1 - r_t/a_1 \simeq 1 - 1.8(M_{\rm bin}/M_\bullet)^{1/3} \gtrsim 0.98$ for values typical of the GC.

We now consider the implications of 3-body exchange interactions of the MBH with old ($t_\star \gtrsim t_H$) binaries and massive young ($t_\star < 5 \times 10^7$ yr) binaries.

The properties of binaries in the inner GC are at present poorly determined. We use the period distribution of Solar neighborhood binaries for old low mass binaries [7] and young massive binaries [19]. The total binary fraction of these binaries is estimated at $f_{\rm bin} \sim 0.3$ for low mass [20] binaries, and $f_{\rm bin} \sim 0.75$ for massive binaries [19]. Adopting these values for the GC, the total binary disruption rate by the MBH can then be calculated by integrating $dN_{\rm bin}/da$ over the binary a distribution and over the power-law stellar density distribution of the GC from the minimal radius where such binaries exist (for young binaries we assume an inner cutoff at 1.5 pcs, where such young stars are not observed [29]), up to 100 pc [11]. Table (2) lists the capture rates for the different perturber models, assuming a typical old equal-mass binary of $M_{\rm bin} = 2\,M_\odot$, or young equal-mass binary of $M_{\rm bin} = 15\,M_\odot$.

The old, low-mass binary disruption rate we derive for stellar relaxation alone is $\sim 5 \times 10^{-7}\,{\rm yr}^{-1}$, ~ 5 times lower, but still in broad agreement with the result of Yu & Tremaine [36]. Their rate is somewhat higher because they assumed a constant binary fraction and a constant semi-major axis for all binaries, even close to the MBH, where these assumptions no longer hold.

MPs increase the binary disruption and high-velocity star ejection rates by factors of $\sim 10^{1-3}$ and effectively accelerate stellar migration to the center. Thus we expect an increase in the number of stars captured close to the MBH and consequently a higher event rate of single star processes such as tidal disruption, tidal heating and GW emission from compact objects, in particular from compact objects on zero-eccentricity orbits [25]. MPs may be implicated in the puzzling presence of a cluster of main sequence B-stars ($4 \lesssim M_\star \lesssim 15\,M_\odot$) in the inner $\sim 1''$ (~ 0.04 pc) of the GC. This so-called "S-cluster" is spatially, kinematically and spectroscopically distinct from the young, more massive stars observed farther out, on the $\sim$0.05–0.5 pc scale, which are thought to have formed from the gravitational fragmentation of one or two gas disks [29]. There is however still no satisfactory explanation for the existence of the seemingly normal, young massive main sequence stars of the S-cluster, so close to a MBH (see review of proposed models by Alexander [1]; also a recent model by Levin [21], and in these proceedings).

Here we revisit an idea proposed by Gould & Quillen [13], that the S-stars were captured near the MBH by 3-body exchange interactions with infalling massive binaries. Originally, this exchange scenario lacked a plausible source for the massive binaries. We suggest that MP-driven 3-body exchanges can serve as a source for binaries, as they increase the rate of infall of young field binaries to the MBH. Such young field binaries should exist in the GC, taking into account the ongoing star formation in the central ~ 100 pc, which implies the presence of a large reservoir of massive stars there. Such stars are indeed observed in the central few $\times$ 10 pc both in dense clusters and in the field [10, 27]. It is plausible that a high fraction of them are in binaries.

We assume star formation at a constant rate for 10 Gyr with a Miller-Scalo IMF [24], and use a stellar population synthesis code [33] with the Geneva stellar evolution tracks [32] to estimate that the present day number fraction of stars in the S-star mass range is 3.5×10^{-4} (and less than 0.01 of that for $M_\star > 15\ M_\odot$ stars). Note that if star formation in the GC is biased toward massive stars [34], this estimate should be revised upward. We adopt the observed Solar neighborhood distribution of the semi-major axis of massive binaries, which shows that massive binaries are thus typically harder than low-mass binaries [19], and will be tidally disrupted closer to the MBH and leave a more tightly bound captured star.

We represent the massive binaries by one with equal mass stars in the mid-range of the S-stars masses, with $M_{\rm bin} = 15\ M_\odot$ and $t_\star(7.5\ M_\odot) \simeq 5{\times}10^7$ yr, and integrate over the stellar distribution and the binary a distribution as before, to obtain the rate of binary disruptions, Γ, the mean number of captured massive stars in steady state, $N_\star = \Gamma t_\star$, and their semi-major axis distribution (Eq. 3). Table (2) compares the number of captured young stars in steady state, for the different MP models, on the $r < 0.04$ pc scale (the S-cluster) and $0.04 < r < 0.4$ pc scale (the stellar rings) with current observations [8, 29].

The number of captured massive stars falls rapidly beyond 0.04 pc (table 2) where the S-cluster is observed because wide massive binaries are rare. This capture model thus provides a natural explanation for the central concentration of the S-cluster (Fig 2). The absence of more massive stars in the S-cluster ($M_\star > 15\ M_\odot$, spectral type O V) is a statistical reflection of their much smaller fraction in the binary population. Figure (2) and table (2) compare the cumulative semi-major axis distribution of captured B-stars, as predicted by the different MP models, with the total number of young stars observed in the inner 0.04 pc (~ 35 stars [8, 12, 29]. Of these, only ~ 10 have full orbital solutions (in particular a and e) at present. The numbers predicted by the MP models are consistent with the observations, unlike the stellar relaxation model that falls short by two orders of magnitude.

The binary capture model predicts that captured stars have very high initial eccentricities. Most of the solved S-star orbits do have $e > 0.9$, but a couple have $e \sim 0.3$–0.4 [8]. Normal stellar relaxation is too slow to explain the decrease in the eccentricity of these stars over their relatively short lifetimes. However, the much faster process of resonant relaxation [31] may be efficient enough to randomize the eccentricity of a fraction of the stars, and could thus possibly explain the much larger observed spread in eccentricities [18].

The companions of these captured stars are ejected from the GC at high velocities [15]. Consequently, our model predicts the number of young hyper velocity stars ejected from the GC (such as observed [4]) to be comparable to that of observed B-stars in the S-cluster, in agreement with current observation based estimations [4].

5. Summary

We presented here the results of a study of the effect of massive perturbers near the MBH in the GC. We have shown that current observations of MPs such as GMCs and clusters indicate that they dominate relaxation process in the GC, where they exist, and they are much more important than stellar two-body relaxation processes by stars which are usually considered. We have used the loss cone formalism in order to analyze the importance of this mechanism in generating low angular momentum stars and binaries to interact with the MBH. We have computed the rates of these interactions, and showed that they can be highly important for the process of binaries disruption by the MBH and its consequent outcomes. The origin of the young massive B-stars at the central arcsecond of the GC can be explained

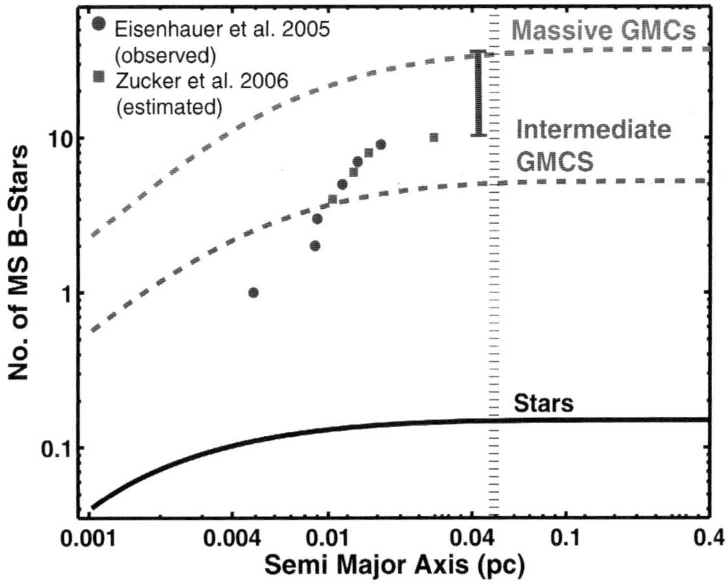

Figure 2. Cumulative number of young B-stars in the GC as predicted by the MP models and by stellar two-body relaxation (listed in table 2).The circles, squares and vertical bar represents the cumulative number of observed young stars inside 0.04 pc [8, 38] . The dotted vertical line marks the approximate maximal distance in which captured B-stars are expected to be observed.

by the capture of stars from binaries disrupted by the MBH after being scattered by MPs. Our calculations also show that some of the companions of these captured stars could be ejected at high velocities, thus explaining the observations of young massive high velocity stars observed recently. In addition MPs increase the number of compact stars captured close to the MBH, and thus increase the emission of zero-eccentricity GWs. We also suggest that such MPs may also help solve the last parsec problem of coalescing BMBHs. Although we focused on the GC of the milky-way, our results can be easily extended to MPs near other MBHs.

Acknowledgments
We thank the organizers of the GC2006 meeting for a very enjoyable and stimulating conference.

References

[1] Alexander, T. 2005, Physical Reports, 419, 65
[2] Binney, J. & Tremaine, S. 1987, Galactic Dynamics (Princeton, NJ: Princeton University Press)
[3] Borissova, J. et al., 2005, Astronomy & Astrophysics, 435, 95
[4] Brown, W. et al. 2006, ApJ Letters, 640, L35
[5] Christopher, M. H. et al., 2005, ApJ, 622, 346
[6] Cohn, H. & Kulsrud, R. M. 1978, ApJ, 226, 1087
[7] Duquennoy, A. & Mayor, M. 1991, Astronomy & Astrophysics, 248, 485
[8] Eisenhauer, F. et al. 2005, ApJ, 628, 246
[9] Ferrarese, L. & Merritt, D. 2000, ApJ Letters, 539, L9
[10] Figer, D. F. et al. 2002, ApJ, 581, 258
[11] Genzel, R. et al. 2003, ApJ, 594, 812
[12] Ghez, A. M. et al. 2005, ApJ, 620, 744
[13] Gould, A. & Quillen, A. C. 2003, ApJ, 592, 935

[14] Güsten, R. & Philipp, S. D. 2004, in The Dense Interstellar Medium in Galaxies, ed. S. Pfalzner, C. Kramer, C. Staubmeier, & A. Heithausen, 253–+
[15] Hills, J. G. 1988, Nature, 331, 687
[16] Hills, J. G. 1991, AJ, 102, 704
[17] Hopman, C. & Alexander, T. 2005, ApJ, 629, 362
[18] Hopman, C. & Alexander, T. 2006, ArXiv, available at http://xxx.lanl.gov/abs/astro-ph/0606161
[19] Kobulnicky, H. A. et al. 2006, ArXiv, available at http://xxx.lanl.gov/abs/astro-ph/0605069
[20] Lada, C. J. 2006, ApJ Letters, 640, L63
[21] Levin, Y. 2006, MNRAS, ArXiv, available at http://xxx.lanl.gov/abs/astro-ph/0603583
[22] Magorrian, J. & Tremaine, S. 1999, MNRAS, 309, 447
[23] Merritt, D. 2006, ArXiv, available at http://xxx.lanl.gov/abs/astro-ph/0605070
[24] Miller, G. E. & Scalo, J. M. 1979, ApJ Supplements, 41, 513
[25] Miller, M. C., Freitag, M., Hamilton, D. P., & Lauburg, V. M. 2005, ApJ Letters, 631, L117
[26] Mouawad, N. et al. 2005, Astronomische Nachrichten, 326, 83
[27] Muno, M. P. et al. 2006, ApJ, 638, 183
[28] Oka, T. et al. 2001, ApJ, 562, 348
[29] Paumard, T. et al. 2006, ApJ, 643, 1011
[30] Perets, H. B., Hopman, C. & Alexander, T. 2006, ArXiv, available at http://xxx.lanl.gov/abs/astro-ph/0606443
[31] Rauch, K. P. & Tremaine, S. 1996, New Astronomy, 1, 149
[32] Schaller, G. et al. 1992, Astronomy & Astrophysics, 96, 269
[33] Sternberg, A., Hoffmann, T. L., & Pauldrach, A. W. A. 2003, ApJ, 599, 1333
[34] Stolte, A. et al. 2005, ApJ Letters, 628, L113
[35] Vollmer, B., Zylka, R., & Duschl, W. J. 2003, Astronomy & Astrophysics, 407, 515
[36] Yu, Q. & Tremaine, S. 2003, ApJ, 599, 1129
[37] Zhao, H., Haehnelt, M. G., & Rees, M. J. 2002, New Astronomy, 7, 385
[38] Zucker, S. et al. 2006, ApJ Letters, 639,L21

Institute of Physics Publishing Journal of Physics: Conference Series **54** (2006) 301–305
doi:10.1088/1742-6596/54/1/047 Galaxy Center Workshop

Ejection of Hyper-Velocity Stars by Intermediate-Mass Black Holes

Holger Baumgardt[1], Alessia Gualandris[2,3] and Simon Portegies Zwart[2,3]

[1] Argelander Institute for Astronomy, University of Bonn, Auf dem Hügel 71, 53121 Bonn, Germany

[2] Astronomical Institute "Anton Pannekoek,", University of Amsterdam, Kruislaan 403, Netherlands

[3] Section Computational Science, University of Amsterdam, Kruislaan 403, Netherlands

E-mail: holger@astro.uni-bonn.de

Abstract. We have performed N-body simulations of the formation of hyper-velocity stars (HVS) in the centre of the Milky Way due to inspiralling intermediate-mass black holes (IMBHs). We find that due to dynamical friction, IMBHs sink into the centre of the Galaxy where they deplete the central cusp of stars. Some of these stars become HVS and are ejected with velocities sufficiently high to escape the Galaxy.

Our simulations show that HVS are generated in short bursts which last only a few Myrs until the IMBH is swallowed by the supermassive black hole (SMBH). After the HVS have reached the galactic halo, their escape velocities correlate with the distance from the Galactic centre in the sense that the fastest HVS can be found furthest away from the centre. Finally, our simulations show that the presence of an IMBH in the Galactic centre changes the stellar density distribution inside $r < 0.02$ pc into a core profile, which takes at least 100 Myrs to replenish.

1. Introduction

[9] was the first to show that the ejection of stars with velocities > 1000 km/sec is a natural consequence of galaxies hosting supermassive black holes and he named these stars 'hyper-velocity stars' (HVS). Recently, several HVS have been discovered in the Galactic halo [4, 9, 5]. Except for one star which might have been ejected from the LMC [6], the travel times of all HVS are short enough that the stars could have been ejected from the Galactic centre within the lifetimes of the stars, confirming Hills' predictions.

The exact formation mechanism of HVS is however still a matter of debate. The ejection of stars by supernova explosions in close binary systems [3] and dynamical encounters [12] cannot produce main-sequence stars with velocities exceeding a few hundred kilometers per second [7], leaving the interaction of stars in galactic nuclei around super-massive black holes (SMBHs) as the only possible source for HVS.

[13] considered three processes which could eject stars with high velocities from the vicinity of SMBHs: (1) close encounters between two single stars, (2) encounters between stellar binaries and the central SMBH and (3) encounters between single stars and a massive black hole binary. For the SMBH in the Galactic centre they found that close encounters between single stars can

create HVS with a rate of only $10^{-11}\,\mathrm{yr}^{-1}$ which would create less than one star during the lifetime of the Milky Way. The other two processes were found to eject stars with rates of up to $10^{-4}\,\mathrm{yr}^{-1}$, sufficiently high to explain the observed number of HVS in the halo of the Milky Way.

A distinction between the two scenarios might come from a detailed analysis of the spatial and kinematical distribution of HVS: the ejection of stars due to the interaction of stellar binaries with an SMBH should be nearly constant with time since the reservoir of binary stars in the Galactic centre is depleted only slowly. Furthermore, the distribution of binary orbits should be nearly isotropic in sufficiently relaxed nuclei, implying that the resulting HVS distribution will also be isotropic. In contrast, the ejection of stars from an SMBH-IMBH binary should show characteristic variations with time and spatial direction (see e.g. [10]).

In the present work we have performed collisional N-body simulations of the dynamics of inspiralling IMBHs in stellar cusps around supermassive black holes. The aim of our simulations was to study whether the ejection of stars by IMBHs leads to observable consequences in the distribution of HVS which might help to distinguish between different ejection scenarios and which could give an indirect hint for the presence of one or more IMBHs in the centre of the Milky Way.

2. Description of the N-body runs

All runs were performed with the collisional N-body code NBODY4 [1] on the GRAPE6 computers [11] of Bonn and Tokyo University. Our runs contained three different components: a central super-massive black hole (SMBH), an IMBH and 10^5 stars. In all simulations the SMBH was initially at rest at the origin and had a mass of $M_{\mathrm{SMBH}} = 3 \cdot 10^6 M_\odot$, similar to the mass of the Galactic centre SMBH. The mass of the IMBHs was varied in the different runs. In total we performed 3 runs, using IMBH masses of $M_{\mathrm{IMBH}} = 10^3 M_\odot$, $3 \cdot 10^3 M_\odot$ and $10^4 M_\odot$ respectively. All IMBHs moved initially in circular orbits at a distance of 0.1 pc from the SMBH.

The stars had masses of $m_* = 30 M_\odot$ and were initially distributed in a power-law cusp around the SMBHs. Stars were merged with the black holes (both SMBH or IMBH) if their separation became smaller than their tidal radius. The mass of disrupted stars was added to the mass of the black holes. IMBHs were merged with the central SMBH if they passed within the radius of the last stable orbit, assumed to be 3 Schwarzschild radii. We did not include the effects of gravitational radiation into our runs. Neither was any softening used in calculating the gravitational forces between the particles in our calculation.

Simulations were stopped when an IMBH merged with the SMBH or the runs reached 15 Myrs, whichever happened first.

3. Results

Fig. 1 depicts the inspiral of the 3 studied IMBHs. As predicted by theory, more massive IMBHs spiral in faster. Inside $r = 0.003$ pc the stellar cusps contain only few stars, (for the cusp profile chosen in our runs only $M(< r) = 2 \cdot 10^3 M_\odot$ in stars were inside this radius), so the mass in stars becomes comparable to the mass of the inspiralling IMBHs. In this case dynamical friction becomes inefficient and the inspiral stalls since the IMBHs cannot displace enough stars to lose further orbital energy.

Fig. 2 depicts the evolution of orbital eccentricity of the IMBHs. During the inspiral phase, the orbits of all IMBHs stay nearly circular, while the eccentricities can reach values as large as $1 - e = 10^{-5}$ once the IMBHs have reached the centre. This is large enough that the IMBHs can merge with the central SMBH due to the emission of gravitational waves (dotted lines). The lifetime of an IMBH in the galactic centre is therefore limited to a few Myrs.

Fig. 3 depicts the escape rate of all ejected stars as a function of time. The overall evolution is very similar for the three IMBH masses: The escape rate rises with time and reaches a maximum

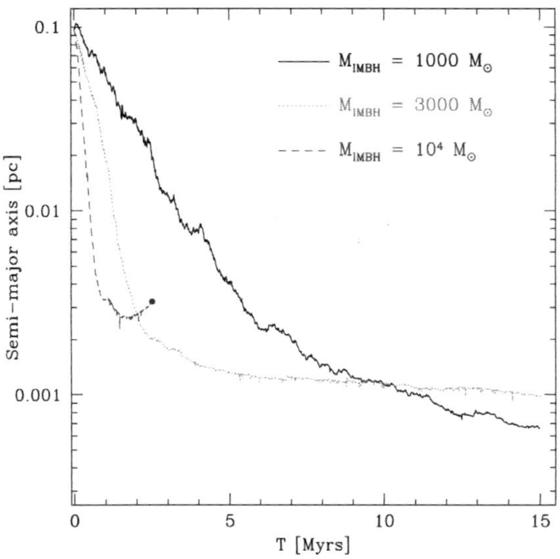

Figure 1. Semi-major axis of the IMBHs as a function of time. IMBHs spiral in as a result of dynamical friction. Inside $r = 0.003$ pc, the central cusp contains too few stars and the inspiral stalls.

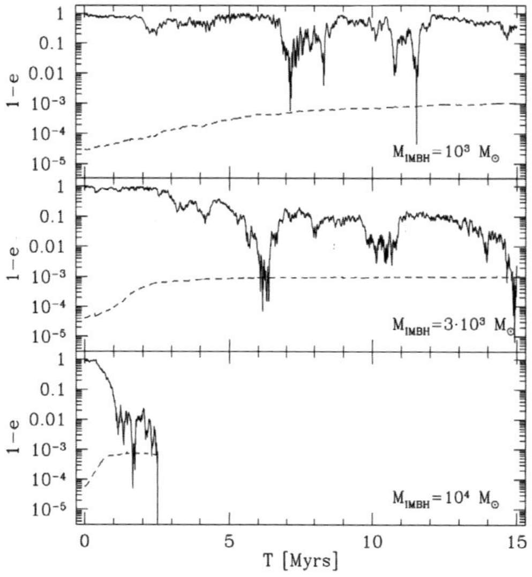

Figure 2. Evolution of the orbital eccentricity of the inspiralling IMBHs. The orbits remain nearly circular during the inspiral phase and become highly eccentric in the stalling phase.

when the IMBHs have spiralled into the centre. It drops on a similar timescale during which the IMBHs scatter away all stars from the inner cusp region. As shown in Fig. 2, the orbital evolution of the IMBH around the SMBH may be terminated by the emission of gravitational waves before the end of the calculation. We have therefore shown the further evolution of the escape rate in Fig. 3 with dotted lines. For the higher mass IMBHs ($M_{\rm IMBH} \geq 3 \times 10^3 M_\odot$), the majority of escapers are ejected well before the two black holes merge due to GW emission.

In order to find the HVSs in our simulations, we followed the orbits of all escapers in the potential of the Galaxy after they left the Galactic centre and considered as HVSs only those stars which acquired large enough velocities to be unbound and escape the Milky Way potential. Fig. 4 shows the spatial distribution of HVS for the $M_{\rm IMBH} = 3000 M_\odot$ run 100 Myrs after the start of the simulation, assuming that the inspiral plane of the IMBH agrees with the plane of the galactic disc (z=0). For the $M_{\rm IMBH} = 3000 M_\odot$ run, the distribution of HVS is nearly isotropic and would thus be indistinguishable from a HVS distribution created by encounters of stellar binaries with a single SMBH. Also shown in Fig. 4 are the velocity vectors of the HVS. As expected, they all point away from the galactic centre. Since HVS are created in a short burst whose duration is much smaller than the time required to travel into the galactic halo, the escape velocities of HVS also increase in magnitude with the distance from the centre. In case HVS are created by an IMBH, we therefore expect a similar correlation between escape velocity and galactocentric distance with the fastest HVS being found at the largest distances.

Fig. 5 shows the density profile of the stellar cusp for the $M_{\rm IMBH} = 10^4 M_\odot$ and $M_{\rm IMBH} = 3 \times 10^3 M_\odot$ runs at the start and by the time the runs were stopped. The initial density profile follows an $\alpha = 1.75$ power-law cusp for radii $r < 1$ pc down to about $r = 3 \times 10^{-4}$ pc, at which radius the cusp runs out of stars. Due to the merger of cusp stars with the central SMBH and the ejection of stars by the IMBHs, the initial cusps are turned into core profiles with a core radius of about $r = 0.02$ pc. The core radii stay nearly constant in time since the ejection rate of stars is low after the initial peak.

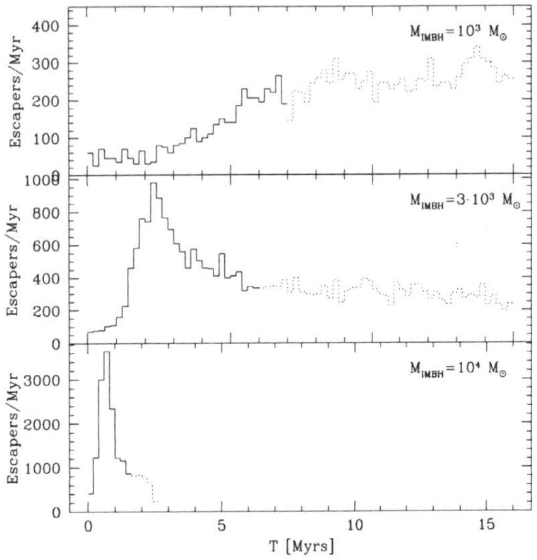

Figure 3. Escape rate of stars ejected from the central cusp as a function of time (solid and dotted lines). The stellar escape rate rises once the IMBHs have reached the centre and drops as the IMBHs deplete the central cusps. Dotted lines show the part of the inspiral which would not have been reached if our runs had included GW emission.

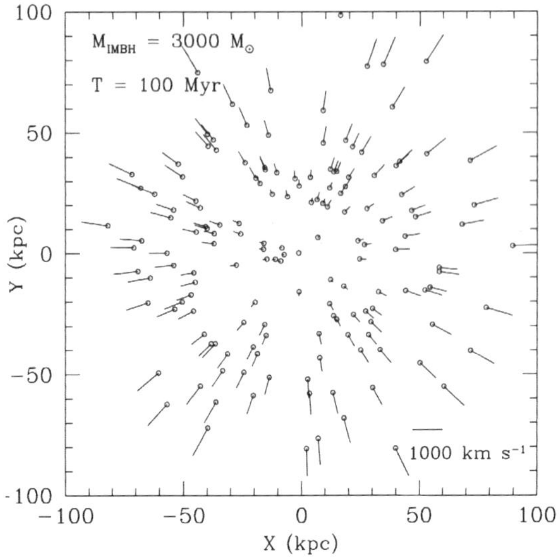

Figure 4. Spatial distribution of HVS created in the $M_{\rm IMBH} = 3000 M_\odot$ run 100 Myrs after the start of the simulation. The orbits of all escapers were followed in a 3D Galactic potential, with the galactic disc assumed to be at z=0. The distribution of HVS after 100 Myrs is isotropic in space and the fastest HVS can be found at the largest distances.

In order to test how quickly the core is replenished, we continued the $M_{\rm IMBH} = 10^4 M_\odot$ run after the IMBH merged with the SMBH. Even after running the simulation for 10 Myrs, there was only a slight increase in central density. Since the relaxation time in our runs is about a factor of 10 smaller than in real Galactic nuclei, replenishing the cusp in the Galactic centre should take at least $T \approx 100$ Myrs and possibly even longer. If observations show that the power-law cusp in the Galactic centre extends down to radii much smaller than 0.02 pc, the presence of massive IMBHs in the Galactic centre within the last 100 Myrs could be excluded. In this case the observed HVS would not have been ejected by IMBHs, since typical travel times of HVS are of order 10^8 years or less [5].

4. Conclusions

We have performed simulations of the inspiral of massive black holes into the centres of galaxies and of the subsequent ejection of hyper-velocity stars. We found that the spatial distribution of HVS is nearly isotropic and would be difficult to distinguish from a HVS distribution created by interactions of stellar binaries with an SMBH if only few HVS were found, as is presently the case. A better indication comes from the escape times of HVS: our simulations confirm that most HVS are ejected in short bursts, lasting only a few Myrs for typical IMBH masses, as soon as the IMBH reaches the galactic centre. The ejection ends when the IMBH merges with the central SMBH, which should happen within a few Myrs for typical IMBH masses. Even if merging can somehow be avoided, the ejection rate of HVS is a factor of 30 to 100 lower than during the burst maximum. The currently observed HVS show a broad distribution of escape times [4], which argues against ejection due to a single IMBH, but would still be consistent with

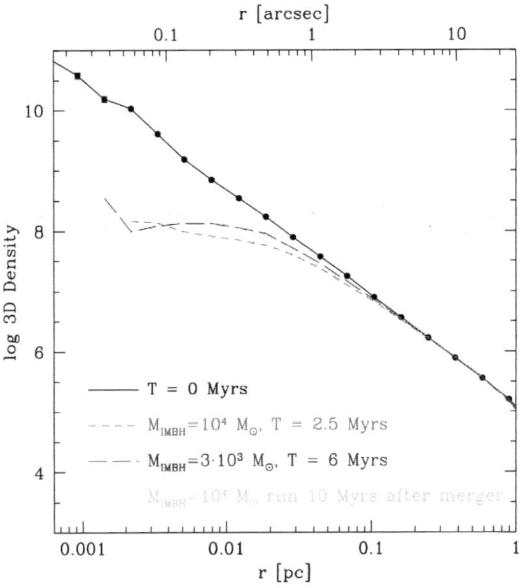

Figure 5. Density distribution of stars at three different times for the $M_{\rm IMBH} = 10^4 M_\odot$ and $M_{\rm IMBH} = 3 \cdot 10^3 M_\odot$ runs. In both runs, the initial density distribution follows a $\rho \sim r^{-1.75}$ power-law cusp which is depleted in the central parts and turned into a core profile after the IMBHs have spiralled into the centre. After the $M_{\rm IMBH} = 10^4 M_\odot$ IMBH merged with the SMBH, the central cusp is replenished only very slowly (dotted lines). The density profile in the Galactic centre should show a similar core if it contained an IMBH within the last 100 Myrs.

the inspiral of several IMBHs. In case HVS are ejected by IMBHs, we expect a strong correlation of escape velocity with galactocentric distance in the sense that the fastest HVS can be found at the largest distances. A more detailed discussion of our results can be found in [2].

Acknowledgments

The authors would like to thank Sverre Aarseth for helping us with NBODY4 and Yuri Levin for useful discussions.

References

[1] Aarseth, S. J., 1999, PASP, 111, 1333
[2] Baumgardt, H., Gualandris, A., Portegies Zwart, S., 2006, MNRAS submitted
[3] Blaauw, A., 1961, Bull. Astron. Inst. Netherlands, 15, 265
[4] Brown, W., et al. 2005, ApJ, 622, L33
[5] Brown, W., et al. 2006, ApJ, 640, L35
[6] Edelmann, H., et al. 2005, ApJ, 634, L181
[7] Gualandris, A., Portegies Zwart, S., Sipior, M. S., 2005, MNRAS, 363, 223
[8] Hills, J. G., 1988, Nature, 331, 687
[9] Hirsch, H. A., et al. 2005, A&A, 444, L61
[10] Levin, Y., 2005, ApJ submitted, astro-ph/0508193
[11] Makino, J., Fukushige, T., Koga, M., & Namura, K., 2003, PASJ, 55, 1163
[12] Poveda, A., Ruiz, J., Allen, C., 1967, Bol. Obs. Tonantzintla Tacubaya, 4, 86
[13] Yu, Q., Tremaine, S., 2003, ApJ, 599, 1129

Institute of Physics Publishing
doi:10.1088/1742-6596/54/1/048

X-ray constraints on the number of stellar mass black holes in the inner parsec

Patrick Deegan and Sergei Nayakshin
University of Leicester, University Road, Leicester, LE1 7RH, UK

Abstract. Due to dynamical friction stellar mass black holes should form a cusp in the inner parsec. Calculations [5, 6] show that approximately 20 thousand black holes would be present in a sphere with radius of about a parsec around Sgr A*. The presence of these objects opens up the possibility that they might be accreting "cool" gas (i.e. the Minispiral) as discussed by Morris [6]. Here we calculate the X-ray emission expected from these black holes as a method to constrain their population. We find that the data limits the total number of such black holes to around 10 – 20 thousand. Even a much smaller number of such black holes, i.e. 5 thousand, is sufficient to produce several sources with X-ray luminosity above $L_X \sim 10^{33}$ erg s^{-1} at any one time. We suggest that some of the discrete X-ray sources observed by Muno [7] with *Chandra* in the inner parsec may be such "fake X-ray binaries".

1. Introduction

Morris [6] noted that dynamical friction of massive stellar remnants on a population of less massive background stars will initiate transfer of these remnants closer to the nucleus of the Galaxy. In particular, he estimated that for stellar mass black holes near our Galactic Centre, the radius from which these will be collected into a central cluster is about 4 parsecs, and the total mass of the black holes is as high as $10^6\,M_\odot$. More recently, Miralda-Escud & Gould [5] updated and quantified Morris's [6] estimates and showed that black holes will force most of low mass stars out of the central region with size $R_{\rm bh} \approx 0.7$ parsec around Sgr A*. In their model, black holes form a central cluster of this size and number about 24 000. The total mass of these black holes could be as high as 5 – 10% of Sgr A*'s mass. In addition to this, more recent observations of young massive stars near Sgr A* suggest another potentially important channel that can contribute to this population of stellar remnants in the central parsec of Sgr A*: in-situ star formation that seems to be heavily biased towards forming many more massive stars per unit stellar mass than in normal galactic environment [8, 9, 11]. Formation of a cluster of black holes surrounding the SMBH thus appears not just plausible but inevitable.

Morris [6] also discussed the possibility that compact remnants will be accreting gas from molecular clouds that may be unusually dense near our Galactic Centre. He noted that if gas density within the central parsec of Sgr A* were much higher than that estimated from the current observations, then these black holes may be collectively rather bright. In this paper we set out to explore this idea in more detail and also speculate on the implications for observations of AGN in general. In particular, we improve on the Bondi-Hoyle accretion rate estimates of Morris [6] in two ways. First of all, we include in the model the formation of small scale discs around accreting black holes due to non-zero residual angular momentum of the accreting gas.

In addition to this, we use a more sophisticated form for the radiative efficiency motivated by the observations of radiatively inefficient accretion flows.

2. Numerical approach

2.1. Bondi-Hoyle gas capture rate estimates

Consider a stellar mass black hole ploughing through a gas cloud or a disc. The black hole will be accreting, or rather capturing, gas at the Bondi-Hoyle accretion rate [3]:

$$\dot{M}_{\rm capt} \sim 4\pi\rho \frac{(GM_{\rm bh})^2}{(\Delta v^2 + c_s^2)^{3/2}} , \tag{1}$$

where ρ, c_s and Δv are the ambient gas density, sound speed and the relative velocity between the black hole and the gas, respectively. The above picture is complicated by the presence of the SMBH. The area of influence of the stellar mass black holes is limited by the Hill radius, $r_H = R(M_{\rm bh}/3M_{\rm smbh})^{1/3}$, where R is the distance between the stellar mass black hole and the SMBH. This imposes a limit onto the capture rate, given by the Hill accretion rate, $\dot{M}_H = 4\pi r_H^2 \rho c_s$.

2.2. Time-dependent disc accretion

The gas captured by a stellar mass black hole will in general have a net angular momentum, resulting in the formation of a small scale accretion disk. We will assume that a disc of size r_c is formed, where $r_c = \zeta r_{\rm capt}$. The capture radius is

$$r_{\rm capt} = min\left[r_H, \frac{GM_{\rm bh}}{\Delta v^2 + c_s^2}\right] , \tag{2}$$

and ζ is a parameter ≤ 1. In fact, below we only consider values of $\zeta \leq 0.1$ because larger values of ζ are possible but not very realistic as it would imply that the captured flow has a maximum attainable asymmetry. For the same reason very small values of $\zeta \ll 10^{-3}$ would also seem quite unlikely.

The viscous time scale in the small scale disk about the stellar mass black hole is $t_{\rm visc} = \alpha^{-1}\Omega_d^{-1}(r/h)^2$, where α is the viscosity parameter, Ω_d is the angular velocity at $r = r_c$ and h is the height of the disk. This can be expressed as,

$$t_{\rm visc} = 1.5 \times 10^3 \text{ year } \alpha_{0.01}^{-1} \mu_d r_{\rm c,12}^{1/2} \, T_{d,3}^{-1} , \tag{3}$$

where $T_{d,3}$ is the disk temperature in units of 10^3 K, the viscosity parameter is $\alpha_d = 0.01\alpha_{0.01}$, μ_d is the mean molecular mass in units of Hydrogen mass, and $r_{\rm c,12}$ is the circularisation radius in 10^{12} cm. This is to be compared with period (the orbital time) about the SMBH. At the distance of 0.1 pc, $P \sim 3000$ year $M_6^{-1/2}$. Thus, the gas captured in the small scale disk about a stellar mass black hole accretes on the black hole after a delay of perhaps a fraction to a few orbital times.

We model the small scale disc evolution in the following way. We estimate the time-dependent accretion rate onto the black hole as $\dot{M}_{\rm acc} = (M_d/t_{\rm visc}) \exp^{-t_{\rm visc}/12t}$. The mass of the disc then evolves according to the addition and accretion of mass as $\dot{M}_d = \dot{M}_{\rm capt} - \dot{M}_{\rm acc}$. The luminosity of the accretion flow is assumed to be given by $L_X = \epsilon \dot{M}_{\rm acc} c^2$, where ϵ is an analytical "fit" to the radiative efficiency in both non-radiative and standard radiative regimes. $\epsilon = 0.01\dot{M}/(\dot{M}_0 + \dot{M})$, where $\dot{M}_0 = 0.01$ is the critical accretion rate where the switch from one regime to the other occurs. We also assumed that X-ray emission visible in the *Chandra* band constitutes 10% of the bolometric efficiency, which would be a lower limit for typical spectra of X-ray binaries in their hard state.

2.3. Orbital evolution of accretors

We model the velocity and space distribution of stellar mass black holes as a cusp that follows the Bahcall-Wolf [2] distribution for heavier species in a mass-segregated cusp. This distribution results in the black hole number density obeying a power law of the form $R^{-7/4}$. Both space and velocity distributions are isotropic. The most recent Monte-Cralo simulations [4] broadly support these classical results.

Generating a series of orbits consistent with this distribution, we randomly prescribe the initial phases of the black holes, and then we follow their spatial motion. When one of these orbits intersects the disc of the Minispiral, the black hole in question starts capturing gas and builds up a disc around it as described in §2.2.

2.4. The model for the Mini-Spiral

In our simple model, the Minispiral is modelled as a disk in circular rotation around Sgr A* with the total gas mass of 50 Solar masses, in accord with estimates in [10]. It extends from a radius of 0.1 pc from the SMBH to a radius of 0.5 pc. [10] suggests that the Minispiral is a dynamical feature almost in free fall onto Sgr A*. The dynamical age of the feature is a few thousand years. Therefore, we ran our calculations for 3 000 thousand years with these assumptions, and then we "remove" the Minispiral instantaneously. This is done as a very rough model of time evolution of the system. In a future work a more complicated, but unavoidably model-dependent, dynamic of the Minispiral might be included.

3. Few tests

The upper panels in Figure 1 display the total X-ray luminosity of a cluster of 5 000 black holes as a function of time for two values of the circularisation radius parameter, ζ, as labelled in the Figure. The lower panels show the number of X-ray sources with luminosity higher than 10^{33} erg s^{-1}. Such sources could be observed by *Chandra*. Several conclusions can be made. With a larger value of $\zeta = 0.1$, the accretion discs around black holes are larger, and thus viscous times are long. As a result, the X-ray emission varies smoothly with time, first increasing as the discs are built up, and then decreasing on $\sim$ a thousand years time scale. Thus the sources are rather steady in time, and are rather dim.

For the smaller value of $\zeta = 10^{-3}$, viscous times in small scale discs are much shorter. Therefore, the X-ray emission from the sources vary on much shorter time scales, i.e., of few years to tens of years. The sources are also brighter as the peak accretion rates are higher – each individual source shines much brighter for a shorter time, of course, as compared with the larger ζ case. Figure 2 shows the same experiments as Figure 1 but for 20 000 black holes. Comparison between the two different values of ζ shows similar trends as before.

It is interesting to compare the two Figures. While the results depend significantly on the apriori unknown value of ζ, both low and high ζ tests show same tendency of a significant luminosity increase with increase in the number of black holes. In fact, the luminosity increased by a larger factor than the black hole number did. The number of sources above the chosen luminosity threshold also increased. This suggest that by performing tests across all reasonable parameter space for ζ we should be able to find the maximum allowed number of stellar mass black holes in the cusp.

4. Results

Following this idea, we ran a number of models for a range of values of ζ and for the total black hole numbers of $N = 5, 10, 20$ and 40 thousand. During the time period modelled, the results vary considerably in each test. For example, it is possible for just one single source to dominate the X-ray luminosity output of the cluster. Thus we have chosen to average several

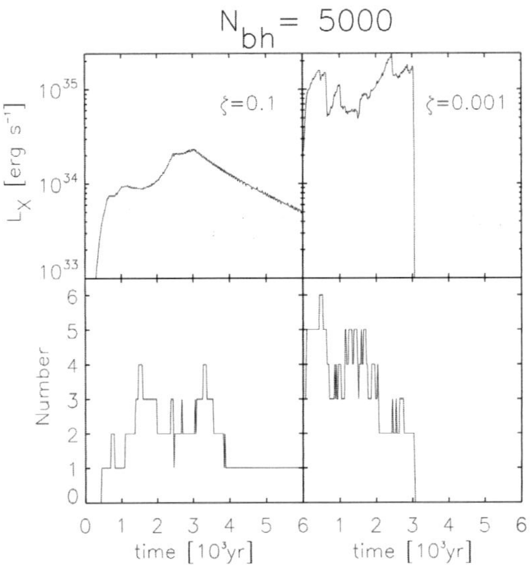

Figure 1. X-ray light curves (upper panels) and number of individual sources (bottom panels) where $L_X < 10^{33}$ erg s^{-1}, for 5 000 black holes in the inner parsec. The left and right panels correspond to $\zeta = 0.1$ and $\zeta = 0.001$ respectively.

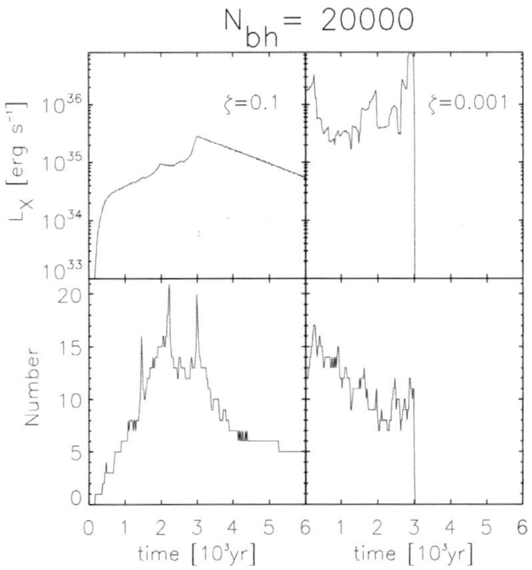

Figure 2. X-ray light curves and number of individual sources where $L_X < 10^{33}$ erg s^{-1}, when the total number of black holes in the inner parsec is 20 000. The left and right panels correspond to $\zeta = 0.1$ and $\zeta = 0.001$ respectively.

Table 1. Characteristics of black hole cusp averaged between 2 000 and 3 000 years.

Number[a] $[10^3]$	ζ[b]	$< L_X >$[c] $[10^{35}$erg s$^{-1}]$	P($L_X > 10^{36}$)[d]	P($L_X > 10^{35}$)[e]	$< N >$[f]	P(N>20)[g]	P(N>10)[h]
5	0.001	5.39 ± 3.06	0.16 ± 0.11	0.58 ± 0.19	4.28 ± 0.97	0.0 ± 0.0	0.05 ± 0.003
5	0.010	1.33 ± 0.64	0.02 ± 0.02	0.27 ± 0.15	3.04 ± 0.75	0.0 ± 0.0	0.0± 0.0
5	0.100	1.52 ± 0.39	0.0 ± 0.0	0.62 ± 0.22	6.13 ± 1.36	0.0± 0.0	0.15 ± 0.11
10	0.001	7.06 ± 4.31	0.59 ± 0.17	1.0 ± 0.01	9.79 ± 1.59	0.001 ± 0.001	0.35± 0.21
10	0.010	7.53 ± 2.11	0.27 ± 0.15	1.0 ± 0.0	10.69 ± 1.30	0.001 ± 0.001	0.50 ± 0.17
10	0.100	1.74 ± 0.69	0.01 ± 0.004	0.59 ± 0.21	11.82 ±0.65	0.002 ± 0.001	0.72 ± 0.12
20	0.001	20.8 ± 5.3	0.55 ± 0.12	1.0 ± 0.0	12.96 ± 0.24	0.0± 0.0	0.92 ± 0.02
20	0.010	20.2 ± 2.65	0.96 ± 0.02	1.0 ± 0.0	18.55 ± 0.48	0.23 ± 0.06	1.0 ± 0.0
20	0.100	8.89 ± 2.20	0.32 ± 0.23	1.0 ± 0.0	19.79 ± 1.15	0.42 ± 0.11	1.0 ± 0.0
40	0.001	84.1 ± 23.9	1.0 ± 0.0	1.0 ± 0.0	29.32 ± 0.37	1.0 ± 0.0	1.0 ± 0.0
40	0.010	101 ± 16.3	1.0 ± 0.0	1.0 ± 0.0	34.20 ± 2.90	1.0 ± 0.0	1.0 ± 0.0
40	0.100	12.0 ± 4.30	0.42 ± 0.211	1.0 ± 0.0	44.32 ± 0.74	1.0 ± 0.0	1.0 ± 0.0

[a] Total number of black holes in the cusp
[b] Circularisation parameter
[c] Time-averaged luminosity of the cusp
[d] Probability that the total luminosity of the cusp is greater than 10^{36} erg s^{-1}
[e] Probability that the total luminosity of the cusp is greater than 10^{35}erg s^{-1}
[f] Average number of sources with X-ray luminosity greater than 10^{33}erg s^{-1}(N)
[g] Probability that N is greater than 20
[h] Probability that N is greater than 10.

observables and then present their mean values. In particular, we calculated the average total X-ray luminosity and the number of black holes brighter than 10^{33} erg s^{-1}, as such sources would have been resolved by *Chandra* into separate point sources. The averaging is done between $2000 < t < 3000$ years as to look at a state that may be similar to the present state of the mini-spiral.

A summary of the results is presented in Table 1, where a number of time-averaged quantities are presented. In order to reduce and estimate statistical noise of the results, for each of the values of ζ and the total black hole number considered in Table 1, the tests were repeated 3 times, each time generating a different random black hole orbit distribution. The mean values of the quantities and their deviation are listed in Table 1.

Observations of the inner parsec by Chandra have placed upper limits on the total luminosity of sources of approximately 10^{35} erg s^{-1} [1] and the number of individual X-ray sources with a luminosity greater than 10^{33} erg s^{-1} of a dozen or so (F. Baganoff, private communication at the meeting). With these constraints in mind, we can immediately rule out the possibility that the cusp contains 40 000 black holes. For any reasonable value of ζ, the total luminosity and the number of individual sources with $L_X > 10^{33}$ erg s^{-1} are too large compared to observations. A cusp containing 5 000 or 10 000 black holes is well within the limits imposed by observations. A cusp containing 20 000 black holes is less likely but cannot be ruled out completely at this time.

5. Conclusions

The presence of stellar mass black holes in a cusp around a SMBH is a robust theoretical prediction [5, 6]. We describe an approach that can begin to constrain the numbers of these objects. Thus far the results presented suggest an upper limit of 20 000 for the number of stellar mass black holes in the inner parsec, consistent with theoretical predictions. Further refinements of the model of the X-ray emission from the black hole cusp, combined with updated observational data, may significantly improve the limits we put on these objects.

We suggest that at least some of the X-ray sources visible in the central parsec may be isolated black holes and neutron stars accreting gas from the Minispiral. Such sources should be preferentially located close to the Minispiral if viscous time is short (ζ is small). In addition, binary systems containing a black hole and a normal low mass star can also accrete gas in roughly the same way as we calculated here. In the case of low values of circularisation parameter, ζ, the size of the disc around the primary (the black hole) can be smaller than the size of the binary itself. Thus, these systems may appear as "fake X-ray binaries", where the gas supply comes from outside rather than from the low mass secondary. Observational signatures of such systems might be warped and out of binary plane accretion discs, "too short" or "too weak" accretion outbursts for the size of the binary.

References

[1] Baganoff, F. K., et al.: ApJ, 591, 891 (2003)
[2] Bahcall, J. N., Wolf, R. A.: ApJ, 209, 214 (1976)
[3] Bondi, H., Hoyle, F.: MNRAS, 104, 273 (1944)
[4] Freitag, M., et al.: These Proceedings
[5] Miralda-Escudé, J., Gould, A.: ApJ, 545, 847 (2000)
[6] Morris, M.: ApJ, 408, 496 (1993)
[7] Muno, M. P., et al.: ApJ, 622, L113 (2005)
[8] Nayakshin, S., to appear in MNRAS (astro-ph/0512255)
[9] Nayakshin, S., Sunyaev, R.: MNRAS, 364, L23 (2005)
[10] Paumard, T., et al.: A&A, 426, 81 (2004)
[11] Paumard, T., et al.: ApJ, 643, 1011 (2006)
[12] Proga, D., Begelman, M. C.: ApJ, 582, 69 (2003)

Institute of Physics Publishing Journal of Physics: Conference Series **54** (2006) 311–315
doi:10.1088/1742-6596/54/1/049 Galaxy Center Workshop

Thin filaments at the Galactic Center: identification and proper motions

K Mužić[1], A Eckart[1,2], R Schödel[1], L Meyer[1] and A Zensus[2,1]

[1] I. Physikalishes Institut, Universität zu Köln, Zülpicher Str. 77, 50937 Köln, Germany
[2] Max-Planck-Institut für Radioastronomie, Auf dem Hügel 69, 53121 Bonn, Germany

E-mail: muzic@ph1.uni-koeln.de

Abstract. L'-band (3.8 μm) images of the Galactic Center show a large number of thin filaments in the mini-spiral, located close to the mini-cavity, along the inner edge of the northern arm and in the vicinity of some stars. We interpret them as shock fronts formed by the interaction of a central wind with the mini-spiral or, in some cases, extended dusty stellar envelopes. The observations have been carried out using the NACO adaptive optics system at the ESO VLT, in 5 subsequent epochs from 2002 to 2006. We present a proper motion study of the thin filaments observed in the central parsec around Sgr A*, obtained using the cross-correlation technique. Our interpretation is consistent with a collimated outflow model from the central few arcseconds. Two possible mechanisms could produce the postulated outflow: stellar winds originating from the high-mass-loosing He-star cluster as well as a wind from Sgr A* due to accretion from the surrounding disk of stars.

1. Introduction

The central cavity of the Galactic Center is surrounded by a dense clumpy molecular ring (also called circum nuclear disk, CND) of warm dust (Zylka et al. 1995) and neutral gas (Güsten et al. 1987; Jackson et al. 1993; Wright et al. 2001; Herrnstein & Ho 2002). The central cavity itself has much lower mean gas density and contains the so called mini-spiral, which consists of mostly ionized gas and dust, and connects the CND to the center of the stellar cluster. The three-dimensional shape of mini-spiral is studied in detail by Vollmer & Duschl (2000) and Paumard et al. (2004).

A cluster of massive stars resides in the central parsec of the Galaxy (Blum et al. 1995; Krabbe et al. 1995; Genzel et al. 1996, 2000; Eckart et al. 1999; Clénet et al. 2001; Paumard et al. 2006). Within that cluster the 7 most luminous ($L>10^{5.75}$ $L_\odot$), moderately hot ($T<10^{4.5}$ K) blue super giants contribute half of the ionizing luminosity of the region (Najarro et al. 1997; Krabbe et al. 1995; Blum et al. 1995). These stars, mostly HeI emission line stars, undergo high mass loss through strong stellar winds, the total mass loss rate being $\sim 10^{-3} M_\odot$ yr^{-1} (Najarro et al. 1997). There is strong evidence for the fact that most of the young and hot stars orbit the very center of the Milky Way in at least one disk like structure well within the central stellar cluster (Genzel et al. 2000, 2003; Levin & Beloborodov 2003; Beloborodov et al. 2006; Paumard et al. 2006).

Adaptive optics L'-band (3.8μm; Fig. 1) images of the central parsec of the Milky Way show large amounts of gas and dust that belong to the mini-spiral. Additionally, one can distinguish between a large number of thin filaments (some have already been reported by Clénet et al.

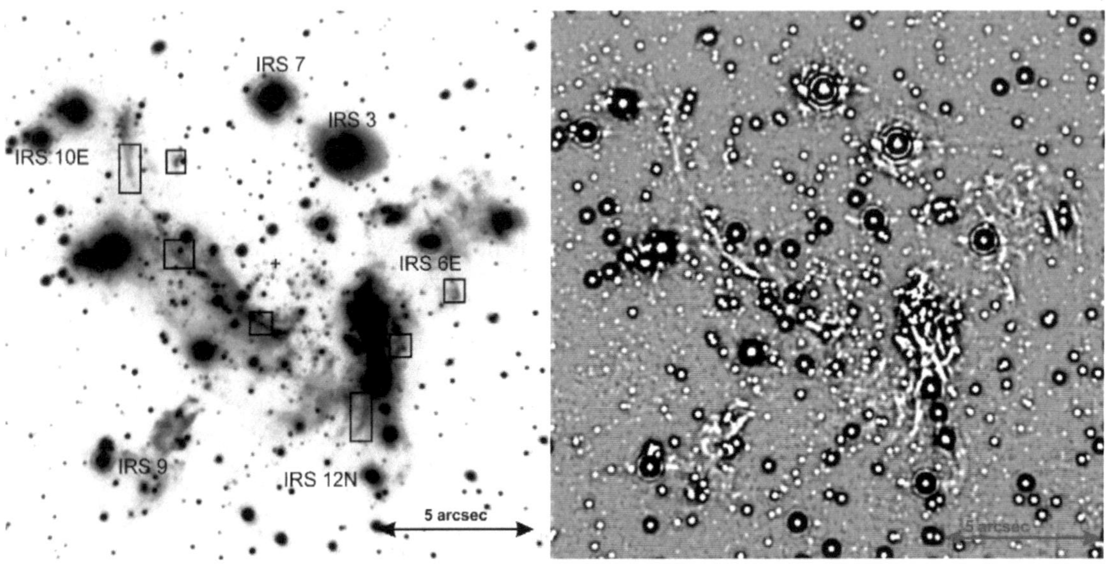

Figure 1. Left: L'-band image of the central part of the Milky Way ($\sim 17 \times 17$ arcsec^2). The image contains a large number of dust embedded stellar sources, and shows the emission from the mini-spiral and a number thin filaments that are associated with this emission. The boxes mark thin filaments with measurable proper motions. Right: High-pass filtered version of the image shown on the left. Here all compact features are enhanced. The cross in the left image marks the position of Sgr A*.

2004) Here, we identify them and present proper motion measurements. We argue that they are most likely shock fronts formed by the interaction between an outflow from the central arcseconds with the ambient ISM.

2. Observations and Data Reduction

The observations were performed using the NAOS/CONICA adaptive optics assisted imager/spectrometer (Lenzen et al. 1998; Rousset et al. 1998; Brandner et al. 2002) at the UT4 (Yepun) at the ESO VLT. The data set includes L'-band images from 5 epochs (2002.660, 2003.356, 2004.320, 2005.364 and 2006.408). Data reduction and formation of final mosaics were performed using the DPUSER software for astronomical image analysis (T. Ott; see also Eckart & Duhoux 1990). The images were transformed to a common coordinate system (that of the epoch 2003.356), using IDL image transformation routines. The transformation was was performed to second order and provides correction of all translations, rotations and possible warps between two images.

In order to highlight the above mentioned filaments, we produced high-pass filtered maps (Fig. 1, right) by subtracting a smoothed version of the images from themselves and smoothing the final images. As a smoothing function we used a 4 pixel Gaussian (the images have a pixel scale of 27 mas/pixel and a resolution of $\sim$100 mas). This procedure enhances all structures that have significant power at spatial frequencies that correspond to the diffraction limited beam size.

3. Proper motions of the thin filaments

3.1. Identification and description of the thin filaments

Fig. 2 shows a schematic representation of the thin filaments observed in the central parsec of the Milky Way. The features located west and south from Sgr A* are denominated SW, those

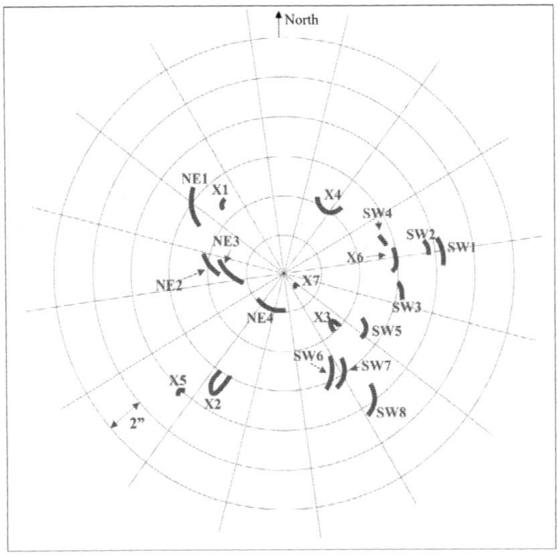

Figure 2. Identification of the thin features observed in the the L'-band. Sgr A* is located at the center of the image.

associated with the northern arm and located east from Sgr A* NE, and the features most probably associated with stars are marked with X. The shocks labeled NE are located along the inner rim of the northern arm and are curved with their convex sides eastwards. Shocks to the west and southwest of Sgr A* are curved with their convex sides westward and are positioned almost perpendicular to the Bar. Exceptions are X3, X4 and X7. X4 is associated with IRS 3, whose bow-shock-like appearance leads to a picture in which the extended dust shell of the mass-losing star is being pushed northwest by a wind from the IRS 16 cluster or Sgr A* (Viehmann et al. 2005). Additionally, IRS 3 has a proper motion component to the southeast, approximately into the direction in which we observe the shock. X3 and X7 are located along the line to Sgr A* and could be due to the postulated central outflow (see below). Features X1, X5 and possibly X6 appear to be associated with individual stars and are probably formed by the interaction between a stellar wind, probably in combination with the stellar motion, and the interstellar medium. X1 is directly associated with a star, both moving in the same direction on the sky, with the comparable plane of the sky velocity. For X5 no significant proper motion was detected, but taking into account the significantly higher brightness of this feature with respect to the other filamentary structures, we suppose that there is an embedded or background star present that cannot be separated from the emission of the bow-shock. On the west side of the mini-cavity, at the location of the IRS13/IRS2 complex or just south of it, we can identify several luminous shocks that cannot be associated with stars in a straightforward way (SW5, SW6, SW7, SW8).

3.2. Results

In order to determine the relative offsets between the positions of the filaments in different epochs, the cross-correlation technique was used. All offsets were calculated with respect to the chosen reference epoch (2003.356). Uncertainties in the proper motions are due to the linear regression uncertainty as well as to the uncertainties of the image transformation and the cross-correlation routine. Table 1 summarizes the proper motions of thin filaments obtained in our study, assuming a distance to Sgr A* of 8 kpc. The velocities of all features, except that of X1 and SW7, are measured approximately perpendicularly to the feature of interest. This results in minimizing the error since many of the filaments are parts of more extended features that could not be completely extracted (e.g. because of the proximity to stellar sources), which basically means that the motion along the feature does not result in a measurable proper motion. Feature X1 is clearly a stellar bow shock, caused by motion of the star (offset from Sgr A* by 2.95″ in

Table 1. Proper motions of thin filaments

Name	$\Delta\alpha$ (arcsec)	$\Delta\delta$ (arcsec)	$size(\Delta x \times \Delta y)$ (arcsec× arcsec)	v (km/s)	Δv (km/s)
NE1	4.76	3.42	0.32× 0.46	-29	16
NE3	2.62	-0.09	0.38× 0.41	167	25
NE4	0.32	-1.95	0.43× 0.43	130	18
SW3	-6.08	-0.77	0.35× 0.35	-229	19
SW5a	-4.20	-2.31	0.35× 0.11	-165	15
SW5b	-4.13	-2.75	0.16× 0.24	-123	22
SW7(R.A.)	-3.07	-5.10	0.73× 1.22	-186	17
SW7(Dec.)	-3.07	-5.10	0.73× 1.22	-80	21
X1(R.A.)	3.22	3.40	0.30×0.38	113	16
X1(Dec)	3.22	3.40	0.30× 0.38	-78	21

Table 2. Positions in right ascension and declination are relative to Sgr A *refer to the center of the box used to extract the filament. All velocities in the table are measured in direction perpendicular to the shock feature except for X1 and SW7. The proper motions of X1 and SW7 are given in R.A. and Dec

R.A. and 3.46″ in Dec) into the medium of the mini-spiral. The proper motion of the bow shock calculated in this work is in very good agreement with the proper motion of the associated star.

3.3. Discussion

There is strong evidence for the existence of at least one stellar disk at the Galactic center. 11 He emission line stars are located within a plane of this (clockwise-rotating) disk, most of them within a radius of 0.1 pc (2.5 arcsec) from Sgr A* (Genzel et al. 2000, 2003; Levin & Beloborodov 2003, Paumard et al. 2006). Strong stellar winds from the He stars cause high mass-loss rate: in total they lose $\sim 10^{-3} M_\odot\ yr^{-1}$ (Najarro et al. 1997). Estimates of the rate at which the material is accreted onto the SMBH at the position of Sgr A* vary depending on the model used (see e.g. Quataert 2003, Baganoff et al. 2003, Narayan et al. 1998, Blandford & Begelman 1999, Yuan et al. 2002), but it is important to note that all of them deal just with a small fraction (less than 1%) of the material available from the high mass-losing stars. This means that, independent of the exact processes that happen on the scale of the Bondi radius, most of the material supplied by the stars never gets close to the BH, i.e. it is blown from the center and therefore can interact with the material of the mini-spiral.

H_2(1-0) S(1) molecular hydrogen line emission from the CND is concentrated on the northern and southern lobes (Gatley et al. 1984, 1986; Burton & Allen 1992; Yusef-Zadeh et al. 2001). The extinction-corrected peak line intensity from the lobes is about a factor 4 higher than the mean surface brightness of the CND (Yusef-Zadeh et al. 2001). The source of excitation of the H_2 emission associated with the CND is unclear. Although the bright appearance of the lobes is usually interpreted as due to the limb brightening of the 0.5 pc thick CND torus, inclined $\sim 70^o$ to the line of sight (Jackson et al. 1993), the limited size of the lobes and their symmetrical position with respect to Sgr A* suggest that the H_2 line emission from the lobes could be dominated by the interaction with a collimated wind.

An alternative source of the outflow could be a jet-like structure directly associated with the black hole at the position of Sgr A*. Then the collimation would have to be seen as a consequence of the yet unexplained accretion process onto the black hole, but could still be linked to the plane the mass-losing He-stars are moving in. The motion within a single plane implies that the viscous friction of gas being blown into that plane is larger than that for gas being blown out perpendicularly to it. On the one hand that results in a pressure gradient perpendicular to the plane, on the other hand it will result in an accretion flow within that plane onto the central black hole and may give rise to a corresponding outflow perpendicular to it.

This scenario would explain the overall geometry shown in Fig.1 and would allow the northern and southern lobes within the CND to be associated with the outflow. Which of the two mechanisms is dominant or whether both mechanisms in combination with the deep central potential provided by the $3.6\times10^6 M_\odot$ black hole represent an efficient collimation process has to be investigated via detailed hydrodynamical model calculations.

References

Baganoff, F. K., et al., ApJ 591, 891 (2003)

Beloborodov, A. M., Levin,Y., Eisenhauer, F., Genzel, R., Paumard, T., Gillessen, S., Ott, T., ApJ in press (2006)

Blandford, R.D., Begelman, M.C., MNRAS 303, L1-L5 (1999)

Blum, R. D., DePoy, D. L., Sellgren, K., ApJ 441, 603 (1995)

Brandner, et al. Messenger 107, 1 (2002)

Burton, M., Allen, D., PASA 10, 55 (1992)

Clénet, Y., Rouan, D., Gendron, E., Montri, J., Rigaut, F., Léna, P., Lacombe, F., A&A 376, 124 (2001)

Clénet, et al. A&A 417, L15 (2004)

Eckart, A., Duhoux, P. R. M., ASP Confrence Series, Volume 14, Astrophysics with Infrared Arrays (San Francisco: ASP), ed. by Richard Elston, p. 336 (1990)

Eckart, A., Ott, T., Genzel, R., A&A 352, L22 (1999)

Gatley, I., Jones, T.J., Hyland, A.R., Wade, R., Geballe, T.R., MNRAS 222, 299 (1986)

Gatley, I., Beattie, D.H., Lee, T.J., Jones, T.J., Hyland, A.R. MNRAS 210, 565 (1984)

Genzel, R., Thatte, N., Krabbe, A., Kroker, H., Tacconi-Garman, L. E., ApJ 472, 153 (1996)

Genzel, R., Pichon, C., Eckart, A., Gerhard, O. E., Ott, T., MNRAS 317, 348 (2000)

Genzel, et al., ApJ 594, 812 (2003)

Güsten, R., Genzel, R., Wright, M. C. H., Jaffe, D. T., Stutzki, J., Harris, A. I., ApJ 318, 124 (1987)

Herrnstein, R.M.; Ho, P.T.P., ApJL 579, L83 (2002)

Jackson, J.M., Geis, N., Genzel, R., Harris, A.I., Madden, S., Poglitsch, A., Stacey, G.J., Townes, C.H., ApJ 402, 173 (1993)

Krabbe, A., et al., AJ 447, L95 (1995)

Levin, Y., Beloborodov, A.M., ApJL 590, L33 (2003)

Lenzen, R., Hofmann, R., Bizenberger, P., Tusche, A., Proc. SPIE Vol. 3354, 606 Infrared Astronomical Instrumentation, ed. Albert M. Fowler (1998)

Najarro, F., Krabbe, A., Genzel, R., Lutz, D., Kudritzki, R. P., Hillier, D. L., A&A 325, 700(1997)

Narayan, R., Mahadevan, R., Grindlay, J. E., Popham, R. G., Gammie, C., ApJ 492, 554 (1998)

Paumard, T., Maillard, J.-P. & Morris, M., A&A 426, 81 (2004)

Paumard, T., et al., ApJ 643, 1011 (2006)

Quataert, E., Astron. Nachr. 324, No. S1, 3-11 (2003)

Rousset, G., et al. Proc. SPIE Vol. 3353, 516, Adaptive Optical System Technologies, ed. Domenico Bonaccini & Robert K. Tyson (1998)

Viehmann, T., Eckart, A.,Schödel, R., Moultaka, J., Straubmeier, C., Pott, J.-U., A&A 433, 117 (2005)

Vollmer, B., Duschl, W.J., New Astronomy 4, 581 (2000)

Wright, M.C.H., Coil, A.L., McGary, R.S., Ho, P.T.P., Harris, A.I., ApJ 551, 254 (2001)

Yuan, F., Markoff, S., Falcke, H., A&A 383, 854 (2002)

Yusef-Zadeh, F., Stolovy, S. R., Burton, M., Wardle, M., Ashley, M. C. B., ApJ 560, 749, (2001)

Zylka, R., Mezger, P.G., Ward-Thompson, D., Duschl, W. J., Lesch, H., A&A 297, 83 (1995)

Institute of Physics Publishing Journal of Physics: Conference Series **54** (2006) 316–320
doi:10.1088/1742-6596/54/1/050 Galaxy Center Workshop

Photometric Stellar Variability in the Galactic Center

M. Rafelski[1], A. M. Ghez[1,2], S. D. Hornstein[1], J. R. Lu[1], M. Morris[1]

[1] Division of Astronomy and Astrophysics, UCLA, Los Angeles, CA 90095-1547

[2] Institute of Geophysics and Planetary Physics, University of California, Los Angeles, CA 90095-1565

E-mail: `marcar, ghez, seth, jlu, morris@astro.ucla.edu`

Abstract. We report the results of a diffraction-limited, photometric variability study of the central 5″ × 5″ of the Galaxy conducted over the past 10 years using speckle imaging techniques on the W.M. Keck I 10 m telescope. Within our limiting magnitude of $m_K < 16$ for maps made from a single night of data, we find a minimum of 25 variable stars out of 131 monitored stars. Among 46 stars brighter than $m_K < 14$ which have roughly uniform photometric uncertainties, there are 16 variable stars. This suggests a minimum variable star frequency of 34%. We see no evidence of flares or dimming of the 7 stars that have known 3-dimensional orbits in our study, which greatly limits the possibility of a cold, geometrically-thin, inactive accretion disk around the supermassive black hole, Sgr A*. While large populations of binaries have been posited to exist in this region both to explain the presence of young stars in the vicinity of a black hole and because of the high stellar densities, only one eclipsing binary is identified. The only periodic source in our sample is the previously identified variable IRS 16SW (P=19.448 ± 0.002 days). In contrast to recent results, our data show an asymmetric phased light curve with a much steeper fall-time than rise-time. IRS 29N shows variability on time scales of ≈ 5 years and has a known spectral type of WC9. This variation is likely due to episodic dust production, which may suggest that this source is a binary star system. Only 2 of the LBV candidates in our sample (16NW, 16SW) show variability and none of the 4 show the characteristic large increase or decrease in luminosity. However, our time baseline is too short to rule them out as LBVs. Our study has shown that photometric variability provides a useful handle on the unusual massive star population surrounding our Galaxy's supermassive black hole and its local environment.

1. Introduction

The cluster of stars at the Galactic center (GC) presents a unique opportunity to study the evolution and properties of stars within the influence of the $\sim 3.7 \times 10^6 M_\odot$ supermassive black hole (SMBH), Sgr A* [1, 2]. These stars are composed of a mix of older giants and young massive stars in transient phases of evolution whose variability characteristics and mechanisms are not well understood. Some of the massive young stars have been spectroscopically identified as blue supergiants (O & B spectral type), luminous blue variables (LBVs), and Wolf-Rayet (WR) stars [3], while some of the older population is composed of asymptotic giant branch (AGB) and cool supergiant stars [4, 5]. The LBVs, WR, and AGB stars are expected to show photometric variability. Determining which stars are variable and the characterization of their time variability can improve our understanding of the circumnuclear material of the SMBH, the rare transitional stars associated with regions of star formation, and the Galactic center stars.

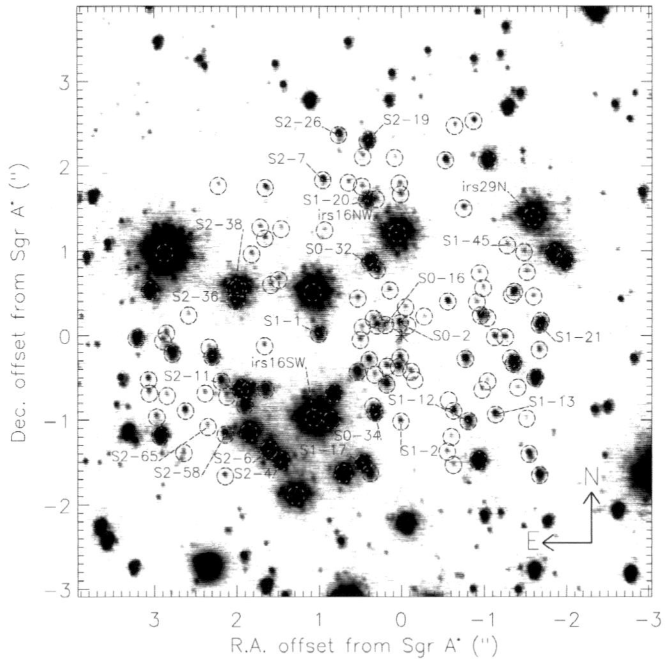

Figure 1. Identification of sources from this study overlaid on a 6″ x 6″ region of an LGS K′-band (λ_o = 2.1μm) image from Ghez et al. taken on June 30, 2005 [7]. All 131 sources are circled, but only variable sources are labeled. The location of Sgr A* is marked with an asterisk.

2. Identifying Variable Stars

Over the last 10 years, K-band ($\lambda_o = 2.2\ \mu m$, $\Delta\lambda$=0.4 μm) speckle imaging observations within the central 5″ × 5″ of the Galaxy's central stellar cluster were obtained with the W. M. Keck I 10 m telescope. For this study, data from each night of observations were combined to produce 50 diffraction limited images. StarFinder [8] was run on all the images to extract the relevant photometric information.

There are a wide range of methods for testing photometric variability and the challenge for these various approaches is to avoid false identifications on the basis of a few outlying data points [6]. We therefore have chosen to use the Kolmogorov-Smirnov (K-S) test, to calculate the probability that a *distribution* of data points is consistent with a model of a distribution of measurements for a non-variable source. This approach is less sensitive to outlying data points than the commonly used χ^2 test. Stars are classified as variable if they lie more than 3σ from a gaussian distribution of a non variable.

Among the 131 stars in our sample, 25 are identified as photometric variables in this study (see Figure 1). Since the relative photometric uncertainties are roughly uniform down to a m_K magnitude of $\sim$ 14 and then increase at fainter magnitudes, we limit our statistical study to those stars brighter than 14 mag. Within this brighter sample of 44 stars, there are 15 variable stars, suggesting a minimum frequency of variable stars of 34%. There is no evidence for radial, magnitude, or type dependence of the variable stars.

3. Variability of stars near closest approach

The photometry of stars near Sgr A* constrain the properties of a possible cold geometrically-thin inactive accretion disk around Sgr A* [9, 10]. In the presence of such a disk we would expect to see nearby stars significantly flaring in the near infrared NIR at times and eclipsed at others due to their interaction with the material in the disk. The time-scales vary based on the geometry of the disk, but are on the order of a year for the eclipses and months for flares. Due to the high stellar densities and the flaring of Sgr A* in the infrared [11, 12], the photometry of the central sources sometimes vary due to stellar confusion with each other or with Sgr A* during their closest approach. Observations averaging multiple nights of data also show additional stars close enough to cause confusion with some of the stars identified in this survey [1].

An optically thin disk may not fully eclipse stars and our gaps in observations would allow different geometries of the disk to account for any one star not showing eclipses or flares in the light curves as was done for S0-2 [12]. We have not observed any flares in the 7 central arcsecond sources that have known 3-dimensional orbits (S0-2, S0-16, S0-19, S0-20, S0-1, S0-4, S0-5). Also, no noticeable eclipses are detected in any of these sources, although three of these (S0-16, S0-19, S0-20) can not be used to constrain the flares since they are all fainter than the others and have missing measurements due to insufficient map sensitivity (see Figure 2). S0-2 and S0-16 are still considered variable stars according to their K-S probability after discarding points affected by well understood confusion, however we attribute this low level variation to confusion with faint unrecognized stars. While optically thin disks may not fully eclipse stars and flares can be missed between observations, the lack of such effects indicate that such a disk probably does not exist around Sgr A*.

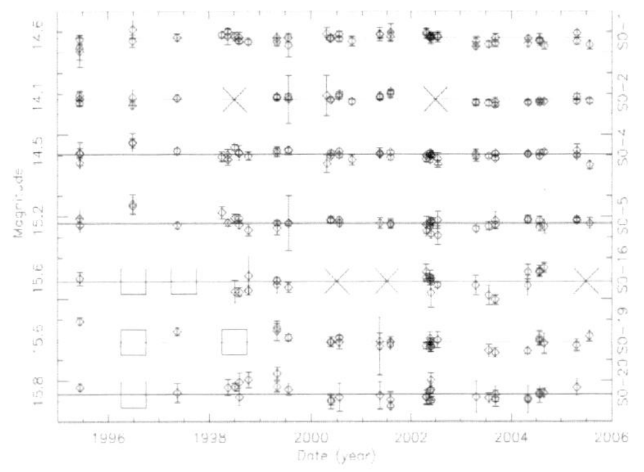

Figure 2. Light curves of 7 central arcsecond sources that have known 3-dimensional orbits. Large tick-marks on y-axis designate 1 magnitude in range. The X's mark observations that are rejected due to confusion with other sources, and the boxes are areas where the nightly maps have missing measurements that are due to insufficient map sensitivity and are detected in monthly averaged images.

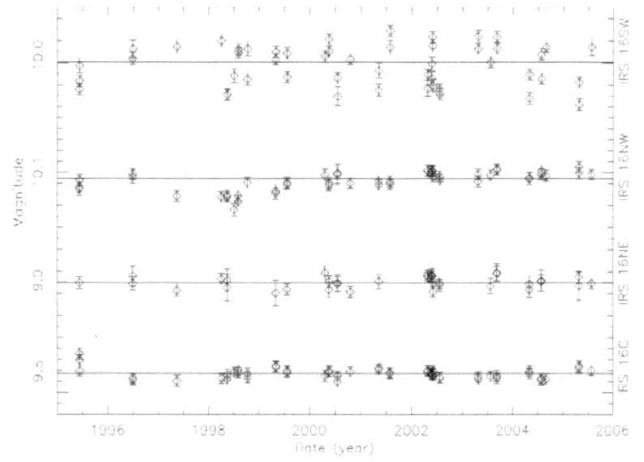

Figure 3. Light curves of the four stars in our sample that are classified as candidate LBVs (IRS 16SW, 16NW, 16NE, 16C). Large tick-marks on y-axis designate 1 magnitude in range. Lines indicate the median magnitude.

4. Luminous Blue Variables

Luminous Blue Variables (LBVs) are rare high mass stars in a transient phase of evolution. Massive O stars go through the LBV phase before becoming Wolf-Rayet stars [13]. During this phase, they are well known for giant eruptions and obscuration events [14]. The presence of LBVs along with Wolf-Rayet stars in the Galactic center would suggest two distinct star formation events occurred due to their differing evolutionary time-scales. Our sample includes 4 of the 23 candidate LBV stars in our Galaxy (IRS 16NE, 16C, 16NW, 16SW) [15, 16]. These 4 stars are cited as possible LBVs based on IRS 34W's (a similar nearby star) categorization as an LBV, their brighter luminosity, and narrower emission lines [15]. If these stars are LBVs, we expect to see eruptions of $\Delta M_V \simeq 1 - 2$ mag occurring on time scales of 10-40 years with minima and maxima that may last several years and have smaller variations superimposed [14].

Two of these stars (16NW and 16SW) show variability (see Figure 3), but these variations are not characteristic of LBV variations; 16NW is an Ofpe/WN9 star with a flat light curve and a decrease in brightness from 1997 to 1999 of $\Delta m_K \simeq 0.2$ and 16SW has periodic variability (see §5). The apparent dimming in 16NW can be explained by ejected circumstellar material obscuring the star, with the smaller amplitude representing a smaller ejection than an LBV, and does not require it to be an LBV. In the case of 16SW, it has been suggested that it might not be an LBV due to its periodic nature, although its spectrum resembles that of the other IRS 16 cluster sources that are Ofpe/WN9 stars [15, 3]. If 16SW consists of two eclipsing LBVs then they show no significant variation evidenced by the smooth phased light curve. The other two LBV candidates(16NE and 16C) are non-variable over a ten-year time frame. LBVs may look like Of/WN9 stars during a quiescent state [14], so these stars may be such LBVs. While none of these stars show the classic characteristics of LBVs, our time baseline is too short to rule them out completely as LBVs.

5. Periodic Variables

IRS 16SW is the only star in our sample with a detectable period. We find a period of 19.448 ± 0.002 days, which is consistent with other studies' second harmonic for this star [17, 18]. As noted by Depoy et al., if the star is an eclipsing binary with two equal mass systems then the actual period would be the second harmonic as the star would have equal primary and secondary minima [18]. Figure 4 shows the phased light curve of IRS 16SW at 19.448 days. The light curve is asymmetric with a shorter fall-time than rise-time, a short minimum, and a possible period of constancy at maximum brightness. Our light curve is similar to the light curve presented by Ott et al., although asymmetry was not reported there [17]. This is in contrast to Depoy et al. who see an almost symmetric light curve with a slightly shorter rise-time than fall-time and no evidence of any part of the light curve remaining constant [18]. The most recent results presented in these proceedings show a period of 19.4 days in the radial velocity data, which matches our photometric period and suggests that 16SW is an eclipsing binary [19]. We suggest that the asymmetry may be due to hot spots on the stars.

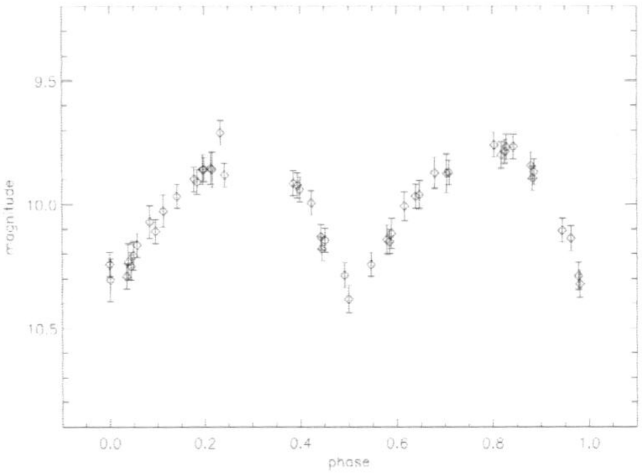

Figure 4. The phased light curve of IRS 16SW at its period of 19.448 ± 0.002 days.

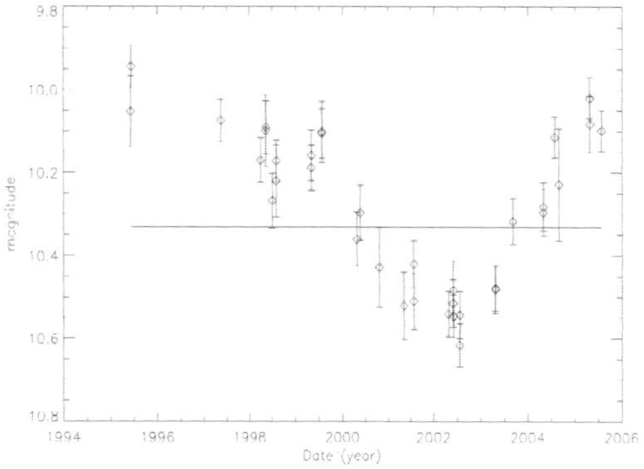

Figure 5. Light curve of IRS 29N.

6. Wolf-Rayet type WC stars

Wolf-Rayet stars of spectral type WC9 are dust producers although the mechanism for this production is still not well understood due to the hostile environment of the stellar winds at the dust condensation radius [20, 21]. One mechanism for dust formation is wind colliding binaries (WCB), and all seven observed (prior to this study) WC stars that variably produce dust are believed to be WCBs [22, 23]. When the dust forms, the stars exhibit enhancements in their infrared fluxes followed by fading emission when dust formation stops and is dispersed by the stellar winds [22]. Light curves of these variable dust producers are presented by van der Hucht in L′-band ($\lambda_o = 3.8\mu m$) [24] and are similar to the light curve of IRS 29N (see Figure 5). IRS 29N is spectroscopically identified as a WC9 star [3], while we show it photometrically varying with a gradual drop and then rise in brightness of $\Delta m_K \simeq 0.7$ over a time scale of ≈ 5 years. We propose that IRS 29N is a periodic or episodic dust producer and probably a WCB based on its light curve and spectral classification.

Acknowledgments

The W. M. Keck Observatory is operated as a scientific partnership among the California Institute of Technology, the University of California and the National Aeronautics and Space Administration. The Observatory was made possible by the generous financial support of the W. M. Keck Foundation. Support for this work was provided by NSF grant AST-0406816.

References

[1] Ghez, A. M., Salim, S., Hornstein, S. D., Tanner, A., Lu, J., Morris, M., Becklin, E. E., Duchêne, G. 2005, *ApJ*, **620**, 744
[2] Schödel, R., Ott, T., Genzel, R., Eckart, A., Mouawad, N., & Alexander, T. 2003, *ApJ*, **596**, 1015
[3] Paumard, T., et al. 2006, *ApJ*, **643**, 1011
[4] Blum, R. D., Ramírez, S. V., Sellgren, K., & Olsen, K. 2003, *ApJ*, **597**, 323
[5] Figer, D.F. et al. 2003, *ApJ*, **599**, 1139
[6] Welch, D.L., & Stetson, P.B. 1993, *AJ*, **105**, 1813
[7] Ghez, A. M., et al. 2006 (in prep)
[8] Diolaiti, E., Bendinelli, O., Bonaccini, D., Close, L., Currie, D., & Parmeggiani, G. 2000, *A&AS*, **147**, 335
[9] Nayakshin, S., & Sunyaev, R. 2003, *MNRAS*, **343**, L15
[10] Cuadra, J., Nayakshin, S., & Sunyaev, R. 2003, *A&A*, **411**, 405
[11] Genzel, R., Schödel, R., Ott, T., Eckart, A., Alexander, T., Lacombe, F., Rouan, D., & Aschenbach, B. 2003, *Nature*, **425**, 934
[12] Ghez, A.M., Wright, S.A., Matthews, K., Thompson, D., Le Mignant, D., Tanner, A., Hornstein, S.D., Morris, M., Becklin, & Soifer, B. T. 2004, *ApJ*, **601**, L159
[13] Massey, P. 2003, *ARA&A*, **41**, 15
[14] Humphreys, R. M., & Davidson, K. 1994, *PASP*, **106**, 1025
[15] Clark, J. S., Larionov, V. M., & Arkharov, A. 2005, *A&A*, **435**, 239
[16] Paumard, T., Genzel, R., Maillard, J. P., Ott, T., Morris, M. R., Eisenhauer, F., & Abuter, R. 2004, *Proc. of XXXIXth Rencontres de Moriond*, Eds: A. Chalabaev, T. Fukui, et al., (Paris: Frontieres), p 377
[17] Ott, T., Eckart, A., & Genzel, R. 1999, *AJ*, **523**, 248
[18] DePoy, D. L., Pepper, J., Pogge, R. W., Stutz, A, Pinsonneault, M., & Sellgren, K. 2004, *ApJ* **617**, 1127
[19] Ott, T. et al. 2006, these proceedings
[20] Williams, P. M., van der Hucht, K. A., & Rauw, G. 2005, *Massive Stars and High-Energy Emission in OB Associations*, p 65
[21] Williams, P. M., van der Hucht, K. A., & The, P. S. 1987, *A&A*, **182**, 91
[22] Williams, P. M., & van der Hucht, K. A. 2000, *MNRAS*, **314**, 23
[23] Lefèvre, L., et al. 2005, *MNRAS*, **360**, 141
[24] van der Hucht, K. A., Williams, P. M., & Morris, P. W. 2001, *ESA SP-460: The Promise of the Herschel Space Observatory*, p 273

Institute of Physics Publishing
doi:10.1088/1742-6596/54/1/051
Journal of Physics: Conference Series **54** (2006) 321–327
Galaxy Center Workshop

Resonant Relaxation near the Massive Black Hole in the Galactic Center

Clovis Hopman and Tal Alexander
Faculty of Physics, Weizmann Institute of Science, POB 26, Rehovot 76100, Israel
E-mail: clovis.hopman@weizmann.ac.il

Abstract. The coherent torques between stars on orbits near massive black holes (MBHs) lead to resonant angular momentum relaxation. Due to the fact that orbits are Keplerian to good approximation, the torques efficiently change the magnitude of the angular momenta and rotate the orbital inclinations. As a result the stars are rapidly randomized. The galactic MBH is a good system for the observational study of resonant relaxation. The age of the young B-stars at a distance of ~ 0.01 pc from the MBH is comparable to the resonant relaxation time, implying that resonant relaxation may have played an important role in their dynamical structure. In contrast, the O-stars in the stellar disks at ~ 0.1 pc are younger than the resonant relaxation time, as required by their dynamical coherence. Resonant relaxation dynamics dominates the event rate of gravitational wave (GW) emission from inspiraling stars into MBHs of masses comparable to the Galactic MBH. Resonant relaxation leads to rates $\lesssim 10$ times higher than those predicted by 2-body relaxation, which would improve the prospects of detecting these events by future GW detectors, such as *LISA*.

1. Introduction

In "collisional" stellar dynamics the potential in which a star moves is considered to be smooth to first order, and the fact that the potential is in fact made out of discrete stars is treated as a perturbation (e.g. Chandrasekhar [11]; Binney & Tremaine [10]). Stars have orbital parameters such as angular momentum and energy that are conserved in the smooth potential, and these parameters indeed remain constant for many dynamical times t_d. Only the weak encounters allow for changes in these quantities. For example, two stars which interact with each other, can exchange energy and angular momentum, so that after their encounters their orbits are described by slightly different quantities. In this way not only the orbits of individual stars are modified, but the distribution function (DF) of the entire system can change, and evolve towards a steady state. The time-scale over which a system evolves is the *relaxation time* t_r. In most systems $t_d \ll t_r$, and relaxation can indeed be treated as a second order effect.

Analyses of the evolution of the DF near a MBH have almost exclusively relied on the assumption that the mechanism through which stars exchange angular momentum and energy is dominated by *uncorrelated two-body interactions*[1]. Any encounter is assumed to be unrelated to previous and future encounters, and changes in energy and angular momenta are considered to be drawn from a specified random distribution. Relaxation can therefore in a meaningful way

[1] N-body simulations form an exception; for N-body simulations near MBHs see e.g. Baumgardt, Makino & Ebisuzaki [7], [8]; Preto, Merritt & Spurzem [44]; Merritt & Szell [37]

be considered to be a random walk process. In the context of stellar dynamics near MBHs, the assumption of uncorrelated encounters is made in Fokker-Planck models (e.g. Bahcall & Wolf [5]; Bahcall & Wolf [6]; Cohn & Kulsrud [12]; Murphy, Cohn & Durisen [39]), where the microscopic interactions are expressed by the diffusion coefficients, and in Monte Carlo simulations (e.g. Shapiro & Marchant [49]; Marchant & Shapiro [35], [36]; Freitag & Benz [15], [16]). Stars around MBHs are described as moving in the smooth average potential of the MBH and the stars, and the scattering by the fluctuating part of the potential is modeled as a hyperbolic Keplerian interaction between a passing star and a test star.

The (non-resonant) relaxation time $T_{\rm NR}$ can be defined as the time it takes for the energy $\mathcal{E}$ of a typical star to change by order unity. This is also the time it takes for its specific angular momentum J to change by an amount of order $J_c(\mathcal{E})$, the maximal angular momentum for that energy. On Keplerian orbits $J_c = \sqrt{GM_\bullet a}$, where a is the semi-major axis. The "non-resonant" relaxation time $T_{\rm NR}$ of stars of mass $M_\star$ can be written in the Keplerian regime as

$$T_{\rm NR} = A_\Lambda \left(\frac{M_\bullet}{M_\star}\right)^2 \frac{P(a)}{N(<a)} \qquad (M_\bullet \gg M_\star)\,, \tag{1}$$

where $P = 2\pi\sqrt{a^3/(GM_\bullet)}$ is the orbital period and A_Λ is a dimensionless constant which includes the Coulomb logarithm. For some stellar systems (like galaxies), $T_{\rm NR}$ is much longer than the age of the system, implying that the system cannot evolve significantly towards steady state by two-body interactions. For other systems, like very dense stellar clusters, the relaxation time can be as small as a few Myr, and such systems may even evaporate within a Hubble time. In our GC, the relaxation time is somewhat smaller than the age of the system, $T_{\rm NR} \sim$ few Gyr (e.g. Alexander [1], [4]), implying that the system has evolved considerably, and two-body relaxation effects such as mass-segregation have occurred (Bahcall & Wolf [6]; Freitag, Amaro-Seoane & Kalogera [19]; Hopman & Alexander [28]). At the same time, the relaxation time is much longer than some other relevant times in the GC, in particular the age of the youngest stars.

The assumption of uncorrelated two-body interactions is well-justified in many systems, such as globular clusters, where stellar orbits are not closed. However, the special symmetry of a Keplerian potential leads to closed, elliptical orbits. The fact that the orbits are closed can be exploited in numerical treatment, and also leads to unique dynamical features (see also the contribution of Touma in this volume). Since $t_r \gg t_d$, the orbits remain closed for many dynamical times, the system may be thought of as a set of "wires" with the mass of the star smeared out over the orbits. In this picture, it is the wires that interact and cause the evolution of the system, rather than point particles interacting at given locations. The idea is reminiscent of the Kozai mechanism in triple stars (Kozai [30]).

Rauch & Tremaine [45] first used this approach in the context of many body stellar dynamics near MBHs, and coined the term *resonant relaxation* (RR), after the $1:1$ resonance between the radial and azimuthal frequencies in a Keplerian potential. The wire approximation is only relevant for times $\ll t_\omega$, where t_ω is the time for the orbit to precess. Precession may be caused by the fact that the potential is not entirely determined by a point mass, and there is still some extended component to the potential due to the stellar mass; this is especially the case far away ($\gtrsim 0.1\,{\rm pc}$) from the MBH. Closer to the MBH ($\lesssim 0.01\,{\rm pc}$), precession may be dominated by effects of General Relativity[2].

1.1. Scalar resonant relaxation

Scalar relaxation results in changes in *both* the direction and the magnitude of the angular momenta. The RR time $T_{\rm RR}$ is estimated by evaluating ΔJ_ω, the coherent change in the

[2] For the parameters of interest here, Lense-Thirring precession is much less efficient than mass and GR precession even for a maximally spinning MBH, and we do not consider it here.

magnitude of the specific angular momentum up to a time t_ω. The change ΔJ_ω is then the step size for the non-coherent growth of the angular momentum over times $t > t_\omega$. Two nearby stars with semi-major axes a exert a mutual specific torque $\sim GM_\star/a$. To zeroth order the torques of the stars on a test wire cancel, so that within a distance a from the MBH the net torque on a test star is determined by the Poissonian excess torque $\dot{J} \sim \sqrt{N(<a)}GM_\star/a$ and

$$\Delta J_\omega \sim \dot{J} t_\omega = \sqrt{N(<a)}(GM_\star/a)t_\omega \,. \tag{2}$$

For $t > t_\omega$ the torques on a particular star-wire become random, and the change in angular momentum grows in a random walk fashion with a timescale $T_{\rm RR} \sim (J_c/\Delta J_\omega)^2 t_\omega$, defined as

$$T_{\rm RR} \equiv A_{\rm RR} \frac{N(>\mathcal{E})}{\mu^2(>\mathcal{E})} \frac{P^2(\mathcal{E})}{t_\omega} \simeq \frac{A_{\rm RR}}{N(<a)} \left(\frac{M_\bullet}{M_\star}\right)^2 \frac{P^2(a)}{t_\omega} \,, \tag{3}$$

where $\mu \equiv NM_\star/(M_\bullet + NM_\star)$, $A_{\rm RR}$ is a numerical factor of order unity, to be determined by simulations, and the last approximate equality holds in the Keplerian regime.

Over most of the relevant phase space the precession is due to the deviations from pure Keplerian motion caused by the potential of the extended stellar cluster. This occurs on a timescale $t_\omega = t_M = [M_\bullet/N(<a)M_\star]P(a)$, assuming $N(<a)M_\star \ll M_\bullet$. The J-averaged RR timescale can then be written as

$$T_{\rm RR}^M = A_{\rm RR} \frac{M_\bullet}{M_\star} P(a) = \frac{A_{\rm RR}}{A_\Lambda} \frac{M_\star}{M_\bullet} N(<a) T_{\rm NR} . \tag{4}$$

Since $T_{\rm RR}^M \ll T_{\rm NR}$ for small a where $N(<a)M_\star \ll M_\bullet$, the RR rate of angular momentum relaxation is much higher than the rate of energy relaxation in the resonant regime. This qualitative analysis has been verified by detailed numerical N-body simulations by Rauch & Tremaine [45] and by Rauch & Ingalls [46].

For most of parameter space, orbital precession is dominated by the mass of the stellar cluster and the RR timescale is well approximated by $T_{\rm RR} \sim T_{\rm RR}^M$. However, very close to the MBH, or on wide orbits with very low angular momentum, so that the periapse is close to the Schwarzschild radius of the MBH, precession is dominated by GR effects. In this case the timescale for precession is given by $t_\omega = t_{\rm GR} = (8/3)(J/J_{\rm LSO})^2 P$; here $J_{\rm LSO} \equiv (4GM_\bullet/c)$ is the angular momentum of the last stable orbit (LSO). When $t_{\rm GR} \ll t_M$ and GR precession dominates, the RR timescale is (Eq. 4)

$$T_{\rm RR}^{\rm GR} = \frac{3}{8} A_{\rm RR} \left(\frac{M_\bullet}{M_\star}\right)^2 \left(\frac{J_{\rm LSO}}{J}\right)^2 \frac{P(a)}{N(<a)} . \tag{5}$$

Generally, GR precession and mass precession occur simultaneously, and the scalar RR timescale $T_{\rm RR}^s(\mathcal{E}, J)$ is given by substituting $1/t_\omega = |1/t_M - 1/t_{\rm GR}|$ in Eq. (3), where the opposite signs reflect the fact that mass precession is retrograde whereas GR precession is prograde. Thus, the scalar RR timescale is

$$T_{\rm RR}^s = \frac{A_{\rm RR}}{N(<a)} \left(\frac{M_\bullet}{M_\star}\right)^2 P^2(a) \left|\frac{1}{t_M} - \frac{1}{t_{\rm GR}}\right| . \tag{6}$$

We use the relation $\mathrm{d}(J^2)/J_c^2 = \mathrm{d}t/T_{\rm RR}^s(\mathcal{E}, J)$ (Eqs. 6) to define the J-averaged time it takes a star to random-walk from $J = J_c(\mathcal{E})$ to the loss-cone $J = J_{lc}$ as

$$\bar{T}_{\rm RR}^s(\mathcal{E}) = \frac{1}{J_c^2} \int_{J_{lc}^2}^{J_c^2} dJ^2 T_{\rm RR}^s(\mathcal{E}, J) \,. \tag{7}$$

1.2. Vector resonant relaxation

For time scales much larger than the dynamical time, orbits precess and describe a rosette shape. One can then consider the torques between different rosettes rather than between different wires. Since the rosettes describe planar rings to good approximation, they cannot modify the magnitude of the angular momentum of the star, but they can change the direction of the angular momentum vector. This process is known as "vector resonant relaxation" (Rauch & Tremaine [45]). Vector RR grows coherently ($\propto t$) on timescales $t \ll t_\varphi$, where t_φ is the timescale for a change of order unity in the total gravitational potential φ caused by the changes in the stellar potential $\varphi_\star$ due to the realignment of the stars as they rotate by π on their orbit,

$$t_\varphi = \frac{\varphi}{\dot{\varphi}_\star} \simeq \frac{N^{1/2}}{\mu}\frac{P}{2} \simeq \frac{1}{2}\frac{M_\bullet}{M_\star}\frac{P}{N^{1/2}}\,, \tag{8}$$

the last approximate equality holds for $NM_\star \ll M_\bullet$. In analogy to scalar RR (Eq. 2), the maximal coherent change in $\mathbf{J}$ is $|\Delta \mathbf{J}_\varphi| \sim \dot{J} t_\varphi \sim J_c$, that is, $\mathbf{J}$ rotates by an angle $\mathcal{O}(1)$ already at the coherent phase. On timescales $t \gg t_\varphi$, $|\Delta \mathbf{J}_\varphi|$ cannot grow larger, as it already reached its maximal possible value, but the orbital inclination angle is continuously randomized non-coherently ($\propto t^{1/2}$) on the vector RR timescale (Eq. 3),

$$T^v_{\rm RR} = 2A^v_{\rm RR}\frac{N^{1/2}(>\mathcal{E})}{\mu(\mathcal{E})}P(\mathcal{E}) \simeq 2A^v_{\rm RR}\left(\frac{M_\bullet}{M_\star}\right)\frac{P(a)}{N^{1/2}(<a)}, \tag{9}$$

where the last approximate equality holds for $NM_\star \ll M_\bullet$.

It is that while the torques driving scalar and vector resonant relaxation are the same, vector RR is much more efficient than scalar RR, $T^v_{\rm RR} \ll \bar{T}^s_{\rm RR}$, due to the much longer coherence time $t_\varphi \sim N^{1/2} t_M \gg t_M$. Furthermore, vector RR proceeds irrespective of any precession mechanisms that limit the efficiency of scalar resonant relaxation.

2. The origin of the young stellar population in the Galactic center

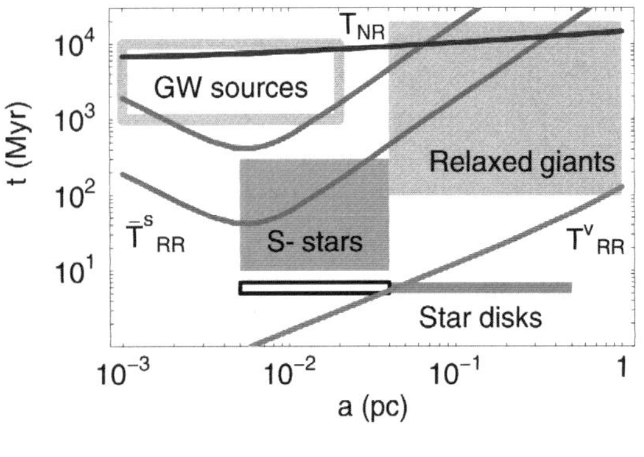

Figure 1. Stellar components, timescales and distance scales in the GC. The NR timescale $T_{\rm NR}$ (top straight line); the timescale $\bar{T}^s_{\rm RR}$, estimated for $1\,M_\odot$ stars (top curved line) and $10\,M_\odot$ stars (bottom curved line); the timescale $T^v_{\rm RR}$ (bottom straight line); the position and estimated age of the young stellar rings in the GC (filled rectangle in the bottom right); the position and age of the S-stars if they were born with the disks (empty rectangle in the bottom left); the position and maximal lifespan of the S-stars (filled rectangle in the middle left). Reprinted with permission from the Astrophysical Journal

Figure (1) compares the distance scales and the ages or lifespans of the various dynamical structures and components in the inner pc of the GC with the relaxation timescales. The NR timescale in the GC, which is roughly independent of radius, is $T_{\rm NR} \sim {\rm few} \times 10^9$ yr (Eq. 1). The scalar RR time $\bar{T}^s_{\rm RR}$ is shown for $M_\star = 1, 10\,M_\odot$. At large radii the RR time decreases towards the center, but for small radii, where GR precession becomes significant, it increases again. The vector RR timescale $T^v_{\rm RR}$, in contrast, decreases unquenched with decreasing radius. Structures with estimated ages exceeding these relaxation timescales must be relaxed.

Two distinct young stellar populations exists in the GC. At distances of 0.04–0.5 pc from the MBH there are about ~ 70 young massive OB stars ($M_\star \gg 10\,M_\odot$, lifespan of $t_\star = 6 \pm 2$ Myr), which are distributed in two nearly perpendicular, tangentially rotating disks (Levin & Belobodorov [32]; Genzel et al. [20]; Paumard et al. [43]). It appears that these stars were formed by the fragmentation of gas disks (Levin & Belobodorov [32]; Levin [31]; Nayakshin & Cuadra [40]; Nayakshin & Sunyaev [41]; Nayakshin [42]). Inside the inner 0.04 pc the population changes. There is no evidence for old stars, and the young stars there (the "S-stars") are main-sequence B-stars ($M_\star \lesssim 15\,M_\odot$, lifespans of $10^7 \lesssim t_\star \lesssim 2\times10^8$ yr; Ghez et al. [21]; Eisenhauer et al. [13]) on randomly oriented orbits with a random (thermal) J-distribution. There is to date no satisfactory explanation for the presence of the S-stars so close to the MBH (see Alexander [4] for a review).

The existence of coherent dynamical structures in the GC constrains the relaxation processes on these distance scales, since the relaxation timescales must be longer than the structure age $t_\star$ to avoid randomizing it. Figure (1) shows that the observed systematic trends in the spatial distribution, age and state of relaxation of the different stellar components of the GC are consistent with, and perhaps even caused by RR. The star disks are young enough to retain their structure up to their inner edge at 0.04 pc, where $t_\star \sim T^v_{\rm RR}$ and vector RR can randomize the disk (Hopman & Alexander [27]). It is tempting to explain the S-stars as originally being the inner part of the same disks that are currently present in the GC. However, this scenario is somewhat problematic. First, we note that vector relaxation can only change the inclinations of the orbits, and not their eccentricities, while many of the S-stars have high ($e > 0.9$) eccentricities; the scalar resonant relaxation time is larger than the age of the disks. Second, resonant relaxation alone cannot explain why the S-stars are systematically less massive than the disk stars. An alternative (Levin [33]) would be that the S-stars were perhaps formed in previous accretion disks of which the dynamical signatures have now disappeared.

If the S-stars were *not* formed in the disk, but captured by either a tidal binary disruption (Gould & Quillen [22]; see also contribution from Perets et al. in this volume) or an exchange interaction with a stellar mass black hole (Alexander & Livio [3]), they may be much older than the disks, and in particular their age may be comparable to the local scalar RR time (see figure 1). In this case, RR will redistribute their orbits within their life-time. This may be an essential element of these formation mechanisms: both scenarios lead to rather eccentric orbits (especially tidal binary disruption), whereas not all the orbits of the S-stars are very eccentric: star S1 has eccentricity $e = 0.358 \pm 0.036$, and S31 has $e = 0.395 \pm 0.032$ (Eisenhauer et al. [13]). Since the age of these stars may well exceed the RR time, RR may have redistributed the eccentricities to the current DF, which is consistent with a thermal DF.

Regardless of the origin of the S-stars, their random orbits are consistent with the effect of RR. Vector RR can also explain why the evolved red giants beyond 0.04 pc, in particular the more massive ones with $t_\star \ll \min(T_{\rm NR}, \bar{T}^s_{\rm RR})$ are relaxed, since $T^v_{\rm RR} < t_\star$ out to ~ 1 pc.

3. Gravitational wave sources

MBHs with masses $M_\bullet \lesssim 5 \times 10^6 M_\odot$ have Schwarzschild radii $r_S = 2GM_\bullet/c^2$, such that a test mass orbiting at a few r_S emits gravitational waves (GWs) with frequencies $10^{-4}{\rm Hz} \lesssim \nu \lesssim 1{\rm Hz}$, detectable by the planned space based *Laser Interferometer Space Antenna*[3] (*LISA*). Such GW sources, for which the mass of the inspiraling object is many orders of magnitude smaller than the mass of the MBH are known as *extreme mass ratio inspiral sources* (EMRIs). GW inspiral events are very rare (of the order of $10^{-7} - 10^{-8}\,{\rm yr}^{-1}$ per galactic nucleus; e.g. Hils & Bender [23];

[3] http://lisa.jpl.nasa.gov/

Sigurdsson & Rees [50]; Ivanov [29]; Freitag [17]; Alexander & Hopman [2]; Hopman & Alexander [26], [27], [28]), and it is unlikely that we will observe GWs from our own Galactic center (GC), although such a possibility is not entirely excluded (Freitag [18]; Rubbo, Holley-Bockelmann & Finn [48]). The galactic MBH plays nevertheless a role of importance in understanding the dynamics of EMRIs, since its mass is very close to the mass of the "optimal" *LISA* EMRI target, and as a consequence one may use the GC to model extra-galactic nuclei.

Hopman & Alexander [26] used a model based on the GC to analyze the dynamics of EMRIs. One of the main results was that inspiraling stars always originate very near the MBH, within a distance of $\sim 0.01\,\mathrm{pc}$: due to the relatively short relaxation time in galactic nuclei, stars that start to spiral in from larger distances are very likely to plunge into the MBH before becoming observable as GW emitters. This result was confirmed qualitatively by N-body simulations by Baumgardt et al. [9] of tidal capture of MS stars by an intermediate mass black hole (Hopman, Portegies Zwart & Alexander [24]; Hopman & Portegies Zwart [25]).

The fact that only stars within $\sim 0.01\,\mathrm{pc}$ spiral in successfully, implies that it is the stellar content and dynamics of that region which determine the rate of GW inspiral events for the different populations in the system. This means, for example, that mass-segregation is likely to play an important role (Hopman & Alexander [28]; Freitag et al.; [19]; see also contribution by Marc Freitag in this volume). Since the resonant relaxation time is very short ($T_{\mathrm{RR}} \ll T_{\mathrm{NR}}$) near $\sim 0.01\,\mathrm{pc}$, it also implies that RR will dictate the rate at which stars are driven towards low J orbits, where energy dissipation is efficient and stars spiral in.

Hopman & Alexander [27] used a Fokker-Planck method in energy space with a sink term due to RR losses in J-space to calculate the GW inspiral rate. At every time-step, stars redistribute in energy-space due to (non-resonant) two body scattering, and stars are accreted by the MBH with some specified rate per energy bin. In the relevant regime, this rate is assumed to be of order $\sim N(E)/T_{\mathrm{RR}}$, i.e., within one RR time all stars in the bin would be accreted if they were not replaced by new stars that flow to higher energies (tighter orbits). In spite of the fact that stars are drained very efficiently near the MBH, Hopman & Alexander [27] found that the rate at which stars are replenished by two-body scattering is sufficiently high that the stellar distribution will not be depleted near the MBH, unless the efficiency of RR is more than an order of magnitude larger than exploratory N-body simulations (Rauch & Tremaine [45]) have indicated. Modifications of the stellar DF due to RR are too small to be observable. Some of the stars that are captured by the MBH are swallowed directly without giving a GW signal, but the stars closed to the MBH (within $\sim 0.01\,\mathrm{pc}$) will spiral in rapidly enough to give obtain an orbit of period $P \lesssim 10^4\,\mathrm{s}$ for more than a year. Such sources would be observable to *LISA* to distances up to a few Gpc, depending on the mass of the inspiraling star. Since the enhanced rate at which stars flow to the loss-cone in angular momentum space is sustained by the larger flow due to two-body scattering in energy space, the rate at which EMRIs are produced is increased. The analysis by Hopman & Alexander [27] indicates that the rate at which *observable* EMRIs are formed in galactic nuclei is ~ 8 times higher than that for the case in which RR was neglected.

4. Conclusions

Resonant relaxation is a relatively unexplored dynamical mechanism. Vector RR, which only affects the orientation of the orbit but not the eccentricity, operates in many stellar systems, while scalar RR, which does affect the eccentricity, is unique for MBH systems. This mechanism becomes important at distances $\lesssim 0.1\,\mathrm{pc}$ from the MBH. It may have played an important role in redistributing the orbits of the S-stars, and enhances estimates of the GW inspiral rate by nearly an order of magnitude. Our own GC provides a unique case study for resonant relaxation.

Acknowledgments
We thank the organizers of the GC2006 meeting for a very stimulating conference, and Yuri Levin for discussions on resonant relaxation.

References

[1] Alexander, T., 1999, ApJ, 520, 137
[2] Alexander, T., & Hopman, C., 2003, ApJ, 590, L29
[3] Alexander, T., & Livio, M., 2005, ApJ, 606, L21
[4] Alexander, T., 2005, Physics Reports, 419(2–3), 65
[5] Bahcall, J. N., & Wolf, R. A., 1976, ApJ, 209, 214
[6] Bahcall, J. N., & Wolf, R. A., 1977, ApJ, 216, 883
[7] Baumgardt, H., Makino, J., & Ebisuzaki, T., 2004a, ApJ, 613, 1133
[8] Baumgardt, H., Makino, J., & Ebisuzaki, T., 2004b,ApJ, 613, 1143
[9] Baumgardt, H., Hopman, C., Portegies Zwart, S. F., & Makino, J. 2005 (astro-ph/0511752)
[10] Binney, J. & Tremaine, S., 1987, Galactic Dynamics (Princeton: Princeton Univ. Press)
[11] Chandrasekhar, S., 1943, Reviews of Modern Physics, 15, 1
[12] Cohn, H., & Kulsrud, R. M. 1978, ApJ, 226, 1087
[13] Eisenhauer, et al., 2005, ApJ, 628, 246
[14] Frank, J., Rees, M. J., 1976, MNRAS, 176, 633
[15] Freitag, M., & Benz, W, 2001, A&A, 375, 711
[16] Freitag, M. & Benz, W., 2002, A&A, 394, 345
[17] Freitag, M., 2001, Class. Quantum Grav., 18, 4033
[18] Freitag, M., 2003, ApJ, 583, L21
[19] Freitag, M., Amaro-Seoane, P., & Kalogera, V., 2006, ArXiv Astrophysics e-prints, astro-ph/0603280
[20] Genzel, R. et al., 2003, ApJ, 594, 812
[21] Ghez, A. M., et al., 2003, ApJ, 586, L127
[22] Gould, A., & Quillen, A., 2003, ApJ, 592, 935
[23] Hils, D., & Bender, P. L., 1995, ApJ, 445, L7
[24] Hopman, C., Portegies Zwart, S.F., & Alexander, T., 2004, ApJ, 604, L101
[25] Hopman, C., & Portegies Zwart, S. F., 2005, MNRAS, 363, L56
[26] Hopman, C., & Alexander, T., 2005, ApJ, 629, 362
[27] Hopman, & Alexander, T., 2006a, pre-print: astro-ph/0601161
[28] Hopman, & Alexander, T., 2006b, pre-print: astro-ph/0603324
[29] Ivanov, P. B., 2002, MNRAS, 336, 373, 2002
[30] Kozai, Y., 1962, AJ, 67, 591
[31] Levin, Y., 2003, (astro-ph/0307084)
[32] Levin, Y., & Belobodorov, A. M., 2003, ApJ, 596, 314
[33] Levin, Y., 2006, MNRAS, submitted
[34] Magorrian, J., Tremaine, S., 1999, MNRAS,309, 447
[35] Marchant, A.B., & Shapiro, S.L., 1979, ApJ, 234, 317
[36] Marchant, A.B., & Shapiro, S.L., 1980, ApJ, 239, 685
[37] Merritt, D., & Szell, A., 2005 (astro-ph/0510498)
[38] Miralda-Escudé, J., & Gould, A., 2000, ApJ, 545, 847
[39] Murphy, B. W., Cohn, H. N., & Durisen, R. H., 1991, ApJ, 370, 60
[40] Nayakshin, S., & Cuadra, J., 2004, A&A, 437, 437
[41] Nayakshin, S., Sunyaev, 2005 (astro-ph/0507687)
[42] Nayakshin, S., 2006 (astro-ph/0512255)
[43] Paumard, T., et al., 2005, ApJ, submitted
[44] Preto, M., Merritt, D., Spurzem, R., 2004, ApJ, 613, L109
[45] Rauch, K. P., & Tremaine, S., 1996, New Astronomy, 149
[46] Rauch, K. P., & Ingalls, 1998, MNRAS, 299, 1231
[47] Rauch, K. P., 1999, ApJ, 514, 725
[48] Rubbo, L. J., Holley-Bockelmann, K., & Finn, L. S. 2006, ArXiv Astrophysics e-prints, astro-ph/0602445
[49] Shapiro, S.L., & Marchant, A.B., 1979, ApJ, 225, 603
[50] Sigurdsson, S., and Rees, M. J., 1997, MNRAS284, 318
[51] Syer, D., Ulmer, A., 1999, MNRAS, 306, 35

Institute of Physics Publishing   Journal of Physics: Conference Series **54** (2006) 328–334
doi:10.1088/1742-6596/54/1/052   Galaxy Center Workshop

Sub-Milliarcsecond Imaging of Sgr A* and M 87

T.P. Krichbaum[1], **D.A. Graham**[1], **M. Bremer**[2], **W. Alef**[1], **A. Witzel**[1], **J.A. Zensus**[1], **and A. Eckart**[3]

[1] MPIfR, Auf dem Hügel 69, 53121 Bonn, Germany
[2] IRAM, 300 rue de la Piscine, 38406 St. Martin d'Heres, France
[3] University of Cologne, I. Physikalisches Institut, Zülpicher Str. 77, 50937 Köln, Germany

E-mail: `tkrichbaum@mpifr-bonn.mpg.de`

Abstract. We present and discuss new result from mm-VLBI observations of M 87 and Sgr A*. The imaging of these sources with a spatial resolutions of a few to a few ten Schwarzschild radii offers new possibilities to study the immediate environment of super-massive black holes.

1. Introduction

In astronomy, the highest angular resolution is obtained with Very Long Baseline Interferometry (VLBI). VLBI observations at short millimeter wavelengths (mm-VLBI) provide angular resolutions in the ten micro-arcsecond range (i.e. 14 μas for a 10,000 km baseline at $\lambda = 1.3$ mm ($\nu = 230$ GHz)). While VLBI observations at 2 mm and 1 mm wavelength are still in a stage of technical development, 3 mm VLBI experiments are performed on a regular basis.

The Global mm-VLBI Array (GMVA[1]) is operational since early 2000. At 86 GHz ($\lambda = 3.5$ mm), it combines the European antennas (Effelsberg 100 m, Pico Veleta 30 m, the phased Plateau de Bure interferometer 6x15 m, Onsala 20 m, Metsähovi 14 m) with the VLBA. With the participation of the two sensitive IRAM telescopes and the 100 m Effelsberg telescope, the array sensitivity is improved by a factor of $3 - 4$, when compared to the VLBA alone. For compact galactic and extragalactic radio sources, the GMVA provides VLBI images with an angular resolution of down to 40 μas.

In the following we present and discuss results from mm-VLBI observations of the nuclei of M 87 and of Sgr A*. When expressed in terms of Schwarzschild radii ($R_S = 2GM_{\rm BH}/c^2$), the spatial resolution obtained for M 87 and Sgr A* become very comparable. M 87 is about $\sim 10^3$ times more distant than Sgr A*, but has a putative black hole at its center, which is also $\sim 10^3$ times more massive than that of Sgr A*. Present day 3 mm-VLBI therefore allows to image both sources with a spatial resolution of only $10 - 20$ R_S. In view of the morphological difference between Sgr A* and M 87 (the latter exhibits a long relativistic jet), a comparison of the sub-milliarcsecond structure and kinematics of both objects may lead to a better understanding of the astro-physical processes acting in the vicinity of such super-massive black holes.

2. 3 mm VLBI observations of M 87

The giant elliptical galaxy M 87 (NGC 4486, 3C 274, Virgo A) is one of the closest radio galaxies with a prominent radio jet. Its relative brightness at millimeter wavelengths makes it ideally

[1] web link: http://www.mpifr-bonn.mpg.de/div/vlbi/globalmm

suited for high angular resolution mm-VLBI studies, which address the still not answered question, of how the powerful jets in radio-galaxies and quasars are launched and collimated.

At a distance of D=16.75 Mpc [1] an angular scale of 0.1 mas corresponds to a spatial scale of $2.5 \cdot 10^{16}$ cm (0.0081 pc). This translates to 27.8 R_S Schwarzschild radii for an assumed $3 \cdot 10^9 M_\odot$ super-massive black hole at the center of M 87 e.g. [2]. Global 3 mm VLBI observations, performed at the edge of the accessible VLBI observing bands, have revealed new images of the inner-most region of M 87. The 3 mm VLBI images shown in Figure 1 were made from observations performed during 2002 – 2004 and show a one-sided core-jet structure, extending ~ 2 mas to the west. The complexity of the source structure and the only moderate dynamic range of the maps (lower sensitivity in earlier observations) do not yet allow to relate the individual structural components seen at the different epochs with each other and by this determine the jet kinematics. Despite these limitations, which will be overcome in the near future, a number of basic statements can already be made:

(i) In the maps, an unresolved compact component at the eastern end of the jet is seen. An upper limit to its size is given by the (elliptical) VLBI observing beam. The uniformly weighted and un-tapered beam size of the observations of April 2004 is $(193 \times 54)\mu$as, which gives an upper limit to the size of the jet base (VLBI core) of $(54 \times 15)R_S$. Circular Gaussian model fitting yields a typical flux density of ~ 0.5 Jy and a size of $\lesssim 50\mu$as ($\lesssim 14R_S$) for the VLBI core. This results in a brightness temperature of $\gtrsim 10^{10.5}$ K, indicative of non-thermal (synchrotron) radiation.

(ii) The 86 GHz VLBI map of April 2004 looks remarkably similar to two previously observed 43 GHz VLBI maps, which indicate conical jet-expansion on sub-milliarcsecond scales [3, 4]. The new 86 GHz maps, which have a ~ 2 times higher resolution, reveal a bifurcated (possibly edge-brightened) expanding one-sided jet, with the southern edge of the emission cone being brighter and more compact than the northern part. The fact that the northern edge of the cone (at $r \leq 0.5$ mas) is almost invisible in the maps of 2002 and 2003 can be explained by the more limited dynamic range of these two maps.

(iii) The large morphological differences seen between the three 86 GHz maps indicate structural variations on time-scales of less than one year. On angular scales ≤ 0.5 mas, all three maps show some south-west oriented jet emission oriented along a position angle of $\sim -110°$. In April 2003, the source was $\sim 1.5 - 2$ times brighter than usual, and a bright component was seen at the map center (at zero position). Since in hybrid VLBI imaging with closure phases, the absolute position information is lost, this bright component most likely should not be identified with the jet base (VLBI-core). Instead, the fainter component located ~ 0.2 mas in the north-east, may be identified with the VLBI-core. We note the tentative detection of a possible counter-jet at 43 GHz [4], which however was not seen earlier [3], nor at longer wavelengths [5], nor in our 86 GHz VLBI maps of 2002 and 2004.

(iv) Owing to the sparse time coverage at 86 GHz, a detailed kinematic study of the component motion in the inner jet of M 87 is not yet possible. Similar to the case of many other VLBI jets, we may assume that the bright central component seen in April 2003 was ejected from the VLBI core after a flux density outburst. This would give an apparent rate of change of $\gtrsim 0.2$ mas/yr, corresponding to an apparent velocity of $\gtrsim 0.05$ c. This speed would be consistent with the observed motion of 0.04 ± 0.02 c seen at larger core separations with VLBI at 15 GHz [6], but not with a lower limit of $0.2 - 0.4$ c from VLBI at 43 GHz [7]. On the other hand, if we compare the relative positions of the two main secondary components seen in April 2002 and April 2004 (at $r \simeq 0.25$ mas & 0.50 mas, respectively) stationarity of some of the inner jet components can also not be excluded. More frequent VLBI observations are clearly necessary to determine the jet kinematics.

We like to point out that the determination of VLBI core size and jet transverse width as a function of core separation, are very important parameters for theoretical jet models. The class of models, which are based on the magnetic sling shot mechanism [8] predict a minimum transverse width of the jet after initial acceleration to be of order $\geq \gamma \cdot 2R_L$, (with bulk Lorentz factor γ and light cylinder radius R_L, e.g. [9]. Notably such jets are expected to be hollow (edge-brightened), just as observed here. After collimation, the light cylinder has a radius of at least $> (10 - 50)\ R_S$, see [10, 11, 12], which sets a lower bound to the observable jet width. With jet Lorentz-factors in the range $3 \leq \gamma \leq 6$ [13], we expect a minimum jet width of $\geq (60 - 600) R_S$. The observed diameter of the jet (at $0.5 \leq r \leq 2$ mas) is $2R_L = (0.5 - 0.7)$ mas, or $(140 - 200) R_S$ (see Figure 1). This is at the lower end of the expected range, favoring a moderately low bulk Lorentz-factor ($\gamma < 2 - 3$) in this region.

At its origin near the black hole, the diameter of the jet depends on the size of the region, where it is initially accelerated. In the Blandford & Payne type models [8], the twisted magnetic field lines are anchored in the inner part of the rotating accretion disk. The minimum jet width is then determined by the diameter of the least stable orbit, which is at $6R_S$ for a Schwarzschild BH, and $2R_S$ for a rotating Kerr BH [28]. The observed small size of the VLBI core ($< 15R_S$) is already within a factor of $2 - 3$ near these values, suggesting that the width of the 'jet-nozzle' is indeed of order of the light cylinder diameter. Since the radiating jet boundary cannot be indefinitely thin, one could expect that the observed jet width is a bit larger than the light cylinder. In this case, the observed small size of the VLBI core would point more towards jet models, in which the jet gains energy directly from BH-rotation [14, 15]. Better VLBI images and more specific theoretical modeling of jet opening angle and speed will help to discriminate between the different types of models (e.g. Blandford & Payne versus Blandford & Znajek), and may tell to what degree black hole rotation plays a role in jet formation.

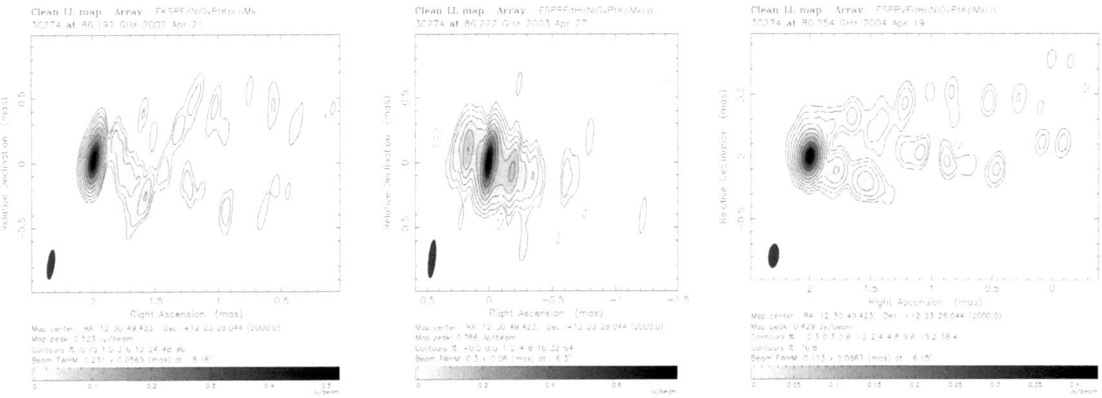

Figure 1. 86 GHz VLBI maps of the jet in M 87. The observations were performed (from left to right) in April 2002, 2003, and 2004. The maps result from global VLBI observations using the GMVA, which combines European radio telescopes with the VLBA.

3. VLBI observations of Sgr A* at 3, 2, and 1 mm wavelength

3 mm: From the early 1990's onwards, Sgr A* was repeatedly observed with VLBI at 3.5 mm (86 GHz) in various array configurations, using European, American and global VLBI arrays [16, 17, 18]. In Figure 2 we show one of the early maps from global 3 mm VLBI, which reveals a partially resolved point source, similar to the recently published VLBA image [19]. At mm-wavelengths, the accurate determination of flux density and size is often affected by calibration uncertainties (low source elevation on the northern hemisphere, atmospheric opacity variations). This requires special observing and calibration methods. In parallel to the ongoing research with the VLBA, we have continued to observed Sgr A*, using the sensitive and large antennas

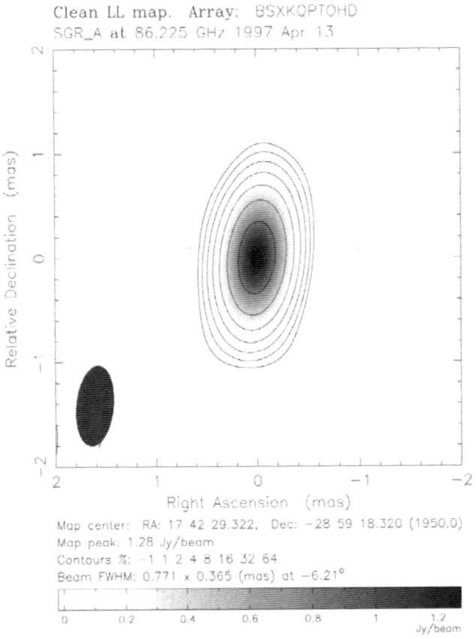

Figure 2. 86 GHz VLBI image of Sgr A* observed in April 1997. The participating antennas are: Effelsberg (100 m), Pico Veleta (30 m), Haystack (37 m), Quabbin (14 m), Pie Town (25 m), Kitt Peak (12 m), Owens Valley (5 x 10 m). A circular Gaussian parameterizes Sgr A* with flux density $S = 1.73 \pm 0.25$ Jy and size $\theta = 0.28 \pm 0.08$ mas.

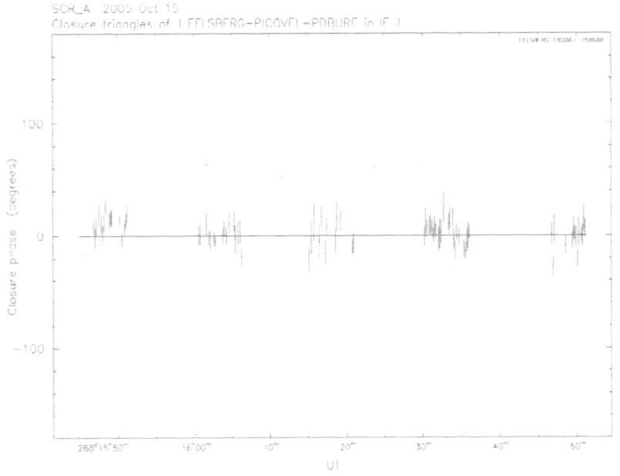

Figure 3. Closure phase of Sgr A* plotted versus time. Data are from a 3 station VLBI experiment performed in October 2005 at 86 GHz using the telescopes at Effelsberg, Pico Veleta and Plateau de Bure (phased). The closure phase is zero within $\pm 10°$, suggesting a point-like or almost perfectly symmetrical brightness distribution.

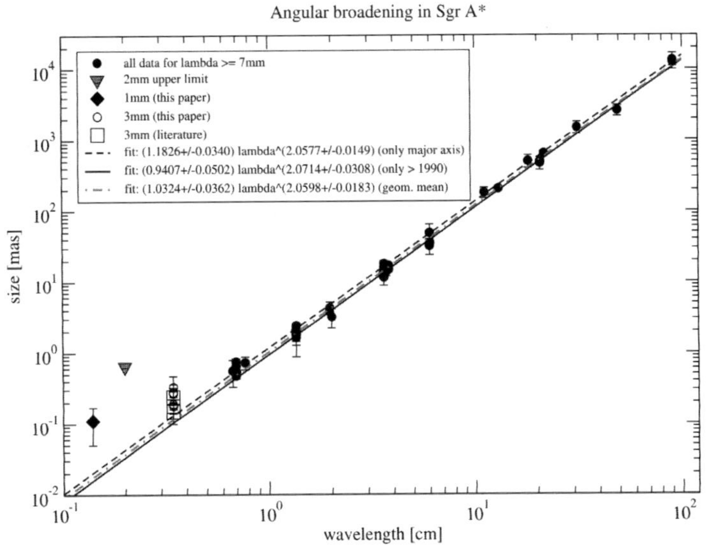

Figure 4. Angular size of Sgr A* plotted versus wavelength. Lines denote fits ($a\lambda^{\beta}$), excluding the short wavelengths $\lambda \leq 7$ mm, which are affected by interstellar scattering. The fit parameters a, β are given inside the figure. Published 3 mm measurements are plotted as open squares, new 3 mm data (this paper) by open circles. New measurements at 2 mm (shaded triangle, an upper limit) and 1.4 mm (diamond) are also shown.

at MPIfR (Effelsberg 100 m telescope) and at IRAM (Pico Veleta 30 m, Plateau de Bure 6 x 15 m). With baseline lengths of $\sim 750 - 1700$ km, and a detection sensitivity of $\gtrsim 50$ mJy (7σ, 512 Mbit/sec) per baseline, this array measures the visibility amplitudes and (closure-) phases with particular high SNR. In April and October 2004, Sgr A* was moderately bright with flux densities of 2.40 ± 0.45 Jy and 2.10 ± 0.23 Jy, respectively. In October 2005, however, the source was detected with a much higher signal-to-noise ratio than before (SNR = 96 on PV-PdB). A preliminary analysis of the new data yields (via circular Gaussian model fitting) a flux density of $S = 3.6 \pm 0.5$ Jy and a size of $\theta = 0.27 \pm 0.05$ mas. The relatively high source flux is confirmed independently by ATCA measurements, performed 3 days after our VLBI experiment ($S = 2.0 - 3.5$ Jy, Z.-Q. Shen, priv. comm.). The source size appears marginally larger (by $2 - 3\sigma$), when compared to our previous size measurements in 2004 ($\theta = 0.19 \pm 0.05$ mas and 0.18 ± 0.01 mas), and in comparison to the VLBA measurements of [19], who obtained $\theta = 0.18 \pm 0.02$ mas with the VLBA. As in previous observations, the closure phase (see Figure 3) is close to zero (within $\pm 10°$), showing no evidence for a deviation from the known point-like (or symmetric) structure. In Figure 5, we plot all available 86 GHz VLBI size measurements, using the data from the literature and including own unpublished old and new data. If one excludes the new data point of October 2005, the weighted mean average size is 0.18 ± 0.01 mas. This appears significantly smaller than the October 2005 measurement. For comparison we also show in Fig. 5 the sizes at 43 GHz (weighted mean: 0.55 ± 0.03 mas) and the flux density of Sgr A* at 86 GHz. It will need a more dense in time sampled 3 mm VLBI monitoring to check, if the brightness distribution of Sgr A* indeed is variable. In view of the observed flux density variations at millimeter wavelengths (see Miyoshi et al., Miyazaki et al., Marrone et al., this conference), variability of the VLBI structure of Sgr A* would be not unexpected (see [20] for possible size variations at 7 mm).

2 mm: VLBI observations at wavelengths shorter than 3 mm are still difficult and are limited by the number of available radio telescopes and their sensitivity. In a pilot VLBI experiment at 147 GHz (2 mm), which aimed at first fringe detection across the Atlantic at this short wavelength, Sgr A* was observed among a dozen other AGN [21]. Several AGN were detected on the long (~ 8500 km) baseline between Spain (Pico Veleta) and the Arizona telescopes (Kitt Peak (KP) 12 m, Heinrich-Hertz telescope (HHT) 10 m), which limits the size of their compact emission regions to $\leq 25 - 40\,\mu$as. Sgr A* was detected with a signal-to-noise ratio of ~ 7 on the short (85 Mλ) baseline between KP and HHT. With a total flux density of ~ 1.9 Jy and a correlated flux density of $S_c = 1.6 \pm 0.25$ Jy an upper limit to its size of ≤ 0.7 mas can be derived, fully consistent with the measured sizes at 3 mm and 1.4 mm.

1 mm: In April 2003, our attempt to observe Sgr A* with global VLBI at 230 GHz failed due to weather and technical problems. In this experiment, which was motivated in a similar manner as the 2 mm VLBI experiment described before, several bright AGN were observed. Finally two blazars (3C 454.3, 0716+714) were detected with SNR $\gtrsim 7$ on the 6.4 Gλ baseline between Pico Veleta and HHT [22]. Although limited by receiver performance and stability of local oscillators at HHT, this experiment demonstrates the technical feasibility of global VLBI at 1.3 mm, and sets a new record in angular resolution. For the detected compact radio sources we obtained sizes of $\lesssim 30\,\mu$as. With such a high angular resolution, which would correspond to a spatial resolution of ~ 3 Schwarzschild radii in Sgr A*, the direct (VLBI-) imaging of the immediate vicinity of a SMBH should become possible within the next few years.

We now like to make a remark with regard to the previous 215 GHz VLBI observations of Sgr A* published in 1998 by Krichbaum et al. [23]. Recently, Eckart et al. [24] (also Eckart et al., this conference) detect an emission component D1 of ~ 170 mas size, located 80 ± 50 mas west of Sgr A*. With single dish radio telescopes operating at short mm-/sub-mm wavelengths, in-beam confusion between Sgr A* and D1 leads to an over-estimate of the total flux density of Sgr A* in this spectral band (overestimate by $\Delta S_{215\,\mathrm{GHz}} \simeq 0.75$ Jy for $T_e \sim 10^3$ K), at least partially removing the so called sub-mm excess, e.g. [25, 26]. In fact, a less pronounced or even not existing sub-mm excess, leads to a flat or only slightly inverted synchrotron-self-absorbed radio spectrum, extending up to at least several 100 GHz, which is typical for the more luminuous AGN. This then would easily allow to interprete the high energy data of Sgr A* via the synchrotron-self-Compton process [27, 24].

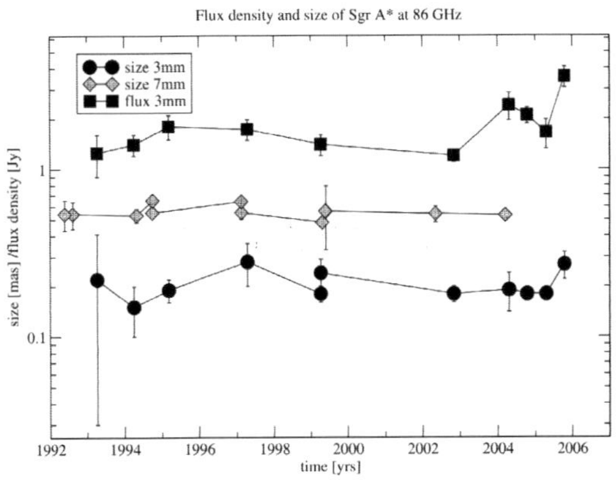

Figure 5. Measured VLBI sizes of Sgr A* at 86 GHz (filled circles) and at 43 GHz (open diamonds) plotted versus time. Data are taken from the literature, including our new and unpublished 86 GHz data. All sizes were calculated from the geometric mean of major and minor axis. We also show the VLBI flux density (filled squares) at 86 GHz.

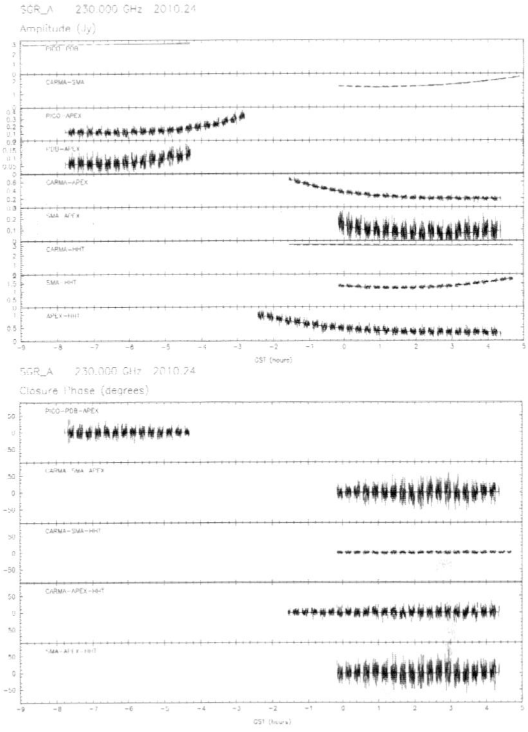

Figure 6. Right: Simulated visibility amplitudes (top) and closure phase (bottom) for a future 1.3 mm VLBI experiment on Sgr A* with the following telescopes: Pico Veleta (Spain), Plateau de Bure (France), CARMA (California), SMA (Hawaii), HHT (Arizona) and APEX (Chile). For the simulation, a circular Gaussian of 2.5 Jy and $\sim 25\,\mu$as FWHM size was assumed.

The existence of such an extended emission region, which is fully resolved and therefore invisible to VLBI, was already considered in 1998 by Krichbaum et al. [23] as a possibility to explain the apparent discrepancy between observed total flux (from single dish measurements) and the correlated VLBI flux, which originates from the compact component. Relating the sub-mm excess of the total spectrum mainly to the flux contribution of component D1, allows to redetermine the total flux of Sgr A* at 215 GHz and reject 'case 1' in the discussion of [23]. This leads to anew and refined estimate of the observed size for Sgr A* at 215 GHz: $\theta = (110 \pm 60)\mu$as, which corresponds to a linear size of $R_S = 11 \pm 6$ Schwarzschild radii and a robust lower limit of $R_S \geq 5$.

4. Intrinsic Source Size

The observed frequency dependence of the source size of Sgr A* (Fig. 4) is commonly interpreted as being due to scatter broadening by the intervening interstellar medium. The intrinsic size is determined by the deconvolution of the observed size $\theta_{\rm obs}$ with the extrapolated scattering size $\theta_{\rm scat}$: $\theta_{\rm int} = \sqrt{\theta^2_{\rm obs} - \theta^2_{\rm scat}}$. The exact form of the assumed scattering law ($\theta_{\rm scat} = a\lambda^{\beta}$, with fit parameters a, β) determines $\theta_{\rm int}$. In Figure 4 we show a few illustrative least-square fit lines, which were obtained from all at present available size measurements, selecting those above $\lambda \geq 1.3\,{\rm cm}$[2] and performing a fit to (i) only the major axis, (ii) the geometric mean of major and minor axis, and (iii) the most recent (> 1990) data. The resulting mean slopes in

[2] these wavelengths should be least affected by a finite intrinsic source size

the range of $\beta = 2.06 \pm 0.01$ deviate slightly from the slope of $\beta = 2$ other authors recently have used ([20, 19], and references therein). Depending on the number of data points included in the fit, the used weighting scheme, and the frequency range over which the fit is performed, one obtains slightly different estimates of the intrinsic size of Sgr A*. Since the size determination at the shortest wavelengths ($\lambda \leq 3$ mm) is still restricted to a circular Gaussian component, we determine the de-convolved intrinsic size of Sgr A* using the geometric mean size from major and minor axis at all wavelengths. Formally we obtain the following sizes[3]: $25 \pm 7R_S$ at 7 mm, $15 \pm 3R_S$ at 3.4 mm, $< 65R_S$ at 2 mm, and $11 \pm 6R_S$ at 1.4 mm.

5. Future Outlook

Global mm-VLBI performed at the shortest wavelengths (≤ 3 mm) will allow to study Sgr A* (and other AGN like, e.g. M 87) with a resolution of only a few Schwarzschild radii, facilitating the direct imaging of the immediate vicinity of super-massive black holes. Global mm-VLBI observations of Sgr A* are clearly a great challenge and demand coordinated efforts. In Figure 6 we show simulations of visibility amplitudes and closure phases for a realistic near future global mm-VLBI observation of Sgr A* at 230 GHz. New telescopes like APEX located in the southern hemisphere (and later supplemented by ALMA), and phased interferometers operating as single VLBI antennas (PdB, CARMA, SMA) are crucial to provide the necessary sensitivity on the longest baselines. When these telescopes become available for mm-VLBI, we may hope to image the environment of the SMBH in Sgr A* and detect possible deviations from circular symmetry, which are predicted by the theory of General Relativity for a rotating accretion disk black hole system.

References

[1] Whitmore B.C., Sparks W.B., Lucas R.A., et al., 1995, ApJ 454, L73.
[2] Machetto F.D., Marconi A., Axon D.J., et al., 1997, ApJ 489, 597.
[3] Junor W., Biretta J.A., Livio M., 1999, Nat 401, 891.
[4] Ly C., Walker R.C., & Wrobel J.M., 2004, AJ 127, 119.
[5] Dodson R., Edwards P.G., Hirabayashi H., 2006, PASJ 58, 253.
[6] Kellermann K.I., Lister M.L., Homan D.C., et al., 2004, ApJ 609, 539.
[7] Ly C., Walker R.C., & Junor W., 2004b, AAS 205, #110.01
[8] Blandford R.D., & Payne D.G., 1982, MN 199, 883.
[9] Tomimatsu A., & Takahashi M., 2003, ApJ 592, 321.
[10] Camenzind M., 1990, RvMA 3, 234.
[11] Fendt C., & Memola E., 2001, A&A 365, 631.
[12] Casse F., & Keppens R., 2004, ApJ 601, 90.
[13] Biretta J.A., Sparks W.B., Machetto F., 1999, ApJ 520, 621.
[14] Blandford R.D., & Znajek R.L., 1977, MN 179, 433.
[15] McKinney J.C., 2006, MN 368, 1561.
[16] Krichbaum T.P., et al., 1994, IAU Symp. No. 159, ed. T. Courvoisier & A. Blecha, Kluwer, p. 187.
[17] Rogers A.E.E., Doeleman S.S, Wright M.C.H., et al., 1994, ApJ 434, L59.
[18] Doeleman S.S, Shen Z.-Q., Rogers A.E.E., et al., 2001, ApJ 121, 2610.
[19] Shen Z.-Q., Lo K.Y., Liang M.-C., et al., 2005, Nat 438, 62.
[20] Bower G.C., et al. 2004, Sci 304, 704.
[21] Krichbaum T.P., et al. 2002, in: *6th EVN Symp., ed. E. Ros et al., MPIfR, Bonn*, p. 125 (astro-ph/0207022).
[22] Krichbaum T.P., et al. 2004, in: *7th EVN Symp., ed. R. Bachiller et al., OAN, Toledo*, p. 15 (astro-ph/0411487).
[23] Krichbaum T.P., Graham D.A., Witzel A., et al., 1998, A&A 335, 106.
[24] Eckart A., Baganoff F.K., Schödel R., et al., 2006, A&A 450, 535.
[25] Serabyn E., Carlstrom J., Lay O., et al., 1997, ApJ 490, 77.
[26] Falcke H., Goss W.M., Matsuo,H., et al., 1998, ApJ 499, 731.
[27] Eckart A., Baganoff F.K., Morris M., et al., 2004, A&A 427, 1.
[28] Meier D.L., Koide S., Uchida Y., 2001, Sci 291, 84.

[3] For $d = 8.0$ kpc and $M_{\rm BH} = 4 \cdot 10^6 M_\odot$ an angle of 0.1 mas corresponds to $1R_S$

Institute of Physics Publishing
doi:10.1088/1742-6596/54/1/053
Journal of Physics: Conference Series **54** (2006) 335–341
Galaxy Center Workshop

Radio spectra and radio-loudness of low-luminosity AGNs

Akihiro Doi[1], **Seiji Kameno**[2], **and Makoto Inoue**[3]

[1] Faculty of Science, Yamaguchi University, 1667-1 Yoshida, Yamaguchi, Yamaguchi 753-8512, Japan

[2] Faculty of Science, Kagoshima University, 1-21-30 Korimoto, Kagoshima, Kagoshima 890-0065, Japan

[3] National Astronomical Observatory of Japan, 2-21-1 Osawa, Mitaka, Tokyo 181-8588, Japan

Abstract. We investigated the empirical relation between black hole mass ($m_{\rm BH}$), Eddington ratio ($L_{\rm bol}/L_{\rm Edd}$), and radio loudness ($R_{\rm RL}$: a ratio of radio to optical luminosity) of nearby low-luminosity active galactic nuclei (LLAGNs). A best-fit plane was found in the three-dimensional space using a sample of 48 nearby LLAGNs: $R_{\rm RL} = m_{\rm BH}^{0.52\pm0.14}(L_{\rm bol}/L_{\rm Edd})^{-0.39\pm0.08}$. This suggests that spectral energy distributions of LLAGNs are controlled by both the black hole mass and accretion rate.

1. Introduction

The Palomar optical spectroscopic survey has detected nuclear emission lines in nuclear regions of 86 per cent of 486 nearby bright galaxies with $B_{\rm T} < 12.5$ mag ($B_{\rm T} < 12.5$ is the total apparent B-band flux), mainly in the northern sky [12]. Active galactic nuclei (AGNs) reside in about one half of these nuclear emission line galaxies [13]. Almost all of them are low-luminosity AGNs (LLAGNs), which are usually formally defined as $L_{\rm H\alpha} < 10^{40}$ ergs s^{-1}. Their low luminosities are thought to be caused by low accretion rates and low radiative efficiencies in the accretion flows. Such accretion has often been explained by an optically-thin accretion flow such as the advection-dominated accretion flow (ADAF, [31]) rather than an optically thick disc ('standard disc'; [38]), which applies to luminous AGNs. Although the ADAF model successfully explains the broadband spectra of LLAGNs, there is a significant departure from observations in the radio regime (e.g., [25, 35, 10]).

From an observational point of view, the connection between radio properties and accretion phenomena in AGNs has not been clear for a long time. Radio loudness, conventionally defined as the ratio of radio at 5 GHz to B-band flux densities [42, 40, 18], spreads more than four orders of magnitude at the same optical luminosities (e.g., [15], and cf. [41]). The jet powers are not well correlated with the radiation powers from accretion discs. On the other hand, there are many reports about radio–optical continuum (or line) flux correlations (e.g., [43, 15]). The coincidences of the jet producing events and soft X-ray flux variations in 3C 120 [26] are one of the phenomenological evidence for the jet–disc connection in the AGNs. The jet–disc connection seems to be weak but real. However, the principal parameters that control jet activities in accretion discs have so far been hardly identified by observations. Recently, several implications have been reported that jet activities are related to black hole masses [5, 27, 32, 19, 11] and/or accretion rates [20, 11, 19], both of which are thought to be two principal parameters that describe the accretion phenomena. For example, radio-loud quasars have been found exclusively in very high-mass black hole systems [20]; radio-louder AGNs tend to show lower accretion rates [11]. A 'fundamental plane' has been found in the three-dimensional space spanned by the radio and X-ray luminosities of the core component and by the black hole mass that applies to stellar-mass as well as supermassive black hole systems ([28], see also [4]). Recently developed methods for estimating black hole masses allow to make such studies. These results suggest that the black hole mass and accretion rate may be principal parameters for the jet activities.

In the present paper, we suggest a unified interpretation that the radio powers and radio loudness can be described using only the two principal parameters (black hole mass and accretion rate) for nearby AGNs, most of which are LLAGNs.

2. Sample

We compiled two samples for nearby AGNs: sample A and sample B. Sample A consists of 48 radio sources that had been significantly detected (>1 mJy) in systematic surveys with the VLA A-configuration at 15 GHz from [29, 30] and archival data. Sample B consists of the other 22 nearby AGNs whose reliable black hole masses have been compiled by [28]; uniform radio data are not available for these sources. All of the AGNs were selected from 486 nearby bright galaxies of the Palomar optical spectroscopic survey [12]. The optical survey collects galaxies with $B_{\mathrm{T}} < 12.5$ mag mainly in the northern sky from the Revised Shapley-Armes Catalogue of Bright Galaxies [37].

Properties of observed optical emission lines were taken from [12, 14], and were extinction-corrected using the mean extinction law of [3], assuming the intrinsic $\mathrm{H}\alpha/\mathrm{H}\beta$ flux ratio = 3.1.

Consequently, most of the AGNs in our samples are LLAGNs. The formal definition of LLAGNs, $L_{\mathrm{H}\alpha} < 10^{40}$ ergs s^{-1} for the narrow-line component, is not adopted in our selections, so that nine nearby AGNs are included in the samples.

3. Estimates of several parameters

3.1. Bolometric luminosities

The nuclear continuum emission from most LLAGNs is dominated by the stellar light of their host galaxies in almost the entire part of their spectral energy distributions (SEDs). Only if sensitive, high-angular resolution observations with instruments like the *Hubble* space telescope and the *Chandra* X-ray observatory are used, the non-stellar continuum emission can be extracted [16, 41]. Since complete data were not available for all objects in our samples, we estimated their bolometric luminosities from Balmer line luminosities, according to [21]. An empirical Hβ line–*B*-band continuum relation in AGNs including LLAGNs [15] and an intrinsic ratio of $\mathrm{H}\alpha/\mathrm{H}\beta = 3.1$ provide $\log \nu L_{\nu}(B) = 1.176 \log L_{\mathrm{H}\alpha} - 6.33$. Furthermore, a bolometric correction factor that has been derived from an averaged SED of LLAGNs [10] is provided by equation (13) in [21]:

$$\log L_{\mathrm{bol}} = 1.176 \log L_{\mathrm{H}\alpha} - 4.91, \qquad (1)$$

where $\log L_{\mathrm{bol}}$ is the bolometric luminosity in ergs s^{-1}. This nonlinearity is consistent with that of the line-continuum correlation in quasars [33]. With this relation, the operational definition of LLAGNs, $L_{\mathrm{H}\alpha} < 10^{40}$, corresponds to $L_{\mathrm{bol}} < 10^{42.1}$ in ergs s^{-1}. The resultant bolometric luminosities, not listed in the present paper, will be used for estimating mass accretion rates.

3.2. Black hole masses

The black hole mass is thought to be one of the fundamental parameters that drive the various activities in all accreting systems; the other one is the accretion rate. We examine in the present paper whether the observed properties of the nearby LLAGNs can be described with the two fundamental parameters.

We compiled black hole masses, mainly provided by the relation between the central stellar velocity dispersion and the black hole mass [8, 7]. We used the relation that has been found by [6] as $m_{\mathrm{BH}} = 1.66 \times 10^8 (\sigma_\star/200)^{4.58}$, where m_{BH} is the black hole mass in $M_\odot$ and $\sigma_\star$ is the central stellar velocity dispersion in km s^{-1}. The HYPERLEDA catalogue [39, 34] provides the values of velocity dispersion. This method can estimate black hole masses with an uncertainty of ~20–30 per cent.

3.3. Accretion rates

The actual dimensionless accretion rate, $\dot{m} \equiv \dot{M}/\dot{M}_{\mathrm{Edd}}$ where $\dot{M}$ is the physical accretion rate and $\dot{M}_{\mathrm{Edd}}$ is the Eddington rate, cannot be measured by observing the radiation from the nucleus because the radiative efficiency in the accretion disc is unknown. Instead of the actual accretion rate, a ratio of bolometric luminosity to Eddington luminosity, $L_{\mathrm{bol}}/L_{\mathrm{Edd}}$, is generally used. We calculated the ratio for our samples. Theoretical studies propose three main solutions: the slim disc ($L_{\mathrm{bol}}/L_{\mathrm{Edd}} > 1$; [1]), the standard disc (or 'thin disc'; $L_{\mathrm{bol}}/L_{\mathrm{Edd}} \sim 0.1$; [38]), and the ADAF ($L_{\mathrm{bol}}/L_{\mathrm{Edd}} \ll 0.1$). Most of the nearby AGNs in our samples show very low $L_{\mathrm{bol}}/L_{\mathrm{Edd}}$, $< 10^{-3}$, where the solution of ADAF is presumably applicable.

Therefore, we assume a radiative efficiency of the ADAF type for the nearby AGNs of our samples. In the ADAF regime, $L_{\mathrm{bol}} \propto \dot{m}^2$, while $L_{\mathrm{bol}} \propto \dot{m}$ in the standard-disc regime; the difference between them is small at $\dot{m} \sim 10^{-2}$, but significant at $\dot{m} \ll 10^{-2}$. Although the conversion factor between L_{bol} and the actual accretion rate may be different for different objects, we adopt a uniform scaling law for the ADAF case (eq. (58) of [24]);

$$\dot{m}_{\mathrm{ADAF}} = \alpha_{\mathrm{vis}} \left(\frac{L_{\mathrm{bol}}}{0.20 L_{\mathrm{Edd}}} \right)^{\frac{1}{2}}, \tag{2}$$

where $\dot{m}_{\mathrm{ADAF}}$ is the dimensionless accretion rate in the ADAF. We assume a viscosity parameter of $\alpha_{\mathrm{vis}} = 0.1$ in the present paper. Note that the ADAF is established only for accretion rates lower than a critical value $\dot{m} < \dot{m}_{\mathrm{crit}} \sim \alpha_{\mathrm{vis}}^2$, i.e. $\sim 10^{-2}$ [36]. In our samples, five objects with relatively high accretion rates do not fulfill this criterion; these objects may be dominated by the standard disc. Note that $L_{\mathrm{bol}}/L_{\mathrm{Edd}}$ is a model-independent parameter, while $\dot{m}_{\mathrm{ADAF}}$ is a model-dependent parameter.

3.4. Radio Loudness

Conventionally, the radio loudness has been defined as the ratio between the 5-GHz radio and the B-band optical flux densities [42, 40, 18]. In the present paper, we define the radio loudness of the core component as $R_{\mathrm{RL}} \equiv P_{15}^{\mathrm{VLA-A}}/P_B$, where $P_{15}^{\mathrm{VLA-A}}$ is the monochromatic flux density that was measured with the VLA in the A-configuration at 15 GHz with a resolution of ~0.15-arcsec and P_B is the B-band monochromatic flux density, which was estimated from the extinction-corrected $H\beta$ luminosity as described in section 3.1.

4. Results

4.1. ($P_{15}^{\mathrm{VLA-A}}$, m_{BH}, $\dot{m}$) plot

The three-dimensional plot of radio power, black hole mass and accretion rate for the sample A is presented in Fig. 1. We found a best-fit plane in the three-dimensional space,

$$\begin{aligned} \log P_{15}^{\mathrm{VLA-A}} &= (1.52 \pm 0.14) \log m_{\mathrm{BH}} \\ &\quad + (0.61 \pm 0.08) \log (L_{\mathrm{bol}}/L_{\mathrm{Edd}}) \\ &\quad + (11.4 \pm 0.97) \\ &= (1.52 \pm 0.14) \log m_{\mathrm{BH}} \end{aligned} \tag{3}$$

$$+(1.22 \pm 0.16)\log \dot{m}_{\rm ADAF}$$
$$+(12.20 \pm 0.94). \qquad (4)$$

This is a 'fundamental plane of radio power'. The radio powers are well-correlated with a function of both the black hole mass and accretion rate, although the radio powers are little correlated individually with the black hole mass or accretion rate, as seen in the two-dimensionally projected plots of $m_{\rm BH}$ vs. $P_{15}^{\rm VLA-A}$ and $\dot{m}_{\rm ADAF}$ vs. $P_{15}^{\rm VLA-A}$ in Fig. 1. The correlation coefficient between $\log m_{\rm BH}$ and $\log P_{15}^{\rm VLA-A}$ is 0.61; the standard deviation of $\log P_{15}^{\rm VLA-A}$ from a regression line is 0.67. The correlation coefficient between $\log \dot{m}_{\rm ADAF}$ and $\log P_{15}^{\rm VLA-A}$ is 0.10; the standard deviation of $\log P_{15}^{\rm VLA-A}$ from a regression line is 0.84. In a two-dimensional plot projected from an edge-on view ($\log P_{15}^{\rm VLA-A}$ vs. $1.52\log m_{\rm BH} + 1.22\log \dot{m}_{\rm ADAF}$, not illustrated in the present paper), the correlation coefficient is 0.85; the standard deviation from the best-fit plane is 0.45.

This best-fit plane suggests that AGNs with higher mass black holes and higher accretion rates tend to have cores with more powerful radio emission. The dependence of the radio power on the black hole mass and its dependence on the accretion rate are nearly the same: $P_{15}^{\rm VLA-A} \propto m_{\rm BH}^{1.52\pm0.14}\dot{m}_{\rm ADAF}^{1.22\pm0.16}$. It is very important that the dependence of radio powers is slightly but significantly different from that of the bolometric luminosities, $L_{\rm bol} \propto m_{\rm BH}\dot{m}_{\rm ADAF}^2$, cf. equation (1). This difference should make another best-fit plane for the radio loudness, which will be described in the next section.

4.2. ($R_{\rm RL}$, $m_{\rm BH}$, $\dot{m}$) plot

As mentioned in the previous section, the dependence of radio powers is different from that of the bolometric luminosities on black hole mass and accretion rate: $P_{15}^{\rm VLA-A} \propto m_{\rm BH}^{1.52}\dot{m}_{\rm ADAF}^{1.22}$ vs. $L_{\rm bol} \propto m_{\rm BH}\dot{m}_{\rm ADAF}^2$. This indicates that also the radio loudness depends on the black hole mass and accretion rate, because of its definition as $R_{\rm RL} \equiv P_{15}^{\rm VLA-A}/P_B \propto P_{15}^{\rm VLA-A}/L_{\rm bol} \propto m_{\rm BH}^{0.52}\dot{m}_{\rm ADAF}^{-0.78}$. Such a relation can be seen in the three-dimensional plot of radio loudness, black hole mass and accretion rate (Fig. 2). The best-fit plane in equation (4) is also expressed as

$$\begin{aligned}\log R_{\rm RL} &= (0.52 \pm 0.14)\log m_{\rm BH} \\ &\quad -(0.39 \pm 0.08)\log(L_{\rm bol}/L_{\rm Edd}) \\ &\quad -(3.48 \pm 0.97) \qquad (5) \\ &= (0.52 \pm 0.14)\log m_{\rm BH} \\ &\quad -(0.78 \pm 0.16)\log \dot{m}_{\rm ADAF} \\ &\quad -(3.98 \pm 0.94). \qquad (6)\end{aligned}$$

This is the fundamental plane of radio loudness. The radio loudness is correlated with both the black hole mass and accretion rate. The correlation coefficient between $\log m_{\rm BH}$ and $\log R_{\rm RL}$ is 0.55; the standard deviation of $\log R_{\rm RL}$ from a regression line is 0.71. The correlation coefficient between $\log \dot{m}_{\rm ADAF}$ and $\log R_{\rm RL}$ is -0.75; the standard deviation of $\log R_{\rm RL}$ from a regression line is 0.51. In a two-dimensional plot projected from an edge-on view ($\log R_{\rm RL}$ vs. $0.52\log m_{\rm BH} - 0.78\log \dot{m}_{\rm ADAF}$ plot, not illustrated in the present paper), the correlation coefficient is 0.82; the standard deviation from the best-fit plane is 0.45. This best-fit plane of radio loudness suggests that AGNs with higher mass black holes and lower accretion rates tend to be radio-louder AGNs.

4.3. Two-dimensional plot of ($m_{\rm BH}$, $\dot{m}$)

A two-dimensional plot of the black hole mass vs. the accretion rate is shown in Fig. 3. It is the projection of the three-dimensional plot onto the ($m_{\rm BH}$, $\dot{m}$) plain, but objects of sample B are also included. Almost all of the nearby AGNs in our samples lie at $\dot{m}_{\rm ADAF} < 10^{-2}$, where

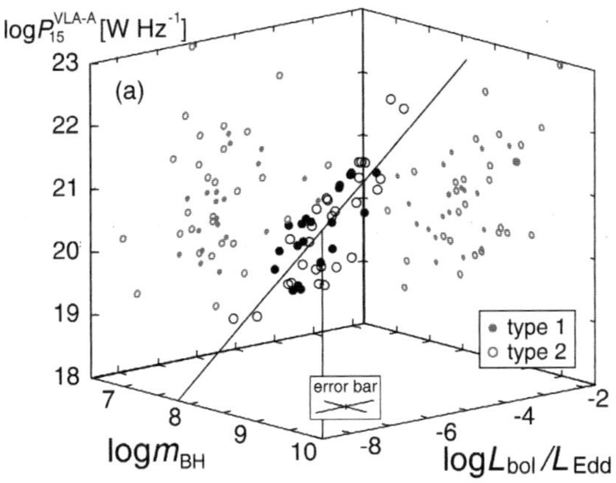

Figure 1. Three-dimensional plot of black hole mass, Eddington ratio, and monochromatic radio power at 15 GHz (*black and white*) and its two-dimensional projected plots (*color*) for sample A. Diagonal lines are the edge-on view of the best fit plane (Section 4.1); we chose such a view point on purpose. Filled and open circles represent type 1s and type 2s, respectively. Error bars are illustrated at the bottom of the plot.

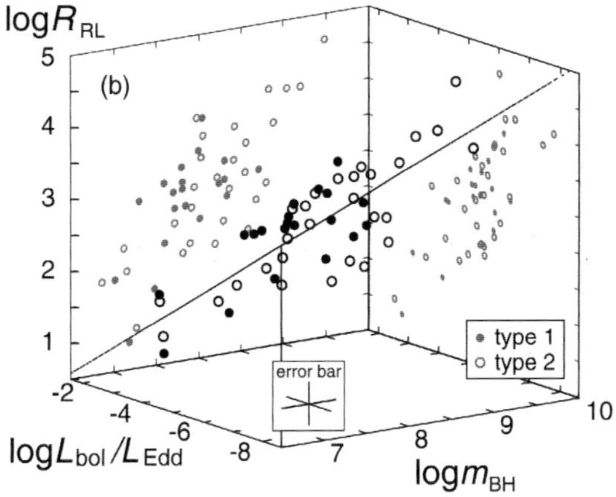

Figure 2. Three-dimensional plot of black hole mass, Eddington ratio, and radio loudness (*black and white*) and its two-dimensional projected plots (*color*) for the same sample as in Fig. 1. Diagonal lines are the edge-on view of the best fit plane (Section 4.2). Symbol descriptions are the same as in Fig. 1

accretion flows can be advection-dominated. Black hole masses are distributed over a range of 10^6–$10^{9.5}$.

In this plot, we illustrate three kinds of lines as follows. The dashed lines represent constant values of bolometric luminosities, which were derived from the extinction-corrected Hβ luminosities of the narrow-line components, on the basis of the hypothesis that photoionization is induced by the radiation of the accretion flow. Blue dot-dashed lines represent constant values of radio power on the best-fit plane (see section 4.1). Nearby AGNs are expected from the best-fit plane to have the radio powers of $\sim 10^{19}$–10^{22} W Hz^{-1}. Red solid lines represent constant values of radio loudness on the best-fit plane of radio loudness (Section 4.2). The nearby AGNs from our samples are expected from the best-fit plane to show the radio loudness of 10^1–10^4, i.e. to be almost always radio-loud.

Type 1s (S1.2–1.9, L1.9) and type 2s (S2, L2, T2) are described with different symbols in Fig. 3. We made statistical tests between the two types in terms of extinction-corrected Hα luminosities, black hole masses, accretion rates, radio powers, and radio loudness. When the parameter includes upper- or lower-limit data, we made a Gehan's generalized Wilcoxon test using the ASURV package for two sample test [17, 22]. When the parameter includes only detected data, we made a Wilcoxon test. We did not found any significant differences between type 1s and 2s for Hα luminosities, black hole masses, radio powers, or radio loudness. On the other hand, the statistical test indicates that type 1s and 2s are not from the same parent population of accretion rate with the significance level of 95 per cent; type 2s show systematically significantly lower accretion rates than type 1s.

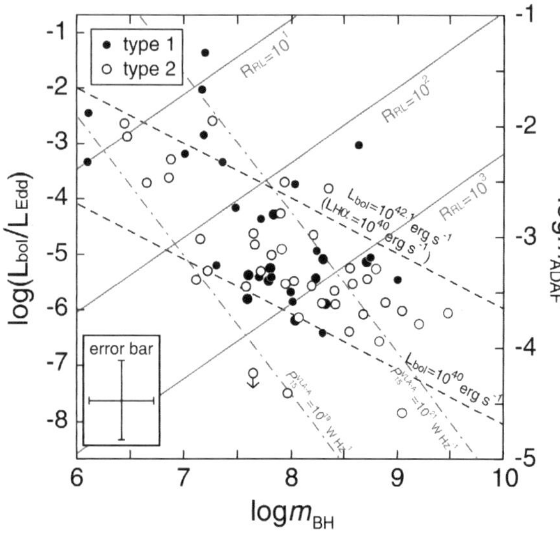

Figure 3. Relation between black hole mass and dimensionless accretion rate. Filled and open circles represent type-1 and type-2 AGNs, respectively. Error bars are illustrated in the lower left-hand corner of the plot. Line descriptions are as follows: black dashed lines represent constant bolometric luminosities, which were estimated from extinction-corrected Hα luminosities (equation (1)); blue dot-dashed lines represent constant monochromatic radio powers of the best-fit plane (equation (4) in Section 4.1); red solid lines represent constant radio loudness of the best-fit plane of radio loudness (equation (6) in Section 4.2).

5. Discussion

The best-fit plane of radio loudness may explain why quasars are found to be both radio-loud and radio-quiet objects, why LLAGNs are exclusively radio-loud, and why narrow-line Seyfert 1s (NLSY1s) are exclusively radio-quiet. The objects with higher luminosities should tend to be radio-quieter as shown in Fig. 3. In the accretion regime of $L_{\rm bol}/L_{\rm Edd} > 10^{-1}$, where NLSY1s and quasars should exist, their radio loudness is expected from the best-fit plane of radio loudness to be 10^{-1}–10^{+2}. The quasars with low mass black holes ($m_{\rm BH} \simeq 10^6$–10^8) are expected to be radio quiet, while those with high mass black holes ($m_{\rm BH} \simeq 10^8$–10^{10}) are expected to be radio loud. In fact, radio-loud quasars have been found exclusively in high mass black hole systems in elliptical host galaxies, while the quasars with $m_{\rm BH} < 3\times10^8$ are practically all radio-quiet [20]. NLSY1s, which have been found exclusively as radio-quiet objects [44], are thought to be luminous Seyferts with very high-accretion rates ($L_{\rm bol}/L_{\rm Edd} \sim 1$) on to low mass black holes ($m_{\rm BH} \sim 10^6$–10^7; [23, 2]). Thus, the observed properties of quasars, NLSY1s and LLAGNs are consistent with the best-fit plane of radio loudness. We suggest 'a unified interpretation of radio loudness': the dependence of bolometric luminosity and radio power on black hole mass and accretion rate provides the best-fit plane of radio loudness, which can phenomenologically explain the radio loudness across the whole luminosity range of AGNs.

However, the best-fit plane of radio loudness was established only with the nearby AGNs, almost all of which are low-accretion rate systems, where the ADAF model is applicable. For quasars and NLSY1s, the standard-disc (or slim-disc) model is more plausible. The radiative efficiency of a standard disc is $L_{\rm bol} = \eta_{\rm eff}\dot{M}c^2$, where $\eta_{\rm eff}$ is the mass-to-energy conversion factor. The radio power of nonthermal jets from the standard discs are theoretically predicted to be $P_{\rm radio} \propto m_{\rm BH}^{1.42}$ [9]. As a result, we find that $R_{\rm RL} \equiv P_{\rm radio}/P_B \propto P_{\rm radio}/L_{\rm bol} \propto m_{\rm BH}^{0.42}\dot{m}^{-1.00}$ is expected in AGNs with a standard disc. Such a dependence is similar to that of our best-fit plane of radio loudness and the model of radiatively inefficient accretion. In fact, radio powers seem to follow a single best-fit plane among stellar-mass black hole systems, LLAGNs and quasars [28]. Therefore, the best-fit plane for radio loudness could be also valid in the quasars and NLSY1 regimes. It is important to examine the best-fit plane also for high accretion-rate AGNs.

Acknowledgments

The authors would like to thank the Yukio Hayakawa Fundation for the travel expense support to participate in the workshop.

References

[1] Abramowicz M. A., Czerny B., Lasota J. P., Szuszkiewicz E., 1988, *ApJ*, **332**, 646
[2] Boroson T. A., 2002, *ApJ*, **565**, 78
[3] Cardelli J. A., Clayton G. C., Mathis J. S., 1989, *ApJ*, **345**, 245
[4] Falcke H., Körding E., Markoff S., 2004, *A&A*, **414**, 895
[5] Franceschini A., Vercellone S., Fabian A. C., 1998, *MNRAS*, **297**, 817
[6] Ferrarese, L. (astro-ph: 0203047)
[7] Ferrarese L., Merritt D., 2000, *ApJ*, **539**, L9
[8] Gebhardt K. et al., 2000, *ApJ*, **539**, L13
[9] Heinz S., Sunyaev R. A., 2003, *MNRAS*, **343**, L59
[10] Ho L. C., 1999, *ApJ*, **516**, 672
[11] Ho L. C., 2002, *ApJ*, **564**, 120
[12] Ho L. C., Filippenko A. V., Sargent W. L. W., 1997a, *ApJS*, **112**, 315
[13] Ho L. C., Filippenko A. V., Sargent W. L. W., 1997b, *ApJ*, **487**, 568
[14] Ho L. C., Filippenko A. V., Sargent W. L. W., 2003, *ApJ*, **583**, 159
[15] Ho L. C., Peng C. Y., 2001, *ApJ*, **555**, 650
[16] Ho L. C. et al., 2001, *ApJ*, **549**, L51
[17] Isobe, T., Feigelson, E. D. 1990, *BAAS*, **22**, 917
[18] Kellermann K. I., Sramek R. A., Schmidt M., Green R. F., Shaffer D. B., 1994, *AJ*, **108**, 1163
[19] Lacy M., Laurent-Muehleisen S. A., Ridgway S. E., Becker R. H., White R. L., 2001, *ApJ*, **551**, L17
[20] Laor A., 2000, *ApJ*, **543**, L111
[21] Laor A., 2003, *ApJ*, **590**, 86
[22] LaValley, M., Isobe, T., Feigelson, E. D., 1992, in ASP Conf. Ser. 25, Astronomical Data Analysis Software and Systems I, ed. D. M. Worrall, C. Biemesderfer, & J. Barnes (San Francisco: ASP), 245
[23] Leighly K. M., 1999, *ApJS*, **125**, 297
[24] Mahadevan R., 1997, *ApJ*, **477**, 585
[25] Manmoto T., Mineshige S., Kusunose M., 1997, *ApJ*, **489**, 791
[26] Marscher A. P., Jorstad S. G., Gómez J., Aller M. F., Teräsranta H., Lister M. L., Stirling A. M., 2002, *Natur*, **417**, 625
[27] McLure R. J., Kukula M. J., Dunlop J. S., Baum S. A., O'Dea C. P., Hughes D. H., 1999, *MNRAS*, **308**, 377
[28] Merloni A., Heinz S., di Matteo T., 2003, *MNRAS*, **345**, 1057
[29] Nagar N. M., Falcke H., Wilson A. S., Ho L. C., 2000, *ApJ*, **542**, 186
[30] Nagar N. M., Falcke H., Wilson A. S., Ulvestad J. S., 2002a, *A&A*, **392**, 53
[31] Narayan R., Yi I., 1994, *ApJ*, **428**, L13
[32] Nelson C. H., 2000, *ApJ*, **544**, L91
[33] Netzer H., Laor A., Gondhalekar P. M., 1992, *MNRAS*, **254**, 15
[34] Paturel G., Petit C., Prugniel P., Theureau G., Rousseau J., Brouty M., Dubois P., Cambrésy L., 2003, *A&A*, **412**, 45
[35] Quataert E., di Matteo T., Narayan R., Ho L. C., 1999, *ApJ*, **525**, L89
[36] Rees M. J., Phinney E. S., Begelman M. C., Blandford R. D., 1982, *Natur*, **295**, 17
[37] Sandage, A. R., & Tammann, G. A. 1981, A Revised Shapley-Ames Catalog of Bright Galaxies (Washington: Carnegie Institution of Washington)
[38] Shakura N. I., Sunyaev R. A., 1973, *A&A*, **24**, 337
[39] Simien F., Prugniel P., 2002, *A&A*, **384**, 371
[40] Stocke J. T., Morris S. L., Weymann R. J., Foltz C. B., 1992, *ApJ*, **396**, 487
[41] Terashima Y., Wilson A. S., 2003, *ApJ*, **583**, 145
[42] Visnovsky K. L., Impey C. D., Foltz C. B., Hewett P. C., Weymann R. J., Morris S. L., 1992, *ApJ*, **391**, 560
[43] Xu C., Livio M., Baum S., 1999, *AJ*, **118**, 1169
[44] Zhou H., Wang T., 2002, *ChJAA*, **2**, 501

Institute of Physics Publishing
doi:10.1088/1742-6596/54/1/054

Journal of Physics: Conference Series **54** (2006) 342–348
Galaxy Center Workshop

The AGN Population in Nearby Galaxies

Mercedes Filho[1], Peter Barthel[2], and Luis Ho[3]

[1] Centro de Astrofísica Da Universidade do Porto, Rua das Estrelas, 4150 – 762, Porto, Portugal
[2] Kapteyn Astronomical Institute, P. O. Box 800, 9700 AV Groningen, The Netherlands
[3] Observatories of the Carnegie Institution of Washington, 813 Santa Barbara Street, Pasadena, CA 91101, USA

E-mail: mfilho@astro.up.pt

Abstract. In order to determine the incidence of black hole accretion-driven nuclear activity in nearby galaxies, we have compiled radio data for the LINERs, composite LINER/HII and Seyfert galaxies from a complete magnitude-limited sample of bright nearby galaxies (Palomar sample). Our results show an overall radio detection rate of 54% (22% of all bright nearby galaxies) and we estimate that at least ~50% (~20% of all bright nearby galaxies) are true AGN. By comparing the radio luminosity function of the LINERs, composite LINER/HII and Seyferts galaxies in the Palomar sample with those of selected moderate-redshift AGN, we find that our sources naturally extend the radio luminosity function of powerful AGN down to powers of about 10 times that of Sgr A*.

1. Introduction

The search for low-luminosity active galactic nuclei (LLAGN) in nearby galaxies has been the subject of many optical surveys. The Palomar survey [10,11,12] has been very useful in this regard by providing a sensitive magnitude-limited (B_T <12.5 mag) sample of almost 500 bright nearby galaxies. About half of the sources are emission-line nuclei, classified as Seyferts, LINERs or composite LINER/HII galaxies, the last category displaying both LINER and HII properties. However, characterizing the powering mechanisms of the sources is not straightforward, particularly in low-luminosity sources. Many of these galaxies possess circumnuclear star-forming regions which blend with and may even drown out the presence of a weak active galactic nucleus (AGN).

Optimally it is necessary to pick spectral regions where the contrast between any hypothetical LLAGN component and circumnuclear stellar component is maximized. X-rays are very useful in this regard as shown by the hard X-ray studies of LLAGN [13,23,24,25]. In the absence of spectral AGN signatures such as a Seyfert or quasar-type continuum or broad emission lines, radio observations can offer an alternative method for determining the LLAGN incidence in nearby galaxies. Measurements of radio flux, compactness, radio spectral index[1] and brightness temperatures provide the necessary diagnostic tools for determining the nature of the radio emission.

Several important radio surveys have been conducted on the magnitude-limited Palomar bright, nearby galaxy sample [10,11,12], revealing a large fraction of radio cores (Figure 1). In

[1] $F_\nu \propto \nu^{-\alpha}$ throughout.

a recent Very Large Array (VLA) 5 and 1.4 GHz, 1" resolution survey of the low-luminosity Seyferts of the Palomar sample [14], it was found that over 80% of the sources harbour a radio core. A VLA, 15 GHz, 0.25" resolution survey of a distance-limited sample of low-luminosity Seyferts, LINERs and composite LINER/HII galaxies [18,19,20] revealed that ~40% of the objects harbour subarcsecond-scale compact radio cores. A recent Multi-Element Radio-Linked Interferometer Network (MERLIN) 5 GHz, 0.1" resolution survey of a subsample of LINERs [5] detected several new subarcsecond-scale radio cores. A similar study with the VLA, at 8.4 GHz, 2.5" resolution of all the composite LINER/HII galaxies in the Palomar sample [7,8], revealing radio cores in ~25% of the sample sources.

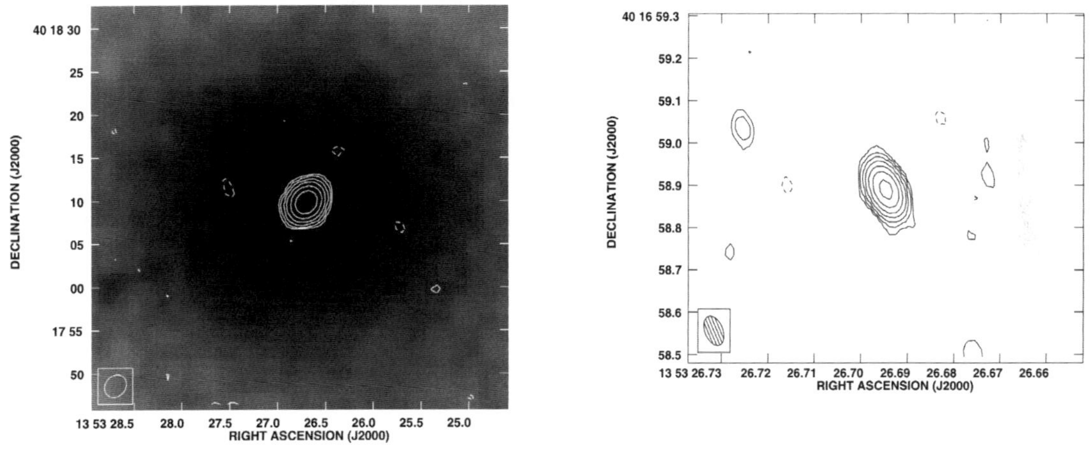

Figure 1. Examples of the radio observations *Left:* The radio contours of the composite galaxy NGC 5354 obtained with the VLA at 8.4 GHz and 2.5" resolution, superimposed on the DSS image [8]. *Right:* The radio contours of the LINER galaxy NGC 5353 obtained with MERLIN at 5 GHz and 0.1" resolution [5].

However, although the radio core emission in these sources is consistent with the presence of a LLAGN, we cannot exclude a stellar origin from the brightness temperature figures ($T_B < 10^5$ K; [3]) obtained at these resolutions. As conclusive judgement requires Very Long Baseline (VLBI)-resolution, multi-wavelength Very Long Baseline Array (VLBA) and European Very Long Baseline Interferometer Network (EVN) observations have been obtained for selected subsamples of low-luminosity Seyferts, LINERs and composite LINER/HII galaxies that showed arcsecond- or subarcsecond-scale radio cores [1,4,6,9,18,19,26]. In sources with subarcsecond- or arcsecond-scale radio peak emission above 2.5 mJy, results reveal a 100% detection rate of high-brightness temperature ($T_B > 10^8$ K), compact, flat spectrum ($\alpha < 0.5$) radio cores, enforcing the LLAGN scenario for the radio emission [1,6,18,26]. Their low radio luminosities suggest we are probing the very faint end of the AGN population.

Unambiguously determining the physical nature of the nearby galaxy radio cores is more than of mere phenomenological interest. If they truly contain an accretion-powered nucleus, then they obviously need to be included in the AGN population. Their non-trivial numbers impact on several astrophysical problems ranging from the cosmological evolution of the AGN luminosity function to their contribution to the X-ray background.

2. The Radio Data

In order to construct a representative AGN radio luminosity function (RLF) of the local Universe, we have used the emission-line (excluding HII) sources in the Palomar sample. All of these sources have now been observed at 2.5" resolution or better. We shall refer to the 196 LINERs, composite sources and Seyferts which satisfy both the magnitude (B_T <12.5 mag) and declination criteria (δ >0 degrees) of the Palomar survey as the 'LTS sources' for brevity (LTS meaning LINER-Transition-Seyfert); these sources constitute the present sample.

Ideally, we would like to have a homogeneous set of radio observations. But lacking such a survey, we have assembled a list of radio measurements used to derive the RLF for the LTS sample sources. Radio observations at different frequencies have been converted to 5 GHz assuming a spectral index of 0.7 and also corrected for different cosmologies if necessary. When multiple observations of the same galaxy were available, by order of preference we choose HU01 for Seyferts, the radio observations in [7,8], and [18,19,20] for composite galaxies, [18,19,20] and MERLIN observations in [5] for LINERs.

We caution that the radio-detection rate will depend strongly on observing frequency, resolution and sensitivity. At higher frequencies, many sources may escape detection because spectral indices are not always flat and instrumental sensitivities will also be lower. Furthermore, the source may suffer resolution effects. With this in mind, the [18,19,20] samples, and [5,7,8] observations provide only lower limits to the radio-detection rate in LTS sources.

3. Statistics

3.1. Radio Detection Rate

Galaxies are considered detected if their radio flux density is above 5σ, where σ is the typical noise associated with the respective survey. 82%, 43%, and 49% of the Seyferts, LINERs, and composite galaxies were detected in the radio at 2.5" or less. This is equivalent to an overall radio-detection rate of 54% in all LTS sources or 22% of all bright nearby galaxies (Palomar sources). For Type 1 and 2 LTS sources, the radio detections are 89% and 46%, respectively. Of the Type 1 radio-detected sources (which must be genuine AGN), 59% are classified as Seyferts and 35% are LINERs. Both broad-lined composite galaxies (Type 1.9 NGC 1161 and NGC 2985) were detected. 100% of Type 1 Seyferts and 67% of Type 1 LINERs were detected. 46%, 68%, and 94% of the Type 2 radio-detected sources are Seyferts, LINERs, and composite galaxies, respectively.

3.2. AGN Fraction

We can also estimate the fraction of these radio-detected LTS sources that are likely to be genuine AGN. In the absence of unambiguous optical spectral signatures of AGN activity (e.g., broad emission lines), the radio regime provides an alternative and complementary diagnostic. Although not a necessary condition for the presence of an AGN, a compact, flat-spectrum radio core is indicative of synchrotron self-absorption, which is associated with jet emission from AGN. However, there are several caveats. First, ground-based optical spectra and VLBI-resolution radio images typically sample very different spatial, and presumably temporal, scales. Thus, optical and radio signatures of AGN activity need not occur concurrently. Second, because the mechanism of jet formation is still uncertain, radio emission cannot be regarded as an inevitable by-product of AGN activity. Finally, empirical evidence suggests that there is a radio flux density threshold of 2 mJy below which the sources become difficult to detect at milliarcsecond-scale resolution using current facilities [4,9,18]. Known sources with submilliJansky radio cores and/or hard X-ray detections (e.g., NGC 660 and NGC 7331; [7,9]), which are likely to be genuine AGN, or sources that are highly radio variable can be missed by these relatively shallow, milliarcsecond-scale resolution observations. We therefore caution that our estimate based on radio detection is a lower limit to the true AGN fraction in LTS sources. For the sake of homogeneity, we will restrict ourselves to observations of <1" resolution. With the above caveats in mind, we conclude that at least 80% of the Seyferts [14], 40% of the LINERs [5,18,19,29] and 20% of the composite sources [5,18,19,20] are likely AGN. Based on the presence of compact radio emission, the total fraction of LTS sources and bright nearby galaxies harbouring AGN is therefore ~50% and ~20%, respectively.

According to classical AGN unification schemes [2], Type 2 objects are simply Type 1 AGN seen edge-on, whereby the molecular torus blocks the direct view of the broad-line region (BLR). However, in the case of LLAGN it is not entirely clear that unification schemes are readily applicable. There is growing evidence that not all low luminosity Seyferts, LINERs and composite sources possess a BLR and when they do the BLR is weak; there are only twelve sources in the Palomar sample [11] classified as Type 1.0-1.8 and all are Seyferts. We can then conservatively argue that all Type 1 sources, which exhibit broad-line emission in their spectra, are genuine AGN. Therefore, based solely on the presence of broad-line emission, we can estimate that at least 20% of the LTS sources and 10% of all bright nearby galaxies harbour an AGN.

4. The Local Radio Luminosity Function

Because the sample sources are nearby (median D = 17 Mpc), we have considered a flat, Euclidean Universe with q_0=0.5 and H_0=75 km s^{-1} Mpc^{-1} for the subsequent calculations. The V/V_{max} method [22] was applied in order to construct the RLF at 5 GHz. The main constraints arise from the magnitude limit of the Palomar survey, which is taken to be B_{limit}=12.5 mag (Ho, Filippenko & Sargent 1995), and the radio flux limit of the survey from where the radio luminosity of the LTS source was obtained. Moreover, only galaxies with positive declination were observed, which restricts the survey area covered to 2π. The calculation of the RLF is then performed over equal bins in log of radio power (0.4 dex). Statistical errors associated with the space densities were assigned assuming Poisson statistics.

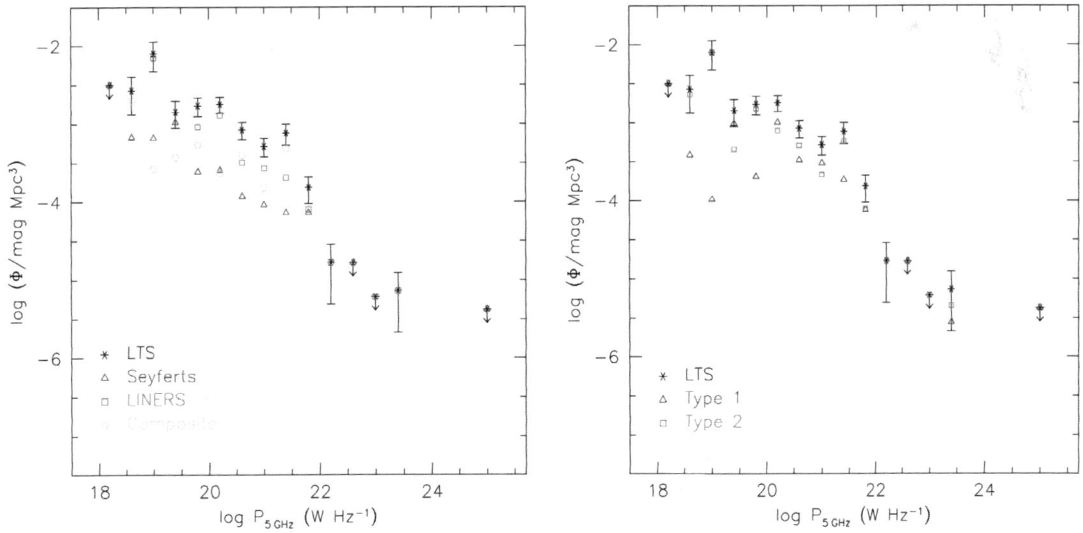

Figure 2. *Left:* The radio luminosity function for the LTS, Seyferts, LINERs and composite galaxies in the Palomar sample. *Right:* The radio luminosity function for the LTS, Type 1 and Type 2 sources in the Palomar sample.

5. Discussion

Figure 2 contains the derived RLF for the radio-detected LTS sample sources, together and separately for Seyferts, LINERs, composite sources and then also for Type 1 and Type 2 objects (with and without broad lines, respectively). It is worth mentioning that, particularly on the low-power end, the RLF is irregular; this is the most likely due to density inhomogeneities in our local volume.

Inspection of the plots shows that space densities of LTS sources continue to rise with decreasing radio power, with some evidence of flattening below 10^{20} W Hz^{-1}, partly due to incompleteness of the radio survey (see also discussion in [18]). At all radio powers the space densities are clearly dominated by the LINER galaxies, in particular at the high radio power end ($> 10^{22}$ W Hz^{-1}). Both Seyfert and LINER galaxies contribute to the steady rise in space densities. On the other hand, the space densities of composite sources appear relatively flat in the range $19 < \log P_{5\,GHz} < 21$ W Hz^{-1}.

For Type 1 and Type 2 sources, at all radio powers the space densities are dominated by the more numerous Type 2 sources, although 56% of the Type 2 sources were not detected on scales <2.5". They do, however, span the same radio power range as Type 1 sources and appear to turn over at a power of 10^{21} W Hz^{-1}. Type 2 sources roughly mimic the composite source and Seyfert behaviour in its flattening below $\log P_{5\,GHz} = 21$ W Hz^{-1}.

It is possible that Type 2 or (equivalently composite sources and some LINERs) are a mixed case, whereby only a small fraction of these are genuine AGN. However, various lines of evidence, as stated above, suggest that we are underestimating the number of LLAGN in the LTS sample. Long integration X-ray and radio observations should prove useful in this regard to provide a complete survey with uniform sensitivity and resolution. If there is a significant population of submillijansky LLAGN that we cannot detect with the present observations, then by including them in the RLF, we should expect higher space densities, in particular at the low-luminosity end of the RLF.

Furthermore, because we have compiled radio measurements from surveys with different resolutions, is it likely that in some cases we are overestimating radio flux densities. Many of these sources are known to suffer from resolution effects [7,8,9,19,20]; the radio power of the underlying AGN may be an order of magnitude or so lower than given by the arcsec-scale observations. The overall effect would be to shift the RLF to lower radio luminosities.

It is interesting to compare our results with available published surveys. In [18] the authors have derived a RLF for a distance-limited sample of Palomar sources. Comparison between our analysis and theirs shows that the results are in rough agreement, within the errors. In particular, the larger sample of the LTS sources presented in this study (our 106 versus their 68 radio-detected sources) allows us to confirm the low-power turnover seen in the RLF.

We have also derived a RLF for the Markarian [17] and CfA Seyferts [15] (2" resolution VLA radio data from [16]), converting flux densities to 5 GHz assuming $\alpha = 0.7$ (Fig. 3; left). We caution that the Markarian Seyferts have been observed with the Westerbork Radio Synthesis Telescope (WRST) and nuclear radio flux densities may be over-estimated. The RLFs are consistent with the Palomar Seyfert RLFs for powers above 10^{21} W Hz^{-1}. The rising of the LTS and Palomar Seyfert RLF towards lower powers demonstrates that the LTS sample contains fainter and more local sources than those in the Markarian and CfA samples. This result is in agreement with that found in VLA/N05 and in HU01, considering the difference in RLF frequency and binning.

Similarly, we can compare our RLF with the AGN sources in the 2dF Galaxy Redshift Survey [21] (2dF/NVSS AGN sample). Galaxies in this sample have been classified as AGN according to their spectral characteristics; they show either an absorption-line spectrum like that of a giant elliptical, an absorption spectrum with weak LINER-type emission lines or stellar continuum dominated by nebular emission lines of [OII] or [OIII], which are strong compared to any Balmer-line emission. The sample has been cross-correlated with the NVSS catalog. We have converted the flux densities to 5 GHz assuming an $\alpha = 0.7$ and corrected for different cosmologies. The resulting RLF is plotted along with the LTS RLF in Figure 3 (right).

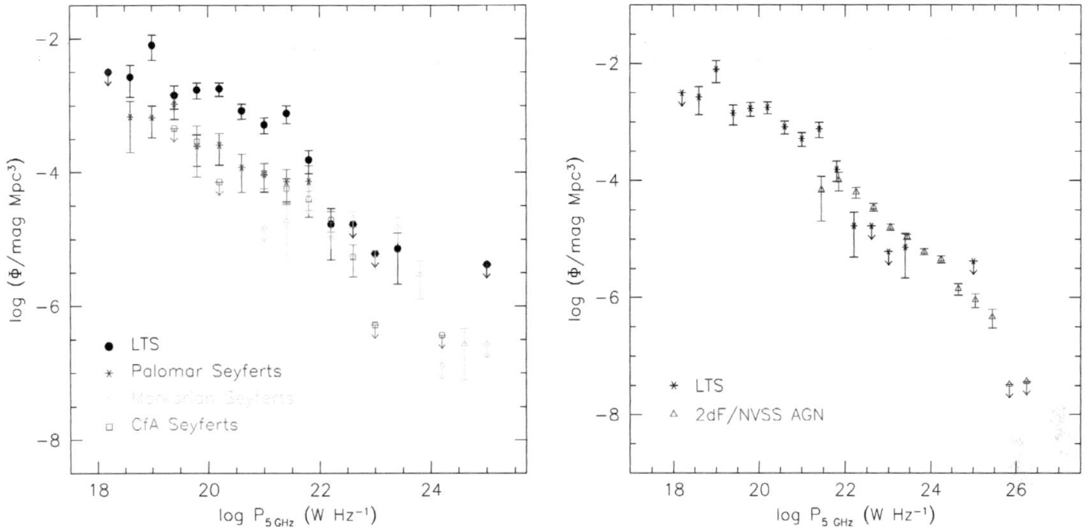

Figure 3. *Left:* The radio luminosity function for the LTS sources compared to the Palomar, Markarian [17] and CfA [15,16] Seyferts. *Right:* The radio luminosity function for the LTS sources compared to the 2dF/NVSS AGN sample [21].

Because the 2dF/NVSS AGN sample is relatively nearby (median z=0.2), it allows a direct comparison with the LTS source RLF. We caution, however, that because this sample has been cross-correlated with NVSS data, the radio flux densities may be slightly over-estimated. The plot shows that there is an overlap in radio luminosities for the LTS and 2dF/NVSS AGN RLF in the regime log $P_{5\,\mathrm{GHz}} \approx 21$ and 23 W Hz^{-1}. In this region of overlap, both the normalization and slope of the two RLFs are roughly similar, within the errors. The LTS sources naturally extend the 2dF/NVSS AGN RLF to lower luminosities. To emphasize the extreme low powers sampled by our RLF, we note that the lowest power LTS sources are only ~10 times more powerful than Sgr A*. The overall shape and the smooth transition from the 2dF/NVSS AGN RLF to the LTS RLF, suggest a luminosity continuation between these two source populations. It is natural to view the LTS sources as the low-redshift, low-luminosity counterparts of the AGN as sampled by the 2dF/NVSS AGN survey.

6. Conclusions

We have compiled radio observations for the non-stellar emission-line sources – low-luminosity Seyferts, LINERs and composite LINER/HII galaxies – in the magnitude-limited Palomar survey. Our results reveal a radio-detection rate of 54% (or 22% of all bright nearby galaxies), with a more than 50% detection rate (or 20% for all bright nearby galaxies) of low-luminosity active nuclei. The radio detection of the Seyferts, LINERs and composite LINER/HII sources in the Palomar survey allow the construction of a local radio luminosity function. Our results show that the Seyferts, LINERs and composite LINER/HII sources form a smooth luminosity transition from higher redshift, more luminous AGN as sampled by the Markarian and CfA Seyfert and 2dF/NVSS AGN survey.

Acknowledgments

M. E. F. acknowledges support from the Fundação para a Ciência e Tecnologia, Ministério da Ciência e Ensino Superior, Portugal through the grant PRAXIS XXI/BD/15830/98 and SFRH/BPD/11627/2002. We are grateful to Jim Ulvestad, Mike Garrett, Simon Garrington, Jim Condon, Naveen Reddy, Marco Spaans and Filippo Fraternali for useful suggestions. Thanks also to Simon Garrington, Anita Richards and Peter Thomasson for valuable help with the data reduction.

References

[1] Anderson, J. M., Ulvestad, J. S., & Ho, L.C. 2004, ApJ, 603, 42
[2] Antonucci, R. 1993, ARA&A, 31, 473
[3] Condon, J. J. 1992, ARA&A, 30, 575
[4] Falcke, H., Nagar, N. M., Wilson, A. S., & Ulvestad, J. 2000, ApJ, 542, 197
[5] Filho, M. E., Barthel, P. D., & Ho, L. C., 2006, A&A, 451, 71
[6] Filho, M. E., Barthel, P. D., & Ho, L. C. 2002b, A&A, 385, 425
[7] Filho, M. E., Barthel, P. D., & Ho, L. C. 2002a, ApJS, 142, 223
[8] Filho, M. E., Barthel, P. D., & Ho, L. C. 2000, ApJS, 129, 93
[9] Filho, M. E., Fraternali, F., Markoff, S., Nagar, N. M., Barthel, P. D., Ho, L. C., & Yuan, F. 2004, A&A, 418, 429
[10] Ho, L. C., Filippenko, A. V., & Sargent, W. L. W. 1997b, ApJ, 487, 568
[11] Ho, L. C., Filippenko, A. V., & Sargent, W. L. W. 1997a, ApJS, 112, 315
[12] Ho, L. C., Filippenko, A. V., & Sargent, W. L. W. 1995, ApJS, 98, 477
[13] Ho, L. C. et al. 2001, ApJ, 549, L51
[14] Ho, L. C., & Ulvestad, J. S. 2001, ApJS, 133, 77
[15] Huchra, J. & Burg, R. 1992, ApJ, 393, 90
[16] Kukula, M. J., Pedlar, A., Baum, S. A., & O'Dea, C. P. 1995, MNRAS, 276, 1262
[17] Meurs E. J. A. & Wilson A. S. 1984, A&A, 136, 206
[18] Nagar, N. M, Falcke, H., & Wilson, A. S. 2005, A&A, 435, 521
[19] Nagar, N. M, Falcke, H., Wilson, A. S., & Ulvestad, J. S. 2002, A&A, 392, 53
[20] Nagar, N. M, Falcke, H., Wilson, A. S., & Ho, L. C. 2000, ApJ, 542, 186
[21] Sadler, E., et al. 2002, MNRAS, 329, 227
[22] Schmidt, M. 1968, ApJ, 151, 393
[23] Terashima, Y., Ho, L. C., & Ptak, A. F. 2000, ApJ, 539, 161
[24] Terashima, Y., Ho, L. C., Ptak, A. F., Mushotzky, R. F., Serlemitsos, P. J., Yaqoob, T., & Kunieda, H. 2000, ApJ, 533, 729
[25] Terashima, Y., & Wilson, A. S. 2003, ApJ, 583, 145
[26] Ulvestad, J. S., & Ho, L. C. 2001b, ApJ, 562, L133

Institute of Physics Publishing
doi:10.1088/1742-6596/54/1/055

Journal of Physics: Conference Series **54** (2006) 349–353
Galaxy Center Workshop

The millimeter variability of M81*

Rainer Schödel[1] **and Andreas Eckart**[1,2]

[1] I.Physikalisches Institut, Universität zu Köln, Zülpicher Str. 77, 50937 Köln, Germany
[2] Max-Planck-Institut für Radioastronomie, Auf dem Hügel 69, 53121 Bonn, Germany

E-mail: rainer@ph1.uni-koeln.de

Abstract. M81*, the nucleus of the nearby spiral galaxy M81 (NGC 3031) is an LLAGN with a luminosity $< 10^{-5}$ times the Eddington luminosity of the related supermassive black hole. It appears to have very similar characteristics to Sagittarius A (Sgr A*), the radiative counterpart of the black hole at the center of the Milky Way. Here, we present simultaneous observations at 3 and 1 mm that were obtained within the framework of a coordinated, multi-wavelength campaign on M81*. We find that at mm-wavelengths M81* is a continuously variable source with the higher variability observed at the shorter wavelength. The variability at 3 and 1 mm appears to be correlated. The data show that M81* is indeed a system with very similar physical properties to Sgr A*. The observed variability time scales point to an upper size limit of the emitting region of the order 25 Schwarzschild radii. The obtained data clearly demonstrate the usefulness and, above all, necessity of simultaneous multi-wavelength observations of LLAGN.

1. Introduction

Until about a decade ago, models of accretion onto black holes (BHs) and the related emission were largely concentrated on thermal emission from thin disks. However, the vast majority of galactic nuclei (at least in the local universe) radiate many orders of magnitude below the Eddington limit. These systems have become observable only since high-resolution, high-sensitivity instrumentation has become available in the (sub)mm, the infrared, and the X-ray domains.

Sub-Eddington accretion flows onto BHs in Low Luminosity AGN, LINERs, and almost quiescent nuclei are clearly distinguished from classical AGN by the absence of the so-called "blue bump" in the optical/UV, that is attributed to thermal emission from an optically thick, geometrically thin accretion disk [17]. In highly sub-Eddington systems (in the following called simply "LLAGN", for brevity), on the other hand, accretion is modeled by radiatively inefficient accretion flows (RIAFs), such as the ADAF [24, 25, 30, 31]. Combined RIAF+outflow models are highly successful in describing the SED of LLAGN, in fact, jets appear to become increasingly important with decreasing overall luminosity of the sources [12, 29, 16]. LLAGN appear to be up-scaled versions of the low/hard state (LHS) of X-ray binaries, which is characterized by the presence of fairly constant, mildly relativistic jets. A universal correlation between X-ray and radio luminosity appears to hold across a range of $> 10^7$ in BH mass [16, 5, 14, 22, 13]. However, the regime of high masses and low Eddington luminosities is still poorly explored and needs to be investigated in order to close the gap in the claimed scaling relation. It seems that synchrotron and SSC/IC from a region very close to the BH is responsible for the submm to X-ray regime in LLAGN. Whether this region is the inner parts (likely outflowing winds) of an RIAF or the base of a jet, or the interface region including both, is still under debate.

Start	Stop	ν_3 [GHz]	ν_1 [GHz]	$\sigma_{\Phi,3}$ [deg]	$\sigma_{A,3}$ [%]	$\sigma_{\Phi,1}$ [deg]	$\sigma_{A,3}$ [%]
24-02 01:11	24-02 19:45	115.3	230.5	10-20	5-8	20-40	11-18
14-07 06:50	15-07 13:51	80.5	241.4	15-35	8-11	40-60	25-30
19-07 23:17	20-07 16:07	86.2	218.2	12-28	~3	30-70	11-18

Table 1. Observations of M81* with the PdBI during February and July 2005. Start and stop times in UTC. ν_3 is the exact frequency used around 3 mm, ν_1 the one at a wavelength of 1 mm. The last four columns list the ranges of the rms values of the phase, Φ, and amplitude, A, between all baselines for each observing epoch and wavelength ($\sigma_{\Phi,3}$ and $\sigma_{A,3}$ for 3 mm and $\sigma_{\Phi,1}$ and $\sigma_{A,1}$ for 1 mm).

Only simultaneous observations with a chance to determine the relationship in time between the various wavebands can constrain the emitting geometry. Invariably we know that RIAFs likely feed jets [20], and while LLAGN often appear as unresolved radio cores at lower resolutions, high resolution observations show that there may be jets present in all LLAGN [23].

With a luminosity $\sim 10^{-9 \ldots -10} \times L_{Edd}$, Sgr A* is – in terms of Eddington luminosity – the weakest accreting black hole with observational statistics good enough to fit models to its spectrum. It is the primary testbed for theoretical models of extreme sub-Eddington accretion, which rely largely on the available radio/submm/NIR/X-ray observations of Sgr A* [21].

The nearby spiral galaxy M81 (NGC 3031) is an Sb spiral galaxy similar to the Milky Way. It is located at a distance of 3.63 ± 0.34 Mpc [15]. From spectroscopic measurements of the $H\alpha + [NII]$ emission, probably emitted from a rotating gas disk inclined at an angle of $14° \pm 2°$, a mass of $7.0^{+2}_{-1} \times 10^7\,M_\odot$ could be determined for the central black hole in M81 [8]. Due to its proximity and due to its luminosity M81* is the ideal bridge between the extremely faint Sgr A* at the one extreme and AGN/Quasars on the other extreme of the luminosity scale. Intriguingly, M81* shows many similarities to Sgr A*, such as a slightly inverted radio spectrum [26] where circular polarization dominates over linear polarization at radio frequencies of 4.8 – 15 GHz [3, 4]. VLBI observations have revealed a stationary core with a variable one-sided jet of ~3600 AU length [2].

Sgr A* is the best example that a highly variable source can only be understood well through simultaneous measurements at different wavelengths [11, 18, 9, 10]. An international, coordinated, multi-wavelength campaign was set up in 2005-2006 in order to observe M81* simultaneously from radio wavelengths to the X-ray domain, involving instruments such as the GMRT, the VLA, the PdBI, the SMA, and the Chandra X-ray observatory (S. Markoff et al., in preparation). In this contribution we describe the result of three epochs of simultaneous observations of M81* at wavelengths of 3 and 1 mm with the Plateau de Bure interferometer that were obtained during the coordinated campaign.

2. Observations

M81* was observed with the PdBI on 24 February, 14-15 July, and 19-20 July 2005 (see Tab. 1). Phase calibration was performed with the sources 1044+719 and 0836+710. Primary flux calibrators to determine the efficiencies of the antennae were the sources 1044+719 for February 23-24 (1.6 Jy at 3 mm/1.1 Jy at 1 mm), 1044+719 (1.8 Jy at 3 mm) and 2200+420 (8.7 Jy at 1 mm) for July 14/15, and MWC349 (1.0 Jy at 3 mm) and 3C454.3 (33.0 Jy at 1 mm) for July 20. The phase calibrators 1044+719 and 0836+710 were used to fit the time-dependent fluctuations of the amplitude for all baselines. The quality of the February data is highest as concerns stability and rms of phase and amplitude (see Tab, 1). A detailed description of the data and their reduction can be found in Schödel et al. (2006, submitted to A&A).

Individual scans of 20 min duration were extracted from the calibrated data. The flux of the

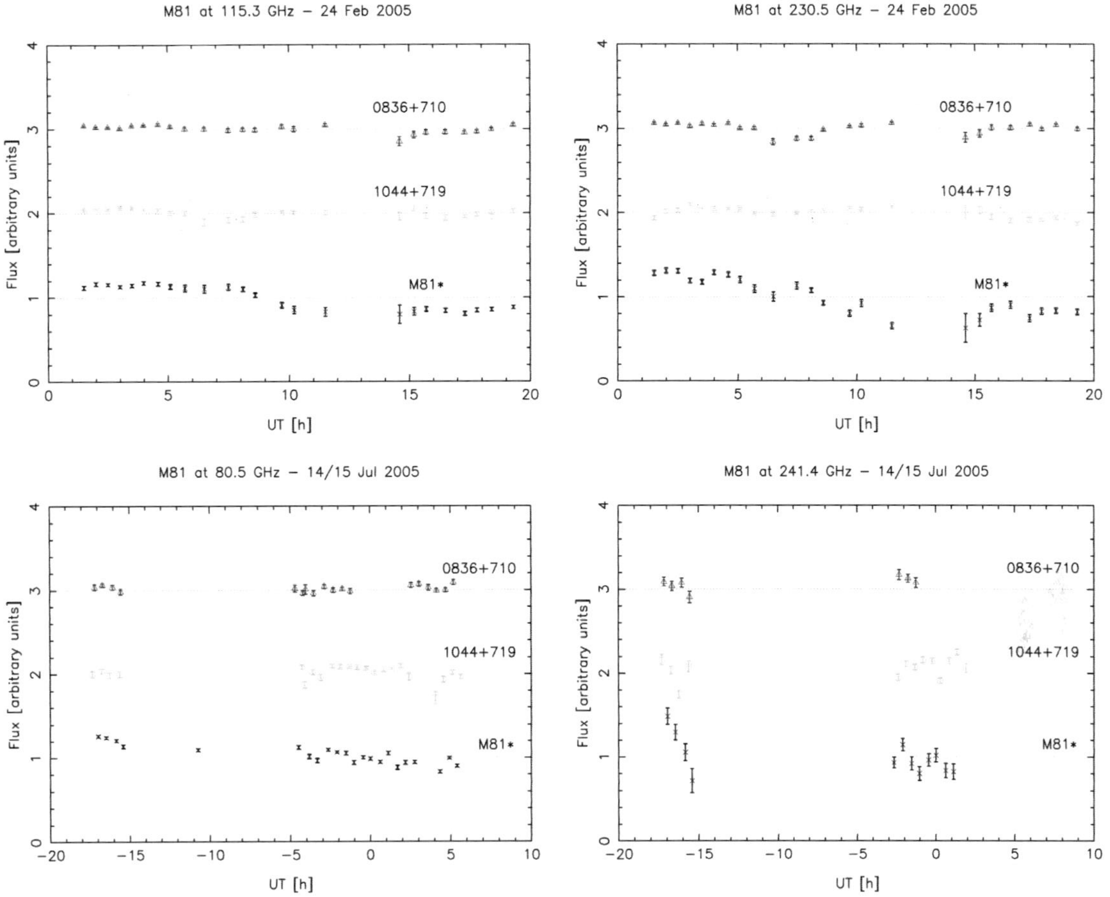

Figure 1. Normalized light curves of M81* and the calibrators 1044+719 and 0836+710 from the PdBI observations at 3 and 1 mm. The x-axes show UTC in hours. The fluxes were scaled to an average value of 1 (curves of the calibrators shifted for better comparison).

sources was determined via fitting of a point source (fitting a Gaussian resulted in very similar results). The resulting light curves for M81* and the two phase calibrators are shown in Figs. 1 and 2.

3. Results

The light curves in Figs. 1 and 2 show that M81* is a continuously varying source at the observing wavelengths of 3 and 1 mm, as has been found before at 3 mm [27]. The three epochs of simultaneous measurements show in summary

- Continuous and strong ($\geq$ 30%) variability of M81*, with the largest variability at the higher frequency.
- Variability at the two wavelengths appears to be correlated.
- With increasing flux the spectral index appears to become more negative, i.e. the general trend of the spectrum becomes steeper with higher flux density at 3 mm.
- $\geq 5\,\sigma$ variability of the flux density occurs on time scales of 5 hours (data from February 24), or even faster (data from July 14, which are unfortunately of much lower quality).

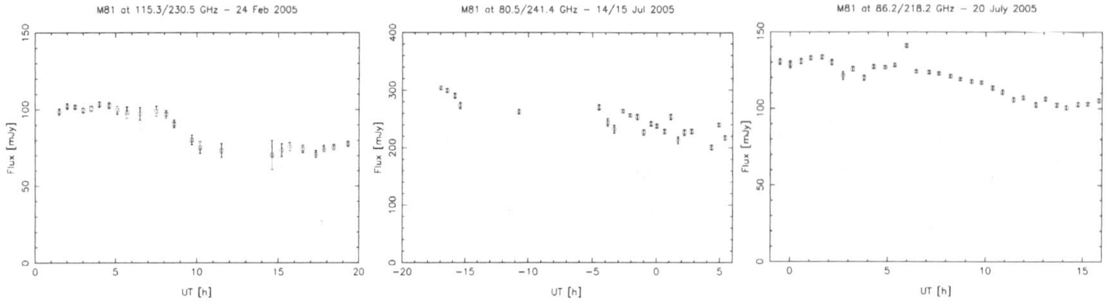

Figure 2. Flux density of M81* on 24 February, 14/15 July, and 20 July 2005. The black boxes mark the flux density at 3 mm and the green (gray) crosses mark the flux density at 1 mm. The exact observing frequencies are indicated in the titles of the individual panels.

As discussed in Schödel et al. (2006, submitted to A&A), the detected variability cannot be due to polarization of the source, which shows absence or at least a very low degree of linear polarization at the observing wavelengths [4].

4. Discussion

The simultaneous flux density measurements of M81* at 3 and 1 mm, obtained at three different epochs indicate that the flux at 1 mm is generally lower than at 3 mm, with the exception of the first ~4 hours of the 24 February light curve. This is in good agreement with previous mm-observations [26], that were, however, not acquired simultaneously.

When combining our measurements with all other radio/mm observations of M81* that have been acquired so far, we find that the highest flux densities were usually measured at a wavelength of ~3 mm [27, 6, 7, 17, 1]. The flux density measurements at shorter wavelengths by [26] and in this work may therefore indicate a decrease towards higher frequencies.

However, the evidence for a turnover in the SED of M81* between 3 and 1 mm is not unambiguous. In fact, observations with the SMA at 345 Ghz that were obtained during the coordinated campaign on M81* indicate that the flux density of M81* increases toward the submm regime, in agreement with the theoretical predictions (Markoff et al., in preparation). Unfortunately, there was only one epoch (24 February 2005), where measurements with the PdBI and the SMA were simultaneous.

A jet is known to exist in M81* [2]. In a simplified jet model the turnover frequency, ν_t, scales with the jet power, Q_j, and mass of the black hole, $M_{\rm BH}$, like $\nu_t \propto Q_j^{2/3} M_{\rm BH}^{-1}$ [13]. When using this relationship to compare M81* with Sgr A* one expects a ~50 times higher ν_t for M81*. Hence, while the SMA observations and theory indicate a ν_t located at submm frequencies (see also Markoff et al., in preparation), the discussion above shows that the 3 and 1 mm data by itself in combination with older data indicate a turnover at lower frequencies. The situation appears therefore somewhat complex.

Possibly, the variability of M81* at radio to mm-wavelengths is related to the ejection of plasma blobs and their subsequent adiabatic expansion [28], which would lead naturally to a stronger variability at shorter wavelengths, in agreement with the observations presented in this work. Events of adiabatic expansion of plasma blobs, that may flow outwards in a jet, would explain a negative 3-to-1 mm spectral index when the 3 mm flux is near its peak emission. In fact, the coordinated observations of M81* show some evidence for variability events that move toward longer wavelengths. Therefore, the overall peak of the emission of M81* may well be located at wavelengths shorter than 1 mm, while observations at individual epochs can indicate turnover at greater λ because of "waves" of variability that move along an expanding outflow

(Markoff et al., in prep.).

The drop-off between 07:00 UT and 12:00 UT in the highest quality data set from 24 February 2006 shows that $\sim 5\sigma$ variability of the flux at 3 and 1 mm occurs on a time scale of 5 hours. This corresponds to an upper limit on the size of the source of merely ~25 Schwarzschild radii, when a mass of 7.0×10^7 $M_\odot$ is assumed for the black hole in M81*. The first four data points on 14 July at 1 mm indicate a possibly even faster variability. The facts that the phase calibrators show only apparently random variations at this time and that the 3 mm light curve decreases, too, support this view. However, due to the low quality of the 14 July data, the existence of such rapid variability needs to be confirmed by additional observations. This rapid variability may be a phenomenon similar to the mm-flares seen in observations of Sgr A* [19] emphasizing again the similarity between the two sources.

The discussion above demonstrates the necessity of observing highly variable sources, such as M81*, simultaneously at different wavelengths. There is clearly an urgent need for more data of this kind. Expanding this work to other LLAGN will provide the necessary constraints for models to describe the dominant mechanisms at work in black holes that accrete in the highly sub-Eddington regime and show whether there exists indeed a scaling law for the sub-Eddington accretion from XRBs to high-mass LLAGN.

Acknowledgments
The authors would like to thank all people that helped in organizing the Galactic Center Workshop 2006.

References

[1] Bartel, N., I.I.Shapiro, B.E. Corey et al. 1982, *Astrophys. J.*, **262**, 556
[2] Bietenholz, M.F. & N. Bartel 2000, *Astrophys. J.*, **532**, 895
[3] Brunthaler, A., G.C. Bower, H. Falcke & R.R. Mellon 2001, *Astrophys. J.*, **560**, L123
[4] Brunthaler, A., G.C. Bower, & H. Falcke 2006, *Astron. Astrophys.*, **451**, 845
[5] Corbel, S., M.A. Nowak, R.P.Fender, A.K. Tzioumis, & S. Markoff 2003, *Astron. Astrophys.*, **400**, 1007
[6] Crane, P.C., T.S. Giuffrida & J.B. Carlson 1976, *Astrophys. J.*, **203**, L113
[7] de Bruyn, A.G., P.C. Crane, R.M. Price & J.B. Carlson 1976, *Astron. Astrophys.*, **46**, 243
[8] Devereux, N., H. Ford, Z. Tsvetanov, & G. Jacoby 2003, *Astron. J.*, **125**, 1226
[9] Eckart, A., F.K. Baganoff, M. Morris et al. 2004, *Astron. Astrophys.*, **427**, 1
[10] Eckart, A., F.K. Baganoff, R. Schödel et al. 2006, *Astron. Astrophys.*, **450**, 535
[11] Falcke, H., W.M. Goss,H. Matsuo et al. 1998, *Astrophys. J.*, **499**, 731
[12] Falcke, H. & S. Markoff 2000, *Astron. Astrophys.*, **362**, 113
[13] Falcke, H., E. Körding, & S. Markoff 2004, *Astron. Astrophys.*, **414**, 895
[14] Fender, R.P., T.M. Belloni, & E. Gallo 2004, *M.N.R.A.S.*, **355**, 1105
[15] Freedman, W.L., S.M. Hughes, B.F. Madore, et al. 1994, *Astrophys. J., Part 1*, **427**, 628
[16] Gallo, E., R.P. Fender, & G.G: Pooley 2003, *M.N.R.A.S.*, **344**, 60
[17] Ho, L.C: 1999, *Astrophysical Journal*, **516**, 672
[18] Markoff, S., H. Falcke, F. Yuan, & P.L. Biermann 2001, *Astron. Astrophys.*, **379**, L13
[19] Mauerhan, J.C., M. Morris, F. Walter & F.K. Baganoff 2005 *Astrophys. J.*, **623**, L25
[20] Meier, D.L. 2001, *Astrophys. J.*, **548**, L9
[21] Melia, F. & H. Falcke 2001, *Annu. Rev. Astron. Astrophys.*, **39**, 209
[22] Merloni, A., S. Heinz, & T. di Matteo 2003, *M.N.R.A.S.*, **345**, 1057
[23] Nagar, N.M., H. Falcke, A.S. Wilson, & J.S. Ulvestad 2002, *Astron. Astrophys.*, **392**, 53
[24] Narayan, R. & I. Yi 1994, *Astrophysical Journal*, **428**, L13
[25] Quataert, E. 2003, *Astronomische Nachrichten Supplement*, **324**, 435
[26] Reuter, H.-P. & H. Lesch 1996, *Astron. Astrophys.*, **310**, L5
[27] Sakamoto, K., H. Fukuda, K. Wada & A. Habe 2001, *Astronom. J.*, **122**, 1319
[28] van der Laan, H. 1966, *Nature*, **211**, 1131
[29] Yuan, F., S. Markoff, & H. Falcke 2002, *Astron. Astrophys.*, **383**, 854
[30] Yuan, F., E. Quataert, & R. Narayan, *Astrophysical Journal*, **598**, 301
[31] Yuan, F., E. Quataert, & R. Narayan 2004, *Astrophysical Journal*, **606**, 894

Institute of Physics Publishing
doi:10.1088/1742-6596/54/1/056

Journal of Physics: Conference Series **54** (2006) 354–362
Galaxy Center Workshop

The Submillimeter Polarization of Sgr A*

Daniel P. Marrone[1], James M.Moran[1], Jun-Hui Zhao[1] and Ramprasad Rao[2]

[1] Harvard-Smithsonian Center for Astrophysics, 60 Garden St., Cambridge, MA 02138, USA
[2] Inst. of Ast. and Astrophys., Academia Sinica, P.O. Box 23-141, Taipei 10617, Taiwan

E-mail: dmarrone@cfa.harvard.edu

Abstract. We report on the submillimeter properties of Sgr A* derived from observations with the Submillimeter Array and its polarimeter. We find that the spectrum of Sgr A* between 230 and 690 GHz is slightly decreasing when measured simultaneously, indicating a transition to optically thin emission around 300 – 400 GHz. We also present very sensitive and well calibrated measurements of the polarization of Sgr A* at 230 and 345 GHz. With these data we are able to show for the first time that the polarization of Sgr A* varies on hour timescales, as has been observed for the total intensity. On one night we find variability that may arise from a polarized "blob" orbiting the black hole. Finally, we use the ensemble of observations to determine the rotation measure. This represents the first statistically significant rotation measure determination and the only one made without resorting to comparing position angles measured at separate epochs. We find a rotation measure of $(-5.6 \pm 0.7) \times 10^5$ rad m^{-2}, with no evidence for variability on inter-day timescales at the level of the measurement error. The stability constrains interday fluctuations in the accretion rate to 8%. The mean intrinsic polarization position angle is $167^\circ \pm 7^\circ$ and we detect variations of 31^{+18}_{-9} degrees. This separation of intrinsic polarization changes and possible rotation measure fluctuations is now possible because of the frequency coverage and sensitivity of our data. The observable rotation measure restricts the accretion rate to the range 2×10^{-7} $M_\odot$ yr^{-1} to 2×10^{-9} $M_\odot$ yr^{-1}, if the magnetic field is near equipartition and ordered.

1. Introduction

The linear polarization of Sgr A* was first detected by Aitken et al. [1] above 100 GHz, after unsuccessful searches at lower frequencies [2, 3]. Through subsequent observations it has been established that the polarization varies in position angle [4, 5] and fraction [6], with variability occurring on timescales comparable to those of the previously observed total intensity variations (see § 3). The variability may be intrinsic to the source or due to propagation effects, but the short timescales suggest processes at work very close to the black hole and thus polarization should be a useful tool for the study of Sgr A*.

The ability to exploit polarization information is significantly sensitivity-limited for this faint source. Until very recently polarimetric observations have required full nights to obtain reliable measurements, obscuring changes on shorter intervals that might be expected based on total intensity variations. Perhaps more important has been the limited instantaneous frequency coverage. Multiple frequencies must be observed in order to separate the local changes in the emission region from the external changes in the propagation medium (such as rotation measure changes). In the absence of spectral information, these changes can at best be separated statistically through comparisons of observations from different frequencies and different times.

As a result of the poor frequency coverage available to date, Faraday rotation has yet to be convincingly detected in this source. Because of its potential to constrain the accretion rate close to the black hole, measurement of Faraday rotation has been a chief goal of polarimetric observations.

Here we present several results obtained with the Submillimeter Array (SMA). The SMA is a sensitive new instrument at a good submillimeter site and favorable latitude for observing the Galactic center. Its has large instantaneous frequency coverage; single-receiver observations provide data in two sidebands 10 GHz apart and in dual-receiver observations the frequency span can be a factor of three (230 to 690 GHz). All of these advantages make the SMA a powerful instrument for the study of the Galactic center, and Sgr A* in particular. We report on observations of the instantaneous spectrum of Sgr A* and the peak in its spectral energy distribution (§ 2), the polarization variability on short timescales (§ 3), and the first measurement of Faraday rotation in this source (§ 4). Other SMA Galactic center observations are presented by Montero-Castaño, Qin, and Winnberg in this volume.

2. Submillimeter Spectrum

Sgr A* is known to have a rising spectrum across three decades in frequency from 300 MHz to 300 GHz. A few detections have been made at frequencies up to 900 GHz [1, 7, 8, 9, 10], but the flux densities are scattered and Sgr A* is occasionally undetectable at the same telescope and frequency [8, 11]. This suggests variability in the source and may also indicate the difficulty of extracting a flux density for Sgr A* from observations with $10 - 20''$ angular resolution in the presence of contaminating dust and free-free emission. There are only upper limits on the flux density in the far and mid-infrared, while (transient) counterparts in the near-infrared and X-ray show quiescent flux densities that are several orders of magnitude below that at 300 GHz [12, 13, 14]. These data suggest an unobserved peak in the spectral energy distribution (SED) between 300 GHz and the mid-infrared.

Using the unprecedented submillimeter angular resolution afforded by the SMA, as well as its ability to observe simultaneously at 230 and 690 GHz, we have made several observations of the submillimeter emission of Sgr A*. The radio spectrum is shown in Figure 1, including four epochs of SMA 230/690 GHz observations. The SMA data were obtained on three nights in 2005,

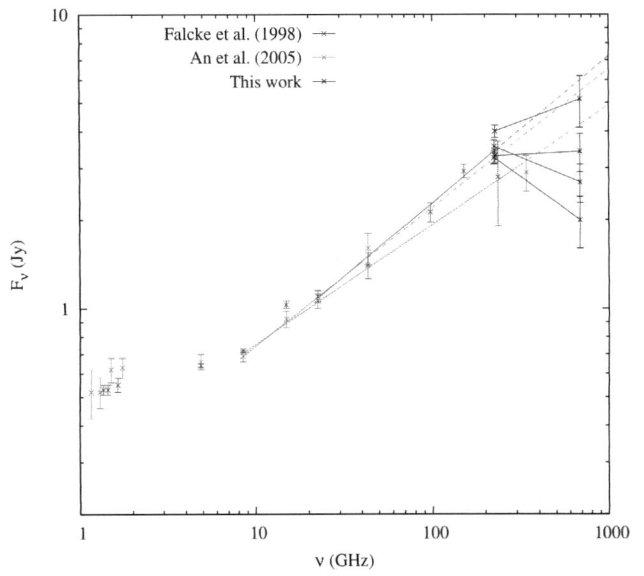

Figure 1. The radio SED of Sgr A*. Points from two multi-wavelength campaigns [15, 16] are shown with power-law fits to the "submillimeter bump". Extrapolated spectra are continued with dashed lines. We provide two fits to the An et al. [16] points, with and without the 340 GHz point, as this frequency is expected to be near the SED peak (and thus below the extrapolation from lower frequencies) based on our observations. The 230 and 690 GHz points in each epoch of SMA data are connected by a solid line.

with one night split into two intervals due to noticeable variability. The data were obtained in excellent weather, 225 GHz zenith opacities of $0.034-0.045$ and corresponding 690 GHz opacities of $0.55-0.75$. Through careful calibration against quasars and multiple primary flux standards (planets), we were able to determine the flux density scale to 15% precision at 690 GHz on the best nights. We find that the spectral index (α, for $S \propto \nu^{\alpha}$) between the two frequencies is flat, ranging between -0.4 and $+0.2$, with a variance-weighted average value of $\alpha = -0.13$.

We contrast these observations to the spectral index measured at lower frequencies. Below approximately 10 GHz, Sgr A* shows a slowly rising spectrum [15, 16, 17] typical of an inhomogeneous optically thick synchrotron source. Closer to our frequencies, observations have found a steeper spectrum in a "submillimeter bump" [18] extending to at least 200 GHz. Spectral measurements at the highest frequencies must be undertaken with care because interday variations of $20-40\%$ are observed above 100 GHz [19], with larger variations observed on few hour timescales [6, 20, 21]. Two multi-facility campaigns to measure the simultaneous spectrum of Sgr A* to ~ 300 GHz are available to provide calibration for our results. In the 1996 observations of Falcke et al. [15] the average spectral index from 22 to 236 GHz is 0.50, while the 2003 campaign of An et al. [16] shows an average spectral index of 0.41 from 8.4 to 340 GHz, or 0.47 if the 340 GHz point is excluded. Yusef-Zadeh et al. [10] also have extensive wavelength coverage but unfortunately their flux densities are contaminated by emission from a radio transient [22] and by flaring. The fits to the simultaneous spectra are plotted in Figure 1 and demonstrate that most of our 230 GHz points correspond well to the extrapolated lower-frequency spectra. The exceptional epoch has a high flux density in both bands and is unusual for other reasons, as shown in Section 3. We clearly observe a deviation from these extrapolations at 690 GHz, with flux density decrements of up to 4 Jy, indicating that we are straddling the peak in the SED. In models of the spectrum of this source the turnover originates from the transition to optically thin emission, typically in a compact component of synchrotron emission corresponding to the inner regions of an accretion flow or the base of a jet [23, 24]. We therefore infer that at frequencies above approximately 300 GHz we have reached this transition and are observing the emission closest to the black hole. This result is particularly important for planned 345 GHz VLBI observations of the relativistically warped image of the black hole. In the case of emission from an accretion flow, VLBI images near the transition frequency should show more structure than would be expected from observations at very optically thin or optically thick frequencies [25]. Something similar might be expected for a jet model since the compact jet nozzle should be the dominant emission source at high frequencies [24], rather than larger scale jet emission that would not appear warped by strong gravity.

3. Polarization Light Curves

The SMA observations are the first to allow short time interval sampling of the millimeter/submillimeter linear polarization of Sgr A* (see Eckart et al. in this volume for polarimetric data in the near infrared). Our 2005 data sample the 230 GHz polarization of Sgr A* in 4 minute integration periods, yielding a typical precision of $0.5-1.0\%$, while the 345 GHz polarization is measured over 5.5 minute periods at a precision of $1.3-1.8\%$. For calibration details see Marrone [26]. This temporal resolution is at least a factor of thirty better than the only previous time-resolved data [6]. As shown by Broderick and Loeb [27] (see also, these proceedings), the polarization changes during flares should be an essential signature of strong gravity. With a favorable flare, our temporal resolution may allow us to measure the black hole spin and properties of the flaring plasma.

A sample polarized light curve from the SMA is shown in Figure 2. The data are from the same night as the highest 230/690 GHz points in Figure 1. Although the total intensity (Stokes I) does vary during these observations, the changes visible in the polarization are much more dramatic. The light curve shows a 50° swing in polarization position angle over a period of just two hours

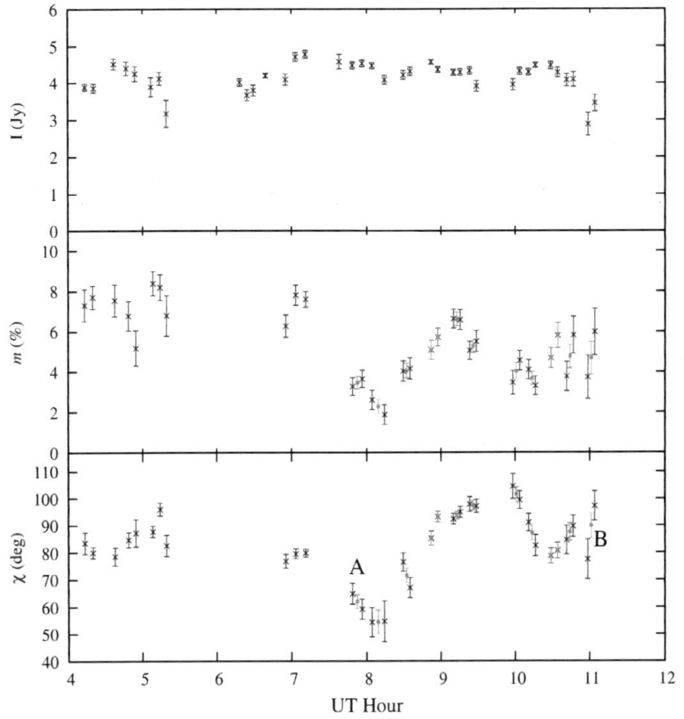

Figure 2. The variability of Sgr A* in total intensity (Stokes I), polarization fraction (m), and polarization position angle (χ) over one night of observations. Each black point corresponds to a four minute sample of the polarization of Sgr A*. In the lower two panels the grey points show the same data rebinned to increase the signal-to-noise ratio; it is these points that are plotted in Figure 3. The grey points labeled A and B in the lower panel are similarly labeled in Figure 3.

and a decrease in polarization fraction from nearly 8% to 2% over just one hour. Maintaining relative flux calibration at the level of a few percent is difficult in the submillimeter, making the $10-20\%$ variability observed in Stokes I difficult to measure to high significance. However, the polarization information is much easier to measure reliably because the observations can be self-calibrated on the relatively strong point source emission from Sgr A* [26]. This makes polarimetry particularly useful for monitoring observations.

The polarization variability near the end of the observations in Figure 2 looks vaguely periodic when plotted in m and χ coordinates. Examination of more natural coordinates, Stokes Q and U, confirms that we do observe a polarization orbit. The grey points from Figure 2 are plotted in the Q$-$U plane in Figure 3, with time proceeding from point A to point B. The polarization appears to circle the mean polarization (marked by the black point) multiple times over the last hours of the light curve. We can understand this track with a simple picture of a submillimeter flare in Sgr A*. We attribute the flaring emission to a newly excited blob of synchrotron-emitting plasma orbiting the black hole in the inner regions of an accretion flow. As the blob orbits the black hole its apparent linear polarization direction will rotate, so when this is added vectorially to the polarization of the surrounding quiescent emission it appears to orbit the mean polarization. The roughly four hour period (twice the period in the Q$-$U plane) would correspond to an orbital radius of around 13 Schwarzschild radii ($r_{\rm S}$), well outside the marginally stable orbit for a black hole of any spin parameter. Although other mechanisms for variability in Sgr A* are certainly possible, this simple picture matches well with our observations. More complicated effects such as synchrotron losses, general relativistic light paths, and radial motion of the blob can also be considered to allow better fits of the observed polarization track, as discussed by Broderick and Loeb in this volume. Nevertheless, the correspondence between this picture and our observations gives us hope for both the utility of short-timescale polarimetry in understanding Sgr A* flares and the potential for modeling polarized flares.

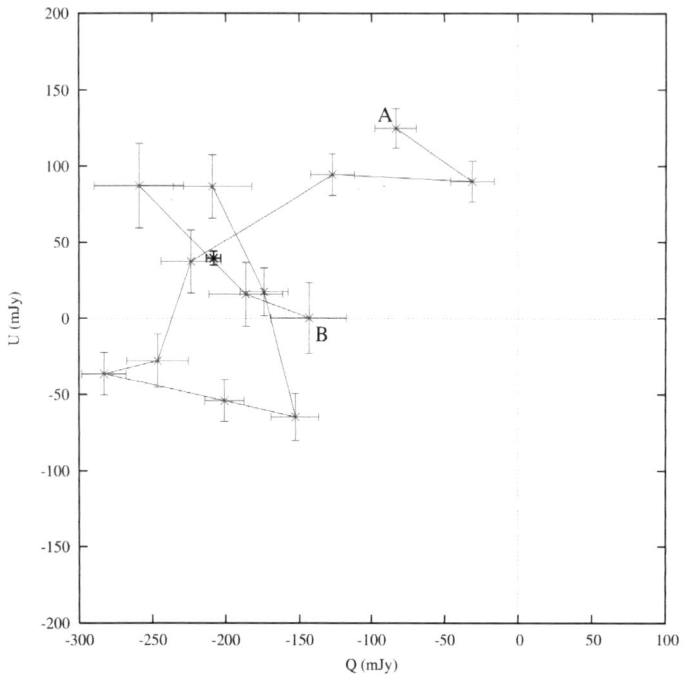

Figure 3. The path traced in the Q–U plane over the last three and a half hours of the observation shown in Figure 2. Time increases from point A to point B. The mean polarization for the entire night is shown with a bold cross near Q= −210 mJy, U=40 mJy. The polarization makes nearly two full orbits around the mean polarization.

4. Rotation Measure and Intrinsic Polarization

Despite great interest, a detection of the Faraday rotation measure (RM) has remained elusive since the discovery of linear polarization in 1999 [1]. The presence of linear polarization was used immediately after its detection to argue that the infalling plasma must be tenuous and the mass accretion rate low (less than $10^{-6}\ M_\odot\ \mathrm{yr}^{-1}$), as larger accretion rates would depolarize the emission through extreme Faraday rotation angles [28, 29]. Further progress on determining the accretion rate and its variability has awaited an RM measurement, which is made difficult by the polarization variability and the diminished effect of Faraday rotation at high frequencies. To date there have been three claims of a RM determination through non-simultaneous observations, although none of them is robust. Bower et al. [4] used the few available measurements to determine an RM of -4.3×10^5 rad m^{-2}, but the discovery of polarization variability rendered the result uncertain [5]. Marrone et al. [6] compared average 230 and 345 GHz position angles to find -5×10^5 rad m^{-2}. Most recently, Macquart et al. [30] found an RM of -4.4×10^5 rad m^{-2} from their 83 GHz polarization and the average of all previous data. However, their analysis allows a much lower RM at a probability of 10% or more due to the 180° degeneracy of the polarization position angle.

Simultaneous measurements are a much more secure way to determine the RM, but the instantaneous frequency coverage of available instruments has not been adequate to show Faraday rotation. The lowest instantaneous RM upper limit is presently -7×10^5 rad m^{-2} from early SMA 345 GHz polarimetric observations [6], comparable to the RM expected from non-simultaneous measurements. Multi-frequency data allow for the determination of both the RM and intrinsic polarization direction, allowing separation of their fluctuations. Isolating these changes is crucial to understanding the conditions in the inner accretion flow. The RM can be used to determine upper and lower limits on the accretion rate, while variations with frequency and time can be used to examine the structure of the flow. In this section we use SMA measurements to obtain a highly significant RM detection and strong evidence that the intrinsic polarization is variable.

4.1. Observations and Calibration

The results in this section depend crucially on careful calibration and thorough examination of possible systematic effects. Measurements of the RM in Sgr A* with the SMA require convincing detections of inter-sideband position angle differences of 1.5° at 345 GHz and 5° at 230 GHz. Here we present a summary of the calibration, with details reserved for a paper (Marrone et al., submitted to *ApJL*).

The data presented here were obtained on ten nights in the summer of 2005 during polarimeter commissioning and as part of regular science programs. There were six tracks in the 230 GHz band and four in the 345 GHz band. The weather was very good, with a typical zenith opacity at 225 GHz of 0.06 – 0.08, or 0.19 – 0.27 when scaled to 343 GHz. All data were obtained with five to seven antennas in the compact configuration (7 – 70 meter baselines), resulting in a synthesized beam width of 2$''$.0×4$''$.0 at 230 GHz and 1$''$.6×3$''$.2 at 345 GHz on Sgr A*. For the purposes of self-calibration and polarization extraction, baselines shorter than $20k\lambda$ were removed to exclude the extended emission around Sgr A*. Polarization was extracted through point-source fits to the visibilities. To obtain full polarization information we used the SMA polarimeter [6], which will be described in a forthcoming instrument paper (D. P. Marrone et al., 2006, in prep.). In 2005 new wave plates were added to the polarimeter to allow simultaneous polarimetry in the SMA 230 GHz and 690 GHz bands. These represent the first scientific data obtained with the new capabilities.

Measurement of polarization relies on precise determination of the fractional contamination ("leakage") of each polarization state by the cross-handed polarization. Uncalibrated leakage contaminates the linearly polarized Stokes parameters (Q and U) with I. Leakages were measured each night by observing polarized quasars (3C 279 or 3C 454.3) over a large range of parallactic angle and simultaneously solving for the source and instrumental polarization. The leakages are stable, with an r.m.s. variability of 0.3% at 230 GHz and 0.4% at 345 GHz. There are three potential sources of variability: (1) real instrumental polarization changes between nights, (2) finite signal-to-noise ratio, and (3) polarization leakage that is not constant across the sky, resulting in leakage determinations that depend on the hour angle coverage. The first two sources are likely present at a low level. The third effect is also present, at a level of around 0.2%, due to the SMA optical configuration. Fortunately, the resulting polarization changes are small. Leakage errors of 0.3%–0.4% contribute 0.2% fractional contamination to the instantaneous polarization measured with seven antennas, significantly less when averaged over a full track due to the parallactic angle rotation. For Sgr A*, which we measure to be 5%–10% polarized, this results in at most 1° of position angle error. More importantly for this purpose, the uncorrected instrumental polarization (source 3, above) is very nearly constant across the sidebands. Therefore, although the absolute position angle varies by up to 1°, the inter-sideband difference varies only by 0.1°–0.2°.

4.2. Rotation Measure and Intrinsic Polarization

Faraday rotation changes the observed polarization position angle (χ) as a function of frequency according to:

$$\chi(\nu) = \chi_0 + \frac{c^2}{\nu^2}\mathrm{RM}, \tag{1}$$

where χ_0 is the intrinsic position angle. The rotation measure (RM) is proportional to the integral of the electron density and parallel magnetic field component along the line of sight. From the observed LSB and USB position angles we derive a RM and χ_0 for each observation (see Figures 4 and 5). The larger errors on 345 GHz points in these figures result from the much smaller inter-sideband difference in λ^2 at the higher frequency. The average RM from all 10 epochs is $(-5.6 \pm 0.7) \times 10^5$ rad m^{-2}, while the 230 GHz and 345 GHz points alone yield $(-5.4 \pm 0.7) \times 10^5$ rad m^{-2} and $(-13 \pm 5) \times 10^5$ rad m^{-2}, respectively, consistent within their

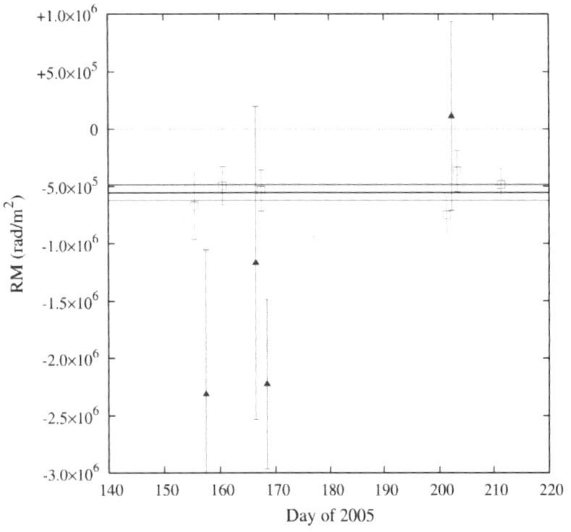

Figure 4. Rotation measure derived from ten nights of observations in the summer of 2005. Measurements at 230 GHz are shown as squares and those at 345 GHz are shown as triangles. The mean RM (-5.6×10^5 rad m^{-2}) and 1σ range are shown with horizontal lines.

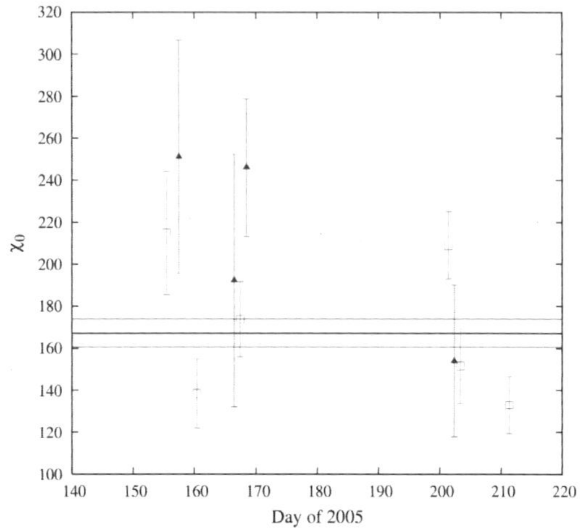

Figure 5. Measurements of the intrinsic polarization position angle (χ_0) from the summer of 2005. The mean value is $\chi_0 = 167° \pm 7°$. Symbols as in Figure 4.

errors. The ten single-night RM values are consistent with a constant value ($\chi_r^2 = 1.09$, for 9 degrees of freedom). We therefore place an upper limit of 7×10^4 rad m^{-2} on the RM dispersion.

This represents the first statistically significant measurement of the RM of Sgr A*, and the only measurement made from simultaneous observations at multiple frequencies and therefore able to isolate source polarization changes. Our ten measurements show an average intrinsic polarization of $167° \pm 7°$, or $162° \pm 7°$ and $210° \pm 21°$ from 230 and 345 GHz observations, respectively. The χ_0 values vary by more than our measurement errors predict, suggesting intrinsic polarization changes. Assuming a constant χ_0, we obtain $\chi_r^2 = 2.8$ (0.3% probability) using all data points, or $\chi_r^2 = 3.9$ (0.16%) for the 230 GHz points only. The variability in the full data set suggests an intrinsic χ_0 dispersion of 31^{+18}_{-9} degrees, with very similar results obtained from 230 or 345 GHz points separately.

*4.3. Implications for the Nature of Sgr A**

The RM we observe is too large to be produced by material beyond the Bondi accretion radius of Sgr A* (approximately $1''$ or 0.04 pc). Using the density measurements of Baganoff et al. [14], the RM in the inner $10''$ is just 8×10^3 rad m^{-2} assuming a 1 mG ambient field. Observations of nearby radio sources have found RMs up to 70% of this estimate [31].

The RM can be used to determine the accretion rate at small radii around Sgr A* if assumptions are made about the nature of the accretion flow. The procedure is outlined in Marrone et al. [6] and assumes a power-law radial density profile ($n \propto r^{-\beta}$) and ordered, radial, equipartition-strength magnetic fields. The accretion rates allowed by our RM detection are shown in Figure 6. The most important parameter in determining the range of accretion rate upper limits is $r_{\rm in}$, the radius at which electrons become relativistic and their RM contribution begins to be suppressed [28]. In general, shallower density profiles ($\beta \to 1/2$) approaching that of the convection-dominated accretion flow (CDAF) [32] yield hotter central temperatures and thus larger $r_{\rm in}$ [33]. Simulations favor β values closer to this type of flow (i.e., $\beta \leq 1$) [23, 34, 35, 36]. Simulations with published temperature profiles show $r_{\rm in}$ between 30 and $100 r_S$ [23, 34], yielding

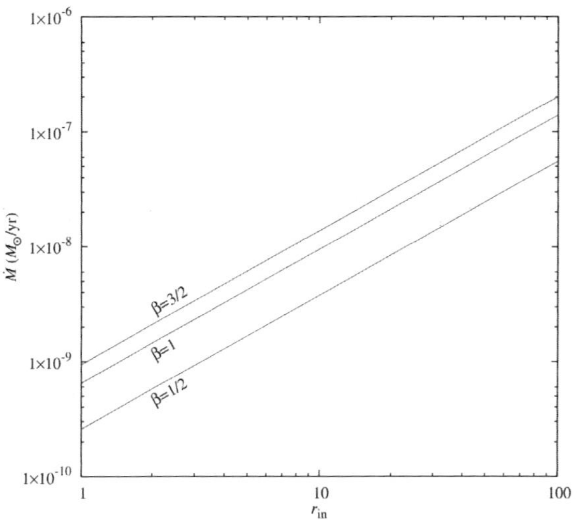

Figure 6. Accretion rate constraints from the RM detection, based on the Marrone et al. [6] formalism. We plot the accretion rate as a function of r_{in}, the radius at which electrons become relativistic, as this parameter strongly influences the derived limits. The three lines trace three density profiles bounded by the ADAF ($\beta = 3/2$) and CDAF (1/2) models. We asssume a coherent magnetic field ($r_{\mathrm{out}} = \infty$). If we instead choose $r_{\mathrm{out}} = 3r_{\mathrm{in}}$, all three lines are within a factor of 2 of the $\beta = 3/2$ line in this figure.

accretion rate upper limits of 2×10^{-7} to 5×10^{-8} $M_\odot$ yr^{-1}. Steeper profiles ($\beta \to 3/2$), as in Bondi-like or advection-dominated accretion flows (ADAF) [37] are marginally relativistic to small radii and may have r_{in} of a few r_{S}, indicating an accretion rate upper limit of 1.5×10^{-8} for $r_{\mathrm{in}} = 10r_{\mathrm{S}}$. This has been used to rule out the original ADAF model, which has an accretion rate comparable to the Bondi rate (10^{-5} $M_\odot$ yr^{-1} for this source, [14]).

It is very important to note that the assumptions of the Marrone et al. [6] formalism, which were taken from previous related works [28, 38], are not observationally constrained. In particular, the assumption of equipartition-strength fields is not well justified, although simulations may show a tendency to develop field strengths of a few percent of equipartition [35]. Magnetic fields that are a fraction ϵ of the equipartition strength will raise the accretion rate limits by $\epsilon^{2/3}$ (a factor of 10 for $\epsilon = 3\%$). The assumption of an ordered field is also potentially suspect. The direction of position angle change with frequency, and thus the sign of the RM, appears unchanged since the Aitken et al. [1] measurements in 1999. This sign stability in the RM and stability of the sense of circular polarization may be evidence of a highly reversed field with a small bias superimposed in an accretion flow or jet [39, 40]. Sign stability would not be expected from turbulent accretion flows unless a special geometry is invoked or a rigid field is assumed [35]. If the stochastic field picture is correct, the accretion rate could be much higher than the limits we have derived.

Our RM also allows us to place lower limits on the accretion rate; these are not subject to the caveats above since all of the uncertainties act to raise the minimum accretion rate. If we take r_{in} to be around $10r_{\mathrm{S}}$ or $3r_{\mathrm{S}}$ (smaller r_{in} yields smaller lower limits, so this is conservative for hot flows), we find that the accretion rate must be greater than $1 - 2 \times 10^{-8}$ $M_\odot$ yr^{-1} or $2 - 4 \times 10^{-9}$ $M_\odot$ yr^{-1}, respectively. This rules out a thin disk in this source, as accretion rates of 10^{-9} $M_\odot$ yr^{-1} or less are required to match the luminosity of Sgr A* with a thin disk [41]. Moreover, if the field is toroidal, reversed, or sub-equipartition, these lower limits are raised and may pose problems for very low accretion rate models. Our limit on RM variability (12%) also limits the fractional accretion rate variability to 8% over two months.

5. Conclusions

We have shown several new observations of Sgr A*, made possible by the Submillimeter Array and its polarimeter. In particular, we report the first significant detection of the RM in this source and discuss its implications. Significant work remains in polarimetric studies of Sgr A*.

In particular, these observations cannot be used to detect deviations from the ν^{-2} dependence expected for Faraday rotation. Such deviations, arising from internal Faraday rotation, would be evidence for cooler than expected gas at small radii, which would favor steep density profiles. Future dual-receiver SMA observations or joint SMA-CARMA polarimetry could yield measurements at three or more frequencies and resolve this question.

Acknowledgments
We thank the Submillimeter Array staff, particularly Ken Young, for vital assistance in developing the polarimetry system. We also thank the organizing committee for a well-managed and useful conference.

References

[1] Aitken D. K. et al. 2000 *ApJL* **534** L173–L176
[2] Bower G. C., Backer D. C., Zhao J.-H., Goss M. and Falcke H. 1999 *ApJ* **521** 582–586
[3] Bower G. C., Wright M. C. H., Backer D. C. and Falcke H. 1999 *ApJ* **527** 851–855
[4] Bower G. C., Wright M. C. H., Falcke H. and Backer D. C. 2003 *ApJ* **588** 331–337
[5] Bower G. C., Falcke H., Wright M. C. and Backer D. C. 2005 *ApJL* **618** L29–L32
[6] Marrone D. P., Moran J. M., Zhao J.-H. and Rao R. 2006 *ApJ* **640** 308–318
[7] Zylka R., Mezger P. G., Ward-Thompson D., Duschl W. J. and Lesch H. 1995 *A&A* **297** 83–97
[8] Serabyn E., Carlstrom J., Lay O., Lis D. C., Hunter T. R. and Lacy J. H. 1997 *ApJL* **490** L77
[9] Pierce-Price D. et al. 2000 *ApJL* **545** L121–L125
[10] Yusef-Zadeh F. et al. 2006 *ApJ* **644** 198–213
[11] Dent W. R. F., Matthews H. E., Wade R. and Duncan W. D. 1993 *ApJ* **410** 650–662
[12] Genzel R. et al. 2003 *Nature* **425** 934–937
[13] Ghez A. M. et al. 2004 *ApJL* **601** L159–L162
[14] Baganoff F. K. et al. 2003 *ApJ* **591** 891–915
[15] Falcke H., Goss W. M., Matsuo H., Teuben P., Zhao J.-H. and Zylka R. 1998 *ApJ* **499** 731
[16] An T., Goss W. M., Zhao J.-H., Hong X. Y., Roy S., Rao A. P. and Shen Z.-Q. 2005 *ApJL* **634** L49–L52
[17] Zhao J.-H., Bower G. C. and Goss W. M. 2001 *ApJL* **547** L29–L32
[18] Melia F. and Falcke H. 2001 *ARA&A* **39** 309–352
[19] Zhao J.-H. et al. 2003 *ApJL* **586** L29–L32
[20] Miyazaki A., Tsutsumi T. and Tsuboi M. 2004 *ApJL* **611** L97–L100
[21] Mauerhan J. C., Morris M., Walter F. and Baganoff F. K. 2005 *ApJL* **623** L25–L28
[22] Bower G. C. et al. 2005 *ApJ* **633** 218–227
[23] Yuan F., Quataert E. and Narayan R. 2003 *ApJ* **598** 301–312
[24] Falcke H. and Markoff S. 2000 *A&A* **362** 113–118
[25] Broderick A. E. and Loeb A. 2006 *ApJL* **636** L109–L112
[26] Marrone D. P. 2006. *PhD thesis* Harvard University
[27] Broderick A. E. and Loeb A. 2006 *MNRAS* **367** 905–916
[28] Quataert E. and Gruzinov A. 2000 *ApJ* **545** 842–846
[29] Agol E. 2000 *ApJL* **538** L121–L124
[30] Macquart J.-P., Bower G. C., Wright M. C. H., Backer D. C. and Falcke H. 2006 *ApJL*. (*Preprint* astro-ph/0606381)
[31] Yusef-Zadeh F. and Morris M. 1987 *ApJ* **322** 721–728
[32] Quataert E. and Gruzinov A. 2000 *ApJ* **539** 809–814
[33] Quataert E. 2003 *Astronomische Nachrichten Supplement* **324** 435–443
[34] Hawley J. F. and Balbus S. A. 2002 *ApJ* **573** 738–748
[35] Igumenshchev I. V., Narayan R. and Abramowicz M. A. 2003 *ApJ* **592** 1042–1059
[36] Pen U.-L., Matzner C. D. and Wong S. 2003 *ApJL* **596** L207–L210
[37] Narayan R. and Yi I. 1995 *ApJ* **452** 710
[38] Melia F. 1992 *ApJL* **387** L25–L28
[39] Ruszkowski M. and Begelman M. C. 2002 *ApJ* **573** 485–495
[40] Beckert T. and Falcke H. 2002 *A&A* **388** 1106–1119
[41] Narayan R. 2002 *Lighthouses of the Universe* , Gilfanov M., Sunyeav R. and Churazov E. ed, page 405. Berlin: Springer

Institute of Physics Publishing
doi:10.1088/1742-6596/54/1/057

Journal of Physics: Conference Series **54** (2006) 363–369
Galaxy Center Workshop

Flux monitoring of Sagittarius A* at mm-wavelengths

Atsushi Miyazaki[1,4], Zhi-Qiang Shen[1], Makoto Miyoshi[2], Masato Tsuboi[3] and Takahiro Tsutsumi[2]

[1] Shanghai Astronomical Observatory, Chinese Academy of Sciences, 80 Nandan Road, Shanghai 200030, P.R.China

[2] National Astronomical Observatory of Japan, 2-21-1 Osawa, Mitaka, Tokyo 181-8588, Japan

[3] Nobeyama Radio Observatory, National Astronomical Observatory of Japan, Minamimaki, Minamisaku, Nagano 384-1305, Japan

E-mail: amiya@shao.ac.cn

Abstract.
We performed the monitoring observations of the flux density toward the Galactic center compact radio source, Sagittarius A* (Sgr A*), which is associated with a supermassive black hole, since 1996 using the Nobeyama Millimeter Array (NMA). The monitoring observations of Sgr A* were carried out in the 3- and 2-mm (100 and 140 GHz) bands over one to several months on each NMA observable season. We have detected several flares of Sgr A* with duration of, roughly, one month. The flux density at the flare peak increases 100%–200% at 100 GHz band and 200%–400% at 140 GHz band, respectively, while the averaged quiescent flux density was about 1 Jy. We also found some intraday variations (IDVs) of Sgr A* at both 2- and 3-mm bands. The shortest twofold increase timescale of the IDV is estimated to be about 1.5 hr at 140 GHz. This short timescale variability suggests that the physical size of emitting region is compact on a scale at or below about 12 AU ($\sim$ 150 R_S). The IDV at mm-wavelengths has a similar increase timescale as those in the X-ray and infrared flares but has a smaller amplitude.

1. Introduction

Sagittarius A* (Sgr A*) is a unique compact radio source located at the dynamical center of the Galaxy, and is widely believed to be associated with a supermassive black hole (SMBH) of mass $\sim 4 \times 10^6\ M_\odot$(see, e.g., [1]). Since its discovery, Sgr A* has been observed at many wavelengths, and temporal flux variations were reported, e.g. from cm- to sub-mm wavelengths in the radio, infrared, and X-rays. Thus, Sgr A* is a source with time-varying luminosity. Because this source is embedded in thick thermal material, it is practically difficult to observe its detailed structure. Generally, the properties of temporal flux variations can give indirect constraints on the source emission geometry and the emission mechanisms.

At centimeter wavelengths, the time variability of Sgr A* has been studied over the past decades (e.g., [2, 3, 4]). Very Large Array (VLA) cm-wavelength monitoring suggests a periodicity of about 110 days [3, 5]. However, at cm-wavelengths, Sgr A* is dominated by effects of refractive interstellar scintillation, and it is difficult to measure all of its intrinsic properties. These effects are usually weaker at higher frequencies, so we can better ascertain the intrinsic nature of Sgr A* at millimeter wavelengths. Earlier, in 1990, significant flux variations of Sgr A*

[4] Present address : Mizusawa VERA Observatory, National Astronomical Observatory of Japan, Mizusawa, Oshu, Iwate 023-0861, Japan; amiya@miz.nao.ac.jp

at 89 GHz were reported by [6]. Since 1996, we have been monitoring Sgr A* 3- & 2-mm flux densities using the Nobeyama Millimeter Array (NMA), the Nobeyama Radio Observatory[5], Japan (see figure 1; [7, 8, 9, 10, 11]). We detected several Sgr A* flares with typical durations of a few weeks [7, 8]. Similar flares of Sgr A* were also detected at 1-mm wavelength using the Sub-Millimeter Array (SMA) [12]. Moreover, the intraday variations (IDVs) in the Sgr A* flux density at mm-wavelengths were detected [10]. In recent years, X-ray and infrared flares have also been detected, indicating very short timescales and violent intensity increases (X-rays: [13, 14]; IR: [15, 16]).

Sgr A* at mm-wavelengths is a relatively weak ($\sim$ 1 Jy) compact component embedded in the extended and strong HII region of Sgr A West. The contribution of the free-free emission from the extended structure is significant even at mm-wavelengths. It is necessary to observe with higher angular resolution to discriminate the compact component from the extended components. Therefore, we conducted the flux monitoring experiments of Sgr A* at short mm-wavelengths using the NMA.

2. Observations

We have performed the flux monitoring observations toward Sgr A* in the 100 and 140 GHz bands (λ=3 and 2-mm) since 1996 using the NMA, a six element 10-m dish interferometer, at the NRO. The observations were carried out using multiple array configurations of the NMA over a period of a few months to half a year for each observable season (winter to spring). Each epoch consists of a set of sequential observations over about 2 days. The epochs were separated by several days to 2 weeks. For each observing day, the observing time was 2–4 hrs within the maximum observable time of about 4 hrs.

We used double sideband (DSB) SIS receivers in the 3- and 2-mm bands as the front-ends. Almost all the data were obtained using the Ultra Wide Band Correlator (UWBC) with 1 GHz bandwidth. With this UWBC system, simultaneous observations of both the lower and upper sidebands, separated by 12 GHz, are allowed. Thus, these observed frequencies were 90 and 102 GHz for the 3-mm band, and 134 and 146 GHz for the 2-mm band. The instrumental gain and phase were calibrated by alternating observations of Sgr A* and NRAO530 at about 20 min intervals. We also used an additional phase calibrator, 1830-210, a known QSO, after 2000. The flux densities of the calibrators are determined from Uranus or Neptune, which were used as the primary flux calibrator. The absolute uncertainties of the flux scaling are about 15% and 20% in the 100 and 140 GHz bands, respectively. Because the observations for Sgr A* in the 2-mm band requires the best weather conditions and phase stability, these observations were made less frequently.

Most of the observations, including the detections of the flares of Sgr A*, were performed by the array configuration with intermediate baselines, C-configuration, of the NMA. The projected baselines in this array range over $\sim 7 - 55$ kλ in the 100 GHz band and $\sim 10 - 77$ kλ in the 140 GHz band. We made spatially filtered maps from the visibilities with projected baselines exceeding 25 kilo-wavelengths ($(U^2 + V^2)^{1/2} \geq 25k\lambda$) to suppress the contamination from the extended components surrounding Sgr A* (17 kλ for the data taken with the compact array configuration, D-configuration, in the 100 GHz band). Typical synthesized beam sizes (HPBW) for the C-configuration were about $3'' \times 6''$ and $2'' \times 4''$ at 100 and 140 GHz, respectively. From the flux densities of the calibrators measured on the maps, the fractions of the decorrelation due to atmospheric phase fluctuation are 10–20% at 100 GHz and 20–40% at 140 GHz. We corrected the measured flux densities of Sgr A* for the decorrelation, and averaged two measured flux densities of Sgr A* which were individually calibrated by each of the two phase calibrators. A

[5] Nobeyama Radio Observatory (NRO) is a branch of the National Astronomical Observatory, National Institutes of Natural Sciences, Japan.

more detailed description of the observations and the data calibrations have also been presented in our other papers [9, 10, 11].

3. Results & Discussion

*3.1. Flare of Sgr A**

Figure 1 shows the light curves of Sgr A* in the 100 and 140 GHz bands constructed from all monitoring data from 1996 to 2005. *Red circles* and *blue squares* indicate the flux densities in the 100 and 140 GHz bands, respectively. The total number of observations is about 102 days. Mean flux densities of Sgr A* in quiescent phase are 1.1 ± 0.2 Jy and 1.2 ± 0.2 Jy at 90 and 102 GHz, respectively. These light curves show that Sgr A* may distinguish between the active phase and the quiescent phases [9]. There are at least four flares in March 1998, March 2000, April 2002, and March 2004 (Flare I, Flare II, Flare III, and Flare IV, respectively, indicated in figure 1) with duration of, roughly, one month during the whole observing period [11]. The peak flux density in these flares increases by 100%–200% at 100 GHz, and by 200%–400% at 140 GHz ($\Delta S/S$), respectively, while the averaged quiescent flux density was about 1–1.5 Jy. For near-simultaneous millimeter spectra made from our observations, it appears that the variability of the flux density in the flare increases with frequency [8, 9, 11].

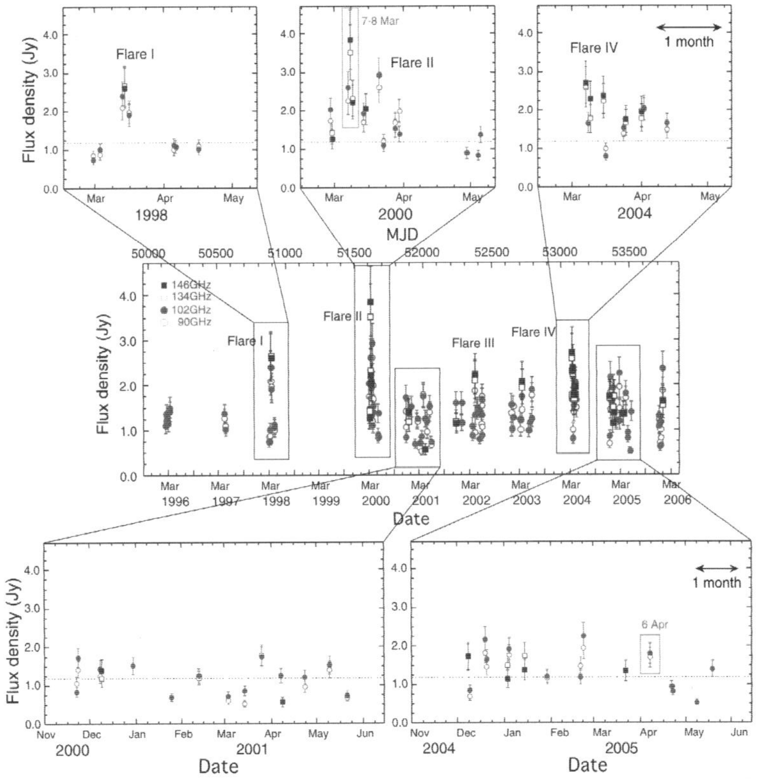

Figure 1. 1996-2005 light curve (*center*) of Sgr A* in the 100 & 140GHz bands. Open circles - 90GHz, filled circles - 102GHz, open squares - 134GHz, filled squares - 146GHz. Top and bottom panels show light curves in 1998, 2000, 2004, 2000-2001, and 2004-2005 in detail. The dotted lines indicate the averaged quiescent flux level.

3.2. Intraday Variation

We detected the intraday variation (IDV) of Sgr A* on the 7th March, 2000 [10]. As shown in the upper panel of figure 1, the light curves of Sgr A* in March 2000 indicate particularly strong variability in the 100 and 140 GHz bands in the active phase of Flare II. The most prominent flare was observed on March 7, 2000 in the 140 GHz band, and the peak flux density at 146 GHz was 3.9 ± 0.8 Jy. On the 7th March, 2000, the Sgr A* flux density have strongly increased during observing session, thus it indicates obviously the IDV [10]. The flux density increases about 30% during 30 minutes, and the twofold timescale of the IDV is estimated to be about 1.5 hours assuming that the increase has a constant gradient [10].

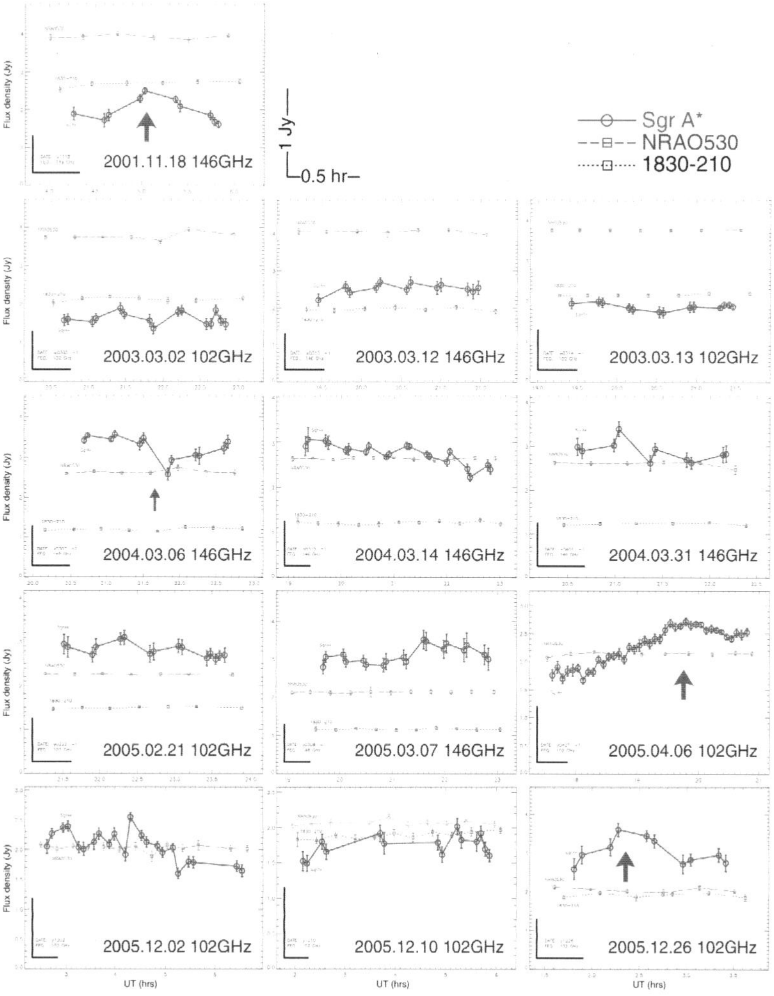

Figure 2. Light-curves of Sgr A* and the calibrators during each observation at 102 or 146 GHz. The light-curves in each panel indicate Sgr A* (red circles with the straight-line), and the calibrators, NRAO 530 (blue squares with the broken-line) and/or 1830-210 (black squares with the dashed-line). The arrows indicate significant flux variations.

Then, we have performed to search for the intraday variability (IDV) based on our NMA monitoring data of Sgr A* flux density. In the analysis, we adopted the monitoring data in good condition from November in 2001 to December in 2005 to exclude the data taken with the compact array configuration. Date of the analyzed data are summarized in Table 1, and total number of the usable data is ~22 days. We measured the flux densities averaged every 4–5 minutes bin on visibility data self-calibrated for phase to search for shorter timescale variability. The flux measurements of Sgr A* were estimated to fit a point source model in visibility plane with longer uv distance of ≥ 25 kλ in order to suppress the contamination from the surrounding extended components around Sgr A*.

In Figure 2, each panel shows some light-curves of Sgr A* at 102 or 146 GHz during each observing session from 2 to 4 hours. Red circles with the straight-line indicate the light-curve of Sgr A*, and blue squares with the broken-line and black squares with the dashed-line indicate the light-curve of the calibrators, NRAO 530 and/or 1830-210, respectively. Some light-curves look like to indicate significant flux variation (see the arrows in figure 2). The most prominent flux variation in the analysis was shown in the light-curves in the 100 GHz band on the 6th April, 2005. The peak-to-peak amplitude (difference between maximum and minimum flux) in the IDV on April 6, 2005 was $\sim$ 1.1 Jy at 102 GHz.

Table 1. Log of IDV searching with the NMA

Frequency	Observing Date
100 GHz	2002.4.6, 2003.1.9, 2003.3.2, 2003.3.13, 2005.2.19, 2005.2.21, 2005.4.6, 2005.12.2, 2005.12.3, 2005.12.5, 2005.12.10, 2005.12.26
140 GHz	2001.11.18, 2002.4.6, 2003.3.12, 2004.3.6, 2004.3.8, 2004.3.14, 2004.3.24, 2004.3.31, 2005.2.20, 2005.3.7

Figure 3 shows the light-curve (*bottom panel*) at 90 and 102 GHz on April 6, 2005. Red circles and blue squares indicate the light-curve of Sgr A* and NRAO530 as the calibrator, respectively. The variation of the flux densities of Sgr A* was much larger than the flux scatters of the calibrator, NRAO530, and thus it indicates obviously the IDV. The peak flux densities were 2.5 Jy at 90 GHz, and 2.7 Jy at 102 GHz. The typical relative uncertainties were at or below about 0.1 Jy. The flux density of Sgr A* at 102 GHz increased from $\sim$1.6 to $\sim$2.7 Jy between 18:15 to 19:45 UT on April 6, 2005. The flux density increases about 70% during 1.5 hours, and the two-fold timescale of the increase is estimated to be about 2 hours assuming that the increase has a constant gradient. Top panel of Figure 3 shows the variability of the spectral indices estimated between the flux at 90 and 102 GHz.

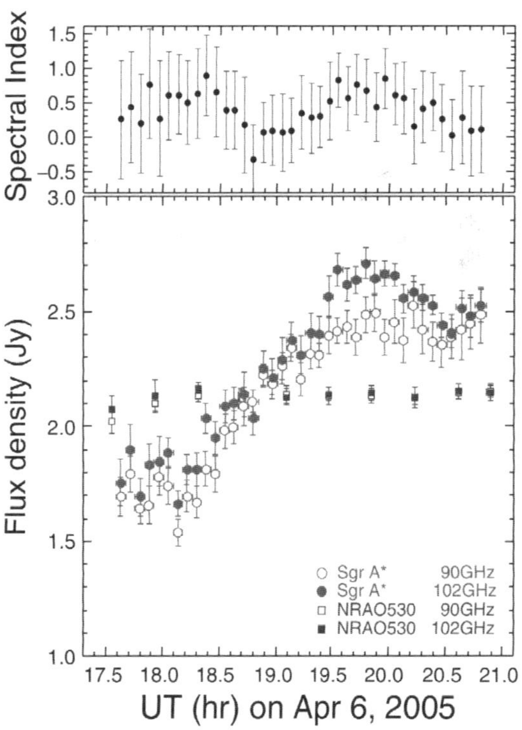

Figure 3. Light-curves (*bottom panel*) of Sgr A* (*red circles*) and the calibrator (*blue squares*), NRAO530, at 90 and 102 GHz on April 6, 2005. In this plot, the data were divided into 5 mins bin. The flux variation of Sgr A* was significantly larger than the scatters of the flux densities of the calibrator. Moreover, top panel shows the spectral indices between the flux at 90 and 102 GHz.

Recently, the other radio IDVs of Sgr A* were reported by some investigators (e.g., VLA at cm-wavelengths [17]; OVRO mm-interferometer at 3-mm [18]; VLA & ATCA at 7- & 12-mm [19]; SMA at 880μm (340 GHz) [20]). The variation of Sgr A* has probably the period of a few hours according to periodicity analyses in these reports [18], and our NMA results are consistent with the results of those. However, as for its amplitude, the IDV at mm- or submm-wavelengths is much larger than that at lower frequency, and these variability of the flux density increases with frequency. Considering these results, it seems that radio IDV in Sgr A* are not rare, especially at shorter wavelengths. The intraday flare at mm-wavelengths has a similar increase timescale as those known in the X-ray and infrared flares, but has a smaller amplitude ([13, 14, 15, 16]). The shortest increase timescale, 1.5 hr, suggests that the physical size of the emitting region in the accretion disk is compact on a scale at or below about 12 AU ($\sim$ 150 R_S; the Schwarzschild radius $R_S = 2GM/c^2$, assuming a black hole mass of 4×10^6 $M_\odot$) [10].

As shown in the top panel of Figure 3, we estimated the spectral indices, α ($S_\nu \propto \nu^\alpha$), between 90 and 102 GHz of each measured bin on the 6th April, 2005. Figure 4 shows the relations between the flux density and the spectral index estimated between 90 and 102 GHz (*left panel*), or 134 and 146 GHz (*right panel*) in each measured bin on each observing session. The typical uncertainties of these spectral indices are $\sim$0.8 in the 100 GHz band and $\sim$1.4 in the 140 GHz band, which is slightly large because of a small frequency span (12 GHz). The straight lines in each panel indicate a best fit for the linear function, and these relations have probably the positive correlation though the uncertainties in the index are large. Each best fit for the linear function are $\alpha_{100\text{GHz}} = (0.81 \pm 0.11) \cdot S_{102\text{GHz}} - (1.11 \pm 0.24)$ in the 100 GHz band, $\alpha_{140\text{GHz}} = (0.15 \pm 0.17) \cdot S_{146\text{GHz}} + (0.67 \pm 0.46)$ in the 140 GHz band. It appears that the scatter of the index in lower flux density is large. As mentioned in the previous subsection, the flux increase during the flare in the 140 GHz band is larger than that in the 100 GHz band [8, 9, 10]. From these results, these demonstrate that the spectral index of Sgr A* was large during the flux increase in the IDV. The spectral variation suggests that the energy injection to photons was occurred in the higher frequency regime and the emitting frequency was shifted to millimeter wave regime with time.

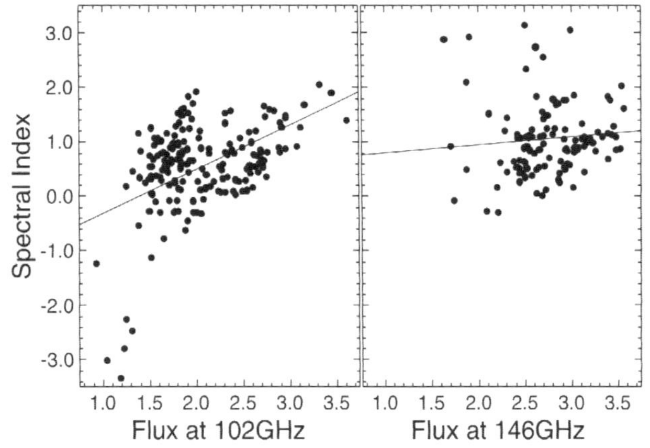

Figure 4. Correlation plots between the flux density and the spectral index estimated between upper- and lower-sidebands (90 and 102 GHz, or 134 and 146 GHz) of each measured bin on each observation. Left-panel and right-panel are in the 100 GHz and 140 GHz bands, respectively. The straight lines each panel indicate a best fit for the linear function.

On the other hand, we have also performed the 3-mm observations of Sgr A* on October 18, 2005 and June 9, 2006 using the the Australia Telescope Compact Array (ATCA), the Australia Telescope National Facility (ATNF) [21]. The ATCA in southern hemisphere has a longer observing window of the Galactic center region for about 10 hrs in a day, and so it is very useful instrument to search for a IDV of Sgr A*. For the preliminary result of our observations with the ATCA, we detected a significant flux variation of Sgr A*, which seem to be the IDV, on the 18th October, 2005 [21]. The Sgr A* flux density first decreased from about 3 Jy to 2 Jy, and then increased to 3.5 Jy on October 18. It appear that the light curve on the 9th June, 2006 don't have significant flux variation.

4. Conclusions

We have performed the monitoring observations of the flux density of Sgr A* in the 3-mm (100 GHz) and 2-mm (140 GHz) bands using the NMA at the NRO since 1996.

(i) We detected several flares of Sgr A* with duration from a few weeks to a month. The flux density at the flare peak increases 100%–200% at 100 GHz band and 200%–400% at 140 GHz band ($\Delta S/S$), respectively, while the averaged quiescent flux density was about 1-1.5 Jy.

(ii) We have searched for the intraday variations (IDV) on the NMA monitoring data of Sgr A* flux density, and detected some IDVs of Sgr A* at both 100 and 140 GHz bands. The amplitude of variation is over 1 Jy during a few hours.

(iii) The twofold increase timescale in these IDVs is estimated to be a few hours. The shortest increase timescale, 1.5 hrs, suggests that the physical size of the emitting region (light crossing time) in the IDV is a scale at or below ~ 12 AU ($\sim 150\ R_{\rm S}$).

Acknowledgments

We thank the staff of NMA group of the NRO for support in the observation.

References

[1] Ghez, A.M., Salim, S., Hornstein, S.D, et al. 2005 *ApJ* **620** 744
[2] Brown, R.L., & Lo, K.Y. 1982 *ApJ* **253** 108
[3] Zhao, J.-H., Bower, G.C., & Goss, W.M. 2001 *ApJL* **547** L29
[4] Herrnstein, R.M., Zhao, J.-H., Bower, G.C., & Goss, W.M. 2004 *AJ* **127** 3399
[5] Zhao, J.-H. 2003a, *Astron. Nachr.* **324** S1 355
[6] Wright, M.C.H., & Backer, D.C. 1993 *ApJ* **417** 560
[7] Miyazaki, A.,Tsutsumi, T., & Tsuboi, M. 1999 *Advances in Space Research* **23** 5/6 977
[8] Tsuboi, M., Miyazaki, A., & Tsutsumi, T. 1999, in *The Central Parsecs of the Galaxy*, ASP Conf. Ser. 186 (San Francisco: ASP) 105
[9] Miyazaki, A.,Tsutsumi, T., & Tsuboi, M. 2003 *Astron. Nachr.* **324** S1 363
[10] Miyazaki, A.,Tsutsumi, T., & Tsuboi, M. 2004 *ApJL* **611** L97
[11] Tsutsumi, T., Miyazaki, A., & Tsuboi, M. 2006 *AJ*, submitted
[12] Zhao, J.-H., Young, K.H., Herrnstein, R.M. et al. 2003b *ApJL* **586** L29
[13] Baganoff, F.K., Bautz, M.W., Brandt, W.N., et al. 2001 *Nature* **413** 45
[14] Porquet, D., Predehl, P., Aschenbach, B. et al. 2003 *A&A* **407** L17
[15] Genzel, R., Schödel, R., Ott, T., et al. 2003 *Nature* **425** 934
[16] Ghez, A.M., Wright, S.A., Matthews, K., et al. 2004 *ApJL* **601** L159
[17] Bower, G.C., Falcke, H., Sault, R.J., & Backer, D.C. 2002 *ApJ* **571** 843
[18] Mauerhan, J.C., Morris, M., Walter, F., & Baganoff, F.K. 2005, *ApJL* **623** L25
[19] Yusef-Zadeh, F., Bushouse, H., Dowell, C.D., et al. 2006 *ApJ* **644** 198
[20] Marrone, D.P., Moran, J.M., Zhao, J.-H., & Rao, R. 2006, *ApJ* **640** 308
[21] Miyazaki, A., Shen, Z.-Q., Fletcher, A.B., Sault, R.J., Miyoshi, M., Tsutsumi, T., & Tsuboi, M. 2006, in preparation

Institute of Physics Publishing
doi:10.1088/1742-6596/54/1/058

Journal of Physics: Conference Series **54** (2006) 370–376
Galaxy Center Workshop

High Resolution Imaging of Sagittarius A*

Geoffrey C. Bower

[1] Astronomy Department & Radio Astronomy Lab, 601 Campbell Hall, UC Berkeley, Berkeley, CA 947220 USA

E-mail: gbower@astro.berkeley.edu

Abstract. High resolution radio imaging of Sgr A* provides an opportunity to study the accretion and outflow environment of a black hole that is not available for any other system. Strong interstellar scattering of radio waves, however, complicate the interpretation of radio images. High quality determination of the intrinsic size and shape require high quality images as well as accurate calibration of the scattering law. New results indicate that the intrinsic size is determined from 3.5 cm to 0.35 cm with a size at the shortest wavelength of $\sim 13R_s$ and a wavelength-size power law index of 1.5. Millimeter and submillimeter observations have the promise of constraining the mass density of the black hole on a smaller radius than any other technique.

1. Introduction

Imaging the radio emitting region surrounding the massive black hole in the Galactic Center, Sagittarius A*, has been a goal since its discovery [1]. Turbulent electrons along the line of sight to Sgr A*, however, scatter radio wavelength photons and produce an image that is an elliptical Gaussian with a major axis size of ~ 0.5 arcsec at 20 cm and a λ^2 dependence [2]. Separating the effects of the small intrinsic source from the effects of scattering has required observations at short wavelengths, careful calibration, and the use of closure amplitude techniques, which reduce sensitivity but remove uncertainty due to calibration error [3, 4]. These efforts have recently resulted in the first robust detections of intrinsic structure in Sgr A* at wavelengths of 1.3 cm, 0.7 cm and 0.35 cm. The intrinsic source has a size that scales with $\lambda^{1.1}$ or $\lambda^{1.6}$ to a minimum of ~ 10 Schwarzschild radii at 0.35 cm (assuming $M_{bh} = 4 \times 10^6 M_\odot$ and $d = 8$ kpc ; [5, 6]).

These detections of the intrinsic size of Sgr A* have a number of consequences. The brightness temperature of 10^{10} K strongly excludes advection dominated accretion flows [7] and Bondi-Hoyle accretion models [8]. These size measurements, however, cannot differentiate between jet models [9], generic radiatively inefficient accretion flows [10], and hybrids of these models [11]. This limitation is principally due to the limited sensitivity in the minor axis size of the scattering ellipse. Coupled with measurements of the proper motion of Sgr A* [12], the assumption that the black hole is smaller than the emission region implies a lower limit to the mass density of the black hole $\sim 10^5 M_\odot \mathrm{AU}^{-3}$, which strongly excludes alternative models for dark mass objects.

The scattering medium itself is a system of intense interest [13, 14, 15]. The λ^2 size dependence of Sgr A* is a strict consequence of the strong scattering and the short projected baselines at the distances of the scattering medium [16].

We wish to clarify here the numerous axes and steps involved in translating observations of Sgr A* into a measurement of the intrinsic size. Very Large Array and Very Long Baseline

Array observations of Sgr A* are obtained with a synthesized beam that is extended North-South. Deconvolution of the observed image with the synthesized beam gives the apparent, scatter-broadened image of Sgr A*. This image is predominantly a two-dimensional Gaussian with the major axis oriented $\sim$ 80 degrees East of North. Typically, when we refer to the major and minor axes, we refer to the orientation of the scattering ellipse. Finally, to obtain the intrinsic image, we deconvolve the apparent image with a model of the scatter-broadened image, which is determined from long wavelength apparent sizes.

2. New VLA Observations at Long Wavelengths

We recently used long wavelength measurements of Sgr A* to determine the scattering properties to unprecedented accuracy with detailed measurements of the source size at wavelengths that range from 17.4 to 23.8 cm using the Very Large Array and the Pie Town Very Long Baseline Array antenna [17]. The addition of the PT antenna to the VLA A configuration improves the East-West resolution by a factor of two.

Sgr A* is clearly resolved in both axes but with considerably more accuracy in the East-West axis than in the North-South axis. Fitting a point source to the data produced very poor quality fits, while fitting an elliptical Gaussian produced a residual image with no obvious systematic errors (Figure 1). The North-South resolution was considerably less accurate due to the presence of a transient due South of Sgr A*.

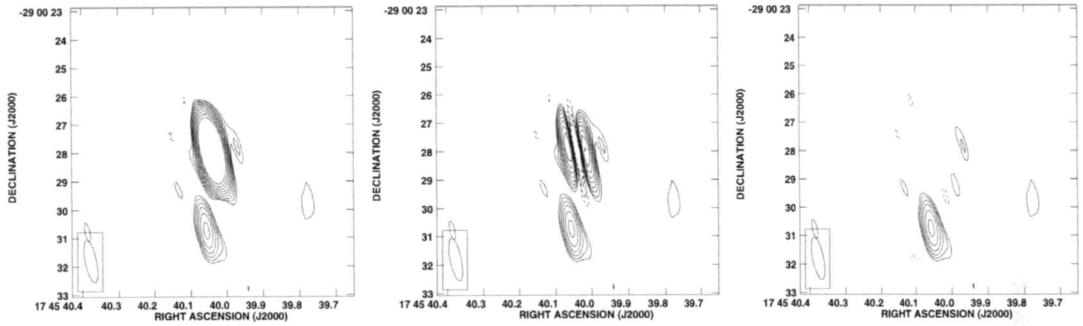

Figure 1. (*left*) Image of Sgr A* at 17.5 cm from the VLA+PT observations. Sgr A* is the bright source at the center of the image. The source to the South of Sgr A* is the transient. The synthesized beam size is indicated in the lower part of the image. Contours begin at 5 mJy/beam and increase by factors of $\sqrt(2)$ to half the peak intensity of Sgr A*. The synthesized beam size is 1.69 $\times$ 0.56 arcsec. (*center*) Residual image of Sgr A* after subtraction of best-fit point source. This shows that the source is clearly resolved East-West and at best marginally resolved North-South. (*right*) Residual image of Sgr A* after subtraction of best-fit Gaussian source.

3. Theoretical Expectations for Scattering

The λ^2 dependence of the scattering law is strongly favored for theoretical reasons. The maximum baseline length projected to the scattering region is $b_{proj} = D_{scattering}/D_{SgrA} \times b_{max} \sim$ 1km, where $D_{SgrA} = 8$ kpc is the distance to Sgr A*, $D_{scattering} \approx 100\,\text{pc}$ is the distance of the scattering region from Sgr A*, and $b_{max} \approx 70\,\text{km}$ is the maximum baseline between the VLA and PT. b_{proj} is substantially smaller than the expected and measured inner scales (b_{inner}) for the power spectrum of turbulent fluctuations (10^2 to $10^{5.5}$ km; [18]). This result holds for the long VLBA baselines involved in imaging at shorter wavelengths, as well, where $b_{proj} \sim 25$ km at

0.7 cm wavelength. For the case of $b_{proj} << b_{inner}$, the resulting image is very heavily averaged and must be Gaussian in shape with size $\propto \lambda^2$ [16].

This expectation of strong scattering is supported by previous measurements of the shape of the image of Sgr A*. Fitting the closure amplitudes of Sgr A* at 0.7 cm with a functional form for the visibilities of $\propto e^{-b^{(\beta-2)}}$, where b is the baseline length, revealed $\beta = 4.00 \pm 0.03$ [3]. That is, the best evidence indicates that the image of Sgr A* is a Gaussian. Following scattering theory, where the size is proportional to λ^α, then $\alpha = \beta - 2 = 2.00 \pm 0.03$ [16]. For the case of the VLA+PT data, we find $\alpha = 1.98 \pm 0.11$, which is consistent with the expectation of $\alpha = 2$.

A final caveat is required. The scattering medium is dynamic. The minimum time scale for a change in the medium is the refractive time scale, which is $0.5\,\mathrm{yr}\,(v/100\,\mathrm{km/s})(\lambda/1\,\mathrm{cm})^2$ for Sgr A*, where v is the velocity of the scattering material relative to Sgr A* and the Sun [16]. The data presented here were obtained in a span of roughly a decade. The long-wavelength scattering properties are very unlikely to change on this time scale. However, at wavelengths as long as 4 cm, the refractive time scale is < 10 year. The many observations at 0.7 cm in this period, however, appear to produce a source of stable size, despite a refractive time scale less than one year. We conclude it is unlikely that the scattering size has changed significantly over this period.

4. High Resolution Observations at Short Wavelengths

Since the realization that the image of Sgr A* is dominated by scattering, there has been a strong motivation to observe Sgr A* at ever shorter wavelengths. The λ^2-scaling for scattering size, the $< \lambda^2$-scaling for intrinsic source, and the λ-scaling for telescope resolution combine to offer the promise of detection of the intrinsic size of Sgr A* at short wavelengths. Poor telescope performance, decreased interferometric coherence, and decreased gain stability conspire, however, to limit the accuracy of amplitude calibration and prevent high quality results at millimeter wavelengths.

To deal with these limitations, closure amplitudes and closure phases have been used to determine the size from VLBA measurements at wavelengths from 6 cm to 3.5 mm (Figure 2). The closure techniques trade reduced sensitivity for improved accuracy that is independent of amplitude and phase calibration. At short wavelengths, the major axis sizes are clearly larger than the expectation of the scattering law determined with longer wavelength observations and indicate sizes of $\sim 25R_s$ at 7 mm and $\sim 10R_s$ at 3.5 mm. The actual intrinsic sizes depend on the scattering law with increasing sensitivity at longer wavelengths.

5. The Intrinsic Size

Ideally, we would fit the intrinsic size and the scattering law simultaneously. A fit of the size proportional to $\sqrt{a^2\lambda^4 + b^2\lambda^{2\gamma}}$ is not sufficiently constrained to set reasonable limits on the parameters a, b, and γ. If we fix γ and search for a and b, we find that χ^2 for $\gamma = 1$ is four times the value for γ unconstrained, indicating that $\gamma = 1$ is strongly excluded. Without the assumption that the second term is negligible for wavelengths longer than 6 cm, therefore, we cannot determine the scattering law or the intrinsic size of Sgr A*.

With the assumption that the scattering law is determined accurately at wavelengths longer than 17 cm, we can determine the intrinsic size. We subtract in quadrature the scattering law size from the measured size (Figure 3). We compute the results for the best-fit major axis scattering law ($b_{sc} = 1.32$ mas/cm^2), and $\pm 3\sigma$ of the best-fit value. For the best-fit case, the intrinsic size is accurately determined from 0.35 cm to 3.6 cm. Over this range, the intrinsic size is well-fit by a power-law λ^γ, where $\gamma = 1.6 \pm 0.1$. For the smaller scattering size, we find a steeper power-law and measure the intrinsic size from 0.35 cm to 6 cm. For the larger scattering size, we cannot measure the intrinsic size at wavelengths longer than 1.3 cm and find a shallower power-law index of $\gamma = 1.3 \pm 0.2$.

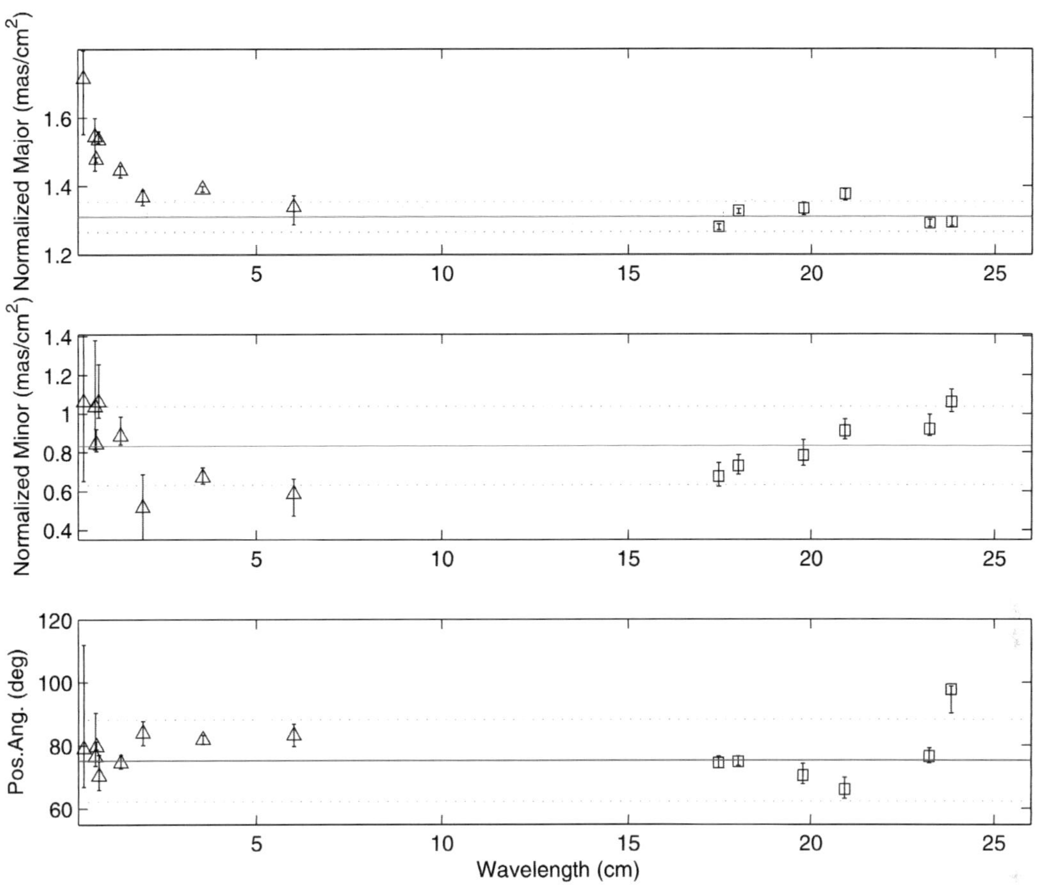

Figure 2. Measured major axis size, minor axis size, and position angle as a function of wavelength. Triangles are VLBA measurements determined through closure amplitude analysis from [3] and [4]. Squares are the new VLA+PT measurements. The major and minor axis sizes have been normalized by λ^2. The solid red line in each curve is the best-fit scattering value determined from the VLA+PT data alone. Dotted red lines are $\pm 3\sigma$ of the best-fit scattering law.

At short wavelengths, the result is consistent with the conclusions of recent efforts by [3] and [4]. The size of Sgr A* at 0.35 cm is $13.3^{+6.7}_{-3.1} R_s$, where $R_s = 1.2 \times 10^{12}$ cm is the Schwarzschild radius for $M_{bh} = 4 \times 10^6 M_\odot$ and the Galactic Center distance $d = 8$ kpc. This compact size confirms tight restrictions on accretion models and black hole alternatives previously claimed and stated in §1.

The wavelength dependence of the source size agrees with that found by [3] and is steeper than that found by [4], who found $\gamma \approx 1.1$. The steeper dependence indicates that the brightness temperature decreases as λ^{-1}, assuming that the size in the second dimension is proportional to the major axis size. The peak brightness temperature at 0.35 cm is $\sim 10^{10}$ K for a flux density of 1 Jy. The power-law dependence of the size as a function of wavelength indicates a stratified, smoothly varying emission region.

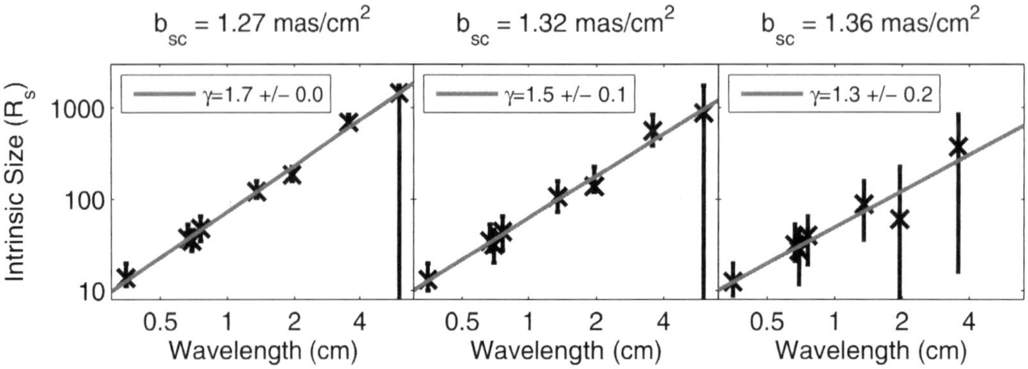

Figure 3. Intrinsic size in Schwarzschild radii for the East-West dimension using three different estimates of the major axis scattering law. We assume a $4 \times 10^6 M_\odot$ black hole at a distance of 8 kpc. In the central panel, we show results for the best-fit scattering law. In the left and right panels, we show the results for the -3σ and $+3\sigma$, respectively. The solid red lines are the best fit curves for size $\propto \lambda^\gamma$.

6. Source Modeling

Detailed jet models for Sgr A* predict $\gamma \approx 1$ [19]. Generalized jet models, however, allow a range of γ, depending on the details of the magnetic field and particle energy density distributions [20]. A jet with $B \propto r^{-1}$, electron density decreasing as r^{-1}, and optically thin power-law index of 1 will show a size $\propto \lambda^{1.4}$.

[21] model the size of Sgr A* for a radiatively inefficient accretion flow. They fit sizes at 0.35 cm and 0.7 cm that are fit with a power law of index $\gamma = 1.1$. Variations in the nonthermal electron distribution or deviations from equipartition, however, could alter γ in their model.

The future of modeling the size of Sgr A* lies in direct comparison between observed and modeled closure quantities. Deconvolution of the size of Sgr A* into a two-dimensional Gaussian component becomes a poor approximation as the size measurements become more accurate. In Figure 4 and 5, we show preliminary results from direct comparison of jet models with 7mm closure amplitudes and closure phases (Markoff & Bower, in preparation).

7. Future Goals for Imaging of Sgr A*

The critical remaining observational goals for understanding the image of the radio emission of Sgr A* are a measurement of the two-dimensional size and detection of structural variability. The simple one-dimensional deconvolution that we have performed here only gives schematic information on the size of Sgr A*. With a more accurate two-dimensional scattering model, future analysis will directly compare the observed image with non-Gaussian source models convolved with the imaging and scattering constraints. Astrometric observations may indicate a shift in the centroid of the image with frequency. A heterogeneous jet will exhibit such a shift due to changing location of the optically thick surface of the source [20].

8. Approaching the Event Horizon

At mm and submm wavelengths, the gravity of the black hole will distort the image of Sgr A* [22, 23]. These effects are apparent on a scale of $\sim 5R_s$. The actual images that will be seen at these short wavelengths depend on the nature of the emission region. Detailed characterization of the shape of the longer wavelength image will provide the background emission region that will be transformed by the black hole's gravity. For the simple case of a homogeneous spherical

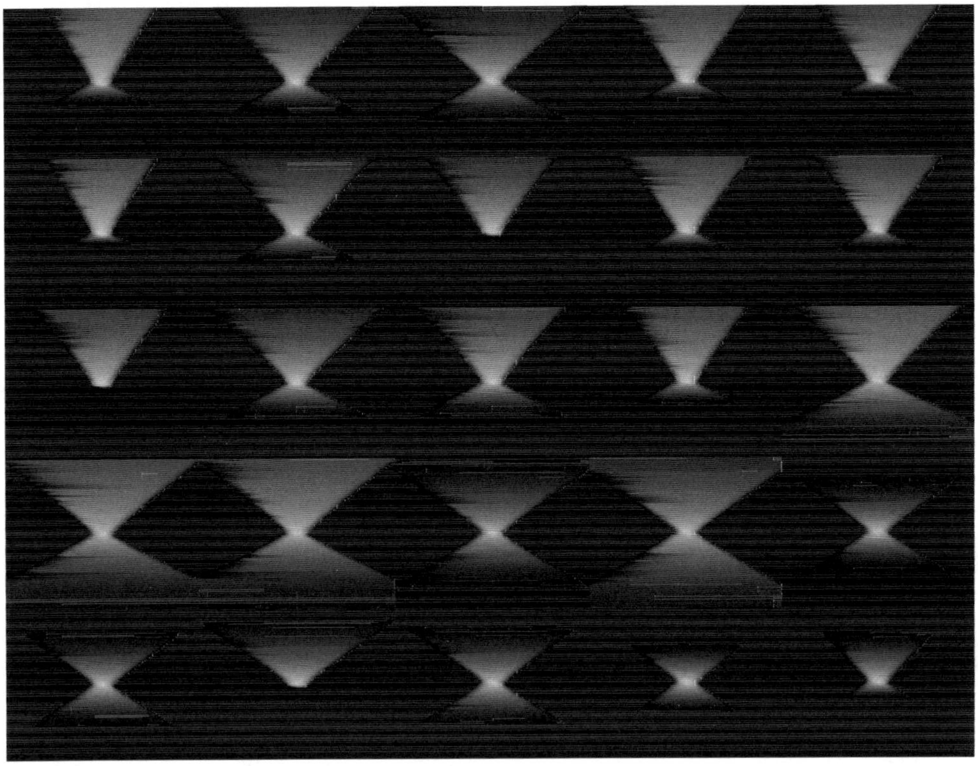

Figure 4. Jet models for comparison with 7mm VLBA observations of Sgr A*. The logarithm of the intensity is shown. The models are required to reproduce the spectrum of Sgr A* using the constraint of polarization on the accretion rate.

emission region, the apparent image will be a ring of radius $5R_s$. Detection of this ring, or of the "shadow" that the black hole casts on the emission region, will constrain the mass of the black hole to the smallest radius yet, providing one of the strongest tests of the existence of black holes.

The observations to achieve this result are technically demanding. They will rely on an *ad hoc* array of telescopes including CARMA, SMA, and ALMA, make use of high bandwidth data recorders, and require high fidelity in time standards and frequency systems. Additionally, the image of Sgr A* may change on a timescale of an orbital period, which is 10s of minutes, preventing traditional synthesis imaging. Phase-referenced imaging may be required to track these changes. This requires rapid switching between Sgr A* and a background source and/or tropospheric phase-correction hardware. Phase-referencing between polarized and total intensity images is a potentially simpler technique, although there have not been theoretical predictions of the relationship between polarized and total intensity images in these regions.

References

[1] Balick, B. & Brown, R. L. 1974, ApJ, 194, 265

[2] Backer, D. C. 1978, ApJ, 222, L9

[3] Bower, G. C., Falcke, H., Herrnstein, R. M., Zhao, J., Goss, W. M., & Backer, D. C. 2004, Sci., 304, 704

[4] Shen, Z.-Q., Lo, K. Y., Liang, M.-C., Ho, P. T. P., & Zhao, J.-H. 2005, Nature, 438, 62

[5] Ghez, A. M., Salim, S., Hornstein, S. D., Tanner, A., Lu, J. R., Morris, M., Becklin, E. E., & Duchêne, G. 2005, ApJ, 620, 744

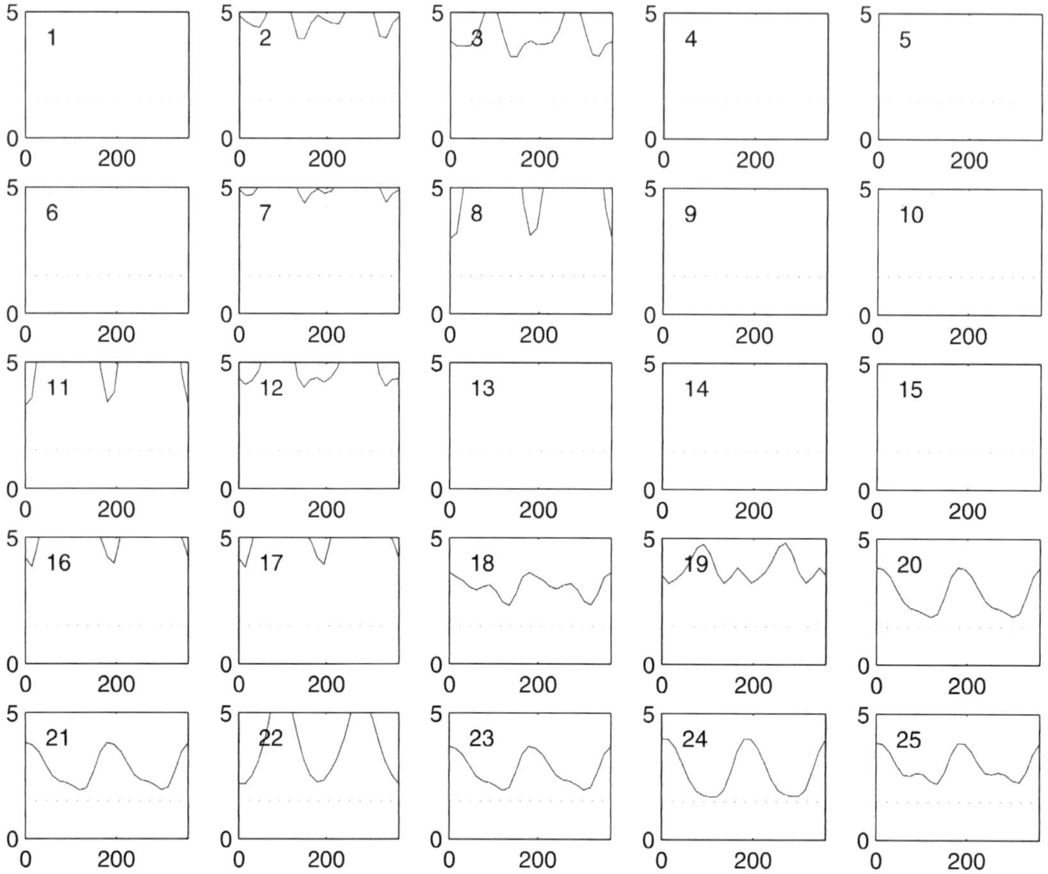

Figure 5. Reduced χ^2 as a function of jet position angle for the models shown in Figure 5. Where no curve is present, the reduced χ^2 is off the scale. Some models and position angles are clearly favored. Detailed characterization of the sensitivity of the technique is necessary to determine which models and position angles are excluded.

[6] Eisenhauer, F., Schödel, R., Genzel, R., Ott, T., Tecza, M., Abuter, R., Eckart, A., & Alexander, T. 2003, ApJ, 597, L121

[7] Narayan, R., Mahadevan, R., Grindlay, J. E., Popham, R. G., & Gammie, C. 1998, ApJ, 492, 554

[8] Melia, F. 1994, ApJ, 426, 577

[9] Falcke, H., Mannheim, K., & Biermann, P. L. 1993, A&A, 278, L1

[10] Quataert, E. & Gruzinov, A. 2000, ApJ, 539, 809

[11] Yuan, F., Markoff, S., & Falcke, H. 2002, A&A, 383, 854

[12] Reid, M. J. & Brunthaler, A. 2004, ApJ, 616, 872

[13] Lazio, T. J. W. & Cordes, J. M. 1998, ApJ, 505, 715

[14] Bower, G. C., Backer, D. C., & Sramek, R. A. 2001, ApJ, 558, 127

[15] Goldreich, P. & Sridhar, S. 2006, ApJ, 640, L159

[16] Narayan, R. & Goodman, J. 1989, MNRAS, 238, 963

[17] Bower, G. C., Goss, W.M., Falcke, H., Backer, D.C, & Lithwick, Y., 2006, ApJ, submitted

[18] Wilkinson, P. N., Narayan, R., & Spencer, R. E. 1994, MNRAS, 269, 67

[19] Falcke, H. & Markoff, S. 2000, A&A, 362, 113

[20] Konigl, A. 1981, ApJ, 243, 700

[21] Yuan, F., Shen, Z.-Q., & Huang, L. 2006, ApJ, 642, L45

[22] Falcke, H., Melia, F., & Agol, E. 2000, ApJ, 528, L13

[23] Broderick, A. E. & Loeb, A. 2005, MNRAS, 363, 353

Institute of Physics Publishing
doi:10.1088/1742-6596/54/1/059
Journal of Physics: Conference Series **54** (2006) 377–380
Galaxy Center Workshop

High-Resolution Millimeter-VLBI Imaging of Sgr A*

Zhi-Qiang Shen[1]

Shanghai Astronomical Observatory, 80 Nandan Road, Shanghai 200030, China

E-mail: `zshen@shao.ac.cn`

Abstract. We present the highest resolution VLBI imaging observations of Sgr A* made at both 7 and 3.5 mm. These data reveal wavelength-dependent intrinsic sizes with an intrinsic emitting region at 3.5 mm of about 1 AU (at a distance of 8 kc to the Galactic Center). When combined with the lower limit on the mass of Sgr A*, these size measurements provide strong evidence that Sgr A* is a super-massive black hole. We also detected a structural variation which results in an intrinsically symmetrical structure that increases in its intrinsic size by more than 25% at 7 mm.

1. Introduction

Sgr A*, the extremely compact non-thermal radio source at the galactic center, is a best candidate for a single super-massive black hole (SMBH) from both the observational studies and the theoretical models (cf. [1]). The determination of the orbital motions of those early-type stars within the vicinity of Sgr A* reveal a central dark mass about 4×10^6 $M_\odot$ within a radius of about 45 AU [2, 3]. On the other hand, the upper limit to the intrinsic proper motion of Sgr A* itself sets a lower limit of 0.4×10^6 $M_\odot$ to the mass directly associated with Sgr A* [4]. Now we can say that we have known the mass of Sgr A* to a great extent, the main uncertainty being the exact distance to the galactic center. But how big it is, and what it looks like, are different questions.

VLBI observations can provide the highest angular resolution achievable with any astronomical instruments [5]. At a distance of 8 kc to Sgr A* at the galactic center [6], an angular separation of 1 mill-arcsecond (mas) corresponds to a linear size of 8 AU. The Schwarzschild radius (R_{sc}) of a 4×10^6 $M_\odot$ SMBH is 1.2×10^{12} cm, or 0.08 AU or 10 micro-arcsecond (μas) in angular size. Indeed, Sgr A* is the closest SMBH candidate with the largest angular size of its Schwarzschild radius on the sky, and thus one of the best prime targets for the VLBI observational study.

2. Millimeter-VLBI of Sgr A*

Attempts to measure the Sgr A* structure with high-resolution VLBI observations at centimeter (cm) wavelength have suffered from the angular broadening caused by the diffractive scattering by the ionized interstellar medium, which dominates the resultant images with a λ^2-dependence apparent size, that is, $\Theta_{obs} = \Theta_{obs}^{1cm}\lambda^2$, where Θ_{obs} is the observed full-width at half-maximum (FWHM) in mas at a wavelength λ in cm, and equals Θ_{obs}^{1cm} at 1 cm. This tells us that the cm radio emission of Sgr A* comes from a region that extends to observers a much smaller

[1] for collaborators see acknowledgments

angle than the scattering angular size. However, following the λ^2-dependent scattering law, the scattering angle decreases very fast with the observing wavelength. So it can be expected that VLBI observations at millimeter (mm) wavelength, where the pure scattering angular size could become small enough to be comparable to the intrinsic (finite) size of Sgr A*, would show deviations of the observed apparent size from the scattering law.

Unlike the cm-VLBI observations, the mm-VLBI observations are severely limited by the atmosphere due to the fact that Sgr A* has a southerly declination ($\sim -30°$) and most existing VLBI antennas are located in the Northern Hemisphere. Therefore, most of the observational data of Sgr A* were taken at low elevation angles ($10° - 20°$) where atmospheric effects are substantial. Furthermore, the short and variable coherence time at mm wavelength, combined with the compromised sensitivity due to the high system temperature and the lower antenna efficiency of mm VLBI antennas, seriously limits the high signal-to-noise-ratio (SNR) detections of mm-VLBI observations of Sgr A*. As a consequence, there are large uncertainties in the results obtained from the conventional imaging process, i.e. the self-calibration technique for VLBI imaging, whose biggest drawback is non-uniqueness when the SNR is poor.

2.1. Revision of Scattering Law

To circumvent the large uncertainty in the commonly used amplitude calibration of VLBA observations of Sgr A*, we have since 2001 developed a model fitting program in which the amplitude closure relation is applied to minimize the calibration errors and thus to improve the accuracy of the measurements [7]. We then applied this procedure to refine the apparent size measurements of the first near-simultaneous VLBA+Y1 data [8] observed at five wavelengths (6, 3.6, 2, 1.35 cm, and 7 mm). By performing the weighted least-squares fits to these sizes at different wavelengths, we conclude that the best-fit two-dimensional scattering structure is [9]

$$\Theta_{\rm sca}^{\rm major} = (1.39 \pm 0.02)\lambda^2 \ , \qquad (1)$$

and

$$\Theta_{\rm sca}^{\rm minor} = (0.69 \pm 0.06)\lambda^2 \ , \qquad (2)$$

with a position angle of 80°. Here, $\Theta_{\rm sca}^{\rm major}$ and $\Theta_{\rm sca}^{\rm minor}$ are FWHM of major and minor axes in mas. This newly revised wavelength-dependent two-dimensional scattering structure is consistent with that from directly fitting to the closure amplitudes of somewhat different databases [10]. Compared with previous one (cf. [8]), new scattering law shows an even smaller scattering angular size along the East-West major axis direction.

2.2. Detections of Intrinsic Size

On November 3, 2002, we successfully carried out the first VLBA imaging observation of Sgr A* at its shortest wavelength of 3.5 mm [9]. Compared to the past 3.5 mm VLBI observations with the CMVA [11], our VLBA observation, with its dynamic scheduling, the highest possible recording rate and more frequent pointing calibration, has for the first time determined its elliptical structure in the radio emission at 3.5 mm, consistent with the morphology seen at longer wavelengths. It should be pointed out that dynamic observing is crucial to minimize the atmospheric effects in the mm-VLBI observation of Sgr A*. The exact observing date is not fixed, but depends on the weather condition at most antenna sites which usually are separated by a few thousand kilometers. While observing, a frequent antenna pointing check observations of compact and strong SiO maser sources, roughly in every 10 minutes, is necessary to maintain the best reference (offset) pointing and calibration.

The best fitted apparent structure from November 2002 VLBA observations is $0.21^{+0.02}_{-0.01}$ mas by $0.13^{+0.05}_{-0.13}$ mas with a position angle 79^{+12}_{-33} degrees. The major axis size is well determined while the fitted size of the minor axis has a quite big error mainly due to the poor resolution along the North-South direction. By subtracting in quadrature the scattering angle from Eq. (1), we obtain, for the first time, an intrinsic size of 0.126±0.017 mas along the major axis at 3.5 mm. In September 2003, we performed the second 3.5 mm VLBA imaging observation of Sgr A*, which confirms the results in November 2002 (Shen et al. 2006 in preparation).

Similarly, two 7 mm VLBA+GBT (Green Bank Telescope) observations with the highest recording rate were dynamically scheduled in March 2004 with the resultant averaged sizes of the major and minor axes of 0.724±0.001 mas and 0.384±0.013 mas, respectively with position angle $80.6^{+0.5}_{-0.6}$ degrees. The inclusion of GBT can improve the resolution in North-South direction by a factor of 3 [7], and actually this is the first VLBI observations to make use of the GBT at 7 mm. The results are consistent with the previous ones, but with much better accuracies. A significant deviation of measured size from the scattering angle along the major axis indicates an intrinsic size of 0.268±0.010 mas for the major axis [9], consistent with the previously reported detection of intrinsic size at 7 mm [11].

*2.3. Temporal Variation in the Structure of Sgr A**

On May 31, 1999, Sgr A* was observed simultaneously at three 7 mm bands (39, 43 and 45 GHz). It shows a larger deviation in the apparent source size from scattering angle than that observed a week agon on May 23, 1999 [9]. Such a deviation is significant at about 3σ along the minor axis direction. Thus, for the first time we can estimate an intrinsic size along the minor axis of 0.359±0.095 mas. The derived intrinsic size for the major axis is 0.334±0.042 mas. Therefore, within the error bars, the intrinsic structure detected at 7 mm on May 31, 1999 could be treated as a circular. Compared with the previously detected major axis size of 0.268 mas, such an intrinsic structure has increased in its size by at least 25% at 7 mm.

3. Discussion

We have sampled a zone of the SMBH closer to the event horizon than ever before, by detecting the intrinsic size of Sgr A* to be only 1 AU, or 12.6 R_{sc}. Assuming an intrinsically spherical structure of Sgr A* and using the lower limit to the mass of Sgr A* of 4×10^5 $M_\odot$ from the absolute proper motion work [4], we can easily derive a lower limit to the dark mass density of Sgr A* of 6.5×10^{21} $M_\odot$ pc^{-3}. This is the highest mass density ever measured in any SMBH candidates, and thus strongly supports the SMBH nature of Sgr A*.

It is intriguing that the detected intrinsic size at 3.5 mm is about twice the diameter of the shadow caused by the strong gravitational bending of light rays [12]. Regardless of the exact emission model, the characteristic diameter of shadow is always about $5R_{sc}$, which is nearly 50 μas for Sgr A*. So, Sgr A* will be the most important target for the future sub-mm VLBI experiment to test the general relativity in the strong field regime. Such kind of shadow, if confirmed, would be the most direct evidence for the existence of SMBH. The success of earlier single baseline 1.3 mm VLBI experiments [13] has already demonstrated the feasibility of capturing an image of the shadow around the edge of Sgr A* at sub-mm wavelengths in the near future.

From the two well determined intrinsic size at 7 [9, 11] and 3.5 mm [9], we can derive a λ^β-dependence of the intrinsic size with $\beta = 1.09^{+0.34}_{-0.32}$ using the two-point fit. This intrinsic size versus wavelength relation is consistent with the two lower limits of 0.02 and 0.008 mas at 1.3 and 0.8 mm, respectively, from the absence of refractive scintillation [14].

Such a wavelength dependence of the intrinsic major axis size provides a strong constraint on emission and accretion models for Sgr A*. It explicitly rules out explanations other than those models with stratified structure. It has been shown that the radiatively inefficient accretion flow (RIAF) model, after taking into account the interstellar scattering, can predict the observed sizes at both 7 and 3.5 mm quite well [15]. The extrapolated size of emitting region at 1 mm will reach the last stable orbit (LSO) radius of $3R_{sc}$ for a non-rotating SMBH. For a rotating

SMBH, the radius of LSO could be only $0.5R_{sc}$. Therefore, a break in the wavelength-dependent intrinsic size is inevitable with the decreasing wavelength, which can be used to constrain the spin of SMBH.

The radio flux density of Sgr A* is known to be variable at all the observable wavelengths on time scales of days to months. The variability appears to be more pronounced at shorter wavelengths with a relatively large amplitude fluctuation. For example, an intra-day variability (IDV) in the 2 mm flux density of Sgr A* has been found by the Nobeyama Millimeter Array (NMA) observations, indicating a very compact emitting region [16]. The correlation between flux density at 7 mm and spectral index suggests that the variation at the short radio wavelengths is intrinsic rather than due to the interstellar scintillation [17]. As such, it seems inevitable that the variability in the radio flux density of Sgr A* would be accompanied by the structural change. Our simulation seems to suggest that the detected structural variability on May 31, 1999 could be related to an outburst/flare occurred at about $50R_{sc}$ from the central SMBH (Shen et al. in preparation).

4. Conclusions

We have revised the scattering law, which gives a smaller scattering size along the East-West major axis direction. With the dynamic scheduling, we successfully carried out two 3.5 mm VLBA imaging observations, which reveal a consistent East-West elongated apparent structure. We also performed two-epoch new 7 mm VLBA+GBT observations with a better sensitivity and resolution. These new data show that the intrinsic size has come to play with the scattering effect at both 3.5 and 7 mm. The inferred intrinsic size is about 1 and 2.1 AU at 3.5 and 7 mm, respectively. The derived extraordinarily high (dark) mass density of Sgr A* strongly argues that Sgr A* is indeed a super-massive black hole.

Acknowledgments

This work was supported in part by the National Natural Science Foundation of China (grant 10573029). Z.-Q. Shen acknowledges support by the One-Hundred-Talent Program of the Chinese Academy of Sciences. In addition to the author, people who are involved in the project are P. T. P. Ho (ASIAA), M.-C. Liang (Caltech), K. Y. Lo (NRAO), M. Miyoshi (NAOJ) and J.-H. Zhao (CfA).

References

[1] Melia F & Falcke H 2001 *ARA&A* **39** 309
[2] Schödel R et al. 2002 *Nature* **419** 694
[3] Ghez A M et al. 2005 *ApJ* **620** 744
[4] Reid M J & Brunthaler A 2004 *ApJ* **616** 872
[5] Kelllermann K I & Moran J M 2001 *ARA&A* **39** 457
[6] Reid M J 1993 *ARA&A* **31** 345
[7] Shen Z-Q et al. 2003 *Astron. Nachr.* **324(S1)** 383
[8] Lo K Y et al. 1998 *ApJ* **508** L61
[9] Shen Z-Q et al. 2005 *Nature* **438** 62
[10] Bower G C et al. 2004 *Science* **304** 704
[11] Doeleman S S et al. 2001 *AJ* **121** 2610
[12] Falcke H Melia F & Agol E 2000 *ApJ* **528** L13
[13] Krichbaum T P et al. 1998 *A&A* **335** L106
[14] Gwinn C R et al. 1991 *ApJ* **381** L43
[15] Yuan F Shen Z-Q & Huang L 2006 *ApJ* **642** L45
[16] Miyazaki A et al. 2004 *ApJ* **611** L97
[17] Herrnstein R M et al. 2004 *AJ* **127** 3399

Institute of Physics Publishing
doi:10.1088/1742-6596/54/1/060
Journal of Physics: Conference Series **54** (2006) 381–385
Galaxy Center Workshop

Simultaneous Multi-Wavelength Spectrum of Sgr A* and Long Wavelength Cutoff at $\lambda > 30$ cm

An T.[1,2], Goss W.M.[3], Zhao Jun-Hui[2], Hong X.Y.[1], Roy S.[4], Rao A.P.[5] and Shen Z.-Q.[1]

[1] Shanghai Astronomical Observatory, China
[2] Harvard-Smithsonian CfA, the USA
[3] NRAO-AOC, Socorro, the USA
[4] ASTRON, the Netherlands
[5] National Center for Radio Astrophysics, Pune University Campus, India

E-mail: antao@shao.ac.cn

Abstract. We reviewed the multiple-wavelength spectrum of Sgr A*. The recent continuum spectrum was determined using data obtained from the Very Large Array (VLA), the Giant Metrewave Radio Telescope (GMRT), the Submillimeter Array (SMA) and the Keck II 10m telescope on the same date (2003 June 17). The nearly simultaneous spectrum ($S \propto \lambda^{-\alpha}$) covers a broad wavelength (λ) range from 90 cm to 3.8 micron. From about 2 cm to 1 mm, the source shows a rising spectrum with a spectral index of $\alpha = 0.43 \pm 0.04$. However, the IR measurements suggested that the flux density of Sgr A* must drastically drop towards the shorter wavelengths and the turnover likely occurs in the submillimeter regime. The spectrum at wavelengths between 47 and 3.6 cm is flat with $\alpha = 0.11 \pm 0.03$. The flux density of Sgr A* is 0.22 ± 0.06 Jy at 90 cm on this date in 2003. Compared with measurement at 47 cm, the flux density at 90 cm has decreased by a factor of two (or $\sim 4\sigma$), suggesting a cutoff at wavelength longer than 47 cm. The fits using free-free absorption opacity to the spectrum suggest a cutoff wavelength at ~ 100 cm. The cutoff wavelength appears to be three times longer than that of ~ 30 cm determined by Davies, Walsh, & Booth [1] three decades ago based on observations in 1974 and 1975. A model interpreting the variation of the long wavelength cutoff using stellar winds (An et al. [2]) is summarized in this paper.

1. Introduction

The compact radio source Sgr A* in the Galactic center (GC) is now believed to be associated with a supermassive black hole (SMBH) with a mass of $M \sim 4 \times 10^6 M_\odot$ ([3, 4, 5, 6]). Owing to its proximity to the earth (1" corresponds to ~0.04 pc at a distance of ~8 kpc, [7, 6]), Sgr A* can be studied in a great detail. Soon after the discovery of Sgr A* ([8]), Davies, Walsh, & Booth ([1]) observed the Galactic center at 0.408, 0.960 and 1.660 GHz using the early MERLIN interferometer. Sgr A* was only detected at the two higher frequencies. The measurements showed a low frequency cutoff around 1 GHz in the spectrum of Sgr A* ([1] and references therein). The authors attributed the decrease in flux densities below 1 GHz to free-free absorption from ionized gas in Sgr A West. Recently, Roy & Rao ([9]) measured the Galactic center region at 620 MHz using the GMRT, and detected Sgr A* with a flux density of 0.5±0.1 Jy (angular resolution ~9"); while at 330 MHz, Nord et al. ([10]) detected Sgr A* using the VLA with a flux density of 0.33±0.12 Jy. The two measurements indicate that the spectrum below 1 GHz may indeed have a pronounced cutoff. However both the measured flux densities of Sgr A* show a significant excess compared to the expected values from Davies et al.'s free-free absorption model.

On 2003 June 17, we observed Sgr A* at multiple wavelengths using the VLA and the GMRT with the purpose of discerning the low frequency spectrum of Sgr A*. The observations and data reduction were reported in An et al. ([2]). We reviewed the spectrum covering wavelength range from 90 cm to 3.8 μm, with the data obtained from VLA, GMRT, SMA ([11, 12, 13]) and Keck II 10m telescope observations on the same date (2003 June 17). In this paper, we will further discuss the model interpreting the variation of the long wavelength cutoff.

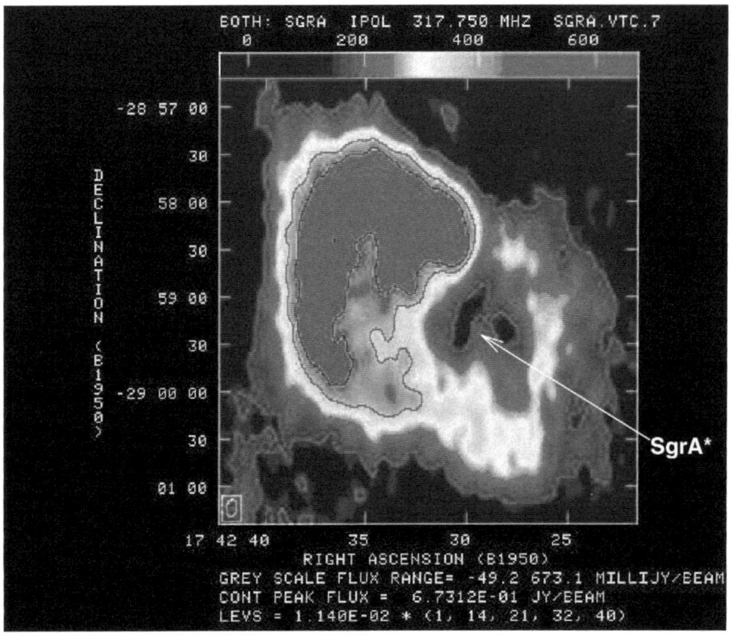

Figure 1. The continuum image of Sgr A* at 90 cm. The resolution is $10.9'' \times 6.8''$ (PA=$-10°$). The *r.m.s.* noise in the image is $\sim$ 12 mJy/beam. The position and expected scattering size of Sgr A* at 90 cm are marked with a red cross in the image center ([14, 15].

2. Flux Densities Measurements and Spectrum of Sgr A*

The measurements of the flux densities for Sgr A* at wavelengths shorter than 20 cm with the VLA in the A-configuration are straight-forward. The measurements have been published by An et al. ([2]). Because the apparent size increases as λ^2 and the confusion of the non-thermal emission from the surrounding sources also increases towards longer wavelengths, the measurements of the flux density at 90 cm are more difficult. Figure 1 shows the 90 cm continuum map in which Sgr A* is marked as a red cross in the image center with an expected scattering size of $\sim 11'' \times 6''$ (e.g. [15, 16, 17]). The supernova Sgr A East dominates the total flux density at this wavelength. The emission from the shell to the east appears to be significantly brighter than that to the west as pointed out earlier by other authors (e.g. [18]). However, at the position of Sgr A* a weak source is detected. The region surrounding this source has a lower flux density to the east, north and west. The deficiency in emission likely arises from absorption by ionized gas in Sgr A West. To determine the flux density at 90 cm, we fit two intensity slices using Gaussians along the expected major ($PA = 80°$) and minor ($PA = -10°$) axes after subtracting the diffuse emission from the background. The measured flux density of Sgr A* at 90 cm is

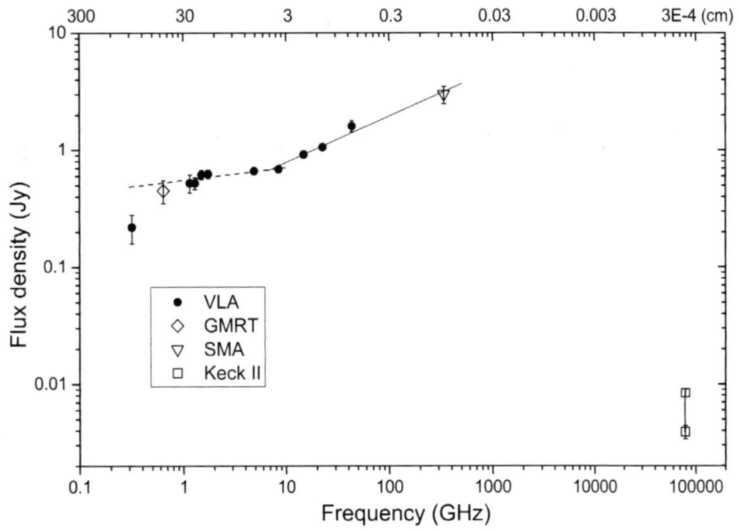

Figure 2. Quasi-simultaneous spectrum of Sgr A* from 90 cm to 3.8 μm on 2003 June 17 ([2]). *solid circle*: VLA; *diamond*: GMRT; *triangle*: SMA; *square*: Keck II 10m.

0.22±0.06 Jy. The result is consistent with the measurement of Nord et al. ([10]) within 1σ at the same wavelength.

Figure 2 shows the quasi-simultaneous spectrum of Sgr A* at wavelengths ranging from 90 cm to 3.8 μm, observed on 2003 June 17 ([2]). A number of interesting results are outlined here:

(i) The source shows a rising spectrum in the range of wavelengths from $\sim$ 2 cm to $\sim$ 0.89 mm. High angular resolution measurements of the simultaneous flux densities at 230 and 690 GHz show that the mean value of 690 GHz flux density observed from multiple epochs is less than that of 230 GHz, suggesting a turnover in the spectrum between these frequencies (Marrone et al., these proceedings).

(ii) A break in the spectrum appears at $\sim$3.6 cm. The spectrum at shorter centimeter to millimeter wavelengths can be described by a power-law: $S_\nu \propto \nu^\alpha$ with $\alpha_{3.6cm}^{0.89mm} = 0.43 \pm 0.04$. The spectrum between 3.6 and 47 cm is flat with a spectral index of $\alpha_{47cm}^{3.6cm} = 0.11 \pm 0.03$. The difference in spectral indices between the two parts of power-law spectra is $\alpha_{3.6cm}^{0.89mm} - \alpha_{47cm}^{3.6cm} = 0.32 \pm 0.05$, suggesting that the break is significant ($> 6\sigma$).

(iii) At 90 cm, the measured flux density of 0.22±0.06 Jy is well below the inferred 0.5 Jy extrapolated from power-law fitting ($S_\nu \propto \nu^{0.11\pm0.03}$). This significant decrease in flux density at 90 cm (larger than 4σ) suggests that the spectrum of Sgr A* may well have a long wavelength cutoff at $\lambda > 47$cm on this date in 2003. In fact, Davies, Walsh, & Booth ([1]) detected a long wavelength cutoff at $\sim$ 30 cm in the spectrum of Sgr A* three decades ago.

Figure 3 compares the observed spectra for Sgr A* at epochs 1975 ([1]) and 2003 ([2]) at wavelengths longer than 3.6 cm. The spectra at both epochs were fitted using a model of slowly rising power-law ($S_\nu \propto \nu^\alpha$) along with a free-free absorption screen ($S_\nu \propto e^{-\tau_{ff}}$) between the observer and Sgr A*, as proposed by Davies et al. ([1]). The comparison of the fits to the 1975 (Figure 3, *dashed line*) and 2003 data (Figure 3, *solid line*) suggest that the cutoff wavelength

($\tau_{\rm ff}(\lambda_{cutoff}) = 1$) due to free-free absorption has changed from $\lambda_{cutoff} \sim 100$ cm in 1975 to $\lambda_{cutoff} \sim 30$ cm in 2003.

If the free-free absorption model is correct, the variation in the long wavelength cutoff would suggest that the column density of the free-free absorption screen must have decreased significantly in the past 28 years. Quantitatively, the free-free absorption opacity τ_{ff} would need to decrease by a factor of nine from 1975 to 2003. Assuming that the electron temperature of the ionized gas is constant, the inferred decrease in τ_{ff} would then correspond to a decrease by a factor of nine in emission measure (EM). Such a large variation in EM would likely occur in a compact region, i.e. 0.1" –10" displaced from Sgr A*. In this region, stellar winds originating from massive stars may have modified the local environment. The stellar wind mass flux captured by the SMBH potential is highly variable owing to orbital motions of stars as well as wind-wind collisions (a number of contributors to this volume have shown the nature of the complex region in the central parsec of the Galactic center). Most likely the source accounting for the long wavelength cutoff is free-free absorption from cool gas close to Sgr A*. The emission line measurements of the star S2 have shown that the surface temperature of this star is about 30,000K; the temperature was measured when S2 was only 100 AU from Sgr A* ([4]). This indicates that free-free absorption gas with temperature $\sim 10^4$ K can exist very close to the SMBH in the Galactic center. In addition, recent simulations ([19, 20] and Cuadra et al. in these proceedings) have shown two phases of gas co-existing in the Galactic center, i.e., hot gas with a typical temperature of $\sim 10^7$ K and cold gas with a temperature of $\sim 10^4$ K. The cold but dense gas clumps originated from the stellar winds can be captured by the potential of the SMBH as have been suggested in the numerical simulations ([19, 20]). If a cold ionized gas component moves to the front of Sgr A*, it would attenuate the flux density of Sgr A* at long wavelengths. As an example, for a cold ionized gas component with a temperature of 10^4 K and a size of 0.004 pc, the free-free absorption opacity is about unity at a wavelength 79 cm for an electron density of 10^4 cm^{-3}, while the cutoff wavelength will be about 30 cm if the emission measure has increased by a factor of nine. In this scenario, the change of the long wavelength cutoff due to free-free absorption toward Sgr A* might be sensitive to fluctuations of stellar wind mass flux captured by the SMBH potential.

Acknowledgments

The work is supported by the NSFC (10503008, 10328306). The VLA is operated by the National Radio Astronomy Observatory, a facility of the National Science Foundation operated under cooperative agreement by Associated Universities, Inc. The GMRT is a component of the National Centre for Radio Astrophysics of the Tata Institute of Fundamental Research.

References

[1] Davies, R.D., Walsh, D., & Booth, R.S., 1976, MNRAS, 177, 319
[2] An, T., Goss W.M., Zhao, J.-H., Hong, X.-Y., Roy, S. & Rao, A.P., Shen, Z.-Q., 2005, ApJ, 634, L49
[3] Schödel, R., Ott, T., Genzel, R., et al. 2002, Nature, 419, 694
[4] Ghez, A.M., Duchene, G., Matthews, K., et al., 2003, ApJ, 586, L127
[5] Reid, M.J., & Brunthaler, A., 2004, ApJ, 616, 872
[6] Eisenhauer, F., Genzel, R., Alexander, T., et al., 2005, ApJ, 628, 246
[7] Reid, M.J., 1993, ARA&A, 31, 345
[8] Balick, B, & Brown, R.L., 1974, ApJ, 194, 265
[9] Roy, S., & Rao, A.P., 2004, MNRAS, 349, L25
[10] Nord, M.E., Lazio, T.J.W., Kassim, N.E., et al. 2004, ApJ, 601, L51
[11] Moran, J.M., 1998, Proc. SPIE, 3357, 208
[12] Blundell, R., 2004, in 15th International Symposium on Space Terahertz Technology, ed. G.Narayanan (Amherst: U Mass), 3 (astro-ph/0508492)
[13] Ho, P.T.P., Moran, J.M., & Lo, K.Y., 2004, ApJ, 616, L1
[14] Rogers, A.E.E., et al. 1994, ApJ, 434, L59

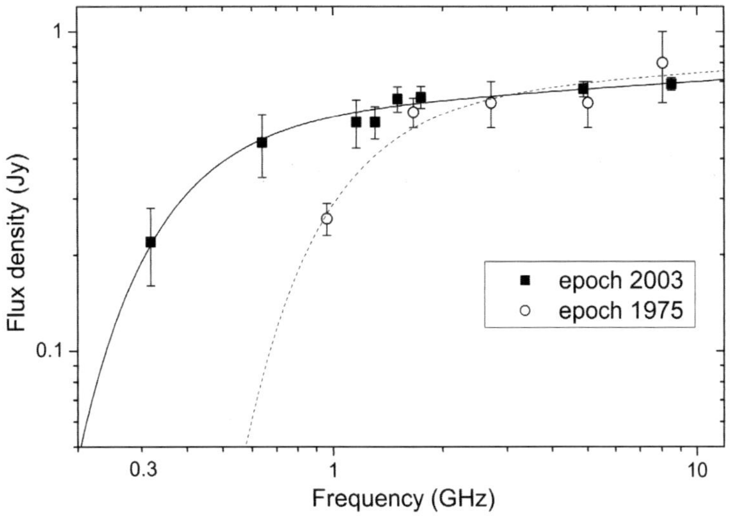

Figure 3. Comparison of the long wavelength spectra of Sgr A* observed between epochs 2003 (*solid square*, [2]) and 1975 (*open circle*, [1]). The *dashed* and *solid* lines represent fits using a free-free absorption opacity model applied to the spectrum at epochs 1975 and 2003, respectively. The free-free opacity decreases by a factor of ~nine at 30 cm (or 1 GHz) from 1975 to 2003.

[15] Lo, K.Y., Shen, Z.-Q., Zhao, J.-H., & Ho, P.T.P., 1998, ApJ, 508, L61
[16] Bower, G.C., Falcke, H., Herrnstein, R.M., Zhao, J.-H., Goss, W.M., & Backer, D.C., 2004, Science, 304, 704
[17] Shen, Z.-Q., Lo, K.Y., Liang, M.-C., Ho, P.T.P., & Zhao, J.-H., 2005, Nature, 438, 62
[18] Pedlar, A., Anantharamaiah, K.R., Ekers, R.D., et al., 1989, ApJ, 342, 769
[19] Cuadra, J., Nayakshin, S., Springel, V., Di Matteo, T., 2006, MNRAS, 366, 358
[20] Cuadra, J., Nayakshin, S., Springel, V., Di Matteo, T., 2005, MNRAS, 360, L55

Institute of Physics Publishing
Journal of Physics: Conference Series **54** (2006) 386–390
doi:10.1088/1742-6596/54/1/061
Galaxy Center Workshop

The infrared emission of the dust clouds close to Sgr A*

Y. Clénet[1], D. Rouan[1], D. Gratadour[2], P. Léna[1] and O. Marco[3]

[1] LESIA, Observatoire de Paris, 5 place Jules Janssen, 92195 Meudon cedex, France;
[2] Gemini Observatory, Gemini Observatory, Hilo, HI 96720, USA
[3] ESO, Alonso de Córdova, Casilla 3107, Vitacura, Casilla 19001, Santiago 19, Chile

E-mail: yann.clenet@obspm.fr

Abstract. Since about an half-decade, adaptive optics (AO) on 8-10m class telescopes has been a key instrumentation for our understanding of Sgr A*. In this article, we first remind the performances of the AO-fed NACO/VLT, with its infrared wavefront sensor, comparing them to other systems worldwide. We then present results obtained in imaging with NACO, on the dust clouds close to Sgr A*, more particularly on the closest one, Sgr A*-f, whose emission could result from a jet from Sgr A*. We finally describe the spectroscopic observations of these dust clouds, performed very recently, using the low resolution prism mode of NACO.

1. Available modes and compared performances of NACO/VLT

NACO has been the first AO-assisted spectro-imager installed at VLT, in late 2001. It is made of the AO system NAOS [1] and the camera CONICA [2]. The main advantage of NAOS for Galactic Centre (GC) observations is its infrared (IR) wavefront sensor (WFS), which is locked on IRS 7 (K=6.5), located 5.5" from Sgr A*. When equipped with visible WFS, AO systems have to be locked on a R∼14 star located about 20" away from Sgr A*, which limits their performances. Even laser-guide star AO systems need an additional guide star for tip-tilt correction, with lower constraints than the classical guide star of visible WFS AO, and thus face the paucity of bright stars in the optical around Sgr A*. As a conclusion, under comparable atmospheric conditions, NAOS-CONICA provides the best correction and image quality for Sgr A* observations (Table 1).

Table 1. Performances of the different AO systems on 8m class telescopes for GC observations

AO systems	Strehl ratio at K (2.1 μm)	Strehl ratio at L′ (3.8 μm)
NAOS+IRWFS	∼50%	∼80%
Visible WFS (e.g. Keck [3])	<10%	∼35%
Laser guide star (e.g. Keck [3])	∼30%	∼70%

The CONICA camera is provided with several observing modes: imaging from J (1.2 μm) to M′ (4.8 μm), spectroscopy from J to M′, Fabry-Perot at K, polarimetry and coronography.

2. Infrared imaging of the dusty clouds close to Sgr A* [4]

2.1. Observations and results

We have performed L′-band (3.8 μm, 0.0271"/pixel) observations of the GC region during five 2004 nights (hereafter Night 1, 2, 3, 4 and 5 respectively) with NACO.

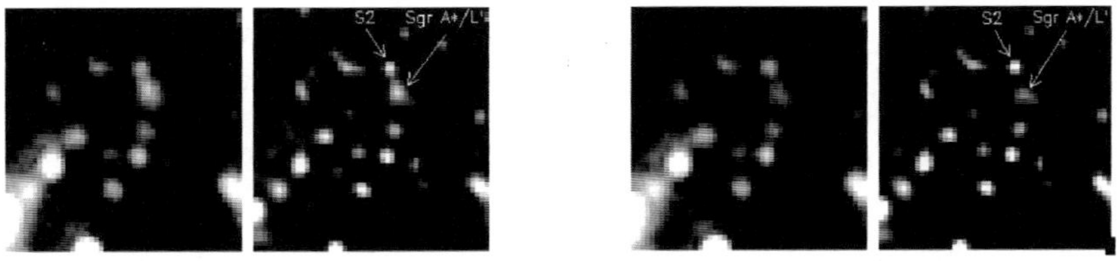

Figure 1. 1.4"×1.4" Night 3 images of the central stellar cluster during Sgr A*/L′ flare (pair of images on the left) and quiescent (pair of images on the right) states. Each couple of images is made of a 105 s integration time reduced image and the corresponding deconvolved one, obtained with 40 iterations of the maximum likelihood deconvolution IDL procedure.

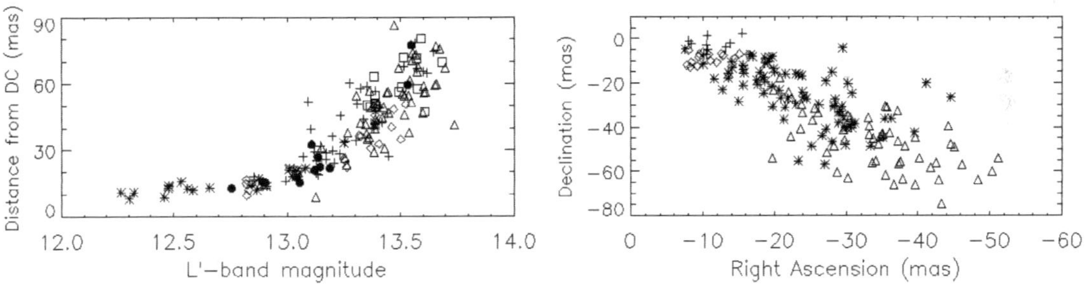

Figure 2. Left: Distance of Sgr A*/L′ from the dynamical centre (DC) of the S2 orbit as a function of the Sgr A*/L′ magnitude. Crosses, stars, diamonds, triangles, squares and filled circles are Night1, 2, 3/part1, 3/part2, 4, 5 data points respectively. Right: Sgr A*/L′ positions with respect to DC. Crosses are positions for 12.2≤L′(Sgr A*/L′)<12.6, diamonds for 12.6≤L′(Sgr A*/L′)<13.0, stars for 13.0≤L′(Sgr A*/L′)<13.4, triangles for 13.4≤L′(Sgr A*/L′)<13.8.

We observe a clear, reproducible, correlation between the Sgr A*/L′ flux and the Sgr A*/L′ position (Fig. 1, 2). This modulation is actually due to the combined and spatially unresolved emissions of SgrA*/IR (the IR counterpart of Sgr A* itself) and Sgr A*-f, a nearby dusty cloud. Hence, we point at a dual mechanism for the Sgr A*/L′ emission (the 3-5 μm emission of Sgr A*): one traced by the flaring state, unresolved and associated to Sgr A*/IR itself and another one associated to the quiet state due to Sgr A*-f. Sgr A*-f has the following characteristics: extended (~130 mas), distance from Sgr A*/IR: 75 mas, unobserved at K and $K - L' > 3$.

2.2. Origin of the Sgr A-f emission*

We consider here the different possible emission mechanisms for Sgr A*-f and their likelihood:

– *dust cloud on the line of sight, heated by a nearby/embedded star:* this mechanism has been invoked by [3, 5] and is of course a possible explanation;

– *stellar emission:* this mechanism is unlikely given the large observed $K - L'$ colour, the extended emission and the lack of detection at K;

– *dust cloud close to Sgr A*, heated by the accretion disk:* the colour temperature T_c of the Sgr A*/L′ quiescent emission, derived from the optically thin case equation $F_\nu(K)/F_\nu(L') = Q_{abs}(K)/Q_{abs}(L') \times B_\nu(T_c, K)/B_\nu(T_c, L')$ (where Q_{abs} is the absorption cross section coefficient,

$B_\nu(T_c, \lambda)$ the Planck function, F_ν(K)=1.9 mJy [6] and F_ν(L′)=3 mJy), is comprised, for dust grain sizes from 0.01 to 0.1 μm, between 1110 K and 1130 K for silicate grains, between 870 K and 930 K for graphite grains. On the other hand, the maximum equilibrium temperature T_e of dust grains heated by Sgr A* itself can be derived from: $\int_0^\infty \frac{L_\lambda}{4\pi r^2}\pi a^2 Q_{abs}(\lambda)\,d\lambda = \int_0^\infty 4\pi a^2 Q_{abs}(\lambda)\,\pi B_\lambda(T_e, \lambda)\,d\lambda$, where L_λ is the Sgr A* flux density, a the dust grain size and r=600 UA, ie about 75 mas. Since the smallest wavelengths dominate the first term, we approximate L_λ by the Sgr A* quiescent state power law found by [7] and usually adopted to constrain the Sgr A* emission models: $L_\lambda = 7.1 \times 10^{-35}\left(\frac{\lambda}{3.1 \times 10^{-10}}\right)^{-0.8}$. For dust grain sizes ranging from 0.001 to 0.1 μm, we obtain an equilibrium temperature between 70 K and 90 K for silicate grains, between 80 K and 110 K for graphite grains. About one order of magnitude is then found between the quiescent IR color temperature T_c of Sgr A*/L′ and the temperature T_e expected if the accretion disk emission was the dust heating source at the Sgr A*-f location. This latter process is therefore excluded.

– *dusty cloud close to Sgr A*, heated by a jet:* the kinetic power delivered by a jet from Sgr A*, assuming an equipartition between ions and electrons, is given by $P_{jet}=\dot{M}_{jet} \times m_e/m_p \times (\beta c)^2$, where m_e, m_p, β are the mass of electron, the mass of proton and the electron to light speed ratio, respectively, β=0.95 [8]: hence, $P_{jet}=1.2\times10^{35}$ erg/s=32 $L_\odot$. On the other hand, the total luminosity corresponding to the observed L′-band luminosity $L_{L'}$, given by: $L_{tot} = \frac{\int_0^\infty Q_{abs}(\lambda) B_\lambda(T_c, \lambda)\,d\lambda}{\int_{L'} Q_{abs}(\lambda) B_\lambda(T_c, \lambda)\,d\lambda} \times L_{L'}$, is L_{tot}=12 $L_\odot$ for silicate grains at T_c=1120 K and L_{tot}=6 $L_\odot$ for graphite grains at T_c=900 K, whatever the grain size from from 0.001 to 0.1 μm. These luminosities are of the same magnitude order as the kinetic power delivered by the jet P_{jet}. The energy transfer of this jet power to Sgr A*-f through collisions can therefore be a possible mechanism to explain the Sgr A*-f emission.

2.3. Connection between IR dusty clouds and gaseous HII regions

Radio observations have shown the presence of a ridge of emission made of a chain of gaseous clouds from Sgr A* to the mini-cavity [9]. These HII coulds could have been formed by the gravitational collimation of the IRS 16 cluster stellar winds by Sgr A* [10, 11]. Alternatively, a jet from Sgr A*, similarly to our scenario, has been proposed as the wind source [12].

These gaseous HII clouds, taking into account their estimated proper motions and their fairly extended sizes [13, 14], are close to red IR extended sources (Fig. 3), which could suggest a common origin, if not an identification.

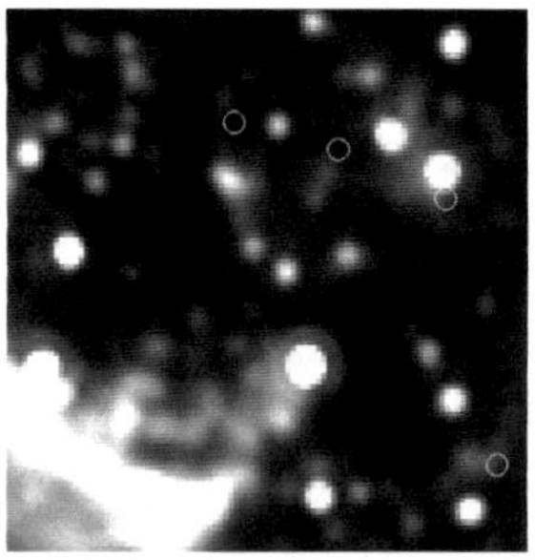

Figure 3. Encircled HII cloud positions on a 2.4"×2.4" L′ NACO image. From left to right: HII 5D, HII 5C, source ϵ, source ξ [13, 14]. The circle diameter is the 0.1" radio resolution.

3. Infrared spectroscopy of the dusty clouds close to Sgr A*

3.1. Observations

During the night of 30 March 2006, we have performed with NACO spectroscopic observations of few dust red clouds close to Sgr A*. For that purpose we have used the NACO spectroscopic mode made of a low resolution prism which allows one to cover at once from J to M′ (1 to 5.4 μm). Its resolution varies from ~60 at 3 μm to 200 at 5 μm. Actually, two configurations are available when using this mode: either with a filter introducing a cut-off at 2.5 μ to cover only from J to K, or a full J-to-M′ coverage, the latter having been used for our observations.

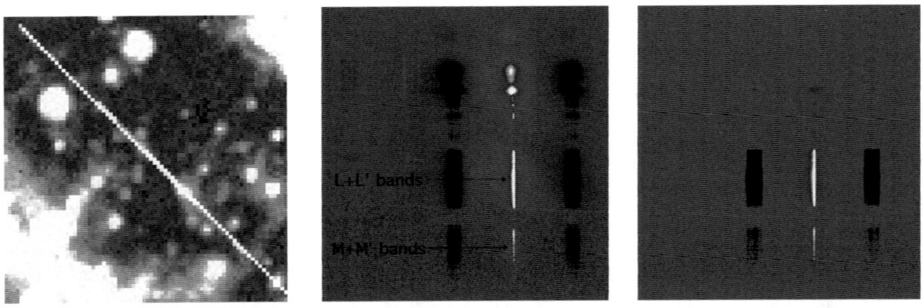

Figure 4. Left: Slit position. Middle and right: raw "spectra" of a B2V and G0V stars.

3.2. Calibration and preliminary results

This observing mode, with its large covered spectral domain, is tricky to use: apart from the lack of signal at the shortest wavelengths for red sources when preventing from saturating the largest wavelengths, calibration is complex, in particular wavelength calibration (given the lack of calibration lines for the available calibration lamps) and telluric feature correction.

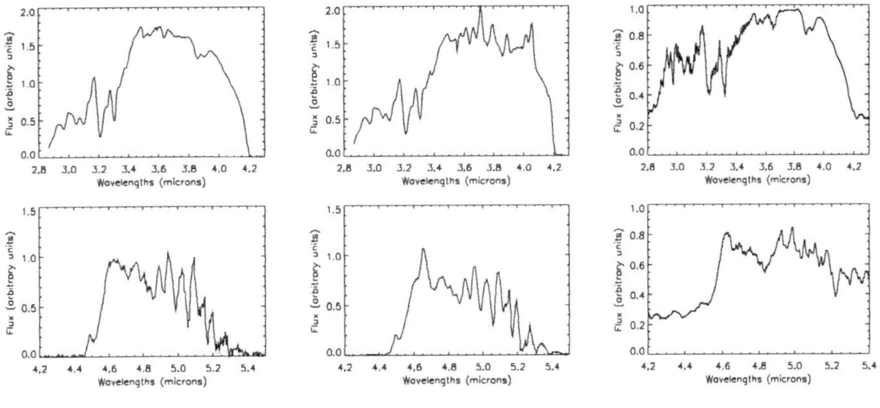

Figure 5. Telluric features. Left column: from the G0V star. Middle: from the B2V star. Right: from [15]

Wavelength calibration and telluric correction have been done simultaneously using the 1-5 μm atmospheric transmission from [15]. Figure 5 presents this calibration of the L′ and M′ atmospheric transmissions, with the observed telluric features from the B2V and G0V calibration stars.

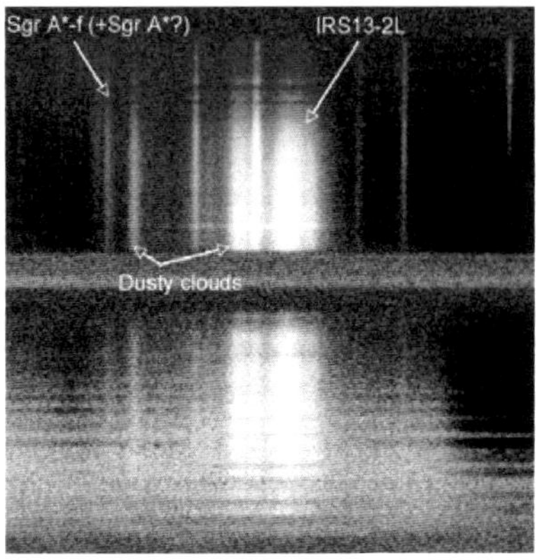

Figure 6. Close-up of the "science" spectra of the different dust clouds close to Sgr A*. The upper part is the L+L′ part of the spectra, the lower part the M+M′ part.

The data is still under data reduction/analysis processes. Still, we can already infer few results (cf. Fig. 6):

– the signal to noise ratio we have obtained on Sgr A*-f is low;

– a Sgr A* flare may have occurred during the observations (from control acquisition images) but it will be probably difficult to extract a spectrum of Sgr A* itself;

– we have achieved a good signal to noise ratio for the other dusty clouds

4. Conclusions

We will most probably be unable to derive a 3-5.4 μm spectrum of Sgr A* during the flare that has occurred during our observations. Though, the derivation and comparison of the dust temperatures and possible spectral features (hydrogen recombination lines, PAH) of the different dusty clouds should help us to test the nature of Sgr A*-f and our shock/jet hypothesis.

Acknowledgments

The authors would like to thank Chris Lidman and Markus Hartung for their helpful assistance in using the NACO prism mode.

References

[1] Rousset G., Lacombe F., Puget P., et al. 2000, *Proc. SPIE*, **4007**, 72
[2] Lenzen R., Hofmann R., Bizenberger P., Tusche A., et al. 1998, *Proc. SPIE*, **3354**, 606
[3] Ghez A., Hornstein S., Lu, J. et al. 2005, *ApJL*, **635**, 1087
[4] Clénet Y., Rouan D., Gratadour D. et al. 2005, *A&A*, **439**, L9
[5] Eckart A., Baganoff F., Schödel R., et al. 2006, *A&A*, **450**, 535
[6] Eckart A., Baganoff F., Morris M., et al. 2004, *A&A*, **427**, 1
[7] Baganoff F., Bautz M., Brandt W., et al. 2001, *Nature*, **413**, 45
[8] Yuan, F., Markoff, S., Falcke, H. 2002, *A&A*, **383**, 854
[9] Yusef-Zadeh F., Morris M., Ekers R. 1990, *Nature*, **348**, 45
[10] Wardle M., & Yusef-Zadeh F. 1992, *Nature*, **357**, 308
[11] Melia F., Coker R., Yusef-Zadeh F. 1996, *ApJL*, **460**, 33
[12] Lutz D., Krabbe A., Genzel R. 1993, *ApJ*, **418**, 244
[13] Yusef-Zadeh F., Roberts D., Biretta J. 1998, *ApJL*, **499**, 159
[14] Zhao J., Goss W. 1999, in ASP Conf. Ser. 186, The Central Parsecs of the Galaxy, ed. H. Falcke et al., 224
[15] Wallace L., Livingston W. 2003, in "An atlas of the solar spectrum in the infrared from 1850 to 9000 cm-1 (1.1 to 5.4 micrometer), revised", National Solar Observatory Technical Report

Institute of Physics Publishing
Journal of Physics: Conference Series **54** (2006) 391–398
doi:10.1088/1742-6596/54/1/062
Galaxy Center Workshop

Multi-wavelength and polarimetric observations [1] of Sagittarius A*

A. Eckart[1] R. Schödel [1] L. Meyer [1] S. Trippe [2] T. Ott [2] R. Genzel [2,3] K. Muzic [1] J. Moultaka [1] C. Straubmeier [1] F. K. Baganoff [4] M. Morris [5] G.C. Bower [6]

[1] I. Physikalisches Institut; University of Cologne, Zülpicher Straße 77; 50937 Köln, Germany
[2] Max Planck Institut für extraterrestrische Physik, Giessenbachstraße, 85748 Garching, Germany
[3] Physics Department, University of California at Berkeley, Le Conte Hall, Berkeley, CA 94720, USA
[4] Kavli Institute for Astrophysics and Space Research, Massachusetts Institute of Technology, Cambridge, MA 021 39-4307, USA
[5] Department of Physics & Astronomy, UCLA, Los Angeles, CA 90095-1547, USA
[6] Department of Astronomy and Radio Astronomy Laboratory, University of California at Berkeley, Campbell Hall, Berkeley, CA 94720, USA

E-mail: `eckart@ph1.uni-koeln.de`

Abstract. We summarize the results of some of the latest NIR/sub-millimeter/X-ray observing campaigns. Those include the latest simultaneous observations as well as the most recent results from VLT NACO observations of polarized NIR flare emission of Sgr A*. We interpret the new NIR polarimetry results using a model in which spots are on relativistic orbits around Sgr A*, which is associated with the massive 3.6 million solar mass black hole at the Galactic Center. In the NIR the observations have been carried out using the NACO adaptive optics (AO) instrument at the European Southern Observatory's Very Large Telescope. In the X-ray and radio domains we used the ACIS-I instrument aboard the Chandra X-ray Observatory and the Submillimeter Array on Mauna Kea, Hawaii, as well as the Very Large Array in New Mexico, respectively.

1. Introduction

The compact radio source Sagittarius A* (Sgr A*) has been convincingly identified in the radio and infrared with a massive black hole (MBH) of mass $3.6\pm0.3\times10^6 M_\odot$at the center of our Galaxy (Schödel et al. 2003; Ghez et al. 2005; Eisenhauer et al. 2005) at a distance of only 7.6±0.3 kpc (Eckart & Genzel 1996, Genzel et al. 1997, 2000, Ghez et al. 1998, 2000, 2003a, 2003b, 2005, Eckart et al. 2002, Schödel et al. 2002, 2003, Eisenhauer 2003, 2005). Especially in the case of the center of the Milky Way, progress has been made through the investigation of stellar dynamics (see references above). Sgr A* is the nearest supermassive black hole. It is about one hundred times closer than the nucleus of the nearest similar galaxy in M31*. It therefore provides us with a unique opportunity to understand the physics and possibly the evolution of these objects.

[1] Based on observations with CHANDRA and ESO VLT observations 271.B-5019(A) and 271.B-5019(A).

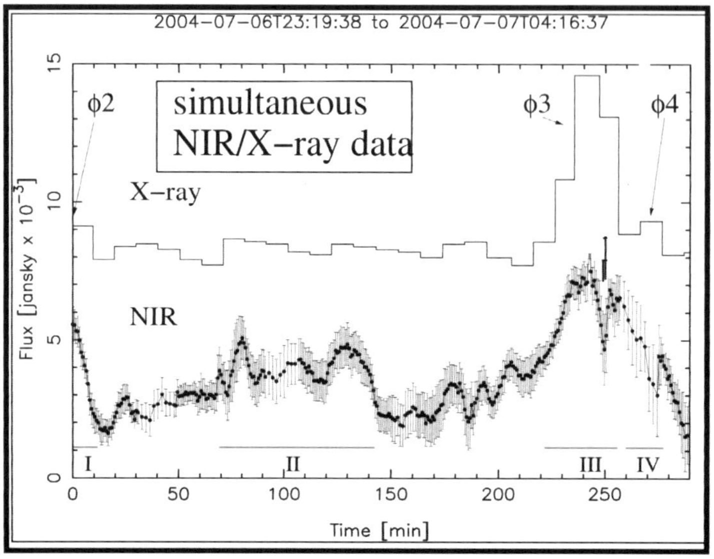

Figure 1. The X-ray and NIR light curves of the 2004 observations plotted with a common time axis.

Compelling evidence for a massive black hole at the position of Sgr A* is also provided by the observation of variable emission from that position both in the X-ray and recently in the near-infrared (Baganoff et al. 2001, 2003, Eckart et al. 2004, 2006a, Porquet et al. 2003, Goldwurm et al. 2003, Genzel et al. 2003, Ghez et al. 2004a, Eisenhauer et al. 2005, Belanger et al. 2005, and Yusef-Zadeh et al. 2006a). The temporal correlation between rapid variability of the near-infrared (NIR) and X-ray emission (Eckart et al. 2004, Eckart et al. 2006a) suggests that the emission showing 10^{33-34} erg/s flares arises from a compact source within a few ten Schwarzschild radii (R_S) of the MBH. This fact points to a common physical origin of the observed phenomena and is probably linked to the variability at radio through sub-millimeter wavelengths (Herrnstein et al. 2004, Mauerhan et al. 2005 and references therein). For a black hole of its size, Sgr A* is extremely underluminous, at about $10^{-9...-10}L_{Edd}$. This surprisingly low luminosity has motivated many theoretical and observational efforts to explain the processes that are at work in the immediate vicinity of Sgr A*.

2. Simultaneous NIR/X-ray Observations of Sgr A*

Simultaneous observations of Sgr A* from the radio to the X-ray regime are an ideal tool to learn about the emission mechanisms responsible for the radiation from the immediate vicinity of the central black hole. The first successful experiment during which simultaneous X-ray and near-infrared flare emission has been detected was presented by Eckart et al. (2004). They detected a weak 6×10^{33} erg/s X-ray flare and covered its decaying flank in the NIR. The flux density variability of Sgr A* at radio through submillimeter wavelengths has been studied extensively, showing that variations occur on time scales from hours to years (e.g. Bower et al. 2002, Herrnstein et al. 2004, Zhao et al. 2003). Some of the radio variability is probably due to interstellar scintillation. So far the connection to variability at NIR and X-ray wavelengths had not been clearly elucidated. Zhao et al. (2004) showed that there is a probable link between the brightest X-ray flare ever observed and flux density at mm- and short cm-wavelengths on a timescale of less than 24 hours (see also Mauerhan et al. 2005).

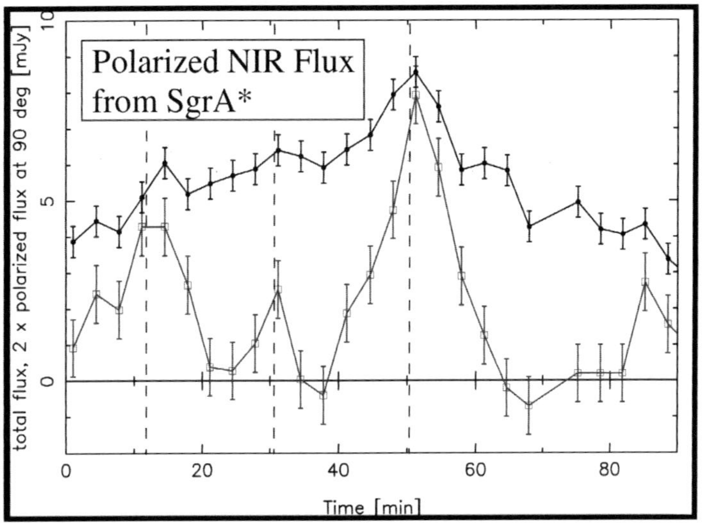

Figure 2. Polarized de-reddened emission from the NIR counterpart Sgr A*. The total flux density light curve (black) and the de-reddened flux density at polarization angle 90° (blue) corrected for the flux at PA 0° at which the sub-flares cannot be seen. The vertical dashed line indicate the times at which sub-flares occurred.

New simultaneous NIR/sub-millimeter/X-ray observations of the Sgr A* counterpart were recently presented by Eckart et al. (2006a). The authors closely investigate the physical processes that may be responsible for the variable emission from Sgr A*. The observations were carried out using the NACO adaptive optics (AO) instrument at the European Southern Observatory's Very Large Telescope and the ACIS-I instrument aboard the *Chandra X-ray Observatory* as well as the Submillimeter Array on Mauna Kea, Hawaii, and the Very Large Array in New Mexico. Eckart et al. (2006a) detected one moderately bright flare event in the X-ray domain and 5 events at infrared wavelengths. At 2 - 8 keV the X-ray flare had an excess luminosity of about 33×10^{33} erg/s. For its entire duration the flare was covered in the X-ray domain and was detected as a simultaneous NIR event with no time lag larger than an upper limit of $\leq$10 minutes - which is mainly given by the required binning width of the X-ray data.

The time-lag between the NIR and X-ray flare emission is very small and in agreement with a synchronous evolution (see Fig.1). This is indicated by all flares covered simultaneously in 2003 and 2004. Cross-correlation shows that the time lag between the flares at different wavelengths is less than 10 minutes and therefore consistent with zero. There was no extensive overlap between radio and the NIR/X-ray measurements in 2004 resulting in no simultaneous flare detections between the NIR/X-ray data and the VLA and SMA data. However, the excess flux densities detected in the radio and sub-millimeter domain may indeed be linked with the flare activity observed at shorter wavelengths. Combined with the information that the NIR flare spectra are very red with variable spectral indices (Eisenhauer et al. 2005, Hornstein et al., this issue and Ghez et al. private communication) we can successfully describe the flares by a SSC model in which a substantial fraction of the NIR emission is due to a truncated synchrotron spectrum. The flaring state can be explained with a synchrotron self-Compton (SSC) model involving

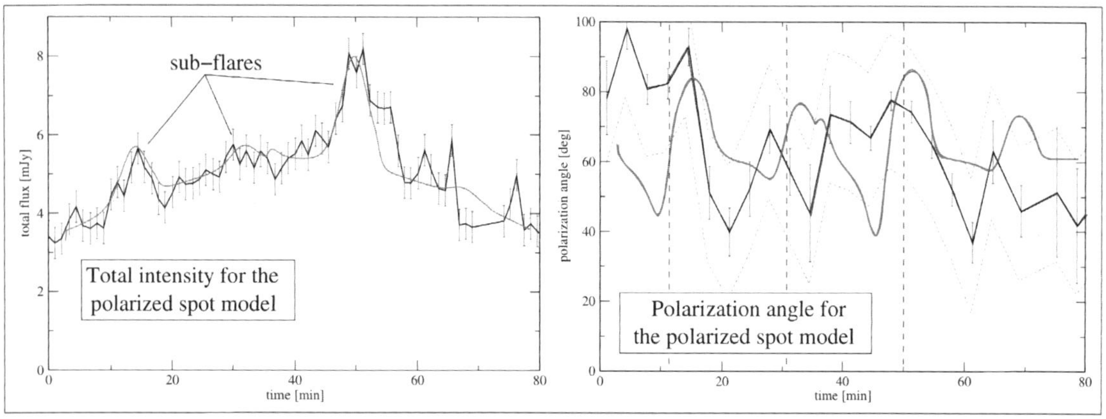

Figure 3. Comparison between model results of an orbiting spot model (red lines) for July 2005 and the measured total flux density (left) and polarization angle (right). The thin dashed lines in the right panel indicate the $\pm 3\sigma$ uncertainties of the data. The vertical dashed lines indicate the times at which sub-flares occurred. For details see Eckart et al. (2006b). The model is shown in red. The model calculations show the compatibility of the orbiting spot model with the NIR polarization data.

up-scattered sub-millimeter photons from a compact source component. Inverse Compton scattering of the THz-peaked flare spectrum by the relativistic electrons then accounts for the X-ray emission. This model allows for NIR flux density contributions from both the synchrotron and SSC mechanisms. Indications for an exponential cutoff of the NIR/MIR synchrotron spectrum allow for a straightforward explanation of the variable and red spectral indices of NIR flares (see also Gillessen et al. 2006).

There are a number of detailed items linked to the variability and the overall spectrum of Sgr A* that have to be mentioned. They are a direct consequence of the model described above and the most recent MIR/NIR measurements.

The NIR flares may contain both synchrotron and SSC flux density contributions: The total number of detectable flares can be obtained by integrating over the amplitude dependent flare rate. The model presented by Eckart et al. (2006a) suggests that in the NIR domain the observed flares can be produced by a mixture of both synchrotron and SSC emission. Here we can assume that the X-ray flares are predominantly produced by SSC emission rather than synchrotron emission. As a consequence - and in good agreement with the observations - the total number of detected X-ray flares is smaller than that in the NIR. However, the NIR flares may have contributions from both the synchrotron and the SSC part the flare spectrum. The contribution of the latter depends on the properties of the relativistic electron spectrum reflected in parameters like the spectral index of the optically thin radio continuum and the exact location of the high and low energy cutoff frequencies of the scattered SSC spectrum. Therefore it may be difficult to discriminate between the SSC and synchrotron dominated flare activity. It appears to be plausible that the SSC dominated NIR flares are bluer than the synchrotron dominated ones.

The NIR flare power spectrum: The description of the flare activity as a power-law under the assumption of a characteristic flare time implies that the 'NIR quiescent phase' of Sgr A* can be regarded as a sequence of frequent low amplitude flares of Sgr A*. Such models have been shown by Eckart et al. (2004) and have been described in detail in Eckart et al. (2006a). The observed flares may be the consequence of a clumpy or turbulent accretion and the flare power spectrum may be coupled to the power spectrum of accreted clumps or turbulence in the accretion flow. In addition - as a consequence of the possible IR turn over of the synchrotron spectrum (see Eckart et al. 2006a) - the flare rates at longer IR wavelengths may be higher than at shorter NIR wavelengths.

The linkage to radio variability: The radio and submillimeter data show clear indications for variability that appear to occur on somewhat longer timescales than the X-ray and IR variations. The exact relation between the radio/sub-mm domain and the NIR/X-ray domain still remains uncertain, due to the lack of sufficient simultaneous coverage. However, the amplitudes and time scales indicated are consistent with a model in which the emitting material is expanding and cooling adiabatically (see also the recent work by Yusef-Zadeh et al. 2006b and the recent interferometric measurements of variable 340 GHz linear polarization in Sagittarius A* by Marrone et al. 2006).

A dust component close to the line of sight toward Sgr A:* Finally Eckart et al. (2006a) have shown that most of the MIR flux density seen towards the position of Sgr A* is due to dust emission (see Fig.5). Ghez et al. (2005) as well as Eckart et al. (2006b) have shown the existence of a dust component close to the line of sight toward Sgr A*. This component may be part of the mini-spiral material, is likely located behind Sgr A* and the blue high velocity stars and may account for as much as $10^{-2}M_{\odot}$in gas and dust. In case the material is not typical of the mini-spiral and if the dust temperature is substantially higher than material associated with the mini-spiral (i.e. higher than ~200-400 K) then the overall mass of this component can be considerably smaller (see also Ghez et al. 2005). This implies that the overall spectral shape of Sgr A* is likely to be significantly less peaked in the FIR wavelength domain, as suggested by the the upper limits that are usually inferred. Combined with the results from the SSC modeling one can expect that the intrinsic spectrum of Sgr A* is peaked at frequencies of a few THz.

3. The Polarized NIR Flare Emission from Sgr A*

Using the NACO adaptive optics (AO) instrument at the ESO VLT Eckart et al. (2006b) have obtained new polarization data of the variable NIR emission of the Sgr A* counterpart. The authors find that the variable NIR emission of Sgr A* consists of a contribution of a non- or weakly polarized main flare with highly polarized sub-flares (see Fig.2). The flare activity shows a quasi-periodicity of 20±3 minutes consistent with previous observations (Genzel et al. 2003). The highly variable and polarized emission clearly shows that the NIR emission is of non-thermal origin.

In Fig.3 we compare the model results of an orbiting spot model (red lines) for July 2005 with the measured total flux density (top) and polarization angle (bottom). As a demonstration of the consistence between the model and the data we have chosen a solution that gives a satisfactory representation of the total flux density. For the chosen inclination of 55° the position angle of the E-vector resulting from the model and interstellar polarization does not wrap over 2π and agrees with the measurements to within ±3 times the mean 1σ uncertainties of the measurements - as indicated by the two thin dashed lines. This solution has to be compared to scenarii with different inclinations as shown in the model results by Eckart et al. (2006b). Details of the modeling are given in Dovciak, Karas & Yaqoob (2004). The calculations show the compatibility of the orbiting spot model with the NIR polarization data. Within the measurement uncertainties the model represents the observed light curve reasonably well. Regarding the fact that we neglected possible tilts and warps of the accretion disk (e.g. Bardeen & Petterson 1975, Lubow, Ogilvie & Pringle 2002) and used a highly idealized and therefore somewhat unphysical intrinsic polarization model, the discrepancy of the polarization angle in detail is not surprising.

Currently the observations cannot discriminate between a jet or a model of a temporarily existing

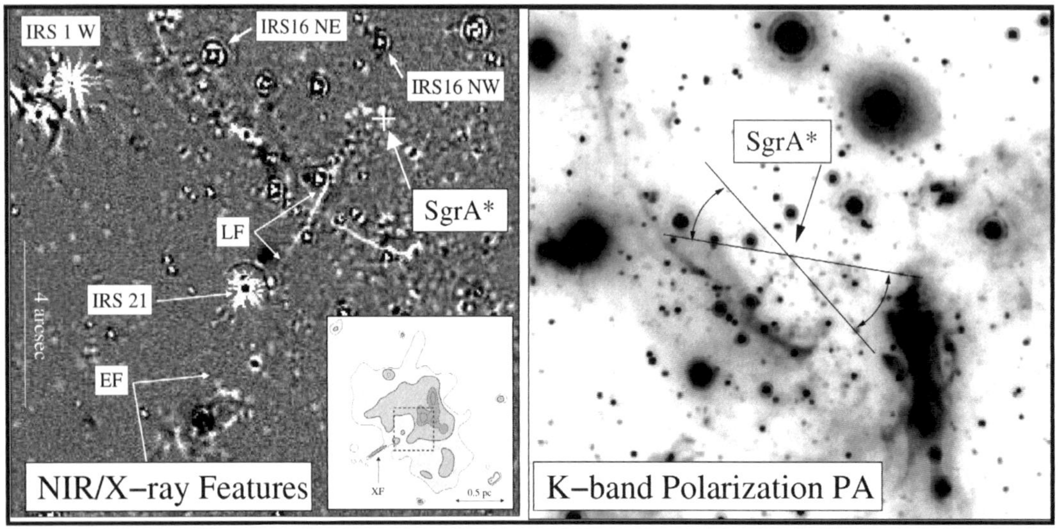

Figure 4. Left: High pass filtered L'-band image. In addition to sharp features in the mini-spiral there are the linear NIR feature (LF) and in the inset the X-ray feature XF (Morris et al. 2004) pointing approximately into the direction of Sgr A*. **Right:** An image of the central 0.5×0.5 pc^2 of the Galactic Center at λ=3.8μm (see Moultaka et al. 2004, 2005; Viehmann et al. 2005, 2006). The two lines centered on the position of Sgr A* limit the range over which the polarization angle varied on the sky. For more details see Eckart et al. (2005, 2006b).

or temporarily bright disk (see Fig.3). In the disk model the quasi-periodic flux density variations can be explained due to polarized spots on relativistic orbits around the central MBH. In the jet model the variable polarization could also be due to a helical magnetic field along a short jet and the variable sub-flare emission could be explained by temporal instabilities in the jet (see elongated features in the NIR and X-ray that are possible - but unlikely jet candidates in Fig.4). However, near the last stable orbit (LSO) a short (possibly only of the order of a few Schwarzschild radii) jet emerging from a disk may look almost indistinguishable from a case involving a pure disk or orbiting spots (see also Fig.4 in Bower et al. 2004). Recent model calculations by Meyer et al. (submitted to A&A) show that the current data indicate spin parameters a>0.5, inclinations of 50°<i<70° and orbital radii slightly larger that that of the last stable orbit. The observed rapid variability - be it related to the jet or the disk model - suggests now that alternative explanations for the high central mass concentration involving boson or fermion balls are increasingly unlikely.

4. Summary and discussion

Our investigation also shows that the NIR K-band is the ideal wavelength band to study the flare emission from Sgr A*. In combination with adaptive optics systems it provides the highest angular resolution at the lowest amount of contamination by dust emission. At wavelengths shorter than the K-band, little emission is found because most flares are expected to be red (Eisenhauer et al. 2005). At longer wavelengths the angular resolution is lower and the dust contamination is high. Observations in the K-band are - in the framework of the presented physical model - ideally suited to observe both synchrotron and SSC flare emission. In addition the model also gives us the opportunity to perform polarization measurements that could provide additional information to study the relevant emission mechanisms.

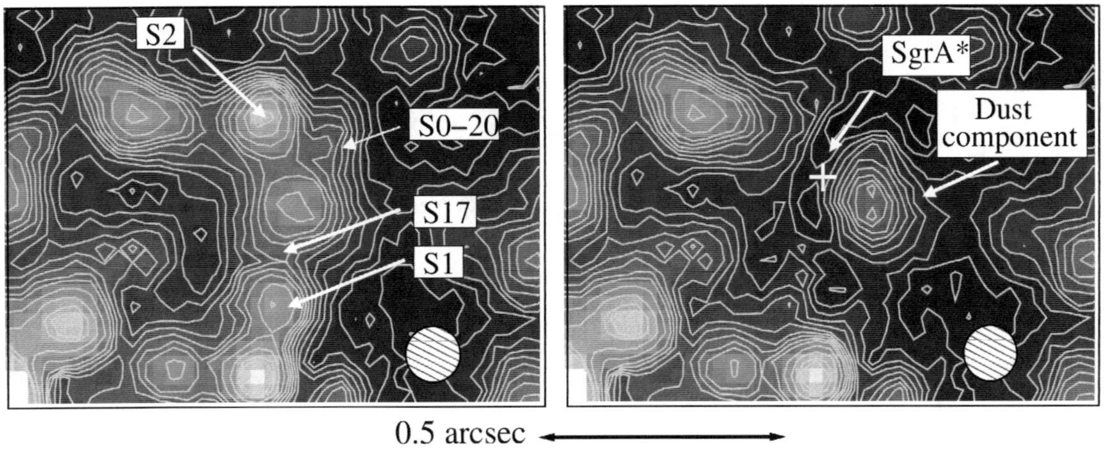

Figure 5. The dust component close to the line of sight toward Sgr A*. On the left we show a portion of an L-band image of the vicinity of Sgr A*. On the right point sources S1, S2, S17, and S0-20 have been subtracted. In both images we show a single resolution element of 104 mas diameter. The scale of 0.5 arcsec corresponds to 19.5 mpc or 4017 AU. More details are given in Eckart et al. 2006a. See also Ghez et al. 2005.

Future progress will mainly depend on further successful simultaneous observing campaigns, - especially between the NIR and (sub)mm-domains to see whether flares are related to extending plasma components. Here extensive simultaneous data are not available so far. Further polarization data from the NIR to the radio as well as (sub-)mm-VLBI and NIR interferometric experiments are also highly desirable to study the details of the accretion process in Sgr A*.
A strong emphasis will also be put on measurements of the Galactic Center using the *Large Binocular Telescope* (LBT) and the *Very Large Telescope Interferometer* (VLTI). High angular resolution interferometric observations of the Galactic Center in the NIR/MIR will provide key information on the central massive black hole and the stellar cluster it is embedded in. These observations have already started by observing the luminous dust enshrowded star IRS3 using MIDI at the VLTI (Pott et al. 2005a, 2005b). As a future NIR wide field interferometric imager offering an angular resolution of about 10 milliarcseconds LINC/NIRVANA at the Large Binocular Telescope will be an ideal instrument for imaging galactic nuclei including the center of the Milky Way.

Acknowledgments: This work was supported in part by the Deutsche Forschungsgemeinschaft (DFG) via grant SFB 494. We are grateful to all members of the NAOS/CONICA and the ESO PARANAL team, especially N. Ageorges for support in setting up the polarization measurements.

Baganoff, F.K., Bautz, M.W., Brandt, W.N., et al. 2001, Nature, 413, 45
Baganoff, F. K., Maeda, Y., Morris, M., et al. 2003, ApJ 591, 891
Bardeen, J.M. & Petterson, J.A., 1975, ApJL 195, L65
Belanger, G., et al., 2005, ApJ 635, 1095, astro-ph/0508412
Bower, G.C., Falcke, H., Sault, R.J. and Backer, D.C., 2002, ApJ, 571, 843
Dovciak, M., Karas, V., & Yaqoob, T. 2004, ApJS 153, 205
Eckart, A., Genzel, R., Ott, T. and Schödel, R. 2002, MNRAS, 331, 917
Eckart, A. & Genzel, R. 1996, Nature 383, 415
Eckart, A., et al., 2004, A&A 427, 1

Eckart, A., et al., 2006a, A&A 450, 535
Eckart, A., Schödel, R., Meyer, L., Trippe, S., Ott, T., Genzel, G. , 2006b in press by A&A
Eisenhauer, F., et al., 2003, ApJ 597, L121
Eisenhauer, F. et al. 2005, ApJ 628, 246
Genzel, T., Eckart, A., Ott, T., Eisenhauer, F., MNRAS 1997, 291, 219
Genzel, R., Pichon, C., Eckart, A., Gerhard, O.E., Ott, T. 2000, MNRAS 317, 348
Genzel, R., Schödel, R., Ott, T., et al. 2003, Nature, 425, 934
Ghez, A., Klein, B.L., Morris, M., Becklin, E.E. 1998, ApJ, 509, 678
Ghez, A., Morris, M., Becklin, E.E., Tanner, A., Kremenek, T. 2000, Nature 407, 349
Ghez, A. M., Duchéne, G., Matthews, K., et al. 2003a, ApJ, 586, L127
Ghez, A.M., et al., 2005, ApJ 635, 1087
Ghez, A.M., Wright, S.A., Matthews, K., et al. 2004a, ApJ 601, 159
Ghez, A.M., et al., 2005, ApJ 620, 744
Gillessen, S., et al. 2006; 2006, ApJ 640, L163
Goldwurm, A., Brion, E., Goldoni, P. et al. 2003, ApJ, 584, 751
Herrnstein, R.M., Zhao, J.-H., Bower, G.C., Goss, W.M., 2004, AJ, 127, 3399
Lubow, S.H., Ogilvie, G.I., Pringle, J.E., 2002, MNRAS 337, 706
Marrone, Daniel P.; Moran, James M.; Zhao, Jun-Hui; Rao, Ramprasad, 2006, ApJ 640, 308
Mauerhan, J.C.; Morris, M.; Walter, F.; Baganoff, F.K., 2005 ApJ 623, L25
Moultaka et al., 2005, A&A 443, 163
Moultaka et al., 2004, A&A 425, 529
Porquet, D., Predehl, P., Aschenbach, et al. 2003, A&A 407, L17
Pott et al. 2005a, AN 326, 569
Pott et al. 2005b, ESO Messenger 119, 43
Schödel, R., Ott, T., Genzel, R. et al. 2002, Nature, 419, 694
Schödel, R., Genzel, R., Ott, et al. 2003, ApJ, 596, 1015
Viehmann et al., 2005, A&A 433, 117
Viehmann et al., 2006, ApJ 642, 861
Yusef-Zadeh, F., et al., 2006a, ApJ 644, 198
Yusef-Zadeh, F., et al., 2006b, ApJ in press, astro-ph/0603685
Zhao, J., et al., 2003, ApJ, 586, L29
Zhao, J.-H., et al., 2004, ApJL, 603, L85

Institute of Physics Publishing Journal of Physics: Conference Series **54** (2006) 399–405
doi:10.1088/1742-6596/54/1/063 Galaxy Center Workshop

Infrared/X-ray intensity variations and the color of Sgr A*

S D Hornstein[1], K Matthews[2], A M Ghez[1,3], J R Lu[1], M Morris[1], E E Becklin[1], F K Baganoff[4] and M Rafelski[1]

[1] Department of Physics and Astronomy, University of California, Los Angeles, CA 90095-1547, USA

[2] Caltech Optical Observatories, California Institute of Technology, MS 320-47, Pasadena, CA 91125, USA

[3] Institute for Geophysics and Planetary Physics, University of California, Los Angeles, CA 90095-1565, USA

[4] Kavli Institute for Astrophysics and Space Research, Massachusetts Institute of Technology, Cambridge, MA 02139-4307, USA

E-mail: seth@astro.ucla.edu

Abstract. We report the first time-series measurements of Sgr A*-IR's broadband infrared color. Using the newly commissioned laser guide star adaptive optics (LGS AO) system on the Keck II telescope, we imaged Sgr A*-IR in the broadband filters H (1.6 μm), K' (2.1 μm), and L' (3.8 μm) every 3 minutes over the course of 120 minutes, during which time the Chandra X-ray Observatory was also monitoring the Galactic center. Complementary measurements of Sgr A*'s L'- and M_s (4.7 μm)-band flux densities were obtained on a separate night with the natural guide star AO system. During our observations, Sgr A*-IR's flux density showed a wide range of values (2 to 12 mJy at 2.1 μm), which are associated with at least 4 peaks in the infrared emission and are among its highest infrared flux density measurements. However, all our near-infrared color measurements are consistent with a constant spectral slope of α = -0.9 $\pm$ 0.2 ($F_\nu \propto \nu^\alpha$), independent of intensity, wavelength, time, or outburst. Assuming that the infrared wavelengths probe synchrotron emission, we interpret the lack of variation in the infrared spectral index as an indication that the acceleration mechanism leaves the distribution of the bulk of the electrons responsible for the infrared emission unchanged. During our coordinated infrared observations, no elevated X-ray emission was detected. While the less frequent X-ray outbursts have shown correlated emission in previous studies, the lack of X-ray variation during the significant infrared variations reported here indicates that one may not be able to connect the infrared and X-ray emission to the *same* electrons. We suggest that while the acceleration mechanism leaves the bulk of the electron energy distribution unchanged, it generates a variable high-energy tail. It is this high-energy tail that gives rise to the less frequent X-ray outbursts.

1. Introduction

Since its discovery, the compact source at the center of our Milky Way Galaxy [1], now associated with a 3.7 $\times$ 10^6 $M_\odot$ supermassive black hole [2, 3], has perplexed scientists with its remarkably low luminosity of only 10^{36} erg s^{-1} or, equivalently, 10^{-9} L_{Edd}. The recent detection of variable emission from Sagittarius (Sgr) A* at wavelengths from radio [4, 5, 6] to X-ray [7, 8, 9] has offered a new view into the central engine of our Galaxy and provides possible clues to its behavior. Models attempting to explain the cause of this emission have yet to converge on a physical cause for the emission but generally rely on a highly-energetic population of electrons producing synchrotron emission at radio and infrared (IR) wavelengths with the X-ray component coming from either a continuation of this synchrotron emission or synchrotron self-Comptonization of the radio/IR emission [10, 11, 12].

In this paper, we present a time series of diffraction-limited, high signal-to-noise ratio images of Sgr A*-IR taken with both the laser guide star (LGS) and natural guide star (NGS) adaptive optics (AO) systems on the Keck II telescope. The LGS AO observations were taken as part of a coordinated campaign, which included observations spanning radio to X-ray wavelengths.

2. Keck Near-IR Adaptive Optics Data

2.1. NIR Observations

On 2005 July 31 (UT), Galactic Center observations in the H- ($\lambda_o = 1.63\mu$m, $\Delta\lambda$=0.30μm), K'- ($\lambda_o = 2.12\mu$m, $\Delta\lambda = 0.35\mu$m), and L'- ($\lambda_o = 3.78\mu$m, $\Delta\lambda$=0.70μm) photometric bandpasses were conducted using the facility near-infrared camera, NIRC2 (K. Matthews 2006, in prep) behind the laser guide star AO system on the W. M. Keck II 10-meter telescope [13, 14]. Observations were made by cycling through the H, K', and L' filters repeatedly for 113 minutes, with a three-filter cycle completed every 3 minutes. Complementary adaptive optics images of the Galactic Center were obtained on 2005 July 16 (UT) using NIRC2 and the AO system in its natural guide star mode [15] by interleaving observations in the L'- and M_s- ($\lambda_o = 4.67\mu$m, $\Delta\lambda = 0.24\mu$m) photometric bandpasses.

2.2. NIR Data Analysis

Identification and characterization of point sources was accomplished using StarFinder, an IDL package developed for astrometry and photometry in crowded stellar fields [16]. StarFinder was first run on an average map made from all the images taken in a night to detect the sources and then run on each individual map with the positions detected in the average map as an input and held fixed, fitting only for source brightness.

Images were photometrically calibrated using the apparent magnitudes of several of the brightest stars in each image [17, 18]. Absolute calibration uncertainties are estimated to be ~4%, 4%, 5%, and 9% at H, K', L', and M_s, respectively. Determination of the uncertainties in the relative photometry of Sgr A*-IR, at all wavelengths, was carried out as in Ghez et al. [19] by calculating the RMS flux density variations between maps for non-variable stars of comparable magnitude. The observed magnitudes were dereddened assuming a visual extinction of 29±1, an extinction law derived by Moneti et al. (A_H=5.14, $A_{K'}$=3.21, $A_{L'}$=1.56, A_{M_s}=1.46 mag), and flux zero points of 1050, 686, 249, 163 Jy for H, K', L', M_s, respectively [18, 20, 21].

Spectral indices are estimated from near-simultaneous apparent magnitude measurements in pairs of filters. To avoid systematics associated with uncertainties in the extinction law, the spectral index calculations are based on apparent colors that are dereddened using the difference of the apparent colors of S0-2 obtained from the average maps ($H - K = 2.05 \pm 0.03, K - L = 1.25 \pm 0.04, L - M = -0.11 \pm 0.13$) and the intrinsic color of a main sequence B0 star ($H - K = -0.05 \pm 0.08, K - L = -0.16 \pm 0.09, L - M = 0.13 \pm 0.13$; [22, 23, 24]).

Systematic uncertainties from this method include the photometric uncertainty of our measurement of S0-2 and the measurement errors for the colors of B0 stars [22]. The colors of early-type stars are relatively insensitive to spectral type over the small range of uncertainties in the type of S0-2, so there is negligible uncertainty associated with its specific spectral type. When combined quadratically, these errors result in an absolute uncertainty of 0.3, 0.2, and 0.8 for $\alpha_{H-K'}$, $\alpha_{K'-L'}$, and $\alpha_{L'-M_s}$, respectively. The spectral index of several nearby stars were calculated in a similar manner and had the expected stellar indices.

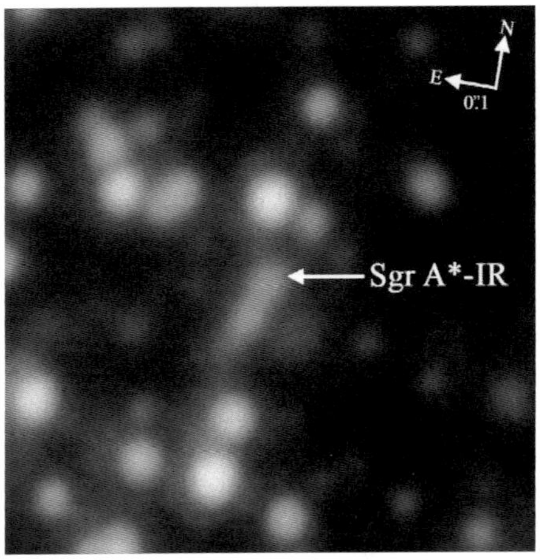

Figure 1. Central 1″ × 1″ region of a diffraction-limited false-color image made from the average H (blue), K' (green) and L' (red) images of the Galactic Center taken on July 31 with the LGS AO system. (See the electronic edition for an animation of this image.)

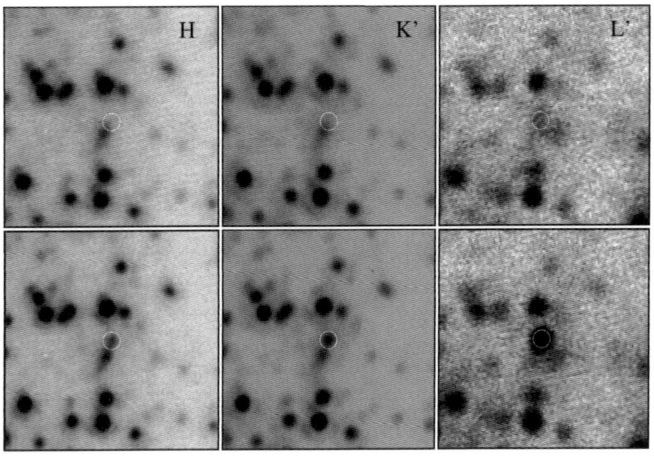

Figure 2. 1″ × 1″ images showing Sgr A* (white circle) at its lowest (top) and highest (bottom) states in H, K', and L' on July 31 taken at 08:32 and 07:57 (UT), respectively.

2.3. NIR Results

Figure 1 shows a color composite of the central 1″ × 1″ H, K', L' maps from 2005 July 31. From this image, one can see the high density of stars surrounding Sgr A*-IR necessitating that the flux from all stars of comparable brightness to Sgr A*-IR be accounted for. This is especially important when the emission from Sgr A*-IR is at its lowest levels, as shown in the top panel of Figure 2. Even at these faint levels, a point source is detected by StarFinder at the location of Sgr A*-IR with correlation values of >0.7.

Figure 3 displays the dereddened light curves of Sgr A*-IR and comparison stars for the LGS AO H-, K-, and L' observations and the same for the NGS AO L' and M_s observations. During our LGS AO observations on July 31, Sgr A*-IR's light curve shows two clear minima that have values of 2-3 mJy in all three filters and that separate emission that can be associated with three distinct peaks in the IR emission. However, only one maximum is observed in its entirety with peak levels of 9, 12, and 15 mJy in the H, K', and L' filters, respectively (see also the bottom panel of Figure 2). During our NGS AO observations, Sgr A*-IR starts out in a low state similar to the LGS AO observations at 3 and 2 mJy at L' and M_s, respectively, and rises to 7 mJy in both filters by the middle of the observations.

Sgr A*-IR's spectral index appears to be independent of flux density. This is most clearly seen in the spectral index derived from the dereddened K'-L' color (Figure 4). At shorter wavelengths, the spectral indices derived at flux densities below 5 mJy are somewhat bluer. Since this effect is not seen at the longer wavelengths, we attribute this effect to contamination from the underlying stellar population. Since the correlation values of the PSF are still quite high (> 0.7), even at these low emission levels the point source contribution from Sgr A*-IR is still being detected. This background contamination is therefore removed by subtracting off the minimum level observed during the observations, which is taken to be the average level detected during a 10 minute period in the first observed minimum beginning at ~07:20 (H = 3.2 ± 0.2, K = 2.8 ± 0.1, L' = 3.6 ± 0.2 mJy) for the LGS AO data set and the average of

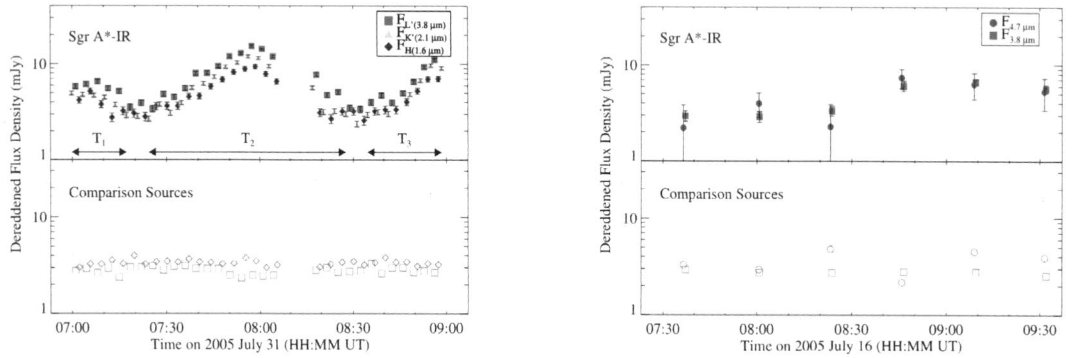

Figure 3. Light curve for Sgr A*-IR and nearby comparison stars of similar brightness at each wavelength for 2005 July 31 (LGS AO; left) and 2005 July 16 (NGS AO; right). The comparison sources are chosen to match the measurements of Sgr A* at its minimum values (S0-37 at H & K', S0-17 at L', and S0-1 at M_s).

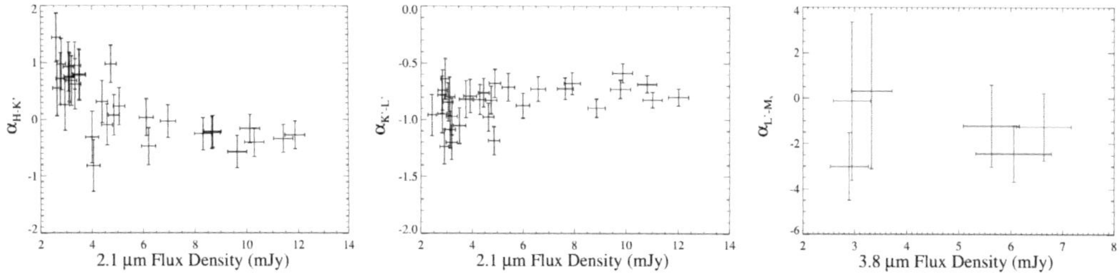

Figure 4. Spectral index ($F_\nu \propto \nu^\alpha$) vs. Sgr A*-IR flux density derived from H-K' (left), K'-L' (center), and L'-M_s(right; not simultaneous with the previous two.) The K'-L' spectral index (middle) shows that the spectral index is independent of flux density.

the first three points for the NGS AO data set (L' = 3.1 ± 0.2, M_s = 2.8 ± 0.9 mJy). With this approach, the spectral indices, as shown in Figure 5, are not only constant with respect to outburst intensity, but also consistent across multiple wavelengths ($\alpha_{H-K'}$ = -1.2 ± (0.4±0.3), $\alpha_{K'-L'}$ = -0.8 ± (0.2±0.2), $\alpha_{L'-M_s}$ = -2.4 ± (0.7±0.8), where the two sources of uncertainty stem from the random photometric errors and the absolute calibration errors, respectively). It should be noted that the K-L' and L'-M_s spectral indices do not change significantly as a result of this subtraction technique, consistent with the assumption that the contaminating background is from the blue stellar population.

Furthermore, as can be seen from Figure 6, we detect no difference in spectral indices within the three separate outburst events. Motivated by this result, we recalculate the spectral index from one pair of K'/L' observations in 2004 [19] using the S0-2 comparison method outlined above and find $\alpha_{K'-L'}$ = -0.8 ± (0.4±0.2) Combining all three 2005 measurements with the 2004 measurement, we infer a near-infrared spectral index for Sgr A*-IR during these observations of α = -0.9 ± 0.2 that is independent of intensity, wavelength, time, or outburst.

Several comments can be made about these results compared to previous results. While photometrically derived spectral indices of Sgr A*-IR are obtained in an inherently different manner than those obtained directly from spectra, they agree within two sigma with OSIRIS K-band (2.02–2.38 μm) spectra obtained at a comparable emission level (i.e. α = -2.6 ± 0.9 when $F_{SgrA*-IR}$ = 6.1 mJy; [25]). Other results, from SINFONI K-band spectra (1.95–2.45 μm),

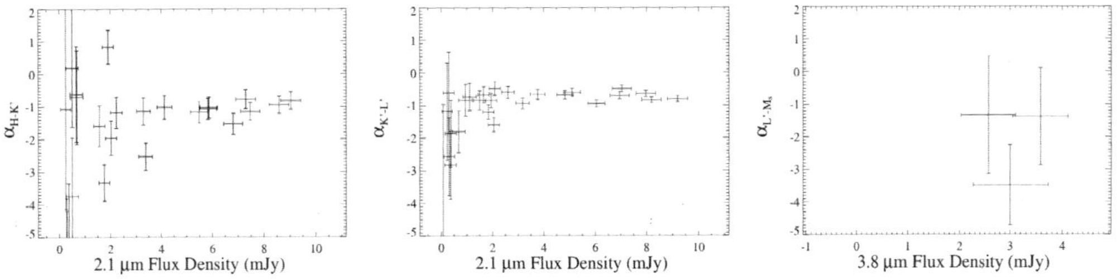

Figure 5. Same as Figure 4 after the background contamination is removed by subtracting the average level of emission during the first minimum for the LGS AO data set and the average of the first three points in the NGS AO data set. The spectral indices across all wavelengths are consistent with each other and have a weighted average of α = -0.9 ± 0.2.

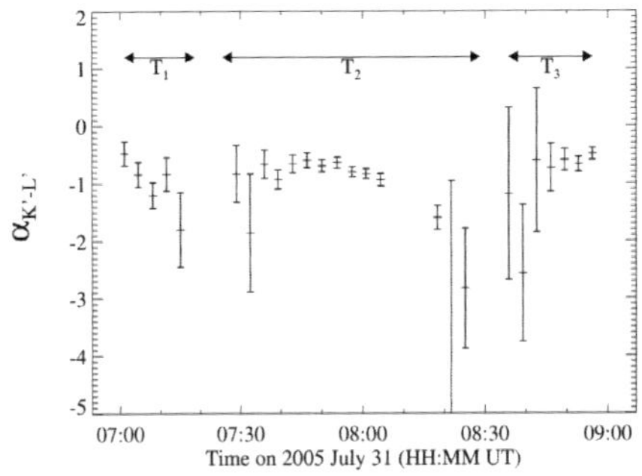

Figure 6. Spectral index derived from K'-L' plotted as a function of time showing no difference in the spectral index within the three separate outburst events (α_{T_1} = -0.9±0.2, α_{T_2} = -0.8±0.3, α_{T_3} = -0.6±0.3).

suggesting a correlation of spectral index with emission intensity [26] are not as discrepant as it may seem. Of the three methods presented by the authors, the 'off state subtraction' method, which most closely resembles the method presented here, shows virtually no correlation down to 2 mJy with $\alpha_{bright\ state} = -0.6 \pm 0.2$. The correlation seen from the other two methods may be explained by the contamination seen in these data below 5 mJy. A local aperture background at K could predominantly sample the contamination from the blue population of surrounding stars and the subtraction of this background would lead to a redder appearance for Sgr A*-IR at low levels (where the background emission becomes comparable to Sgr A*-IR's intrinsic emission.) We therefore believe that the apparent variation in the IR color of Sgr A* in previous experiments arises simply as a consequence of background contamination.

3. Chandra X-ray Data

3.1. ACIS X-ray Observations & Data Analysis

X-ray observations of the Galactic Center were conducted with the *Chandra X-Ray Observatory* [27] using the imaging array of the Advanced CCD Imaging Spectrometer (ACIS-I; [28]) from 2005 July 30 19:44 to 2005 July 31 09:10 (UT). The last two hours of these observations overlapped with the Keck LGS AO observations from the ground. While full details about these observations will be published elsewhere (F. K. Baganoff 2006, in prep), data acquisition and reduction was similar to that reported in [7, 29] (see also [30, 31]).

3.2. X-ray Results

Figures 7 and 8 show the results of the X-ray observations as well as the overlap with ground-based Keck observations. While one moderate-strength outburst is detected, it is not

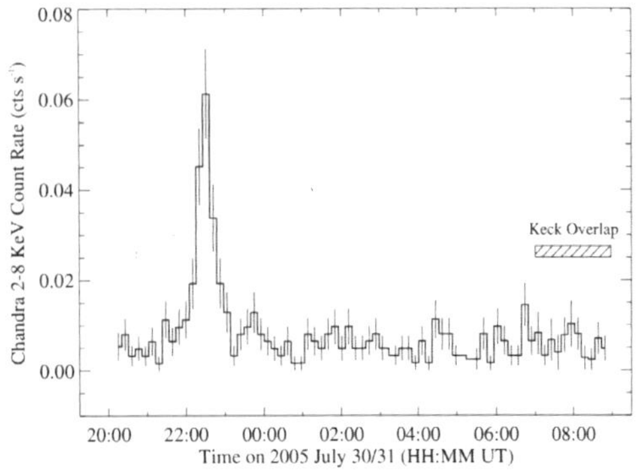

Figure 7. Chandra 2-8 keV light curve of 2005 July 30/31 including the overlap with ground-based Keck observations. The Chandra observations begin on 2005 July 30 20:15:12.0 and end on 2005 July 31 08:49:04.0.

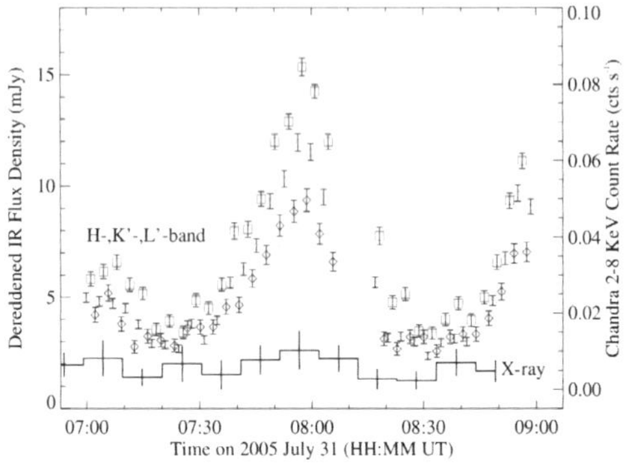

Figure 8. Expanded view of the overlapping X-ray and NIR light curves (See Figure 3 for NIR symbol description). During the entire Keck observations, the X-ray luminosity did not change by more than ~16%.

simultaneous with the Keck NIR observations. Here we restrict our analysis to the period overlapping the Keck observations. A Bayesian blocks analysis, as described in Eckart et al. [30], was performed on the X-ray light curve without background subtraction. This analysis indicates that Sgr A*'s X-ray light curve is consistent with no variability at the 90% confidence level during the entire remaining period after the outburst. The mean count rate of the background-subtracted light curve from ~23:00 - 07:00 is 4.29 ± 0.68 cts ks^{-1} corresponding to a 2–8 keV luminosity of 1.8 ± 0.3 × 10^{33} erg s^{-1}. During the Keck overlap, the mean count rate is 4.94 ± 1.54 cts ks^{-1}, which lies within the 90% confidence interval for the period cited above. We therefore adopt an upper limit during the Keck observations of 2.1 × 10^{33} erg s^{-1}. Thus, the X-ray luminosity could not have changed by more than ~16% during the NIR outburst observed with Keck.

4. Discussion

Comparing the emission presented here with the 12 previously reported 2 μm outburst events, reveals that the peak emission detected on July 31 ranks in the top quartile of all 2 μm outbursts [19, 24, 25, 26, 30, 31, 32]. Furthermore, the low point of infrared emission is comparable to some of the lowest measurements. It is therefore quite notable that, in spite of the wide range of infrared intensities measured, the infrared spectral index is constant. We assume that the infrared probes synchrotron emission from a distribution of relativistic electrons and interpret the lack of variation in the infrared spectral index as an indication that the acceleration mechanism leaves the distribution of the bulk of the electrons responsible for the infrared emission unchanged.

In contrast to the uniformity of the infrared colors, the ratio of infrared to X-ray emission appears to be quite variable. While no elevated X-ray emission was seen during the observations presented here, several previous infrared outbursts (both of higher and lower peak emission levels) have associated X-ray flares [30, 31]. The lack of X-ray variations during our observations indicates that one cannot connect the infrared and X-ray emission to the *same* electrons without invoking a precise tuning of the magnetic field in reaction to a change in the volume of the emitting region such that the X-ray emission could vary substantially while the IR spectral index and intensity remain constant. We therefore suggest that while the acceleration mechanism leaves the bulk of the electron energy distribution unchanged, it generates a variable high-energy tail. It is this high-energy tail that gives rise to the less frequent X-ray outbursts, which have shown correlated infrared activity in previous studies.

Acknowledgments

Support for this work was provided by NSF grant AST 04-06816 and the NSF Science and Technology Center for Adaptive Optics, managed by UCSC (AST 98-76783). FKB was supported by NASA through Chandra award G05-6093X. The data presented herein were obtained at the W. M. Keck Observatory, which is operated as a scientific partnership among the California Institute of Technology, the University of California and the National Aeronautics and Space Administration. The Observatory was made possible by the generous financial support of the W. M. Keck Foundation.

References

[1] Balick, B., & Brown, R. L. 1974, *ApJ* **194** 265

[2] Ghez, A. M., Salim, S., Hornstein, S. D., Tanner, A., Lu, J. R., Morris, M., Becklin, E. E. and Duchêne, G. 2005a, *ApJ* **620** 744

[3] Schödel, R., Ott, T., Genzel, R., Eckart, A., Mouawad, N. and Alexander, T. 2003 *ApJ* **596** 1015

[4] Falcke, H. 1999, ASP Conf. Ser. 186: The Central Parsecs of the Galaxy, **186** 113

[5] Zhao, J.-H., Young, K. H., Herrnstein, R. M., Ho, P. T. P., Tsutsumi, T., Lo, K. Y., Goss, W. M. and Bower, G. C. 2003 *ApJ* **586** L29

[6] Mauerhan, J. C., Morris, M., Walter, F. and Baganoff, F. K. 2005, *ApJ* **623** L25

[7] Baganoff, F. K. et al. 2001, *Nature* **413** 45

[8] Goldwurm, A., Brion, E., Goldoni, P., Ferrando, P., Daigne, F., Decourchelle, A., Warwick, R. S. and Predehl, P. 2003, *ApJ* **584** 751

[9] Porquet, D., Predehl, P., Aschenbach, B., Grosso, N., Goldwurm, A., Goldoni, P., Warwick, R. S. and Decourchelle, A. 2003, *A&A* **407** L17

[10] Markoff, S., Falcke, H., Yuan, F. and Biermann, P. L. 2001, *A&A* **379** L13

[11] Liu, S. & Melia, F. 2002, *ApJ* **566** L77

[12] Yuan, F., Quataert, E. and Narayan, R. 2004 *ApJ* **606** 894

[13] Wizinowich, P., et al. 2006 *PASP* in press

[14] van Dam, M. A., et al. 2006 *PASP* in press

[15] Wizinowich, P., et al. 2000 *PASP* **112** 315

[16] Diolaiti, E., Bendinelli, O., Bonaccini, D., Close, L. M., Currie, D. G. and Parmeggiani, G. 2000, *Proc. SPIE* **4007** 879

[17] Blum, R. D., Sellgren, K., & Depoy, D. L. 1996, *ApJ* **470** 864

[18] Wright, S. A., et al. 2006, in prep

[19] Ghez, A. M., et al. 2005, *ApJ* **635** 1087

[20] Moneti, A., Stolovy, S., Blommaert, J. A. D. L., Figer, D. F. and Najarro, F. 2001, *A&A* **366** 106

[21] Tokunaga, A. T. and Vacca, W. D. 2005 *PASP* **117** 1459

[22] Ducati, J. R., Bevilacqua, C. M., Rembold, S. B. and Ribeiro, D. 2001, *ApJ* **558** 309

[23] Ghez, A. M., et al. 2003, *ApJ* **586** L127

[24] Eisenhauer, F., et al. 2005, *ApJ* **628** 246

[25] Krabbe, A., Iserlohe, C., Larkin, J. E., Barczys, M., McElwain, M., Weiss, J., Wright, S. A. and Quirrenbach, A. 2006, *ApJ* **642** L145

[26] Gillessen, S., et al. 2006, *ApJ* **640** L163

[27] Weisskopf, M. C., O'dell, S. L. and van Speybroeck, L. P. 1996 *Proc. SPIE* **2805** 2

[28] Garmire, G. P., Bautz, M. W., Ford, P. G., Nousek, J. A., & Ricker, G. R., Jr. 2003 *Proc. SPIE* **4851** 28

[29] Baganoff, F. K., et al. 2003, *ApJ* **591** 891

[30] Eckart, A., et al. 2004, *A&A* **427** 1

[31] Eckart, A., et al. 2006, *A&A* **450** 535

[32] Genzel, R., Schödel, R., Ott, T., Eckart, A., Alexander, T., Lacombe, F., Rouan, D. and Aschenbach, B. 2003, *Nature* **425** 934

Institute of Physics Publishing
doi:10.1088/1742-6596/54/1/064
Journal of Physics: Conference Series **54** (2006) 406–410
Galaxy Center Workshop

Diffraction Limited Imaging Spectroscopy of a Sgr A* Flare with OSIRIS

A. Krabbe[1], C. Iserlohe[1], J. E. Larkin[3,], M. Barczys[3,], M. McElwain[3,], J. Weiss[3,], S. A. Wright[3,] and A. Quirrenbach[4]

[1] I. Physikalisches Institut, Universität zu Köln, 50937 Köln, Germany
[2] Division of Astronomy, University of California, Los Angeles, CA, 90095-1562, USA
[4] Landessternwarte Königstuhl, 69117 Heidelberg, Germany

E-mail: `krabbe@ph1.uni-koeln.de`

Abstract. We present diffraction limited K-band integral field spectroscopy of a flare associated with SgrA*. From the spectrum we determined the K-band spectral index of the pure flare emission to be ($F(\nu) \propto \nu^{\alpha}$) of $\alpha = -2.6 \pm 0.9$. If we do not subtract the quiet state emission of SgrA*, then our spectral index is consistent with earlier observations of Ghez et al. 2005 [8]. We compare our observations with other data already published and discuss the implications.

1. Introduction

The infrared variability or flaring of SgrA*, first reported by [5] and [7], has become an important observable to investigate the physics in the immediate vicinity of the black hole in the center of our Galaxy. Flares are fairly frequent and they are also readily observable with adaptive optics at 8-10m class telescopes [2, 8]. The infrared variability is closely linked with the X-ray variability detected a few years earlier [1, 16, 10]. In fact, flares are related multiwavelength phenomena [4]. The reported short timescales of the variability of a few minutes imply that the emission arises from an environment close to the Schwarzschild radius. Flares thus have become a key in our understanding of mass accretion processes towards the Galactic Center black hole and the related relativistic and high energy phenomena. In particular, the spectral indices of flare emissions and their possible variations from flare to flare and during a single flare hold important clues about the emission processes involved [19, 15]. We observed SgrA* with OSIRIS, the new Keck NIR integral field spectrometer during commissioning of the instrument. These data are the first laser guide star (LGS) assisted spectra ever taken of the Galactic Center region.

2. Observations and Data Reduction

The OSIRIS (OH Suppressing InfraRed Imaging Spectrograph) instrument is a new facility near infrared (NIR) integral field spectrograph designed for the Keck Observatory's Adaptive Optics (AO) system. It utilizes an array of microlenses and a HAWAII-2 detector (2048×2048 pixels) to simultaneously obtain about 1020 full IR band spectra at a spectral resolution of R = 3700 over a rectangular field of view with about 16×64 spatial positions. More information about OSIRIS can be found in [14, 17, 20].

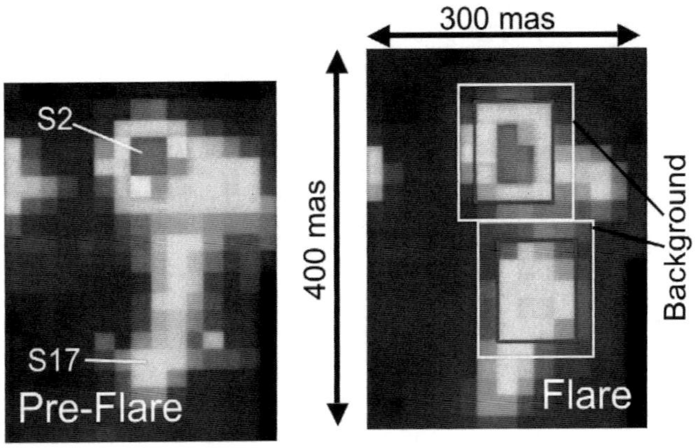

Figure 1. These wavelengh collapsed K-band images centered on SgrA* were obtained with OSIRIS on Keck II. The left data set shows SgrA* at the beginning of a flare still at a low level of activity. The image on the right was obtained 5 minutes later and shows a typical flare at the location of SgrA*. Two field stars are labeled. The rectangles indicate the pixels used for extracting the on-source and background spectra on S2 and SgrA*.

We observed the Galactic Center on April 29, 2005 at Keck II during the first deployment of OSIRIS with the LGS AO system (Wizinowich et al., in prep.). Two consecutive K-band frames, each with 300 s integration time were obtained at an airmass of 1.65 with SgrA* in the field of view (FOV). Overhead between the frames was about 37 s. A sky field was observed immediately after the 2nd frame.

The data was reduced using the dedicated OSIRIS data reduction pipeline, described in more detail in [12, 11]. After sky subtraction, individual spectra in each frame were straightened and corrected for crosstalk from adjacent spectra. Arc line spectra were used for wavelength calibration. Atmospheric differential dispersion effects were also corrected by tracing the peak emission of the stellar continuum point spread function (PSF) through the wavelength slices of the cube. Telluric and instrumental transmission, and also the foreground extinction to the Galactic Center were corrected by dividing all spectra by the average spectrum of star S2 and multipying them by a 30,000K black-body, representing S2 of type O8V or B0V [6]. Due to the low signal/noise we did not attempt to model the stellar Brγ absorption line at 2.166 μm. For the same reason, OH residuals have been ignored in the spectra.

3. Results

Basic results have been published in [13]. A more detailed view on SgrA* is presented in Figure 1, which shows the vicinity of SgrA* including the stars S2 and S17 (according to [3]) and excludes the rest of OSIRIS' field of view. The angular resolution on the sky is 60 mas and both images have been obtained at a pixel scale of 20 mas. The angular resolution achieved is lower than the diffraction limit of 46 mas at 2.18 μm, probably due to the low telescope elevation. The two panels display the images produced by collapsing all of the wavelength channels from 2.02 μm through 2.38 μm. The left image, obtained first, shows SgrA* close to its quiet state (pre-flare). The right image shows clear indication of a flare at the position of SgrA* (flare).

The flare at the position of SgrA* is located 145$\pm$5 mas south and 22$\pm$5 mas west of S2 at the epoch of the observations. Inspecting the identical position in the pre-flare frame reveals a weaker but still notable emission at the location of SgrA*. The spectrum of SgrA* was determined from the flare frame using a 5 $\times$ 4 pixel^2 box centered on the position of the flare. The general background was determined by measuring the average pixel value in the 22 pixels that form a circumference around the 5$\times$4 aperture, also indicated in Figure 1. The spectrum of S2, used for calibration as described in the previous section, was determined in an identical fashion from the same frame. An identical procedure was applied to the pre-flare frame. Assuming that the flare is unresolved in our data, the extraction regions for SgrA* and star S2 cover the same fraction of the PSF for each source.

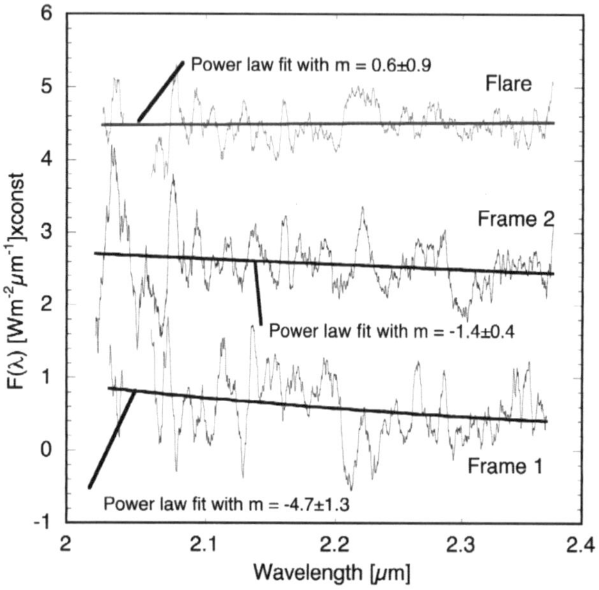

Figure 2. Spectra of SgrA* obtained during pre-flare (bottom) and flare (middle, shifted up by 1 unit). The upper spectrum (shifted up by 4 units) is the difference of the lower two spectra and represents the pure-flare spectrum. The spectra have been smoothed with a 30 pixel wide boxcar filter for display. The lines are power law fits to the unsmoothed continua of the spectra. The derived values for the slopes($F(\lambda) \propto \lambda^m$) are indicated. Adopted from [13].

The resulting spectra are shown in Figure 2. The lowest spectrum represents the pre-flare emission. The middle spectrum, shifted up by one unit, is the extracted spectrum of the flare. The upper spectrum, shifted up by 4 units, is the difference between the flare spectrum and the pre-flare spectrum, and we will refer to this as the pure-flare spectrum. All three of the spectra have been smoothed by a 30 pixel wide boxcar filter and were then fitted with a power law ($F(\lambda) \propto \lambda^m$). The resulting fits and slopes (m) are indicated in Figure 2 along with 1σ errors. These slopes were converted into a frequency power law index α ($F_\nu \propto \nu^\alpha$) listed in Table 1. We note here that the spectral index of the flare is quite red (α=-2.6 ± 0.9).

4. Discussion

From the spectral fits, the K-band flux of the flare and the pre-flare were determined (Table 1) assuming that S2 has a magnitude of m_K = 13.9 mag (24 mJy) [6]. The flux in in the flare spectrum is 9.6 mJy, which is 6.1 mJy stronger than the pre-flare and corresponds to an increase of the K-band flux by a factor of 2.8 within 5.6 minutes. Photometric K-band data by [4] indicate that the lowest activity level of SgrA* can be 12 times fainter than S2 corresponding to a K-band flux density of 2 mJy. Comparing this factor with Table 1 suggests that the K-band emission in the pre-flare frame is within a factor of 2 of the lowest activity level measured. Our observations thus very likely record the beginning of a flare brightening up to a typical level of activity. We also conclude that the intrinsic K-band flux of the flare is probably best represented by the difference between the peak value of 9.6 mJy and the lowest background of 2 mJy, yielding ($7.6 \pm 34\%$) mJy. This value has been used in Figure 3 (see below).

[3] were first to report measurements of the spectral index of the emission of SgrA* in the NIR. Their spectral indices α for the flux density F_ν lie in the range between -3.3 and -4.8, with an averaged index of $\alpha_E = -4 \pm 1$ for fluxes < 2 mJy. Their value has been obtained in a similar fashion to ours, in that two spectra of different activity levels were subtracted from each other. However, their result is different from our value of $\alpha_K = -2.6 \pm 0.9$ (see Table 1 and Figure 3).

Figure 3 summarizes all published spectral indices for SgrA*'s K-band flares as a function of their 2 μm flux density. In a recent paper, [8] report spectral indices based on K′ and L′ imaging. Their method of determining α is directly based on K′-L′ PSF fitting photometry.

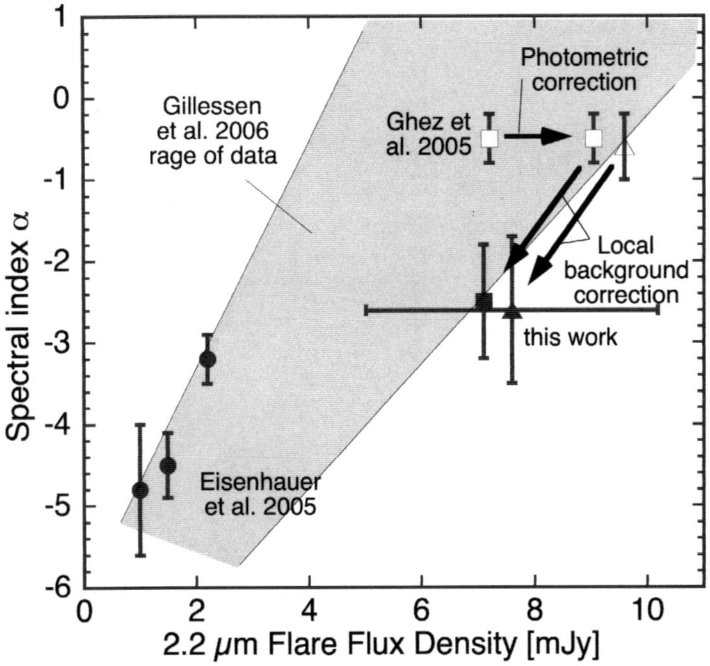

Figure 3. SgrA*'s K-band spectral index (α, where $F(\nu) \sim \nu^{\alpha}$) as a function of its 2 μm flux density. The results of [3], [8], and [9] are indicated. The result of this work is denoted by the triangles. The open triangle marks the spectral index and flux density derived from the flare-spectrum (Figure 2 middle), the filled triangle represents the spectral index of the pure-flare spectrum (Figure 2 top). For the arrows see text. The rage of data points obtained by [9] is indicated by the shaded area between the two boundary lines.

However, different from our result and from [3], their result does not account for flux from the source during its quiet phase. Thus their value of α = -0.5 has a different meaning than our flare index. It is more correctly compared to the slope of the flare spectrum in Figure 2 (middle) which includes the local background and has an index of $\alpha = -0.6 \pm 0.4$. We find that our value for the spectral slope is in fact very consistent with the observations by [8]. In addition, their K-band flux densities are close to our results, especially if we correct for the difference between K′ and K bands (as indicated by the arrow). Using their spectral index value of α = -0.5 and K′ flux density of 7.2 mJy, we calculate a K-band flux density of 9.2 mJy, again very comparable to our measurement of 9.6 mJy for frame 2.

We thus confirm the findings of [8] for the slope of the flare plus local background within the 1 σ errors. It is interesting to note that both spectral indices were obtained by very different techniques and using different wavebands: K′ and L′ versus K band only. The spectral indices reported by [3] are clearly different but were also taken at a significantly lower level of activity. Our data and the data by [3] imply a dependency of the spectral index on the flare intensity, another suggestion first made by [8].

However, new observations by Ghez et al. (private communication and this volume), let them conclude that they do not find any significant dependence of the flare spectral index on the flare intensity. Such finding is not only inconsistent with [3] but does also contradict the conclusion reached in a recent paper by [9] based on near infrared spectra taken with the SINFONI instrument at the VLT. More observations are clearly needed to clarify the possible dependence of the flare spectral index on it's (K-band) flux.

To further complicate the picture there also is an interesting twist with the observations by [9]. Although their version of Figure 3 covers a range of K-band flare flux densities which look statistically significant, their distribution of values in the diagram does not comply well with the data published by [8] and by us [13]. In order to illustrate this, Figure 3 also shows the [9] range of data values indicated by the shaded area between the boundaries. Here, since [9] assumes the S2 flux to be 19 mJy instead of 24 mJy (this paper), we already accounted for the different calibration in order to be able to directly compare the results. The reason for the discrepancy

Table 1.

Target	F_ν [mJy]	Spectral Index α ($F_\nu \propto \nu^\alpha$)
Star S2 (both frames)	24	
SgrA* (pre-flare)	$3.5 \pm 20\%$	2.7 ± 1.3
SgrA* (frame 2)	$9.6 \pm 20\%$	-0.6 ± 0.4
SgrA* flare	$6.1 \pm 34\%$	-2.6 ± 0.9

between [9] and our results or those by [8] remains unclear.

The argument has been made that the flare's spectral index as presented by [13] might need to be corrected slightly. Since they have subtracted a strong flare from a weak flare (see above), they might have subtracted two flares with different spectral indices. The intrinsic spectral index of the strong flare could thus be different from the value they determined. This argument can easily be checked by adopting the spectral indices determined by [3] in Figure 3 and by assuming that the weak 1.5 mJy flare included in our pre-flare observation might have a spectral index as low as α = -4.5. Adding such a spectrum to the pure-flare spectrum in Figure 2 would slightly pull the spectral index of the resulting pure-flare spectrum towards more negative values. Such possible correction would rather increase the discreapancy between our data and the observations by [9].

Both the infrared and the X-ray variability are probably due to transiently heated relativistic plasma accelerated at $\sim$ 10 times the Schwarzschild radius [15, 18]. The synchrotron emission from the high-energy electrons within this plasma can account for the observed X-ray flares. The models can also be tuned to explain the infrared flux, its variability, and also the range of spectral indices [15, 19]. If the flare spectral index is indeed a function of the flare activity level, stronger flares involve more energetic electrons, making the infrared spectrum bluer. Such a finding would considerably narrow the range of process parameters describing the origin of the flares.

References

[1] Baganoff, F.K., et al. 2001, Nature, 413, 45
[2] Clénet Y., Rouan D., Gratadour, D., Marco, O., Léna, P., et al. 2005, A&A, 439, L9
[3] Eisenhauer, F., Genzel, R., Alexander, T., Abuter, R., Paumard, T., et al. 2005, ApJ, 628, 246
[4] Eckart, A., Baganoff, F.K., Morris, M., Bautz, M.W., Brandt, W.N., et al. 2004, A&A, 427, 1
[5] Genzel R., Schödel, R., Ott T., Eckart, A., Alexander, T., et al. 2003, Nature , 425, 934
[6] Ghez, A.M., Duchêne, G., Matthews, K., Hornstein, S.D., Tanner A., et al. 2003, ApJ, 586, L127
[7] Ghez, A.M., Wright, S.A., Matthews, K., Thompson, D., Le Mignant, D., et al. 2004, ApJ, 601, L159
[8] Ghez, A.M., Hornstein, S.D., Lu, J., Bouchez, A., Le Mignant, et al. 2005, ApJ, 635, in print
[9] Gillessen, S., Eisenhauer, F., Quataert, E., Genzel, R., Paumard, et al. 2006, ApJ, 640, L163
[10] Goldwurm, A., Brion, E., Goldoni, P., Ferrando, P., Daigne, F., et al. 2003, ApJ, 584, 751
[11] Krabbe, A., Gasaway T., Weiss J., Larkin J., Barczys M., et al. 2002, SPIE, 4847, 448
[12] Krabbe, A., Gasaway, T., Song, I., Iserlohe, C., Weiss, J., et al. 2004, SPIE, 5492, 1403
[13] Krabbe, A., Barczys, M., Larkin, J.E., LaFreniere, D., Quirrenbach, A., et al. 2006, ApJL, 642, L145
[14] Larkin, J.E., Quirrenbach, A., Krabbe, A., Aliado, T., Barczys, M., et al. 2003, SPIE, 4841, 1600
[15] Liu, S., Petrosian, V., & Melia F. 2004, ApJ, 611, L101
[16] Porquet, D., Predehl, P., Aschenbach, B., Grosso, N., Goldwurm, A., et al. 2003, A&A, 407, L17
[17] Quirrenbach, A., Larkin, J.E., Krabbe, A., Barczys, M., & LaFreniere, D. 2003, SPIE, 4841, 1493
[18] Yuan, F., Quataert, E., & Narayan R. 2003, ApJ, 598,301
[19] Yuan, F., Quataert, E., & Narayan R. 2004, ApJ, 606, 894
[20] Weiss, J., Barczys, M., Larkin, J.E., LaFreniere, D., Quirrenbach, A., et al. 2002, SPIE, 4848, 519

Institute of Physics Publishing
doi:10.1088/1742-6596/54/1/065

Variations in the spectral slope of Sgr A* during a NIR flare

Stefan Gillessen[1], **F. Eisenhauer**[1], **E. Quataert**[2], **R. Genzel**[1,3], **T. Paumard**[1], **S. Trippe**[1], **T. Ott**[1], **R. Abuter**[1], **A. Eckart**[4], **P. O. Lagage**[5], **M. D. Lehnert**[1], **L. J. Tacconi**[1], **F. Martins**[1]

[1]Max-Planck-Institut für extraterrestrische Physik, 85748 Garching, Germany
[2]Astronomy Department, University of California, Berkeley, CA 94720, USA
[3]Physics Department, University of California, Berkeley, CA 94720, USA
[4]1. Physikalisches Institut, Universität zu Köln, 50937 Köln, Germany
[5]UMR 7158, CEA-CNRS-Université Paris 7, DSM/DAPNIA/Service d'Astrophysique, CEA/Saclay, France

E-mail: ste@mpe.mpg.de

Abstract. We have observed a bright flare of Sgr A* in the near infrared with the adaptive optics assisted integral field spectrometer SINFONI. Within the uncertainties, the observed spectrum is featureless and can be described by a power law. The associated power law index is subject to systematic effects, namely the determination of the background level. We explore these effects and can show that while the absolute value of the spectral power law index is relatively uncertain, our data nevertheless suggest that the spectral index is correlated with the instantaneous flux. Both quantities experience significant changes within less than one hour. We argue that the near infrared flares from Sgr A* are due to synchrotron emission of transiently heated electrons, the emission being affected by orbital dynamics and synchrotron cooling, both acting on timescales of ≈ 20 minutes.

1. Introduction

The detection of stellar orbits [1, 2, 3, 4] close to Sgr A* has proven that the center of our Galaxy hosts a massive black hole (MBH) with a mass of $(3.6 \pm 0.2) \times 10^6\,\mathrm{M}_\odot$. Sgr A* appears rather dim at all wavelengths, which is explained by accretion flow models [5, 6, 7, 8]. In the near infrared (NIR) it was detected after diffraction limited observations at the 8-m class telescopes had become possible [9, 10]. Usually the emission is not detectable. However, every few hours Sgr A* flares in the NIR for 60 to 90 minutes, reaching up to $K \approx 15\,\mathrm{mag}$. During these flares the lightcurves show characteristic quasi-periodic oscillations [9] with a timescale of $\approx 17\,\mathrm{min}$. The position of the flares coincides with the dynamical center of the stellar orbits.

Recent progress in understanding the Galactic Center (GC) has been made by integral field spectroscopy using the 2004 commissioned instrument SINFONI at the VLT [11, 12]. One main result [2] is the finding that the S-stars visibly orbiting Sgr A* are mostly B dwarfs, intensifying the so-called "paradox of youth" [13]. Furthermore, using SINFONI it becomes possible to determine radial velocities for these stars on a regular basis. Also, a first flare spectrum was obtained by [2], showing a featureless, red spectrum (power law index in $\nu S_\nu \sim \nu^\beta$ of $\beta \approx -3$).

The basic consensus model which explains the overall, multiwavelength spectral energy distribution (SED) of Sgr A* is that the mm-IR emission from Sgr A* is synchrotron emission from relativistic electrons close to the last stable orbit (LSO) [14, 15, 16, 17]. Radiatively inefficient accretion flow (RIAF) models with a thermal electron population ($T_e \approx 10^{11}$ K) produce the observed peak in the submm but fail to produce enough flux at $2\,\mu$m. The NIR emission requires transiently heated or accelerated electrons as proposed by [18]. If a few % of the electrons are heated up to $\approx 10^{12}$ K significant NIR flux emerges. This basic scenario is illustrated in Fig. 1.

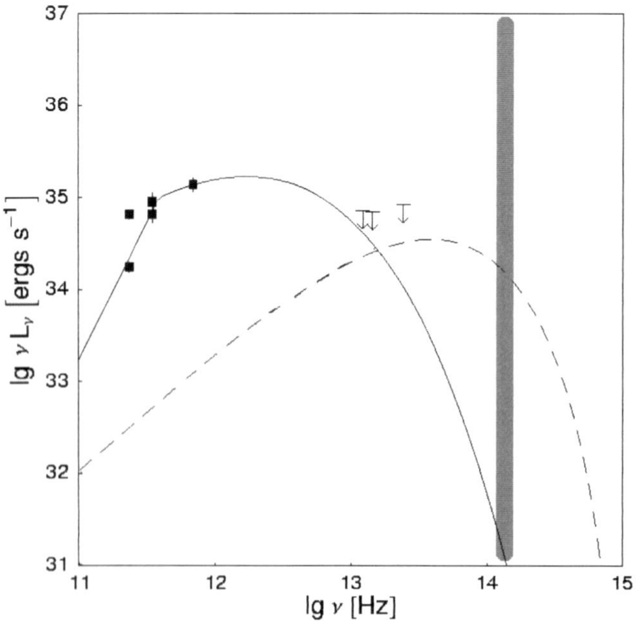

Figure 1. Basic scenario for the spectral energy distribution of Sgr A*. The sub-mm data from [19, 20] and the MIR data from [21] are shown together with a toy model for the emission of Sgr A*. The solid line (commonly known as the sub-mm bump) is the synchrotron emission of a thermal electron population with $T_e \approx 10^{11}$ K. These electrons do not produce significant flux in the NIR (grey shaded area). If a few % of the electrons are heated up to 10^{12} K (dashed line), NIR flux emerges. It has a red spectral slope ($\beta < 0$).

2. New observations

We observed the GC region on 2005 June 18 from 2:40 to 7:15 UT with SINFONI. The field of view was 0.8"×0.8" for individual exposures, mapped onto 64×32 spatial pixels. We observed in K-band, spectrally dispersed at a resolution of FWHM 0.5 nm. The first 12 integrations lasted 5 min each. During those we noticed NIR activity of Sgr A* (see Fig. 2) and we switched to 4 minute exposures. We followed Sgr A* for another 32 exposures. In total we interleaved nine integrations on a specifically chosen off field (712" W, 406" N of Sgr A*). The seeing was $\approx$0.5" and the optical coherence time $\approx$ 3 ms, some short-time deteriorations excluded. The AO was locked on the closest optical guide star (m_R = 14.65, 10.8" E, 18.8" N of Sgr A*), yielding a spatial resolution of $\approx$ 80 mas FWHM, close to the diffraction limit of UT4 in K-band ($\approx$ 60 mas).

Our detection triggered immediate follow-up observations with VISIR, a mid-infrared (MIR) instrument mounted at the ESO-VLT unit telescope Melipal (UT3). From 5:25 UT onwards VISIR was pointing to the GC. At the position of Sgr A* no significant flux was seen. A conservative upper limit of 40 mJy (not dereddened) at 8.59 μm is reported.

The reduction of the SINFONI data followed the standard steps: From all source data we subtracted the respective sky frames to correct for instrumental and atmospheric background. We applied flatfielding, bad pixel correction, a search for cosmic ray hits, and a correction for the optical distortions of SINFONI. We calibrated the wavelength dimension with line emission

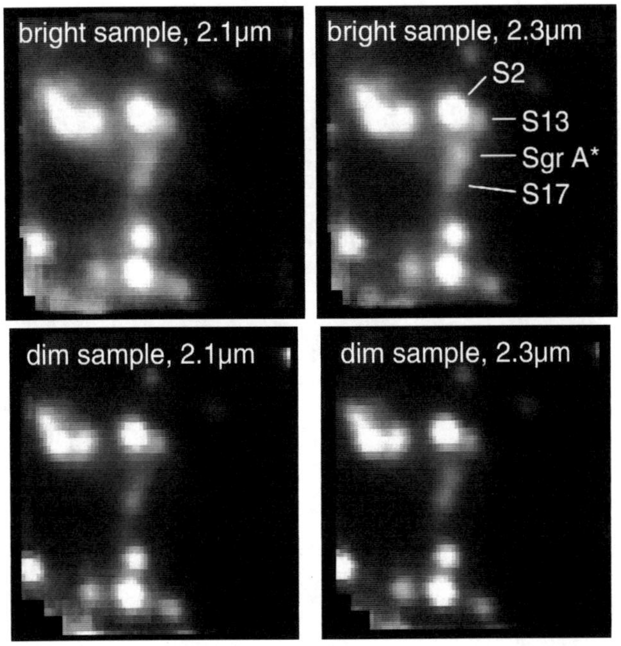

Figure 2. The Galactic Center as seen with SINFONI. Channel maps from the bright and the dim sample obtained by taking the mean in spectral dimension from $2.05\,\mu$m to $2.15\,\mu$m and from $2.25\,\mu$m to $2.35\,\mu$m. All images are scaled the same way and use an identical color map. Sgr A* is brighter in the longer wavelength maps, indicating that it is is redder than the field stars. Furthermore, the effect is more pronounced in the dim sample, meaning that Sgr A* is redder therein than in the bright sample.

lamps and tuned on the atmospheric OH-lines of the raw frames. Finally we assembled the data into cubes with a spatial grid of 12.5 mas/pix.

3. Analysis

3.1. Flux determination

For all 44 cubes we extracted a collapsed image (median in spectral dimension) of a rectangular region (0.25"×0.5") centered on Sgr A* and containing the three S-stars S2, S13, and S17. We determined the flux of Sgr A* from a fit with five Gaussians to each of these images. Four Gaussians with a common width describe the four sources. The fifth Gaussian (with a width 3.5× wider, typical for the halo from the imperfect AO correction) accounts for the halo of the brightest star S2 ($K \approx 14$ mag). The halos of the weaker sources (all $K < 15$ mag) could be neglected for the flux measurement. We fixed the positions of all sources (known a-priori from a combined cube) and the amplitude ratios for the stars. Five parameters were left free: An overall amplitude, the background, the width, the flux ratio halo/S2, and $\mathcal{F}$, the flux ratio Sgr A*/S2. This procedure disentangles real variability from variations in the background, the Strehl ratio, and the seeing.

As a crosscheck we determined $\mathcal{F}$ in a second way for all images; for both Sgr A* and S2 we measured the flux difference between a signal region centered on source and a reasonable, symmetric background region. The such determined values agreed very well with the fits. For further analysis we used the fitted ratios and included the difference between the two estimates in the errors.

3.2. Determination of the spectral slope

The procedure adopted to obtain the spectral power law β index for a given cube was as follows:

(i) Extract the spectrum of Sgr A* from the signal region, namely a ring with radius 3 pixels centered on source.

(ii) Extract the spectrum of the background from the background region. For its choice see the following subsection.

(iii) The raw spectrum is obtained as the difference of signal and background spectrum

(iv) Repeat the same steps for S2, using identically chosen signal and background regions

(v) Divide the raw spectrum by the S2 spectrum in order to correct for instrumental effects and interstellar extinction

(vi) Multiply the result with $\nu\, S_\nu$ of a black body with the temperature of S2 (T=25000 K)

(vii) Clean and rebin the such obtained spectrum into 60 bins

(viii) Fit the result with a power law $\alpha \cdot \nu^\beta$.

(ix) In order to estimate the error of the result, repeat the determination of β for a sample of slightly differently chosen signal and background regions.

While the definition of the signal region is straight forward, it is less clear how a proper estimate of the background can be achieved, as the background flux is varying spatially due to the bright neighbouring S-stars and temporally due to variable AO correction. The value of the spectral power law index β crucially depends on the background subtraction. Subtracting too much light would artificially make the signal look redder than it is, underestimating the background would make Sgr A* artificially too blue.

3.3. Choice of the background

For a reasonable choice of the background region the nearby sources S2 and S17 have to be excluded explicitely. Furthermore, the background flux is varying spatially. Actually Sgr A* lies close to a saddle point in the background light distribution, caused by S2 and S17 (see Fig. 1). In the East-West direction the background has a maximum close to Sgr A*, in the North-South direction it has a minimum near the position of Sgr A*. A proper estimate of the background can be achieved in two ways: a) working with small enough, symmetric apertures, and b) subtracting from the signal an off state obtained at the position of Sgr A* from cubes in which no emission is seen. We used the first method as well as two variants of the second.

Small apertures: The local background at a position $\vec{x}$ can be estimated by averaging over a small, symmetric region centered on $\vec{x}$. Given the background geometry we have chosen a ring with inner radius 3 pix and outer radius 7 pix. The circular symmetry was only broken since we explicitely excluded those pixels with a distance to S17 and S2 smaller than 3 and 7 pixels respectively. Unfavorable for this method is that the local background is only approximated, since a sufficiently large region has to be declared as signal region.

Off state subtraction: Another way to estimate the local background is to extract it from cubes in which no signal is seen. Since the seeing conditions change from cube to cube, one still has to correct for the varying amount of stray light in the signal region. We estimate this variation by measuring the difference spectrum between signal and off cube in a stray light region. The latter must not contain any field stars and should be as far away from the nearby sources as Sgr A*. We used two stray light regions to the left and to the right of Sgr A*, between 5 and 10 pixels away. The disadvantage of this method is that one needs a suitable off state. The latter point is critical for our data. Even though Sgr A* has not been detected directly in the first three cubes, the light at its position appears redder than the local background[1]. Assuming that this light is due to a very dim, red state of Sgr A*, we would subtract too much red light and artificially make the flare too blue. In this sense, this method strictly speaking yields an upper limit for β.

Constant subtraction: The off state method can be varied to obtain a lower limit for β.

[1] Inspecting older (non-flare) SINFONI data we found cases similar to the new data and other cases in which the light was identical to the local background emission. This is consistent with the L-band observations by [10, 22].

Assuming that the true off state spectrum has the color of S2, one can demand that it is flat after division by S2. In our data, the S2 divided off state spectrum is rising towards longer wavelengths. Hence, in this third method we estimate the background at blue wavelengths and use that constant as background for all spectral bins.

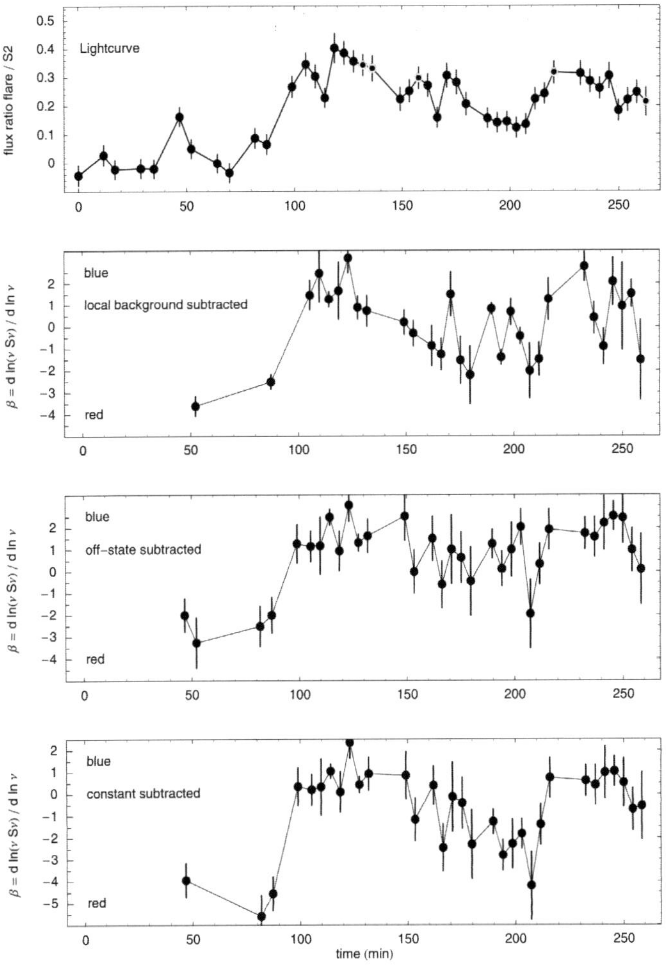

Figure 3. Lightcurve and variation of the spectral power law index β during the flare. Time is counted from 2:40 UT. Top: Flux ratio flare/S2. Thin dots are exposures affected by bad seeing. Middle/top: β using the small apertures background. Middle/bottom: β using the off state subtraction background. Bottom: β using the constant subtraction background.

4. Results

We observed a strong (flux density up to 8 mJy or $\nu L_\nu \approx 10^{35}$ ergs/s), long (more than 3 hours) flare which showed significant brightness variations on timescales as short as 10 minutes (Fig. 3, top). While the data is not optimal for a periodicity search (poorer sampling than our previous imaging data), it is worth noting that the highest peak in the periodogram (significance of $\approx 2\,\sigma$) lies at a period of $\approx$ 18 min. This is also the timescale found by [9], who identify the quasi-periodicity with the orbital time close to the LSO of the MBH.

We divided the data into three groups: a) the cubes of the first peak near $t = 50$ min ("pre-flare"), b) the cubes at $t > 100$ min with $\mathcal{F} < 0.25$ ("dim state"), and c) the cubes at $t > 100$ min with $\mathcal{F} > 0.25$ ("bright state"). For the three sets we created combined cubes in which we determined β, using all three background estimates. In all cases we obtained the correct spectral index 2.9 ± 0.5 for S17 (a star with a spectrum similar to S2 but a flux comparable to Sgr A*).

For Sgr A* we get:

β	preflare	dim state	bright state
off state subtr.	-1.4 ± 0.4	-0.7 ± 0.4	$+0.4 \pm 0.2$
small apertures	-1.8 ± 0.3	-1.7 ± 0.4	-0.1 ± 0.3
constant subtr.	-3.4 ± 0.4	-2.3 ± 0.3	-0.3 ± 0.2

The values obtained from the small apertures lie between the values from the other two methods, consistent with the idea that they yield upper and lower limits. The absolute values for β vary systematically according to the chosen background method. However, independent from that, it is clear that the preflare is redder than the dim state which in turn is redder than the bright state. An obvious question then is whether there is an instantaneous correlation between flux and β.

Hence, we applied the spectral analysis to the individual cubes. We kept the data in which Sgr A* is detected, the error $\Delta\beta < 1.5$ and the spectral index for S17 does not deviate more than 1.5 from the expected value. The resulting spectral indices appear to be correlated with the flux (Fig. 3, 4). The values match the results in [2] and [22]. Bright flares are indeed bluer than weak flares, as suspected by [22] and consistent with the earlier multi-band observations of [9]. Our key new result is that this even holds within a single event.

For all three background methods it is clear that the main event was preceded by a weak, red event. For the small apertures and the constant subtraction method instantaneous flux and spectral index are correlated. For the off state subtraction method one could instead group the data into a preflare at the beginning and a bluer, brighter main event.

We checked our data for contamination effects. If stray light would affect the flare signal, β should be correlated with the seeing (measured by the width from the multiple fits). Since we did not find such a correlation, we exclude significant contamination.

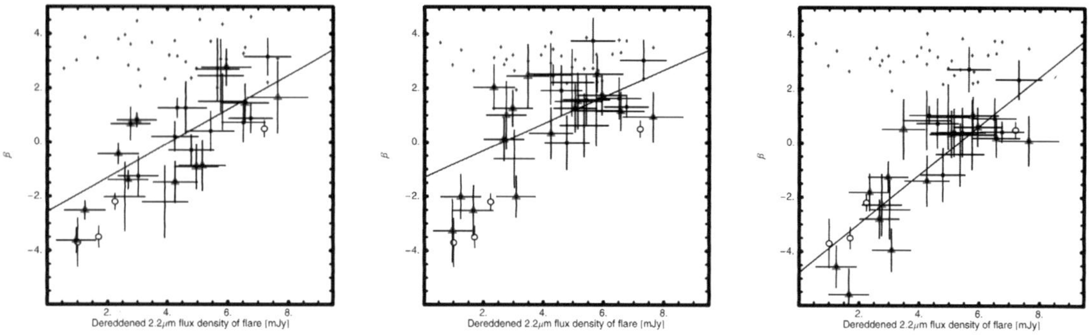

Figure 4. Correlation plots between the flare flux and the spectral index β. Points with error bars represent the flare, blue dots are β_{S17}. Thick black dots mark the points with an error $\Delta\beta < 1$, red triangles mark the data with good seeing (FWHM < 75 mas). The line is a fit to the thick black dots. Open circles denote the data from [2] - near 2 mJy - and [22] - near 7 mJy. Left: Small apertures method; middle: off state subtraction method; right: constant subtraction method.

5. Interpretation

A conservative interpretation of our data is suggested by Fig. 4, middle. There was a weak, red event before, and possibly independent from, the main flare which was a much bluer event.

Plausibly the preflare is then due to the high-energetic tail of the submm peak [9]. The main flare requires nevertheless a population of heated electrons.

A more progressive interpretation follows from the instantaneous correlation between flux and β (Fig. 4, left & right). Then our data suggest that we observed a single event and that the NIR variability is caused by the combination of transient heating with subsequent cooling and orbital dynamics of relativistic electrons. In the following subsection we will exploit this idea.

5.1. Synchrotron emission

In the absence of continued energy injection, synchrotron cooling will suppress the high energy part of the electron distribution function. This results in a strong cutoff in the NIR spectrum with a cutoff frequency decreasing with time. At a fixed band one expects that the light becomes redder as the flux decreases. This can cause the correlation between flux and β. The observed flare apparently needs several heating and cooling events. In Fig. 5 we show variants of our thermal synchrotron model that demonstrate explicitely that suitable electron distribution functions exist: One can reproduce the fluxes and spectral slopes of Sgr A* in the NIR without violating any other observation, namely in the MIR and submm regime.

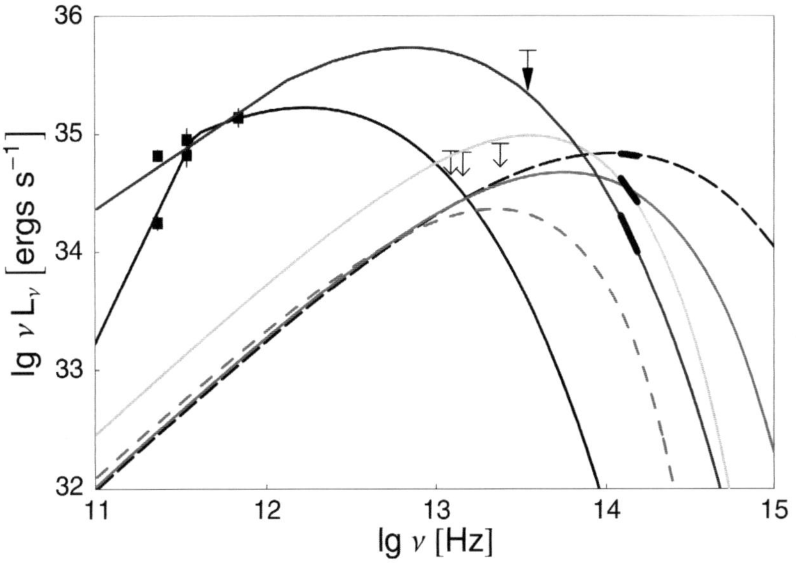

Figure 5. Measured fluxes of Sgr A* from the submm to the NIR and synchrotron models. The submm data and MIR upper limits are the same as in Fig. 1. The upper limit at 8.6 μm is from the simultaneous VISIR observation (dereddened). Our NIR data are shown as short bars for typical flux density values of 1.5, 3.5, and 7 mJy. Solid black line: Thermal synchrotron model for an emitting sphere with $R = R_S$, $B = 30\,\mathrm{G}$, $n_e = 2\times10^7\,\mathrm{cm}^{-3}$ and $T_e = 1.5\times10^{11}\,\mathrm{K}$. Dashed black line: 1% of the electrons are heated up to $T_e = 2\times10^{12}\,\mathrm{K}$. This model can account for the bright state but a parameter set that reproduces the dim state requires a large electron density, yielding high MIR- and FIR fluxes (blue line). Red lines: The effect of synchrotron cooling of the heated electrons, assuming 4-minute exposures taken 4 minutes (solid) and 16 minutes (dashed) after the heating event. The observed range of spectral indices can be reproduced, but larger flux variations than observed are predicted. Green: A variation of the cooling model that mildly increases T, B and n_e while the electrons cool for 8 minutes. This model seems more likely given its lower flux predictions for the MIR, FIR and submm.

The synchrotron cooling time is comparable to the observed timescale of decaying flanks as in Fig. 3 (top) or [9]. In a RIAF model [23] the magnetic field B is related to the accretion rate $\dot{M}$. For disk models with $\dot{M} \approx 10^{-8}\,M_\odot/\mathrm{yr}$ [24, 25] one has $B \approx 30\,\mathrm{G}$ at a radial distance of $3.5\,R_\mathrm{S}$ (Schwarzschild radii). The cooling time for electrons emitting at $\lambda = 2\,\lambda_2\,\mu\mathrm{m}$ for $B = 30\,B_{30}\,\mathrm{G}$ is

$$t_\mathrm{cool} \approx 8\,B_{30}^{-3/2}\,\lambda_2^{1/2}\,\mathrm{min}\ . \tag{1}$$

This is similar to the orbital timescale, making it difficult to disentangle flux variations due to heating and cooling from dynamical effects due to the orbital motion.

5.2. Orbital dynamics

The timescale of $\approx$ 20 minutes for the observed variations suggests orbital motion close to the LSO as one possible cause. Any radiation produced propagates through strongly curved space-time. Beaming, multiple images, and Doppler shifts have to be considered [26, 27]. Recent progress has been made in simulating these effects when observing a spatially limited emission region orbiting the MBH [28, 29, 30, 31].

That the emission region is small compared to the accretion disk can be deduced from X-ray observations [32, 33, 34]. In order to explain the X-ray luminosity the emission region has to be a small spot.

For such a small emitting region the Doppler effect can lead to a time-dependent variation of the spectral slope. In this dynamical model the observed light corresponds to different rest frame frequencies depending on the orbital phase. If the source spectrum is curved, flux and spectral index appear correlated. Following this interpretation, the emission during the brightest part originates from a rest-frequency with larger β than the dimmer state emitted at shorter wavelengths. Note that such a concavely curved spectrum is naturally expected from the synchrotron models.

6. Conclusions

The observation of a long, bright flare from Sgr A* with adaptive optics assisted integral field spectroscopy at the VLT allowed us to measure for the first time the temporal evolution of the spectral power law index of the emission. While the absolute values of the index have a large systematic uncertainty caused by the difficulty to establish a clean background estimate in the crowded region, it is clear that the index changes with time. We adopted three different background estimates and found consistently that the spectral index is correlated with the instantaneous flux of Sgr A*. This can be interpreted in terms of synchrotron emission models as well as in purely dynamical models. Most probably the observed flux and power law index are affected by both phenomena.

References

[1] Schödel, R. et al. 2002, Nature, 419, 694
[2] Eisenhauer, F. et al. 2005, ApJ, 628, 246
[3] Ghez, A., Becklin, E., Duchene, G., Hornstein, S., Morris. M., Salim, S. & Tanner, A. 2003, ANS, 324, 527
[4] Ghez, A., Salim, S., Hornstein, S., Tanner, A., Lu, J., Morris, M. & Becklin, E. 2005, ApJ, 620, 744
[5] Narayan, R., Yi, I., & Mahadevan, R. 1995, Nature, 374, 623
[6] Quataert, E. & Narayan, R. 1999, ApJ, 520, 298
[7] Melia, F., Liu, S. & Coker, R. 2000, ApJ, 545, L117
[8] Melia, F., Liu, S. & Coker, R. 2001, ApJ, 553, 146
[9] Genzel, R. et al. 2003, Nature, 425, 934
[10] Ghez, A. et al. 2004, ApJ, 601, L159
[11] Eisenhauer, F. et al. 2003, Proc. SPIE, 4841, 1548
[12] Bonnet, H. et al. 2004, ESO Messenger, 117, 17
[13] Ghez, A. et al. 2003, ApJ, 586, L127

[14] Liu, S. & Melia, F. 2001, ApJ, 561, L77
[15] Quataert, E. 2003, ANS, 324, 435
[16] Liu, S., Petrosian, V. & Melia, F. 2004, ApJ, 611, L101
[17] Liu, S., Melia, F. & Petrosian, V. 2006, ApJ, 636, L798
[18] Markoff, S., Falcke, H., Yuan, F. & Biermann, P. 2001, A&A, 379, L13
[19] Marrone, D., Moran, J., Zhao, J., Rao, R., ApJ, 640, 308
[20] Zhao, J., Young, K., Herrnstein, R., Ho, P., Tsutsumi, T., Lo, K., Goss, W., Bower, G. 2003, ApJ, 586, L29
[21] Morris, M., Tanner, A., Ghez, A., Becklin, E., Cotera, A., Werner, M. & Ressler, M. 2001, BAAS, 33, 841
[22] Ghez, A. et al. 2005, ApJ, 635, 1087
[23] Yuan, F., Quataert, E. & Narayan, R. 2003 ApJ, 598, 301
[24] Agol, E. 2000, ApJ, 538, L121
[25] Quataert, E. & Gruzinov, A. 2000, ApJ, 539, 809
[26] Hollywood, J. & Melia, F. 1997, ApJS, 112, 423
[27] Hollywood, J., Melia, F., Close, L., McCarthy, D., Dekeyser, T. 1999, ApJ, 448
[28] Paumard, T. et al. 2005, AN, 326, 568
[29] Broderick, A. & Loeb, A. 2005, MNRAS, 363, 353
[30] Broderick, A. & Loeb, A. 2005, ApJ, 636, L109
[31] Broderick, A. & Loeb, A. 2006, MNRAS, 367, 905
[32] Eckart, A. et al. 2004, A&A, 427, 1
[33] Eckart, A. et al., A&A 2006, 450, 535
[34] Gillessen, S. et al., ApJ, 640, L163

Institute of Physics Publishing
doi:10.1088/1742-6596/54/1/066

Journal of Physics: Conference Series **54** (2006) 420–426
Galaxy Center Workshop

Periodic Modulations in an X-ray Flare from Sagittarius A*

G. Bélanger[1,2], R. Terrier[2], O.C. de Jager[3], A. Goldwurm[1,2], F. Melia[4]

[1] Service d'Astrophysique, DAPNIA/DSM/CEA, 91191 Gif-sur-Yvette, France
[2] UMR Astroparticule et Cosmologie, 11 place Berthelot, 75005 Paris, France
[3] Unit for Space Physics, North-West University, Potchefstroom 2520, South Africa
[4] Physics Dept. and Steward Observatory, University of Arizona, Tucson, AZ 85721, USA

E-mail: belanger@cea.fr

Abstract. We present the highly significant detection of a quasi-periodic flux modulation with a period of 22.2 min seen in the X-ray data of the Sgr A* flare of 2004 August 31. This flaring event, which lasted a total of about three hours, was detected simultaneously by EPIC on *XMM-Newton* and the NICMOS near-infrared camera on the HST. Given the inherent difficulty in, and the lack of readily available methods for quantifying the probability of a periodic signal detected over only several cycles in a data set where red noise can be important, we developed a general method for quantifying the likelihood that such a modulation is indeed intrinsic to the source and does not arise from background fluctuations. We here describe this Monte Carlo based method, and discuss the results obtained by its application to other *XMM-Newton* data sets. Under the simplest hypothesis that we witnessed a transient event that evolved, peaked and decayed near the marginally stable orbit of the supermassive black hole, this result implies that for a mass of $3.5 \times 10^6\, M_\odot$, the central object must have an angular momentum corresponding to a spin parameter of $a \approx 0.22$.

1. Introduction

Detecting Keplerian motion in the accretion flow orbiting around a black hole (BH), in the case where an event, localized to a portion of the flow, retains its coherence over several orbits, entails being able to detect periods of the order of 10^{-4} s (or frequencies of ~10 kHz) for solar mass objects. The problem is entirely different when considering a supermassive black hole (SMBH), and in particular Sgr A*. For a non-rotating BH of $3.5 \times 10^6\, M_\odot$, the period at the marginally stable orbit (MSO, $r(\textsc{mso}) = 3r_s = 6GM/c^2$) is 1592 s or about 26.5 min. So as much as detecting kHz quasi-periodic oscillations may be a delicate matter for certain reasons, as we explore long periodic signals, low frequency noise and windowing effects become important, and must be taken into account in a proper characterization of the signal.

In general, to test for a periodic signal, we must know what is the probability distribution function (PDF) of the test statistic in the absence of such a periodic signal, so that a reliable significance can be assigned to the claimed detection. In Fourier analysis, a PDF of the form e^{-Z}, where Z is the Fourier power gives the classical probability measure. For time series with important fluctuations or flares (colored noise), one usually obtains a distribution that is broader than e^{-Z} resulting in overestimated significances. Similarly, windowing effects introduced by the testing of non-integral periods with respect to the total observation time also affect the probability distribution. These effects can be taken into account in order to derive an accurate probability for a given peak in the power spectrum.

For Sgr A*, the MSO—where we expect the minimum fundamental Keplerian period—ranges from $P_{\rm min} \sim 300$ to $P_{\rm max} \sim 1600$ s, for a mass of $3.5 \times 10^6\,M_\odot$ and spin parameter $0 \geq a = \frac{Jc}{GM^2} < 0.999$. Here, J is the angular momentum, c is the speed of light, and G the gravitational constant. A number of X-ray observations of Sgr A* have revealed flares with durations between $\approx 3\,000$ and $10\,000$ s. We adopt a standard minimum quality factor, Q, of 4–5 cycles to claim a periodicity, and thus find that for the longest period of 1600 s, the minimum flare duration must be 8000 s. The number of Independent Fourier Spacings (IFS or trials) in our full interval (8752 s) is given by: $\text{IFS} = T_{\rm obs}(1/P_{\rm min} - 1/P_{\rm max}) = 24$.

Aschenbach et al. (2004) claimed the detection of 5 periods ranging from 100 to 2250 s each of which were identified with one of the gravitational cyclic modes associated with accretion disks. The work of these authors was based on X-ray data from *XMM-Newton.* Genzel et al. (2003) claimed the detection of an ≈ 17 min period detected during two near-IR flares from Sgr A*. Clearly, a confirmed detection of a periodic or quasi-period signal present in a flare from the central BH would be very useful for constraining the nature of the flares, and more importantly, as evidence for the geometry of the space-time around Sgr A* through indirect BH spin measurements. In the context of this paper, we use the term *quasi-period* to refer to the average period of a signal that exhibits a periodic quality, while possibly comprising an evolution from longer to shorter periods, which would leads to a broader peak due to a spread in the power over several frequencies.

2. Observations and Methods

2.1. Observations

In the last few years, observations of the central BH with the *Chandra* and *XMM-Newton* X-ray telescopes have revealed several flares with different durations and spectral indices [2, 3, 11, 14]. Recently, two more flares were detected by *XMM-Newton*, and their properties are discussed by Bélanger et al. (2005).

Here, we restrict our analysis to *XMM-Newton* data, and pay particular attention to two data sets: 0516-0111350301 and 0866-0202670701. These observations of Sgr A* were respectively performed on 2002 October 3 and 2004 August 31, and each comprises one major flare, that lasted ~3 and 9 ks respectively. Details on the observation and characteristics of the first flare are presented and discussed by Porquet et al. (2003), and those of the other event by Bélanger et al. (2005). In addition, we analysed the data of pointing 0789-0202670601, which also comprised a flare from Sgr A*, but solar proton contamination caused a large number of data gaps during the flaring events thus making these data unreliable for period analysis. As described in Bélanger et al. (2005), the light curve can be reconstructed under the condition that the binning is large enough to be insensitive to the presence of very short good time intervals that would otherwise lead to an overestimate of the count rate. In this case, 500 s was found to be a reasonable bin size, but this does not allow a meaningful period analysis for an event that lasted ~3 ks.

For each data set, we used the PN event list constructed by selecting good X-ray events in the 2–10 keV energy range, within a radial distance of 10″ from the radio position of Sgr A* [17]. The PN instrument has a larger collecting surface, better sensitivity and most importantly, a much higher time resolution than the MOS cameras—73 ms compared to 2.6 s in *FullWindow* mode[1]. This makes the PN camera the *XMM-Newton* instrument of choice for timing analysis. Figures 1 and 2 show the light curves of the 2004 August 31 flare and of the 2002 October 3 observations respectively, both with 120 s resolution.

[1] PN's time resolution can be as high as 0.3 ms in the standard *Timing* mode with a duty cycle close to 100%, and 7 μs in the special *Burst* mode. The duty cycle, however, is only 3% in the latter mode of operation.

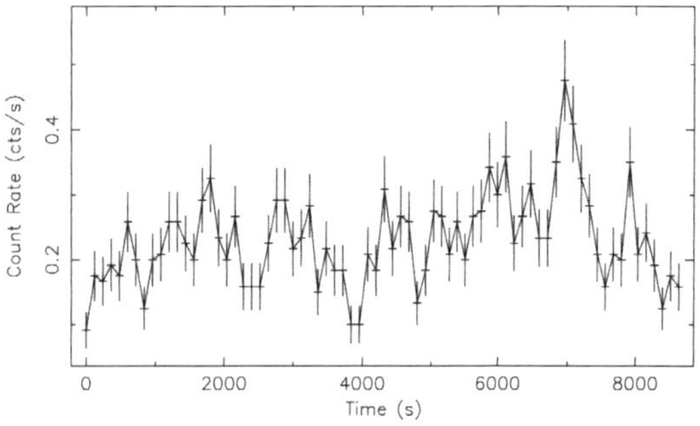

Figure 1. Light curve of the 2004 August 31 flare ($T \sim 9$ ks).

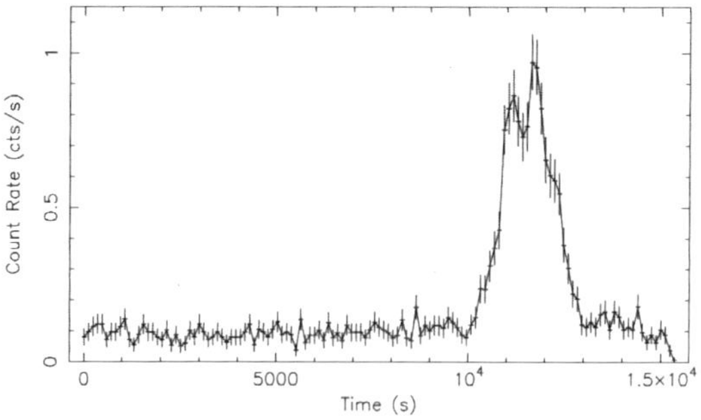

Figure 2. Light curve of complete ~15 ks 2002 October 3 observation.

2.2. Test-statistic and Periodogram

The Z^2-periodogram is constructed as follows. For a given test period P_j, the phase ϕ_i, of each arrival time t_i, is calculated as $\phi_i = 2\pi t_i/P_j$. These phases are then used to calculate the Z_m^2 value defined by Buccheri et al. (1983) as

$$Z_m^2 = \frac{2}{N}\sum_{k=1}^{m}\left[\left(\sum_{i=1}^{N}\cos(k\phi_i)\right)^2 + \left(\sum_{i=1}^{N}\sin(k\phi_i)\right)^2\right] \tag{1}$$

Here, N is the number of events and m is the number of harmonics over which the sum is performed. We used the Rayleigh statistic ($m=1$) and thus performed the sum over the first harmonic only, since this is the most powerful test for sinusoidal signals [12].

The Rayleigh statistic, Z_1^2, is distributed as a χ^2 with two degrees of freedom (dof), or in compact form: $\chi^2_{\rm dof} = \chi^2_2$. This is so because $\cos\phi$ and $\sin\phi$ are Gaussian distributed, the square of a Gaussian variable is χ^2_1 distributed, and the sum of χ^2 variables is also a χ^2 variable for which the number of dof is given by the sum of the dof from the individual variables. Thus, the Z_m^2 statistic is χ^2_{2m} distributed. de Jager (1994) showed that the predicted significance for a sinusoid from the Rayleigh power is

$$Y = -\log_{10}(\text{Prob}) = 0.434(Np^2/4 + 1), \tag{2}$$

where N is the number of events, and p is the pulsed fraction.

As mentioned above, the range of periods related to the MSO we consider physically interesting is 300–1600 s. Nonetheless, periodograms are shown from 100 s to $T_{\rm obs}/3$.

2.3. Background

In the simplest case, the background is composed of white or Poisson noise: fluctuations for which the range of values is consistent with counting statistics. A more general case considers the background as a combination of white and red noise. Red noise (sometimes referred to as $1/f$ noise) arises from the inherent source variability and has a power spectral distribution that goes as $f^{-\alpha}$, where f is the frequency and α is the slope of the power spectrum in log-log space—the larger the amplitude of the variations, the steeper the slope.

Clearly, it is essential to determine the relative importance of the component of red noise with respect to the Poisson noise in order to accurately estimate the probability associated with a given peak in the periodogram. This can be done in different ways, one of which was recently suggested by Vaughan (2005) and entails determining the index α by performing a linear fit on the log-log periodogram and then calculating in an analytical way the confidence levels, conveniently represented as lines parallel to, and lying above the fitted power-law. This method is simple, does not require Monte Carlo (MC) simulations, and is convenient for data that has an important component of red noise and whose duration is long enough to allow for a suitably large range of frequencies with a well defined power spectrum over which to perform the fit.[2]

Another way to characterize the component of red noise in an observation is to construct the power spectrum of the appropriately smoothed light curve. The power spectrum will then be dominated by the features associated with the inherent source variability, and will exhibit the features of the underlying red noise component, which can then be characterized as described above. This procedure should give results closely compatible with the method of de-trending—fitting the light curve with a polynomial and subtracting it—which leaves the flattened light curve free of the inherent source variability. Here, however, the stochastic nature of the red noise is not accounted for.

We have used a method that allows us to work directly with the event list and to take into account all noise components present in the data in a natural and transparent manner that also takes into account the random nature of the noise. This was done with the use of MC simulations to generate event lists having the same statistical properties as the data.

2.4. Pseudo-Random Event Lists

Most analyses aimed at estimating the significance of a periodic signal use MC simulations. We have used such a technique to estimate the probability of the null hypothesis, that is, the probability that a peak in the periodogram is caused by a background fluctuation. However, several different methods can be employed, and thus we detail the one we have used.

In counting photons from a given astrophysical source or radioactive decays from an unstable isotope, each event occurs independently of the previous but with a certain regularity given by the average count rate. These processes can be described by the same statistical law, characterised by an exponential PDF that defines the distribution of time intervals between two consecutive events and for which the mean corresponds to the average count rate. Therefore, in order to construct a true pseudo-random light curve of total duration T and average count rate r, one must draw $T \times r$ numbers from an exponential PDF with a mean of r. Each of these corresponds to the Δt between two consecutive events. Arrival times are calculated by summing these Δt, and the light curve is finally constructed from an event list that has the same statistical properties (Poissonian for $r = \text{const}$), as the data.

In the case where the mean is not constant, we can generate event lists with equivalent statistical properties as the data set by drawing the Δts from different exponential PDFs with means determined by the sliding average count rate calculated from the data over an appropriately sized window.

The size of the smoothing window determines the range of periods accessible, (period longer than this size are smoothed out). By performing the simulations using different window sizes determined on the basis of the length of the observation as $T/4$, $T/8$, $T/16$, $T/32$, etc. down to a specified minimum, the probability associated with each trial period is determined taking into account fluctuations in the base level of the light curve that occur on timescales of the smoothing window size. Therefore, in each range of accessible trial periods, the significance of each peak is properly assessed. A period will be picked out only if it is significant compared to general trends in the light curve. Drawing numbers from PDFs with means that follow the variation in the count rate takes into account the overall noise properties of the data by simultaneously characterizing the inherent variations in the base level of the light curve, and allowing fluctuations about this level that are proportional to the amplitude of the variation. This is the method we used to estimate the probability associated with each value in the Rayleigh periodogram.

[2] A proper fitting technique must be used in order to determine the power-law index (see [10]).

In practice, for a given window size, we determine a set of mean rates and errors. For a mean rate $r \pm \sigma_r$, a Δt is constructed in two steps: one pseudo-random number, $\hat{r}$, is drawn from a Gaussian $N(r,\sigma_r)$, and another, $\Delta\hat{t}$, is drawn from an Exponential $E(\hat{r})$. The phases are constructed from the arrival times as described above. The full range of testable periods (chosen as 100–$T/3$) is subdivided in 100 logarithmically spaced period bins, and logarithmically sampled by 150 trial periods, P_j, to ensure that each bin contains the same number of trials. For each P_j, we obtain a Z^2 value histogrammed in the corresponding period bin. Finally, the probability is determined directly from the Z^2 values by finding the quantile inverse for the Z^2 value of P_j in the data set.

3. Results

Figures 3 and 4 show the Z^2-periodograms, and their associated probability as a function of period derived from the MC simulations for the flare data sets of 2004 August 31 and and 2002 October 3 respectively. The In Fig. 5, we present the periodogram and probability of the data obtained on 2002 October 3 before the flare. The vertical scale is the same on all three Figures.

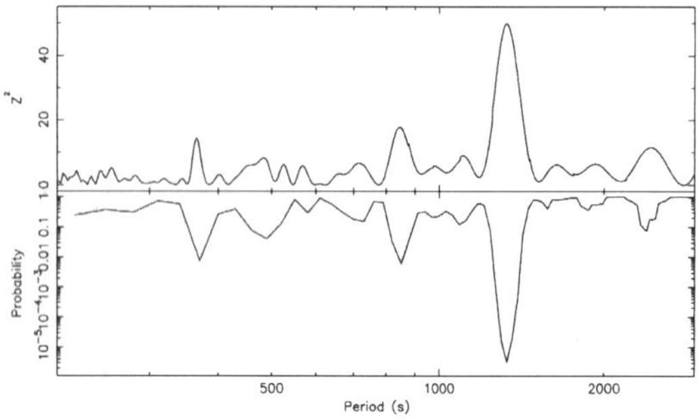

Figure 3. Flare of 2004 August 31: Z^2-periodogram and associated probabilities derived from MC simulations of 10^6 event lists. Total flare duration was ~9000 s and thus the period range is 100–3000 s. The peak is at $P = 1330$ s and rises to a Z^2 value of 46, and has an associated probability of $\sim 10^{-6}$.

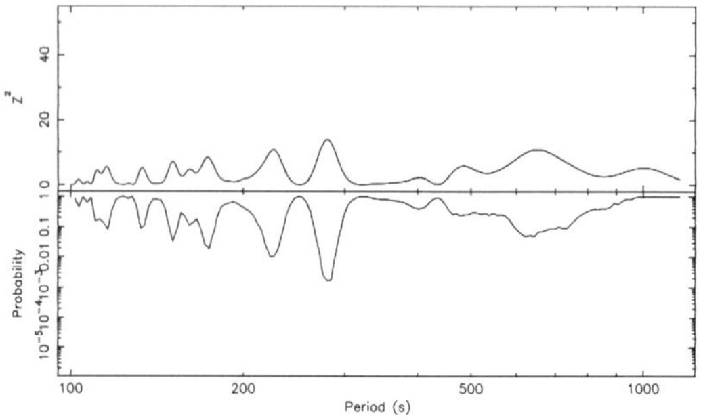

Figure 4. Flare of 2002 October 3: Z^2-periodogram and associated probabilities derived from MC simulations of 500 event lists. Total flare duration is ~3000 s and period range spans 100–1000 s. No significant peaks are detected

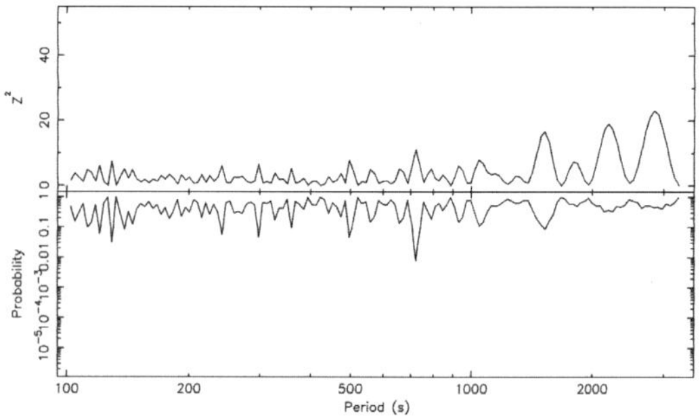

Figure 5. Data of 2002 October 3 before flare: Z^2-periodogram and associated probabilities derived from MC simulations of 500 event lists. Total duration of observation is ~10000 s and period range spans 100–3000 s. No significant peaks are detected

As expected, the probability function is perfectly anticorrelated with the value of Z^2 over most of the range. Deviations are apparent towards to end of the range, where P_j becomes commensurable with T_{obs}. This effect is most noticeable in Fig. 5 where the last three peaks in the periodogram grow in height whereas their probabilities hover around one, indicating that the MC simulations indeed model the overall statistical properties of the data accurately and thus reproduce features associated with the noise. Notice that similar behaviour, although not as marked, is also seen in the 2004 August 31 as well as 2003 October 3 events.

In Fig. 3, the strongest peak at 1330 s has an associated probability of 3×10^{-6}. The other two noticeable peaks at 370 s and 850 s have respective probabilities of 0.006 and 0.007, and are therefore not significant. Peaks with probabilities greater than 10^{-3} cannot be considered significant if we apply a 3σ requirement to claim a detection and thus, no peaks in the periodograms other than the one at 1330 s can be taken as significant. The after-trial significance for this signal is $Y = 4.14$, and using Eq. 2 we find a pulsed fraction of 0.13 ($N = 2023$).

Fig. 6 presents three phasograms resulting from folding the flare light curve with a phase of 1330 s. These were constructed in the 2–4 keV, 4–10 keV and 2–10 keV bands. There is a clear asymmetry in the overall light curve, and a marked peak present at low energies, making the 2–4 keV band radically different from the much smoother 4–10 keV light curve. This peak is shifted by ~266 s from the centre of the sinusoid, corresponding to a distance of ~0.5 AU for light. This intriguing feature will be investigated elsewhere.

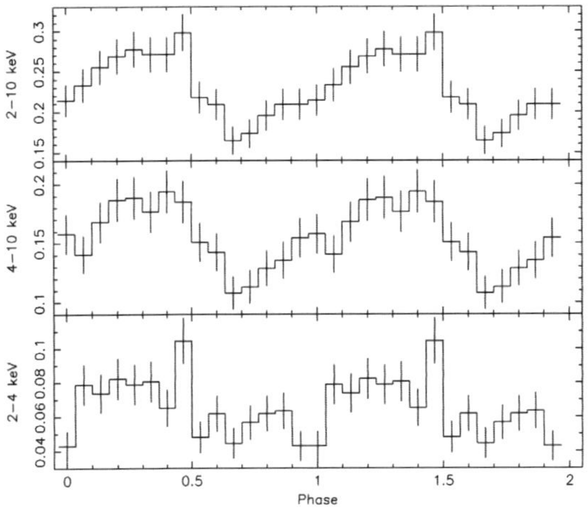

Figure 6. Light curve of 2004 August 31 flare folded with a phase of 1330 s, in three energy bands: 2–10 keV (top), 2–4 keV (middle) and 4–10 keV (bottom).

Finally, we also analyzed the pre-flare period of the 2002 October 3 observation to search for the periods reported by Aschenbach et al. (2004) in these data. The periodogram is shown in Fig. 5 and is remarkably flat over most of the range of physically interesting periods. As we can see from the associated probabilities, there are no significant peaks present.

4. Summary and Conclusion

We detected the presence of a 22.2 min quasi-period in the X-ray data of the flaring event in Sgr A* that occurred on 2004 August 31 while simultaneous *XMM-Newton* and HST observations were under way (see Yusef-Zadeh et al. 2006, Bélanger et al. 2005 and Bélanger et al. 2006 for details on the results of the campaign). The probability that such a signal arises from statistical fluctuations was determined to be 3 × 10–6 from MC simulations and is conservatively estimated to be less than $\sim 7.2 \times 10^{-5}$ if we consider the 24 IFS in the period range 300–1600 s [7]. It is important to clarify that although an evolution of the period as a function of time may be present (as expected from the spiraling motion), and that it can be qualitatively seen in the light curve as alluded to by Liu et al. (2006), our statistics do not permit us to claim such a detection.

We considered and analysed two other *XMM-Newton* observations of Sgr A* using the same method for the probability estimation: the one that contains the 2004 March 31 flare, and the one of the 2002 October 3 flaring event. The former was found to be unreliable due to solar proton contamination, and no significant signals were detected in the latter, in contrast with the results obtained by Aschenbach et al. (2004) using the same data set.

The method we have developed to detect a periodic signal, and evaluate the associated probability is well suited for searches involving long periods with respect to the total observation time, and in which fluctuations in the average count rate do not follow Poisson statistics. This MC based method is just as powerful in the absence of red noise but clearly not necessary since the probability can be computed analytically. This straightforward and reliable method for quantitatively estimating the significance of a given periodic or semi-periodic modulation will be useful in future applications of this kind, and hopefully help, at least in part, relieve some of the ambiguities that arise in this type of analysis.

The radius of the MSO in units of $r_g = GM/c^2$ is given by:

$$r_{\rm mso} = [3 + z_2 - \sqrt{(3 - z_1)(3 + z_1 + 2z_2)}], \tag{3}$$

where $z_1 = 1+(1-a^2)^{1/3}[(1+a)^{1/3}+(1-a)^{1/3}]$, and $z_2 = (3a^2+z_1^2)^{1/2}$. The Keplerian frequency as a function of a is:

$$\omega = \frac{c^3}{2\pi GM}(r^{3/2} + a)^{-1}. \tag{4}$$

Taking a mass of $3.5 \times 10^6\, M_\odot$ and assuming that the periodic signal originates from orbital motion at the MSO, the 1330 s period implies that the BH is spinning at a rate given by $a \approx 0.22$ (in the prograde case, the only considered).

Thus, the main result presented here can be interpreted as the signature of a transient event that evolved over about 10 ks, simultaneously giving rise to an X-ray and near-infrared flare. The emitting region spiraled around the mildly spinning SMBH, near or at the last stable orbit. In this simple scenario, we would have directly witnessed the accretion of matter through the accretion disk in the Kerr metric around the central dark mass (see [15]).

References

[1] Aschenbach, B. et al. 2004, *A&A*, **417**, 71
[2] Baganoff, F.K. et al. 2001, *Nature*, **413**, 45
[3] Baganoff, F.K. et al. 2003, *ApJ*, **591**, 891
[4] Bélanger, G. et al. 2005, *ApJ*, **635**, 1095
[5] Bélanger, G. et al. 2006, *ApJ*, **636**, 275
[6] Buccheri, R., et al. 1983, *A&A*, **128**, 245
[7] de Jager, O.C., Swanepoel, J.W.H. & Raubenheimer, B.C. 1989, *A&A*, **221**, 180
[8] de Jager, O.C. 1994, *ApJ*, **436**, 239
[9] Genzel, R. et al. 2003, *Nature*, **425**, 934
[10] Goldstein, M.L., Morris, S.A. & Yen, G.G. 2004, (cond-mat/0402322)
[11] Goldwurm, A. et al. 2003a, *ApJ*, **584**, 751
[12] Leahy, D.A., Elsner, R.F. & Weisskopf, M.C. 1983, *ApJ*, **272**, 256
[13] Liu, S. Melia, F. & Petrosian, V. 2006, *ApJ*, **636**, 789
[14] Porquet, D. et al. 2003a, *A&A*, **407**, L17
[15] Tagger, M., & Melia, F. 2006, *ApJ*, **636**, L33
[16] Vaughan, S. 2005, *A&A*, **431**, 391
[17] Yusef-Zadeh, F., Choate, D., & Cotton, W. 1999, *ApJ*, **518**, L33
[18] Yusef-Zadeh, F. et al. 2006, *ApJ*, **644**, 198

Institute of Physics Publishing
Journal of Physics: Conference Series **54** (2006) 427–435
doi:10.1088/1742-6596/54/1/067
Galaxy Center Workshop

Accretion models of Sgr A*

Feng Yuan[1,2]

[1] Shanghai Astronomical Observatory, Chinese Academy of Sciences, 80 Nandan Road, Shanghai 200030, China

[2] Joint Institute for Galaxy and Cosmology (JOINGC) of SHAO and USTC

E-mail: `fyuan@shao.ac.cn`

Abstract.
The supermassive black hole in the center of our Galaxy, Sgr A*, is unique because the angular size of the black hole is the largest in the sky thus providing detailed boundary conditions on, and much less freedom for, accretion flow models. In this paper I review advection-dominated accretion flow (ADAF; another name is radiatively inefficient accretion flow) models for Sgr A*. This includes the dynamics and developments of ADAFs, and how to explain observational results including the multi-waveband spectrum, radio polarization, IR and X-ray flares, and the size measurements at radio wavebands.

1. Introduction

The center of our Galaxy provides the best evidence to date for a massive black hole [1, 2], associated with the compact radio source, Sgr A* [3]. Since the original discovery of Sgr A* in 1974, there have been intensive efforts in both observational and theoretical aspects, with dramatic progresses in the past few years. The reason why we are so interested in this object is because of its proximity, which allows us to determine observationally the dynamics of gas quite close to the BH, providing unique constraints on theoretical models of accretion flows.

Before introducing the model, I first briefly review the main observational results of Sgr A*. As shown by the data points in Fig. 2, its radio spectrum consists of two components. The component below 86 GHz has a spectrum $F_\nu \propto \nu^{0.2}$, while the high frequency component, the "submm bump", has a spectrum $F_\nu \propto \nu^{0.8}$ up to $\sim 10^3$ GHz [4, 5]. Variability at centimeter and millimeter wavelengths is detected with a timescale ranging from hours to years with an amplitude of less than 100% [5, 6, 7, 8]. The high level of variable linear polarization ($\sim 2\% - 10\%$) at frequencies higher than $\sim$ 150 GHz sets an upper limit on the rotation measure of 7×10^5rad m^{-2}, which argues for a low density at the innermost region of the ADAF [9, 10, 11, 12, 13].

At IR wavelength, the source is highly variable. Genzel et al. [14] detected Sgr A* at 1.6-3.8 μm, with a factor of $\sim 1-5$ variability on timescales of $\sim 10-100$ min. Similarly, at 3.8 μm, Ghez et al. [15] found that the flux changes by a factor of 4 over a week, and a factor of 2 in just 40 min. If describing the IR spectrum with a power-law, Gillessen et al. [16] found that the spectral index is correlated with the instantaneous flux but Hornstein et al. [17] found it remains constant during the flare process.

Sgr A* has been convincingly detected in the X-rays [18, 19, 20, 21]. The X-ray emission has two distinct components. In "quiescence," the emission is soft and relatively steady, with a large fraction of the X-ray flux coming from an extended region with a diameter $\approx 1.4''$ [18]. Several times a day, however, Sgr A* has X-ray "flares" in which the X-ray luminosity increases by a factor of a few – 50 for roughly an hour. For the most flares, the spectrum is hard, with a photon index of $\Gamma = 1.3^{+0.5}_{-0.6}$. *XMM*, however, detected a very bright and soft flare with $\Gamma = 2.5^{+0.3}_{-0.3}$ [21]. Recent several multiwavelength campaigns found that there is no time lag between the IR and X-ray flares [22, 23, 24]. This strongly suggests a common physical origin. The short timescales indicate that the emission arises quite close to the BH, within $\sim 10R_S$ (where R_S is the Schwarzschild radius). Sgr A* is extremely dim overall, with a bolometric luminosity of only $L \approx 10^{36}\,\mathrm{erg\,s^{-1}} \approx 3\times10^{-9}\mathrm{L_{Edd}}$.

2. Accretion flow models: from ADAF to RIAF

Consider a steady axisymmetric accretion flow. The dynamics of the flow are described by the following height-integrated differential equations, which express the conservation of mass, radial momentum, angular momentum, and the energy of electrons and ions [25, 38, 44]:

$$\dot{M} = -4\pi R H \rho v = \dot{M}_{\rm out}\left(\frac{R}{R_{\rm out}}\right)^s, \tag{1}$$

$$v\frac{dv}{dr} = -\Omega_{\rm k}^2 r + \Omega^2 r - \frac{1}{\rho}\frac{dp}{dr}, \tag{2}$$

$$v(\Omega r^2 - j) = \alpha r \frac{p}{\rho}, \tag{3}$$

$$\rho v\left(\frac{d\varepsilon_e}{dr} - \frac{p_e}{\rho^2}\frac{d\rho}{dr}\right) = \delta q^+ + q_{ie} - q^-, \tag{4}$$

$$\rho v\left(\frac{d\varepsilon_i}{dr} - \frac{p_i}{\rho^2}\frac{d\rho}{dr}\right) = (1-\delta)q^+ - q_{ie} = -(1-\delta)\alpha p r\frac{d\Omega}{dr} - q_{ie}. \tag{5}$$

The quantities have their usual meaning. Specifically s describe the strength of the outflow and δ the fraction of the turbulent energy which directly heats electrons. The radiative processes include synchrotron and bremsstrahlung emission and their Comptonization. After the dynamics of the accretion flow is solved, we can calculate the emitted spectrum and polarization fraction and compare with the observed ones.

The above set of equations hold for any accretion models including, e.g., the standard thin disk, except that the standard thin disk is one-temperature due to the strong coupling between electrons and ions so the two energy equations are unified into one. The ADAF solution is one set of solution in the regime of $\dot{M} \lesssim \alpha^2 \dot{M}_{\rm Edd}$ where $\dot{M}_{\rm Edd} \equiv 10 L_{\rm Edd}/c^2$ is the Eddington accretion rate[1]. An ADAF has many distinct properties compared to the standard thin disk. The gas temperature is much higher, the flow is optically thin and geometrically thick. Perhaps most importantly, the radiative efficiency of an ADAF is very low, $\eta_{\rm ADAF} \sim 0.1\dot{M}/(\alpha^2\dot{M}_{\rm Edd})$ [28, 29, 25]. This is the reason why ADAF models are so successful in explaining the dimness of Sgr A* [30, 31, 32].

There has been a significant change in the theoretical understanding of ADAFs over the past few years. First, global, time-dependent, numerical simulations reveal that only a very small fraction of the mass that is available at large radii actually accretes onto the black hole and most of it is lost to a magnetically driven outflow or circulates in convective motions [33, 34], i.e, $s > 0$ in eq. (1). The physical reason for the presence of an outflow is that, since very

[1] There are four solutions to the eqs. (1)-(5), two "cold" ones and two "hot" ones. The former includes the standard thin disk and slim disk; the latter includes ADAF and LHAF. See Fig. 6 in [26] and Fig. 1 in [27] for details.

little energy is radiated away, the Bernoulli parameter of an ADAF is almost zero[2], much larger than that in the standard thin disk. Therefore the gas escapes more easily to infinity once it is perturbed [28, 36].

In the early version of an ADAF, δ *is assumed* to be very small, $\delta \approx 10^{-2} - 10^{-3}$, i.e., the turbulent energy only heats ions. However, there has been large theoretical uncertainties in how electrons are heated in the accretion flow [37] and a larger δ seems to be physically more plausible since processes like magnetic reconnection is likely to heat electrons directly. A large δ is also required from the observational side. Quataert & Narayan [38] show that when outflows are present, to produce the correct amount of emission of Sgr A*, the electrons in the accretion flow should be hotter than in the early version of ADAF to compensate for the decrease of radiation due to the decrease of density. Stated another way, δ must be larger. Sometimes, the "updated" version of ADAF with $s > 0$ and $\delta \lesssim 1$ is called radiatively inefficient accretion flow (RIAF). In this review we simply use the original term: "ADAF".

When $\delta \approx 10^{-2}$, almost all of the turbulent energy q^+ is stored in the flow as the entropy of ions and very little is transferred into electrons through Coulomb collision q_{ie}, i.e., $q_{ie} \ll q^+$. We generally think this is why the radiative efficiency of ADAF is so low. If $\delta \lesssim 1$, which means that a large amount of turbulent energy will heat electrons directly, what is the radiative efficiency of an ADAF? In this case, as shown by Fig. 1, almost all of the turbulent energy heating electrons is stored in electrons as entropy, i.e., the electrons are advection-dominated. This ensures the low efficiency of an ADAF even though δ is large. We would like to emphasize that with an increase of the accretion rate, the flow is no longer advection-dominated and the radiative efficiency rapidly increases (ref. Fig. 7 in [25] and Fig. 6 in [26]).

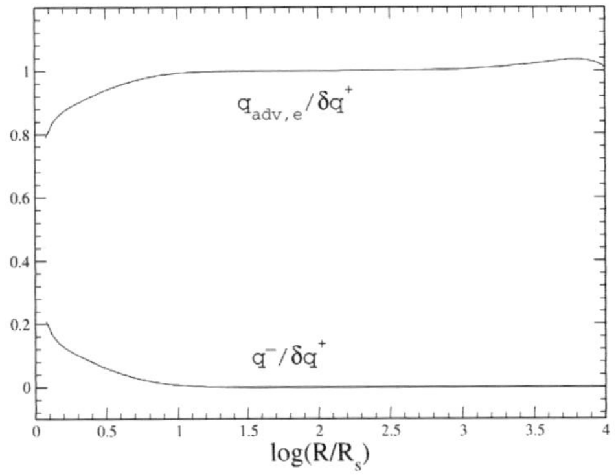

Figure 1. The energy relationship of electrons in the ADAF in Sgr A* (with $\delta \approx 0.55$).

3. Application to Sgr A*

3.1. Outer boundary conditions

The outer boundary conditions include the temperature, density, mass accretion rate, and angular momentum at the outer boundary. The outer boundary is determined by the Bondi radius. For uniformly distributed matter with an ambient density ρ_0 and sound speed c_s, the Bondi radius of a black hole of mass M is $R_{\rm Bondi} \approx GM/c_s^2$ and the accretion rate is $\dot{M}_{\rm Bondi} \approx 4\pi R_{\rm Bondi}^2 \rho_0 c_s$. *Chandra* observations of the Galactic center detect extended diffuse emission within $1'' - 10''$. The inferred gas density and temperature are ≈ 100 cm^{-3} and ≈ 2 keV on $1''$ scales [18]. The corresponding Bondi radius $R_{\rm Bondi} \approx 0.04$pc $\approx 1'' \approx 10^5 R_s$ and the

[2] The self-similar solution of [28] gives a positive Bernoulli parameter of the flow. However, Nakamura [35] argue that the Bernoulli parameter in an ADAF should be mostly negative. This is confirmed by my global solution of ADAFs.

Bondi accretion rate is $\dot{M}_{\rm Bondi} \approx 10^{-5} M_\odot$ yr^{-1}. Given that the accreted gas should have some angular momentum (see below), the accretion rate should be somewhat smaller. The recent 3D numerical simulation for the accretion of stellar winds on to Sgr A* by Cuadra et al. [39] obtains $\dot{M} \approx 3 \times 10^{-6} M_\odot$ yr^{-1}. If gas were accreted at this rate onto the black hole via the standard thin disk, the expected luminosity would be $L \approx 0.1 \dot{M}_{\rm Bondi} c^2 \approx 10^{41}\,{\rm erg\,s^{-1}}$, 5 orders of magnitude higher than the observed luminosity. This is the strongest argument against a thin disk in Sgr A*.

The numerical simulation indicates that the angular momentum is large, with a circularization radius of about $10^4 R_s$ [39]. This result casts doubt on the "small angular momentum" assumption of spherical accretion models [40, 41].

3.2. Explaining the quiescent state

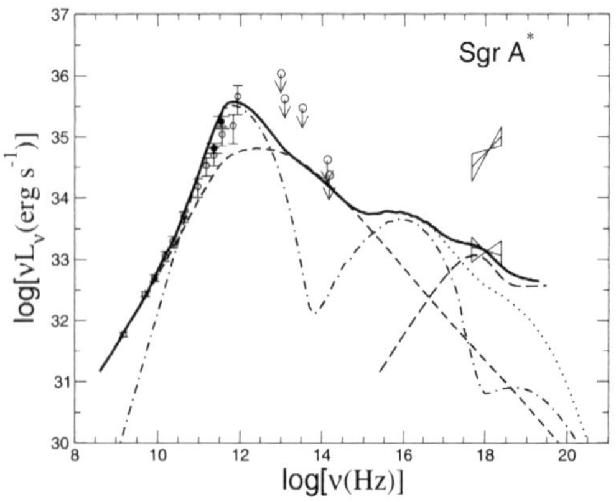

Figure 2. ADAF model for the quiescent state emission from Sgr A*. The dot-dashed line is the synchrotron and SSC emission by thermal electrons; the dashed line is the synchrotron emission by non-thermal electrons. The dotted line is the total synchrotron and SSC emission while the solid line includes the bremsstrahlung emission from the outer parts of the ADAF (shown by the long-dashed line). Taken from YQN03.

A number of authors have applied ADAF models to explain the observations of Sgr A*. In the early applications [30, 31, 32], they neglect outflows and the direct heating of electrons by turbulence, i.e., $s = 0$ and $\delta = 10^{-2}$ are adopted. These works can roughly account for the observed low luminosity and the spectrum of Sgr A*. However, such models can't explain the observed high linear polarization. This is because the predicted density is too high and the rotation measure is $\sim 10^{10}$radm^{-2} in the region where high-frequency radio emission is produced, much higher than the upper limit of 7×10^5rad m^{-2} obtained by Marrone et al. [11]. This problem can be solved if we take into account the presence of outflows, i.e, $s > 0$. In this case, most of the accreted gas is lost to the outflow. Thus the density in the innermost region is much lower.

Another problem with the original ADAF model is that it under-predicts the low-frequency radio emission significantly. This can be solved by introducing another component to the model. One possibility is a jet [42, 43]. Alternatively, some fraction of electrons in the ADAF may be in nonthermal distribution, due to acceleration processes in the accretion flow such as weak shocks and turbulent dissipation. Their synchrotron emission can explain the radio "excess". Fig. 2 shows the modeling result of an ADAF model with outflow and nonthermal electrons, taken from Yuan, Quataert & Narayan [44] (hereafter YQN03). The parameters are $\dot{M}_{\rm out} \approx 10^{-6} M_\odot$ yr$^{-1}, s = 0.27, \delta = 0.55$. The nonthermal electrons are assumed to be in a power-law distribution with a spectral index of $p = 3.5$. The injected energy in nonthermal electrons is equal to a fraction of 1.5% of the energy in thermal electrons. The accretion rate close to the black hole is $\dot{M} \approx 4 \times 10^{-8} M_\odot$ yr$^{-1} \ll \dot{M}_{\rm out}$. As shown in the figure, the submm bump is mainly due to the synchrotron emission from thermal electrons, the low-frequency radio and IR emissions from the synchrotron emission from nonthermal electrons, while the extended X-ray emission is mainly due to the bremsstrahlung emission from the region around the Bondi radius.

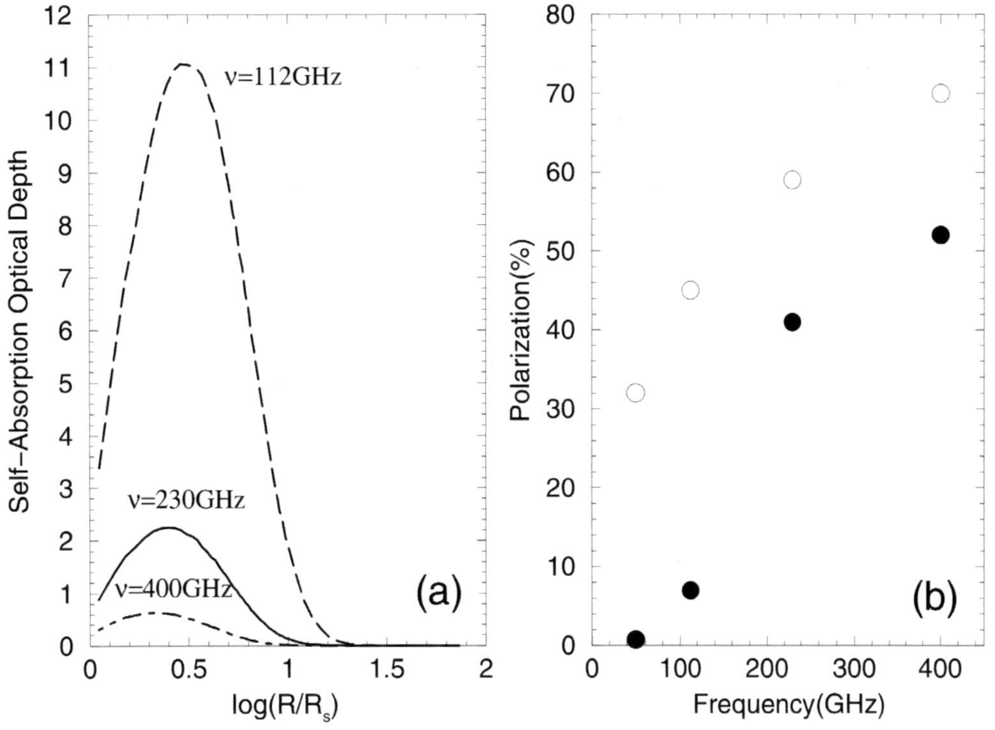

Figure 3. (a) Synchrotron self-absorption optical depth of thermal electrons as a function of radius; (b) Polarization fraction from thermal electrons with (filled circles) and without (open circles) Faraday rotation. These values are upper limits because a uniform magnetic field is assumed. Taken from YQN03.

Since most of the accreted gas is lost to the outflow, the rotation measure is very small, RM $\approx 10^7$ and $\lesssim 5 \times 10^5$rad m^{-2} at 230 GHz if we integrate along the equatorial plane and rotation axis, respectively. Note that these should be regarded as upper limits since we assume that the magnetic field is fully coherent and points along the line of sight while the magnetic field should be predominantly toroidal.

As noted above, the submm emission is due to thermal electrons. Since the linear polarization of optically thick thermal synchrotron emission from a uniform medium is suppressed by $\exp[-\tau]$, where τ is the synchrotron optical depth, one also needs to check whether the model can roughly account for the magnitude of the observed linear polarization ($\lesssim 10\%$). YQN03 have calculated the linear polarization produced by the thermal electrons in their models, including optical depth effects and Faraday rotation. Fig. 3a shows τ as a function of radius for three frequencies and Fig. 3b shows by open circles the degree of linear polarization as a function of frequency when Faraday rotation is neglected. From the figure we see that thermal electrons can readily account for the observed level of linear polarization from Sgr A*.

3.3. Understanding the flares
Markoff et al. [45] showed that the flares are probably due to enhanced electron heating or acceleration, rather than a change in the accretion rate onto the black hole. The IR and X-ray flares in Sgr A* may be analogous with solar flares, in which magnetic energy is converted into thermal energy, accelerated particles, and bulk kinetic energy, due to magnetic reconnection [44]. This speculation is confirmed by the 3D MHD numerical simulation by [46]. Unfortunately, even for solar flares, some important aspects such as the detailed acceleration mechanism and the energy distribution of heated/accelerated electrons remains unclear [47].

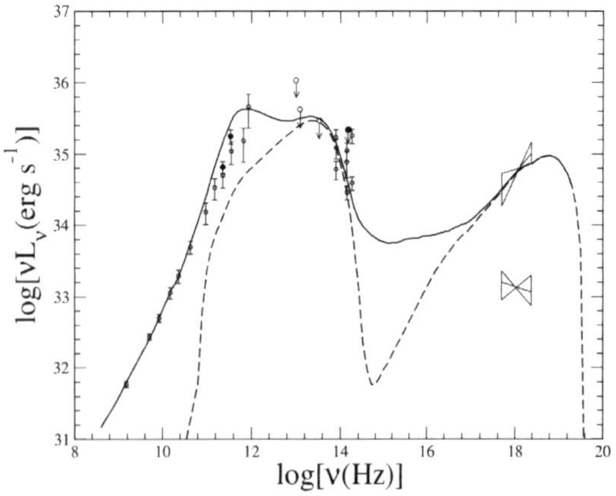

Figure 4. Synchrotron+SSC model for the IR and X-ray flares in Sgr A*. The dashed line shows the synchrotron and SSC emission from power-law electrons, and the solid line shows the total flare spectrum, including the emission from the thermal electrons. Taken from YQN03.

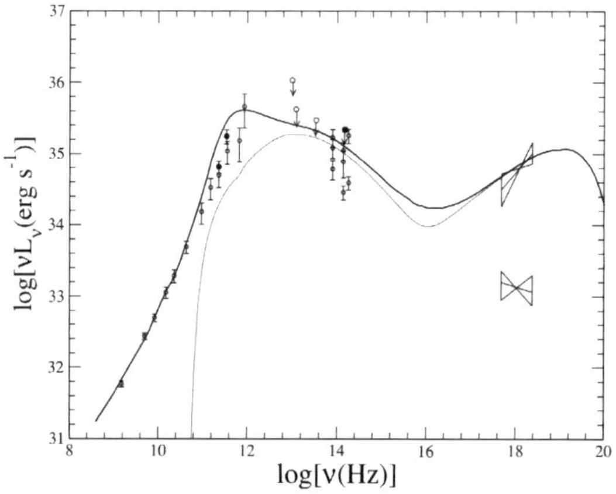

Figure 5. Synchrotron model for the IR and X-ray flares in Sgr A*. The light solid line corresponds to the emission from power-law electrons and the thick solid line to the total emission including the thermal electrons. Taken from [48].

Based on this idea, YQN03 and Yuan, Quataert & Narayan [48] propose that during the flare events, magnetic reconnection occurs in the innermost region of the ADAF. As a result, some electrons are heated/accelerated into a thermal/power-law distribution. Two models are proposed to explain the IR and X-ray flares in Sgr A*. In the first "synchrotron+SSC" model shown in Fig. 4, the electrons are accelerated into a single power-law distribution with a maximum Lorentz factor of $\gamma_{\rm max} \sim$ several hundred. The IR flare is due to the synchrotron emission while the X-ray flare is due to the up-scattering of the IR photons by these electrons. In the second "synchrotron" model shown in Fig. 5, it is assumed that some electrons are *heated* into a higher temperature while some electrons are *accelerated* into a (hard) power-law distribution with $\gamma_{\rm max} \sim 10^6$. The synchrotron emission of these two types of electrons are responsible for the IR and X-ray emissions, respectively[3]. I would like to emphasize that currently both models can explain the observations well and there are no strong arguments against any one given our poor knowledge of particle acceleration.

[3] Liu et al. [49] propose that the electrons are in a *thermal* distribution after some stochastic acceleration process. The synchrotron emission and its inverse Compton scattering produce the observed IR and X-ray flares.

3.4. Using the size measurement to test the models

Recent radio observations by the VLBA at 7 and 3.5 mm produced high-resolution images of Sgr A*, and detected that its intrinsic size is wavelength-dependent [50, 51]. The measured size of images of Sgr A* at 7 mm is $0.712^{+0.004}_{-0.003}$ mas by [50], and 0.724 ±0.001 mas and $0.21^{+0.02}_{-0.01}$ mas at 7 and 3.5 mm, respectively by [51]. This provides an additional test to accretion models of Sgr A*.

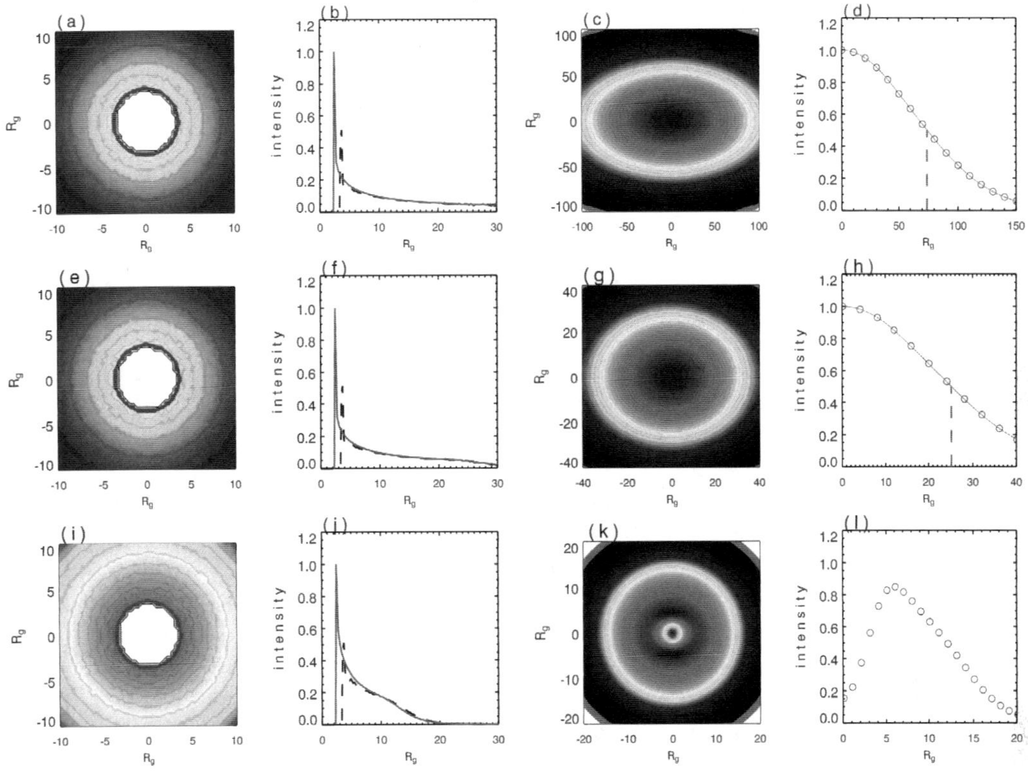

Figure 6. The images and sizes of Sgr A* at 7, 3.5, and 1.3 mm from top to bottom. The first column shows the "input" intensity distribution. The second column shows the intensity profile calculated from the YQN03 model, with (dashed) and without (solid) GR effect taken into account. The third column shows the simulated image after the interstellar scattering is taken into account. The fourth column shows the intensity profile of the simulated image (dots) and the Gaussian fit (line). The vertical dashed line shows the location of the FWHM. At 1.3 mm, the simulated profile can't be fit by a Gaussian, and thus no FWHM is indicated. Taken from [52].

Yuan, Shen, & Huang [52] calculate the predicted size of the YQN03 model. The most important point they realize is that since the intrinsic intensity profile of the ADAF is not a Gaussian, as shown by the dashed lines in the second column of Fig. 6, it is impossible to directly compare the predicted intrinsic size with the observations. Rather, they first calculate the predicted intrinsic intensity profile of the ADAF models, and then take into account the scattering broadening toward the Galactic center to obtain the *simulated* image. They then compare the simulated image with the observed one. The third column in Fig. 6 shows the simulated image at three wavelengths and the fourth column shows the fitting results of the intensity profile by a Gaussian. The obtained sizes of the simulated images are $0.729^{+0.01}_{-0.009}$ mas and $0.248^{+0.001}_{-0.002}$ mas at 7 mm and 3.5 mm, respectively. The simulated size at 7 mm is in good agreement with the observed value by [51] within the error bars but slightly larger than the observed size by [50]; the size at 3.5 mm is a little larger than the observation of [51]. Given that the size of the source may be variable [50] and the uncertainties in the calculations, they conclude that the predictions of the YQN03 model are in reasonable agreement with the observations. They predict that GR effects may be detectable at 1.3 mm.

There are two alternative models of Sgr A*, namely the jet model [42] and the coupled jet-ADAF model [43]. One difference between these two models is that in the former the radio emission above $\sim$ 86GHz is produced by the nozzle of the jet while in the latter the contribution of the ADAF is significant. In the jet-ADAF model, the contribution of the emission from the ADAF can dominate over that from the jet under suitable parameters. Moreover, if the "old" ADAF in this model is "updated" (i.e, including outflow), the only difference between the jet-ADAF model and the YQN03 model is the origin of the radio emission below $\sim$ 86 GHz. In this case, the predicted size of Sgr A* by the jet-ADAF model will be consistent with the observations. On the other hand, the predicted size at 3.5 mm by the jet model is $\gtrsim$ 0.04 mas [42], much smaller than the observed value, so some modifications of the model are required (Markoff et al. in preparation).

4. Summary

The supermassive black hole in our Galactic Center represents a unique opportunity to probe the physics of accretion, especially at extremely low accretion rates. In this review, I first briefly introduce the dynamics and evolution of the ADAF. I then try to argue that this model can provide a reasonable explanation to most of the current observations.

Chandra observations provide us with measurements of the density and temperature of the hot gas at $\sim 1'' \sim 0.04$ pc. This radius happens to be the Bondi radius where the gas is captured by the gravity of the center black hole and starts to be accreted. The Bondi accretion rate, $\dot{M}_{\rm Bondi} \sim 10^{-5} M_{\odot}\ {\rm yr}^{-1}$ provides a good estimation to the real accretion rate. As a comparison, the numerical simulation gives $\dot{M} \sim 3 \times 10^{-6} M_{\odot}\ {\rm yr}^{-1}$. Since the bolometric luminosity of Sgr A* is only $10^{36}\,{\rm erg\,s^{-1}}$, the radiative efficiency should be very low, $\sim 5 \times 10^{-6}$. The standard thin disk model [53] is therefore ruled out immediately.

"Old" ADAF models can naturally explain such a low efficiency [30, 31, 32]. However, these models fail to explain the high linear polarization in the submm waveband because the density and , further, the rotation measure, are too large. On the other hand, theoretical studies of ADAFs also indicate the presence of outflows [33, 34, 36]. This feature is taken into account in the "new" ADAF models (YQN03). The YQN03 model can explain the quiescent state spectrum and the polarization, as shown in Figs. 2 and 3. The accretion rate close to the event horizon is only $4 \times 10^{-8} M_{\odot}\ {\rm yr}^{-1}$. So the low radiative efficiency of the ADAF in Sgr A* is partly due to the outflow, which contributes a factor of $\sim 10^{-2}$, and partly due to the advection of energy, which contributes a factor of $\sim 5 \times 10^{-4}$. The IR and X-ray flares are explained by the synchrotron and inverse-Compton emissions from the heated/accelerated electrons during magnetic reconnection events in the innermost region of the ADAF (Figs. 4 and 5). Finally, the YQN03 model satisfactorily passes the test of recent observations of the size of Sgr A* in the 3.5 and 7 mm wavebands (Fig. 6). Moreover, this model predicts that observations at 1.3 mm should be able to detect GR effects (Fig. 6).

Acknowledgments

This work was supported in part by the One-Hundred-Talent Program and the National Natural Science Foundation of China (grants 10543003)

References

[1] Schödel, R. et al. 2002, Nature, 419, 694
[2] Ghez, A. M., et al. 2003, ApJ, 586, L127
[3] Melia, F., & Falcke, H. 2001, ARA&A, 39, 309
[4] Falcke, H., Goss, W. M., Matsuo, H., Teuben, P., Zhao, J. H., & Zylka, R., 1998, ApJ, 499, 731
[5] Zhao, J.-H., et al. 2003, ApJ, 586, L29
[6] Zhao, J.-H., et al. 2004, ApJ, 603, L85
[7] Herrnstein, R.M., Zhao, J.-H., Bower, G.C., Goss, W.M. 2004, AJ, 127, 3399
[8] Miyazaki, A., Tsutsumi, T., & Tsuboi, M. 2004, ApJ, 611, L97
[9] Aitken, D.K., Greaves, J.S., Chyrsostomou, A. et al., 2000, ApJ, 534, L173
[10] Bower, G.C. et al. 2003, ApJ, 588, 331
[11] Marrone, D.P. et al. 2006, ApJ, 640, 308
[12] Macquart J.P. et al. 2006, APJ, 646, L111
[13] Quataert, E. & Gruzinov 2000, ApJ, 545, 842
[14] Genzel. R. et al. 2003, Nature, 425, 934
[15] Ghez, A. M. et al. 2004, ApJ, 601, L159
[16] Gillessen, S.; et al. 2006, ApJ, 640, L163
[17] Hornstein, S. D. et al. 2006, these proceedings
[18] Baganoff, F. K., Maeda, Y., Morris, M. et al., 2003, ApJ, 591, 891
[19] Baganoff, F. K., Bautz, M. W., Brandt, W. N. et al., 2001, Nature, 413, 45
[20] Goldwurm A., et al. 2003, ApJ, 584, 751
[21] Porquet, D., et al. 2003, A&A, 407, L17
[22] Eckart, A.; Baganoff, F. K.; Morris, M. et al. 2004, A&A, 427, 1
[23] Eckart, A. et al. 2006, A&A, 450, 535
[24] Yusef-Zadeh, F. et al. 2006, ApJ, 644, 198
[25] Narayan, R., Mahadevan, R., & Quataert, E. 1998, in "The Theory of Black Hole Accretion Discs", eds. M.A. Abramowicz, G. Bjornsson, and J.E. Pringle, (Cambridge University Press)
[26] Yuan, F. 2001, MNRAS, 324, 119
[27] Yuan, F. 2003, ApJ, 594, L99
[28] Narayan, R. & Yi, I. 1994, ApJ, 428, L13
[29] Narayan, R. & Yi, I. 1995, ApJ, 444, 231
[30] Narayan, R., Yi, I., & Mahadevan, R., 1995, Nature, 374, 623
[31] Narayan, R., Mahadevan, R., Grindlay, J.E., Popham, R. & Gammie, C., 1998, ApJ, 492, 554
[32] Manmoto, T., Mineshige, S., & Kusunose, M. 1997, ApJ, 489, 791
[33] Stone, J., Pringle, J., & Begelman, M. 1999, MNRAS, 310, 1002
[34] Hawley, J. F. & Balbus, S. A. 2002, ApJ, 573, 738
[35] Nakamura, K. E. 1998, PASJ, 50, L11
[36] Blandford, R.D., & Begelman, M.C. 1999, MNRAS, 303, L1
[37] Quataert, E. & Gruzinov 1999, ApJ, 520, 248
[38] Quataert, E., & Narayan, R., 1999, ApJ, 520, 298
[39] Cuadra, J. et al. 2006, MNRAS, 366, 358
[40] Melia, F., 1992, ApJ, 387, L25
[41] Melia,F., Liu, S., & Coker, R. 2001, ApJ, 553, 146
[42] Falcke, H., Markoff, S. 2000, A&A, 362, 113
[43] Yuan, F., Markoff, S., & Falcke, H., 2002, A&A, 383, 854
[44] Yuan, F., Quataert, E., & Narayan, R. 2003, ApJ, 598, 301 (YQN03)
[45] Markoff, S., Falcke, H., Yuan, F., & Biermann, P. 2001, A&A, 379, L13
[46] Machida, M. & Matsumoto, R. 2003, ApJ, 585, 429
[47] Miller, J.A. 1998, Spac. Sci. Rev. 86, 79
[48] Yuan, F., Quataert, E., & Narayan, R. 2004, ApJ, 606, 894
[49] Liu, S. et al. 2006, submitted to ApJ (astro-ph/0603136)
[50] Bower, G. C., et al. 2004, Science, 304, 704
[51] Shen, Z.-Q. et al. 2005, Nature, 438, 62
[52] Yuan, F., Shen, Z.Q., Huang, L. 2006, ApJ, 642, L45
[53] Shakura, N. I.; Sunyaev, R. A, 1973, A&A, 24, 337

Institute of Physics Publishing
doi:10.1088/1742-6596/54/1/068
Journal of Physics: Conference Series **54** (2006) 436–442
Galaxy Center Workshop

Variable accretion of stellar winds onto Sgr A*

Jorge Cuadra[1] **and Sergei Nayakshin**[2]

[1] Max-Planck-Institut für Astrophysik, D-85741 Garching, Germany
[2] Department of Physics and Astronomy, University of Leicester, LE1 7RH, UK

E-mail: `jcuadra@mpa-garching.mpg.de`, `Sergei.Nayakshin@astro.le.ac.uk`

Abstract.
We report a 3-dimensional numerical study of the accretion of stellar winds onto Sgr A*, the super-massive black hole at the centre of our Galaxy. Compared with previous investigations, we allow the stars to be on realistic orbits, include the recently discovered slow wind sources, and allow for optically thin radiative cooling. We first show the strong influence of the stellar dynamics on the accretion onto the central black hole. We then present more realistic simulations of Sgr A* accretion and find that the slow winds shock and rapidly cool, forming cold gas clumps and filaments that coexist with the hot X-ray emitting gas. The accretion rate in this case is highly variable on time-scales of tens to hundreds of years. Such variability can in principle lead to a strongly non-linear response through accretion flow physics not resolved here, making Sgr A* an important energy source for the Galactic centre.

1. Introduction

Sgr A* is identified with the $M_{\rm BH} \sim 3.5 \times 10^6\,{\rm M}_\odot$ super-massive black hole (SMBH) at the centre of our Galaxy [1, 2]. By virtue of its proximity, Sgr A* may play a key role in the understanding of Active Galactic Nuclei (AGN). Indeed, this is the only SMBH where observations detail the origin of the gas in its vicinity. This information is absolutely necessary for the accretion problem to be modelled self-consistently.

One of Sgr A* puzzles is its low luminosity with respect to estimates of the accretion rate. Young massive stars in the inner parsec of the Galaxy emit in total $\sim 10^{-3}\,{\rm M}_\odot\,{\rm yr}^{-1}$, filling this region with hot gas. From *Chandra* observations, one can measure the gas density and temperature around the inner arcsecond[1] and then estimate the Bondi accretion rate [3]. The expected luminosity is a few orders of magnitude higher than the measured $\sim 10^{36}\,{\rm erg\,s}^{-1}$.

The hot gas, however, is continuously created in shocked winds expelled by tens of young massive stars near Sgr A*, and the stars themselves are distributed in two discs [4]. The situation then is far more complex than in the idealised, spherically symmetric and steady state, Bondi model. An alternative approach is to model the gas dynamics of stellar winds, assuming that the properties of the wind sources are known [5, 6, 7]. Unfortunately, in the calculations just cited the stars were kept fixed in space and radiative cooling was not included. Here we present our numerical modelling of wind accretion onto Sgr A*, the first to include optically thin radiative cooling and allow the wind-producing stars to be on Keplerian orbits.

[1] One arcsecond ($1''$) corresponds to $\sim 0.04\,{\rm pc}$, $\sim 10^{17}{\rm cm}$ or $\sim 10^5 R_{\rm S}$ for Sgr A*.

2. Numerical approach and first results

We use the smooth particle hydrodynamics (SPH) code GADGET-2 [8] to simulate the dynamics of stars and gas in the gravitational field of the SMBH. Because of the Lagrangian nature of SPH, we can easily set the stars on orbits around Sgr A*. To model the stellar winds, new gas particles are continously created around the stars. The SMBH is modelled as a 'sink' particle [9, 10], with all the gas passing within a given distance from it ($0.05''$ in the present simulations) disappearing from the computational domain. In addition, since we are not interested in following the evolution of the gas that leaves the inner region, we eliminate all the gas that goes beyond a $20''$ outer boundary. More details of the numerical method along with validation tests have been presented elsewhere [11, 12].

We first performed several simulations with different configurations for the stellar orbits. We found that the angular momentum of the gas has a strong dependence on the orbital motions. Moreover, only the fraction of the gas with the 'right' – low – angular momentum goes to the inner region and can be accreted. Therefore, when the stars are rotating in a disc, the gas has more angular momentum and the accretion rate is lower than in the case where stars orbit the central black hole isotropically [12]. This result highlights the necessity of using realistic orbits to model the accretion onto Sgr A*.

We then introduced in the simulations the luminous blue variable candidates (LBVs), mass-losing stars with outflow velocities of only $v_{\rm w} \approx 300\,{\rm km\,s^{-1}}$, low enough to be susceptible to radiative cooling. When including this process in the calculations, we found that the winds emitted from these stars indeed cool and form clumps and filaments, creating a complex two-phase medium together with the hotter gas coming from Wolf–Rayet stars (WRs). In addition, the existence of clumps adds variability to the accretion rate. Every time a clump is captured by the black hole, a sharp peak is produced [11, 12].

3. Set-up of the new simulations

In our previous calculations (§ 2), we used an ensemble of stars in circular orbits with roughly the same properties as the observed population. For the new simulations presented here, we use the latest available data for individual stars, going a step forward on the creation of more realistic models.

3.1. Stellar wind data

As wind sources in our calculations, we use the 30 stars identified as WRs or LBVs [4]. We do not include the stars IRS 3E and 7SE2, because of the poor constraints on their orbits.

To set the initial conditions of the stellar winds, we use the wind velocities and mass loss rates calculated for the stars IRS 5SW, 7SE, 7W, 15NE, and 16SE2, and for AFNWNW, AF, and AFNW [13]. For the rest of the stars we use the wind velocities derived previously [14] when available, otherwise we set $v_{\rm w} = 1000\,{\rm km\ s^{-1}}$. The mass loss rate for the stars without data is set to a common value such that the total mass loss rate from all the stars is $10^{-3}\,{\rm M_\odot\,yr^{-1}}$. However, as the slow wind stars ($v_{\rm w} < 500\,{\rm km\ s^{-1}}$) are very important for the resulting gas morphology, and their mass loss rates are not well constrained, we experiment in some of the runs changing their mass loss rate to a lower value, while the fast wind stars values are increased accordingly, to keep the total $\dot{M}$ fixed.

In addition to the WR stars, we include as a wind source the object IRS 13E3. This is supposed to be the core of a small stellar cluster harbouring a few mass-losing stars [15, 16]. We count IRS 13E3 as 2 stars in terms of mass loss rate. As we expect some of the kinetic energy from its winds to dissipate 'inside' the cluster, we set its wind velocity to only $v_{\rm w} = 500\,{\rm km\ s^{-1}}$.

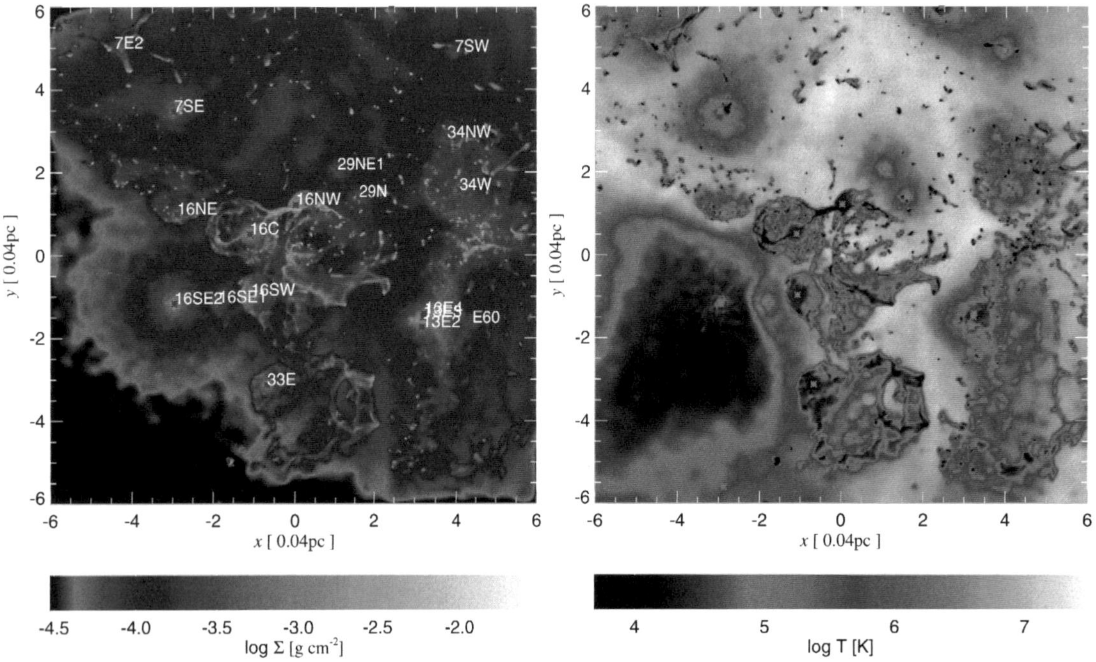

Figure 1. Left panel: Column density of gas in the inner 6″ of the computational domain, as it would be observed from Earth. Stars are shown with green symbols, with labels indicating their identity. Right panel: Averaged temperature of the same region. Notice the dense cold clumps forming around the IRS 16 and 13 groups, where slow winds from different stars collide. Additionally, the winds from the powerful WR star IRS 16E2 that have not collided yet with any other winds remain cold but diffuse.

3.2. Orbital data

We use the published positions and velocities for the mass-losing stars [4]. For the stars without a z coordinate, this is set putting the star in the corresponding disc. For some of the stars (IRS 16NW, 29N, AFNWNW) this would put them in orbits that take them too close to the black hole. To avoid numerical problems, we changed their velocities within the error bars, increasing the pericentre distance for their orbits. It should be noticed that this is a conservative approach when computing the accretion rate, both in terms of quantity and variability.

After setting the current 3d positions and velocities for the stars, we use NEMO's [17] N-body code GYRFALCON [18] to evolve the orbits back in time for $\sim 1200\,\mathrm{yr}$. The positions and velocities at this time were used as initial conditions for the SPH simulations.

4. Two-phase gas

Figure 1 shows the resulting morphology of the gas at the end of one simulation. Cool dense regions in the gas distribution are mainly produced by winds from LBVs. When shocked, these slow winds attain a temperature of only around 10^6 K, and, given the high pressure environment of the inner parsec of the GC, quickly cool radiatively [11]. LBV winds form bound clouds of gas, often flattened into filaments due to the SMBH potential. On the other hand, the WRs by themselves do not produce much structure. The fast winds they emit have temperatures $> 10^7$ K after shocking, and do not cool fast enough to form clumps. This temperature is comparable to

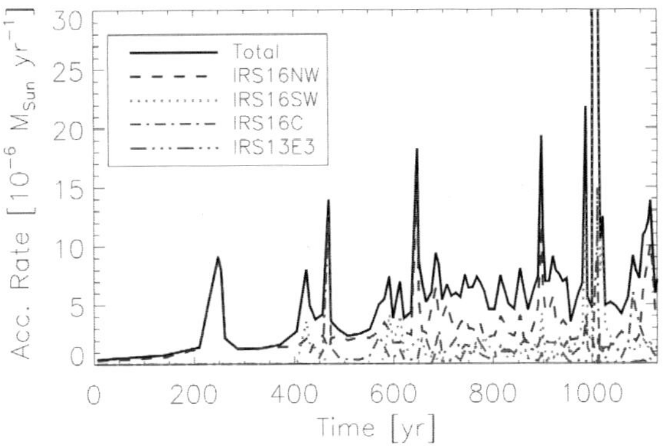

Figure 2. Accretion history from a simulation with the LBVs mass loss rate set to $1.58 \times 10^{-5}\,M_{\odot}\,yr^{-1}$. Different lines show the contribution of material coming from a few selected stars. Notice that most of the accreted matter comes from the stars in the IRS 16 group.

that producing X-ray emission detected by *Chandra*. Gas cooler than that would be invisible in X-rays due to the finite energy window of *Chandra* and the huge obscuration in the Galactic plane.

We find that it is harder to form a disc-like structure like the one found in our previous studies [11, 12]. The main reason for this is that in the present simulations there is not as much cold gas. The LBVs – from whose winds the cold clumps are mostly formed – have mass loss rates in the range 5×10^{-7}–$1.5 \times 10^{-5}\,M_{\odot}\,yr^{-1}$, smaller than the $5 \times 10^{-5}\,M_{\odot}\,yr^{-1}$ used before, while their number went down from 9 to 6. Moreover, while before we put most slow wind stars in the same plane, now we put only two out of the three innermost stars in one plane, so in this case the preference for one plane is lower.

Another reason that prevents the formation of a disc is the presence of the energetic star IRS 16SE2 – its winds blow away any forming disc. Actually, one test where $\dot{M}_{16SE2}$ was decreased by a factor 10 from its measured value appeared to be more favourable for the formation of a disc.

5. Accretion onto Sgr A*

¿From any of our simulations, we can extract the accretion history onto Sgr A*. Accretion is here defined by the quantity of gas entering the inner boundary (0.05″) of our computational domain.

In all the simulations the accretion rate is quite variable, changing by factors $\sim$ 10 on timescales as short as the chosen resolution of 10 yr. The variability is mainly caused by the accretion of clumps originating from material produced by the few innermost slow wind stars in the IRS 16 group (see Fig. 2). But even if the mass loss rates from these stars are reduced by more than a factor 10, a large degree of variability remains (see Fig. 3).

In our previous work [12], we had found that the accretion onto the SMBH could be separated into a quasi-steady hot component and a variable cold one coming from the episodic accretion of clumps. In these new calculations, we find instead that most of the accreted material is actually hot ($T > 10^7$ K), regardless of which star it originates from. This is most likely the effect of eccentric orbits for the stars. Such orbits increase the total velocity of the wind near orbit's pericenter. Hence even winds from narrow line stars become hot when shocked in the inner arcsecond.

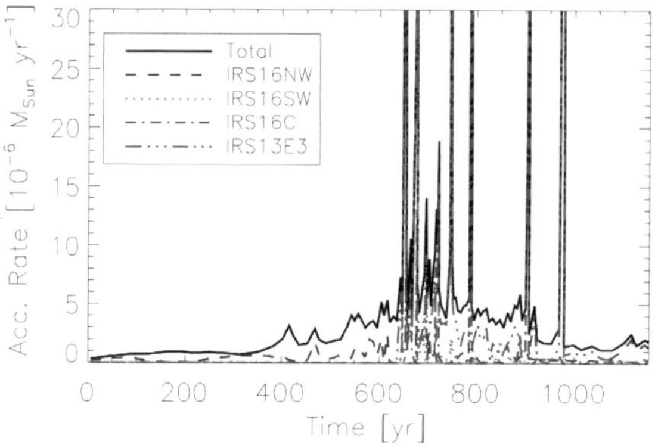

Figure 3. Same as Fig. 2, but with the LBVs mass loss rate set to $10^{-6}\,M_\odot\,yr^{-1}$. Notice that some variability remains in the accretion rate, and that this is caused by both stars in the IRS 16 group and also the IRS 13E3 source.

While we cannot resolve the inner accretion flow to predict the actual accretion rate onto Sgr A*, we estimate that variability by a factor of a few should still reach the black hole. In the extremely sub-Eddington regime of Sgr A*, a small change in the accretion rate produces a non-linear response on the luminosity [19]. The results from our simulations, the observational evidence for higher luminosity in the recent past [20], and the idea of star formation in an AGN-like accretion disc a few million yr ago [21], all suggest that on long time-scales Sgr A* is an important energy source for the inner Galaxy.

6. Line emission

To compare better the outcome of our simulations with actual observations, we create emission maps of atomic lines that are expected from gas at $T \sim 10^4\,$K. Here we concentrate mostly in the inner arcsecond, where the influence of the mini-spiral can be neglected.

In the simulations the minimum temperature is set to $T = 10^4\,$K, otherwise the time-steps could become prohibitively short. It is then questionable whether in reality atomic line emission would arise from the gas in the GC. However, the powerful UV radiation from the stars ensures that most of the gas remains ionised[2].

¿From our simulations we create maps to compare with published images of Paα emission [22]. We calculate the luminosity per unit volume as $4\pi j_{\mathrm{Pa}\alpha} = 6.41 \times 10^{-18}\,\mathrm{erg\,cm^{-3}\,s^{-1}} T_{6000}^{-0.87} n_4^2$. Examples of resulting images are plotted in Fig. 4. Since there is still uncertainty on the stellar orbits, it is not possible to make a direct comparison of these images with observations. However, from the examples it is clear that the total emitted luminosity depends on the winds properties. While the emission on the right panel of Fig. 4 can be easily hidden in the observed maps [22], this might be not the case with the left panel. With this approach, we can constrain the mass loss rate from slow wind stars in the GC in a way complementary to the stellar spectra analysis [13].

7. Conclusions

Here we presented our new numerical simulations of wind accretion onto Sgr A*. Compared with previous works, our methodology includes a treatment of stellar orbital motions and of

[2] For one single star with ionising radiation rate $Q = 10^{48} Q_{48}\,\mathrm{s^{-1}}$, the Strömgren radius is $1.38\,\mathrm{pc}\, Q_{48}^{1/3} n_2^{-2/3}$.

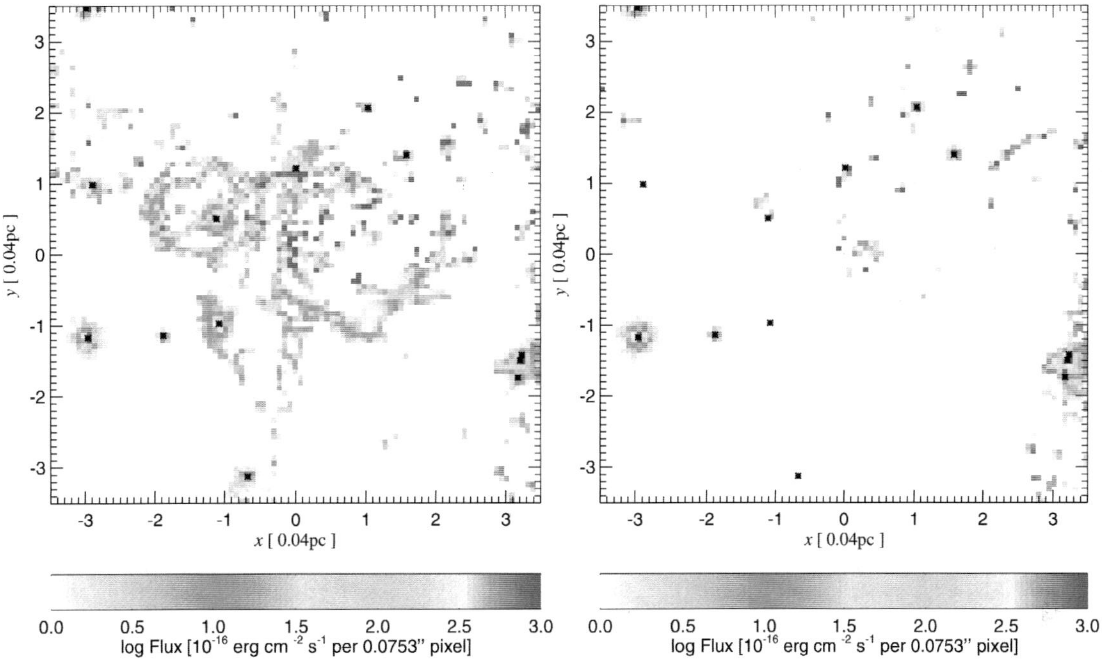

Figure 4. Maps of Paα emission from simulations with the mass loss rate from the LBVs set to either 1.58×10^{-5} (left) or $10^{-6}\,M_\odot\,yr^{-1}$ (right). Notice the strong dependence of the emitted flux on the mass loss rate.

optically thin radiative cooling. While the results depend on the assumptions about stellar mass loss rates, orbits, and wind velocities, some relatively robust conclusions can be made.

Unless mass loss rates of narrow line stars are strongly over-estimated, the gas at $r \sim 1''$ distances from Sgr A* has a two-phase structure, with cold filaments immersed into hot X-ray emitting gas. Both the fast and the slow phase of the winds contribute to the accretion flow onto Sgr A*. The accretion rates we obtain are of the order of a few times $10^{-6} M_\odot$ year^{-1}, consistent with the *Chandra* estimates, although very variable on time-scales as short as tens of years. This implies that the current very low luminosity state of Sgr A* may be the result of a relatively unusual quiescent state. It also means that the real time-averaged output of Sgr A* in terms of radiation and mechanical jet power may be orders of magnitude higher than what is currently observed. The role of Sgr A* for the energy balance of the inner region of the Galaxy may therefore be far more important than its current meager energy output would suggest.

References

[1] A. M. Ghez, M. Morris, E. E. Becklin, A. Tanner, and T. Kremenek. The accelerations of stars orbiting the Milky Way's central black hole. *Nature*, 407:349–351, September 2000.

[2] R. Schödel, T. Ott, R. Genzel, R. Hofmann, M. Lehnert, A. Eckart, N. Mouawad, T. Alexander, M. J. Reid, R. Lenzen, M. Hartung, F. Lacombe, D. Rouan, E. Gendron, G. Rousset, A.-M. Lagrange, W. Brandner, N. Ageorges, C. Lidman, A. F. M. Moorwood, J. Spyromilio, N. Hubin, and K. M. Menten. A star in a 15.2-year orbit around the supermassive black hole at the centre of the Milky Way. *Nature*, 419:694–696, October 2002.

[3] F. K. Baganoff, Y. Maeda, M. Morris, M. W. Bautz, W. N. Brandt, W. Cui, J. P. Doty, E. D. Feigelson, G. P. Garmire, S. H. Pravdo, G. R. Ricker, and L. K. Townsley. Chandra X-Ray Spectroscopic Imaging of Sagittarius A* and the Central Parsec of the Galaxy. *ApJ*, 591:891–915, July 2003.

[4] T. Paumard, R. Genzel, F. Martins, S. Nayakshin, A. M. Beloborodov, Y. Levin, S. Trippe, F. Eisenhauer, T. Ott, S. Gillessen, R. Abuter, J. Cuadra, T. Alexander, and A. Sternberg. The Two Young Star Disks in the Central Parsec of the Galaxy: Properties, Dynamics, and Formation. *ApJ*, 643:1011–1035, June 2006.

[5] R. F. Coker and F. Melia. Hydrodynamical Accretion onto Sagittarius A* from Distributed Point Sources. *ApJL*, 488:L149, October 1997.

[6] G. Rockefeller, C. L. Fryer, F. Melia, and M. S. Warren. Diffuse X-Rays from the Inner 3 Parsecs of the Galaxy. *ApJ*, 604:662–670, April 2004.

[7] E. Quataert. A Dynamical Model for Hot Gas in the Galactic Center. *ApJ*, 613:322–325, September 2004.

[8] V. Springel. The cosmological simulation code GADGET-2. *MNRAS*, 364:1105–1134, December 2005.

[9] M. R. Bate, I. A. Bonnell, and N. M. Price. Modelling accretion in protobinary systems. *MNRAS*, 277:362–376, November 1995.

[10] V. Springel, T. Di Matteo, and L. Hernquist. Modelling feedback from stars and black holes in galaxy mergers. *MNRAS*, 361:776–794, August 2005.

[11] J. Cuadra, S. Nayakshin, V. Springel, and T. Di Matteo. Accretion of cool stellar winds on to Sgr A*: another puzzle of the Galactic Centre? *MNRAS*, 360:L55–L59, June 2005.

[12] J. Cuadra, S. Nayakshin, V. Springel, and T. Di Matteo. Galactic Centre stellar winds and Sgr A* accretion. *MNRAS*, 366:358–372, February 2006.

[13] F. Martins et al. *in preparation.*

[14] T. Paumard, J. P. Maillard, M. Morris, and F. Rigaut. New results on the helium stars in the galactic center using BEAR spectro-imagery. *A&A*, 366:466–480, February 2001.

[15] J. P. Maillard, T. Paumard, S. R. Stolovy, and F. Rigaut. The nature of the Galactic Center source IRS 13 revealed by high spatial resolution in the infrared. *A&A*, 423:155–167, August 2004.

[16] R. Schödel, A. Eckart, C. Iserlohe, R. Genzel, and T. Ott. A Black Hole in the Galactic Center Complex IRS 13E? *ApJL*, 625:L111–L114, June 2005.

[17] P. Teuben. The Stellar Dynamics Toolbox NEMO. In *ASP Conf. Ser. 77: Astronomical Data Analysis Software and Systems IV*, pages 398–+, 1995.

[18] W. Dehnen. A Hierarchical $\mathcal{O}(N)$ Force Calculation Algorithm. *Journal of Computational Physics*, 179:27–42, June 2002.

[19] F. Yuan, S. Markoff, and H. Falcke. A Jet-ADAF model for Sgr A*. *A&A*, 383:854–863, March 2002.

[20] M. G. Revnivtsev, E. M. Churazov, S. Y. Sazonov, R. A. Sunyaev, A. A. Lutovinov, M. R. Gilfanov, A. A. Vikhlinin, P. E. Shtykovsky, and M. N. Pavlinsky. Hard X-ray view of the past activity of Sgr A* in a natural Compton mirror. *A&A*, 425:L49–L52, October 2004.

[21] S. Nayakshin and J. Cuadra. A self-gravitating accretion disk in Sgr A* a few million years ago: Is Sgr A* a failed quasar? *A&A*, 437:437–445, July 2005.

[22] N. Z. Scoville, S. R. Stolovy, M. Rieke, M. Christopher, and F. Yusef-Zadeh. Hubble Space Telescope Paα and 1.9 Micron Imaging of Sagittarius A West. *ApJ*, 594:294–311, September 2003.

Institute of Physics Publishing
Journal of Physics: Conference Series **54** (2006) 443–447
doi:10.1088/1742-6596/54/1/069
Galaxy Center Workshop

A two component hot spot/ring model for the NIR flares of Sagittarius A*

Leonhard Meyer[1], Andreas Eckart[1], Rainer Schödel[1], Wolfgang J. Duschl[2,3], Michal Dovčiak[4] and Vladimir Karas[4]

[1] I.Physikalisches Institut, Universität zu Köln, Zülpicher Str. 77, 50937 Köln, Germany
[2] Institut für Theoretische Physik und Astrophysik, Universität zu Kiel, 24098 Kiel, Germany
[3] Steward Observatory, The University of Arizona, 933 N. Cherry Ave. Tucson, AZ 85721, USA
[4] Astronomical Institute, Academy of Sciences, Boční II, CZ-14131 Prague, Czech Republic

E-mail: `leo@ph1.uni-koeln.de`

Abstract. The supermassive black hole at the Galactic Center, Sgr A*, shows frequent radiation outbursts, so-called flares. In the near-infrared some of these flares were reported to show intrinsic quasi-periodicities of 18 ± 3 min. In 2005, we have carried out polarimetric observations of these QPOs in the K-band. These observations allow for a detailed investigation of Sgr A* within the hot spot model. In this model, inhomogeneities in the accretion flow are represented as confined orbiting material. By simultaneous fitting of the lightcurve fluctuations and the time-variable polarization angle, we address the question whether these changes are consistent with the hot spot model, in which the interplay of relativistic effects plays the major role. We consider all general relativistic effects that imprint on the polarization lightcurves. As the synchrotron mechanism is most likely responsible for the intrinsic polarization, we consider two different magnetic field configurations as approximations to the complex structure of the magnetic field in the accretion flow. Considering the quality of the fit, we think that the spot model is favoured. Finally, our confidence contours give constraints to the spin parameter and the inclination of the supermassive black hole associated with Sgr A*.

1. Introduction

The compact radio, infrared and X-ray source Sagittarius A* (Sgr A*) is the manifestation of a massive ($M = 3.6 \times 10^6 M_\odot$) black hole (MBH) at the center of the Milky Way (e.g. [26], [17], [14]). With Eddington luminosities of $L_{\rm Edd} = 10^{-9} - 10^{-10}$, it is the most extreme sub-Eddington source accessible to observations. Sgr A* shows the feature of so-called 'flares', short bursts of increased radiation that last for about 60-100 min. These flares were first discovered at X-ray wavelengths, where the flux of Sgr A* may rise by factors up to ~100 during such an event (e.g. [2], [25], [11]). At near-infrared (NIR) wavelengths, the flares show very similar timescales, but the flux varies only by factors of ≤ 10, see [16]. Although the exact cause of the flares is still unclear, it is generally accepted that they are caused by synchrotron and synchrotron self-Compton emission processes within $\lessapprox 10$ Schwarzschild radii (R_S) of the black hole [12, 18].

The most intriguing feature related to these flares is the presence of quasi-periodic oscillations (QPOs) with a period of 17-22 min, which have been detected in several of these events [16, 4, 13, 22]; see also Bélanger et al., this volume. The quasi-periodicity manifests itself in the lightcurve as sub-flares superimposed on the underlying main flare. It is not clear yet, how these quasi-periodicities are created. The timescale of the lightcurve variations and the rather small volume around Sgr A*, where these variations originate, indicate that rapid motion in strong gravity is involved. Also, frequencies in microquasar QPOs scale with the inverse mass of the BH, and the frequency of the QPOs in Sgr A* indicates that an extrapolation from the stellar-mass BH case in microquasars to the MBH case is applicable [1, 21, 23].

However, these QPOs of Sgr A* have only been observed unambiguously in a few cases, which has raised some doubts about their nature, in particular whether they may just be related to a red-noise-like process in the source. This aspect is discussed extensively in Meyer et al. [22] for the flare that shows the clearest evidence of quasi-periodicities. They conclude that the ~17 min periodicity reported by Genzel et al. [16] is most probably not due to red noise. The discovery by Eckart et al. [13] that periodicities can show up in NIR polarized light, while at the same time being hardly visible in the total flux, raises the possibility that such periodicities may be present in most flares.

Variations of the degree and angle of polarization can carry valuable information on relatvistic effects (e.g. [8], [5], [10]). Here, we apply the relativistic ray-tracing code by Dovciak et al. [10] to a combined orbiting spot/ring model to fit the recent NIR polarimetric observations by Eckart et al. [13].

2. The model of an orbiting blob

In our model the broad underlying flare is caused by some kind of non-axisymmetric accretion disk instability of a truncated disk/ring, see [22]. The sub-flares are due to a compact, confined emission region on a relativistic orbit around the MBH. Such 'hot spots' can be created in reconnection events and/or at shocks, in which thermal electrons are accelerated into a broken power law [29]. Due to the proximity of the BH horizon, general relativistic effects imprint on the synchrotron radiation of such inhomogeneities [6, 5, 10, 20, 21, 7, 15, 24]. Redshifts, lensing, time delay, change in emission angle and change in polarization angle belong to these. The code by Dovciak et al. [10] takes these special and general relativistic effects into account by using the concept of a transfer function, first utilized in [9]. A transfer function relates the flux as seen by a local observer comoving with an accretion disk to the flux as seen by an observer at infinity. This transfer of photons is numerically computed by integration of the geodesic equation. For the change of polarization angle the method of Connors & Stark [8] has been used.

The disk/ring's time behavior was assumed to be Gaussian and to account for the main flare. The spot is orbiting within the disk and the equatorial plane of a Kerr BH. Its intrinsic luminosity follows roughly a two-dimensional normal distribution with $\sigma \sim 1.5R_S$ ($R_S = 2GM/c^2$), which is jolted in the radial direction so that the azimuthal extent is larger. As the periodicity reported by Eckart et al. [13] is $\sim$ 19min, the spot follows a circular trajectory in the case $a_\star \gtrsim 0.5$. Note that the radius for the corresponding orbit is spin dependent [3] if it is chosen such that this observed periodicity is matched. In the case $a_\star \lesssim 0.5$ the spot is freely falling, i.e. plunging towards the horizon.

To investigate the parameters of Sgr A*, we have fitted for the inclination angle, the dimensionless spin parameter $a_\star$, the brightness excess of the spot with respect to the disk, the initial phase of the spot on the orbit, the orientation of the equatorial plane on the sky and the polarization degree of the disk (constrained to $\leq 15\%$) and the spot ($\leq 70\%$). From these, only the spin parameter and the inclination are needed for the ray-tracing computation. For these two parameters we have chosen a discrete grid with $0 \leq a_\star \leq 1$ and steps of 0.1, and an inclination angle $10^\circ \leq i \leq 70^\circ$ with steps of 5°; $i = 0^\circ$: face-on. For $i > 70^\circ$ multiple images could become important, which are not included in our treatment. The special geometries, where strong lensing – and therefore higher order images – is important, are all characterized by viewing angles very close to edge on.

The observational data include the effects of foreground polarization, so they have to be taken into account in the modeling calculations, too. Following the calibration of Eckart et al. [13], we fix it to be 3.4% at 25°. The geometry of the magnetic field of the spot is completely unknown. As a first order approximation, the fits are done for two different simple field configurations [24]. The first configuration is such that the resulting projected E-Vector is always perpendicular to the equatorial plane. Such preferred orientation could result from perturbations in the disk similar to sunspots (see also

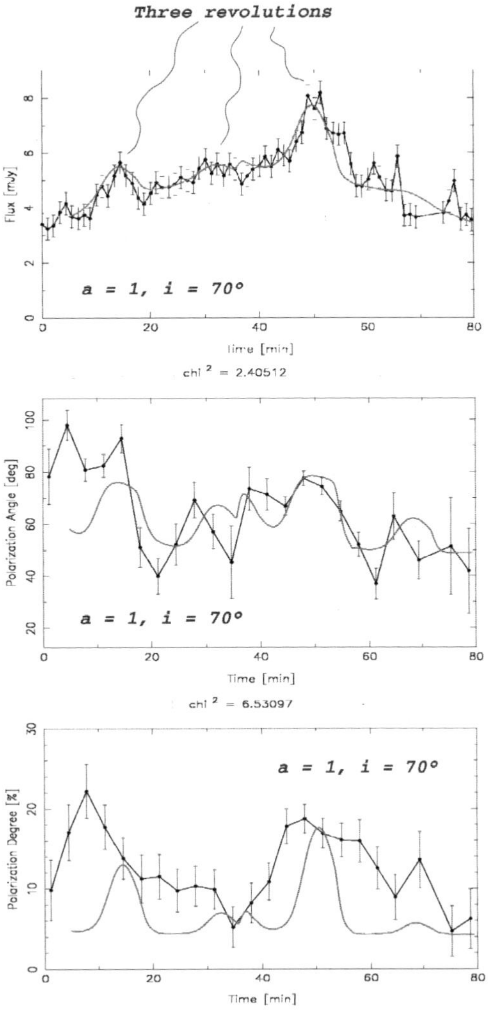

Figure 1. The solution with the least χ^2 values (in red) for the constant E-vector scenario. Shown is the flux (top), polarization angle (middle), and degree of linear polarization (bottom). The parameters of the model are inclination = 70°, $a_\star \approx 1$, an orientation of the projection of the disk on the sky of 0° (east of north) and a polarization degree of the disk (spot) of 9% (60%). The spot is confined during its three revolutions and shearing is not taken into account. As there is no simple physical model for the evolution of the spot within the accretion flow, the changes of its intrinsic lumionsity are incorporated in a discontinuous way.

Shakura & Sunyaev [27]). As a second configuration, we have chosen a global azimuthal magnetic field. This behaviour may be caused by the magneto-rotational instability and is motivated by global MHD simulations (Hirose et al. [19]). In this case, the E-vector would rotate along the orbit, if relativistic effects were switched off.

3. Results

Figure 1 shows the fit with the least reduced-χ^2 value within the discrete grid discussed in the last section. High inclination and high spin give the best solutions. The case of the azimuthal magnetic field leads to slightly higher χ^2 values, i.e. worse fits. There, high inclination and medium spin is preferred. Regarding

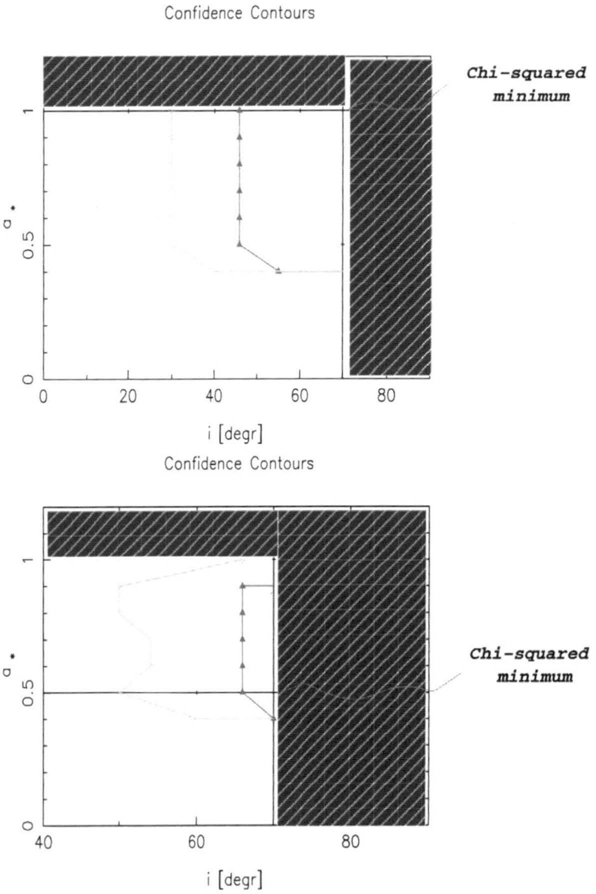

Figure 2. The confidence contours for the constant E-vector (top) and the azimuthal magnetic field case (bottom). The red (green) line is chosen such that the projection onto one of the parameter axes gives the 1σ (3σ) limit for this parameter. The values for the dimensionless spin parameter are constrained to the physical region $0 \leq a_\star \leq 1$. We constrained the inclination to $\leq 70°$, because multiple images are not incorporated in our approach.

the uncertainty of the data and the extremely simple picture of an orbiting spot, the quality of the fits is very good. This favors relativistically orbiting and confined material for NIR flares of Sgr A*.

The confidence contours shown in figure 2 reveal that for both magnetic field cases the spin parameter $a_\star$ cannot be well constrained from the current data due to the relatively large error bars. However, within the 3σ limit, $a_\star$ can be inferred to lie in the range $0.4 \leq a_\star \leq 1$. This is in agreement with previous observations [16, 4]. The range of the inclination is higher for the scenario of the azimuthal magnetic field. Since the χ^2-minimum is lower for the constant E-vector, this case should be given more weight. That means the inclination is $\gtrsim 35°$ on a 3σ level.

Acknowledgments
This work was supported in part by the Deutsche Forschungsgemeinschaft (DFG) via grant SFB 494. We want to thank Avery Broderick for enlightening discussions during this workshop. We are grateful to Thomas Müller, Frank Grave and Andrei Lobanov for their support and advice and especially to Claus Kiefer for the promotion of this project. L.M. is supported by the International Max Planck Research School (IMPRS) for Radio and Infrared Astronomy at the Universities of Bonn and Cologne.

References

[1] Abramowicz, M. A., Kluzniak, W., Stuchlik, Z., Török, G. 2004, in Proceedings of RAGtime 4/5 Workshops on black holes and neutron stars, edited by Hledik, S. ans Stuchlik, Z., Silesian University in Opava, p. 1

[2] Baganoff, F. K., Bautz, M. W., Brandt, W. N., et al. 2001, Nature, 413, 45

[3] Bardeen, J. M., Press, W. H., Teukolsky, S. A. 1972, ApJ, 178, 347

[4] Belanger, G., Terrier, R., De Jager, O., Goldwurm, A., Melia, F. 2006, submitted to ApJL; astro-ph/0604337

[5] Broderick, A. E., Loeb, A. 2006, MNRAS, 367, 905

[6] Broderick, A. E., Loeb, A. 2005, MNRAS, 363, 353

[7] Bromley, B. C., Melia, F., Liu, S. 2001, ApJ, 555L, 83B

[8] Connors, P. A., Stark, R. F., 1977, Nature, 269, 128

[9] Cunningham, C. T., 1975, ApJ, 202, 788

[10] Dovciak, M., Karas, V., Yaqoob, T. 2004, ApJS, 153, 205

[11] Eckart, A., Baganoff, F. K., Morris, M., et al. 2004, A&A, 427, 1

[12] Eckart, A., Baganoff, F. K., Schödel, R., et al. 2006a, A&A, 450, 535

[13] Eckart, A., Schödel, R., Meyer, L., et al. 2006b, A&A, 455, 1

[14] Eisenhauer, F., Genzel, R., Alexander, T., et al. 2005, ApJ, 628, 246

[15] Falcke, H., Melia, F., Agol, E. 2000, ApJ, 528L, 13F

[16] Genzel, R., Schödel, R., Ott, T., et al. 2003, Nature, 425, 934

[17] Ghez, A. M., Salim, S., Hornstein, S. D., et al. 2005a, ApJ, 620, 744

[18] Gillessen, S., Eisenhauer, F., Quataert, E., et al. 2006, ApJ, 640, L163

[19] Hirose, S., Krolik, J. H., DeVilliers, J.-P., Hawley, J. F., 2004, ApJ, 606, 1083

[20] Hollywood, J. M., Melia, F., Close, L. M., McCarthy, D. W. Jr., Dekeyser, T. A. 1995, ApJ, 448L, 21H

[21] Hollywood, J. M., Melia, F. 1997, ApJS, 112, 423

[21] McClintock, J. E., Remillard, R. A. 2004, Chapter 4 in "Compact Stellar X-Ray Sources", eds. Lewin, W. H. G., van der Klis, M., Cambridge University Press.

[22] Meyer, L., Eckart, A., Schödel, R., et al. 2006, A&A, in press

[23] Nowak, M., Lehr, D. 1998, in "Theory of Black Hole Accretion Disks", eds. Abramowicz, M. A., Bjornsson, G., Pringle, J. E., Cambridge University Press, p.233; eprint: astro-ph/9812004

[24] Pineault, S. 1981, ApJ, 246, 612

[25] Porquet, D., Predehl, P., Aschenbach, B., et al. 2003, A&A, 407, L17

[26] Schödel, R., Ott, T., Genzel, R., et al. 2002, Nature, 419, 694

[27] Shakura, N. I., Sunyaev, R. A. 1973, A&A, 24, 337

[28] Yuan, F., Quataert, E., Narayan, R. 2003, ApJ, 598, 301

[29] Yuan, F., Quataert, E., Narayan, R. 2004, ApJ, 606, 894

Institute of Physics Publishing
doi:10.1088/1742-6596/54/1/070

Journal of Physics: Conference Series **54** (2006) 448–455
Galaxy Center Workshop

Testing General Relativity with High-Resolution Imaging of Sgr A*

Avery E. Broderick, Abraham Loeb

Institute for Theory and Computation, Harvard-Smithsonian Center for Astrophysics, Cambridge, MA, 02145, USA

E-mail: abroderick@cfa.harvard.edu, aloeb@cfa.harvard.edu

Abstract. Submilliarcsecond astrometry and imaging of the black hole Sgr A* at the Galactic Center may become possible in the near future at infrared and submillimetre wavelengths. This resolution is sufficient to observe the silhouette the supermassive black hole in the Galactic center casts upon background emission. However, more exciting is the prospect of observing "hot spots" in the accretion flow. Here we discuss how such measurements may be used to test not only the consistency of General Relativity, but also the validity of the Kerr metric in particular.

1. Introduction

Testing strong field gravity remains one of the primary objectives of observational astronomy. Due to their compact nature, black holes provide an ideal environment to do this. Nevertheless, an unambiguous confirmation of strong field relativity has been elusive thus far.

There have been a number of attempts to probe the strong gravity regime, including observations of the relativistically broadened Fe Kα line (see, e.g., [1; 2]), interpretations of quasi-periodic oscillations (QPOs) (see, e.g., [3; 4]), and multiwavelength spectropolarimetric observations (see, e.g., [5–7]). However, the interpretation of each of these are dependent upon unknown accretion physics, making the implications for general relativity ambiguous.

For example, the failure to find an expected correlation between the variability in the Fe Kα line emission and the soft X-ray continuum implies that the simplest emission models for the Fe Kα observations are incomplete (see, e.g., [8–12]), though attempts to rectify this with the inclusion of strong gravitational lensing have been made [13]. In addition, alternative explanations for the formation of the broad iron lines exist (see, e.g., [14; 15]), further complicating their interpretation.

The most commonly discussed black hole QPO's are those observed in the X-ray spectra of stellar-mass black hole candidates. These typically have Q's of 10–100 and are at kilohertz frequencies. Unfortunately, in the absence of a definitive theory for how these are produced, the identification of these with the epicycles of black hole spacetimes is tenuous at best. A second class of QPO's are those observed in supermassive black holes. In the context of Sgr A*, these are observed in the near-infrared (NIR) and X-ray bands, have Q's of 3, (though see [16]), and periods on the order of 20 minutes. If these are interpreted as the Keplerian orbital periods of the innermost stable circular orbit (ISCO), they imply black hole spins as high as 0.5.

More recently, arguments based upon the lack of a thermal peak in the spectra have been used to infer the absence of a surface in stellar-mass black hole candidates (see, e.g., [17]) and Sgr A* [18]. In the latter case, where the putative thermal emission due to the small accretion rate peaks in the near infrared (NIR), this result appears especially robust. However, these arguments present only a qualitative confirmation of the observational consistency of general relativity.

In contrast, it is now technologically feasible to image a black hole directly. A background illuminated black hole will appear in silhouette, with an angular size of roughly twice that of the horizon [19], and may be directly observed. With an expected resolution of $\sim 20\,\mu$as, submillimeter very-long baseline interferometry (VLBI) would be able to image the silhouette cast upon the accretion flow by Sgr A* (with an angular scale of $\sim 50\,\mu$as), and M87 ($\sim 25\mu$as) [20; 21]. In principle, detailed measurements of the size and shape of the silhouette could yield information about the mass and spin of the central black hole. In practice, the interpretation of such an image will likely depend upon the accretion flow model employed (this is discussed in more detail in §2).

Sgr A* has exhibited strong flares in the NIR and X-ray [4; 22–24], and more recently in the submillimeter [25], implying that the innermost portions of the emitting region are strongly variable. A simple model for the flares, motivated by the evidence of periodicity, is that of transient orbiting bright regions, hot spots, which dominate the flaring luminosity. Such hot spots appear inevitable, the product of shocks and magnetic reconnection events within the accretion flow. Due to its dynamical and compact nature, images of such a spot will contain significantly more information about the spacetime. Each imaged spot will allow the measurement of both, the mass and spin of the black hole. Combining observations of many hot spots orbiting at different radii provides a way to test not only the consistency of general relativity, but the validity of the Kerr metric generally, and the no-hair theorems specifically.

In §2 and §3 we discuss the expected images for a radiatively inefficient accretion flow (RIAF) and hot spots, respectively, in the context of the Galactic center. Some concluding remarks are in §4.

2. Silhouettes

For a truly back-lit black hole, strong gravitational lensing produces a silhouette with radius $\sqrt{27}GM/c^2$, roughly twice the size of the horizon of a non-rotating black hole. The size and shape are nearly independent of the black hole spin, though the position of the silhouette relative to the black hole changes substantially (see, e.g., [26]).

However, for a black hole embedded in an accretion flow the silhouette will generally be asymmetric regardless of black hole spin. Even in an optically thin accretion flow asymmetry will result from special-relativistic effects (aberration and Doppler shifting). The degree of asymmetry will be strongly dependent upon the spectral index, vanishing for $\alpha = -2$ (i.e., a thermally emitting, *optically thin* disk). For $\alpha \simeq 0$, the asymmetry can be quite large, as seen in left column of images in Figure 1.

When the opacity of the disk cannot be ignored, as is likely to be the case in Sgr A* at submillimeter wavelengths, this too will alter the shape and position of the silhouette. Indeed, as can be seen in the left two columns of Figure 1, the shape of the photosphere itself is asymmetric, again due to special relativity. Hence, it appears unlikely that it will be possible to completely disentangle the effects of strong gravity from the accretion disk physics. Nonetheless, comparisons of the image centroid positions between the optically thick (where the photosphere is large, and thus unaffected by the spin of the black hole) and thin limits (where the emission is dominated by the inner disk edge, presumably at the ISCO) provides a rough measure of the black hole spin, independent of the details of the accretion flow [21].

A second discriminant between low and high spin may be found in the net polarization. In the

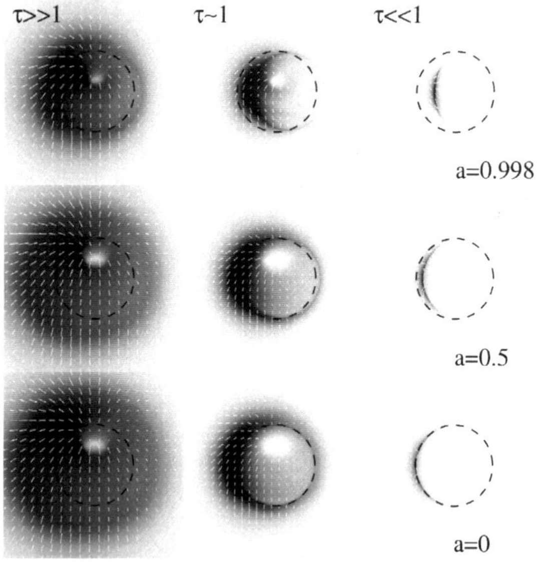

Figure 1. The expected silhouettes for low, moderate, and high spin (bottom to top) and high, unity, and low optical depth (left to right). Each plot is normalized separately, with black being the highest intensity and white corresponding to zero. Overlayed upon the image are white polarization ticks, the magnitudes of which corresponds to size of the polarized intensity. For reference, the silhouette for the non-rotating black hole is shown by the dashed black line. Note that the hole in the center of the optically thick images is due to an evacuated funnel in the accretion flow, *not* the black hole. In each plot

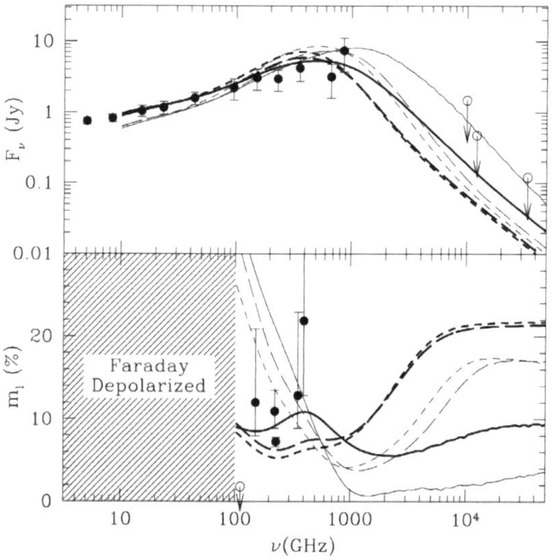

Figure 2. The total and polarization fraction spectra for the non-rotating (short-dash), moderately-rotating (long dash) and maximally rotating (solid) black holes. The thick and thin lines correspond to inclination angles of 45° and 0°, respectively. The data are taken from [27–29].

optically thin limit the emission is restricted to a small arc, and thus the polarimetric properties of the source are dominated by the magnetic field in that region. Conversely, as the optical depth increases the emission is more distributed and thus the net polarization is an average over a wide variety of magnetic field directions (compounded with position angle rotations due to both, the relativistic motion of the disk, and gravitational lensing). As a result, quite generally, the polarization fraction will increase with increasing frequency, plateauing at a value set by the optically thin spectral index and the spin of the black hole. Low asymptotic polarization fractions are then indicative of high spin, as a consequence of frame dragging. This is shown for two magnetic field geometries in Figure 2.

Unfortunately, due to the underlying uncertainty in the accretion model, it will be difficult to make precise measurements of the black hole spin with these observations. Alternatively, these measurements will provide invaluable information on the nature of the accretion flows in low-luminosity active galactic nuclei.

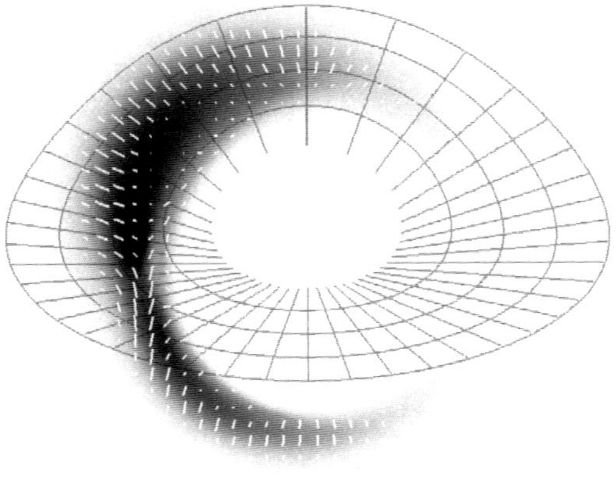

Figure 3. An instantaneous image of a hot spot orbiting a non-rotating black hole at the ISCO. Overlayed upon the intensity are polarization ticks in white. The lensed upper surface of the equatorial plane is shown for reference.

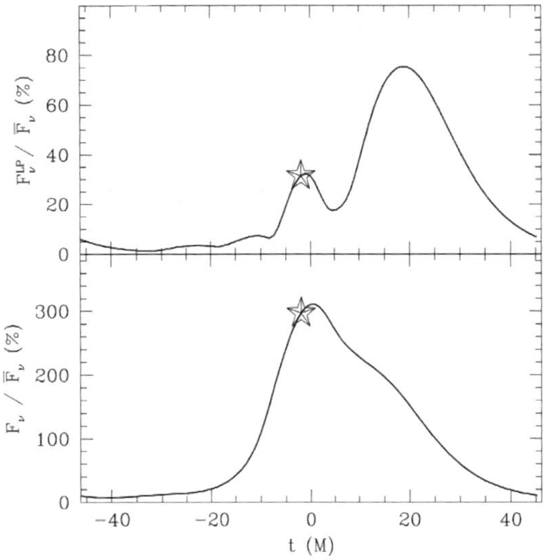

Figure 4. Typical light curves for the total flux, polarized flux and position angle. The star indicates the time at which the image in Figure 3 was produced. Note the features in the light curve that are due to the higher order images.

3. Hot Spots

Observations of Sgr A* show strong variability, with the possibility of periodicity, thus suggesting hot spots in the accretion flow. This is not unexpected theoretically as strong inhomogeneities develop in the innermost regions of general-relativistic magnetohydrodynamic simulations (see, e.g., [30]), although mapping these into inhomogeneities in the emission is non-trivial. This is likely to happen via particle acceleration at strong shocks and/or magnetic reconnection events. Both mechanisms will produce a distribution of compact, non-thermally emitting regions[1].

There are a number of advantages to imaging these "hot spots" as opposed to the underlying quiescent accretion flow. Firstly, due to their compact, and essentially local nature, they are likely to be considerably simpler to model than the global structure of the accretion flow. Hence, the primary difficulty in using the black hole silhouettes to measure the mass and spin is naturally solved in this case. Secondly, encoded in the images of the hot spot is significantly more information. This is because at each instant, the light rays which connect the spot to the observer fill, and thus are sensitive to, a far smaller region of the spacetime. Therefore, as the hot spot orbits the black hole, the observed image is continually sampling new regions of the spacetime. Observing multiple hot spots at different orbital radii completes the sampling of the spacetime.

Figure 3 shows the instantaneous image of a spherically symmetric, synchrotron emitting hot spot (see [7] for details regarding the spot model and the generation of the image). Immediately visible is the fact that two, albeit merging, images of the hot spot are present. Within each image the polarization angles are roughly uniform. However, the polarization angles of the two images are significantly different. The result is that strong lensing is typically associated with

[1] Note that due to the high radiative efficiency of non-thermal electrons in this environment, this is unlikely to have a significant effect upon the accretion flow structure.

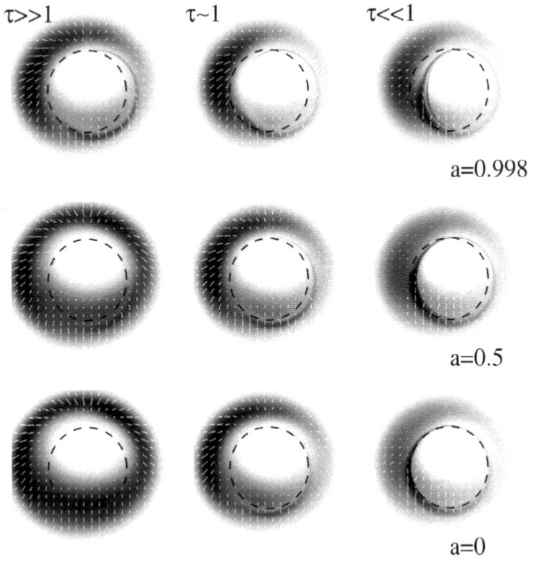

Figure 5. Orbit averaged *background subtracted* images of a hot spot orbiting at a non-rotating, moderately rotating, and maximally rotating black hole (bottom to top) for low, moderate and high optical depth (left to right). Overlayed on the image are polarization ticks in white and the size of the silhouette of a non-rotating black hole (the black dashed circle).

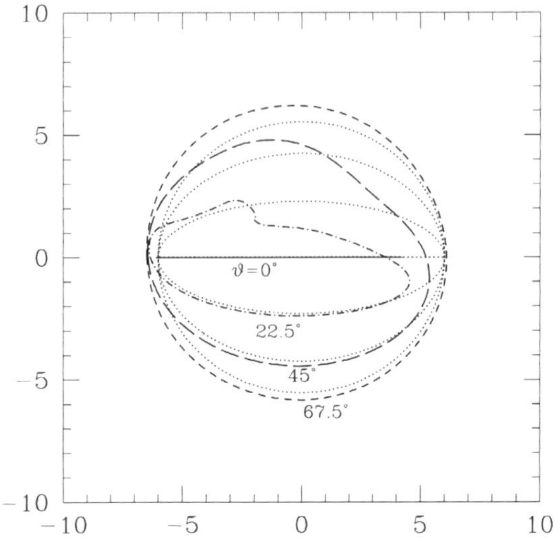

Figure 6. The path of the image centroid for a hot spot orbiting a non-rotating black hole at the ISCO for various inclinations, ϑ. For references, the projected Newtonian orbits are shown by the dotted lines.

a decrease in the polarized flux. This is indeed apparent in the light curve shown in Figure 6. Thus, the polarized flux is sensitive to the presence of higher order images. Since the light rays which produce these images are more strongly lensed than those that produce the primary, and consequently spend more time in the strongly gravitating region, they are more sensitive to the spacetime. Therefore, as mentioned earlier, the polarization evolution is diagnostic of the black hole parameters.

Images of the hot spots are likely to be averaged over many orbits. Figure 5 shows *background subtracted* orbit-averaged hot-spot images for the same cases presented in Figure 1. As with the accretion disk, the hot-spot images are generally asymmetric, becoming more so as the optical depth decreases. However, in this case the images form sets of rings, one for each hot-spot image. As spin increases, these rings are displaced, with those formed by higher order images naturally being displaced further. Therefore, if the hot-spot parameters (orbital parameters, size, density and spectral index) are known, then the spin of the black hole may be determined by measuring the *relative* shift between the primary and secondary images. Since this involves relative astrometry, this is an ideal problem for submillimeter VLBI.

In addition to the submillimeter imaging, the proposed GRAVITY instrument at the very-large telescope interferometer promises to provide $\sim 10\,\mu$as phase-referenced astrometry on few minute timescales in the NIR [31; 32]. Therefore, it will be possible to track the image centroid path taken by a hot spot. A seen in Figure 6, deviations due to strong lensing do occur. These are, in general, functions of the black hole spin. However, the dominant effect is due to the orbital parameters (inclination and radius). Hence, at the least, monitoring the centroid motion provides a method by which to determine the hot-spot orbit.

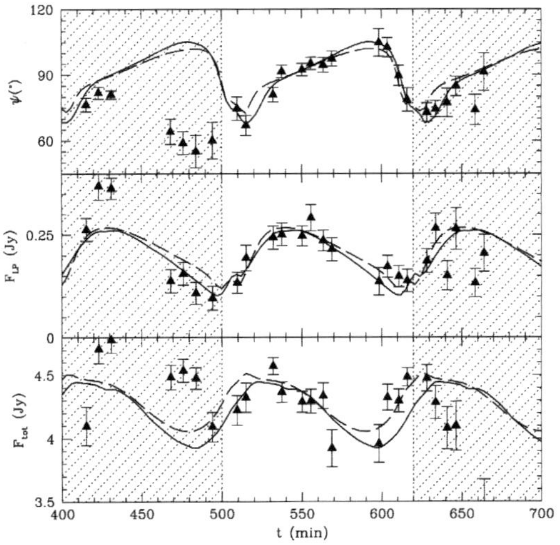

Figure 7. A preliminary fit of a total flux, polarized flux and position angle light curves for a submillimeter flare [25]. The fit was performed in the unhatched region only. For the dashed line the entire region was used. For the solid line only minutes from 500 to 600 were used. In both cases the reduced-χ^2 is of order unity.

Despite its simplicity, the toy hot spot model we have employed here fits actual flare intensity and polarization light curves remarkably well. Two fits to submillimeter observations are shown in Figure 7, differing in the time region over which the fits were attempted. Clearly, the flare is evolving on orbital timescales, with the period decreasing as expected for infall. This was not taken into account in the light curve modeling, and thus the fit was restricted to a single period. Nevertheless, the hot spot model is capable of fitting the total intensity, polarized flux and position angle simultaneously, with a reduce-χ^2 of order unity. While a claim to have measured the flare orbit and/or black hole spin would be premature, the fact that even the simplest model fits implies that such hot spots are indeed simple.

Combining spectopolarimetric observations, imaging, and the centroid paths, it is possible in principle to determine simultaneously the hot-spot orbit and size in physical units, and thus the black hole mass and spin [7; 33]. Because they are compact, the rays connecting the hot spot and an observer at infinity sample only a small portion of the spacetime around the black hole. Thus, the measured mass and spin are actually indicative of the spacetime in a small region. Observations of multiple spots, presumably at different orbital radii, will consequently provide a way in which to measure the mass and spin at a number of distinct points near the black hole. These may then be compared to the general relatavistic prediction that the spacetime is fully described by a mass and spin alone, thus providing a way in which to quantitatively test the Kerr metric.

4. Conclusions

Despite their putative size, imaging the black hole horizons at submillimeter wavelengths is now technologically feasible. In addition, phase-referenced astrometry will be available in the NIR in the coming decade. These capabilities will allow, for the first time, direct observations of gravity in the strongly non-linear regime.

Observations of the size, shape and location of the silhouette cast by the black hole on the surrounding accretion flow may be used in principle to determine its mass and spin. However, in practice this is unlikely to be simple due to uncertainties in accretion physics. Nevertheless, at the very least it will provide a means to learn about accretion onto compact objects.

Of more interest will likely be observations of hot spots. Due to their compact and dynamical nature their images contain more information than those of the underlying quiescent accretion flow. It appears to be possible, even with the simplest models, to constrain the hot spot parameters (e.g., orbit, size and spectral index). If this is indeed the case, then hot-spot observations will provide a method for quantitatively testing general relativity.

Acknowledgments

A.E.B. gratefully acknowledges the support of an ITC Fellowship from Harvard College Observatory. A. L. was supported in part by NASA grants NAG 5-1329 and NNG05GH54G and by the Clark/Cooke fund of Harvard University.

References

[1] Pariev V I, Bromley B C and Miller W A 2001 *ApJ* **547** 649–666

[2] Reynolds C S and Nowak M A 2003 *Phys. Rep.* **377** 389–466

[3] Remillard R A 2005 in L Burderi *et al.*, eds, *AIP Conf. Proc. 797: Interacting Binaries: Accretion, Evolution, and Outcomes* pp 231–240

[4] Genzel R *et al.* 2003 *Nature* **425** 934–937

[5] Connors P A, Stark R F and Piran T 1980 *ApJ* **235** 224–244

[6] Laor A, Netzer H and Piran T 1990 *MNRAS* **242** 560–569

[7] Broderick A E and Loeb A 2006 *MNRAS* **367** 905–916

[8] Wang J X *et al.* 1999 *ApJL* **516** L65–L68

[9] Chiang J *et al.* 2000 *ApJ* **528** 292–305

[10] Lee J C, Fabian A C, Reynolds C S, Brandt W N and Iwasawa K 2000 *MNRAS* **318** 857–874

[11] Wang J, Wang T and Zhou Y 2001 *ApJ* **549** 891–895

[12] Weaver K A, Gelbord J and Yaqoob T 2001 *ApJ* **550** 261–279

[13] Matt G, Fabian A C and Reynolds C S 1997 *MNRAS* **289** 175–184

[14] Elvis M 2000 *ApJ* **545** 63–76

[15] You J H, Liu D B, Chen W P, Chen L and Zhang S N 2003 *ApJ* **599** 164–172

[16] Belanger G *et al.* 2006 (*Preprint* `astro-ph/0604337`)

[17] Garcia M R *et al.* 2001 *ApJL* **553** L47–L50

[18] Broderick A E and Narayan R 2006 *ApJL* **638** L21–L24

[19] Bardeen J M 1973 in C Dewitt and B S Dewitt, eds, *Black Holes* (New York: Gordon and Breach, 1973.)

[20] Falcke H, Melia F and Agol E 2000 *ApJL* **528** L13–L16

[21] Broderick A E and Loeb A 2006 *ApJL* **636** L109–L112

[22] Ghez A M, Wright S A, Matthews K, Thompson D, Le Mignant D, Tanner A, Hornstein S D, Morris M, Becklin E E and Soifer B T 2004 *ApJL* **601** L159–L162

[23] Eckart A, Baganoff F K, Morris M, Bautz M W, Brandt W N, Garmire G P, Genzel R, Ott T, Ricker G R, Straubmeier C, Viehmann T, Schödel R, Bower G C and Goldston J E 2004 *A&A* **427** 1–11

[24] Baganoff F K, Bautz M W, Brandt W N, Chartas G, Feigelson E D, Garmire G P, Maeda Y, Morris M, Ricker G R, Townsley L K and Walter F 2001 *Nature* **413** 45–48

[25] Marrone D 2006 in R Schčhodel *et al.*, eds, *From the Center of the Milky Way to Nearby Low-Luminosity Galactic Nuclei* (Institute of Physics Publishing)

[26] Takahashi R 2004 *ApJ* **611** 996–1004

[27] Aitken D K *et al.* 2000 *ApJL* **534** L173–LL176

[28] Bower G C *et al.* 2002 *ApJ* **571** 843–855

[29] Yuan F, Quataert E and Narayan R 2004 *ApJ* **606** 894–899

[30] De Villiers J, Hawley J F and Krolik J H 2003 *ApJ* **599** 1238–1253

[31] Bougeret J L *et al.* 2006 Phase-A proposal for GRAVITY: An infrared wavefront sensor assisted beam combiner instrument for narrow angle astrometry and phase referenced faint imaging at the VLTI

[32] Eisenhauer F *et al.* 2005 *Astronomische Nachrichten* **326** 561–562

[33] Broderick A E and Loeb A 2005 *MNRAS* **363** 353–362

Institute of Physics Publishing
doi:10.1088/1742-6596/54/1/071

Growth of black holes and dark matter accretion

Faustin Munyaneza[1,2] **and Peter L. Biermann**[1,3,4]

[1] Max-Planck-Institut für Radioastronomie, Auf Dem Hügel 69, D53121 Bonn, Germany
[3]Department of Physics and Astronomy, University of Bonn, Germany
[4] Department of Physics and Astronomy, University of Alabama, Tuscaloosa, USA

E-mail: `munyanez@mpifr-bonn.mpg.de`

Abstract. We investigate the distribution of fermion dark matter in the Milky Way galaxy and find that dark matter could gravitationally condensate in a degenerate core of mass of $3 \times 10^6 M_\odot$ embedded in a dark matter halo of $3 \times 10^{12} M_\odot$ with a size of about 200 kpc. We then show that the galactic black hole of mass of about $3 \times 10^6 M_\odot$ might have grown from a stellar seed black hole by mainly accreting dark matter from the compact degenerate fermion core. This leads to a lower limit on the mass of the fermion dark matter of about (6 − 10) keV. It is then argued that the constrained dark matter could be a sterile neutrino.

1. Introduction

There is mounting evidence that most galaxies harbor supermassive black holes (BHs) of masses from $10^{6.5}$ to $10^{9.5} M_\odot$. The typical case is of the Galactic black hole of mass $(3.1 \pm 0.9) \times 10^6 M_\odot$ [1, 2], which is associated with the radio source Sagittarius A*. It has also been established that the mass of the central black hole is tightly correlated with the velocity dispersion σ of its host bulge, where it is found that $M_{BH} \sim \sigma^{4-5}$ [3]. In spite of the vast and tantalizing work on black hole physics, their genesis and evolution are not well understood. Moreover, the particle nature of dark matter (DM) remains a mystery. Two scenarios have been discussed in modeling the growth of BHs. One is that BHs grow out of a low mass 'seed' BH through accretion of dark matter and baryonic matter [4], and another one is that BHs grow by merging [5]. In this work, we explore the first scenario and consider a seed BH in the mass range from 5 to 10 $M_\odot$ which might have evolved in BH binaries [6]. We will then constrain the mass of the dark matter particle by establishing the relationship between the growth of the black hole in galactic centers and the dark matter accretion.

2. Main equations

We will assume that the dark matter particles are fermions that are described by the Fermi-Diracs distribution

$$f(E) = \begin{cases} \dfrac{g_f}{8\pi^3\hbar^3} \dfrac{1}{\exp\left(\frac{E-\mu}{kT}\right) + 1} + C \ , & E < E_c \\ 0 & , \ E > E_c \ , \end{cases} \tag{1}$$

[2] Humboldt Fellow

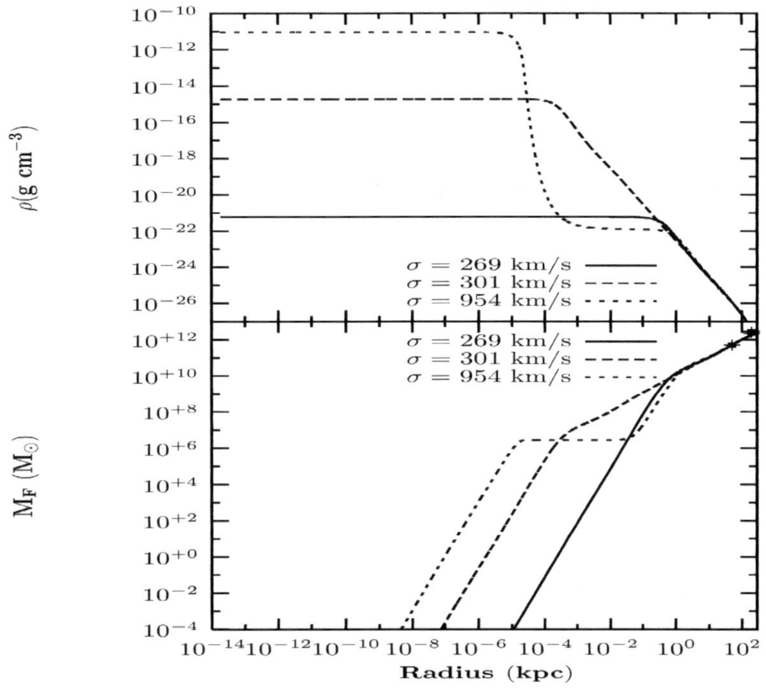

Figure 1. The mass density ρ and the mass enclosed M_F are plotted as a function of the radius r in the upper and lower panels, respectively. The velocity dispersion σ is varied as shown on the plot. In the outer edge of the DM halo, the density ρ scales as $1/r^2$. The mass of fermions used in this plot is $m_f c^2 \approx 12$keV and the size of the fermion ball is about 10^{-2}pc. The data points for the mass within 50 kpc and 200 kpc are taken from [8].

where E is the DM particle total energy, μ is the chemical potential and T is the temperature. The constant C is chosen so that the distribution function vanishes at the cutoff energy $E = E_c$[7]. The mass density $\rho(r)$ and the gravitational potential $\Phi(r)$ are given by

$$\rho(r) = m_f \int_0^{E_c} 4\pi f(p) p^2 dp(E), \quad \Delta\Phi(r) = 4\pi G\rho(r) \ . \tag{2}$$

We then integrate equation (2) to get the density and the total mass of DM [7] as shown in Fig. 1, for solutions with a $3 \times 10^{12} M_\odot$ mass and a radius of 200 kpc. The assumption of a degenerate fermion ball of $3 \times 10^6 M_\odot$ at the center of a DM halo of $3 \times 10^{12} M_\odot$ with a density scaling as $1/r^2$ in the outer edge of the halo constrains the fermion mass to

$$m_f c^2 \gtrsim 12 \text{ keV}. \tag{3}$$

Near the center of the DM halo, fermions become degenerate and can form gravitationally stable objects called fermion balls. The physics of fermion balls has been studied in a series of papers [7, 9, 10, 11, 12, 13, 14]. The mass and radius of non relativistic degenerate fermi balls scale as

$$M_F R_F^3 = 27.836 M_\odot \left(\frac{15\text{keV}}{m_f c^2}\right)^8 \left(\frac{2}{g_f}\right)^2 \text{pc}^3. \tag{4}$$

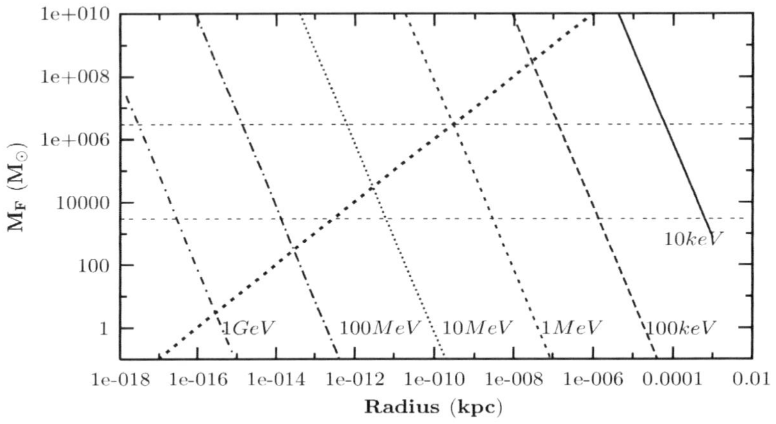

Figure 2. The mass - radius relation for degenerate fermion balls. The total mass scales as $M_F \sim R_F^{-3} m_f^{-8}$. In the same plot, we have shown the BH line and two horizontal lines for the lower and upper limits for the total mass of the degenerate Fermi balls.

The relativistic fermi ball can have a maximum mass, also called the Oppenheimer-Volkoff limit $M_{OV} \sim m_{pl}^3 m_f^{-2}$, where $m_{pl} = (\hbar c/G)^{1/2}$ is the Planck's mass. One could use the virial theorem to express the rotational velocity $v_{rot}^2 = \alpha G M_F / R_F$, where α is chosen so that for $M_F = 3\,10^6 M_\odot$, $v_{rot} = 220$ km/s, we get $m_f = 12$ keV/c^2 and this leads to the following relation

$$m_f \approx 12\text{keV/c}^2 \left(\frac{v_{rot}}{220\text{km/s}}\right)^{3/4} \left(\frac{M_{bh}}{3\,10^6 M_\odot}\right)^{-1/2} \left(\frac{2}{g_f}\right)^{1/4}, \tag{5}$$

where M_F has been replaced by the BH mass M_{bh} as it assumed here that the black hole will eat the whole fermion ball. Using the relation between the velocity dispersion and the rotational velocity for an isothermal sphere, i.e. $v_{rot} = \sqrt{2}\sigma$ and using $\sigma = 100 \pm 20$ km/s [3] and the mass of the black hole of $M_{BH} \sim (2.2-4)10^6 M_\odot$ [1], then we obtain the lower limit on the mass of the DM particle

$$m_f \gtrsim (6-10)\ \text{keV/c}^2. \tag{6}$$

The above limits lie within the range obtained by Kusenko [15]from the study of the origin of the high velocities up to 1000 km/s of pulsars. Moreover, similar limits were also obtained from X-ray background studies [16]. Recently, Biermann & Kusenko [17] showed that the decay of such a sterile neutrino could help initiate star formation in the early Universe.

In Fig. 2 we plot the mass - radius relation for fermion balls for different values of the fermion mass m_f.

3. Black hole growth

The black hole growth law for Eddington baryonic matter accretion is given by

$$M_{BH}(t) = M_{BH}^i \exp\left(\frac{t-t^i}{\tau_{EDD}}\right), \tag{7}$$

where $\tau_{EDD} = 10^{7.69}$years $\times \left(\frac{\epsilon}{0.1}\right)$ is the relevant timescale and ϵ being the efficiency.

In Fig. 3 we illustrate the growth of a seed black hole from Eddington baryonic matter accretion. The mass of the seed black hole is varied from 5 to $10^4 M_\odot$ and the seed formation

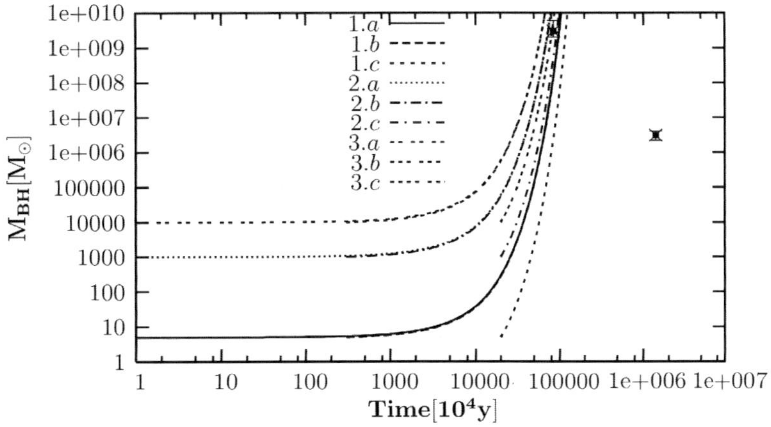

Figure 3. The growth of the black hole mass from Eddington baryonic matter accretion. The curves 1 to 3 have different seed BHs masses of 5, 10^3 and $10^4 M_\odot$. The accretion starting time t^i is also varied.

time is changed from $t = 0$ to $t = 2. \times 10^8$ years. ¿From this plot, we learn that in order to fit the data point of mass $M \sim 3 \times 10^9 M_\odot$, the mass of seed black hole should be greater than $10^3 M_\odot$ and the accretion onto the black hole cannot start later than $z \sim 17$. In order to grow a stellar seed black hole to $10^{3-4} M_\odot$, we use DM accretion to boost its initial growth. In order to study the accretion of DM onto the BH[7], we will assume that dark matter is degenerate near the center of the DM halo and we will then assume a spherical accretion onto the BH. Under the above assumptions, the density of dark matter near the BH scales as

$$n = \frac{g_f}{6\pi^2} \frac{m_f^3}{\hbar^3} \left(\frac{2GM_{BH}}{r}\right)^{3/2} . \tag{8}$$

The BH accretion rate is given by

$$\dot{M}_{BH} = 1.03 \times 10^{-7} g_f \left(\frac{m_f c^2}{15 \text{ keV}}\right)^4 \left(\frac{M_{BH}}{M_\odot}\right)^2 M_\odot \text{yr}^{-1}, \tag{9}$$

which is integrated to yield the following growth law

$$M_{BH} = M_\odot \left[\frac{M_\odot}{M_{BH}^i} - \frac{1}{t_0}\left(t - t^i\right)\right]^{-1}, \tag{10}$$

where M_{BH}^i is the mass of the seed BH at the starting time t^i of accretion, and $t_0 = 9.7 \times 10^6 \text{yr} \left(\frac{15\text{keV}}{m_f c^2}\right)^4 g_f^{-1}$ is the corresponding timescale. In order to fit the Galactic center black hole of $M \sim 3 \times 10^6 M_\odot$ from only DM accretion, the mass of the fermions should be $m_f \approx 1.4$ keV/c^2. Thus, the degeneracy pressure of fermions near the black hole allows to capture a huge amount of DM in a short time. In Fig. 4, we plot the growth of a seed black hole that fully consumes the whole fermion ball of mass of $M \sim 3\ 10^6 M_\odot$ and then grows by Eddington limited baryonic accretion to a supermassive black hole of $M \sim 3 \times 10^9 M_\odot$ at redshift $z \sim 6.41$. We find that DM accretion could be started at $z \sim 10$ which corresponds to the reionisation time from most recent WMAP observations. A fermion mass of $m_f \approx 10$keV/c^2 is used in this plot.

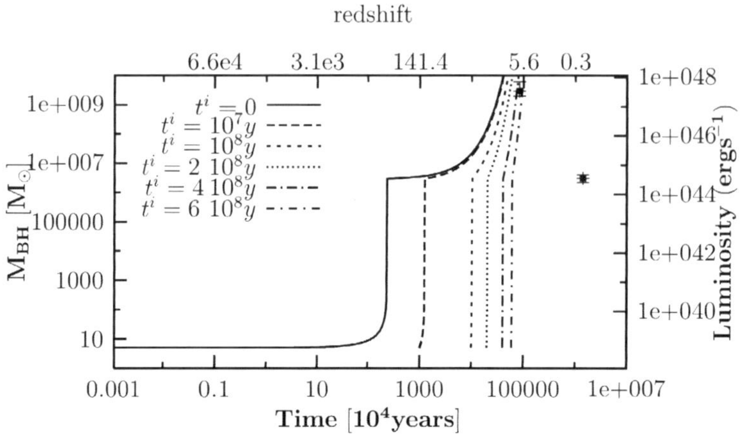

Figure 4. The growth of a seed black mass from DM and Eddington limited baryonic matter accretion.

4. Conclusion

We find that in order to grow the Galactic center black hole of $3 \times 10^6 M_\odot$ and the supermassive black hole in SDSS quasars of $3 \times 10^9 M_\odot$, the DM mass is constrained to a lower limit of about $(6 - 10)$ keV. DM accretion boosts the growth of stellar seed black holes to $10^{3-4} M_\odot$ and a further growth to $10^{6.5-9.5} M_\odot$ is achieved through Eddington baryonic matter accretion. Such DM could be interpreted as a sterile neutrino [15, 16, 17] and it has been shown that its decay could speed up the cooling of the gas and the early star formation [18], which, in turn lead to an early reionisation. A key prediction of this sterile neutrino mass will be the detection of an X-ray line at half the sterile neutrino by XMM-Newton and CHANDRA satellites.

Acknowledgments

Faustin Munyaneza's research is supported by the Alexander von Humboldt Foundation.

References

[1] Schödel R et al. 2003 *ApJ* **596** 1015
[2] Ghez A M et al. 2005 *ApJ* **620** 744
[3] Ferrarese L and Merritt D 2000 *ApJ* **539** L9
[4] Rees M J 1984 *ARA&A* **22** 471
[5] Barkana R , Haiman Z and Ostriker J. P. 2001 *ApJ* **558** 482
[6] Podsiadlowski P , Rappaport S and Hau Z 2003 *MNRAS* **341** 385
[7] Munyaneza F and Biermann P L 2005 *A&A* **436** 805
[8] Wilkinson M I and Evans N W 1999 *MNRAS* **310** 645
[9] Viollier R D 1994 *Prog.Part. Nucl. Phys.* **32** 51
[10] Tsiklauri D & Viollier R D 1998 *ApJ* **500** 591
[11] Munyaneza F & Viollier R D 2002 *ApJ* **564** 274
[12] Munyaneza F, Tsiklauri D and Viollier R D 1998 *ApJ* **509** L105
[13] Munyaneza F, Tsiklauri D and Viollier R D 1999 *ApJ* **526** 744
[14] Bilić N, Munyaneza F and Viollier R D 1999 *Phys. Rev. D* **59** 024003
[15] Abazajian K et al. 2001 *Phys. Rev. D* **64** 023501
[16] Mappelli M and Ferrara A 2005 *MNRAS* **364** 2
[17] Kusenko A 2004 *Int. J. Mod. Phys.* **D13** 2065
[18] Biermann P L and Kusenko A. 2006 *Phys. Rev. Lett.* **96** 091301

Institute of Physics Publishing
Journal of Physics: Conference Series **54** (2006) 461–467
doi:10.1088/1742-6596/54/1/072
Galaxy Center Workshop

News from the year 2006 Galactic Centre workshop

Mark Morris[1] **and Sergei Nayakshin**[2]

[1] Dept. of Physics & Astronomy, Univ. of California, Los Angeles, USA
[2] Dept. of Physics & Astronomy, University of Leicester, LE1 7RH, UK

Abstract. We summarize some of the new results from contributions made to the Galactic Centre workshop that took place in Bad Honnef, Germany, on April 18-22 2006.

1. Introduction

One of the web definitions of "a center" reads: "A place of concentrated activity, influence, or importance" (http://www.answers.com/topic/center). Another adds, "a point of origin from which ideas or influences originate". We believe this conference on the Galactic Center is a great illustration of these definitions for several important fields of research in modern Astrophysics. While it is very hard to summarize all the exciting work presented at this meeting, we shall try to give the reader a glimpse of the state of the art of the field as of April 2006. We shall cite the results by simply the name of the first author on the corresponding proceedings article, except for papers external to the proceedings.

2. The young massive stars in the central parsec

Young massive "He-I" stars dominate the power output of the central half-parsec [6] of our Galaxy. "Standard" models of star formation are not easily applicable here due to a huge tidal field of the central object at $R = 0.1$ pc distances from Sgr A*. The required gas density is $n_H > 10^{11}\mathrm{cm}^{-3}(R/0.1\mathrm{pc})^{-3}$. In the last several years, a significant breakthrough in understanding of these stars has taken place. We now know that ...

- These young stars are not dynamically relaxed (Levin & Beloborodov 2003). They appear to form one (J. Lu, A. Ghez) or arguably two stellar disks (T. Paumard and also Genzel et al. 2003).
- There are inner and outer radii that seem to encompass most of the massive early type stars, $R \sim 0.03$ and $R \sim 0.5$ pc, respectively. (The "S" stars inside of 0.03 pc are discussed separately below.)
- This implies that the stars are very unlikely to have migrated from outside, thus suggesting an *in situ* star formation, despite the tidal shear produced by Sgr A*. The previous debate about whether the massive stars in the central parsec formed *in situ* or whether they are remnant of a tidally disrupted cluster brought into the Galactic center by dynamical friction is not entirely concluded, but the latter hypothesis seems very strongly constrained, requiring a rather heavy burden of extreme assumptions.
- The young stars surrounding Sgr A* are then likely to be the first observational evidence for star formation in massive gaseous disks in the inner parsecs of galactic nuclei (Levin,

Paumard). Recurring *in situ* star formation was suggested to occur in the GC due to a periodic inflow of gas from the Circumnuclear Disk (inner few parsecs) by Morris et al. (1999).

- The inner radius might be indicative of the minimum distance at which star formation can still occur, and is consistent with theoretical predictions (Levin, Nayakshin).
- NIR observations and X-ray constraints suggest that the IMF of young stars is very top-heavy (Paumard, Nayakshin). Such an IMF might be the result of inefficient cooling in the optically thick conditions in the disk (Nayakshin, Levin). The dynamically hot condition of the disk (*i.e.*, likely strong turbulence, as evidenced by the current circumnuclear disk), and the presumably strong magnetic field might also have this effect on the IMF (Morris 1993).
- The exact orbits of the stars in the stellar ring(s) remain a subject of debate. The acceleration limit (J. Lu) places lower limits on the eccentricities of some of the young stars, some of which are rather high. This spells trouble for the simplest version of the *in situ* model, i.e., the model of a gaseous disk in circular rotation about Sgr A*.
- Numerical simulations, however, show that stars may also form via gravitational fragmentation in an eccentric ring of gas (Sanders 1998; Nayakshin). Interestingly, an initially flat eccentric disk of stars becomes geometrically thick much faster than a flat circular disk (Nayakshin et al. 2006) due to orbital precession (Touma). Thus, the fact that the second feature in the stellar velocity distribution of young stars in the GC (the counter-clockwise "disk") is more geometrically thick than the clockwise disk may be naturally explained by the different rates of orbital precession.
- New observations constrain stellar effective temperatures with increasing reliability. There appear to be no stars more massive than about 60 $M_\odot$, which would be consistent with the age of the Sgr A* young star cluster (F. Martins).
- The star cluster IRS13E is a mysterious collection of several very massive young stars and perhaps a couple dozen lighter ones, so it may be the surviving core of a massive cluster which has been almost entirely tidally stripped (T. Paumard). Indeed, its dynamics place it in the counter-rotating stellar disk, so it may provide evidence that that disk has resulted from the disruption of a dense cluster. However, the visible mass of IRS13E is insufficient to bind it gravitationally. An alternative to the rather implausible explanation that the cluster core is only now making its first pass near the central black hole is that it is bound by an intermediate mass black hole. However, the X-ray evidence favors colliding stellar winds over black hole accretion. IRS13E-like objects are in fact not formed naturally in current *in situ* star formation simulations (Nayakshin). Understanding IRS13E is therefore crucial for understanding of the overall star formation event in the GC.
- Other co-moving groups of young stars might be present (R. Schoedel) in the inner parsec.
- IRS16SW appears to be an eclipsing contact binary with two ~ 50 $M_\odot$ stars orbiting with a period of about 18 days (T. Ott, M. Rafelski, F. Martins et al. 2006).

3. S-stars in the central ~ 0.03 parsec

The so-called "S-stars" are the stars observed in the inner arcsecond in projection from Sgr A*, i.e., in the inner region that excludes the He-I stars discussed above. These stars are now spectroscopically resolved to be less massive but nonetheless still quite young B-type stars. The paradox of youth is orders of magnitude more severe for these stars than for the He-I stars. Furthermore, the orientations of the orbits of these stars appear to be consistent with isotropic, so they do not point to any organized formation mechanism such as a disk. A large fraction of the S-stars has relatively eccentric orbits, although this fraction is difficult to determine reliably because of selection effects. There is also a strong selection effect toward studying only the brightest sources, although more stars are emerging all the time, so this situation is improving.

While there were many theoretical suggestions for the origin of these stars, there does not yet appear to be a clear winner among them. Gravitational instability of a massive accretion disk is not predicted to form stars closer in than about $1''$, and hence the in-situ star formation for S-stars is not favored. Some of the new ideas emerged from the meeting on the origin of these stars are:

- Massive perturbers, such as massive molecular clouds or star clusters, can significantly enhance the rate at which massive young binaries diffuse on nearly radial orbits (H. Perets). If these binaries are then disrupted, a population of stars similar to S-stars might be created.
- The population of remnants of disrupted binaries would have very large eccentricities ($e \approx 0.99$; Y. Levin). Ordinary N-body relaxation is too slow to evolve the orbits, which would then be a serious problem for the model, as several of the S-stars have moderated eccentricities ($e < 0.5$). Resonant relaxation (C. Hopman, T. Alexander), might however result in the orbital evolution that is rapid enough to explain those "rouge" S-stars.
- Although this was not discussed at the meeting, we note that the relatively young, ultra-high velocity stars now being found in the galactic halo, whose orbits are consistent with being purely radial, are widely seen to result from binary encounters either with the central black hole or with stellar mass black holes in a central cluster. Such process may be a promising way of producing the S-stars as the former companions of the ultra-high velocity stars (O'Leary & Loeb 2006).
- Spectra of the S-stars are consistent with those of normal B-type stars (F. Martins), possibly ruling out the model in which the S-stars are created by tidal disruption of giants on nearly radial orbits.

4. NIR/X-ray flares from Sgr A*

At the time of the last workshop on the Galactic Center, Sgr A* had not yet been discovered at infrared wavelengths, and the first observations of X-ray flares had only recently been announced (Baganoff et al. 2001). Millimeter-wavelength flares had been reported already at the 1998 Tucson workshop, and continued follow-up work on that was reported here by A. Miyazaki. Since that time, the characteristics of the emission have become clearer, and theories for the spectrum of Sgr A* have been elaborated to account for that (F. Yuan). However, this study is still at a very early stage, both observationally and theoretically, as the frequency, spectrum, location, and cause of the flares all remain widely debated. X-ray flares are well-defined events, rising above the feeble, marginally extended (~ 1.5") quiescent emission almost once per day. The near-infrared intensity variations of more than an order of magnitude appear to be essentially continuous red noise variations, rather than being well-defined "flares". The common use of the term "flare" in the NIR typically refers to a broad intensity maximum in the continuous light curve.

- The locations of NIR flares are offset from the dynamical center of the Sgr A* star cluster by no more than 2 mas $\sim 2 \times 10^{14}$ cm. This implies they do originate from Sgr A* (S. Trippe).
- There is a dusty clump of gas very near the line of sight to Sgr A*, which may influence the mid-infrared fluxes of Sgr A* during flares (Ghez et al. 2005; A. Eckart).
- There is not yet universal agreement about whether the spectral index observed during flares is correlated with the flare luminosity, but much recent effort has gone into measuring the NIR spectrum and its relationship to the X-ray emission properties (S. Hornstein; S.

Gillessen; A. Krabbe; A. Eckart). Hornstein finds no variability of the NIR spectral index of Sgr A* from 1.6 to 4.6 μm.

- Simultaneous observations of Sgr A* at infrared, radio and X-ray wavelengths have been attempted since 2002, and a few events have been captured in multiple bands (A. Eckart; Yusef-Zadeh et al. 2006). The present indications are that X-ray flares are accompanied by NIR maxima, although the statistics on this can only be improved. This strengthens theories in which the same population of electrons is responsible for both the NIR and X-ray events, but probably by virtue of different processes (synchrotron and synchrotron self-Compton, respectively).
- A breakthrough has occurred with the first measurements of the polarization of the NIR emission from Sgr A*and of its variability (A. Eckart). The polarization shows much more time structure than the total intensity, suggesting that it is affected by the varying direction of motion of the nonthermally emitting gas. A similar phenomenon is also evidenced at submillimeter wavelengths (D. Marrone). Polarization variability time scales of $\sim$ 20 minutes are particularly intriguing in view of the occasional periodicities that have been claimed for the total IR (Genzel et al. 2003) and X-ray intensities (G. Belanger), and in view of the coincidence of this time scale with the period of the innermost stable circular orbit around a rotating black hole of 3.6 $\times$ 10^6 $M_\odot$. Theoretical models suggest that polarised emission of a rotating blob near Sgr A* could produce characteristic signatures that would allow observers to constrain the spin of Sgr A*(L. Meyer; A. Broderick).
- A compelling 22-minute periodicity was inferred from the data on one rather long X-ray flare observed with XMM (G. Belanger). The duration of the entire flare was only about 7 times this period, so the analysis of the statistics of this event is an important and delicate issue. The $\sim$15 other flares seen with Chandra and XMM have not yet shown any significant periodicity, so any such modulation of the X-ray flare signal is apparently uncommon.

5. New X-ray results on the GC

- The *Suzaku* telescope, formerly known as *AstroE2*, can now distinguish between rather closely lying X-ray lines. K. Koyama reported *Suzaku* results on the Fe K-*alpha* line at $\sim$ 6.7 keV in the GC diffuse emission. It is found that the centroid energy is 6.680 keV, which favors the thermal collisional origin of that line, rather than its being due to charge exchange with cosmic rays (in which the line energy would be 6.666 keV). The ratio of X-ray line intensities shows that plasma emitting the diffuse emission is in near ionization equilibrium. Also, the emitting medium does not have a large bulk motion with respect to Sgr A*.
- The diffuse X-ray emission presented a puzzle for many years, as the plasma emitting it would be too hot to be confined by the gravity of the GC, and would require an enormous energy source to maintain (R. Belmont). R. Warwick suggests that the diffuse X-ray emission is produced by a population of cataclysmic variables (CVs) rather than by a hot diffuse plasma. This parallels the recent results of Revnivtsev et al. (2006) on the larger scale Galactic ridge X-ray emission, but conflicts with the conclusions of Muno et al. (2004) and Koyama et al. (2006).
- Association of the "neutral" 6.4 keV Fe K-α line with the 4.5 – 6 keV continuum emission argues for origin of most of that emission in local cosmic ray ionization (R. Warwick), rather than photo-ionization.
- Sgr B2 cloud's 6.4 keV Fe K-α X-ray emission, and that in the newly found "X-ray reflection nebula", are however the results of photo-ionization (K. Koyama).
- *Chandra* observations show that low mass stars $M \lesssim 1 - 2\,M_\odot$ seem to be deficient by a factor of up to ten in both the central star cluster (S. Nayakshin) and the Arches star cluster

(D. Wang). The extent to which the apparently top-heavy MF of the Arches cluster is owed to the preferential tidal stripping of its original low mass stars is not yet clear.

6. The central molecular zone and star formation

- Gas in the CMZ is particularly warm, dense, and turbulent, with temperatures ranging up to several hundred degrees K (N. Rodriguez-Fernandez). About 10-30% of the column density of the CMZ can be attributed to a hot (~150 K) photodissociation region. It is not yet possible to decide definitively among the possible heating mechanisms proposed for the rest of the gas, or even whether the right mechanism has yet been proposed. Observations of an unusually high relative abundance of SiO in CMZ clouds imply that shocks affect the chemical abundances there.
- The study of the CMZ is a mature field, with many surveys done in past decades. Therefore, it was remarkable to see in a recent CO survey some altogether new morphological features: very large molecular loops rising far above the Galactic plane (Y. Fukui). The shapes and velocity fields of these features are reasonably interpreted as having been produced by the Parker instability, although it remains unclear why the loop-like magnetic structures resulting from that instability would have a molecular form.
- The Spitzer/IRAC survey has provided a wealth of detail that will be mined for a long time to come (S. Stolovy). Hot dust and PAH emission are ubiquitous throughout the inner few hundred parsecs, pointing the way to sites of star formation activity. The close correspondence between the IRAC images and the radio images of HII regions – notably near the Quintuplet Cluster – illustrate the strong interaction between the luminous, hot stars and gas in the CMZ (A. Cotera).
- Star formation in the CMZ is a currently active process that has been ongoing quasi-continuously since the formation of the Galaxy (Figer 2007). It has been punctuated recently, and very likely at a steady rate in the past, by the formation of massive, compact clusters, three of which have formed within the past 10^7 years. Inevitable tidal dissolution makes it very difficult to identify older clusters than that. One of the important outstanding problems in Galactic center research is to identify the mechanism for the formation of such massive clusters as the Arches and Quintuplet.

7. The Galactic center magnetic field

Two important puzzles that emerge from this workshop are the strength and the dispersion of the magnetic field in the central 100 parsecs (M. Morris). The paradigm suggested some time ago by the presence of numerous, vertically-oriented, nonthermal radio filaments was a pervasive milligauss field having a relatively small dispersion in both strength and direction. The field was hypothesized to be highlighted by synchrotron radiation only at the places where relativistic electrons have been injected. The pervasive milligauss field has been called into question based on recent studies of low-frequency synchrotron radiation (T. LaRosa), and using the minimum energy assumption, which may be inapplicable to this situation. The debate about the field strength hinges also on the relatively short synchrotron lifetime of the emitting electrons in the strong field case, and therefore whether there exists a reacceleration mechanism. Others have argued that the radio filaments represent local concentrations of the field caused by turbulence or by some transitory current structure. However, these proposals encounter difficulties with time scales or geometry. Additional constraints on the magnetic field can perhaps be supplied by the high energy results from INTEGRAL (G. Belanger) and HESS (J. Hinton). What is now needed is a direct measurement of the magnetic field; submillimeter measurements of polarized line and continuum emission may settle this issue before the next workshop. Also, current theoretical efforts to model cosmic rays and their synchrotron emission will soon be informing this debate.

It is an important issue to settle, because a strong magnetic field could play a vital role in many of the phenomena observed in the Galactic center.

8. Other interesting results

- Adaptive optics near-infrared observations of the Arches cluster also suggest that its IMF is top-heavy (A. Stolte, S. Kim). The IMF is mass-segregated as a function of the projected distance from the center of the cluster.
- The prominent mid-infrared source AFGL 5376 appears to have resulted from a strong, ~100-pc shock between two vertically oriented molecular systems. The association of this structure with magnetic radio filaments (J. Staguhn) raises the possibility that the shock is associated with a region of strongly compressed magnetic field.
- The Sgr A East supernova remnant, which encircles the GC in projection, and is probably not very far behind Sgr A* is impacting the dense molecular material surrounding it, giving rise to 1720 MHz OH maser emission (L. Sjouwerman) and to shocked H_2 line emission (S. Lee) at several points around its periphery. The impact upon the 50 km s^{-1} cloud gives rise to a compressed ridge (M. Tsuboi). The energetics and the age of this supernova are still under discussion. While some consider it to be a normal Type II supernova, Others suggest that it may have resulted from a much more energetic hypernova. Similarly, age estimates for the remnant now range between 1700 and 10,000 years.
- The mass of the Circumnuclear Disk (CND) , which orbits Sgr A* and has an inner radius of ~1 pc, has been debated for some time, with estimates ranging from 10^4 to a few $\times$ 10^5 $M_\odot$. With the SMA results of M. Montero-Castano on J = 4-3 HCN emission, it now appears that, while hot gas is characteristic of the CND, the density is not uniformly as high as had previously been estimated (Christopher et al. 2005). Consequently, the CND mass may lie near the lower end of the above range. The CND is a likely reservoir for future accretion activity onto SgrA*, so this result has important implications for the long-term accretion rate. The CND also serves as a test case for the clumpy medium model of T. Beckert for dusty tori in low-luminosity AGNs.
- There appears to be a low extinction channel along the Galactic plane, from NE to SW which points to the minicavity (R. Schödel). If ascribed to an outflow from Sgr A* it agrees with the model of K. Mužić for the proper motions of mid-IR filaments, and with the orientation of the polarization vector found by A. Eckart.
- The GC distance determined using the orbital fits to the proper motions of the S-stars, in concert with measurements of their radial velocities, could be as small as 7.4 kiloparsec (T. Ott), although ongoing efforts are warranted to eliminate astrometric biases occurring near periapse and during close passages by other stars.
- Sophisticated Monte-Carlo modeling of dynamical mass segregation in the GC predicts about 20 thousand stellar-mass black holes in the inner parsec (M. Freitag).
- Horizontal branch/Red clump stars, i.e. stars with $M \lesssim 2\,M_\odot$ and older than 10^9 years, seem to be missing in the inner 9", possibly due to their expulsion from that region by stellar mass black holes (T. Alexander).
- Numerical models of stellar wind accretion onto Sgr A* predict a strong variability of the accretion rate (J. Cuadra) on time scales of tens to hundreds of years. The implications of these results for the structure of the inner radiatively inefficient accretion flow (F. Yuan) are at the moment not clear. Radio evidence gathered over a long term with the VLA indicates that there has been a secular change in the low-frequency turnover frequency (T. An), suggesting a variable mass flux in the stellar winds which feed into the accretion flow. These variations would modulate the opacity of the local absorbing medium.

- Moving toward higher frequencies so that the λ^2 dependence of the size of the elliptical scattering disk is minimized, several VLBI groups are reporting an intrinsic size of for the radio source, as well as its variation with wavelength (G. Bower, Z-Q. Shen, T. Krichbaum). The intrinsic size at 3mm is ~15 Schwarzschild radii, consistent with equating the dominant variability time scales reported at that wavelength with the dynamical time at that radius (Mauerhan et al. 2005). There is still some discrepancy between the suggested wavelength-dependences of the intrinsic source size: $\lambda^{1.1}$ (Shen) to $\lambda^{1.6}$ (Bower).
  There is enthusiasm in this community for the future use of VLBI techniques at submillimeter wavelengths, where the scattering disk will presumably be substantially smaller than the source. In one of the next few meetings, we can perhaps expect to witness images of the shadow of the black hole's event horizon. As A. Broderick shows, the theorists are ready for that.

Acknowledgments

The conference organizers, and especially Rainer Schödel, must be thanked for organizing this workshop, running it amazingly well and problem-free, which allowed the participants to concentrate on the exciting science of the galactic Center.

References

[1] Baganoff, F.K., Bautz, M.W., Brandt, W.N., Chartas, G., Feigelson, E.D., Garmire, G.P., Maeda, Y., Morris, M.R., Ricker, G.R., Townsley, L.K. & Walter, F. 2001, Nature, 413, 45
[2] Christopher, M.H., Scoville, N.Z., Stolovy, S.R. & Yun, M.S. 2005, ApJ 622, 346
[3] Figer, D.F. 2007, in *Massive Stars: From Pop III and GRBs to the Milky Way*, ed: M. Livio, Cambridge Univ. Press
[4] Genzel, R., Schödel, R., Ott, T., Eckart, A., Alexander, T., Lacombe, F., Rouan, D. & Aschenbach, B. 2003, Nature 425, 934
[5] Ghez, A.M. et al. 2005, ApJ 635, 1087
[6] Krabbe, A., et al. 1995, ApJL, 447, L95
[7] Koyama, K. et al. 2006, PASJ, in press, astro-ph/0609215
[8] Levin, Y., & Beloborodov, A. M. 2003, ApJL, 590, L33
[9] Martins, F., Trippe, S., Paumard, T., Ott, T., Genzel, R., Rauw, G., Eisenhauer, F., Gillessen, S., Maness, H., & Abuter, R. 2006, ApJL 649, L103
[10] Mauerhan, J.C., Morris, M., Fabian, W. & Baganoff, F.K. 2005, ApJL 623, L25
[11] Morris, M. 1993, ApJ, 408, 496
[12] Morris, M., Ghez, A.M. & Becklin, E.E. 1999, Adv. Spa. Res. 23, 959
[13] Muno, M.P., Baganoff, F.K., Bautz, M.W., Feigelson, E.D., Garmire, G.P., Morris, M.R., Park, S., Ricker, G.R. & Townsley, L.K. 2004, ApJ 613, 326
[14] Nayakshin, S., Dehnen, W., Cuadra, J., & Genzel, R. 2006, MNRAS, 366, 1410
[15] R.M. O'Leary and A. Loeb 2006, MNRAS, in press.
[16] Revnivtsev, M., Sazonov, S., Gilfanov, M., Churazov, E., & Sunyaev, R. 2006, A&A, 452, 169
[17] Sanders, R. H. 1998, MNRAS, 294, 35
[18] Yusef-Zadeh, F. et al. 2006, ApJ 644, 198

Institute of Physics Publishing
doi:10.1088/1742-6596/54/1/073
Journal of Physics: Conference Series **54** (2006) 468–473
Galaxy Center Workshop

Infall, Fragmentation, and Outflow in Sgr B2

Sheng-Li Qin[1,2], Jun-Hui Zhao[1], James M. Moran[1], Daniel Marrone[1], N. Patel[1], Sheng-Yuan Liu[3], Yi-Jehng Kuan[3,4], Jun-Jie Wang[2]

[1] Harvard-Smithsonian Center for Astrophysics, 60 Garden St, Cambridge, MA 02138, USA
[2] National Astronomical Observatories, CAS, Beijing, 100012, P. R. China
[3] Academia Sinica Institute of Astronomy and Astrophysics, Taipei 106, Taiwan
[4] Department of Earth Sciences, National Taiwan Normal University, Taipei 116, Taiwan

E-mail: `sqin@cfa.harvard.edu`

Abstract. Observations of H_2CO lines and continuum at 1.3 mm towards Sgr B2(N) and Sgr B2(M) cores were carried out with the SMA. We imaged H_2CO line absorption against the continuum cores and the surrounding line emission clumps. The results show that the majority of the dense gas is falling into the major cores where massive stars have been formed. The filaments and clumps of the continuum and gas are detected outside of Sgr B2(N) and Sgr B2(M) cores. Both the spectra and moment analysis show the presence of outflows from Sgr B2(M) core. The H_2CO gas in the red-shifted outflow of Sgr B2(M) appears to be excited by a non-LTE process which might be related to the shocks in the outflow.

1. Introduction

The giant molecular cloud Sgr B2 is located in the Galactic center ($\sim 44'$ from SgrA*). High resolution radio continuum and recombination line observations at centimeter wavelengths showed the two most active star forming cores, Sgr B2(N) and (M), hosting clusters of massive stars associated with UCHII regions [1-6]. Previous observations showed that massive star formation is taking place in both Sgr B2(N) and (M), suggesting that the two hot cores are at different evolutionary stages [7-12]. The H_2CO molecule has a simple chemical formation path and has been proven to be a useful probe of the physical conditions of star formation regions [13,14]. We observed Sgr B2 at 1.3 mm with the SMA including the lines of H_2CO. The new results regarding the ongoing star formation in the Galactic center region are reported in this paper.

2. Observations and Results

Two fields centered on Sgr B2(N) and (M) were observed at 1.3 mm with the SMA[1] in the compact configuration on August 1, 2005. Three H_2CO transitions were included in the 2 GHz sideband, one of them is blended with other molecular lines. The two unblended transitions are H_2CO ($3_{03}-2_{02}$) and ($3_{21}-2_{20}$) with upper level energies (E_u) of 21.0 K and 67.8 K at the rest frequencies 218.222 and 218.760 GHz, respectively. The data reduction including calibration, imaging and analysis was carried out using *Miriad*. The continuum subtraction was made in

[1] The Submillimeter Array is a joint project between the Smithsonian Astrophysical Observatory and the Academia Sinica Institute of Astronomy and Astrophysics and is funded by the Smithsonian Institution and the Academia Sinica.

the UV domain using UVLIN. The mosaic of Sgr B2(N) and Sgr B2(M) was made using a simple linear mosaicing algorithm. The primary beam attenuation was also corrected.

The mosaic of the continuum images of Sgr B2 with an angular resolution of 5$''$.4 × 3$''$.2 is shown in Fig. 1, showing the two bright components Sgr B2(N) (S_p=29.2±0.1 Jy beam^{-1}) and Sgr B2(M) (S_p=20.2±0.1 Jy beam^{-1}). The continuum sources associated with the UCHII regions K1, K2 and K3 in the Sgr B2(N) core are not resolved at this angular resolution. However, a few continuum clumps have been detected, namely, Sgr B2(N)-E (S_p=1.1±0.1 Jy beam^{-1}), Sgr B2(N)-W (S_p=1.0±0.1 Jy beam^{-1}) and K4 (S_p=1.4±0.1 Jy beam^{-1}). We have checked previous observations at longer wavelengths and have found no radio continuum sources to be associated with Sgr B2(N)-E and Sgr B2(N)-W, indicating that the continuum emission at 1.3 mm arises mainly from dust and the two clumps are probably in an early stage of star formation. The strong continuum emission from the unresolved Sgr B2(M) core mainly arises from the UCHII regions F1, F2, F3, and F4. In addition, a continuum component (Sgr B2(M)-W, S_p=1.7±0.1 Jy beam^{-1}) is located west of the Sgr B2(M) core. The position of the continuum peak at 1.3 mm of Sgr B2(M)-W is consistent with that of the source B at 3.8 mm [9] and contains components A1 and A2 at 1.3 cm [2]. Finally, a component, associated with the UCHII region Z10.24, is detected at 1.3 mm (S_p=2.0±0.1 Jy beam^{-1}), revealing a filamentary morphology located between Sgr B2(N) and (M).

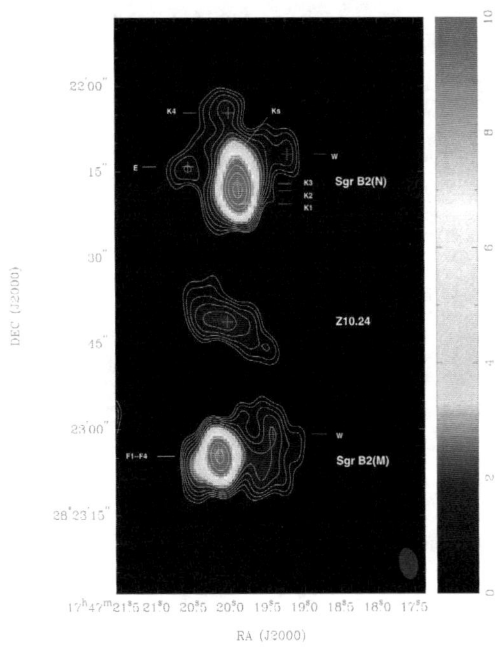

Figure 1. Mosaic image of the Sgr B2 continuum at 1.3 mm, obtained by combining the 218 (LSB) and 228 GHz (USB) data. The synthesized beam is 5$''$.4 × 3$''$.2, PA=12.5° (lower-right corner). The wedge indicates the intensity scale from 0 to 10 Jy beam^{-1}. The contours are -4, 4, 5.66, 8, 11.3, 16, 22.63, 32, 45.25, 64, 90.51, 128, 181.26 σ. The rms (1 σ) is 0.1 Jy beam^{-1}.

The integrated H_2CO line emission flux images are shown in Fig. 2. Most of the emission is distributed around the two cores with peak line emission flux of $F_{3_{03}-2_{02}}$=653 Jy beam^{-1} km s^{-1} and $F_{3_{21}-2_{20}}$=696 Jy beam^{-1} km s^{-1} for Sgr B2(N), and $F_{3_{03}-2_{02}}$ =92.6 Jy beam^{-1} km s^{-1} and $F_{3_{21}-2_{20}}$=94.7 Jy beam^{-1} km s^{-1} for Sgr B2(M). Clearly, the distribution of the H_2CO gas is not spherically symmetric with respect to each of the two star formation cores. There are a few clumps of gas distributed in the vicinity of the Sgr B2(N) region, in addition to the gas concentration on the core. We note that the emission images of the two transitions have a very similar morphology in the Sgr B2(N) region. In Sgr B2(M), the brightest emission of the higher transition gas is located 5$''$ SE of the continuum peak while the peak emission of the lower transition gas coincides with the continuum peak, suggesting that the two transitions in Sgr B2(M) correspond to different physical conditions. Most of the gas traced by the low transition H_2CO line in Sgr B2(M) is located NW of the core, showing filamentary and clumpy structures.

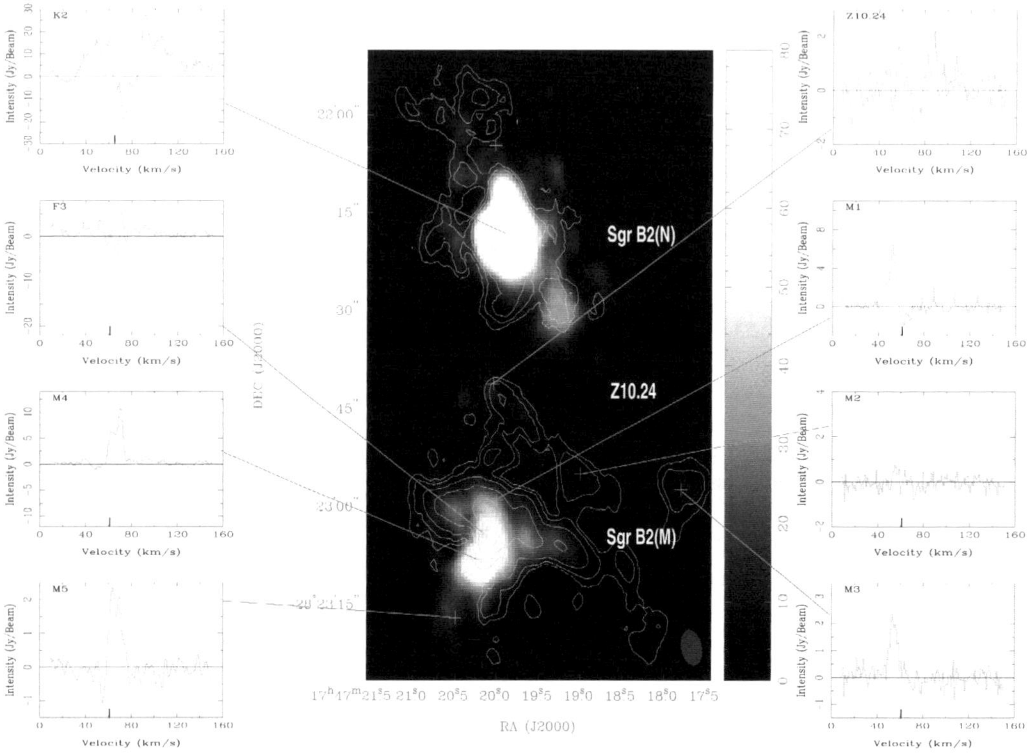

Figure 2. Middle panel is the mosaic images of the integrated intensities of H_2CO ($3_{21}-2_{20}$) (pseudo-color in a scale range from 0 to 80 Jy beam^{-1} km s^{-1}) and H_2CO ($3_{03}-2_{02}$) (contours 4,8,16,32,64,128,256 Jy beam^{-1} km s^{-1}) with FWHM beam of $5''.4 \times 3''.2$, PA=12°.5. The left and right panels are the spectra at the positions of the various components for the low transition $3_{03}-2_{02}$ (green) and the high transition $3_{21}-2_{20}$ (red), respectively. The horizontal line is the zero level after continuum-subtraction, and the vertical thick bars mark the systematic velocity of the sources. The systematic velocities are 65 and 61 km s^{-1} for Sgr B2(N) and Sgr B2(M), respectively.

The typical spectra at several positions are made from channel maps. The spectra towards the two continuum peaks in Sgr B2(N) and (M), show the gas components in either absorption or emission. The absorbing gas associated with the two cores is dominated by gas that is red-shifted with respect to the systemic velocity, suggesting that the gas is moving towards the cores. The spectrum of the higher transition emission at K2 shows three absorbing peaks at velocity 66.0, 73.0 and 80.4 km s^{-1}, respectively, while the absorbing intensities of the lower transition appear to be saturated. The observed line intensity (F_ν), the linewidth (ΔV), the ratios of L/C and the ratio of $F_{3_{03}-2_{02}}/F_{3_{21}-2_{20}}$ give a strong constraint on the physical conditions. Considering the excitation condition and opacity effect of the two transitions, a kinetic temperature of 40-50 K and an H_2 density of 10^7 cm^{-3} are suggested by a large-velocity-gradient (LVG) model for the absorbing gas in the core. The lower transition spectrum at F3 shows two absorbing peaks near 65 and 75 km s^{-1} while the high transition spectrum shows an absorption near 65 km s^{-1} and an emission peak near 75 km s^{-1}. The multiple velocity components of the spectra in both absorption and emission appear to indicate that there are multiple gas clumps located at different depths inside the two cloud cores. High spatial resolution H_2CO line observations will be a critical key to resolve the gas kinematics of the multiple components in the two cores.

The lower transition H_2CO ($3_{03}-2_{02}$) spectrum of Z10.24 shows that most of the gas is seen in absorption with a broad line width (~50 km s^{-1}) and the emission is red-shifted with respect to the absorption, while the spectrum of the higher H_2CO transition shows a double-peaked profile with an absorption dip at 65 km s^{-1} which is consistent with optically thin radio recombination line H66α observations [5]. De Pree et al. (1996) [5] argued that an ionized outflow is centered on Z10.24. The P-cygni profile in the H_2CO ($3_{03} - 2_{02}$) spectrum and the double-peaked profile in the H_2CO ($3_{21} - 2_{20}$) spectrum appear to favor the argument of an outflow from the UCHII region. M1, located within the continuum core, shows an inverse P-cygni profile in the spectra of both H_2CO transitions, providing evidence for the gas moving into the core. M2 located within the filament (30″ in length and 7″ in width) shows significant emission (30 σ in the peak line intensity) in the lower transition while the higher transition line emission is less significant (2-3 σ), suggesting that the gas is cold in this extended arclike structure. Line emission of both transitions is detected in clump M3 ($> 15\ \sigma$).

M4 and M5 reside in the outflow region of Sgr B2(M) [8, 11]. At M4 near the core, the spectra show that the high transition gas is in emission while the low transition gas is in absorption against the continuum. Most of the lower transition gas that is seen in absorption, which appears to be red-shifted with respect to the systemic velocity, provides evidence for an infall of the gas into the core, while the higher transition gas in red-shifted emission suggests an outflow of the gas from the core. M5 is located at the tip of the outflow, where the spectrum shows a very significant line emission (20 σ) of the higher transition gas with a linewidth of $\sim$ 20 km s^{-1}, composed of at least two velocity components while less significant emission (2-3 σ) of the lower transition gas is shown at the same spot, suggesting that the molecular gas does not satisfy the LTE condition. The H_2CO gas in the outflow appears to be excited by a non-LTE process, which might be related to the shocks in the outflow. The shocks could be C-shocks if the gas is weakly ionized and the shock velocity is below the critical value of ~50 km s^{-1} [15]. M5 is far away from the continuum at cm wavelengths and the H_2CO line width is below 20 km s^{-1}. The high transition H_2CO (3_{21}-2_{20}) line in the Sgr B2(M) outflow appears to be excited by a C-shock.

Fig. 3 shows the velocity field of the H_2CO emission. The kinematical structure observed from the lower transition (see Fig. 3(a)) in Sgr B2(M) is relatively simple, consisting of a red spot 5″ SE of the compact core and the NW arch structure. The red spot (red-shifted emission, see also the spectrum M4 in Fig. 2) appears to be a fast moving compact component ($V_{lsr} \sim 100$ km s^{-1}) ejected from the core. The appearance of the NW arch (blue-shifted emission) suggests that the gas is undergoing a complicated infall-process by interacting with the outflow while spiraling into the core rather than simply falling freely. There is a NS velocity gradient in Sgr B2(N) (see also Fig. 3(a)). Based on the HC_3N observations, Lis et al. (1993) [8] argued that the NS velocity gradient in Sgr B2(N) traces the rotation. The kinematics of the lower H_2CO transition in the emission gas of Sgr B2(N) is likely dominated by the gas rotating around the core. The velocity field of the higher transition gas appears to trace the outflow better (see Fig. 3(b)). The red-shifted outflow SE to the core of Sgr B2(M) is clearly present. The kinematics of the NW arch of Sgr B2(M) observed in the high transition gas appears to be similar to that observed in the lower transition gas. The outflow in Sgr B2(N) was observed in SW direction [8]. The SE-NW velocity gradient of the high transition gas in Sg B2(N) could be caused by the combination

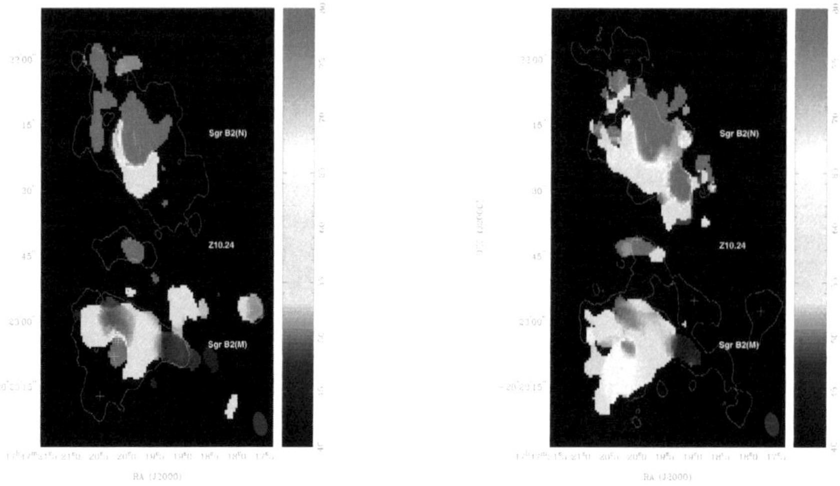

Figure 3. The mosaic images of the intensity weighted velocity, constructed by using a cutoff of 8 σ for the H_2CO $(3_{03} - 2_{02})$ emission (color) overlaid with the contours of H_2CO $(3_{21} - 2_{20})$ intensity in the left panel. The right panel is the velocity field of the H_2CO $(3_{21} - 2_{20})$ emission (color) overlaid with the contours of the H_2CO $(3_{03} - 2_{02})$ intensity. The color wedge scales the velocity range from 40 to 80 km s^{-1}. The contours outline the intensity level of 4 Jy $beam^{-1}$ km s^{-1} for both images.

of the rotation and outflow.

Based on the line ratio of $F_{3_{03}-2_{02}}/F_{3_{21}-2_{20}}$ from the spectra, most of the gas in the Sgr B2 region does not satisfy the LTE condition. Assuming the molecular cloud is in plane-parallel geometry, the molecular excitation in non-LTE has been modelled based on an LVG approximation (e.g. [14],[16]). The non-LTE calculation suggests that the absorbing gas of the compact cores has a kinetic temperature of ~40 K and an H_2 density of $\sim 10^7$ cm^{-3}. A higher kinetic temperature ($\geq$50K) and relatively lower H_2 density ($\sim 10^6$ cm^{-3}) are needed to explain the emission gas both residing in the cores (probably tracing the outflows) and surrounding the cores. The detected larger-scale emission appears to be from a warm envelope, and absorbing gas appears to trace cold dense condensations associated with the massive star forming cores.

3. Conclusions

H_2CO lines and continuum at 1.3 mm were observed with the SMA. The infalling gas was detected through the red-shifted absorbing gas against the continuum in the Sgr B2 compact cores and their associated dusty clumps and the filament Z10.24. The gas clumps in emission were also observed in both H_2CO lines. Based on an LVG model, our analysis suggests that the gas in the Sgr B2 complex consists of cold dense cores and a warm but less dense envelope. The red-shifted outflow in Sgr B2(M) is observed in the H_2CO $(3_{21}\text{-}2_{20})$ line and appears to be excited by a C-shock.

Acknowledgments

We thank all the SMA staff of SAO and ASIAA for making the SMA observations possible.

References

[1] Gaume R A & Glaussen M.J 1990 *ApJ* **351** 538
[2] Gaume R A, Glaussen M J, De Pree C G *et al.* 1995 *ApJ* **449** 663
[3] Mehringer D M, Palmer P, Goss W M 1993 *ApJ* **412** 68

[4] De Pree C G, Gaume R A, Goss W M *et al.* 1995 *ApJ* **451** 284
[5] De Pree C G, Gaume R A, Goss W M *et al.* 1996 *ApJ* **464** 788
[6] De Pree C G, Goss W M, Gaume R A *et al.* 1998 *ApJ* **500** 84
[7] Vogel S N, Genzel R, Palmer P, 1987, *ApJ*, **316** 243
[8] Lis D C, Goldsmith P F, Carlstrom, J E *et al.* 1993 *ApJ* **402** 238
[9] Kuan Y J & Snyder L E 1994 *ApJS* **94** 651
[10] Kuan Y J & Snyder L E 1996 *ApJ* **470** 981
[11] Kuan Y J, Mehringer D M, Snyder L E 1996 *ApJ* **459** 619
[12] Liu S Y & Snyder L E 1998 *ApJ* **523** 683
[13] Mangum J G & Wootten A 1993 *ApJ* **89** 123
[14] Evans N J, Davis J H, Plambeck R L 1979 *ApJ* **227** L25
[15] McKee C F, Chernoff D F and Hollenbach D J 1984, in Proc. 16th ESLAB Symp. Galactic and Extragalactic infrared Spectroscopy, ed. M. Kessler & J.Phillips (Dordrechi:Reidel), 103
[16] Scoville N Z & Solomon P M 1974 *ApJ* **187** L71

Institute of Physics Publishing
doi:10.1088/1742-6596/54/1/074

Journal of Physics: Conference Series **54** (2006) 474–480
Galaxy Center Workshop

Radio Linear and Circular Polarization from M81*

Andreas Brunthaler[1], Geoffrey C. Bower[2], and Heino Falcke[3,4]

[1] Max-Planck-Institut für Radioastronomie, Auf dem Hügel 69, 53121 Bonn, Germany
[2] Radio Astronomy Laboratory, University of California, Berkeley, CA 94720, USA
[3] ASTRON, Postbus 2, 7990 AA Dwingeloo, The Netherlands
[4] Department of Astrophysics, Radboud Universiteit Nijmegen, Postbus 9010, 6500 GL Nijmegen, The Netherlands

E-mail: `brunthal@mpifr-bonn.mpg.de`

Abstract. We present results from archival Very Large Array (VLA) data and new VLA observations to investigate the long term behavior of the circular polarization of M81*, the nuclear radio source in the nearby galaxy M81. We also used the Berkeley-Illinois-Maryland Association (BIMA) array to observe M81* at 86 and 230 GHz. M81* is unpolarized in the linear sense at a frequency as high as 86 GHz and shows variable circular polarization at a frequency as high as 15 GHz. The spectrum of the fractional circular polarization is inverted in most of our observations. The sign of circular polarization is constant over frequency and time. The absence of linear polarization sets a lower limit to the accretion rate of $10^{-7} M_{\odot} y^{-1}$. The polarization properties are strikingly similar to the properties of Sgr A*, the central radio source in the Milky Way. This supports the hypothesis that M81* is a scaled up version of Sgr A*. On the other hand, the broad band total intensity spectrum declines towards milimeter wavelengths which differs from previous observations of M81* and also from Sgr A*.

1. Introduction

The nearby spiral galaxy M81 (NGC 3031) shares many properties with the Milky Way. It is similar in type, size and mass and it also contains a nuclear radio source, M81*, that is most likely associated with a supermassive black hole. M81* has been studied extensively using Very Long Baseline Interferometry (VLBI) in the past. M81* was resolved into a stationary core with a one sided jet [1]. Multi-wavelength [2] and sub-millimeter observations [3] showed many similarities between M81* and Sgr A*, the central radio source in our Milky Way [4]. A jet model of Sgr A* has been applied to M81*, where it can reproduce the radio flux density and the size of the radio core by changing the accretion rate [5]. The sizes of both radio sources show a $\sim 1/\nu$ dependency on the frequency (e.g. [6] for M81*, and [7, 8] for Sgr A*).

M81* is an apparent transitional object between Sgr A* and high luminosity AGN. As the brightest of the nearby low luminosity AGN (LLAGN), it is 5 orders of magnitude brighter than Sgr A* at radio wavelengths. M81* is substantially underluminous at X-ray wavelengths ($L \sim 10^{-5} L_{edd}$), yet not as much as Sgr A* ($L \sim 10^{-10} L_{edd}$). Still, it is the faintest LLAGN we can study.

Furthermore, the polarization properties of M81* and Sgr A* are very similar. Sgr A* shows circular polarization in absence of linear polarization [9, 10, 11] and we detected the same behaviour in M81* [12]. The polarization properties of Sgr A* and M81* are in contrast to the properties of most radio jets in active galactic nuclei where linear polarization often exceeds circular polarization by a large factor (e.g. [13, 14]). The absence of linear polarized emission in Sgr A* can be explained as a consequence of the accretion flow.

The long term behavior of the circular polarization in Sgr A* was investigated from archival VLA data that showed that *i)* the circular polarization is variable on timescales of days to months, *ii)* the sign of the circular polarization stayed constant over the entire time range of almost 20 years, and *iii)* the average spectrum of circular polarization is inverted [15].

After the discovery of circular polarization in M81* we used the VLA to investigate the variability of the circular polarization on short timescales. We used additional archival VLA data to investigate the long term behavior of the circular polarization of M81*. This work is also published in [16].

2. Observations

2.1. Archival VLA Data

M81* and the supernova SN1993J in M81 were observed many times with the Very Long Baseline Array (VLBA) and the phased Very Large Array (VLA) over the last decade (e.g. [17, 1]). The observations were typically 16 hours in duration, were made at different frequencies and involved rapid switching between M81*, SN1993J and scans roughly every hour on the extragalactic background source 0954+658 for calibration purposes. We used the VLA data from the observations on 5 November 1993 (ν = 8.4 and 15 GHz), 16 December 1993 (8.4 and 15 GHz), 29 January 1994 (4.8 and 8.4 GHz), 21 April 1994 (4.8 GHz), 29 August 1994 (4.8 and 8.4 GHz), 31 October 1994 (8.4 GHz), 23 December 1994 (8.4 GHz), and 7 April 1996 (8.4 GHz). The VLA data had 50 MHz of bandwidth in two sidebands in right (RCP) and left (LCP) circular polarization modes.

Data reduction was performed with the Astronomical Image Processing System (AIPS). 3C 48 was used as primary amplitude calibrator. Then amplitude and phase self-calibration was performed on 0954+658. This forces 0954+658 to have zero circular polarization. The amplitude calibration was transfered to M81* and SN1993J before we performed phase self-calibration on M81* and SN1993J. Finally, we mapped all three sources in Stokes I and V. Flux densities were determined by fitting an elliptical Gaussian component to the sources.

2.2. New VLA observations

In addition to the archival VLA data, we used the VLA to observe M81* on 02 March 2001 and 9 dates between 15 June and 30 August 2001. During the latter period, the minimum and maximum separations between observing dates were 2 and 15 days, respectively. The VLA was in B configuration for the first observation and C array for the remaining observations. Observations were made at 5.0, 8.4, 15 and 22 GHz with 50 MHz of bandwidth in two sidebands in RCP and LCP modes.

Each observation was between two and four hours in duration. Observing and analysis was performed with AIPS and followed the procedures outlined in [12]. The extragalactic point source J1044+719 was used as a phase, amplitude and polarization calibrator as well as reference pointing source. The extragalactic point source J1053+704 was used to check for polarization calibration errors. 3C 286 was used as an absolute amplitude calibration source. Finally, we mapped all three sources in Stokes I, U, Q, and V. Flux densities were determined by fitting an elliptical Gaussian component to the sources.

We also used the VLA to observe M81* on 09 August 2003 at 15 GHz, 22 GHz, and 43 GHz in polarimetric mode. The VLA was in A configuration with a resolution of about 50 milliarcseconds at 43 GHz. The total integration time on M81* was 8, 48, and 52 minutes at 15, 22 and 43 GHz respectively and spread over a time range of 11 hours. Phase and amplitude calibration were performed using the nearby compact source, J1056+714. Phase self-calibration was also performed on M81* to eliminate the effects of atmospheric decorrelation. Polarization

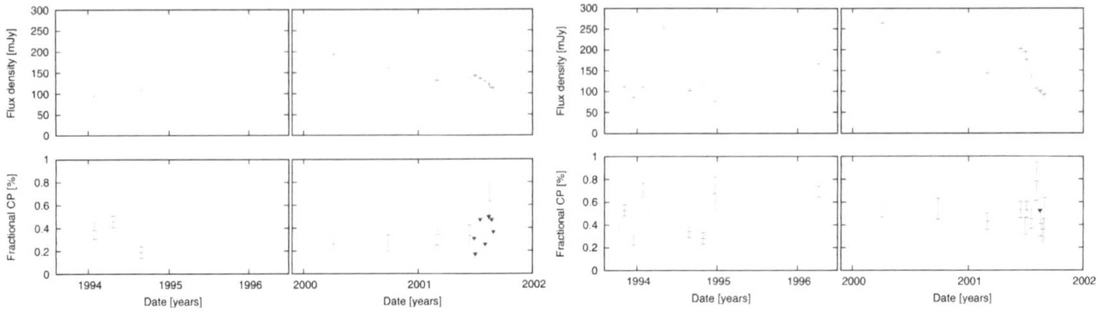

Figure 1. Light-curve in total intensity (top) and fractional circular polarization of M81* at 4.8 (left) and 8.4 GHz (right). The filled triangles are 3σ upper limits. The two data points in 2000 are taken from [12]

leakage calibration was performed using simultaneous full track observations of J1056+714 and J1048+701.

2.3. BIMA observations

We used the Berkeley-Illinois-Maryland Association (BIMA) array to observe M81* at 3 mm wavelength [18]. Observations were made in the multiplexed polarimetric mode described in [11]. The receivers were tuned to sky frequencies of 82.8 GHz (lower sideband) and 86.2 GHz (upper sideband). The compact source 0954+658 was used for phase calibration. Leakage calibration was determined from observations of 3C 279 on 11 November 2003. Observations of M81* were made on 6 dates in September and October 2003. BIMA observations at 230 GHz on M81* were also performed in November 2003. These data were also obtained in polarimetric mode with similar observing parameters to the 3 mm observations. Each observation was an 8 hour track.

2.4. Error Analysis

The Stokes parameter V is measured as the difference between the left- and right-handed parallel polarization correlated visibilities. Errors in circular polarization measurements with the VLA have numerous origins: thermal noise, gain errors, beam squint, second-order leakage corrections, unknown calibrator polarization, background noise and radio frequency interference. The errors caused by amplitude calibration errors, beam squint, and polarization leakage scale with the source strength, and therefore the fractional circular polarization is a more relevant indicator for the detection of circular polarization. A detailed discussion of these errors is given in [9] and [15]. We calculated the systematic errors based on the model for the VLA for circular polarization from [15].

3. Results

3.1. Circular and Linear Polarization

The results for the VLA data at 4.8, 8.4, and 15 GHz are shown in Fig. 1 and 2. We consider circular polarization as detected if the measured flux density exceeds the combined statistical and systematic errors (added in quadrature) by a factor of three. 0954+658 showed no circular polarization as expected. M81* showed circular polarization in all observations except one (16 December 1993 at 15 GHz). SN1993J showed circular polarization only in one epoch (29 August 1994 at 4.8 GHz) and no circular polarization in all other observations. The upper limits on the circular polarization of SN1993J are not very meaningful at 15 GHz due to the low flux density and unexpected high noise. At 4.8. 8.4, and 15 GHz, 1044+719 showed no circular polarization

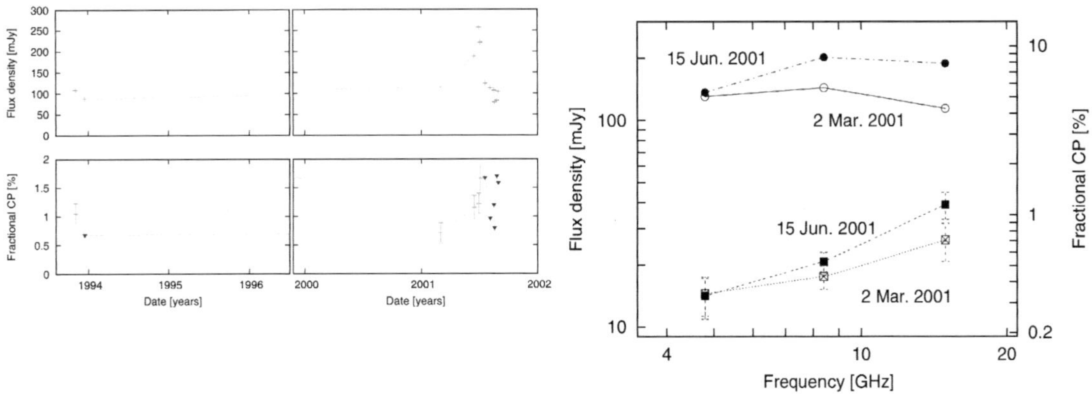

Figure 2. Left: Light-curve in total intensity (top) and fractional circular polarization of M81* at 15 GHz for the archival data (left) and the new observations (right). The filled triangle is a 3σ upper limit. Right: Spectra of the total intensity (circles) and the fractional circular polarization (squares) from 2 March 2001 (open) and 15 June 2001 (filled).

as expected. M81* showed circular polarization in three epochs at 4.8 GHz, all except one epoch at 8.4 GHz and four epochs at 15 GHz. The check source 1053+704 showed only circular polarization in two epochs at 15 GHz. The 22 GHz data and the high frequency observations on 9 August 2003 gave no useful limits on the circular polarization, mainly because of the short integration time and higher systematic errors.

The three *detections* of circular polarization in the check sources SN1993J and 1053+704 are caused by either remaining amplitude calibration errors or by a small level of circular polarization in the sources. Since both sources show no circular polarization at 8.4 GHz, the frequency band with the highest sensitivity, it is most likely that the *detected* circular polarization in SN1993J and 1053+704 comes from residual amplitude calibration errors. The fractional circular polarization in M81* at 15 GHz is higher by a factor of 2 and 4 than in 1053+704 in the observations on 15 June 2001 and 4 July 2001 respectively. In these two cases, the measured circular polarization of M81* is probably only partly caused by the amplitude calibration errors.

In the BIMA observations at 86 and 230 GHz neither linear nor circular polarization is detected for M81* in any individual epoch. Mean linear polarization at 86 GHz is 1.2 mJy, or 1.6% (3σ upper limit). Mean circular polarization at 86 GHz is 2.8 ± 0.4 mJy, or 3.9 ± 0.5%. Although this is formally a detection, it is not clear whether systematic errors are significant. Due to the low flux density, limits on the polarized flux density are not significant at 230 GHz.

The 8.4 GHz data seems to be the most reliable data since M81* showed circular polarization in all epochs except one while the check sources SN1993J and 1053+704 never showed circular polarization. The light-curve of total intensity and fractional circular polarization is shown in Fig. 1. The fractional circular polarization shows significant variability on timescales of a few weeks which is not correlated with the variability in the total intensity. Between 4 August 2001 and 16 August 2001, the fractional circular polarization at 8.4 GHz dropped from 0.78% to less than 0.35% while the total intensity showed no significant change. However it is remarkable that, despite the strong variability, the sign of the circular polarization is always positive. At the other two frequencies, the sign is also positive when circular polarization is detected in M81*.

The spectrum of the fractional circular polarization is inverted ($\alpha > 0$ for $m_c \propto \nu^{\alpha}$) with values between 0.4 and 2 in most epochs when it was detected at more than one frequency. The mean spectral index between 4.8 and 8.4 GHz is 0.51, while the mean spectral index between 8.4 and 15 GHz is 1.52. Only the observation on 18 August 2001 shows a steep spectrum of

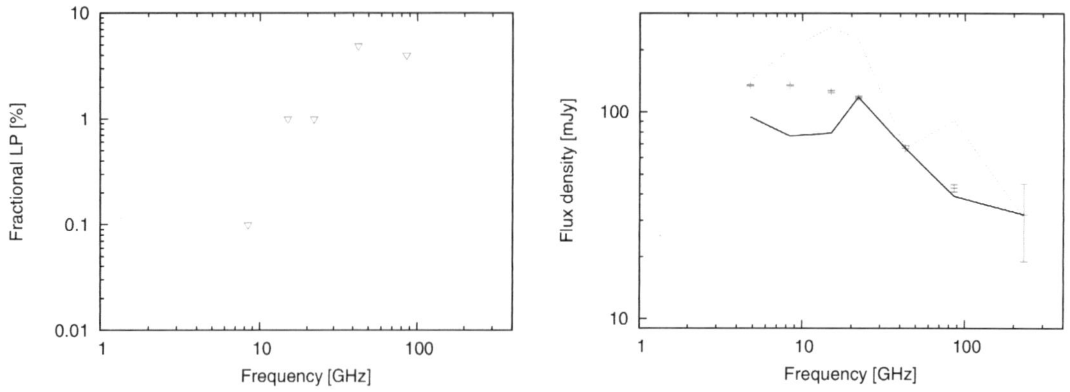

Figure 3. Left: 3σ upper limits on the linear polarization of M81* up to 86 GHz. The 8.4 GHz data point is taken from [37]. Right: Non simultaneous broad-band spectrum of M81*. Also shown are the maximal (dashed line) and minimal (solid line) measured values at each frequency.

α=-0.77 between 4.8 and 8.4 GHz.

Linear polarization was not detected in the VLA observations on 9 August 2003 and the BIMA observations at 86 GHz (Fig. 3). The archival data was not searched for linear polarization.

3.2. Total intensity

The high frequency VLA data taken on 9 August 2003 give a simultaneous total intensity spectrum of M81* that shows the flux density decreasing from 15 to 43 GHz with a spectral index of ~ -0.6. In the BIMA observations M81* is clearly detected at 86 GHz in total intensity and is strongly variable on a time scale $\sim$ 10 days. The mean total intensity is 71 mJy. Due to low sensitivity, M81* is only marginally detected at 230 GHz in total intensity. The mean flux density for M81* from this observation is $31 \pm 4 \pm 15$ mJy, where the first error is the statistical error and the second error is the systematic error due to decorrelation. Atmospheric decorrelation may be serious and these results may significantly underestimate the total flux density of M81*.

Fig. 3 shows a non simultaneous broad-band spectrum of M81*. The 15 – 43 GHz data points are from the VLA observations on 9 August 2003. The 4.8 and 8.4 GHz data points are from the observation on 19 July 2001, where the 15 GHz flux was comparable to the 15 GHz flux in the 9 August 2003 observation. The 86 GHz and 230 GHz data points are from the BIMA observations on 7 September 2003 and November 2003 respectively. Also shown are the maximal and minimal measured values at each frequency in our observations. Although the spectrum is not simultaneous it is clear that the spectrum declines towards higher frequencies.

M81* underwent a flare in total intensity during the new VLA observations between June and August 2001. The peak was reached at 8.4 GHz before 15 June 2001, while the flux density continued to rise at 4.8 until 4 July 2001. At 15 GHz, the peak was reached in 30 June 2001. The spectrum at the lower frequencies shows a smooth transition from an inverted ($\alpha \sim +0.7$) to a steep ($\alpha \sim$ - 0.4) spectrum. At higher frequencies, the spectral index does not follow a trend and is scattered between -0.4 and +0.5. The fast change in spectral index between 4.8 and 8.4 GHz could be caused by a drop in the turnover frequency of a syncrotron self-absorpted jet from above 8.4 to below 4.8 GHz and should be accompanied by a fast expansion of the jet. This behaviour is known in other active galactic nuclei (e.g. III Zw 2: [19, 20]). The scatter of the spectral index between 8.4 and 15 GHz could be caused by multiple sub-flares that occur at

15 GHz.

4. Discussion

The origin of circular polarization in AGN is still not known. Several mechanisms have been proposed in the literature. Interstellar propagation effects predict a very steep spectrum [21] which is not consistent with our observations. One possible mechanism could be Faraday conversion [22, 23] of linear polarization to circular polarization caused by the lowest energy relativistic electrons. [9] proposed a simple model for Sgr A* in which low-energy electrons reduce linear polarization through Faraday de-polarization and convert linear polarization into circular polarization. Faraday conversion can also affect the spectral properties of circular polarization and may lead to a variety of spectral indices, including inverted spectra [23]. In inhomogeneous sources, conversion can produce relatively high fractional circular polarization [24]. Gyro-synchrotron emission, can also lead to high circular polarization with an inverted spectrum and low linear polarization [25]. However, this mechanism is to some degree related and also requires that M81* and Sgr A* both contain a rather large number of low-energy electrons. Faraday conversion is also favored by [26] and [27].

The long-term stability of the sign of circular polarization suggests that the Faraday conversion is connected to fundamental properties of the source and the material which is responsible for the conversion. It requires uniformity in the magnetic pole and accretion conditions over the observed timescales. One possible scenario is described by [28] where the sign of the circular polarization is connected to the sense of rotation of the central engine. In this scenario M81* is expected to rotate counter-clockwise.

The size of the radio emission of M81* at 8.4 GHz is $\sim$ 0.45 mas, or $\sim$ 1800 AU at 4 Mpc [29]. The fact that M81* is depolarized at at level of $< 0.1\%$ at the same frequency requires that the depolarizing material is also present at a scale of $\sim$ 1800 AU.

While the polarization properties of M81* and Sgr A* are strikingly similar, the total intensity spectrum seems to be different. In Sgr A*, the radio flux density rises towards the sub-mm regime (e.g. [30, 31]). Our measurements indicate a different trend for M81*. Although the spectrum in Fig. 3 is not simultaneous and M81* shows strong variability our measured flux densities at 86 and 230 GHz are lower than the typical flux densities at centimeter wavelengths. This is different from the results in [3] who find an inverted spectrum up to $\sim$ 100 GHz. We cannot tell whether we observed M81* in a phase with unusual low millimeter emission or the [3] observations were made during an outburst at millimeter wavelengths. A simultaneous monitoring project from centimeter to millimeter wavelengths is needed to decide this question ([32, 33, 34]).

Linear polarization is not detected for M81* at any wavelength longward of 3.6 mm. The presence of a jet at VLBI resolution, radio synchrotron emission, and circular polarization suggest that M81* is intrinsically linearly polarized but depolarized during propagation through a magnetized plasma. The case is similar to that of Sgr A*, which is detected in linear polarization only at wavelengths shortward of 3.6 mm. For Sgr A*, the detection of linear polarization at short wavelengths provides an upper limit to the rotation measure of a few times 10^6 rad m^{-2}. This provides an upper limit to the accretion rate of $\sim 10^{-7} M_\odot y^{-1}$. For M81*, we find a lower limit to the rotation measure of $\sim 10^4$ rad m^{-2} for the case of beam depolarization and $\sim 4 \times 10^5$ rad m^{-2} for the case of bandwidth depolarization, under the assumption that the intrinsic source is polarized. The lesser value could originate in the dense interstellar medium but the larger value exceeds that seen anywhere in the ISM. ADAF and Bondi-Hoyle accretion models will depolarize the source unless the accretion rate falls below $10^{-9} M_\odot y^{-1}$ [35]. However, for radiatively inefficient accretion flows, the larger of the RM limits implies a lower limit to the accretion rate of $10^{-7} M_\odot y^{-1}$. The accretion rate necessary for the X-ray luminosity is $10^{-5} M_\odot y^{-1}$. Since bandwidth depolarization effects decrease as λ^3, measurement of the linear polarization at a wavelength of 0.8 mm would increase the accuracy of the accretion rate constraint by nearly two orders of magnitude.

5. Summary & Conclusion

We have presented VLA observations of M81* from 1994 until 2002 that show that circular polarization is present at 4.8, 8.4, and 15 GHz in absence of linear polarization. The fractional circular polarization is variable on timescales of days and months and not correlated with the total flux density of the source. The sign of the circular polarization was, if detected, at all frequencies and times always positive. The polarization properties are strikingly similar to the properties of Sgr A*, the central radio source in the Milky Way. This supports the hypothesis that M81* is a scaled up version of Sgr A*. [36] and [37] detected linear polarization at 230 GHz and higher frequencies that also shows variability [38, 39]. Given the similarity between M81* and Sgr A* we expect to see also linear polarization in M81* at higher frequencies.

References

[1] Bietenholz M. F., Bartel N., Rupen M. P., 2000, ApJ, 532, 895
[2] Ho L. C., Filippenko A. V., Sargent W. L. W., 1996, ApJ, 462, 183
[3] Reuter H.-P., Lesch H., 1996, A&A, 310, L5
[4] Melia F., Falcke H., 2001, ARA&A, 39, 309
[5] Falcke H., 1996, ApJ, 464, L67
[6] Bietenholz M. F., Bartel N., Rupen M. P., 2004, ApJ, 615, 173
[7] Bower G. C., Falcke H., Herrnstein R. M., et al., 2004, Science, 304, 704
[8] Shen Z.-Q., Lo K. Y., Liang M.-C., Ho P. T. P., Zhao J.-H., 2005, Nature, 438, 62
[9] Bower G. C., Falcke H., Backer D. C., 1999, ApJ, 523, L29
[10] Bower G. C., Backer D. C., Zhao J., Goss M., Falcke H., 1999, ApJ, 521, 582
[11] Bower G. C., Wright M. C. H., Backer D. C., Falcke H., 1999b, ApJ, 527, 851
[12] Brunthaler A., Bower G. C., Falcke H., Mellon R. R., 2001, ApJ, 560, L123
[13] Wardle J. F. C., Homan D. C., Ojha R., Roberts D. H., 1998, Nature, 395, 457
[14] Rayner D. P., Norris R. P., Sault R. J., 2000, MNRAS, 319, 484
[15] Bower G. C., Falcke H., Sault R. J., Backer D. C., 2002, ApJ, 571, 843
[16] Brunthaler A., Bower G. C., Falcke H., 2006, A&A, 451, 845
[17] Bartel N., Bietenholz M. F., Rupen M. P., et al., 1994, Nature, 368, 610
[18] Welch W. J., Thornton D. D., Plambeck R. L., et al., 1996, PASP, 108, 93
[19] Brunthaler A., Falcke H., Bower G. C., et al., 2000, A&A, 357, L45
[20] Brunthaler A., Falcke H., Bower G. C., et al., 2005, A&A, 435, 497
[21] Macquart J.-P., Melrose D. B., 2000, ApJ, 545, 798
[22] Pacholczyk A. G., 1977, Oxford Pergamon Press International Series on Natural Philosophy, 89
[23] Jones T. W., Odell S. L., 1977, ApJ, 214, 522
[24] Jones T. W., 1988, ApJ, 332, 678
[25] Ramaty R., 1969, ApJ, 158, 753
[26] Beckert T., Falcke H., 2002, A&A, 388, 1106
[27] Ruszkowski M., Begelman M. C., 2002, ApJ, 573, 485
[28] Enßlin T. A., 2003, A&A, 401, 499
[29] Bietenholz M. F., Bartel N., Rupen M. P., et al., 1996, ApJ, 457, 604
[30] Zylka R., Mezger P. G., Ward-Thompson D., Duschl W. J., Lesch H., 1995, A&A, 297, 83
[31] Falcke H., Markoff S., 2000, A&A, 362, 113
[32] Markoff S., et al. in prep.
[33] Schödel R., Eckart A., 2006 J. Phys.: Conf. Ser. 54 349-353
[34] Schödel, R., Krips, M., Markoff, S.,Neri, R., Eckart, A., 2007, A&A, in press, astro-ph/0611844
[35] Quataert E., Gruzinov A., 2000, ApJ, 545, 842
[36] Aitken D. K., Greaves J., Chrysostomou A., et al., 2000, ApJ, 534, L173
[37] Bower G. C., Wright M. C. H., Falcke H., Backer D. C., 2003, ApJ, 588, 331
[38] Bower G. C., Falcke H., Wright M. C., Backer D. C., 2005, ApJ, 618, L29
[39] Marrone D. P., Moran J. M., Zhao J.-H., Rao R., 2006, ApJ, 640, 308

AUTHOR INDEX

AUTHOR INDEX